第八届中国国际管道会议（CIPC）论文集

《第八届中国国际管道会议（CIPC）论文集》编委会　编

中国石化出版社

·北京·

图书在版编目(CIP)数据

第八届中国国际管道会议(CIPC)论文集 /《第八届中国国际管道会议(CIPC)论文集》编委会编. 北京：中国石化出版社, 2025.4. —ISBN 978-7-5114-7933-4

Ⅰ. U172-53

中国国家版本馆 CIP 数据核字第 20257P5Q58 号

未经本社书面授权，本书任何部分不得被复制、抄袭，或者以任何形式或任何方式传播。版权所有，侵权必究。

中国石化出版社出版发行

地址：北京市东城区安定门外大街 58 号
邮编：100011 电话：(010)57512500
发行部电话：(010)57512575
http://www.sinopec-press.com
E-mail:press@sinopec.com
北京鑫益晖印刷有限公司印刷
全国各地新华书店经销

*

880 毫米×1230 毫米 16 开本 40.75 印张 1044 千字
2025 年 4 月第 1 版 2025 年 4 月第 1 次印刷
定价：298.00 元

序

当今世界正面临百年未有之大变局，能源转型与工业革命的浪潮激荡交融：碳中和目标引领全球可再生能源蓬勃发展，新型能源体系加速构建；新一轮科技革命和产业变革向纵深推进，人工智能、大数据、物联网、云计算、区块链等新技术快速普及，其与能源工业及交通运输业深度融合，正在重塑能源行业、交通运输系统乃至整个人类社会。在这场深刻的变革中，管道运输作为能源供应的"生命线"及交通运输系统向地下延伸的载体，以其绿色、高效、安全、经济等优势，在油气、氢能、甲醇、液氨等能源介质及二氧化碳、矿物浆体等非常规介质的输送中具有不可替代的作用，成为保障能源供应安全、维护经济稳定的"压舱石"，未来更有望成为综合立体交通运输网络的重要一环，在提高物流运输效率、降低物流运输成本中创造管道价值。

近年来，全球管道科技和产业发展突飞猛进，为能源保供和经济社会发展做出了重要贡献。众多能源管道企业、科研机构、高等院校聚焦能源安全需求，以高质量发展为首要任务，以科技创新为驱动引擎，推动行业向高端化、智能化、绿色化迈进：高钢级管材及制管技术为大规模管道建设提供了保障，智能化设计与机械化施工技术提高了管道设计建造水平，仿真优化技术解决了复杂管网系统稳定经济高效运行难题，智能感知与数字孪生技术实现了管道全生命周期动态监测及预测性管理，氢能、甲醇、液氨等新能源介质的管道输送及二氧化碳捕集封存等前沿领域取得了重要进展。这些创新成果不仅有效保障了能源输送安全，而且为社会生产方式低碳转型提供了新赛道。

2025年4月8-10日在中国北京举办的"第八届中国国际管道会议（CIPC）暨技术装备与成果展"，以"绿色、智能、融合、发展"为主题，重点交流管道绿色与智能发展背景下的科技创新成果，深入探讨管道与新型能源体系及交通运输系统的融合路径，进一步明确管道运输在新时代、新需求下的发展任务。

本论文集围绕大会主题组织稿件，汇聚了众多管道工业领域专家学者及科技工作者近年的学术成果，内容上具有创新性和可读性，期望能够为管道行业科技创新和成果应用提供参考，推动全球管道行业在新时代需求下实现跨越式发展。

中国工程院院士
2025 年 3 月 28 日

前　言

在全球能源转型加速、地缘政治博弈加剧、能源安全需求升级、工业革命持续深化的复杂背景下，作为"能源大动脉"的油气管道行业正处于"传统与新兴并存、风险与机遇交织"的关键发展阶段。应对挑战、把握机遇、推动高质量可持续发展，成为全球能源管道行业的发展任务。

作为引领行业潮流、推动创新发展的国际性学术交流与商务合作平台，中国国际管道会议(CIPC)已成功举办七届，成为全球管道行业的盛事，吸引了众多国内外知名专家学者及管道从业者参加，在推动油气与新能源储运技术进步、促进能源管输转型升级、实现高水平国际交流合作中发挥了重要作用。

"第八届中国国际管道会议(CIPC2025)暨技术装备与成果展"将于2025年4月8-10日在中国北京举办。本次会议以"绿色、智能、融合、发展"为主题，将全面总结上届会议以来全球油气管道行业的研究进展与技术进步，不仅分享管材研制与应用、管道设计与施工、管道输送与储存、管道监检测与抢修、管道风险管理、管道数字化与智能化等重要领域的创新成果与工程经验，而且探讨在能源转型及工业4.0背景下油气管网多介质灵活输运等新兴领域的技术挑战与应对策略，以及以人工智能为代表的新一代信息技术在油气管道转型升级进程中的应用潜力、技术瓶颈及研发路线。

本次会议基于会议主题及上述行业热点、难点开展征文，得到全球管道从业者积极响应，共收到486篇来稿，经过专家评审，最终66篇入选论文集。论文内容涉及管道产业链前沿思考及各环节科技创新，希望能够为广大读者提供有价值的参考，并催生更多科研成果，为提升能源供应安全水平、推动管道产业高端化、智能化、绿色化发展提供新动能。借此机会，向全体论文作者和关心支持能源管道行业发展的各界朋友致以崇高的敬意和衷心的感谢！

本书编委会

2025年4月

目 录

Effect of Welding Parameters on Stress Corrosion Cracking Susceptibility of Welded API X-70 Pipeline Steel in a Near-neutral pH Simulated Soil Solution Protected by CP
　　…………………… Reza Sadeghi M. Sc.　Abdoulmajid Eslami　Ali Ashrafi　Khosrow F. Zare（ 1 ）

基于神经网络的典型天然气管网短期单耗预测
　　………………………………… 李　雨　王　瑞　刘屹然　石　峰　张文喆　张天娇（ 14 ）

天然气地下储气库注采方案优化研究 ………… 邹雪晴　张　轶　岳远志　杨　飞　耿金亮（ 21 ）

基于电力制度的天然气管道电驱压缩机组生产运行成本优化应用实践
　　………………………………… 乔　欣　刘屹然　孟　霏　高振宇　刘旭东　张华斌（ 29 ）

绿色发展导向下天然气与电力折标系数的融合应用研究与价值重塑
　　………………………………… 孟　霏　乔　欣　刘麦伦　刘义茜　孙　敏　石咏衡（ 35 ）

双碳目标下深地空间利用及展望 ……………… 熊昊天　王文权　刘晓旭　陈　伟　周练武（ 40 ）

基于黏菌优化算法的全国一张网下的宏观路由自动选择
　　………………………………… 李　娟　刘　亮　李明旭　李　强　孟祥海　李海润（ 49 ）

站场管座角焊缝专用扫查装置开发 …… 路兴才　项小强　赵　岩　赵子峰　周　聪　赵龙一（ 59 ）

基于IMU数据的管道力学响应分析 ……… 余晓峰　黄忠宏　陈翠翠　吕志阳　李云涛（ 67 ）

LNG接收站罐容及其配套外输管道规模确定方法研究
　　………………………………………………… 候宇坤　明　亮　吴国其　陈　彬　于润泽（ 73 ）

国家管网集团新建油气管道与航道、港口相遇相交跨行业系统性协作管理创新与探索
　　…………………………………………………………………………… 王　丽　李　维　刘　垒（ 78 ）

竞合关系在海外油气管道建设EPC项目界面管理中的应用 ……………………………… 杨　洋（ 82 ）

Feasibility and challenges of digital twin technology development in the field of oil and gas pipelines
　　……………… Jiang Yuyou　Kou Bo　Chen Ruolei　Jiang Haibin　Zhu Yanxiang
　　　　　　　　　　　　　　　　　　　　　Wang Can　Cheng Yutao　Wang Bo（ 87 ）

油气管道完整性管理智能化实施方案研究 ……… 薛鲁宁　杨秦敏　田明亮　陈　钻　赵俊丞（ 95 ）

Optimization of water blending in unheated gathering based on low-temperature flow characteristics of heavy oil with high water contents
　　………………………… Lv Yuling　Hu Weili　Shi jiakai　Ma lihui　Song kunlin　Wang jiaxin（101）

阀芯结构对蝶阀内部空化演化及抑制的影响 …… 张　光　吴　璇　张浩钿　林　哲　朱祖超（115）

Integrated process optimization of natural gas pretreatment and pressurized liquefaction based on low temperature packed bed decarburization ……… Fu Juntao　Liu Jinhua　Li Zihe　Zhu Jianlu（131）

Numerical simulation study on pressure reduction characteristics of dense phase CO_2 pipeline leakage
　　…… Yin Buze　Huang Weihe　Li Yuxing　Hu Qihui　Yu Xinran　Zhao Xuefeng　Meng Lan（145）

Effect of Epoxy ResinCoating on Hydrogen Embrittlement Behavior of X80 Pipeline Steel in Hydrogen Blended Methane Environment: Mechanism of Coating Delamination in Aggravating Hydrogen Embrittlement …… Zhu Yuanchen　Yang Hongwei　Zhu Yun　Liu Fang　Chen Lei　Liu Gang (154)

A Synergistic Approach to BOG Generation Calculation Using Horizontal Feature Fitting and Vertical Time Series Prediction …… Wang Yuqian　Liu Gang　Wang Bo　Zhang Chao　Chen Lei (169)

基于 WebGIS 的天然气站场甲烷泄漏监测可视化平台实现
………………………… 詹佳琪　刘　标　洪　啸　宋　成　易　欣　刘　翔(181)

水下气液分离与增压技术及装备研究进展 ……………………… 罗小明　李蔚迪　许欣怡(188)

天然气超声速旋流分离技术现状与应用展望
………………… 边　江　王　颖　王泽润　鞠　淋　邓涵玉　张嘉伟　曹学文　曹恒广(201)

Comparison and selection of decarbonization process based on natural gas liquefaction under pressure
………………………………………………………… Li Zihe　Liu Jinhua　Zhu Jianlu (216)

A Review of Cavitation Problems of Cryogenic Fluids in Gathering Pipelines
………………………………………………………… Ao Di　Li Yuxing　Liu Cuiwei (228)

基于状态空间的天然气管道系统模型预测主动控制 ……… 李　轩　陆洋帆　陈国龙　高　伟
　　　　　　　　　　　　　　　　伍梓文　金　凤　郭雨茜　张炜奇　宫　敬　温　凯(246)

甲醇对 X65 管线钢腐蚀性实验研究
………………… 陈更生　姜子涛　黄梓耕　聂超飞　刘罗茜　刘世贸　常滋茹　刘冠一(255)

含蜡原油管道停输再启动初凝仿真研究 ………… 崔润麒　王　吉　苏　怀　李鸿英　张劲军(263)

A Mapless Path Generation and Control Method for UAV Pipeline Inspection
………………………………………………………………………… Ma Yinghan　Zhao Hong (273)

Research on Obstacle Evasion and Joint Angle-Minimized Trajectory for Pipeline Grinding Robot based on PSO ………………………………………………………… Yan Zhouyu　Zhao Hong (284)

光电+光热耦合相变蓄能供热系统在输油管道站场应用初探 ……… 王思杨　张灿灿　王新宇(293)

国内外成品油管网公平开放机制研究 ………… 张澍原　薛　庆　王建良　马青天　刘　钰(299)

一种音速喷嘴摩尔质量的修正方法 ……………………………… 陈曦宇　王柯栩　裴勇涛(315)

基于双向流固耦合的天然气科氏质量流量计模拟与优化 ………………………………… 裴全斌(320)

Fatigue Failure Analysis of Tensile Armor Layers of Deep-sea Flexible Pipes at Joints
……………… Liu Yu　Chen Yanfei　Zhang Ye　Zhong Ronfeng　Lu Shuntian　Xiang Tao (329)

油气管道智能工地建设质量问题剖析及技术提升策略
………………… 刘海春　郭　旭　苏维刚　李庆生　姜　鹏　陈　群　李增材(339)

综合法天然气管道泄漏监测系统研发与应用 ………… 张　春　曹旦夫　王军防　王浩霖(347)

燃气轮机伺服控制系统的关键技术分析与架构设计 ……… 郑　明　关　睿　刘　超　姚　珺(356)

Optimization analysis of shutdown and restart of multiphase transportation operations in the South China Sea
………………………… Han Dong　Zheng Chengming　Jiang Ling　Zhang Meng　Qi Baobao
　　　　　　　　　　　　　Chen Zhu　Zhang Lin　Guo Yingzhen　Sun Xinghua　Wang Bing (362)

Study on the effect law of microcrystalline wax on EVA to improve the low temperature rheology of model waxy oil …………………… Xia Xue　Yan Feng　Li Qifu　Yu Hongmei　Li Zhengbin (373)

| 掺氢长输管道气体传质规律研究 ………………………… 柴 冲 张瀚文 彭世垚 裴业斌（380）
| 站场典型燃气锅炉对掺氢天然气的适应性研究 …………………… 程 磊 张瀚文 张 扬（386）
| 国外成品油管输定价监管现状及对我国的启示 … 牛国富 石博涵 张仲藜 温 文 阴佳乐（394）
| 弯管相连腐蚀缺陷力学—电化学相互作用规律研究 ……… 张 鹏 赵 明 黄云飞 许 田（402）
| 基于改进门控循环单元的原油储罐关键参数预测方法 ………………………………… 刘鹏涛（410）
| 内压-缺陷尺寸耦合下管道极限弯曲应变特性分析
| ………………………… 裴迎举 王 聪 凌瑜基 蒋程晨 薛喆中（417）
| 油气田在役玻璃钢管道老化规律研究 ………… 熊新强 宫 敬 刘 杰 唐德志 廖丹丹（426）
| 多井水溶造腔排量优化调配及现场应用 ………………………… 秦 垦 任众鑫 廖友强（437）
| 文23储气库完整性协同管理体系
| ………… 许 锋 苏小健 张思远 杨佳坤 施玉霞 周栋梁 高立超 苗 刚 常 帅（443）
| Numerical study on cathodic protection effect of corrosion defects on pipelines under constant load ………
| …………………………………………………… Dongxu Sun Bo Wang Lei Li Yang Yu Yi Ji（450）
| 盐穴储气库智能造腔预测与设计 ………………… 王桂九 陈加松 赵廉斌 李金龙 王卓腾（463）
| Seismic-hazard risk assessment of long-distance pipelines based on an improved unascertained
| measurement model ……………………………………… Ying Wu Qing Peng Yu Tian Xiao You（478）
| 基于数据中台的油气管道行业数据治理模式及实施路径探讨
| ………………………………… 李 梁 姜 辉 张建军 徐加兴 李步伟 刘 峰（499）
| Seismic vulnerability analysis of natural gas distribution station … Wu Ying Tian Yu Meng Bojie（505）
| 基于GA-PSO混合算法和XGBoost算法的城市燃气管道风险评价 ……………… 彭善碧 王鸿扬（522）
| A Panoramic Visualization Platform for Petroleum Pipeline Supply Chain Procurement：Enhancing
| Transparency，Efficiency，and Data-Driven Decision-Making ……………… by Bing HAN（531）
| 设计参数对硬岩储气库稳定性影响规律研究 …… 周小松 黄康康 闫 磊 王颖蛟 刘 卫（539）
| 基于PSO-SVR和熵权-TOPSIS的双金属复合管焊接残余应力预测及优化
| …………………………………………… 彭星煜 蒋海洋 冯梁俊 祝星语（548）
| 油田含硫天然气小口径管道内检测技术与案例分析
| ………………………… 张 佳 薛文明 周智勇 秦 林 李潮浪 孙明楠 文绍牧（564）
| 复杂地质条件下水平定向钻技术在长输油气管道施工中的应用 ……………… 曹子建 胡乾彬（574）
| 油气管道智能阴极保护技术应用及发展 ………… 刘红波 王 钰 高秀宝 郭春雷 高 晓（578）
| 国内外油气行业管材标准现状及发展趋势分析 ……………………………… 田 灿 崔绍华（592）
| 天然气管道投产及运行工艺计算与软件开发
| ………………………… 姜新慧 何国玺 孙 勇 杨 洋 钟瀚宇 廖柯熹（600）
| 基于GWO-BP算法的页岩气集输管道内腐蚀速率预测 …… 祝星语 彭星煜 冯梁俊 蒋海洋（608）
| 气田在线仿真平台架构及仿真模型探讨 ………… 李长俊 廖钰朋 贾文龙 杨 帆 黄巧竞（618）
| Application of Composite Materials in Pipeline Repair and Optimization
| …………………………………………………………………… Casey Whalen Matthew Green（629）

CONTENTS

Effect of Welding Parameters on Stress Corrosion Cracking Susceptibility of Welded API X-70 Pipeline Steel in a Near-neutral pH Simulated Soil Solution Protected by CP
　　　　　……………… Reza Sadeghi M. Sc.　Abdoulmajid Eslami　Ali Ashrafi　Khosrow F. Zare (1)
Short-term unit energy consumption prediction of typical natural gas pipeline network based on neural network ………… Li Yu　Wang Rui　Liu Yiran　Shi Feng　Zhang Wenzhe　Zhang Tianjiao (14)
Research onoptimization of injection and production plan for underground gas storage
　　　　　……………………… Zou Xueqing　Zhang Yi　Yue Yuanzhi　Yang Fei　Geng Jinliang (21)
Application of production and operation cost optimization of electric drive compressor unit for natural gas pipeline based on electric power system
　　　　　………………… Qiao Xin　Liu Yiran　Meng Fei　Gao Zhenyu　Liu Xudong　Zhang Huabin (29)
Integration application research and value reshaping of natural gas and electricity discount factor under the orientation of green development
　　　　　………………… Meng Fei　Qiao Xin　Liu Mailun　Liu Yixi　Sun Min　Shi Yongheng (35)
Utilization and Prospects of Deep Underground Space under the Dual Carbon Goals
　　　　　……………… Xiong Haotian　Wang Wenquan　Liu Xiaoxu　Cheng Wei　Zhou Lianwu (40)
Macro-routing automation under a nationwide network based on Slime Mold Optimization Algorithm
　　　　　…………………… Li Juan　Liu Liang　Li Mingxu　Li Qiang　Meng Xianghai　Li Hairun (49)
Development of special scanning device for fillet weld of pipe seat in station and yard
　　　　　………… Lu Xingcai　Xiang Xiaoqiang　Zhao Yan　Zhao Zifeng　Zhou Cong　Zhao Longyi (59)
Pipeline mechanical response analysis based on the IMU data
　　　　　………………… Yu Xiaofeng　Huang Zhonghong　Chen Cuicui　Lv Zhiyang　Li Yuntao (67)
A Method for Determining the Tank Capacity of LNGTerminal and the Scale of Pipeline
　　　　　…………………………… Hou Yukun　Ming Liang　Wu Guoqi　Chen Bin　Yu Runze (73)
The Pipe China's Innovation and Exploration of Cross-Industry Systematic Collaborative Management for New Oil and Gas Pipelines Meeting and Intersecting with Waterway and Ports
　　　　　………………………………………………………………… Wang Li　Li Wei　Liu Lei (78)
Practice of coopetition relationship on interface management of international oil and gas pipeline EPC project ……………………………………………………………………………………… Yang Yang (82)
Feasibility and challenges of digital twin technology development in the field of oil and gas pipelines
　　　　　……………………… Jiang Yuyou　Kou Bo　Chen Ruolei　Jiang Haibin　Zhu Yanxiang
　　　　　　　　　　　　　　　　　　　　　　　　　　　　Wang Can　Cheng Yutao　Wang Bo (87)

Research on the Intelligent Implementation Plan of Oil and Gas Pipeline Integrity Management
………………………… Xue Luning　Yang Qinmin　Tian Mingliang　Chen Zuan　Zhao Juncheng（95）

Optimization of water blending in unheated gathering based on low-temperature flow characteristics of heavy oil with high water contents
………………………… Lv Yuling　Hu Weili　Shi jiakai　Ma lihui　Song kunlin　Wang jiaxin（101）

Effects of core structure on cavitation evolution and inhibition in the butterfly valve
………………………… Zhang Guang　Wu Xuan　Zhang Haotian　Lin Zhe　Zhu Zuchao（115）

Integrated process optimization of natural gas pretreatment and pressurized liquefaction based on low temperature packed bed decarburization ………… Fu Juntao　Liu Jinhua　Li Zihe　Zhu Jianlu（131）

Numerical simulation study on pressure reduction characteristics of dense phase CO_2 pipeline leakage
…… Yin Buze　Huang Weihe　Li Yuxing　Hu Qihui　Yu Xinran　Zhao Xuefeng　Meng Lan（145）

Effect of Epoxy ResinCoating on Hydrogen Embrittlement Behavior of X80 Pipeline Steel in Hydrogen Blended Methane Environment: Mechanism of Coating Delamination in Aggravating Hydrogen Embrittlement …… Zhu Yuanchen　Yang Hongwei　Zhu Yun　Liu Fang　Chen Lei　Liu Gang（154）

A Synergistic Approach to BOG Generation Calculation Using Horizontal Feature Fitting and Vertical Time Series Prediction ………… Wang Yuqian　Liu Gang　Wang Bo　Zhang Chao　Chen Lei（169）

Implementation of a Visualization Platform for Monitoring Methane Leakage in Natural Gas Stations Based on WebGIS ……………… Zhan Jiaqi　Liu Biao　Hong Xiao　Song Cheng　Yi Xin　Liu Xiang（181）

Research Progress on Technologies and Equipment for Underwater Gas-Liquid Separation and Pressurization
………………………………………………… Luo Xiaoming　Li Weidi　Xu Xinyi（188）

Current status and application prospects of natural gas supersonic separation technology …… Bian Jiang　Wang Ying　Wang Zerun　Ju Lin　Deng Hanyu　Zhang Jiawei　Cao Xuewen　Cao Hengguang（201）

Comparison and selection of decarbonization process based on natural gas liquefaction under pressure
………………………………………………… Li Zihe　Liu Jinhua　Zhu Jianlu（216）

A Review of Cavitation Problems of Cryogenic Fluids in Gathering Pipelines
………………………………………………… Ao Di　Li Yuxing　Liu Cuiwei（228）

Model predictive active control of natural gas pipeline system based on state space
………………………… Li Xuan　Lu Yangfan　Chen Guolong　Gao Wei　Wu Ziwen
　　　　　　　　　　　　　　Jin Feng　Guo Yuqian　Zhang Weiqi　Gong Jing　Wen Kai（246）

Experimental study of methanol corrosion on X65 pipeline steel …… Chen Gengsheng　Jiang Zitao
　　　　Huang Zigeng　Nie Chaofei　Liu Luoqian　Liu Shimao　Chang Ziru　Liu Guanyi（255）

Initial Gelling for the Shutdown and Restart Processes of a Waxy Crude Oil Pipeline
………………………… Cui Runqi　Wang Ji　Su Huai　Li Hongying　Zhang Jinjun（263）

A Mapless Path Generation and Control Method for UAV Pipeline Inspection
………………………………………………… Ma Yinghan　Zhao Hong（273）

Research on Obstacle Evasion and Joint Angle-Minimized Trajectory for Pipeline Grinding Robot based on PSO ……………………………………………… Yan Zhouyu　Zhao Hong（284）

Preliminary Study on the Application of Photovoltaic and Photothermal Coupled Phase Change Energy
    Storage Heating System in Oil Pipeline Stations ...... Wang Siyang   Zhang Cancan   Wang Xinyu (293)
Research on the Open-Access Mechanism of Products Pipeline Networks at Home and Abroad
    .... Zhang Shuyuan   Xue Qing   Wang Jianliang   Ma Qingtian   Liu Yu (299)
Method for molar mass correction of a sonic nozzle .... Chen Xiyu   Wang Kexu   Pei Yongtao (315)
Simulation and Optimization of Coriolis Mass Flowmeters for Natural Gas Based on Bidirectional Fluid-
    Solid Coupling .... Pei Quanbin (320)
Fatigue Failure Analysis of Tensile Armor Layers of Deep-sea Flexible Pipes at Joints
    .... Liu Yu   Chen Yanfei   Zhang Ye   Zhong Ronfeng   Lu Shuntian   Xiang Tao (329)
Analysis of QualityProblems and Technical Improvement Strategies for Intelligent Construction Sites of Oil
    and Gas Pipelines .... Liu Haichun   Guo Xu   Su Weigang   Li Qingsheng   Jiang Peng
                                                                                                    Chen Qun   Li Zengcai (339)
Research and application of comprehensive natural gas pipeline leakage monitoring system
    .... Zhang Chun   Cao Danfu   Wang Junfang   Wang Haolin (347)
Key technology analysis and architecture design of gas turbine servo control system
    .... Zheng Ming   Guan Rui   Liu Chao   Yao Jun (356)
Optimization analysis of shutdown and restart of multiphase transportation operations in the South China Sea
    .... Han Dong   Zheng Chengming   Jiang Ling   Zhang Meng   Qi Baobao
                                        Chen Zhu   Zhang Lin   Guo Yingzhen   Sun Xinghua   Wang Bing (362)
Study on the effect law of microcrystalline wax on EVA to improve the low temperature rheology of model
    waxy oil .... Xia Xue   Yan Feng   Li Qifu   Yu Hongmei   Li Zhengbin (373)
Study on the law of gas mass transfer in long-distance hydrogen pipeline
    .... Chai Chong   Zhang Hanwen   Peng Shiyao   Pei Yebin (380)
Study on the adaptability of typical gas boilers in stations to hydrogen blended natural gas
    .... Cheng Lei   Zhang Hanwen   Zhang yang (386)
The Current Situation of Pricing and Regulatory Supervision of Foreign Oil Pipeline Transportation and Its
    Implications for China .... Niu Guofu   Shi Bohan   Zhang Zhongli   Wen Wen   Yin Jiale (394)
Study onmechano-electrochemical interaction laws of connected corrosion defects on elbows
    .... Zhang Peng   Zhao Ming   Huang Yunfei   Xu Tian (402)
Key Parameter Prediction Method for Crude Oil Storage Tanks Based on Improved Gated Recurrent Unit
    .... Liu Pengtao (410)
Analysis of ultimate bending strain characteristics of pipelines under coupling of internal pressure and
    defect size .... Pei Yingju   Wang Cong   Ling Yuji   Jiang Chengchen   Xue Zhezhong (417)
Research on the main aging factors and influence law of in-service GFRP pipes in oil and gas fields
    .... Xiong Xinqiang   Gong Jin   Liu Jie   Tang Dezhi   Gu Tan   Liao Dandan (426)
Optimization and Application of Multi-Well Cavity Discharge Volume Allocation for Salt Cavern Gas
    Storage .... Qin Ken   Ren Zhongxin   Liao Youqiang (437)

Integrity Collaborative Management System Of Wen 23 gas storage ············ Xu Feng　Su Xiaojian　Zhang Siyuan　Yang Jiakun　Shi Yuxia　Zhou Dongliang　Gao Lichao　Miao Gang　Chang Shuai（443）

Numerical study on cathodic protection effect of corrosion defects on pipelines under constant load ·········
·· Dongxu Sun　Bo Wang　Lei Li　Yang Yu　Yi Ji（450）

Prediction and design of intelligent leaching for salt cavern gas storage
················· Wang Guijiu　Chen Jiasong　Zhao Lianbin　Li Jinlong　Wang Zhuoteng（463）

Seismic-hazard risk assessment of long-distance pipelines based on an improved unascertained measurement model ······················· Ying Wu　Qing Peng　Yu Tian　Xiao You（478）

Research on the Data Governance Model and Implementation Path of the Oil and Gas Pipeline Industry Based on the Data Middle Platform ···················· Li Liang　Jiang Hui　Zhang Jianjun
Xu Jiaxing　Li Buwei　Liu Feng（499）

Seismic vulnerability analysis of natural gas distribution station ··· Wu Ying　Tian Yu　Meng Bojie（505）

Risk assessment of urban gas pipeline based on GA-PSO hybrid algorithm and XGBoost algorithm
·· Peng Shanbi　Wang Hongyang（522）

A Panoramic Visualization Platform for Petroleum Pipeline Supply Chain Procurement: Enhancing Transparency, Efficiency, and Data-Driven Decision-Making ·············· Han Bing（531）

Study on the influence of design parameters on the stability of hard rock gas storage
················ Zhou Xiaosong　Huang Kangkang　Yan Lei　Wang Yingjiao　Liu Wei（539）

Prediction and optimization of welding residual stress in bimetallic composite pipes based on PSO-SVR and entropy weight TOPSIS ········ Peng Xingyu　Jiang Haiyang　Feng Liangjun　Zhu Xingyu（548）

ILI technology and field case analysis of small-diameter pipelines of sulfur natural gas
················· Zhang Jia　Xue Wenming　Zhou Zhiyong　Qin Lin　Li Chaolang
Sun Mingnan　Wen Shaomu（564）

Application of Horizontal Directional Drilling Technology in Long-distance Oil and Gas Pipeline Construction under Complex Geological Conditions ················ Cao Zijian　Hu Qianbin（574）

Application and Development of Intelligent Cathodic Protection Technology
················ Liu Hongbo　Wang Yu　Gao Xiubao　Guo Chunlei　GaoXiao（578）

Application Status and Development Tendency for the Standards of Tubular Goods for Oil and Gas Industries
·· Tian Can　Cui Shaohua（592）

Gas pipeline commissioning and operation process calculation and software development
··········· Jiang Xinhui　He Guoxi　Sun Yong　Yang Yang　Zhong Hanyu　Liao Kexi（600）

Corrosion Rate Prediction in Shale Gas Gathering Pipelines Based on GWO-BP Algorithm
················ Zhu Xingyu　Peng Xingyu　Feng Liangjun　Jiang Haiyang（608）

The architecture of the online simulation platform for gas fields and discussion on simulation models
··········· Li Changjun　Liao Yupeng　Jia Wenlong　Yang Fan　Huang Qiaojing（618）

Application of Composite Materials in Pipeline Repair and Optimization
·· Casey Whalen　Matthew Green（629）

Effect of Welding Parameters on Stress Corrosion Cracking Susceptibility of Welded API X-70 Pipeline Steel in a Near-neutral pH Simulated Soil Solution Protected by CP

Reza Sadeghi M. Sc.   Abdoulmajid Eslami   Ali Ashrafi   Khosrow F. Zare

(Isfahan University of Technology Faculty of Material Engineering)

Abstract Stress Corrosion Cracking is one of the main causes of pipeline failure. This study investigates the effect of welding parameters on stress corrosion cracking(SCC) susceptibility of API X-70 pipeline steel in simulated soil solution(C_2 solution) protected by cathodic polarization(CP). To this end, API X-70 pipeline steel plates were welded via 70-70 degree V-shaped grooves using the shielded metal arc welding process. The SCC susceptibility of welded steel was evaluated using the slow strain rate test. To determine the SCC susceptibility at OCP, the tensile strength drop factor(F_σ), fracture surface area drop factor(F_R), and elongation drop factor(F_E) were calculated. Results showed that API X-70 pipeline specimens welded at the highest heat input of 2.227kJ/mm had the largest HAZ and grain size, while the number of secondary phases like perlite, banded structures, and precipitates decreased with an increase in weld heat input. For the welded API X-70 pipe specimens in solution at OCP, with the increase in weld heat input from 1.912kJ/mm to 2.227kJ/mm, SCC susceptibility decreased. The highest strength was also obtained for specimens welded with the lowest weld heat input of 1.912kJ/mm. The SCC susceptibility decreased for specimens protected by CP of -850mV, and -1200mV vs Ag/AgCl compared to specimens at OCP and increased by increasing the CP from -850mV to -1200mV vs Ag/AgCl. Furthermore, a decrease in SCC susceptibility was observed with the increase in weld heat input for specimens protected by CP.

Keywords Cathodic Polarization; Pipeline Steel; Stress Corrosion Cracking; Weld Heat Input

1 Introduction

Oil and gas transfer pipelines are susceptible to corrosion and cracking. Pipeline failure is catastrophic and can lead to economic and environmental losses. Stress corrosion cracking(SCC) has been recognized as one of the leading causes of pipeline failure and is caused by the simultaneous effects of tensile stress and corrosive environment on the external surface of a pipeline. The type of SCC in these pipelines is generally classified as high pH SCC(the classical SCC) and near-neutral pH SCC. These two types of SCC have similarities and differences. Cracks in both types of SCC are surface cracks and can eventually lead to pipeline leaks or ruptures. High pH SCC often occurs at a pH above 9 and can be expressed as intergranular sharp cracks, with the presence of minor corrosion in the walls of the cracks. Near neutral pH, SCC occurs in aqueous environments with a pH close to 6.5. Cracks in this type of SCC are usually widespread transgranular cracks, with noticeable corrosion on both sides of crack faces. For high pH SCC, anodic dissolution at the grain boundary is described as the primary mechanism of the failure while near-neutral pH SCC is often associated with uniform corrosion. Cracks associated with both types of SCC(i.e., Near-neutral and High pH SCC) are often found close to weld lines.

There have been some researches investigating the effect of welding parameters on corrosion and SCC susceptibility of pipeline steels. In this regard, Kong et al. studied the SCC behavior of X-80 pipeline steel welded by submerged arc welded (SAW) in different environments, i.e., air, NACE solution free from H_2S, and NACE solution saturated with H_2S. Results showed ductile fracture occurred in the air, while the fracture surface was quasi cleavage with some dimples in NACE solution free from H_2S. However, in the NACE, solution saturated with H_2S, the fracture surfaces were brittle, and high susceptibility to SCC was observed. Lu et al. investigated the effect of plastic pre-strain on near-neutral pH SCC of API X-70 welded pipeline steel. Slow Strain Rate Test(SSRT) was used to investigate the SCC susceptibility of the pipeline steel. Results showed a decrease in SCC resistance at weldment of X-70 pipeline steel by an increasing plastic pre-strain. Zhang et al. investigated the microstructure and corrosion resistance of welded X-70 pipeline steel in a near-neutral pH solution (in a pH range between 5.7 to 6.5). Results showed that the microstructure of weld metal was a mixture of acicular ferrite and grain boundary ferrite. Whereas the HAZ consisted of acicular ferrite, bainite, and martensite. The microstructure of base metal was mainly ferrite and perlite. Electrochemical corrosion tests illustrated that the HAZ had the highest corrosion rate upon hydrogen charging. In another study, Zhang et al. investigated the corrosion behavior of welded X-100 pipeline steel exposed to a near-neutral pH solution using the electrochemical scanning vibrating electrode technique (SVE). Results showed that the phase transformations during welding from acicular ferrite and bainite (in base metal) to ferrite and perlite (in weld metal) affected the corrosion resistance of the pipeline steel. Higher corrosion resistance was observed for the weld metal compared to the base metal.

This study investigates the effect of welding parameters on the SCC susceptibility of X-70 pipeline steel in a simulated near-neutral pH solution protected by CP.

2 Solutions and Methods

2.1 Sample Preparation

API X-70 pipeline steel with the chemical composition shown in Table 1 was used in this study. Previous studies have shown this steel is susceptible to SCC. To study the effect of weld heat input on the SCC susceptibility of this steel, API X-70 plates with 15×30 cm dimensions were welded via 70 70-degree V-shaped groove. Welding of the plates was done using a Shielded Metal Arc Welding(SMAW) process through 4 welding passes without preheating. AWS-E6010 penetrating cellulosic electrodes with diameters of 2.5mm at a constant current of 70 were used for the root pass. For the other welding passes, AWS-E8018 electrodes with a diameter of 3.5mm were used. E8018 electrodes were preheated at 250℃ per the manufacturer's instructions prior to the welding process 1. The chemical composition of the welding electrodes is shown in Table 1. Different weld heat inputs were used in this study(Table 2 to 4). Welding heat inputs were calculated using the following Equation.

$$Q = \frac{0.6VI}{S} \quad (1)$$

Where Q is the weld heat input, V is the welding voltage, I is the welding amperage, and S is the welding speed. After each welding pass, the surface of the weld zone was polished using a grinding machine. The schematic of the weld pattern and the image of the prepared welding sections are presented in Figure 1.

Table 1 Chemical composition of used X-70 pipeline steel and weld electrodes.

	V	Cr	Ni	S	P	Ti	Mo	Al	Si	Mn	C	Fe
API X-70	0.03	0.01	0.04	0.005	0.008	0.0096	0.207	0.447	0.327	1.97	0.0759	Balance
E8018	—	—	—	—	—	—	0.4	—	1.4	1.4	0.08	Balance
E6010	—	—	—	—	—	—	—	—	0.25	0.4	0.13	Balance

Table 2 Welding parameters used for specimen No. 1.

Welding Passes	Welding Current/A	Voltage/V	Welding Speed/(mm/s)	Heat Input/(kJ/mm)	Total Heat Input/(kJ/mm)
Root	70	26	3.21	0.340	1.912
1	90	26	2.80	0.501	
2	90	26	2.61	0.538	
3	90	26	2.68	0.524	

Table 3 Welding parameters used for specimen No. 2.

Welding Passes	Welding Current/A	Voltage/V	Welding Speed/(mm/s)	Heat Input/(kJ/mm)	Total Heat Input/(kJ/mm)
Root	70	26	3.15	0.346	1.976
1	110	26	3.11	0.551	
2	110	26	3.20	0.536	
3	110	26	3.16	0.543	

Table 4 Welding parameters used for specimen No. 3.

Welding Passes	Welding Current/A	Voltage/V	Welding Speed/(mm/s)	Heat Input/(kJ/mm)	Total Heat Input/(kJ/mm)
Root	70	26	3.18	0.343	2.227
1	130	26	3.34	0.607	
2	130	26	3.38	0.600	
3	130	26	3.30	0.614	

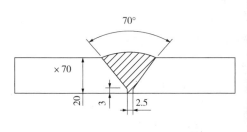 

(a) The schematic of the weld pattern　　　(b) Image of the prepared welding section

Figure 1

2.2 Slow Strain Rate SCC Tests

To investigate the SCC susceptibility of the welded X-70 pipeline steel at different weld heat inputs, SSRT samples were machined from the central part of the welded specimen with dimensions shown in Figure 2. For the welded specimens, the weld and base metal were located in the gage length of the tensile specimens, exposed to soil solution during the SSRT tests as schematically shown in Figure 3. According to previous studies, the strain rate in SSRT tests was fixed at 1×10^{-6} m/s. C2 solution was used as the simulated soil solution in this study. The chemical composition of this solution is shown in Table 6. To maintain an anaerobic environment, a mixture of 95% N_2 and 5% CO_2 was purged in a C2 solution for 48 hours before and during each test. During the SSRT, the pH of the C2 solution remained constant close to 6.3. Cathodic Protection (CP) of -850mV vs Ag/AgCl and -1200mV vs Ag/AgCl were applied to specimens using a three-electrode system via a 2050BEHPAJOOH BHP coulometer device. These CP potentials were chosen to cover a range of variations. Although the NACE recommendation is -850 mv Cu/CuSO4 respective to off potential, there could be a wide variation of CP potentials depending on field conditions. In this system, the specimen was the Working Electrodes, Platinum was the Counter Electrode, and Ag/AgCl was the Reference electrode. After each test, specimens were removed from the solution and dried with cold air. Before microscopic investigations, corrosion products were also removed from the surface of the specimens using a rust remover solution with the chemical composition shown in Table 7.

Table 5 Dimensions of SSRT specimens used in this study.

d_0	5mm
d_1	10mm
L_0	50mm
L_C	60mm
L_t	110mm
H	20mm

Table 6 Chemical composition of C2 solution

Compound	Concentration(g/L)
$MgSO_4 \cdot 7H_2O$	0.0274
$CaCl_2$	0.0255
KCl	0.0035
$NaHCO_3$	0.0195
$CaCO_3$	0.0606

Table 7 Chemical composition of Rust remover sol

Compound	Volume/mL
H_2O	100
Cis-2-Butene-1,4-diol	4
HCl	3

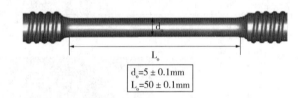

Figure 2 The Schematic of the SSRT specimen used in this study

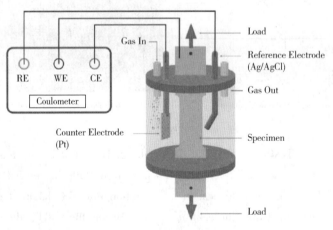

Figure 3 The schematic of the corrosion cell used for the SSRT tests

2.3 Electrochemical Corrosion Tests

Electrochemical corrosion tests were performed in C2 solution using a three-electrode system explained in section 2.2. The Potentiodynamic polarization test was carried out in simulated C2 soil solution purged with (N_2 95% and CO_2 5%) at a stable pH of 6.3 and room temperature. For potentiodynamic polarization tests, the potential range of −250mV to +750mV versus the Open Circuit Potential (OCP) was used. The scan rate of 0.5mV/s was selected for the potentiodynamic polarization tests.

3 Results and Effects

3.1 Microstructure Studies

Figure 4 shows the microstructure of X-70 pipeline steel investigated in this study. As shown in this Figure, the microstructure consists of ferrite grains, perlite colonies, and perlite bands, with grain size in the range of 3 to 4 μm. The average hardness for ferrite grains, perlite colonies, and perlite bonds were 201, 225, and 250 Vickers, respectively. Depending on the chemical composition of the pipeline steel, banded zones could have martensite or perlite microstructures. Elemental analysis showed that the banded zones in X-70 pipeline steel were highly rich in manganese. For banded zones, Ar temperature is reduced, and perlite is formed in regions with high concentrations of manganese. Researchers have shown that perlite bands are more prone zones for nucleation and growth of cracks. Higher heat inputs in the welding process enhanced the size of HAZ and increased the time above the recrystallization temperature. For specimens welded at heat inputs of 1.912, 1.976, and 2.227kJ/mm, the size of HAZ was 2, 3, and 4mm, respectively.

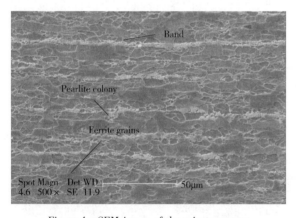

Figure 4 SEM image of the microstructure of X-70 pipeline steel

Researchers have shown that cyclic heat treatment of banded zones can dissolve banded perlite and re-precipitate these zones at ferrite grain boundaries. The average grain size and number of secondary phases formed in HAZ and welded regions for specimens investigated in this study are shown in Figure 5. As shown in this Figure, the percentage of secondary phases has decreased, while grain size has increased with an increase in weld heat input from specimen No. 1 to specimen No. 3. Higher weld heat inputs also caused the better distribution of secondary phases. This could be due to the peak temperature effect and the longer penetration time at the higher welding heat inputs, as reported by other researchers.

3.2 Electrochemical Corrosion Tests

Figures 6 and 7 show potentiodynamic polarization curves and also corrosion current densities, for weld and HAZ of specimens welded at different heat inputs. As can be seen from Figure 7, by the increase in weld heat input, the corrosion rate of pipeline steel has decreased (specimen No. 3, with a weld heat input of 2.227kJ/mm has the lowest corrosion rate in both HAZ and weld line). This could be due to a decrease in grain size and also the percentage of secondary phases, acting as the more active sites for corrosion, by the increase in the weld heat input (see Figure 5).

3.3 Slow Strain Rate SCC Tests

3.3.1 Slow Strain Rate SCC Tests in Air

Air was used as the neutral environment for SCC tests. After SSRT was performed in the air, all specimens were fractured from HAZ. In this regard, a cross-sectional image from the fracture surface of the specimen welded at a heat input of 2.227kJ/mm is shown in Figure 8. As shown in this Figure, cracks have propagated through banded perlite, which is the appropriate region for crack initiation/ growth (see Figure 8). Figure 9 and Table 8 show the results of SSRT in air. As shown in Figure 9, and Table 8, the lowest tensile strength and yield strength after SSRT is for specimen 3, which was welded at the highest heat input (i.e., heat input of 2.22kJ/mm). Reduction in strength with increased weld heat input could be due to grain growth at the weld line and HAZ. Also, for all welded specimens, the yield point phenomenon (Luders Bands) was observed after the SSRT. This was due to the presence of secondary phases such as acicular and polygonal ferrite or the existence of interstitial solid solution phases, as reported in other studies.

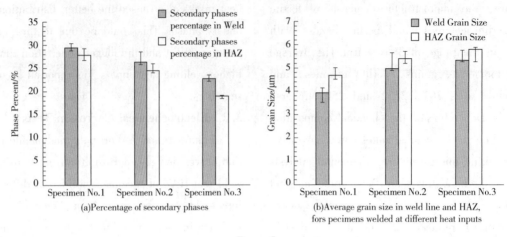

Figure 5

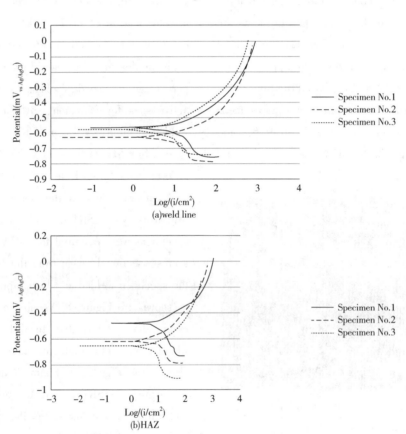

Figure 6 Potentiodynamic polarization curves for specimens welded at different heat inputs measured at a) weld line, and b) HAZ

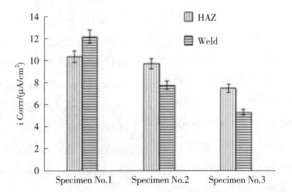

Figure 7 Corrosion current densities from weld and HAZ for specimens welded at different heat inputs

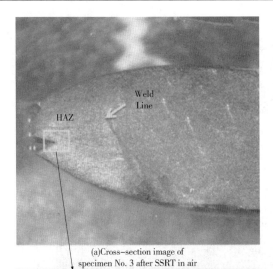

(a) Cross-section image of specimen No. 3 after SSRT in air

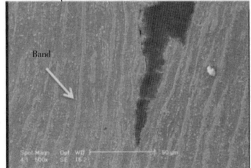

(b) The SEM image showing crack formed in perlite band

Figure 8

The typical fracture surface of specimens welded at different heat inputs after the SSRT in air is shown in Figure 10. As can be seen from this Figure, fracture surfaces contain dimples, which indicates ductile fracture. The highest strength was also obtained for specimen No. 1, which was welded with the lowest weld heat input (i.e., weld heat input of 1.912kJ/mm). Smaller dimples were observed on the fracture surface for this specimen. For the specimen welded with the highest heat input of 2.227kJ/mm, lower strength was obtained, while dimples were larger and stretched on the fracture surface after the SSRT. In addition, as shown in Figure 10(d), the fracture surface of specimens at lower magnification after the SSRT is oval in appearance, which is probably due to the non-homogeneity of the microstructure formed in the pipeline production process.

3.3.2 SSRT in C_2 solution at open circuit potential (OCP)

Figure 11 shows the stress vs. strain curve results for SSRT performed on welded specimens in simulated soil solution (C2 solution) at room temperature. By comparing Figures 9 and 11, it can be seen that the failure strain in a near-neutral pH solution is less than that in air. To determine the SCC susceptibility for specimen tests at OCP, the tensile strength drop factor (F_σ), fracture surface area drop factor (F_R), and elongation drop factor (F_E) were calculated using Equations 2-4.

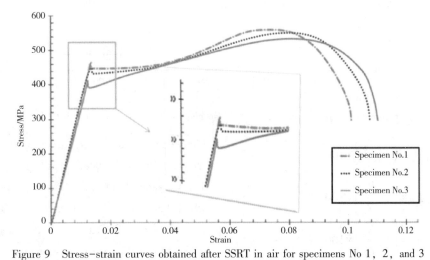

Figure 9　Stress-strain curves obtained after SSRT in air for specimens No 1, 2, and 3

Table 8　Mechanical properties of the specimens welded at different heat inputs after SSRT in air

Specimen Number	Fracture Time/h	Tensile Strength/MPa	Yield Strength/MPa
Specimen No. 1	28.33	558	480
Specimen No. 2	28.42	542	463
Specimen No. 3	31.11	530	422

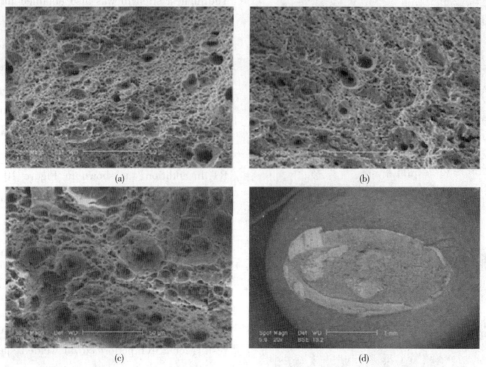

Figure 10 SEM images of the fracture surface of failed specimens after SSRT in air for
a) Specimen No. 1, b) Specimen No. 2, c) Specimen No. 3, and d) Macrostructure of fractured surface.

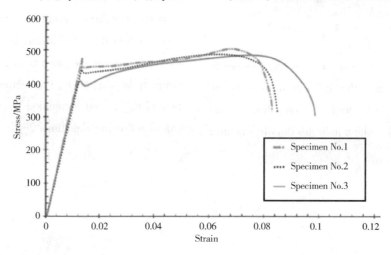

Figure 11 Stress-strain curves of the specimens obtained after SSRT at OCP

$$F_\sigma = \left[1 - \frac{S_E}{S_0}\right] \times 100 \quad (2)$$

$$F_R = \left[1 - \frac{R_E}{R_0}\right] \times 100 \quad (3)$$

$$F_E = \left[1 - \frac{E_E}{E_0}\right] \times 100 \quad (4)$$

In these Equations, R_E, E_E, S_0, R_0, and E_0 are tensile strength, fracture surface area reduction, elongation in solution, tensile strength, fracture surface area reduction, and elongation in the air, respectively. An increase in the mentioned factors indicates higher susceptibility to SCC. These factors were calculated and are shown in Figure 12. As seen from this Figure, specimen No. 3, which had the highest weld heat input (i. e., weld heat input of 2.227kJ/mm), has the lowest susceptibility to SCC, while specimen No. 1, with the lowest weld heat input (i. e., weld heat input of 1.912kJ/mm) exhibits the highest susceptibility to SCC. This could be due to the lower percentage of perlite and martensite phases within the weld line and the heat-affected zone. Considering anodic dissolution is the dominant mechanism of SCC initiation at OCP, these perlite and martensite phases can serve as suitable regions for crack initiation and propagation. These results are in very good agreement with the corrosion test results shown in Figures 6 and 7.

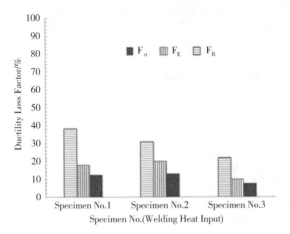

Figure 12　Ductility loss factor for specimens after SSRT test at OCP

Comparison of microscopic images from the fracture surfaces of specimens tested in air and near-neutral pH solution at OCP showed that the relative number of dimples and cavities caused by plastic deformation had significantly dropped for specimens tested in solution at OCP as that of in air (Figure 13). In other words, the fracture surface for specimens after SSRT in solution at OCP was more brittle than the fracture surface of specimens after SSRT in air. This is an indication of an increase in SCC susceptibility in solution. SCC micro-cracks were observed in the necking region of specimens tested in solution at OCP. The cracks had nucleated from the surface and propagated towards the fracture region. These cracks were not observed in specimens tested in air.

Figure 14 shows a typical fractured surface cross-section of the welded specimen after the SSRT in the solution at the OCP. As shown in Figure 14 (a), SCC cracks have nucleated and grown along the perlite bands. As mentioned earlier, these bands are regions more prone to nucleation and growth of cracks. Figure 14(b) shows a crack that has nucleated from a metallurgical discontinuity on a grain boundary and has grown into the ferrite grain. This is an indication of trans-granular SCC. Figure 14(c) shows the nucleation of a crack from the bottom of a corrosion pit, which has a high stress concentration.

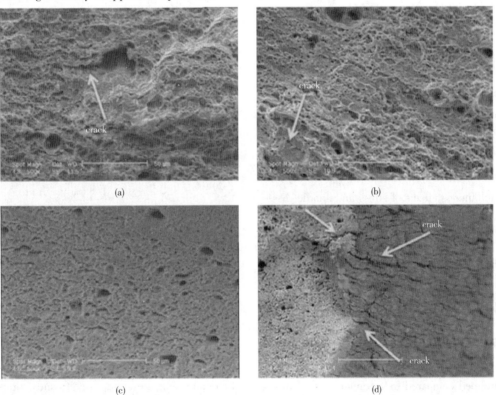

Figure 13　SEM images of the fracture surface of welded specimen after SSRT test in solution at OCP for Specimen No. 1(a & d), Specimen No. 2(b), and Specimen No. 3(c)

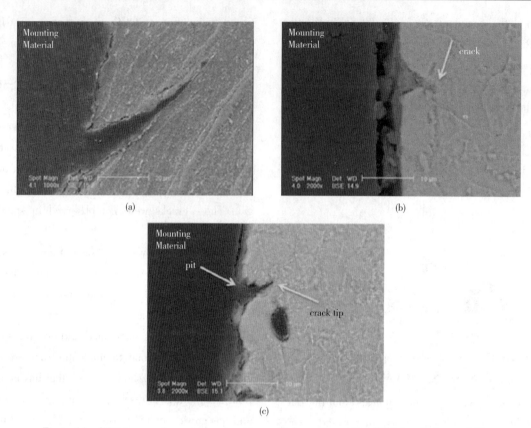

Figure 14 SEM images of a cross-section of a welded specimen after SSRT in solution at OCP.

3.3.3 SSRT at CP Potential of −850mV (vs Ag/Ag/Cl) and CP Potential of −1200mV (vs Ag/Ag/Cl)

For specimens protected at CP potentials of −850mV (vs Ag/Ag/Cl) and CP Potential of −1200mV(vs Ag/Ag/Cl) F_σ, F_R and F_E were calculated after the SSRT and are shown in Figure 15. F_E was negative after SSRT for specimen No. 3. Which had the highest weld heat input (i.e. weld heat input of 2.227kJ/mm). In this regard, Javidi et al. suggested that this phenomenon is probably due to the simultaneous effects of stress and anodic dissolution, which facilitates slippage of dislocations and increases the amount of elongation respective to specimens tested in air. However, the negative value in the F_E factor does not indicate lower susceptibility to SCC by itself, while the two other factors (i.e.: F_σ and F_R) should also be considered. Comparing Figure 15 and Figure 12, it can be concluded that the susceptibility of welded specimens protected by CP has dropped compared to specimens tested in solution at OCP. This could be due to a reduction in anodic dissolution rates, affecting the SCC susceptibility, for specimens protected by CP.

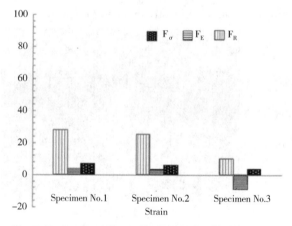

Figure 15 Ductility loss factor obtained from SSRT for specimens protected at CP potential of −850mV vs Ag/AgCl

By comparing the results of SSRT for specimens protected at a CP potential of −1200mV vs Ag/AgCl with results of SSRT for specimens protected at a CP potential of −850mV vs Ag/AgCl (Figures 16 and 17), it can be seen that with an increase in CP potentials from −850 to −1200mVvs Ag/AgCl, failure time and toughness has significantly dropped. This could be due to the destructive role of hydrogen in the increase in SCC susceptibility of over-protected pipeline steel by the Hydrogen Embitterment (HE) mechanism, accelerating SCC crack growth rates. Also, from Figures 16 and 17, it can be seen that specimen No. 3, which had the highest weld heat input (i.e., weld heat input of 2.227kJ/mm), has the lowest susceptibility to SCC. This could be due to the formation of larger grains at higher heat inputs acting as the dominant paths for hydrogen diffusion for pipeline steels it has been observed that an increase in grain boundaries up to 50 micrometers could increase hydrogen penetration rates. In addition to the increase in grain size, the reduction in the percentage of regions more susceptible to crack initiation and propagation, i.e., perlite and martensite phases within the weld line, and the HAZ (see section 3.3.1).

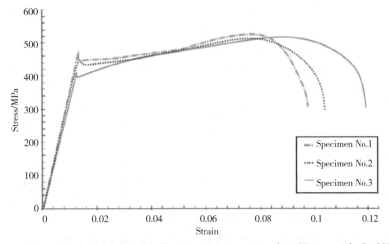

Figure 16 Stress-strain curves obtained from SSRT for specimens protected at CP potential of −850mV vs Ag/AgCl

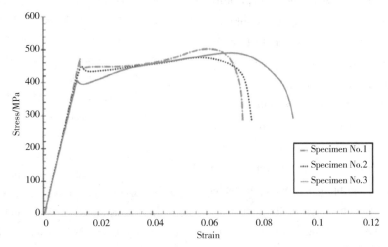

Figure 17 Ductility loss factor obtained from SSRT for specimens protected at CP potential of −1200mV vs Ag/AgCl

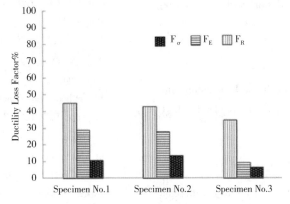

Figure 18 Stress-strain curves obtained from SSRT for specimens protected at CP potential of −1200mV vs Ag/AgCl

4 Conclusion

(1) For weld heat inputs investigated in this study (i.e., heat inputs of 1.912, 1.976, and 2.227kJ/mm), API X-70 pipeline specimens welded with the highest weld heat input (2.227kJ/mm) had the largest HAZ and grain size, while the number of secondary phases like perlite, banded structures, and precipitates decreased with increase in weld heat input.

(2) For the welded API X-70 pipe specimens in solution at OCP, with the increase in weld heat input from 1.912kJ/mm to 2.227kJ/mm, SCC susceptibility decreased. This was explained by reducing the percentage of regions more susceptible to crack initiation and propagation, such as perlite and banded structures. These results correlated nicely with the corrosion test results.

(3) For welded API X-70 pipeline specimens protected by CP potentials of -850mV vs Ag/AgCl and CP of -1200mV vs Ag/AgCl, the susceptibility to SCC dropped as compared to samples in solution at OCP. This was explained by the reduction in anodic dissolution rates for specimens protected by CP compared to samples in solution at OCP.

(4) For welded API X-70 pipe specimens protected by CP, with the increase in weld heat input from 1.912kJ/mm to 2.227kJ/mm, SCC susceptibility decreased. The formation of larger grain size at the higher heat inputs, acting as the dominant paths for hydrogen diffusion is considered as the main reason behind the decrease in SCC susceptibility in addition to the reduction in the percentage of regions more susceptible to crack initiation and propagation within the weld and HAZ.

Acknowledgment

The authors would like to thank Khuzestan Gas Company and the Isfahan University of Technology for their support.

References

[1] S. Vervynckt, "Control of the Non-Recrystallization Temperature in High Strength Low Alloy (HSLA) Steels.," 2010. [Online]. Available: https://api.semanticscholar.org/CorpusID:139028892.

[2] B. Y. Fang, A. Atrens, J. Q. Wang, E. H. Han, Z. Y. Zhu, and W. Ke, "Review of stress corrosion cracking of pipeline steels in 'low' and 'high' pH solutions," *J Mater Sci*, vol. 38, no. 1, pp. 127–132, 2003.

[3] W. Chen and R. L. Sutherby, "Crack Growth Behavior of Pipeline Steel in Near-Neutral pH Soil Environments," *Metallurgical and Materials Transactions A*, vol. 38, no. 6, pp. 1260–1268, Jun. 2007.

[4] A. Contreras, S. L. Hernández, R. Orozco-Cruz, and R. Galvan-Martínez, "Mechanical and environmental effects on stress corrosion cracking of low carbon pipeline steel in a soil solution," *Mater Des*, vol. 35, pp. 281–289, Mar. 2012.

[5] G. F. Li, G. L. Zhang, J. J. Zhou, C. B. Huang, and W. Yang, "Characteristics and Mechanism of High pH Stress Corrosion Cracking of Pipeline Steel X70," *Key Eng Mater*, vol. 353-358, pp. 219–222, Sep. 2007.

[6] Y. Frank Cheng, *Stress Corrosion Cracking of Pipelines*, First. John Wiley & Sons, 2013.

[7] I. V. Ryakhovskikh and R. I. Bogdanov, "Model of stress corrosion cracking and practical guidelines for pipelines operation," *Eng Fail Anal*, vol. 121, p. 105134, Mar. 2021.

[8] Z. LIU, X. LI, Y. ZHANG, C. DU, and G. ZHAI, "Relationship between electrochemical characteristics and SCC of X70 pipeline steel in an acidic soil simulated solution," *Acta Metallurgica Sinica (English Letters)*, vol. 22, no. 1, pp. 58–64, Feb. 2009.

[9] S. Longfei, L. Zhiyong, L. Xiaogang, G. Xingpeng, Z. Yinxiao, and W. Wu, "Influence of microstructure on stress corrosion cracking of X100 pipeline steel in carbonate/bicarbonate solution," *Journal of Materials Research and Technology*, vol. 17, pp. 150–165, Mar. 2022.

[10] L. Zhiyong, C. Zhongyu, L. Xiaogang, D. Cuiwei, and X. Yunying, "Mechanistic aspect of stress corrosion cracking of X80 pipeline steel under non-stable cathodic polarization," *Electrochem commun*, vol. 48, pp. 127–129, Nov. 2014.

[11] D. Kong, Y. Wu, and D. Long, "Stress Corrosion of X80 Pipeline Steel Welded Joints by Slow Strain Test in NACE H$_2$S Solutions," *Journal of Iron and Steel Research International*, vol. 20, no. 1, pp. 40–46, Jan. 2013.

[12] B. Lu, J.-L. Luo, and D. G. Ivey, "Near-Neutral pH Stress Corrosion Cracking Susceptibility of Plastically Prestrained X70 Steel Weldment," *Metallurgical and Materials Transactions A*, vol. 41, no. 10, pp. 2538-2547, Oct. 2010.

[13] G. A. Zhang and Y. F. Cheng, "Micro-electrochemical characterization of corrosion of welded X70 pipeline steel in near-neutral pH solution," *Corros Sci*, vol. 51, no. 8, pp. 1714-1724, Aug. 2009.

[14] C. Zhang and Y. F. Cheng, "Corrosion of Welded X100 Pipeline Steel in a Near-Neutral pH Solution," *J Mater Eng Perform*, vol. 19, no. 6, pp. 834-840, Aug. 2010.

[15] A. Eslami, R. Kania, B. Worthingham, G. V. Boven, R. Eadie, and W. Chen, "Corrosion of X-65 Pipeline Steel Under a Simulated Cathodic Protection Shielding Coating Disbondment," *Corrosion*, vol. 69, no. 11, pp. 1103-1110, Nov. 2013.

[16] A. Eslami et al., "Stress corrosion cracking initiation under the disbonded coating of pipeline steel in near-neutral pH environment," *Corros Sci*, vol. 52, no. 11, pp. 3750-3756, Nov. 2010.

[17] S. Kou, *Welding Metallurgy*. Wiley, 2002.

[18] R. N. Parkins and J. A. Beavers, "Some Effects of Strain Rate on the Transgranular Stress Corrosion Cracking of Ferritic Steels in Dilute Near-Neutral-pH Solutions," *CORROSION*, vol. 59, no. 3, pp. 258-273, Mar. 2003.

[19] NACE TM0497: 2022, "Standard Test Method Measurement Techniques Related to Criteria for Cathodic Protection on Underground or Submerged Metallic Piping Systems," Mar. 2022. Accessed: Nov. 09, 2024. [Online]. Available: https://www.intertekinform.com/en-gb/standards/nace-tm0497-2022-734257_saig_nace_nace_3137306/.

[20] K. Herlambang, Irwan Setyo Wibowo, and Maskuri Junaedi, "Cathodic Protection Online Monitoring Using a Low Power Wide Area Network Communication System," *Natural Sciences Engineering and Technology Journal*, vol. 3, no. 2, pp. 224-231, Jun. 2023.

[21] R. Chu, W. Chen, S.-H. Wang, F. King, T. R. Jack, and R. R. Fessler, "Microstructure Dependence of Stress Corrosion Cracking Initiation in X-65 Pipeline Steel Exposed to a Near-Neutral pH Soil Environment," *Corrosion*, vol. 60, no. 3, pp. 275-283, Mar. 2004.

[22] A. Mustapha, E. A. Charles, and D. Hardie, "The Effect of Microstructure on Stress-Strain Behaviour and Susceptibility to Cracking of Pipeline Steels," *Journal of Metallurgy*, vol. 2012, pp. 1-7, Feb. 2012.

[23] M. Alizadeh and S. Bordbar, "The influence of microstructure on the protective properties of the corrosion product layer generated on the welded API X70 steel in chloride solution," *Corros Sci*, vol. 70, pp. 170-179, May 2013.

[24] S. Kumar and A. S. Shahi, "Effect of heat input on the microstructure and mechanical properties of gas tungsten arc welded AISI 304 stainless steel joints," *Mater Des*, vol. 32, no. 6, pp. 3617-3623, Jun. 2011.

[25] R. Ashari, A. Eslami, M. Shamanian, and S. Asghari, "Effect of weld heat input on corrosion of dissimilar welded pipeline steels under simulated coating disbondment protected by cathodic protection," *Journal of Materials Research and Technology*, vol. 9, no. 2, pp. 2136-2145, Mar. 2020.

[26] S. Y. Shin, B. Hwang, S. Lee, N. J. Kim, and S. S. Ahn, "Correlation of microstructure and charpy impact properties in API X70 and X80 line-pipe steels," *Materials Science and Engineering: A*, vol. 458, no. 1-2, pp. 281-289, Jun. 2007.

[27] T. T. Nguyen, N. Tak, J. Park, S. H. Nahm, and U. B. Beak, "Hydrogen embrittlement susceptibility of X70 pipeline steel weld under a low partial hydrogen environment," *Int J Hydrogen Energy*, vol. 45, no. 43, pp. 23739-23753, Sep. 2020.

[28] J.-J. Xu, S. Wang, Z. Chai, C. Yu, J.-M. Chen, and H. Lu, "Comparison of the Stress Corrosion Cracking Behaviour of AISI 304 Pipes Welded by TIG and LBW," *Acta Metallurgica Sinica (English Letters)*, vol. 34, no. 4, pp. 579-589, Apr. 2021.

[29] H. Alipooramirabad, A. Paradowska, S. Nafisi, M. Reid, and R. Ghomashchi, "Post-Weld Heat Treatment of API 5L X70 High Strength Low Alloy Steel Welds," *Materials*, vol. 13, no. 24, p. 5801, Dec. 2020.

[30] R. de Carvalho Paes Loureiro, M. Beres, M. Masoumi, and H. Ferreira Gomes de Abreu, "The effect of pearlite morphology and crystallographic texture on environmentally assisted cracking failure," *Eng Fail Anal*, vol. 126, p. 105450, Aug. 2021.

[31] B. T. Lu, Z. K. Chen, J. L. Luo, B. M. Patchett, and Z. H. Xu, "Pitting and stress corrosion cracking behavior in welded austenitic stainless steel," *Electrochim Acta*, vol. 50, no. 6, pp. 1391-1403, Jan. 2005.

基于神经网络的典型天然气管网短期单耗预测

李 雨　王 瑞　刘屹然　石 峰　张文喆　张天娇

(国家管网集团油气调控中心)

摘 要　2024年，国家管网集团运营管理的长输油气管道里程已经超过10万公里，并将于2035年增至30万公里。随着国家管网集团所辖天然气管网规模日益扩大全管网单耗也逐步提升。对天然气长输管道单耗进行预测，能够有效控制管网单耗水平，为制定管网单耗目标和制定管网运行优化方案提供依据。本文首先基于某管道三年运行原始数据分析单耗影响因素，再通过相关性分析分析定量确定不同影响因素的重要程度，结合以往运行经验定性评价计算结果，提取出与运行单耗相关性最强的几组特征数据作为模型输入数据。使用BP神经网络预测模型对天然气管道进行日单耗预测。通过遗传算法、粒子群算法、改进后的粒子群算法等优化方法确定BP神经网络隐含层神经元数和初始学习速率最优值，确保模型预测效果。最后对比不同模型预测效果，确定最优预测模型。经过测算，使用改进粒子群算法优化参数的BP神经网络模型预测精度最高，模型的R^2、MAPE、MAE和RMSE分别为0.98、2.60%、3.22kgce/10^7Nm^3·km、10.56kgce/10^7Nm^3·km，预测效果最好。

关键词　混合神经网络模型；单耗预测；天然气管道；改进粒子群算法

Short-term unit energy consumption prediction of typical natural gas pipeline network based on neural network

Li Yu　Wang Rui　Liu Yiran　Shi Feng　Zhang Wenzhe　Zhang Tianjiao

(PipeChina Oil & Gas Control Center)

Abstract　In 2024, the length of long-distance oil and gas pipelines operated and managed by the Pipe China has exceeded 100000 kilometers, and will increase to 300000 kilometers in 2035. With the increasing scale of natural gas pipeline network controled by the PipeChina, the unit energy consumption of the whole pipeline network has gradually increased. Predicting the unit consumption of long-distance natural gas pipeline can effectively control the unit consumption level of pipeline network, and provide a basis for formulating the target of unit consumption of pipeline network and the optimization plan of pipeline network operation. In this paper, the factors affecting unit consumption are analyzed based on the original data of a pipeline in three years, and then the importance of different factors is quantitatively determined through correlation analysis. Combined with the qualitative evaluation results by previous operation experience, several sets of characteristic data with the strongest correlation with unit consumption are extracted as the model input data. BP neural network prediction models are used to forecast the daily unit energy consumption of natural gas pipeline. The number of hidden layer neurons and the optimal initial learning rate of BP neural network were determined by genetic algorithm, particle swarm algorithm and improved particle swarm algorithm to ensure the prediction effect of the model. Finally, the prediction effect of different models is compared, and the optimal prediction model is determined. After calculation, the BP neural network model using the improved particle swarm optimization algorithm to optimize the parameters has the highest prediction accuracy, and the R^2, MAPE, MAE and RMSE of the model are 0.98, 2.60%, 3.22kgce/10^7Nm^3·km and 10.56kgce/10^7Nm^3·km, respectively, showing the best prediction effect.

Key words　hybrid neural network model; Unit energy consumption forecast; Natural gas pipeline; Improved particle swarm optimization

国家管网集团成立后，针对智能调控、管道运行优化、节能降本增效的研究越来越多，而管道单耗预测作为管道优化运行的一部分，对提升管网运行效率有着重要作用。管道单耗计算方式为管道能耗除以管道周转量。近年来，针对管网运行优化的研究热度不减，其中管道周转量、进出站压力等特征都被认为是影响管道单耗的重要因素。由于每条管线运行工况不同，不同因素影响管道单耗的程度亦不同。有的仅利用日输量便可比较准确地预测天然气管道运行能耗，这种情况通常是考虑输量的时序信息，进而开展预测。也有的需要同时考虑多个特征才能实现管道电耗的准确预测。这种情况通常是使用输量、压力、温度等信息，对运行能耗开展预测，但管道系统运行过程中统计的数据种类毕竟有限，常统计的信息也仅有流量、压力、温度等，不包括用于评价流体流动情况的雷诺数、摩阻等信息。因此通常需要在原始数据的基础上进一步扩增新特征，以提升预测效果。为确定一个能够最精确预测管道单耗耗的小特征集合，需要对全部待选特征进行相关性分析。Kraskov 提出基于 K 近邻的互信息估计方法，避免了皮尔逊系数计算非线性关系不敏感和离散数据应用于互信息计算不准确的问题，且本身具有较好的相关性分析效果，适合用于本研究中评价天然气管道相关性。最后本研究以天然气管道运行数据为基础，将 K 近邻互信息估计和 BPNN 结合，提出一种天然气管道电耗预测模型，并利用西气东输一线管段 3 年运行数据验证该模型预测效果。

1 管线数据预处理

数据输入模型前必须经过预处理，该过程包括数据归一化和划分数据集两个过程。本文通过 minmax 归一化将数据规整至[0，1]区间，这样能够避免数据级差别过大导致预测效果不好的问题。因为是线性处理，因此能够在一定程度上保持原始数据的分布类型，避免出现数据分别变化影响预测效果的问题。利用分层抽样的方法，按照 7∶3 的比例划分训练集和验证集。分层抽样能够让训练集和验证集的分布接近原数数据分布，避免数据分布的波动。利用训练集训练机器学习模型，形成拟合效果较好的天然气管网单耗预测模型，再利用验证集评价模型预测效果，避免过拟合情况发生。

2 影响因素分析

如果用于预测单耗的特征数据与单耗几乎毫无关系，那么必然不能准确预测单耗，特征数据与单耗相关性越强，则预测效果通常越好。影响天然气管道运行单耗的因素众多，如管道输量、周转量、出口压力、管道管存等。只有确定影响管道运行单耗的主要因素，才能更好地预测管道单耗。

在统计学中常用相关性分析来定量判断一个因素受另一个因素的影响程度。如果影响程度越高，则相关性越高。考虑到原数数据的离散性和各特征与单耗间的非线性关系，本文利用 K-EMI 定量评价不同特征与运行单耗的相关性程度。运用 K-EMI 确定与单耗相关性强的特征，再根据管道的实际情况验证效果，通常能够得到更准确的相关性分析结果，确定出影响单耗的主要因素。

2.1 K-EMI

K-EMI 在评价非线性相关性时效果较好，同时避免离散型数据计算相关性的问题。李雨曾详细介绍过 K-EMI 计算方式，并将其应用于原油管道能耗预测中。该方法主要通过计算不同数据点的欧氏距离得到各特征间的相关性结果。

2.2 分析结果

经过计算后，各数据相关性结果如图 1 所示。观察图 1 能够发现与管段运行电耗呈强正相关性的特征主要有日输量、周转量、雷诺数；与管道运行电耗呈较强负相关性的特征为出站压力、管存。

其中，日输量增加需要启更多机组或机组提升负荷将更多天然气输送至下游，因此运行单耗增加。如图 2 所示为 2023 年某月西气东输一线日输量和运行单耗示意图，能够看到日输量和单耗呈明显的正相关性，尤其 1~15 日，单耗随输量增加同步增加。当管道上载点与下载点距离较近时，大输量下可能存在不过压缩机组等情况，因此周转量相比输量更直接的影响管道单耗，因此周转量与单耗相关性更高。

天然气在管道中运行时一般处于混合摩擦区，雷诺数越高，运行时能量损失越高，所需能量越多，单耗越高。当输量接近时，管存高天然气在输送过程中通常损失能量更少,而在低管存时损失能量更多,因此管存与单耗呈负相关。由于管存在长输管道中对能耗的影响往往存在延时性,因此用月平均管存和月平均单耗的关系能够更清晰观察出管存对单耗的影响。如图3所示,2月西一线管存与单耗呈现较明显的负相关趋势。出站压力高意味着管存跟高,单耗相对较少。

如图4所示,轮南站进站压力在当月于6~7MPa间波动,轮南站压力较高时,西一线通常可维持较低单耗,尤其是因某些原因,轮南气源压力波动剧烈时,单耗也呈负相关同步波动,更能验证压力与单耗的负相关性。综上所述,通过定性分析发现,相关性分析的结果比较符合运行的实际情况。由于保密原因,部分图标纵坐标数值不能展示。

	能耗	输量	周转量	入口压力	出口压力	管道管存	沿线地温	雷诺数	摩阻系数	沿程摩阻
能耗	1	0.87	0.93	-0.52	-0.73	-0.77	0.22	0.62	-0.45	0.42
输量	0.87	1	0.93	0.71	0.81	0.64	0.14	0.59	-0.71	0.58
周转量	0.93	0.93	1	0.68	0.82	0.51	0.31	0.67	-0.44	0.4
入口压力	-0.52	0.71	0.68	1	0.56	0.66	-0.31	-0.19	0.22	-0.34
出口压力	-0.73	0.81	0.82	0.56	1	0.54	-0.28	0.48	-0.52	0.39
管道管存	-0.77	0.64	0.51	0.66	0.54	1	-0.29	-0.39	0.35	-0.28
沿线地温	0.22	0.14	0.31	-0.31	-0.28	-0.29	1	0.17	0.11	0.16
雷诺数	0.62	0.59	0.67	-0.19	0.48	-0.39	0.17	1	-0.4	0.53
摩阻系数	-0.45	-0.71	-0.44	0.22	-0.52	0.35	0.11	-0.4	1	0.14
沿程摩阻	0.42	0.58	0.4	-0.34	0.39	-0.28	0.16	0.53	0.14	1

图1 相关性分析热力图

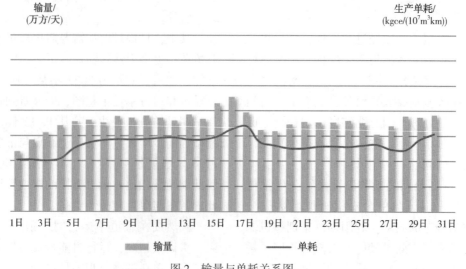

图2 输量与单耗关系图

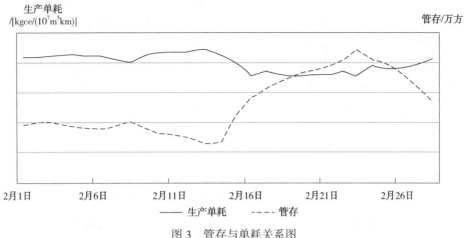

图3 管存与单耗关系图

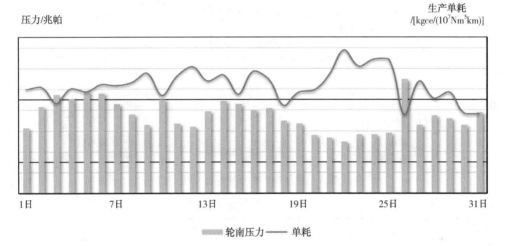

图4 出站压力与单耗关系图

3 模型介绍

建立本预测模型应用了 BPNN、搜索算法等，下面逐一介绍这些方法。

3.1 BPNN 算法介绍

管道运行电耗与日输量、进出站压力等众多特征相关，这造成了电耗特征与其他特征间强烈的非线性关系。BPNN 结构如图 5 所示。其中，X 为导入输入层神经元的数据值，即各项影响管道单耗的数据，W_{mn} 为连接输入层和隐含层神经元的权重数值，f 为隐含层激活函数，H 则代表各个隐含层神经元，W_{pk} 为连接 BPNN 隐含层和输出层的权重数值，最终输出层导出管道单耗的预测值 Y。

3.2 改进粒子群算法

改进粒子群算法（ISPO）通过计算粒子群中每个个体在不同位置下得到的适应度函数值来指导下一代个体的运动，最终全部粒子聚集在适应

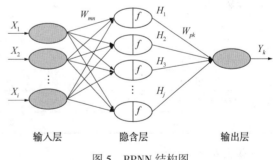

图5 BPNN 结构图

度函数值最高的位置，从而达到寻优目的。李雨在其论文中已详细介绍其计算方法。

3.3 其他搜索算法

本研究也利用粒子群算法（PSO）、引力搜索算法（GSA）、遗传算法（GA）和模拟退火算法（SA）调节 BPNN 的参数，分别建立 PSO-BPNN、GSA-BPNN、GA-BPNN、SA-BPNN 四个预测模型，将其与 IPSO-BPNN 模型进行对比。

3.4 混合 BPNN 模型

混合 BPNN 模型是一个结合了相关特征提

取、BPNN 和搜索算法的天然气管网单耗预测模型。利用 K-EMI 对已有数据做特征分析，提取出相关性强的主要特征，同时实现降维作用；把处理后数据作为 BPNN 输入，天然气管道运行单耗值作为输出，建立单耗预测模型；初始化一个 BPNN 的隐含层神经元数和初始学习速率；设定模型测试集 R^2 作为适应度函数，以评价搜索算法的搜索效果；执行搜索算法得到最大 R^2 对应的 BPNN 的最优参数值；得到精度最高的 BPNN 预测模型；利用误差指标评价模型精度。具体流程如图 6 所示。

先对完整的数据集做相关性分析，确定出与单耗数据相关性最强的特征群，利用确定好的特征作为输入，单耗值作为输出建立 BPNN 模型。其余对比预测模型按照表 1 规则命名。

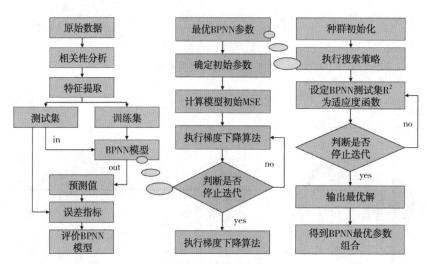

图 6 混合 BPNN 预测模型结构图

表 1 各模型命名表

模型简称	模型介绍
BPNN	特征提取后数据利用神经网络建模
GA-BPNN	特征提取后数据利用神经网络建模 GA 算法调整参数
SA-BPNN	特征提取后数据利用神经网络建模 SA 算法调整参数
GSA-BPNN	特征提取后数据利用神经网络建模 GSA 算法调整参数
PSO-BPNN	特征提取后数据利用神经网络建模 PSO 算法调整参数
IPSO-BPNN	特征提取后数据利用神经网络建模 IPSO 算法调整参数
PSO-BPNN-W	原始数据利用神经网络建模 IPSO 算法调整参数
IPSO-BPNN-W	原始数据利用神经网络建模 IPSO 算法调整参数

3.5 误差指标

采用均方根误差（RMSE）、决定系数（R^2）、平均绝对误差（MAE）和均绝对百分比误差（MAPE）以衡量模型预测效果，各项误差计算公式如下：

$$RMSE = \sqrt{\frac{1}{N}\sum_{i=1}^{N}(y_i - \hat{y_i})^2} \quad (1)$$

$$R^2 = 1 - \frac{\sum_{i=1}^{N}(y_i - \hat{y_i})^2}{\sum_{i=1}^{N}(y_i - \overline{y_i})^2} \quad (2)$$

$$MAE = \frac{1}{N}\sum_{i=1}^{N}|y_i - \hat{y_i}| \quad (3)$$

$$MAPE = \frac{1}{N}\sum_{i=1}^{N}|y_i - \hat{y_i}| \times 100 \quad (4)$$

式中，y_i，$\overline{y_i}$ 和 $\hat{y_i}$ 分别代表真实值、真实值的平均值和预测值。

4 结果分析

图 7 展示了 IPSO-BPNN 模型的收敛情况。模型在前 50 次迭代时，均方误差快速下降，并在后 350 次迭代时处于稳定区间内，说明 400 次

迭代能够让模型达到充分收敛。在测试集中选出15组数据，记录下真实值和IPSO-BPNN预测值，形成表2。能够发现，模型的预测效果较好，误差较低。

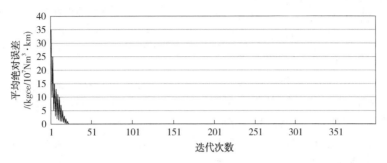

图7　IPSO-BPNN模型迭代收敛情况

表2　模型部分预测结果

真实值	预测值	真实值	预测值	真实值	预测值
131.4	128.616948	138.8	145.401328	147.9	149.583102
140.6	141.594042	139.8	145.317906	141	137.6865
143.8	143.41893	141.2	143.75572	147.1	149.933146
131.1	136.277139	144.6	149.368908	134.3	128.118171
136.1	137.956404	144.2	151.386928	146.7	151.912251

将上述模型的4种误差指标统计形成图6。该图横坐标为模型类型，左侧纵坐标为RMSE、MAE、MAPE和R2数据。观察图6能够发现利用K-EMI提取出重要特征，再利用IPSO算法确定模型参数的BPNN模型预测效果最好。模型的R2、MAPE、MAE和RMSE分别为0.98、2.60%、3.22kgce/107Nm³·km、10.56kgce/107Nm³·km。该模型比未利用搜索算法优化的BPNN模型4项误差指标分别提升15.29%、7.04%、6.82%、4.75%。对比IPSO-BPNN模型与IPSO-BPNN-W模型以及PSO-BPNN模型与PSO-BPNN-W能够发现，开展特征提取确实能够提升模型预测精度，IPSO-BPNN模型的4项误差指标相对IPSO-BPNN-W分别提升2.04%、30.77%、41.61%和2.65%。IPSO-BPNN模型的4项误差指标相对IPSO-BPNN-W分别提升1.05%、4.55%、13.49%和7.49%。

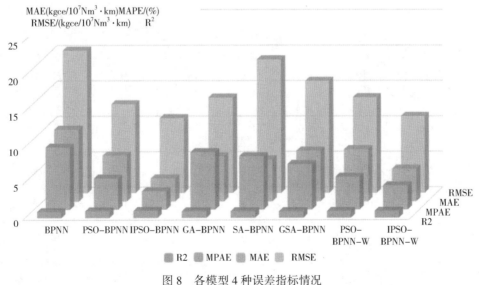

图8　各模型4种误差指标情况

5 总结

(1) 针对已有的管线运行数据将统计方法与经验充分结合,绘制基于 K-EMI 计算的热力图确定与单耗相关的主要特征,提取出相关特征作为模型输入数据,并验证了提取结果的合理性。

(2) 为提高搜索算法的全局寻优能力,对 PSO 进行改进,获得 IPSO 算法,针对西气东输一线管段建立单耗预测模型,通过对比 IPSO-BPNN 和其他模型的精度,验证改正的粒子群能够更准确的找出 BPNN 模型的最优参数,提升模型预测效果。

(3) 为进一步提升模型精度,提取重要特征,针对西气东输一线管段,对比开展特征提取和保留原始数据模型的精度,证明在训练模型前提取出关键特征,确实能够进一步提高模型预测精度。

参 考 文 献

[1] 张对红,杨毅.大型复杂天然气管网离线仿真软件国产化研发及应用[J].油气储运,2023,42(09):1064-1072+1080.

[2] 唐善华,杨金辉,徐春野,等.工业数据驱动技术在大型复杂天然气管网运行中的应用[J].天然气工业,2021,41(09):135-141.

[3] 杨毅,刁洪涛,向敏,等.基于动态规划和黄金分割法的环状天然气管网运行优化[J].天然气工业,2020,40(02):129-134.

[4] 孙晓波.天然气管道自动分输模式及应用[J].天然气技术与经济,2019,13(04):69-73.

[5] 邢同胜,孙晓波,胡善炜,等.港清线和港清复线的并管运行特性研究[J].石油机械,2014,42(07):109-111.

[6] Zeng C, Wu C, Zuo L, et al. Predicting Energy Consumption of Multiproduct Pipeline Using Artificial Neural Networks[J]. Energy, 2014, 66(1): 791-798.

[7] 侯磊,许新裕,崔金山,等.基于 BP 神经网络的输油管道能耗预测方法[J].节能技术,2009,27(5):401-406.

[8] Kraskov A, Stgbauer H, Grassberger P. Estimating Mutual Information [J]. Physical Review E, 2004, 69(6Pt2): 066138.

[9] 李雨,侯磊,徐磊,等.基于 K 近邻互信息估计的原油管道电耗预测[J].节能技术,2021,39(02):144-148+164.

[10] 李雨,侯磊,徐磊,等.基于混合 BP 神经网络的原油管道电耗预测研究[J].石油化工高等学校学报,2022,35(02):68-73.

天然气地下储气库注采方案优化研究

邹雪晴　张　轶　岳远志　杨　飞　耿金亮

（国家管网集团油气调控中心）

摘　要　【目的】随着经济社会的不断发展，天然气需求量逐年增加且具有季节不均匀性，冬季需求量明显高于夏季，平均峰谷差高达2左右。此外还存在区域调峰需求不平衡、基础设施不平衡、储气调峰能力不足等问题。国家管网集团成立后，天然气"全国一张网"已初步形成，与其连接的储气库已有近20座，合理制定储气库的注采方案，对管网平稳运行十分重要。【方法】在供需不平衡的前提下，传统的人工制定储气库注采方案的方法效率较低，难以最大化发挥储气库的调峰能力，使得储气库利用率较低，需建立合理的优化模型，得到合理的注采方案。本文以管网公司为视角，以管网公司收益最大为目标，建立了协同考虑天然气供应和需求情况、用户用气保障、管网的输送能力、地下储气库注采和储气限制等方面的多周期天然气管网运输优化模型，并采用分支定界法对模型进行求解。【结果】基于模型求解结果，在保障管网安全运行和下游用户用气的前提下，能够降低管网公司的输气成本，得到最合理的储气库注采方案。通过注采方案优化，在带来经济效益的同时储气库利用率也得到了显著提升。【结论】基于优化结果把握储气库注采规律，快速掌握储气库储气量变化情况，能够增强管网管控和应急保供能力，对天然气稳定供应具有重要意义。

关键词　天然气管网；地下储气库；运输优化

Research onoptimization of injection and production plan for underground gas storage

Zou Xueqing　Zhang Yi　Yue Yuanzhi　Yang Fei　Geng Jinliang

(PipeChina Oil & Gas Control Center)

Abstract　[Objective] With the development of the economy and society, the demand for natural gas is increasing and seasonal imbalance. The demand in winter is significantly higher than that in summer, and the average peak-valley difference is about 2. In addition, there are also problems such as an imbalance in regional demand, an imbalance in infrastructure, and insufficient capacity for gas storage. After the establishment of Pipe China, the "national network" of natural gas was initially formed, and there are nearly 20 underground gas storage connected with it. Developing a reasonable injection and production plan for gas storage is very important for the operation of the pipeline network. [Methods] Under the premise of the imbalance between supply and demand, the traditional manual method of making gas storage injection and production plan is inefficient, and it is difficult to maximize the peak load balancing capacity of gas storage, which makes the utilization of gas storage inefficient. It is necessary to establish a reasonable optimization model and obtain a reasonable injection and production plan. This paper takes Pipe China as the perspective, aims to minimize operating costs, and establishes a multi-cycle natural gas pipeline network transportation optimization model that considers the supply and demand of natural gas, gas guarantee for users, transportation capacity of pipeline network, gas storage for injection, storage, and production, etc. and uses the branch and boundary method to solve the model. [Results] Based on the model solution results, under the premise of ensuring the gas consumption of downstream users. On the premise of ensuring the gas consumption of users, the cost of gas transportation is reduced, and the most reasonable gas storage injection and production plan is obtained. Through the optimization of injection and production plans, the utilization rate of gas storage has been significantly improved while bringing economic benefits. [Conclusion] Based on the optimization results, the injection production and inventory of gas storage are obtained. It can enhance the ability of pipeline network control and emergency supply assurance, which is of great significance to the stable supply of natural gas.

Key words Natural gas pipeline network; Underground gas storage; Transportation optimization

随着我国经济的不断发展，天然气资源需求量剧增。据国家能源局发布的《中国天然气发展报告(2024)》以及国家发改委公布的数据，2023年全国天然气消费量为3945亿立方米，同比增长7.6%，占我国一次能源消费总量的8.5%。我国天然气资源主要分布在中西部地区和海域，消费区则主要在京津冀、长三角、珠三角等发达地区，导致供需不平衡，必须经过管网或者其他方式运输。天然气资源的不平衡是我国天然气市场遇到的最大问题，其中最大的不平衡是季节不平衡性。由于冬季气温较低，居民取暖需求大幅增加，如北方地区的集中供暖以及部分南方地区也开始使用天然气取暖，使得冬季天然气消费量显著上升，用气的冬夏峰谷差异大。图1为2019-2023年每月天然气表观消费量和增速对比，可知冬季的天然气需求量明显高于夏季，且在每年的12月左右达峰，其需求量峰值可达到夏季6-8月需求量谷值的2~3倍。除了季节不平衡外还存在区域调峰需求不平衡、基础设施不平衡、储气调峰能力不足问题。目前，我国主要调峰方法分别是地上/地下储气库调峰、液化天然气(LNG)、气田调峰和压减可中断用户用量。就储气库调峰而言，它具有建设成本较低、存储容量较大、对环境污染小以及安全可靠性强等诸多优点，不仅可以进行季节调峰和资源保障，也可以作为供应系统的应急储备装置，是保障国家天然气供应安全的关键战略内容之一。据统计，全球约有700个地下储气库，其工作容量占全球天然气消耗量的11.8%。LNG接收站因具有灵活性，在我国储气调峰中发挥着越来越重要的作用。LNG调峰主要通过LNG储气罐进行调峰，通常储存于特制的钢制筒体储气罐中，这些储气罐在用气低谷时储存多余的LNG，而当用气高峰来临、天然气供应紧张时，将储存的LNG气化后输送到管网中，补充天然气的供应量，以此平衡用气的峰谷差异，起到调峰作用，因储气罐罐容有限，其储气调峰能力低于储气库。气田调峰是指在具备一定条件时，可以通过调节气田天然气的开采量和外输量来实现调峰目的，对气田的天然气储备规模、底层能量要求较高，也会在一定程度上影响气田的正常生产，气田配套管道输送能力也有一定的限值。气田调峰存在诸多限制因素，具有一定的不可持续性，往往不能作为最主要的天然气调峰手段。减压可中断用户根据自身实际用气情况以及天然气供应系统整体的调配需求，中断供气措施，进而优化用气需求结构。目前我国的天然气调峰措施还是应以储气设施为主。

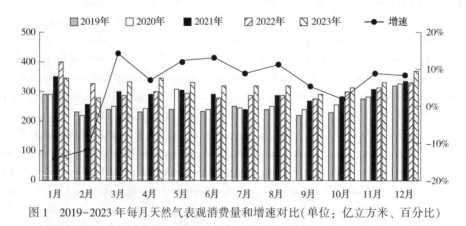

图1 2019-2023年每月天然气表观消费量和增速对比(单位：亿立方米、百分比)

当冬季用气高峰来临时，无法有效储存和调节天然气供应，导致供需矛盾加剧。地下储气库是调节天然气季节不平衡的关键设施，需加大对地下储气库的投资建设，提高储气库的工作气量。如我国在华北、东北等用气集中区域建设大型地下储气库，在夏季将多余的天然气注入储气库，冬季再采出供应市场。通过这种方式，可以有效缓解冬季供气压力。欧美地区的储气库工作气量占全年消费量的比例达20%，且拥有数量众多的储气库，分布广泛，能够较好地满足不同地区的用气需求。我国储气库建设在早期未得到足够重视，"十二五"前，储气库建设、运营主体只有中石油一家，建设速度缓慢，导致储气库数量相对较少、规模有限，难以满足日益增长的天然气消费需求，限制了利用率的提升，我国地下储气库工作气量及LNG接收站罐容占全国消费量的比例远低于国际平均水平，储气能力有限，难以满足季节调峰需求。十四五规划提出统筹推进地下储气库、LNG接收站等储气设施建设，打造华北、东北、西南、西北等数个百亿方级地下储气库群。国家管网集团成立后，以西气东输管道、华北管道系统、川气东送等管道系统为主干的天然气"全国一张网"已初步形成，截至2023年，国内已建地下储气库30座，形成调峰能力230亿立方米，占天然气消费量的5.8%。在大力发展储气设施基础建设的同时，也要最大化储气库的利用率。目前我国储气库的注采方案为人工制定，难以最大化发挥储气库的调峰能力，使得储气库利用率较低。为了最大化发挥储气库的调峰能力，需建立合理的优化模型，得到合理的注采方案。

总之，得到合理的天然气储存和储气库注采方案，调节天然气的峰谷差非常重要。本文以管网公司为视角，以管网公司收益最大为目标，建立了协同考虑天然气供应和需求情况、用户用气保障、管网的输送能力，地下储气库注采气和储气限制等方面的多周期天然气管网注采优化模型。基于模型求解结果，在保障管网安全运行和下游用户用气的前提下，能够降低管网公司的输气成本，得到最合理的储气库注采方案。通过注采方案优化，在带来经济效益的同时储气库利用率也得到了显著提升。

1 研究框架

本研究旨在得到在满足各种生产运输条件及用户需求前提下，得到管网公司效益最大的运输方案和储气库注采方案。对于拟建立的优化模型来说，我们将管网所连接的上游进口和生产、中游运输、下游销售等环节看作一个整体的运输过程，并将这个过程抽象为一个输送网络。主要包括以下几个部分：

① 上游气源：气源包括塔里木油田、西南油气田等国内气田气；俄气、缅气、霍尔果斯进口气等国外进口气和各类进口LNG。

② 天然气管网：天然气通过管道将气源输送到用户终端，多个输气管道互联互通组成输气管网，成为天然气运输的基础设施网络，目前我们已经基本建成"三横三纵"天然气主干管网。

③ 下游用户：天然气需求单位，在本模型中将其表示为一个分输站。

④ LNG接收站：液化天然气接收站，是指储存液化天然气然后往外输送天然气的装置。

⑤ 节点站场：一个能够将不同组件连接起来的连接点，本研究中各组件通过站场相连，形成一个真正的网络系统。天然气也是通过这个节点注入和分输。

⑥ 储气设施：可以储存天然气，并在必要的环节进行注采，我们将其抽象为一个单节点组件。

本技术基于相关供需、管网拓扑结构、管网物理参数等基础数据，以管网公司为视角，以管网公司收益最大为目标，结合天然气管网实际生产运行等约束建立优化模型，求解得到各供应/需求节点的最优流量值和储气设施注采方案，主要研究框架如图2所示。

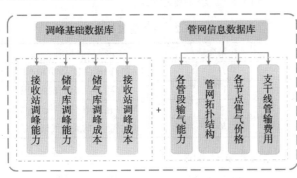

图2 储气库注采优化方法框架

地下储气库包含油藏型、气藏型、盐穴型、高含水型等类型，现阶段我国主要以油气藏型和盐穴型为主。目前，我国油气田位置分布不均，主要气源集中在中西部地区，但主要消费城市在东南部地区,为解决此供需分离问题,国家将"西气东输"战略与地下储气库同步规划,实现了天然气资源和储气设施之间的精准对接。在此过程中,地下储气库发挥了至关重要的作用,有力地保障了主干天然气管网能够安全且平稳地运行。我国主要地下储气库设施概况如下表1所示。

表1 我国主要地下储气库设施概况表

设施名称	所属公司	类型	所在地
储气库1	国家管网	盐穴	江苏省
储气库2	国家管网	油气藏	江苏省
储气库3	国家管网	油气藏	河南省
储气库4	国家管网	油气藏	河北省
储气库5	国家管网	油气藏	天津市
储气库6	合资	油气藏	新疆
储气库7	合资	油气藏	重庆市
储气库8	合资	油气藏	辽宁省
储气库9	中石化	盐穴	河南省
储气库10	中石化	油气藏	河南省
储气库11	中石化	盐穴	湖北省
储气库12	中石油	油气藏	陕西省
储气库13	中石油	油气藏	河北省
储气库14	中石油	油气藏	天津市
储气库15	中石油	油气藏	黑龙江省
储气库16	港华	盐穴	江苏省

与欧美等地区相比,我国天然气行业发展较慢,天然气市场化也较晚,我国储气库收费历史大致经历了三个阶段。在天然气产业发展初期,储气库主要由中石油、中石化等国有油气企业自行建设和运营,主要是为了满足企业自身天然气生产、运输和销售的调峰需求,未形成对外公开的、明确的收费模式和标准,费用通常在企业内部进行核算和分摊,通过收取管输费用收回储气库的建设运营成本。这种一体化的捆绑运营模式并不利于天然气市场的发展。随着我国天然气市场化的推进,油气管道设施逐渐实行第三方准入政策,储气业务与管输业务逐渐分离,也渐渐有了自己的价格机制。2016年,国家发改委发布《关于明确储气设施相关价格政策的通知》,规定储气库不再作为附属设施,可以根据市场竞争情况单独定价。2017年"气荒"后,国家开始推动了储气库建设和运营模式的改革。在天然气市场改革下,天然气储气设施实行"两部制"价格机制,即容量费和注采服务费,各运营商可以通过收取储气费用、库存预定费用等方式将部分库存通过租赁的方式租给其他企业。现如今国家管网集团文23储气库和刘庄储气库储气容量竞价交易,其储气服务费采用两部制计价,包括容量费和注采费。容量费按月滚动支付,每个结算周期结算当月费用,注/采费按实际气量收取。随着天然气市场化改革的继续推进,储气库运营模式将完全独立和市场化,在自由化的天然气市场中,各方参与者可以根据实际情况进行天然气的低买高卖,收益于差价变化。

由于我国完全实现储气设施的市场化运营还需一段时间,本文对地下储气库设施参照"两部制"方法进行建模,收取储气能力占用费用和储气设施运行费用。在本模型中,容量预定费收费方式为 0.6 元/($m^3 \cdot a$),储气设施运行费用为注采费用,按模型中的实际注采量收费,收费方式为 0.1 元/m^3。在模型建立中,我们将储气设施抽象为一个单节点组件,不考虑其内部物理组件和注采气压力约束。

2 储气库注采优化模型

考虑天然气的运输和储存等环节,以管网公司收益最大为目标,基于天然气管网基础物理数据和天然气供需数据,建立储气库注采优化模型,通过最优化方法进行优化,求解得到天然气运输方案和储气设施的注采计划。

在此模型中,下标 N 表示所有的节点,子集 N_P、N_S 和 N_N 分别表示一般气源供应节点、需求节点和中间节点,N_L、N_U 和 N_T 分别表示 LNG 接收站供应节点、储气库和 LNG 接收站储存节点。上标 S 表示需求节点,而 P 表示供应节点。T 为流量分配优化的周期的集合,时间变量 t 代表 1-12 月,二者满足 $t \in T$。模型决策变量主要为节点的注入/分输/注采流量、节点的多气/缺气量、管段流量、储气库的天然气储存量,具体决策变量解释见表2。

表2 决策变量符号与解释

决策变量	解释说明
$Q_{i,t}$	节点 i 在 t 时间的注入/分输/注采流量
$V^{TRA}_{i,j,t}$	节点 i 到 j 在 t 时间的管段流量
$S_{i,t}$	节点 i 在 t 时间的多气/缺气量
$IU_{i,t}$	储气库 i 的天然气储存量

2.1 目标函数

该模型的目标函数是管网公司收益最大化,除了基础的管输费用外,同时也需考虑储气设施的费用,主要包括储气库注气、采气和储气过程中产生的费用和 LNG 接收站的注采气费用。总利润可以表示为管输收益减去注采气费用和惩罚费用。具体数学表示如下：

$$\max Pr = C_1 - C_2 - C_3 - C_4 - C_5 \tag{1}$$

式中，Pr 为管网公司收益；C_1 为管输费用；C_2 为 LNG 接收站装卸的操作费用；C_3 为储气库的注采费用和储存费用；C_4 为 LNG 接收站的注采气费用；C_5 为因上载点多气和用户缺气造成的惩罚成本。具体计算方法如下：

$$C_1 = \sum_{i \in N} \sum_{j \in N} \sum_{t \in T} V_{i,j,t}^{TRA} c_{i,j}^{TRA} L_{i,j}^{TRA} \tag{2}$$

$$C_2 = \sum_{i \in N_L} \sum_{t \in T} c_i^{LU} (Q_{i,t}^P - S_{i,t}^P) \tag{3}$$

$$C_3 = \sum_{i \in N_U} \sum_{t \in T} Q_{i,t}^{GS} c_i^{IP} + \sum_{i \in N_U} \sum_{t \in T} IU_{i,t} c_i^{ST} \tag{4}$$

$$C_4 = \sum_{i \in N_T} \sum_{t \in T} Q_{i,t}^{LNG} c_i^{TA} \tag{5}$$

$$C_5 = \sum_{i \in N_S} \sum_{t \in T} S_{i,t}^S d_i^S + \sum_{i \in N_P} \sum_{t \in T} S_{i,t}^P d_i^P \tag{6}$$

式中，$Q_{i,t}^S$ 和 $Q_{i,t}^P$ 分别为在 t 周期内需求节点 i 的天然气计划需求量和供应节点 i 的天然气计划供应量；$S_{i,t}^S$ 和 $S_{i,t}^P$ 分别为在 t 周期内需求节点 i 的天然气缺气量和供应节点 i 的天然气生产多气量；c_i^{LU} 为接收站 i 的单位 LNG 汽化和装车费用的费用系数；$V_{i,j,t}^{TRA}$ 为 t 周期内从节点 i 到节点 j 的天然气运输量；$c_{i,j}^{TRA}$ 为天然气从节点 i 到节点 j 的单位运输成本；$L_{i,j}^{TRA}$ 为节点 i 到节点 j 的运输距离；$Q_{i,t}^{GS}$ 和 $IU_{i,t}$ 分别为在 t 周期内储气库 i 的注采流量和库存量；c_i^{IP} 和 c_i^{ST} 分别为储气库 i 的单位注采操作费用和单位储存费用；$Q_{i,t}^{LNG}$ 为 LNG 接收站 i 在 t 周期的操作流量；c_i^{TA} 为接收站 i 向储罐注采的操作费用，d_i^S 和 d_i^P 分别表示下游缺货和生产多气的惩罚成本。

2.2 约束条件

（1）节点流量守恒约束

天然气管网中存在很多连接管段的节点，节点的流量守恒约束是保证天然气正常输送的必要条件。我们根据节点类型提出了如下节点流量守恒约束条件：

$$\sum_{j \in N} V_{j,i,t}^{TRA} + S_{i,t}^S = Q_{i,t}^S + \sum_{j \in N} V_{i,j,t}^{TRA}, \quad \forall i \in N_S, \forall t \in T \tag{7}$$

$$\sum_{j \in N} V_{i,j,t}^{TRA} - \sum_{j \in N} V_{j,i,t}^{TRA} + S_{i,t}^P = Q_{i,t}^P, \quad \forall i \in N_P \cup N_L, \forall t \in T \tag{8}$$

$$\sum_{j \in N} V_{i,j,t}^{TRA} = \sum_{j \in N} V_{j,i,t}^{TRA}, \quad \forall i \in N_N, \forall t \in T \tag{9}$$

$$\sum_{j \in N} V_{j,i,t}^{TRA} = Q_{i,t}^{GS} + \sum_{j \in N} V_{i,j,t}^{TRA}, \quad \forall i \in N_U, \forall t \in T \tag{10}$$

$$\sum_{j \in N} V_{i,j,t}^{TRA} = \sum_{j \in N} V_{j,i,t}^{TRA} - Q_{i,t}^{LNG}, \quad \forall i \in N_T, \forall t \in T \tag{11}$$

$$IU_{i,t} = IN_i + Q_{i,t}^{GS}, \quad \forall i \in N_U, \forall t = 1 \tag{12}$$

$$IU_{i,t} = IU_{i,t-1} + Q_{i,t}^{GS}, \quad \forall i \in N_U, \forall t > 1 \tag{13}$$

需求节点的天然气流出量加上缺气量等于计划下载量加上天然气流入量，对于供应节点和中间节点也是类似逻辑。对于 LNG 接收站节点，其输出流量也等于气源点输入量减去储罐注入量。对于储气库，我们将其考虑为一个单节点组件，只与上游站场相连，输送至该节点的气体等于从该节点输送的气体加周期 t 间的注入或提取量，注入为正数，而提取为负数。此外还需要考虑到储气量和注采量。在第一个时间步长中，储气库的库存量等于初始气体加上操作量（注入或提取），当时间步长大于 1 时，库存量等于上一期的库存量加上操作量。其中 IN_i 为储气库 i 的初始库存。

（2）储气库库存、注采情况约束

$$Q_{i,t}^{GS} \leq UC_i, \quad \forall i \in N_U, \forall t \in T \tag{14}$$

$$\gamma_i US_i \leq IU_{i,t} \leq US_i, \quad \forall i \in N_U, \forall t \in T \tag{15}$$

储气库在一段周期内的注采量和库存量应在相应的上下界之间。由于各储气库所处地理条件、站内设备规模不同，各储气库的最大注采能力也不同，所以需建立约束将注采量控制在合理注采值内。其中 UC_i 表示每个时间步长中为储气库的最大注/采气能力；US_i 表示每个时间步长中储气库储存量的上限；由于储气库存在铺底气，γ_i 为各储气库的储存下限因子。

$$MinGS_i \leq \sum_{t \in T} Q_{i,t}^{GS}, \quad \forall i \in N_U, \forall t \in T \tag{16}$$

此外，由于各储气库属于不同的公司，为了保证天然气市场的平衡，促进储气库的建设发展，同时也设置了各储气库的最低注采量，$MinGS_i$ 为各储气库月最少注采量。

（3）LNG 接收站接卸情况约束

与储气库建模方式类似，LNG 接收站约束同样考虑其流量平衡约束和实际接卸操作约束。

$$IU_{i,t} = IN_i + Q_{i,t}^{LNG}, \quad \forall i \in N_L, \forall t=1 \quad (17)$$

$$IU_{i,t} = IU_{i,t-1} + Q_{i,t}^{LNG}, \quad \forall i \in N_L, \forall t>1 \quad (18)$$

$$\alpha UL_i \leq Q_{i,t}^{GS} \leq UL_i, \quad \forall i \in N_L, \forall t \in T \quad (19)$$

$$\beta UD_i \leq IU_{i,t} \leq UD_i, \quad \forall i \in N_L, \forall t \in T \quad (20)$$

式中，UL_i 表示每个时间步长中LNG储罐接卸量的上限；UD_i 表示每个时间步长中LNG储罐的储存量上限；α 和 β 均为下界尺度因子。

（4）管段其他边界约束

除了以上约束外，天然气的下载量、供应量、管道流量等也需受到边界条件的限制。具体约束如下：

$$q_{mini,t}^{S} \leq Q_{i,t}^{S} \leq q_{maxi,t}^{S}, \quad \forall i \in N_S \forall t \in T \quad (21)$$

$$q_{mini,t}^{P} \leq Q_{i,t}^{P} \leq q_{maxi,t}^{P}, \quad \forall i \in N_P, \forall t \in T \quad (22)$$

$$b_{i,j}^{TRA} v_{mini,j}^{TRA} \leq V_{i,j,t}^{TRA} \leq b_{i,j}^{TRA} v_{maxi,j}^{TRA}, \quad \forall i,j \in N, \forall t \in T \quad (23)$$

$$S_{i,t} \leq Q_{i,t}, \quad \forall i \in N, \forall t \in T \quad (24)$$

式(21)对需求节点进行了流量边界约束，其中 $q_{mini,t}^{S}$ 和 $q_{maxi,t}^{S}$ 分别表示时间步长 t 中需求节点 i 的最小和最大下载量。式(22)与其类似，表示供应节点的流量边界约束，其中 $q_{mini,t}^{P}$ 和 $q_{maxi,t}^{P}$ 分别表示时间步长 t 中供应节点 i 的最小和最大供应量。$b_{i,j}^{TRA}$ 是一个二元变量，如果从节点 i 到节点 j 存在一个实际管段，则等于1，否则为0。$v_{mini,j}^{TRA}$ 和 $v_{maxi,j}^{TRA}$ 分别表示从节点 i 到节点 j 的最小和最大输送能力。式(24)表示缺货/多气量要小于实际的供应/下载量。

3 算例分析

本算例以月为周期，年为时间尺度建立优化模型，共设置12个周期。共包含281个节点，其中有24个需求节点，含有9个LNG接收站；21个储气节点，包括12个储气库节点，9个LNG储罐节点；237个需求节点。由包括输气管道和注采气管道的309条管道将各个节点连接起来构成完整的拓扑结构。该模型为MILP模型，我们用Matlab语言进行编程，并调用Gurobi求解器进行求解。

通过对模型的求解，可以得到每个节点的实际供应/需求量、各管段的流量、储气设施的注采情况和库存情况。图3展示了模型优化的流量分配结果，可看出本算例涉及到了我国30个省市地区。其中，蓝色管段为输气管道、蓝色空心圆点为气源点、红色实心圆为需求节点、绿色星星为储气设施。算例的滞留/缺货值均为0，表示该分配计划能够满足天然气需求。气源点附着的彩色圆圈面积大小代表供气量的多少。

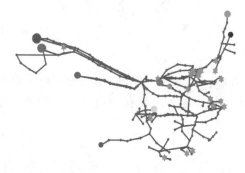

图3 模型优化的流量分配结果示意图

图4展示了各储气库在全年各周期的天然气注采量和库存量，对于注采量，注气为正值，储气为负值。由图可知，为了给冬季供应做准备，大部分储气库从4月份开始转采为注，储气库存量在十月份达到峰值。十一月份天气渐凉，进入冬季保供，由于天然气需求量的迅速增加，储气库开始进行采气，采气最高峰为十二月份，直至三月份采气结束，库存量也渐渐趋于最低值。

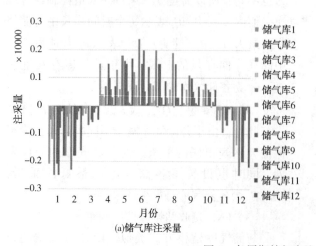

(a) 储气库注采量

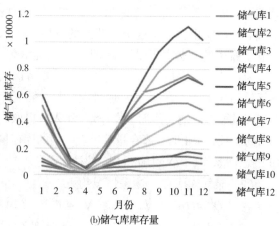

(b) 储气库库存量

图4 各周期储气库注采流量和库存量

据统计，我国储气库利用率普遍不高，为30%左右。为了评估目前我国储气设施的利用率，我们提出了两个指标并进行计算。首先是储气库利用率指标GSU_i，该指标计算方法见式(25)，表示为各循环周期内储气量变化值与各循环周期储气量理论最大变化值(各储气库的最大注采气能力)之比。其次，基于各储气库的储气利用指标GSU_i，提出了储气库平均利用率AGSU来衡量多个储气库的总体利用效率。

$$GSU_i = \frac{\sum_t |IU_{i,t} - IU_{i,t-1}|}{UC_i * T}, \forall i \in N_U, \forall t \in T \quad (25)$$

$$AGSU = \frac{\sum_i GSU_i}{N_U}, \forall i \in N_U \quad (26)$$

图 5 为计算出的各储气库的利用率，不同储气库的利用率差别较大，最大值为57%，最低为17%，其平均利用率为42%，相比优化前有显著提升。从地理位置上看，需求量高且气源稀缺的东部沿海地区或距离气源点丰富地区的储气库利用率较高。此外，从总体上看，储气库的平均利用率距欧美发达地区还有较大差距，还需进一步优化提升。

4 结论

该研究基于我国的天然气需求的季节不均匀性，考虑通过天然气储气设施来解决天然气的需求峰谷差较大的问题。以管网公司收益最大为目标建立了协同考虑天然气供应和需求情况、用户用气保障、管网的输送能力，地下储气库注采气和储气限制等方面的多周期天然气地下储气库注采优化模型，优化了天然气的流量配置和管段输量、在保障管网安全运行和下游用户用气的前提下得到了多个周期下最优的储气库注采量方案，在优化企业运营成本的同时提高了天然气资源及其储存设施的利用率。基于优化结果把握储气库注采规律，快速掌握储气库储气量变化情况，能够增强管网管控和应急保供能力，对天然气稳定供应具有重要意义。

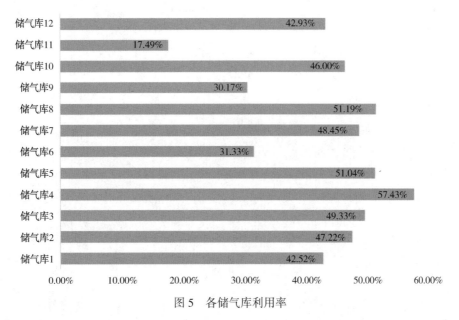

图 5　各储气库利用率

参 考 文 献

[1] 梁永图，邱睿，涂仁福，等．中国油气管网运行关键技术及展望[J]．石油科学通报，2024，9(2)：213-223.

[2] 池立勋，郝迎鹏，苏亮，等．中国区域级新型能源体系建设思考——以某沿海城市为例[J]．油气与新能源，2024，36(05)：1-9.

[3] 王浩，李爽，田佳丽，等．关于进一步完善LNG储气调峰成本疏导机制的探讨——以四川省雅安市为例[J]．天然气技术与经济，2024，18(03)：48-55.

[4] 丁国生, 李春, 王皆明, 等. 中国地下储气库现状及技术发展方向[J]. 天然气工业, 2015, 35(11): 6.

[5] 周淑慧, 孙慧, 梁严, 等. "双碳"目标下"十四五"天然气发展机遇与挑战[J]. 油气与新能源, 2021, 33(3): 27-36.

[6] 周淑慧, 梁严, 王占黎. 中国LNG接收站公平开放实践与展望[J]. 油气与新能源, 2022, 34(3): 1-10.

[7] 何春蕾, 段言志, 张颢, 等. 中国天然气价格改革理论研究进展及其应用回顾与展望[J]. 天然气工业, 2023, 43(12): 121-129.

[8] 李建君. 中国地下储气库发展现状及展望[J]. 油气储运, 2022, 41(7): 780-786

[9] 徐东, 唐国强. 中国储气库投资建设与运营管理的政策沿革及研究进展[J]. 油气储运, 2020, 39(05): 481-491.

[10] 邹才能, 林敏捷, 马锋, 等. 碳中和目标下中国天然气工业进展、挑战及对策[J]. 石油勘探与开发, 2024, 51(02): 418-435.

[11] 高芸, 王蓓, 胡迤丹, 等. 2023年中国天然气发展述评及2024年展望[J]. 天然气工业, 2024, 44(2): 166-177.

[12] 文韵豪, 王秋晨, 巴玺立, 等. 国内外智能化储气库现状及展望[J]. 油气与新能源, 2022, 34(6): 60-64.

[13] 完颜祺琪, 丁国生, 赵岩, 等. 盐穴型地下储气库建库评价关键技术及其应用[J]. 天然气工业, 2018, 38(5): 111-117.

[14] 阳小平. 中国地下储气库建设需求与关键技术发展方向[J]. 油气储运, 2023, 42(10): 1100-1106.

[15] 段言志, 何春蕾, 敬兴胜, 等. 天然气管输经济与政策: 以四川盆地为例[M]. 北京: 石油工业出版社, 2024: 50-53.

[16] 胡奥林, 何春蕾, 史宇峰, 等. 我国地下储气库价格机制研究[J]. 天然气工业, 2010(9): 6.

[17] 瞿静川, 张蓉, 田红英, 等. 我国天然气地下储气库运营与定价模式改进对策研究[J]. 价格理论与实践, 2022, (07): 191-194.

[18] 王胜男. 我国油气管网监管立法研究[D]. 北京: 华北电力大学(北京), 2022.

[19] 段言志, 郭焦锋, 邬宗婧, 等. 天然气管输体制改革成效与展望[J]. 油气储运, 2024, 43(10): 1089-1098.

[20] 曾伟, 肖长久, 梅琦, 等. 电力市场化改革对川渝地区天然气生产企业用电成本影响分析[J]. 天然气技术与经济, 2024, 18(03): 63-68.

基于电力制度的天然气管道电驱压缩机组生产运行成本优化应用实践

乔 欣[1]　刘屹然[1]　孟 霏[1]　高振宇[1]　刘旭东[2]　张华斌[2]

(1. 国家管网集团油气调控中心；2. 国家管网集团甘肃分公司)

摘　要　在天然气管网的运营体系中，电驱压缩机组占据着关键地位，是最为主要的耗电设备。以国家管网集团西部管道公司作为典型范例，其站场的耗电量分布情况显示，电驱压缩机组耗电量的占比极高，达到了站场总耗电量的约90%。深入调研电费构成可知，主要涵盖电度电费、基本电费以及力调电费这三个重要部分。其中，电度电费与基本电费的高低，在很大程度上受到月度最大需量以及峰谷电价的左右。

鉴于此，通过全面且细致地剖析国家电力制度的细则、天然气管道站场的基本运营状况，还有压缩机组的具体运行配置等多方面因素，将燃驱和电驱压缩机组的运行模式，与电费结算方式进行深度的有机融合。一方面，精准控制基本电费的每路外电的月度最大需量，使用减少机组占用外电线路、暖机或切机前降低机组负荷、调整机组暖机窗口期、倒闸前切为燃驱机组、短期启机优选燃驱机组、月末谨慎启停电驱机组等优化方案避免不必要的高额费用支出；另一方面，巧妙借助电度电费中峰谷电价存在的差异，在用电低谷时段加大机组运行负荷，高峰时段适度减少高耗电作业，优化用电策略。通过这样一系列科学合理的举措，每年能够为西部管道公司成功节省数千万元的电力成本，在降本增效方面展现出极为显著的成效，为企业的可持续发展注入强劲动力。

关键词　降本增效；电费成本；电驱机组；峰谷平分时电价；基本电费；最大需量；力调电费

Application of production and operation cost optimization of electric drive compressor unit for natural gas pipeline based on electric power system

Qiao Xin[1]　Liu Yiran[1]　Meng Fei[1]　Gao Zhenyu[1]　Liu Xudong[2]　Zhang Huabin[2]

(1. Oil and Gas Control Center of PipeChina; 2. Gansu Branch of PipeChina)

Abstract　An electric-driven compressor unit is like a "heart" in today's natural gas pipeline network operation system. It steadily occupies a key position and is the undisputed main force of power consumption. It may be worthwhile to explore in depth with the State Pipe Network Group Western Pipeline Company as a typical sample. From its station detailed power consumption distribution data, electric drive compressor unit power consumption accounted for a strikingly high proportion, nearly reached 90% of the total station power consumption. This data visually highlights its decisive influence in the energy consumption segment.

When focusing on the composition of the electricity bill, an in-depth research clearly found that it is mainly interwoven by three key parts, namely, the electricity bill, the basic electricity bill and the power adjustment electricity bill. Among them, the ups and downs of the tariffs and the basic tariffs are closely related to the two core elements of the monthly maximum demand and the peak-valley time-sharing tariffs. They are like a "baton" that largely controls the direction of electricity costs.

In view of this, a comprehensive and detailed analysis of the rules of the national electricity system, the basic operation of the gas pipeline station, and the specific operating configuration of the compressor units was carried out. The operating modes of fuel-driven and electric-driven compressor units are integrated with the electricity billing method in an in-depth manner.

On the one hand, we have made a lot of efforts in controlling the basic electricity cost. By reducing the number of external power lines occupied by the unit, the demand for electricity is "slimmed down" from the source. Before warming up or cutting down, reduce the unit load in an organized manner to avoid instantaneous power peaks. Fine-tuning the window period for unit warm-up to match the rhythm of power supply. Decisively switch to fuel-driven units before the reverse gate operation to avoid the impact of external power. Prioritize the use of fuel-driven units for short-term startups to reduce dependence on external power. At the end of the month, the startup and shutdown operations of electric drive units are handled with caution. Firmly control the monthly maximum demand of each external power, successfully avoiding the "minefield" of unnecessary high expenses.

On the other hand, seize the opportunity of peak and valley tariff differences in electricity tariffs. In the low valley of electricity consumption, make full use of the low price advantage, turn on full power to increase the operating load of the unit, so that every one degree of electricity are to play the maximum efficiency. During the peak hours, the company will take advantage of the situation and moderately cut down the high power-consuming operations, so as to skillfully balance the production and energy consumption.

With this series of interlocking, scientific and reasonable innovative initiatives, Western Pipeline Company can successfully save tens of millions of dollars from the electricity bill every year. The company has made great strides on the road of cost reduction and efficiency enhancement, and has achieved great results. For the sustainable development of the enterprise to inject a steady stream of strong power, more other enterprises in the industry to provide a valuable reference example.

Key words Cost Reduction and Efficiency; Electricity Cost; Electric Drive Unit; Peak-to-Valley Equalization Hourly Tariff; Basic Tariff; Maximum Demand; Force Modulation Tariff

国家管网集团原西部管道公司(以下简称"西部管道公司",现包含国家管网集团西部管道公司、甘肃公司)所辖天然气管道主要包括西气东输一线西段(轮南-中卫)、西气东输二线西段和三线西段(霍尔果斯-中卫,并行敷设)、轮吐线(轮南-吐鲁番)、涩宁兰管道等。在天然气管道沿线压气站中,压缩机组是站场核心设备,根据驱动方式,压缩机组可分为燃机驱动(简称燃驱)和电机驱动(简称电驱),目前西部管道公司共管辖158台压缩机组,其中电驱机组53台,燃驱机组105台。根据机组类型,西部管道公司管辖的34座压气站可分为燃驱站场、电驱站场和燃电混合站场三类,其中电驱站场和燃电混合站场占比35%。据统计,在西部天然气管道生产运行中,电驱机组每年耗电约20亿 kW·h,约占总耗电量的90%。可以看出,电驱机组是管道生产运行中最主要的耗电设备,降低电驱机组的用电成本对管道运营企业降本增效具有重大意义。

1 电费构成

调研发现,西部地区电费构成包括电度电费、基本电费、力调电费,而电费的结算主要受月度最大需量和峰谷平分时电价影响,具体如下。

电度电费是按电能量表计量的有功示数结算的电费,由有功电量和电价决定,电价存在峰谷平分时差异。以新疆精河地区为例,每日时段划分为:峰时段10:00-13:00,19:30-0:30;平时段8:30-10:00,13:00-19:30;谷时段0:30-8:30。峰电价0.5505元/kW·h,平电价0.3360元/kW·h,谷电价0.1215元/kW·h,各时段电价差异明显。

基本电费的缴纳方式分为三种:一按实际最大需量;二按合同需量;三按变压器容量,由用电单位与供电公司协商确定。目前,西部地区各压气站均按实际最大需量缴纳基本电费(基本电费=最大需量×需量价格)。最大需量是指在一定结算期内(一般为一个月),某一段时间(我国现执行15分钟）客户用电的平均功率，保留其最大一次指示值作为这一结算期的最大需量。新疆和甘肃地区110KV外电需量价格2022年分别为33元/千瓦·月和28.5元/千瓦·月，2023年分别为30.4元/千瓦·月和32.8元/千瓦·月。

力调电费是指供电公司根据客户一段时间内所使用的有功和无功电量计算其平均功率因数，并据此收取的相关电费。该功率因数标准要求通常为0.9，当功率因数低于标准时，电力公司会对用电单位进行罚款，反之会进行奖励。由于西部管道公司各站场配备了无功补偿装置，功率因数基本可以满足标准要求，极少产生力调电费。

2 运行现状及存在问题

2.1 运行现状

对于电驱站场和燃电混合站场，外电线路和压缩机组是生产运行关注的重点，与电驱机组相比，燃驱机组耗电量可忽略不计，因此电驱机组是影响电费支出的关键因素。

2.1.1 外电线路方面

西部管道现有的12座电驱站场和燃电混合站场，除永昌站配备四路外电，其余均配备两路外电，机组均匀分配至每路外电。为保障线路可靠性，每年春季各站场需对外电线路开展检修，检修前会对用电设备所连外电线路进行切换，俗称倒闸。外电检修的主要过程为：第一路外电检修前，全站负荷切至第二路；第一路外电检修完成后，站场负荷切至第一路，开展第二路外电检修；第二路外电检修完成后，两路外电恢复至作业前负荷。除春检倒闸外，站场外电出现故障、配合上游电网检修等情况时，也会进行倒闸操作。

2.1.2 压缩机组方面

国家管网集团天然气管道各站场的机组运行情况由国家管网集团油气调控中心（以下简称"油气调控中心"）统一调度指挥，天然气管道调度员根据运行方案以及管道实际工况调整机组配置。为减小外电波动对压缩机组运行的影响，站场启机时通常会均衡分配各路外电上的电驱机组数量。天然气管道站场根据公司规定及运行需求进行暖机、切机、测试等作业，对备用机组每月至少开展一次暖机作业，并对机组进行4000、8000小时运行保养等定期维检修作业。对于上述涉及机组的作业，大部分需要对该站的机组进行启停操作。此外，当机组运行出现异常时，站场需要对在运机组进行停机检查并切换备用机组运行，在检查后需要对该机组进行启机测试。

2.2 存在问题

2.2.1 基本电费方面

基本电费取决于各用电基本单元（站场层级）每条外电线路月度最大需量，电驱机组启停可能导致最大需量增加5~15MW，产生15~50万元基本电费。在通常的管道生产运行和调控指挥过程中，未考虑启机、暖机、切机、测试、倒闸等对站场外电线路月度最大需量的影响，造成不必要的基本电费支出。

2.2.2 电度电费方面

在我国西部地区，电度电费中的电价存在明显的峰谷平分时差异，即峰时段电价高，谷时段电价低，平时段电价居中。在常规的调控指挥中，如管道沿线生产运行平稳或工况变化较小，机组往往以恒定转速运行，未考虑不同时段电价差异，管道运营企业对机组的暖机、切机、测试等作业也未考虑避开电价较高的峰时段。

3 优化建议

为减少非必要的基本电费和电度电费，关键在于控制每路外电月度最大需量以及利用峰谷平分时电价差异，具体优化建议如下。

3.1 控制每路外电月度最大需量

3.1.1 减少机组占用外电线路

西部地区各站场至少配备两路外电，电力公司收取基本电费时，各路外电的最大需量单独计算，电驱机组占用的外电线路越多，机组暖机、切机、测试等作业越容易造成基本电费增加。个别站场电驱机组长期无需运行，但占用两路外电，每月仅开展例行暖机作业，导致两路外电均产生高额基本电费。如将电驱机组连接到同一路外电，可避免另一路外电因暖机产生基本电费的问题。因此，有必要减少电驱机组对外电线路的占用。

3.1.2 暖机或切机前降低机组负荷

西部地区各电驱站场通常一路外电同时与2~4台电驱机组相连，备用机组开展暖机、切机等作业前，若同一路外电有其他电驱机组正在运行，可先降低在运机组转速，再对备用机组进行暖机、切机等作业，维持总用电功率不变或控制功率增幅，进而控制该路外电月度最大需量。

为量化电驱机组运行功率和转速之间的关

系,笔者团队对某站机组进行了测试,结果显示二者呈正相关。根据测试结果,如图1所示,在特定工况下,机组转速每降低100rpm(转/分钟),运行功率降低0.733MW。根据现场经验和监测数据,额定功率为18MW的电驱机组暖机时,防喘阀全开且最低转速运行时功率约5MW。因此,如果要保持暖机、切机等作业前和作业时的同一条外电线路总功率不变,大约需要该外电线路在运的其他电驱机组合计降低转速700rpm。

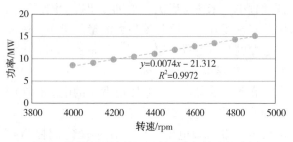

图1 某电驱压缩机组运行功率与转速拟合图

3.1.3 调整机组暖机窗口期

为保证机组可靠性,备用机组每月需进行暖机,暖机时长通常为0.5~1小时,超过了我国现执行的最大需量的结算期(15分钟),即如果不进行其他调整,电驱机组暖机必然导致最大需量增加。在无法调整同路外电所连电驱机组转速的情况下,可通过调整该机组暖机窗口期来控制最大需量。例如,电驱机组每月暖机调整为奇数月暖机两次(月初、月末)、偶数月不暖机,既能保证暖机效果,又能降低偶数月电费支出。

3.1.4 倒闸前切为燃驱机组

春季检修期间,站场需对两条外电交替停运依次开展检修作业,倒闸操作将导致每路外电月度最大需量大幅增加,全站最大需量增加一倍。针对该问题,可通过减少电驱机组运行控制最大需量增幅。对于燃电混合站场,建议先将在运电驱机组切换为燃驱机组,再开展倒闸操作。对于电驱站场,站内没有燃驱机组,可通过适当降低倒闸前后机组转速减小最大需量增幅。此外也可与供电公司协商调整外电检修作业窗口,等到管道输量降低,运行机组较少时再开展作业。

3.1.5 短期启机优选燃驱机组

天然气管网生产运行中难免出现工况调整,存在需临时增启机组的情况,如增启电驱机组可能导致当月最大需量增加10~15MW,产生基本电费30~50万元。遇到此工况,为避免基本电费增加,建议优先考虑增启该站燃驱机组,或不会增加最大需量的电驱机组。如增启该站机组无法避免最大需量增大的情况,可考虑增启相邻站场机组进行替换。

3.1.6 月末谨慎启停电驱机组

电费结算通常以月度为周期,对于某一站场,月末如需增启机组,应充分考虑最大需量问题。如增启电驱机组会大幅增加最大需量,应优先增启燃驱机组,或通过其他调整暂缓增启本站机组。反之,对于某一站场,月末如有电驱机组超配置低效运行,应及时停机,切勿等到次月1日后再停机,避免次月的最大需量过早出现。因此建议月末谨慎增启电驱机组并及时关停低效电驱机组。

3.2 利用峰谷平电价差异

目前,国内多地工业用电实行峰谷平分时计价,各地峰谷平时段略有不同,但峰谷平电价均具有明显差异。因此,在确保管道运行安全的基础上,具备条件的站场可采取"错峰用电"措施,减少电度电费支出,具体建议如下。

3.2.1 峰时段降速谷时段提速

压缩机组通常以恒定转速运行,调度人员调整机组转速大多出于生产运行需要,而非利用峰谷平分时电价差异。对于精河地区,峰时段和谷时段电价相差0.429元/kW·h,如果机组转速峰时段降低100rpm,谷时段提高100rpm,每天可节约电费支出2500元。因此,对于非满负荷运行的电驱机组,应充分利用不同时段电价差异,谷时段提高转速,峰时段降低转速,根据需要动态调整平时段转速。

3.2.2 高耗电作业避开峰时段

由于谷时段为夜间,站场极少开展作业,现场大多数作业在峰时段和平时段开展。以精河地区为例,峰时段和平时段电价相差0.2145元/kW·h,每次暖机耗电约3000kW·h,暖机选择平时段可比峰时段节约600元/次。因此,建议暖机、切机、测试等高耗电作业充分利用电价差异,优先选择平时段和谷时段,尽量避开峰时段。

4 优化措施应用效果

4.1 A压气站应用效果

西一线A压气站配有3台RR机组,驱动类型为1台燃驱和2台电驱,2台电驱机组的额定功率均为22MW,暖机最低功率约为6.5MW。站内1#电驱和2#燃驱机组与第一路外电连接,

3#电驱机组与第二路外电连接,见图2。A压气站每年只有半年时间需要启机运行,且以运行燃驱机组为主,该站生产运行中存在两项问题:(1)站场每月对电驱机组进行例行暖机作业,导致每月产生30余万元基本电费。(2)两台电驱机组分别与不同路外电相连接,导致每路外电均会因暖机产生基本电费。

针对上述问题,调控人员提出两项优化措施:(1)优化暖机时间安排。将每月暖机1次调整为奇数月月初和月末各暖机1次、偶数月不暖机。该措施不影响暖机效果,又能减少偶数月电费支出。(2)暖机前,通过倒闸将电驱机组连接到同一路外电,减少对外电线路的占用,确保另外一路外电月度基本电费不会增加。

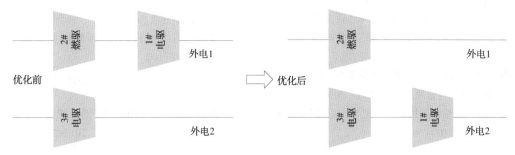

图2 A压气站优化前后压缩机组与外电线路关系图

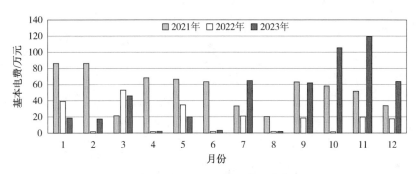

图3 A压气站2021-2023年基本电费支出对比图

根据电费账单统计分析,见图3,两项优化措施实施后,A压气站基本电费支出大幅降低,奇数月基本电费降低一半,偶数月基本电费接近0,2022年和2023年全年基本电费支出相比2021年分别减少440万元和128万元,生产运行成本大幅降低。

4.2 西部管道公司总体应用效果

2022年4月至今,上述各项优化措施在西部管道公司全面落地实施。据统计,在管输量同比增加,压缩机负荷增大的情况下,西部管道公司2022年和2023年相比2021年,基本电费支出减少2800万元和3900万元,电度电费节约数百万元。由于降本增效成果显著,各项优化措施已在国家管网集团北京管道公司、西气东输公司等陆续实施。

5 结论

(1)电驱压缩机组是天然气管网最主要的耗电设备,在管道生产运行中,通过将机组启停和电力制度深度融合可节省高额电费支出,在国家管网集团西部管道公司已经实现每年减少超过3000万元的电力成本支出。

(2)电费成本中的基本电费取决于每条外电线路月度最大需量,可以通过减少机组占用外电线路、暖机或切机前降低机组负荷、调整机组暖机窗口期、倒闸前切为燃驱机组、短期启机优选燃驱机组、月末谨慎启停电驱机组等措施减少基本电费支出。

(3)电费成本中的电度电费存在峰谷平分时电价差异,可以通过峰时段降速谷时段提速、高耗电作业避开峰时段等措施减少电度电费支出。

参 考 文 献

[1] 路毅铭. 西部管道有限责任公司发展战略研究[D]. 兰州大学, 2018.

[2] 徐铁军. 天然气管道压缩机组及其在国内的应用与发展[J]. 油气储运, 2011, 30(05): 321-326+313.

[3] 周大鹏. 天然气长输管道压气站压缩机组选型[J]. 油气田地面工程, 2015, 34(09): 88-89.

[4] 林森, 董昭旸, 康焯, 等. 西气东输二线电驱压缩机组驱动方式比选[J]. 油气储运, 2012, 31(06): 470-472+486-487.

[5] 孙晓宝. 西三线西段工程压缩机组选型及节能措施[J]. 石油石化节能, 2016, 6(10): 43-44+50.

[6] 王浩儒, 赵凯. 大工业用电站场基本电费计费方式优化[J]. 石油石化节能, 2020, 10(04): 36-38+11.

[7] 王其林, 穆立臣. 基本电费的两种计费方式对汽车生产企业电费的影响[J]. 汽车工业研究, 2017(12): 19-21.

[8] 郏海明. 大工业用电企业如何选择基本电费计费方式[J]. 企业改革与管理, 2017(24): 205.

[9] 刘广成. 最大需量控制措施的研究与应用[J]. 科技创新导报, 2012(07): 57.

[10] 陈媛. 电网企业电费管理研究[D]. 华北电力大学(北京), 2011.

[11] 孙国荣. 从电费构成看企业如何合理减少月电费支出[J]. 电力需求侧管理, 2005(05): 50-51.

[12] 周冠. 浅论技术管理措施在控制最大需量电费中的运用[J]. 企业技术开发, 2011, 30(14): 106.

[13] 倪保利, 杨军, 祁生, 等. 最大需量综合控制在高电耗企业中的实践运用[J]. 中国有色金属, 2018(S1): 25-30.

[14] 杨焜, 刘卉圻, 吕玲, 等. 电驱压气站的力调电费平衡分析与探讨[J]. 天然气与石油, 2019, 37(06): 97-102.

[15] 甘秀仕. 电费与功率因数的关系[J]. 电力需求侧管理, 2007(02): 79-80.

[16] 钟秋. 提高功率因数对于企业应缴电费的影响研究[J]. 中国新技术新产品, 2012(22): 180.

绿色发展导向下天然气与电力折标系数的融合应用研究与价值重塑

孟 霏[1]　乔 欣[1]　刘麦伦[1]　刘义茜[2]　孙 敏[1]　石咏衡[1]

（1. 国家管网集团油气调控中心；2. 国家管网集团甘肃分公司）

摘 要　本文以我国当前实施的《综合能耗计算通则》GB/T 2589—2020 为深入研究的核心对象，旨在细致剖析天然气和电力这两种主要能源的折标系数对企业能源消耗可能产生的深远影响。折标系数作为能耗计算的关键参数，其准确性和适用性直接关系到企业能耗统计的精确度与节能措施的制定。

文章首先通过对比当量值与等价值两种折标方法，深入探讨了在能耗核算过程中可能遇到的理论与实践问题。当量值侧重于能源的"热量转换"，而等价值则更侧重于能源的"品味价值"，两者在反映能耗特性上存在显著差异，进而影响到企业的能耗评估与成本控制策略。为进一步说明问题，文章结合了具体实例，通过对比分析不同折标系数选取下天然气和电力的能耗量变化以及其实际成本分析，直观展示了折标系数对企业能耗数据的重要影响。这一分析不仅揭示了当前能耗核算中可能存在的偏差，也为后续提出合理的修正建议提供了有力支撑。在此基础上，文章提出了针对性的修正建议，旨在优化折标系数的选取与应用，以更客观、准确地反映企业的实际能耗水平。这些建议目的是平衡能耗指标与成本指标之间的关系，确保能耗统计既能反映真实情况，又能为企业节能降耗提供科学依据，从而推动能耗统计与节能规划工作的进一步完善，助力企业实现更加高效、可持续的能源管理。

关键词　天然气折标系数；电力折标系数；当量值；等价值

Integration application research and value reshaping of natural gas and electricity discount factor under the orientation of green development

Meng Fei[1]　Qiao Xin[1]　Liu Mailun[1]　Liu Yixi[2]　Sun Min[1]　Shi Yongheng[1]

(1. Oil and Gas Control Center of Pipe China; 2. Gansu Branch of Pipe China)

Abstract　This paper takes China's current implementation of the "General Rules for Comprehensive Energy Consumption Calculation" GB/T 2589-2020 as the core object of in-depth research. It aims to comprehensively and meticulously analyze the far-reaching impact that the discount factor of two major energy sources, namely natural gas and electricity, may have on the energy consumption of enterprises. As an indispensable key parameter in the process of energy consumption calculation, the discount factor is directly related to its accuracy and applicability. Its accuracy and applicability are directly related to the accuracy of the enterprise's energy consumption statistics and the effective formulation of energy-saving measures. Therefore, the in-depth study of the discount factor not only has theoretical value, but also has great significance in practical application.

In the practice of energy consumption accounting, the choice of discount factor is not static, but needs to be flexibly adjusted according to the actual situation. The article firstly compares and analyzes the two discounting methods of equivalent value and equivalent value. Equivalent value discounting method mainly focuses on the "quantity" of energy. That is, according to the calorific value of various energy sources for uniform conversion, so as to derive a standardized energy consumption indicators. This method can intuitively reflect the energy characteristics of energy, but in practical application may ignore the "level" of energy. In contrast, the equivalent value discounting method focuses more on the value of the energy.

In order to more intuitively illustrate the impact of the discount factor on the energy consumption data of enterprises, the article combined with specific examples for comparative analysis. By selecting different discount factors, the changes in energy consumption of natural gas and electricity were calculated. And a detailed analysis was carried out in combination with the actual cost. The results show that the choice of discount factor has a significant impact on the calculation results of energy consumption data. It may even lead to the deviation of energy consumption statistics. This deviation may not only affect the cost control strategy of the enterprise, but also mislead the formulation and implementation of energy-saving measures.

On this basis, the article puts forward targeted correction suggestions. Obviously, the in-depth study and optimization of the discount factor is of profound and immeasurable significance to significantly improve the accuracy of enterprise energy consumption statistics and the actual effect of energy-saving measures. As a bridge connecting actual energy consumption and standard energy consumption, the accuracy and applicability of the discount factor are directly related to the reliability and comparability of energy consumption data. Through scientific and reasonable selection and application of the conversion factor, we can reveal the actual energy consumption of enterprises more objectively, comprehensively and accurately. Thus, we can avoid the problem of distortion of energy consumption data caused by improper statistical methods or irrational selection of coefficients.

The optimization of this process not only helps enterprises to accurately grasp their own energy consumption structure, timely detection of energy consumption anomalies and potential energy saving space. It can also provide solid data support and scientific basis for the enterprise's energy management and energy saving work. On this basis, enterprises can be more targeted to develop and implement energy-saving measures. Optimize the energy use process. Enhance the efficiency of energy utilization. Achieve the goal of energy saving and consumption reduction.

At the same time, the in-depth study and optimization of the application of the discount factor can also promote the further improvement of energy consumption statistics and energy conservation planning. With the deepening of the understanding of the discount factor and the application of continuous optimization, we can gradually establish a more scientific and systematic energy consumption statistics system. Provide strong support for the energy decision-making of enterprises. In addition, this process will also promote the continuous innovation of energy-saving technologies and the transformation and upgrading of energy-saving models, helping enterprises to realize more efficient and sustainable energy management, and laying a solid foundation for the sustainable development of enterprises.

Key words natural gas discount factor; electricity discount factor; equivalent value; equivalent value

1 引言

根据《综合能耗计算通则》GB/T 2589—2020的规定，折标系数为能源单位实物量或者生产单位耗能工质所消耗能源的实物量，折算为标准煤的数量。这一转换过程对于准确评估企业的能耗水平至关重要。

在工业生产实践中，单一能源消耗的企业可以不将能耗转换为标准煤进行评价，因为在纵向（不同年份）或横向（不同所属企业）比较时，采取一致的度量单位即可，即便进行转化，折标系数的取值对评价结果的影响也相对有限。然而，对于多数工业企业而言，能源消耗种类繁多，包括电力、煤炭、天然气、蒸汽、燃油等多种类型。在这种情况下，为了准确开展能耗统计和制定节能计划，就必须将各类能耗通过折标系数换算为统一的标准煤单位，从而建立一个具有可比性和规范性的计量基准。

在能耗折算过程中，折标系数的选择直接决定了能耗统计的精确度和合理性。不同的折标系数可能会导致能耗数据的显著差异，进而影响企业对能耗状况的全面了解和准确判断。因此，科学选择折标系数不仅是确保能耗数据准确性的关键，也是开展企业能耗审计与节能评估工作的重要保障，更是制定有效节能措施和提供可靠决策的依据。

为了科学选择折标系数，企业通常需要参考国家相关标准和行业规范，并考虑自身的生产特点和能源消耗情况。同时，企业还需要定期更新折标系数，以适应能源市场的变化和新技术的应用。

2 当量值与等价值折标分析

电力，作为一种高效且应用广泛的二次能源，在推动社会各个领域的运作中扮演着至关重要的角色，是所有耗能企业或生产流程中不可或缺的能量来源，其潜在的节能减排效应巨大且深远，因此备受社会各界关注。在国内企业的能源统计流程中，针对电力的折标系数处理，明确区分了当量值与等价值两种类型，以适应不同层级与目的的能源管理需求。为了促进国际交流与合作，并保持数据统计的连贯性，国家规定在计算国家、省级及市级层面的能源消费总量时，电力折标需采用等价系数进行核算，以确保宏观能源数据的准确性和可比性。而在基层企业层面进行能源消费量计算时，为了简化操作和保持一致性，电力折标则统一采用当量系数，具体数值设定为 $0.1229 \text{kgce}/(\text{kW} \cdot \text{h})$，即 $1 \times 10^4 \text{kW} \cdot \text{h}$ 电力折合 1.229t 标煤。

能源的当量值（energy calorific value）——按照物理学电热当量、热功当量、电功当量换算的各种能源所含实际能量，又称理论热值（或实际发热值）——是指某种能源本身所含热量，按热量的多少可以折算成的标准煤。即不同形式的能量相互转换时的相当量，按照能量的法定计量单位焦耳，热能、电能、机械能等不同形式的能量，其相互之间的换算系数均为1。

该换算方法只考虑了其"数量"，对其"品位"没有加以重视。按照热力学第二定律，像热能这样的低品位能源是不能充分转换成电能的。$1 \text{kW} \cdot \text{h}$ 的电能，其工作效率比 $1 \text{kW} \cdot \text{h}$ 的其他能量要高得多。在对能耗进行宏观核算的过程中，使用等量值折标系数容易造成"误导"效果。

电力是高品位的二次能源，近50%由火力发电而来，其转换效率很低，如图1所示。且电力系统由发电、输电和配电三个环节组成，每个环节都有电能损失。一般火力发电的能源利用效率只有30%~40%，输配电损耗约5%~10%。如果仅统计其所含能量的数量，会忽视发电过程能量的损失。也将不合理地模糊了节能的重点，且在一定程度上影响了节能工作的客观评价。

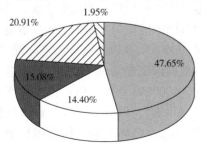

图1 2023年全国主要发电方式占比示意图

能量的等价值（energy equivalent value），是生产单位数量的二次能源或耗能工质所消耗的各种能源折算成一次能源的能量，也就是消耗一个度量单位的某种二次能源，等价于消耗了以热值表示的一次能源量。

以天然气为代表的一次能源，其折标系数，无论是当量还是等价，均旨在将单位量的天然气转换为标准煤的数量，便于统一度量与比较。相比之下，电力作为典型的二次能源，其折标系数分为了当量值与等价值两种。当量值折标系数直接反映了单位电力本身的能量大小，将其换算为标准煤量；而等价值折标系数则深入考量了生产该单位电力所消耗的全部一次能源量及输配电损失，更全面地体现了电力生产过程中的能耗水平。

《综合能耗计算通则》GB/T 2589—2020 中天然气和电力折标系数分别为：天然气 $1.1 \sim 1.33 \text{kgce}/\text{m}^3$，电力（当量值）$0.12293 \text{kgce}/(\text{kW} \cdot \text{h})$，电力（等价值）按上年电厂发电标准煤耗计算。

3 折标系数在国家管网的应用现状

随着中国天然气管道运输业务的蓬勃发展，国家管网集团公司正面临着一系列严峻的挑战，包括用能总量的持续增加、电价成本的不断攀升以及碳减排压力的日益加大。为了应对这些挑战，国家管网必须摒弃粗放的管理模式，转而采取更加精细化的管理手段，不断细化标准应用，深入挖掘能耗潜力，以持续节约能源、减少消耗并降低成本。这对于油气管网的生产运行而言，是一项至关重要的任务。

在天然气管网运行的过程中，压缩机作为关键的"心脏动脉"，其能耗水平直接关系到整个管网的能效表现。因此，对压缩机中的电驱机组与燃驱机组进行能耗分析，成为节能降耗工作的重中之重。在进行能耗分析时，折标系数的取值对于评估结果具有至关重要的影响。目前，国家管网在进行能耗分析时，采用的是天然气折标系数为 $1.33 \text{kgce}/\text{m}^3$，电力折标系数为当量值 $0.12293 \text{kgce}/(\text{kW} \cdot \text{h})$。

在天然气折标系数取值方面，旧标准《综合能耗计算通则》GB/T 2589—2008 规定，油田天然气和气田天然气的折标系数分别为 $1.33 \text{kgce}/\text{m}^3$ 以及 $1.2143 \text{kgce}/\text{m}^3$。而国家管网当前采用的 $1.33 \text{kgce}/\text{m}^3$ 的折标系数，实际上是旧版标准中油田天然气的参考值，这一取值相对偏高。在进行能耗评价时，如果采用这一偏高的折标系数，将会夸大天然气的折标煤量，从而影响评价结果的准确性。

在电力折标系数取值方面，当量值折标系数只能反映所消耗的二次能源电力本身所具备的能量，而无法体现生产电力所消耗的一次能源的数量。此外，它还不考虑输配电过程中的损失。因此，仅仅根据天然气和电力自身具备的能量来直接折算为标准煤量，显然是不合理的。这种做法从根本上忽略了一次能源和二次能源之间的客观差异，无法准确反映能耗的真实情况。

这就导致了这两种折标系数的取值存在显著的差异，相差十余倍。表1和表2分别为国家管网集团西气东输管线Z站的三台机组基础数据和某日该三台机组能耗情况，可以看到单台燃驱机组的能耗是单台电驱机组的4倍，但燃驱机组的运行成本却略低于电驱机组。由此可见，能耗与成本存在矛盾，在燃驱机组和电驱机组能耗对比的问题上，简单地依据上述折标系数来得出结论可能并不准确，生产运行也无法兼顾能耗指标与成本指标，需根据实际情况做出取舍。

表1　Z站三台机组数据

压缩机组编号	驱动方式	进口/国产	压缩机型号	连续运行工作转速范围/rpm	驱动机额定功率/MW
二线3#	燃驱	进口	PCL802	3965~6100	31.4
三线1#	电驱	进口	D16R3S	3380~5200	18.0
三线4#	电驱	国产	PCL804	3120~4800	18.0

表2　某日Z站三台机组能耗情况

压缩机组编号	转速	耗气/电量	折标煤量	单价	成本
二线3#	5525rpm	$13.0\times10^4 m^3$	172.3t	1.236元/m^3	16.0万元
三线1#	4796rpm	$34.6\times10^4 kW\cdot h$	42.5t	0.494元/$kW\cdot h$	17.1万元
三线4#	4598rpm	$36.2\times10^4 kW\cdot h$	44.5t	0.494元/$kW\cdot h$	17.9万元

在"碳达峰"和"碳中和"的战略背景下，电力折标系数的取值问题变得更加复杂而重要。为了更加准确地反映电力的能耗水平和对环境的影响，采用等价值折标系数来计算所消耗的一次能源的数量，比采用当量值折标系数更具合理性和时代意义。等价值折标系数能够更全面地考虑电力生产过程中所消耗的一次能源以及输配电过程中的损失，从而更准确地反映电力的能耗水平和对碳排放的贡献。这对于推动国家管网集团公司的节能减排工作、实现可持续发展目标具有重要意义。

4　结论

在天然气折标系数取值方面，天然气折标系数按实测低位发热量计算。《综合能耗计算通则》GB/T 2589—2020中规定：能源的低位发热量和耗能工质耗能量，应按实测值或供应单位提供的数据折标准煤，无法获得实测值的，其折标准煤系数可参照国家统计局公布的数据或参考附录。根据国标定义，1kg标准煤的低位发热值取29271KJ，可得天然气折标系数公式：

$$\text{天然气折标系数} = \frac{\sum_i Ql\times\varepsilon}{29271} \quad (1)$$

式中，i为气源数量；Ql为低位发热值，kJ；ε为气源含量占比。

根据国家管网生产管理系统气质报表数据，选取黑河、霍尔果斯、长庆等20个站场实测的天然气低位发热量按权重平均后约为33.8MJ，计算可得折标系数约1.153kgce/m^3。

在电力折标系数取值方面，电力折标系数按等价值取值并考虑线路输损。根据陕西省工业和信息化厅对《综合能耗计算通则》GB/T 2589—2020要点解读，对于自发电企业，在计算电力等价值时通常是按照上年供电标煤耗进行计算，而非发电标煤耗。而对于非发电企业，外购电力的等价值按国家能源局公布的上年全国电力工业统计数据（6000kW及以上电厂供电标准煤耗）进行计算。根据国家能源局2023年全国电力工业统计数据，6000kW及以上电厂供电标准煤耗为302.0g/（$kW\cdot h$），因此，电力等价值折标系数应取0.3020kgce/（$kW\cdot h$）。但考虑到全国线路损失率5.62%，对于外购电力企业，电力等价值折标系数取0.3190kgce/（$kW\cdot h$）更为合理。

根据上述天然气折标系数和电力等价值折标系数，计算 Z 站 3 台机组折标煤量分别为 149.9t、110.4t 和 115.5t，显然这种计算方法较好的兼顾了能耗指标与成本指标。

综上所述，通过科学合理地确定天然气和电力折标系数，企业能够更加精准地把握其能源消耗的实际状况，从而实现对能耗水平的准确评估。这一基础性工作不仅为制定针对性强、效果显著的节能减排措施提供了坚实的数据支撑，使得企业能够在能源消耗和排放控制方面做出更加明智的决策，同时也为企业推进可持续发展战略、实现绿色低碳转型奠定了坚实的基础。合理的折标系数确保了能源数据的准确性和可比性，有助于企业挖掘节能潜力，优化能源结构，提升能源利用效率，从而在减少能源消耗和环境污染的同时，增强企业的核心竞争力和社会责任感。

参 考 文 献

[1] GB/T 2589—2020，综合能耗计算通则[S].
[2] 陈向方. 论"折标系数"的罅隙与补丁——基于《能源加工转换与回收利用》表的探索[J]. 内蒙古统计，2022，(05)：47-49.
[3] 杨晓东，孔文胜. 正确计算能源消耗是节能监察能力建设的技术基础[J]. 节能技术，2017，35(06)：569-574.
[4] 李文旦. 基于电力需求的特征分析及其对工业经济发展的影响分析[J]. 现代工业经济和信息化，2023，13(11)：303-305.
[5] 梁荣进，郭广磊. 关于采用当量值能耗还是等价值能耗分析项目新增能耗对所在地影响的探讨[J]. 中国工程咨询，2018，(11)：37-40.
[6] 夏志昊. 火力发电行业碳排放效率分析及碳减排政策评估[D]. 曲阜师范大学，2023
[7] 赵施阳，迟永福，滕征旭，等. 基于自适应蒙特卡罗方法的输电线路损耗计算[J]. 制造业自动化，2023，45(05)：45-49.
[8] GB/T 2589—2008，综合能耗计算通则[S].
[9] 王海荣，张郗郡. 发展新质生产力"碳"索未来之路[N]. 深圳商报，2024-10-29(A01).
[10] http：//gxt.shaanxi.gov.cn/jjnhb/64571.jhtml. 陕西省工业和信息化厅 2020 版《综合能耗计算通则》要点解读.
[11] https：//www.nea.gov.cn/2024-01/26/c_1310762246.htm. 国家能源局 2023 年全国电力工业统计数据.
[12] 董丽. 促进绿色低碳转型提升碳市场活跃度[N]. 中国财经报，2024-11-28(007).

双碳目标下深地空间利用及展望

熊昊天　王文权　刘晓旭　陈 伟　周练武

（国家管网集团工程技术创新有限公司）

摘　要　【目的】2024年《政府工作报告》指出："大力推进现代化产业体系建设，加快发展新质生产力"。深地储能利用深地空间作为能源存储"密封罐"，可以用于储气、储油、储氢、CO_2封存等，既包括传统产业改造提升、又包括新兴产业和未来产业的发展布局，是新质生产力的重要组成部分。新型储能行业发展形式和经济分析，聚焦新型储能行业发展前景，为合理、经济利用深地空间做出战略支撑。【方法】按照储存对象，将深地空间分为储气、储电、储氢、碳封存等利用方法。基于双碳目标下，储能行业发展全产业链研究，归纳总结新型储能行业发展趋势、政策动向和经济效益，提出新型储能行业发展前景分析方法。【结果】文章以传统、新兴和战略储备三大储能业务为例，根据行业前景分析方法，以需求、政策、经济效益等角度，对各行业全产业链进行分析核算，最终获得中国新型储能行业发展前景形式结论。【结论】通过对储能行业发展分析，储气业务市场需求广阔、经济效益好，空气压缩储能技术发展迅速，储氢业务潜在市场需求广阔、应用市场丰富，碳封存业务市场需求广阔、技术发展相对成熟，从"制、运、储、销"产业链四大模块提出发展对策建议，为中国新型储能行业发展前景提供了合理的建议。

关键词　新型储能；规划方法；CO_2封存；电力市场

Utilization and Prospects of Deep Underground Space under the Dual Carbon Goals

Xiong Haotian　Wang Wenquan　Liu Xiaoxu　Cheng Wei　Zhou Lianwu

(Pipe China Engineering Technology innovation Co., Ltd.)

Abstract　Objective: The 2024 Government Work Report emphasizes the need to "vigorously promote the construction of a modern industrial system and accelerate the development of new productive forces." Deep underground energy storage utilizes underground space as a "sealed container" for energy storage, which can be applied to gas storage, oil storage, hydrogen storage, and CO_2 sequestration. This includes both the transformation and upgrading of traditional industries as well as the development layout of emerging and future industries, making it an essential component of new productive forces. The development form and economic analysis of the new energy storage industry focus on the development prospects of the industry, providing strategic support for the rational and economic utilization of deep underground space. 【Methods】The study categorizes the use of deep underground space based on the stored object, including gas storage, electricity storage, hydrogen storage, and carbon sequestration. Under the dual carbon goals, the study conducts a full industrial chain analysis of the energy storage industry, summarizing the development trends, policy directions, and economic benefits of the new energy storage industry, and proposes a method for analyzing the development prospects of the industry. 【Results】The article takes the three major energy storage businesses—traditional, emerging, and strategic reserves—as examples. It analyzes and calculates the entire industry chain from the perspectives of demand, policy, and economic benefits, ultimately drawing conclusions about the development prospects of China's new energy storage industry. 【Conclusion】Through the analysis of the energy storage industry, the gas storage business has broad market demand and good economic benefits, compressed air energy storage technology is developing rapidly, the hydrogen storage business has vast potential market demand and diverse application markets, and the carbon sequestration business has broad market demand with relatively mature technology. The paper puts forward development countermeasures for the four major modules of the industrial chain: production, transportation, storage, and sales, providing reasonable suggestions for the development prospects of China's new energy storage industry.

Key words New Energy Storage; Planning Methods; CO_2 Sequestration; Electricity Market

2020年9月，习近平总书记代表中国政府郑重承诺，中国二氧化碳排放2030年前达到峰值，努力争取2060年前实现碳中和。2024年《政府工作报告》指出："大力推进现代化产业体系建设，加快发展新质生产力"。中国作为世界上最大的能源生产国和消费国，能源消费结构以煤炭为代表的化石能源为主，2023年我国能源消费总量57.2亿吨标准煤，占能源消费总量比重55.3%，清洁能源消费占能源消费总量比重26.4%。天然气、氢能和太阳能等清洁能源的大规模使用与传统化石能源的协同转型是实现"双碳"战略目标的关键之一。然而天然气消费存在强季节性和时段性，风能、氢能等清洁能源存在地域性分布，发展清洁能源可通过储能等手段"削峰填谷"克服地域性和时段性。深地空间作为能源存储工具，可用于储气、储氢和碳封存等，深地空间开发利用既包括传统产业改造提升，又包括新兴产业和未来产业建设布局，对优化我国能源结构，建全能源储备体系，保证能源安全供应有重要作用。

1 深地储能概念及意义

双碳目标下深地储能是指通过将石油、天然气、氢气和二氧化碳等储存在地下深处，实现顶峰消费调节和保障资源供应，减少二氧化碳排放。以储存空间为例，可将深地空间划分为孔隙型和洞穴型两大类。其中孔隙型储气库分为气藏、油藏、含水层储气库；洞穴型储气库分为盐穴、矿坑储气库。结合"全国一张网"现有优势和"X+1+X"发展路径，以油气和电力产业在能源供应和消费上互补为方向，空气压缩储能、碳封存和储氢等深地空间利用业务发展如图1所示。将风光电等可再生能源转化为氢能，通过天然气掺氢发挥"全国一张网"优势大规模运输，即可作为新型储能平抑新能源发电波动，同时利用氢能载体可丰富氢能储运场景，加大储氢深地空间开发利用是实现绿氢高效利用开发的关键之一。电力产业链，截止2024年上半年，全国已建成投运新型储能项目累计装机规模达到440GW，光伏装机610GW，风光电等可再生能源存在供给连续性不足和地域性较强等问题，储电业务随着新能源"强制配储"政策导向，新型深地储能大规模发展，供应大多集中于电力生产链，可支撑新能源发电建设。其他产业链，碳封存业务作为双碳目标下核心业务之一，碳源至封存点可通过大规模管道运输；氢气作为国内战略物资，2023年国内氢气消费量$2565×10^4 m^3$，利用盐穴储气库储存氢气可保障国内氢气资源安全供应。因此，深地空间利用是实现双碳目标和"全国一张网"中重要的组成部分。

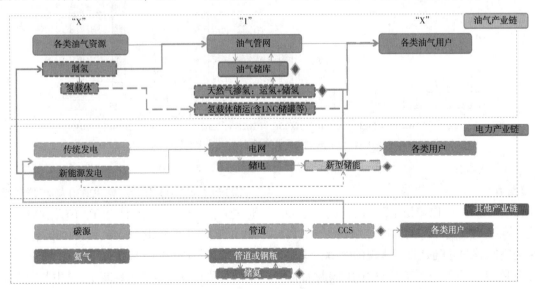

图1 深地空间利用在"X+1+X"发展路径作用示意图

2 我国深地空间利用现状分析

2.1 储气调峰

1999年，我国第一座商业储气库——大港大张坨储气库投入全面建设，标志着我国地下储气库进入了一个新的发展阶段。截止2023年全球已建储气库689座，总工作气量4165亿方，主要分布在北美、欧盟和独联体国家。国内已建枯竭气藏储气库28座，总工作气量约215亿方，盐穴储气库3座，总工作气量约15亿方。根据历年我国不同地区天然气消费量实际数据，以及城市燃气、工业、发电、化工四大行业的用气特性，测算未来分年分区域用气不均匀系数、分区域和全国调峰需求以及占消费量的比例。"十四五"期间，全国天然气表观消费量持续增长，预计2025年全国天然气消费需求4500亿方，2030年达到5700亿方，2040年达到峰值6500~7000亿方。预计2025-2030年调峰需求从492亿方增长到615亿方，2040年达到776亿方，调峰需求持续增长。目前，全国已建储气库31座，总工作气量230亿方，相对实际调峰需求还有很大差距，同时也为储气调峰深地空间利用带来了新机遇。

2.2 压缩空气储能

2024年，《政府工作报告》提出"发展新型储能"积极稳妥推进碳达峰碳中和。盐穴压缩空气储能是以压缩空气为载体实现能量的存储和利用，可以实现风能及太阳能等新能源发电功率的平滑输出，降低新能源并网给系统带来的冲击，降低弃风弃光率、实现电网削峰填谷。预估2030年发电量达到13万亿千瓦时左右，储电需求将显著提高。截至2023年底，中国已投运电力储能项目累计装机规模86.5GW。预测2030年新型储能累计装机规模保守情景和理想情景下将分别达到221.1GW和313.9GW，如图2所示，其中压缩空气储能占比仅0.7%，盐穴液流电池储能尚属空白，盐穴压缩空气储能和盐穴液流电池储能前景广阔。

2.3 储氢

地下储氢作为氢能产业中的重要环节，可以为风能、太阳能等间歇性能源提供电网储能，也可以为发电和运输提供燃料。2022年，我国氢气表观消费量3527万吨，2023年约3570万吨。预测到2030年氢气年需求量将达到4000万吨，到2060年，我国氢气的年需求量将增至1.3亿吨。氢既不仅可以用于短期和小规模的储能需求，也可以用于长期和大规模储能调峰。利用风光发电制氢-储氢是大容量长期储能的有效方式，可储氢及其衍生物(如氨、甲醇)，对于改善我国能源结构，能源绿色可持续发展具有重要意义。

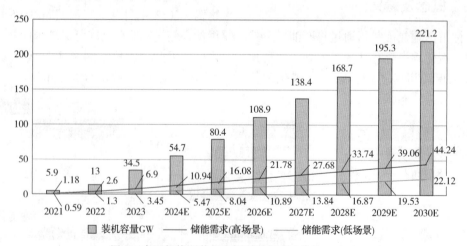

图2 中国新能源装机发展趋势展望(单位：GW)

2.4 碳封存

2015年12月，《巴黎协定》明确指出：必须努力使全球平均气温升高的幅度控制在1.5℃以内。深地空间碳封存既是目标实现的保障，又是新兴行业风口。经预测，2025年中国碳封存减排需求约为2400万吨/年，2030年将增长到近1亿吨/年，2040年预计达到10亿吨/年左右，2050年将超过20亿吨/年，2060年约为23.5亿吨/年。碳达峰碳中和目标下，中国经济生产和消费方式正在发生系统性变革。《中共中央国务院关于完整准确全面贯彻新发展理念做好碳达峰碳中和工作的意见》指出，到2060年，绿色低碳循环发展的经济体系和清洁低碳安全高效的能源体系全面建立。为实现国家"30·60"双碳目标，CO_2地质封存既是目标实现的保障，又是新兴行业风口。

3 我国深地空间利用产业形式发展分析

3.1 储气调峰产业发展分析

（1）支持政策

自2018年开始，发改委、能源局等部门陆续出台加快储气设施建设、完善油气储备体系、提升能源储运能力、加强保供增储等相关政策。2020年，国家发改委联合印发《关于加快推进天然气储备能力建设的实施意见》，加大政策力度，促进储气能力快速提升。2024年，国家能源局印发《2004年能源工作指导意见》，确立能源工作保障能力持续增强。国家政策积极稳健，具有明显的连续性。

（2）经济效益

储气调峰方式包括储气库、LNG、气田、管网等，其中储气库、LNG是最主要的调峰方式，其他方式作为补充和应急手段。储气库的单位工作气量投资低于新建LNG接收站。储气库主要作为管网配套工程，受国家统一调度，未单独核算收益情况。近年随着市场化机制引导，逐步走向独立化运行、商业化运作的模式，从当前已有的库容交易情况看，已经进行商业化运作的储气库盈利能力较佳。不同类型储气调峰优缺点如表1所示。

表1 储气调峰优缺点表

调峰方式	优点	缺点	单位投资
储气库	容量大，可靠性高 削峰填谷，平衡气田生产 提高管网运行效率	受限于地理位置和地质构造，库址选择难度较大 建设周期较长	油气藏储气库2.11~4元/方 含水层储气库5.4~9.3元/方 盐穴储气库5~8.33元/方
LNG	启停灵活，随时可用 建设周期相对较短，各项工艺均较成熟	储存能力受制于储罐个数，通常小于储气库 夏季无法发挥储气库的注气作用，无法填谷 接卸作业受海况和天气影响较大	扩建储罐7.05元/方 新建LNG接收站9.6元/方
气田	依托气田现有产能和地面工程，可动用规模大 短期释放产能强	导致气田冬季生产不平稳，易造成边底水入侵，出砂等 淡季生产设施闲置，降低气田开发效益	—
管网	管存储气，动用灵活	调峰量较小，不能作为季节调峰	—

（3）技术发展

针对我国特殊建库地质条件和储气库多周期运行特点，油气藏型、盐穴型储气库建库理论和技术方面基本成熟，形成了储气库气藏扩容达产、盐穴优快建腔、降本增效工程和安全运行评价等系列配套技术。常规油气藏储气库建库理论和技术方面基本成熟，形成了储气库气藏库址筛选、动态密封性评价、钻采工艺、地面注入及采出处理工艺、安全风险管控体系等技术。随着建库地质条件日趋复杂，油气藏从中高渗、高丰度转向中低渗、低丰度，复杂地质条件油气藏储气库技术尚需攻关。盐穴从层状、高品位盐层转向低品位、多夹层和厚夹层等复杂盐岩条件，复杂条件下盐层建库技术储备不足。盐穴储气库建库技术难点如图3所示。

（4）应用场景分析

地下储气库具有储气调峰和应急供气功能，是"全国一张网"的重要组成部分，是保障天然气产业链平稳运行的重要基础设施。储运应用场景成熟，主要用于提供储气调峰，保证管道安全高效运营，同时探索库容竞拍等多种经营模式，实现深地储气商业化。储气调峰产业分析如图4所示。随着储气调峰能力的增强，将进一步对"全国一张网"高效运营作出贡献，实现"运""储"协调发展。同时需加快储气建设能力，快速弥补短期调峰能力不足的短板，并主动应对远期全国储气能力可能过剩和竞争的风险。

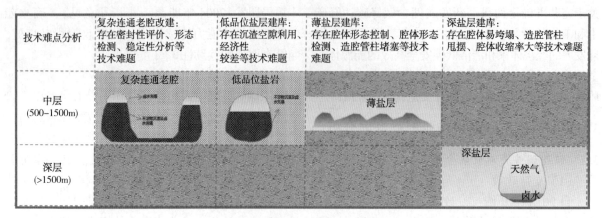

图3 盐穴储气库建库技术难点

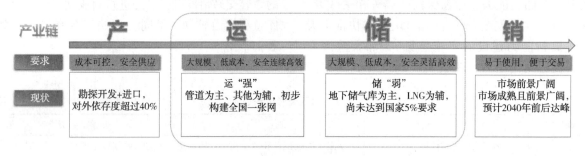

图4 储气调峰应用场景分析

3.2 压缩空气储能产业发展分析

（1）支持政策

自2017年以来国家逐渐重视储电业务，伴随连续密集相关政策出台，推动技术创新和产业升级，呈现出良好的发展态势，截至2023年有15省出台了新能源项目配置储能的要求。在全面落实"碳达峰、碳中和"要求的进程中，随着新能源强制配储及补贴政策带动，储能可与新能源发电、电力系统协调优化运行。

（2）经济效益

各类储能技术参数及成本如表2所示，抽水蓄能的度电成本最低，其次是压缩空气储能和竖井重力储能，液流电池等电化学储能的度电成本仍然较高。压缩空气储能依托于电力系统，当地的上网电价、峰谷电价差、补贴是储能系统外重要变量，对储能收益水平影响较大。发电侧储能应用能够减少新能源弃电场景，储能利用率提高。用户侧固定储能负荷曲线，储能运行方式明确，利用效率较高，峰谷电价差相对较高，两者储能度电成本较低。盐穴压缩空气储能和竖井重力储能较于盐穴液流电池储能成本更低，具有实现经济效益的基础和广泛开发应用的潜力。

表2 主要新型储能技术参数及成本对比

类型	储能时长/h	响应时间	规模/MW	效率/%	系统成本/(元/kW)	度电成本/(元/kWh)	寿命/年
抽水蓄能	4~6	分钟级	100~5000	70~80	5500	0.2~0.3	40~60
压缩空气	4	分钟级	1.5~300	40~70	6000	0.2~0.5	30~40
铅酸电池	4~10	毫秒级	0~20	60~75	1500	0.83~0.98	5
液流电池	4	百毫秒级	0.03~5	65~75	3000	0.79~1.1	5~15
锂电池	2	毫秒	0~20	80~88	1400~2200	0.56~0.94	8~10

（3）技术发展

1978年德国建设全球第一座大型盐穴压缩空气储能电站，目前国外共有2座盐穴压缩空气储能电站投入运行、1座处于建设期，国外已有盐穴压缩空气储能项目建成投产，技术发展成熟。中国第一座盐穴压缩空气储能电站位于金坛，于2022年5月投入商业运行，目前我国有2座盐穴压缩空气储能项目正处于商业运行、1

座处于建设期、5个项目处于可研阶段，国内盐穴压缩空气储能装机规模由 60MW 发展到 660MW，正由示范应用阶段转向初期商业化发展阶段。现已形成非补燃先进绝热压缩空气储能技术，建立研发了高负荷离心压缩机和高参数换热器等核心设备。

（4）应用场景分析

在风光发电大力建设及国家"强制配储"政策推动下，新能源储能建设需求高，可探索峰谷获利等应用场景。

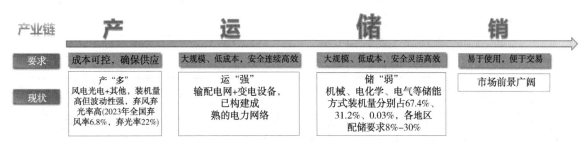

图 5 储电调峰应用场景分析

3.3 储氢产业发展分析

（1）支持政策

2006年以来，国家陆续出台系一系列关于引导和支持氢能发展的政策或者规范性文件，2021年《"十四五"新型储能发展实施方案》出台，将氢能列入"十四五"新型储能核心技术装备攻关重点方向。2022年3月《氢能产业发展中长期规划》指出持续推进绿色低碳氢能制取、储存、运输和应用等各环节关键核心技术研发。目前30个省份已将氢能写入了"十四五"发展规划，从氢能产业的整体发展到具体应用领域，都给予了明确的指导和支持。

（2）经济效益

地下储氢相较于地面储氢、液态储氢成本低，枯竭油气藏储氢最为经济（9.1元/kg），其次是含水层（9.4元/kg）和盐穴（11.7元/kg）。结合氢能全产业链进行预估，结果如图6所示。由于可再生能源电解制氢成本在制氢-储氢产业链中占主导，地下储氢技术能否在工业规模上应用，还需要进一步降低氢能全产业链成本，利用规模化低成本储存提高全产业链经济效益。

（3）技术发展

目前欧洲、北美有5座盐穴储氢库建成投产，用于存储纯氢气，油气藏储氢国外已建成2座，存储人造煤气（氢气含量10%），技术发展相对成熟，国外储氢库建设情况如表3所示。国外技术发展成熟。国内已有盐穴储氢理论研究基础，技术发展处于初期，国内学者研究了氢气对管材腐蚀的影响、分析了储氢过程中水泥降解、氢气泄漏等问题，因氢气密度小、黏度低且渗透性高，且存在与地层矿物发生氧化还原反应等影响国内尚无相关理论研究，关键基础理论与核心技术尚待突破。目前在河南叶县，中科院岩土力学所与平煤盐化开展盐穴储氢先导试验，预计2025年形成150万方的储氢规模。

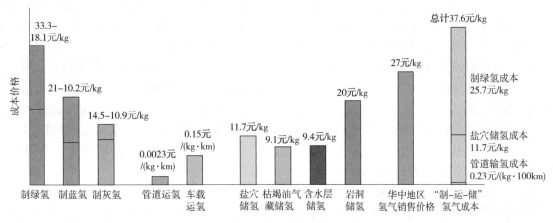

图 6 氢能产业链各环节成本对比

表3 国外储氢库建设情况

类型	国家	运营时间	深度/m	压力/bar	容积/库容/m³	储库中流体组成
盐穴储氢库	德国	1971	1305~1400	80~100	32000	60%左右 H_2
	英国	1972	350	45	3×70000	95%H_2+3%~4%CO_2
	美国	1983	850~1150	70~135	580000	95%H_2
	美国	2007	820~1400	55~152	566000	95%H_2
	美国	2014	1500	68~202	906000	95%H_2
油气藏储氢库	奥地利	2013	500~1000	78bar/30℃-80℃	—	10%H_2
	阿根廷	2015	600	10bar/50℃	—	10%H_2

(4) 应用场景分析

在氢能全产业链中，储氢不仅具有调节电网波动性的作用，还具有代替部分化石能源、充当化工原料的作用。储氢应用场景如图7所示，以盐穴储氢为出发点，为风光电制氢基地建设奠定基础，同时调节新能源发电波动，基于管道掺氢输送和地下储气库用于天然气调峰。

图7 储氢调峰应用场景分析

3.4 碳封存产业发展分析

(1) 支持政策

2007年以来国家累计出台80余项碳减排相关政策，强调碳封存项目的集成化、规模化发展。2017年底，中国全国碳市场正式启动，至今累计出台20余项涉及碳市场相关政策，明确了碳排放权市场交易制度。2024年1月，全国温室气体自愿减排交易市场启动，产生的减排效果经过科学方法量化核证后，通过全国碳排放权交易市场出售获取相应的减排贡献收益，减排量项目必须有对应方法学支持。发布造林碳汇等4项方法学，处于征求意见阶段2项。第一批温室气体自愿减排项目审定与减排量核查，能源产业拟审批4家，林业和其他碳汇类型5家。目前尚未建立碳封存项目相关方法学、碳封存项目缺乏相关政策支持、碳封存项目商业化应用程度低。

(2) 经济效益

国外碳市场交易以欧盟碳排放交易体系为主，2019年其碳交易量为67.8亿吨，占世界总交易量的77.6%，交易额为1689.7亿欧元，占世界总交易额的87.2%。1990-2019年，欧盟温室气体排放下降26%。国内碳市场交易逐渐完善，国内全国碳排放权交易市场(CEA)，实行配额交易，发电行业成为首个且唯一纳入全国碳排放权交易的行业。2023年配额成交量2.12亿吨，总成交额144.44亿元。全球碳市场交易价格如图8所示，未来中国规模化 CO_2 地质封存经济可行性较高。

(3) 技术发展

咸水层碳封存包括陆地咸水层碳封存和海洋咸水层碳封存，中国咸水层碳封存潜力巨大，封存潜力约2.42~4.13万亿吨，占比约98.8%。国内咸水层封存项目多为先导工程，尚处于现场试验与工程探索阶段，当前正在开展世界级规模水层CCS先导试验，技术水平与国外并行，已形成碳封存的场地筛选技术、实施层位选择及评价技术、二氧化碳注入流动与传热模拟分析技术、防腐蚀固井水泥技术等一套全流程CCS示范技术体系。针对枯竭油气藏碳封存，国内已建设多个二氧化碳驱油项目，形成了二氧化碳驱油与封存配套技术体系、高压混相驱油技术、全过程 CO_2 驱前缘数值模拟及动态调控技术，并建立了封存安全性评价、监测及全生命周期评价技术。

(4) 应用场景分析

以咸水层碳封存为出发点，结合自愿减排碳市场可探索咸水层碳封存方法学、自愿减排交易

等模式构建 CO_2"运-封-易"一体化产业链，为大规模发展碳封存产业奠定基础。该场景近期盈利较低，且较难实现纵向一体化，随着碳交易价格增长及碳排放考核、交易政策的调整，远期可能存在盈利机会。

4 总结与展望

在双碳目标下，深地空间利用是大规模储能最有效、最经济的方法之一。通过对各传统及新兴行业政策、经济、技术和应用场景分析，传统业务与新兴业务并重发展战略，以全国一张网为产业发展基础，根据各个储能业务方向特点，明确发展思路和措施，科学布局新业务方向。可加快推动深地空间开发利用产业相关政策制定，如矿权转让机制和新型储能业务的补贴政策等。针对深地空间下储天然气、氢气、二氧化碳和压缩空气储能等应用场景，开展一系列针对新技术研究。持续开展新技术、新材料、新装备、新工艺、新工法等研究攻关，积极开展国际交流，共同推进工艺技术创新，促进深地储能未来行业发展。

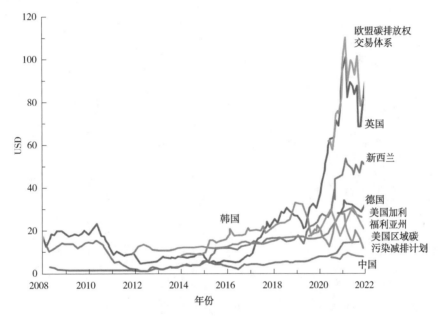

图8 2008年-2022年全球碳市场交易价格图

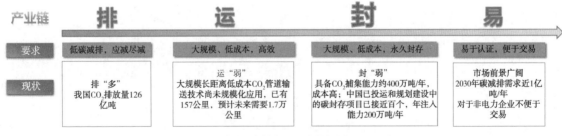

图9 碳封存应用场景分析

参 考 文 献

[1] Baoguo F. Reflections on Cultivating New Quality Productive Forces in the Oil and Gas Industry[J]. China Oil & Gas, 2024, 31(05): 21-25.
[2] 中华人民共和国国务院新闻办公室. 中国的能源转型. 2024.
[3] 中国能源传媒集团有限公司. 中国能源数据大报告. 2024.
[4] 杨春和, 王同涛. 我国深地储能机遇、挑战与发展建议[J]. 科学通报, 2023, 68(36): 4887-4894.
[5] 张怀水. 氢能在我国构建新型能源体系进程中将发挥重要作用[N]. 每日经济新闻, 2023-03-14(004).
[6] 张哲, 黄骞, 王春燕, 等. 中国氢气全产业链发展现状与展望[J]. 油气与新能源, 2024, 36(02): 1-9.
[7] 中石油规划总院.《中国油气与新能源市场发展报告

（2024）》. 2024.
[8] 国家能源局. 2023年全国电力工业统计数据. 2024.
[9] 前瞻产业研究院. 2024-2029年中国可再生能源制氢（绿氢）行业市场前瞻与投资战略规划分析报告. 2024.
[10] 发展改革委财政部自然资源部住房城乡建设部能源局《关于印发关于加快推进天然气储备能力建设的实施意见》. 发改价格〔2020〕567号. 成文日期：2020年04月10日. https://www.gov.cn/zhengce/zhengceku/2020-04-26/content_5506189.htm.
[11] 王晓宇, 马果靖, 李彪, 等. 新型储能发展的场景和必要性分析[J]. 中国设备工程, 2024,（22）：14-17.
[12] 张来斌, 胡瑾秋, 肖尚蕊, 等. 深部地下空间储能安全与应急保障技术现状与发展趋势[J]. 石油科学通报, 2024, 9(03): 434-448.
[13] 张运东, 方辉, 刘帅奇, 等. 深地油气勘探开发技术发展现状与趋势[J]. 世界石油工业, 2023, 30(06): 12-20.
[14] 中国21世纪议程管理中心. 中国二氧化碳捕集利用与封存（CCUS）年度报告（2024）. 2023.
[15] 生态环境部环境规划院. 中国二氧化碳捕集利用与封存（CCUS）年度报告（2024）. 2024.
[16] 杨春和, 王同涛. 深地储能研究进展[J]. 岩石力学与工程学报, 2022, 41(09): 1729-1759.

基于黏菌优化算法的全国一张网下的宏观路由自动选择

李娟[1] 刘亮[1,2] 李明旭[1] 李强[1] 孟祥海[1] 李海润[1]

(1. 国家管网集团工程技术创新有限公司；2. 南京大学计算机科学与技术系)

摘 要 国家管网集团成立，天然气管输与销售分离，油气体制改革迈出关键一步，管输体系从"公司一张网"迈向"全国一张网"。这一转变使天然气管网规划面临全新挑战。作为管网建设首道程序，天然气路由规划的科学性与合理性至关重要。在资源与市场纳入全国管网统一调配的背景下，管网规划布局是节省造价、降低管理维护难度的首要环节，还关乎可持续发展、生态保护及土地资源节约。以天然气管网规划宏观路由最佳路径为目标，借助黏菌觅食时强大的路线规划能力，建立黏菌觅食数学模型求解最优路径。该技术有望为天然气管网、道路、通信等基础设施网络的规划设计提供指导，提升天然气输送效率，降低成本，保障管网安全可靠运行，提高规划效率，助力战略规划业务开展。

关键词 全国一张网；X+1+X；管网规划；黏菌优化算法；Physarealm；OSGEarth

Macro-routing automation under a nationwide network based on Slime Mold Optimization Algorithm

Li Juan[1] Liu Liang[1,2] Li Mingxu[1] Li Qiang[1] Meng Xianghai[1] Li Hairun[1]

(1. PipeChina Engineering Technology Innovation Co. Ltd;
2. NanJing University Department of Computer Science and Technology)

Abstract With the establishment of the National Pipe Network Group, the separation of natural gas pipeline transportation and sales has been achieved. This represents a crucial step in deepening the reform of the oil and gas industry system, marking the transition of the natural gas pipeline transportation system from "company-specific networks" to a "national integrated network". This transformation has presented unprecedented challenges to natural gas pipeline network planning. As the initial procedure in natural gas pipeline network construction, the rationality and scientific nature of natural gas routing planning are of great significance. Against the backdrop of integrating resources and markets into the unified optimization and allocation of the national pipeline network, the planning and layout of the natural gas pipeline network are the first crucial steps in saving project costs and reducing the difficulty of management and maintenance. Moreover, they are also related to sustainable development, ecological protection, and the conservation of land resources. Aiming to find the optimal path for the macro routing of natural gas pipeline network planning, we can leverage the remarkable route planning ability of slime molds during their foraging process. By establishing a mathematical model of slime mold foraging, we can solve for the optimal routing path. This technology holds promise in guiding the planning and design of infrastructure networks such as natural gas pipeline networks, roads, and communication networks. It can enhance the efficiency of natural gas transportation, reduce costs, ensure the safe and reliable operation of the pipeline network, improve the efficiency of network planning, and facilitate the implementation of strategic planning tasks.

Key words A nationwide network; X+1+X; Pipe Network Planning; SMOA; Physarealm; OSGEarth

根据国家发展和改革委员会公布的能源相关数据，2022年我国天然气产量2178亿 m³，占比58.92%；进口量10925万吨，占比41.08%，进口天然气包括进口液化天然气（Liquefied Natural Gas，LNG）和进口管道天然气（Pipeline Natural Gas，PNG），占比依次为57.64%和42.36%。2017年国家发改委和国家能源局发布了《中长期油气管网规划》，未来10年中国油气管网的规划布局将着眼于拓展"一带一路"进口通道，基本构成"西油东送、北油南运、西气东输、北气南下、缅气北上、海气登陆"的油气供应格局。预计到2025年，中国油气管网规模将达到24万km，其中原油管道、成品油管道、天然气管道分别为3.7万km、4万km和16.3万km，而目前我国油气管道总长度仅为12.23万km，其中天然气管道长度约为7.6万km，与国家目标差距较大。国家管网集团加快构建从资源到市场的"X+1+X"体系，进一步完善四大能源战略通道和油气管网布局。其中，第一个"X"表示上游的油气资源，多为油气田、储气库、LNG接收站，第二个"X"代表下游的消费市场，主要受地区人口和GDP分布的影响。而中间的"1"代表了国家管网集团构建的"全国一张网"。从资源与市场平衡的角度来看，"全国一张网"是连接上、下游资源与市场的关键桥梁，其建设和运营受沿途地区的地形、地貌、地质和水文等自然条件的限制，对前期的规划设计提出了更高的要求。许多技术，如：卫星遥感、GIS、人工智能、运筹学、数字孪生以及仿生算法等都被应用到管线路由规划当中。

多头绒泡菌是一种黏液霉菌，简称黏菌，栖息在阴凉、潮湿的地方，在有丝分裂形成的变形体成熟之后，进入营养生长时期，会形成网状形态，且依据食物、水与氧气等所需养分改变其表面积。黏菌是最容易培养的真核微生物之一，在许多涉及变形虫和细胞运动的研究中被用作模式生物。例如：日本北海道大学的Tero等人用燕麦片在地图上标记出东京及其周边城市，经过反复测试，黏菌形成了类似东京铁路网的性能优越的觅食路径。Nakagaki等人将黏菌放在迷宫中，并在起点和终点处均放置食物，结果发现黏菌在觅食过程中表现出了惊人的线路规划能力，在找寻食物的同时会对路径进行优化。尽管这些生物没有大脑，但它们能够在留下的黏液痕迹中储存化学信息，包含对过去事件的"记忆"，方便更快地找到食物来源。黏菌甚至可以学习，然后通过与其他黏菌融合，将学到的信息传递下去。黏菌不会浪费时间向已经探索过的地方发出触角，它伸展形成复杂的管道网，并通过一些简单规则管理网络，那些与食物没有连接的管道逐渐衰退，而找到食物的管道不断强化，并将营养物质送回中心。这一系列实验充分证明了黏菌寻径演化对实际路网的构建具有重要的指导意义。Li等人于2020年提出了一种模拟黏菌在觅食过程中的行为和形态变化的新型群体智能优化算法——黏菌算法（Slime Mould Algorithm，SMA），该算法模拟黏菌搜索、接近和包围食物三种阶段的觅食形态变化来实现路径的智能寻优功能。杰夫·琼斯采用了一种综合的方法和一个移动的多智能体系统，再现了黏菌的生物学行为。王庆基于迷宫模型，建立了完善的黏菌模型，提出了根据流量的变化对道路网络进行动态优化的算法。高金兰提出一种基于改进黏菌算法的配电网重构策略以降低网络总有功损耗和实现负载均衡。任丽莉提出了一种基于黏菌觅食的多元宇宙优化算法，利用黏菌觅食行为在局部最优和全局最优之间寻求最优解。SMA属于元启发算法，具有收敛速度快，寻优能力强的特点，具有广泛的应用前景。但一种算法并不能解决所有的工程问题，因此，针对实际情况对传统SMA进行优化是有必要的。

清华大学的Ma Yidong（https://github.com/maajor/Physarealm）设计了基于Rhino三维建模软件和SMA的黏菌插件（Physarealm），因其模拟黏菌的觅食线路呈现美观的流线型，被广泛应用于建筑设计领域。但该插件存在诸多不足之处，在实际的线路规划领域应用较少。本文在国家管网集团构建"全国一张网"的大背景下，基于Physarealm插件进行C#二次开发，解决SMA的平面避障、多边形边界和多源发射点等问题，并结合我国的社会经济和地理环境等影响因素，将其应用于油气管线宏观路由的自动选择。

1 数据与平台概述

1.1 研究区域和数据

我国陆上油气资源主要分布在松辽、渤海湾、准噶尔、塔里木、柴达木、吐哈、鄂尔多斯、四川、二连、南襄、苏北、江汉、海拉尔等盆地。而消费市场多集中于人口密集、经济发达地区，如：京津冀、长三角、珠三角、成渝等城市群等。油气管道线路的选择，应根据该工程建设的目的和市场需要，结合沿途地区的地形、地貌、地质、水文、气象、地震等自然条件，在施工便利、运营安全和技术经济的前提下，通过综合分析确定线路总走向。一般而言，管道不得通过城市水源区、工矿用地、机场、火车站、海（河）港码头、军事设施、国家重点文物保护单位和国家级自然保护区。本文以中国除港、澳、台地区和南海九段线内的南海诸岛及相关海域的其他30个省市自治区（不涉及领土完整性问题）为研究区域，基于"全国一张网"数据，选取了8个影响管线规划和选址的社会经济和地理环境因子来进行分析，依据《中华人民共和国石油天然气管道保护法》、《中华人民共和国环境保护法》、中华人民共和国国家标准《输气管道工程设计规范（GB 50251—2015)》、《油气输送管道跨越工程设计标准（GB/T 50459—2017）》、《油气输送管道穿越工程设计规范（GB 50423—2007）》和《石油化工管道常用钢管（JIS）材料及使用温度范围》等相关的政策法规和工程规范设置各影响因子的阈值，如表1和图1。

表1 实验数据说明

数据名称	来源	描述
全国一张网	国家管网集团构建的全国石油、天然气管网基础设施数据	陆气分布于油气田、储气库，海气和外来气集中于LNG接收站，以上均为为资源点。管线数据为国家管网集团和各省网公司修建的油气运输管道
GDP	中国科学院资源环境科学与数据中心每隔5年更新的中国GDP空间分布km网格数据集（https://www.resdc.cn/DOI/DOI.aspx?DOIID=33，最新版本2019年	GDP>的地区为下游消费市场
人口分布	来源于中国科学院资源环境科学与数据中心每隔5年更新的中国人口空间分布km网格数据集（https://www.resdc.cn/DOI/DOI.aspx?DOIID=32），最新版本2019年	人口密度>区域为下游消费市场
土地利用类型	中国科学院资源环境科学与数据中心更新的2020年1km栅格数据（https://www.resdc.cn/DOI/DOI.aspx?DOIID=54）	不适宜修建管线的土地利用类型有：厂矿、大型工业区、油田、盐场、采石场、机场、湖泊、滩涂、沼泽地等
海拔	美国地质调查局2009年发布的30m分辨率的ASTGTM2 DEM数据（https://earthexplorer.usgs.gov/）	管道施工处在4000m以上的高海拔地区，极度高寒缺氧，会对参建人员的身体健康构成极大威胁
坡度	由DEM数据计算坡度值得到	坡度>30°为陡坡，地形陡峭，高差大，地质不稳定，容易造成施工安全隐患，且对油气输送的动力系统有极高的要求
风景名胜和自然保护区	国家基础地理信息中心发布的2021年全国1:100万基础地理数据库（http://www.webmap.cn/commres.do?method=result100W）	风景名胜和自然保护区禁止破坏性施工，不适宜修筑管线
地震动峰值加速度	国家强制性标准《中国地震动参数区划图》（GB 18306—2015）	地震动峰值加速度>=0.4g为地震高风险地区，不适宜修筑地下管线
最冷月最低温	WorldClim网站发布的全世界1970~2000年的气候数据（https://www.worldclim.org/data/worldclim21.html）	高寒地区（东北、西北、青藏高原）土壤多为冻土层，气温低于-30°会给施工人员野外作业带来困难

*注：因一个地区的地形、气候、土地利用类型、人口等在短时间内未发生重大变化，故可用所选取年份的数据来代表区域最近的情况。

1.2 Physarealm 插件概述

Rhino软件是由美国Robert McNeel公司于1998年推出的一款基于非均匀有理B样条曲线（Non-Uniform Rational B-Spline, NURBS）的高级三维建模软件，主要应用于工业和建筑设计行业。Grasshopper是一款基于Rhino平台运行的可视化编程语言，以插件形式存在于 Rhino 软件中，它提供了一种基于节点的可视化编程方法，允许用户通过将不同的组件(节点)连接在一起来创建复杂的算法和设计过程。每个节点代表一个特定的功能或操作，可以在设计中进行组合和重新排列以实现各种效果，通过拖拽和连接节点，用户可以创建自定义的设计和生成参数化模型。Grasshopper 还拥有众多插件，如：Anemone、Cocoon、Culebra、ELK、Fabtools、Human、Hummingbird、LunchBox、Mosquito、Octopus、Physarealm 等。本文所使用的黏菌插件(Physarealm)是通过 C#二次开发得到的，其主要功能是通过模拟多头绒泡菌在搜索食物过程中不断向外扩张，以找寻获得食物最有效的路径，其主界面如图 2。

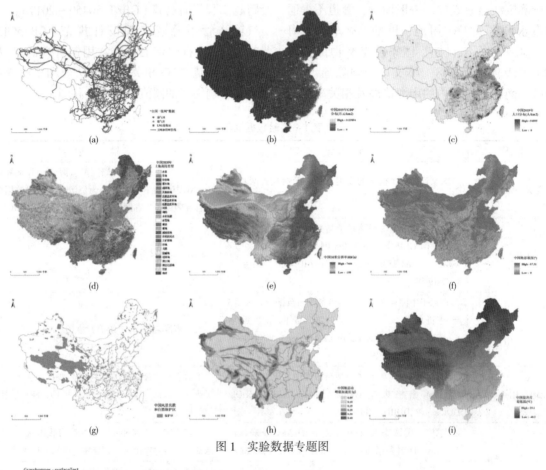

图 1　实验数据专题图

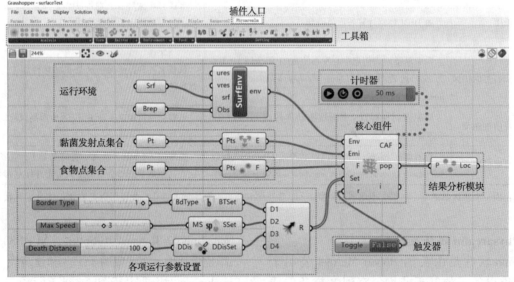

图 2　二次开发的 Physarealm 插件的界面

1.3 OSGEarth 概述

OSGEarth 是一种基于 OpenGL 标准的支持三维地图展示的地景渲染工具箱，采用 C++编写，使用了标准模版库(STL)。如图 3，OSGEarth 使用场景树的方式来管理三维场景，使用逻辑组来构建场景树，以便进行高效的渲染和遍历等。OSGEarth 支持加载多种格式（Shapefile、GeoJson、KML、CAD、TIF、3ds Max、LiDAR 点云等）的数据到三维场景中，并将加载的图层自动进行合并，完成地形、影像、矢量数据的综合展示。OSGEarth 提供了多细节层次模型(Levels of Detail, LOD)，使得海量的三维地形影像数据可按瓦片金字塔方式实现加载，以减小内存负荷，提升浏览速率。OSGEarth 具备跨平台性，可以运行在 Windows、Mac OS X、UNIX 和 Linux 操作系统上，支持手机、平板以及其他嵌入式设备。上述优点使得 OSGEarth 成为地学领域一个高效、表现力好的三维渲染引擎。

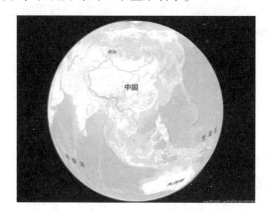

图 3 OSGEarth 三维虚拟地球平台

2 研究方法

2.1 黏菌优化算法

本文设计了一种黏菌优化算法（Slime Mold Optimization Algorithm, SMOA），通过模拟黏菌的觅食行为来实现智能寻优功能。黏菌会根据当前位置的客观条件（适应度函数优劣），决定每个个体所在位置的权重，然后个体会根据权重决定新的位置。算法的实现分如下五个步骤。

2.1.1 布局设置

首先，在 Rhino 软件的绘图界面上设置黏菌的运行环境这里设置为平面，宽度为 width，高度为 height，黏菌发射点、食物点的分布，以及障碍物的位置，发射点、食物点和障碍物可以有多个，如图 4，黏菌由发射点不断向外发射，需要避开障碍物找寻每一个食物点，从而形成觅食路径网络。算法的核心思想是分割平面运行环境，将其横向设置 u 个控制点，纵向设置 v 个控制点，一共细分为 $u \times v$ 个格网，格网的宽度为横轴上相邻两控制点的最小间隔距离 $u_{interval}=width/u$，高度为纵轴上相邻两控制点的最小间隔距离 $v_{interval}=height/v$。初始时，用二维数组 griddata[u][v]为每个格网赋值为-1，如下公式（1）和图 5：

$$griddata[i][j]=-1, 0 \leqslant i < u, 0 \leqslant j < v \quad (1)$$

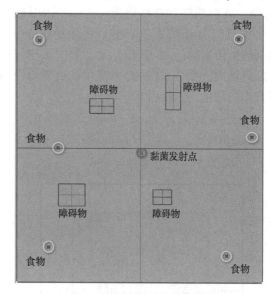

图 4 设置黏菌的运行环境

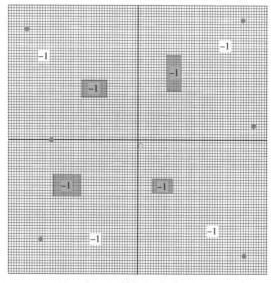

图 5 平面环境的分割与初始化

2.1.2 躲避障碍

避障问题，即从一个点到另一个点，一个区域到另一个区域之间并非一马平川，总会存在一些障碍，如高山、河流、地质灾害点和自然保护区等，因此，在线路规划问题中需考虑如何避开这些障碍物进行线路设计。本文模拟黏菌在觅食过程中绕开障碍物继续寻找食物，如图6，算法思路为：首先，判断平面上是否存在障碍物。计算平面环境的最小外包矩形 EFGH，在 EFGH 的两条平行边 EF 和 HG 上沿横轴按 $u_{interval}$ 向右推进，依次遍历各控制点，连接上下两条边上对应位置处的控制点形成一条曲线，如图中的带箭头的红线，与平面内的障碍物 ABCD 进行相交判断，若无交点，则继续向右推进；若存在交点 M、N，则可判断存在障碍物，进一步沿纵轴方向由 N 点向 M 点按 $v_{interval}$ 推进，将途径的每一个格网的值由初始的-1 变更为-2，表明该格网位于障碍物内，如公式(2)。由此可找出平面内的所有障碍物格网，如图7。

$$\begin{cases} griddata[i][j]=-2, \ inside \ obstacles \\ griddata[i][j]=-1, \ outside \ obstacles \end{cases} \quad (2)$$

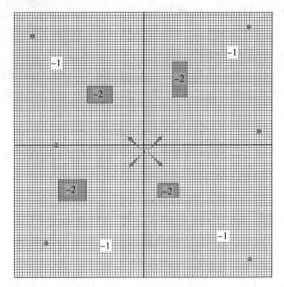

图7 判断平面内所有障碍物格网

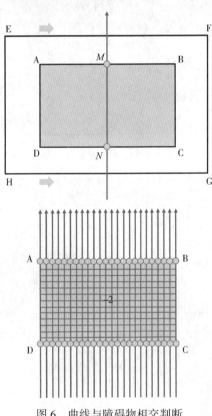

图6 曲线与障碍物相交判断

2.1.3 动态觅食

动态觅食主要包含①搜索食物：黏菌由发射点向外不断发射并在其 $n×n$ 邻域内寻找食物的过程；②接近食物：黏菌通过空气中的气味接近食物，在食物浓度低时更慢地接近食物，食物浓度高时更快地接近食物；③包围食物：当黏菌发现食物时会发生收缩，逐渐包围食物。黏菌静脉接触的食物浓度越高，生物振荡器产生的传播波越强，细胞质流动越快。其位置更新公式如，第一行表示黏菌的①②过程；第二、三行表明黏菌的③过程。

$$X(t+1)=\begin{cases} r×(UB-LB)+LB, \ r<z \\ X_b(t)+vb×(W×X_A(t)-X_B(t)), \\ \quad r<p, \ -a \leq vb \leq a \\ vc×X(t), \ r \geq p, \ -1 \leq vc \leq 1 \end{cases} \quad (3)$$

式中，UB 和 LB 为搜索范围的上、下界；r 为 $[0,1]$ 之间的随机数，$z=0.03$；t 为当前迭代次数；vb 为 $[-a, a]$ 内的随机数；vc 为在 $[-1,1]$ 内振荡并最终趋于零的参数；$X(t)$ 为当前黏菌个体的位置；$X(t+1)$ 为下一次迭代的黏菌个体位置，$X_b(t)$ 表示当前发现食物气味浓度最高位置(也是适应度最优的个体位置)，$X_A(t)$ 和 $X_B(t)$ 表示两个随机个体的位置，W 表示黏菌的权重系数。vb、vc 和 W 参数的微调可以改变黏菌位置。控制参数 p、参数 a 和权重系数 W 的更新公式分别为(4)~(7)：

$$p=\tanh|S(i)-DF|, \ i=1, 2, \cdots, N \quad (4)$$

$$a=arctanh\left(-\frac{t}{\max T}+1\right) \quad (5)$$

$$W(SmellIndex(i))=\begin{cases} 1+r×\log\left(\frac{bF-S(i)}{bF-wF}+1\right), \ i<\frac{N}{2} \\ 1-r×\log\left(\frac{bF-S(i)}{bF-wF}+1\right), \ i \geq \frac{N}{2} \end{cases}$$
$$(6)$$

$$SmellIndex=sort(N) \quad (7)$$

式中，N 为种群规模；$S(i)$ 为第 i 个黏菌个体的适应度；DF 为所有迭代中的最佳适应度；$\max T$ 为最大迭代次数；r 为[0，1]之间的随机数；$i<N/2$ 为 $S(i)$ 排名前一半的黏菌个体；bF 为在当前迭代获得的最佳适应度；wF 为在当前迭代获得的最差适应度；$SmellIndex$ 为适应度序列；$sort(\ )$ 为排序函数。

整个黏菌群的觅食过程是由单细胞黏菌组成的由局部到整体的运动趋势，单个黏菌向外搜索过程如下图所示。原始单细胞黏菌所在格网索引 Index=0，假设搜索半径 radius=4，则在其 80 个邻域内按上述规则进行搜索，避开障碍物(图 8 中的红色格网)，向外增殖到下一格网，Index 递增，并将格网值由初始的-1 变为新的 Index，表明该格网已被黏菌占据(图 8 中的黄色格网)。每一个新生成的黏菌又可向外继续增殖，依此类推，可形成一条觅食路径。对整个黏菌群而言，则可形成多条觅食路径。

$$\begin{cases} griddata[i][j]=-2, & inside\ obstacles \\ griddata[i][j]=-1, & outside\ obstacles \\ griddata[i][j]=Index, & occupied\ grids \end{cases} \tag{8}$$

-1	-1	-1	-1	-1	-1	-1	-1	-1
-2	-1	-1	-1	-1	-2	-2	-1	-1
-2	-1	-1	-1	-1	-2	-2	-1	-1
-1	-1	-1	-1	-1	-1	-1	-1	-1
-1	-1	-1	-1	0	-1	-1	-1	-1
-1	-1	-1	-1	-1	-1	-1	-1	-1
-1	-1	-1	-1	-1	-1	-1	-1	-1
-1	-1	-1	-1	-1	-1	-1	-2	-1
-1	-1	-1	-1	-1	-1	-1	-2	-1

-1	-1	-1	-1	-1	4	-1	-1	-1
-2	-1	-1	-1	3	-2	-2	-1	-1
-2	-1	-1	2	-1	-2	-2	-1	-1
-1	-1	-1	1	-1	-1	-1	-1	-1
-1	-1	-1	-1	0	-1	-1	-1	-1
-1	-1	-1	-1	-1	-1	-1	-1	-1
-1	-1	-1	-1	-1	-1	-1	-1	-1
-1	-1	-1	-1	-1	-1	-1	-2	-1
-1	-1	-1	-1	-1	-1	-1	-2	-1

图 8　单个黏菌向外增殖过程

2.1.4　触碰边界

在实体建模和计算机辅助设计中，边界表示(Boundary representation)，通常缩写为 B-rep 或 Brep，是一种利用极限来表示形状的方法。一个实体被表示为一组连接的表面元素，这些元素定义了内部和外部点之间的边界。Brep 是在 20 世纪 70 年代早期由剑桥大学的 Braid(用于 CAD)和斯坦福大学的 Baumgart(用于计算机视觉)独立开发的。因为黏菌的运行环境和障碍物均为 Brep 对象，当黏菌在运动过程中触碰到 Brep 边界后需设置其反应行为，如图 9，α 为入射角，β 为反射角。第一种情形为触边后方向随机，即 $\alpha \neq \beta$；第二种情形为触边后消亡，即 $\beta=0$；第三种情形为触边后反弹，即 $\alpha=\beta$。

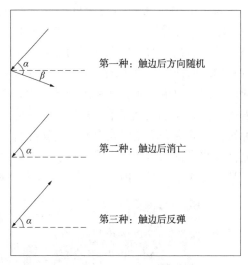

图 9　黏菌触碰到边界后的几种行为方式

2.1.5　路径优选

针对黏菌群形成的多条觅食路径，选择“最优”路径即可，这里的“最优”可以有不同的含义。对于“全国一张网”的管线路由规划问题而言，“最优”可以是距离最短、耗费最少、服务范围最大、惠及人口最多等指标，其公式可表示为(8)，其中，(x_{l_0}, y_{l_0})，(x_{l_1}, y_{l_1})，…，(x_{l_n}, y_{l_n}) 为黏菌觅食的足迹点，L_{best} 为最优路径。

$$L_{best}=\sum_{i=0}^{n-1}\sqrt{(x_{l_{i+1}}-x_{l_i})^2+(y_{l_{i+1}}-y_{l_i})^2},\quad 0 \leq i < n \tag{9}$$

2.2　OSGEarth 中算法的实现

当黏菌在真实地形上飞行时，包含起伏的地形以及障碍物等环境，要保证黏菌觅食成功率，就需要使用到地形回避技术。地形回避技术是指按照地形回避系统引导黏菌在规定高度上前进以达到避开起伏地形效果的一种低空突防技术，在飞行过程中使飞机保持在一定高度之上，从而避免发生飞机在飞行过程中与地形产生碰撞的可能性。根据实际，将飞机飞行高度确定为定值，此时在三维空间中的避障过程将转换为二维平面避障路径的处理。

将黏菌觅食路径与真实平面以网格化的形式抽象，以间距10m对平面进行分割并对得到的网格点以数组进行保存，以两点之间的真实距离作为估算代价运行，执行A*算法并记录所规划的路径上生成的坐标点。

3 实验与分析

3.1 几种常用寻径算法的对比

在Matlab中对几种不同的算法进行测试对比，蝠鲼觅食优化算法(MRFO)算法多为螺旋觅径方式，虽然蝠鲼的螺旋觅食行为能够对空间进行较好的全面搜索，但是翻滚觅食行为并没有很好地产生预期效果，尤其在跳出局部最优解方面仍需优化。海洋捕食者算法(MPA)算法是根据各种研究的规则和要点以及自然界的实际行为模拟捕食者和猎物的运动，虽然此算法具有合理的探索和开发速度，但MPA仍然在局部最优解附近停滞不前，无法实现全局最优解。黏菌优化算法(SMOA)是利用黏菌的扩散和觅食行为，属于元启发算法，具有收敛速度快，寻优能力强的特点，在线性规划方面具备突出的优势。将SMOA算法与往期的蝠鲼觅食优化算法(MRFO)、海洋捕食者算法(MPA)在部分CEC2017测试函数上进行对比测试，取种群规模为30，维度为30，最大迭代次数为10000，计算结果如图10所示。

算法	平均值	标准差	最差值	最优值
SMOA	70	0	14.145	0
MRFO	74	5.714	13.166	6.921
MPA	80	14.286	12.419	12.202

图10 不同算法的运行结果

3.2 黏菌觅食避障的实现

如图11(a)所示，黏菌在觅食过程中不能识别障碍物格网，致使黏菌穿透障碍物，这样的实验结果不能应用于实际线路规划当中。为了解决此问题，采用上述SMOA的"躲避障碍"算法进行Physarealm插件的二次开发，运行结果如图11(b)，红色的点为黏菌，中间的小方块为障碍物，可以实现黏菌成功躲避障碍物功能，表明实验结果具有可行性。

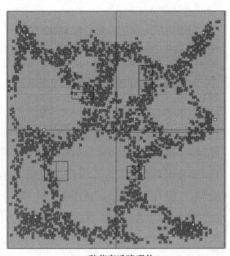
(a)黏菌穿透障碍物

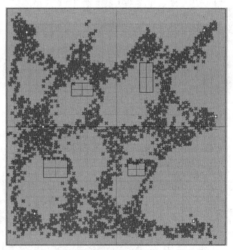
(b)黏菌绕开障碍物

图11 黏菌觅食避障的实现

3.3 全国一张网下的宏观路由自动选择

在解决了黏菌避障问题之后，将SMOA应用于实际的管线宏观路由规划，以全国6个省为研究区域进行实验。如表3所示，第三列列出了各省市实验设定的管线长度，第四列展示了黏菌优化算法计算得到的管线长度。第五列展示了各省市自治区的计算结果和实际情况的匹配度百分比。将黏菌优化算法计算的结果展现在OSGEarth构建的三维地理信息平台上，可直观的了解到各省的实际情况。总体而言黏菌优化算法计算得到的结果精度较高，可应用于未来的管线规划问题的可行性分析，为管网规划及工程建设提供一定的科学决策和支持。

表3 基于黏菌优化算法计算得到的全国6个省、市、自治区的宏观路由与实验设定情况的对比分析

序号	省/市/自治区	事先设定管线长度/km	SMOA计算得到的管线长度/km	匹配度/%
1	重庆市	69	73	94.20%
2	河北省	98	102	95.92%
3	山西省	53	57	92.45%
4	内蒙古自治区	89	83	93.26%
5	宁夏回族自治	35	37	94.29%
6	新疆维吾尔自治区	102	109	93.14%

4 结论与讨论

4.1 结论

通过黏菌优化算法动态觅食模型，通过环境布局设置，黏菌障碍躲避，黏菌动态觅食，黏菌触碰边界等验证，实现黏菌优化算法构建。利用黏菌会根据当前位置的客观条件（适应度函数优劣），决定每个个体所在位置的权重，然后个体会根据权重决定新的位置。针对黏菌觅食避障的实现是利用SMOA的"躲避障碍"算法进行Physarealm插件的二次开发，可以实现黏菌的自动避障。

针对全国一张网下的宏观路由规划问题，黏菌优化算法计算的结果展现在OSGEarth构建的三维地理信息平台上，实现了宏观路由最优路径规划。算例分析验证了本算法在天然气管网规划中的表现，也为黏菌优化算法指导天然气管网规划的有效性与可行性提供了证明。

4.2 讨论

通过Physarealm插件来研究黏菌优化算法在管道规划的应用，提供了一定的便利性。但是，Physarealm插件还存在诸多问题，例如稳定性和持续性等，导致无法满足管道应用场景要求。还需要投入较多的时间和精力解决上述存在的问题。对黏菌插件的科学性以及结合地理数据应用于管道规划的可行性进行深入研究。

参 考 文 献

[1] 国家发展改革委,国家能源局.关于印发《中长期油气管网规划》的通知:发改基础[2017]965号[EB/OL].(2017-05-19)[2019-03-21]. https://www.ndrc.gov.cn/xxgk/zcfb/ghwb/201707/t20170712_962238_ext.html.

[2] 田姗姗.基于多约束因子识别和层次分析法的油气管道路由选择方法研究[J].石油工程建设,2019,45(1):5.

[3] 赵倩维.卫星遥感技术在油气管道规划设计中的应用[J].化工管理,2017(1):1.

[4] Abudu D, Williams M. GIS-based Optimal Route Selection for Oil and Gas Pipelines in Uganda[J]. Advances in Computer Science, 2015, 4(4): 93-104.

[5] 王小龙.浅析人工智能在油气行业中的应用[J].现代信息科技,2017,1(2):3.

[6] Yildirim V, Yomralioglu T, Nisanci R, et al. A spatial multicriteria decision-making method for natural gas transmission pipeline routing[J]. Structure and Infrastructure Engineering, 2017, 13(5): 1-14.

[7] 丛瑞,冯骋,沈晨等.油气管道数字孪生技术应用[J].油气田地面工程,2022,41(10):108-113.

[8] 屈洪春,蹇霜,王平,等.基于多头绒泡菌仿生算法的城市燃气管网自动优化方法:CN201410086412.1[P].CN103886389A[2023-11-06].

[9] Atsushi Tero, Seiji Takagi, Tetsu Saigusa, et al. Rules for Biologically Inspired Adaptive Network Design [J]. Science, 2010, 327: 439-442.

[10] Toshiyuki Nakagaki, Hiroyasu Yamada, ágota Tóth. Intelligence: Maze-solving by an amoeboid organism [J]. Nature, 2000, 407(6803): 470-470.

[11] Li S M, Chen H L, Wang M J, et al. Slime mould algorithm: A new method for stochastic optimization [J]. Future Generation Computer Systems, 2020, 111, 300-323. aliasgharheidari.com: 300-323.

[12] Jeff Jones. From Pattern Formation to Material Computation: Multi-agent Modelling of Physarum Polycephalum[M]. 2015. Berlin: Springer.

[13] 王庆.基于多头绒泡菌的路网优化算法[D].重庆:西南大学,2023.

[14] 高金兰,王良禹,宋爽.基于改进黏菌算法的配电网重构研究[J].吉林大学学报(信息科学版),

2022, 40(5): 759-766.

[15] 任丽莉, 王志军, 闫冬梅. 结合黏菌觅食行为的改进多元宇宙算法[J]. 吉林大学学报(工学版), 2021(006): 051.

[16] 郑民, 李建忠, 吴晓智, 等. 我国主要含油气盆地油气资源潜力及未来重点勘探领域[J]. 地球科学, 2019, 44(3): 833-847.

[17] 张小强, 蒋庆梅. 寒冷地区埋地输气管道最低设计温度选取[J]. 油气田地面工程, 2017, 36(12): 4.

[18] 肖鹏, 刘更代, 徐明亮. OpenSceneGraph 三维渲染引擎编程指南[M]. 北京: 清华大学出版社, 2010.

[19] Ian C. Braid. Designing with volumes[D]. London: University of Cambridge, 1973.

[20] Bruce G. Baumgart. Winged Edge Polyhedron Representation[R]. Palo Alto: Computer Science Department, School Of Humanities And Sciences, Stanford University, 1972, 1-46.

站场管座角焊缝专用扫查装置开发

路兴才 项小强 赵岩 赵子峰 周聪 赵龙一

(国家管网集团工程技术创新有限公司)

摘要 【目的】近几年油气管道站场管座角焊缝失效泄漏事件时有发生,磁粉、渗透仅能检测表面缺陷,常规超声检测(UT)不能完全覆盖且信号干扰多,容易漏检和误判,射线检测(RT)同样存在漏检的现象,而相控阵超声检测(PAUT)能有效发现潜在的内部缺陷,由于管座角焊缝具有复杂的结构和坡口型式,目前PAUT尚无在管座角焊缝检测中的相关研究和应用,尤其是机械扫查装置不适合管座角焊缝,需对站场管座角焊缝专用扫查装置进行研究。【方法】首先结合缺陷分析、管座建模,自行设计了17种试块,根据试块的管座角焊缝主管、支管和管座规格型号,通过3D建模,按照PAUT在主管侧检测和在支管侧检测,分别研发了两种扫查装置。其次使用CIVA软件进行仿真分析,考虑不同管径、壁厚对仿真结果的影响,筛选出适用的扫查装置。并在实验室进行组装和调试,选取典型站场进行专用扫查装置性能试验。【结果】该成果在西三线西峡分输压气站、西四线吐鲁番站和了墩站开展了管座角焊缝超声相控阵技术的验证,从缺陷响应幅值得出,除了横向裂纹缺陷的灵敏度较低,PAUT检出了17个管座角焊缝的未熔合、未焊透、气孔等缺陷,实现了超声相控阵扫查支架配装对站内不同管径管座角焊缝的适应及缺陷的有效检出;PAUT能够有效检出内部缺陷,通过发现潜在的内部缺陷,可以避免缺陷引起的严重安全事故和环境污染,因此通过PAUT检测及时发现并修复管座角焊缝的缺陷,可以有效预防泄漏事故的发生,减少因维修和停产造成的巨大经济损失。【结论】管座角焊缝相控阵超声检测专用扫查装置的设计及开发,实时控制并记录相控阵检测扫查位置,解决了以往管座角焊缝检测中的位置偏差对检测结果的影响,同时管座角焊缝无损检测工艺方案制定,实现管座角焊缝内部缺陷及近表面缺陷的数字化检测和存储。(图13,表3,参8)

关键词 管座角焊缝;相控阵超声检测(PAUT);仿真分析;专用扫查装置;缺陷响应幅值;内部缺陷

Development of special scanning device for fillet weld of pipe seat in station and yard

Lu Xingcai Xiang Xiaoqiang Zhao Yan Zhao Zifeng Zhou Cong Zhao Longyi

(Pipe China Engineering Technology Innovation Co. Ltd)

Abstract [Objective] In recent years, the failure and leakage of fillet welds of pipe seats in oil and gas pipeline stations often occur. Magnetic particle and penetration can only detect surface defects, while conventional ultrasonic detection(UT) cannot completely cover them and has a lot of signal interference, which is easy to miss and misjudge. Radiometric detection(RT) also has the phenomenon of missing detection, while phased array ultrasonic detection(PAUT) can effectively detect potential internal defects. Due to the complex structure and groove type of fillet weld, there is no relevant research and application of PAUT in the detection of fillet weld. In particular, mechanical scanning device is not suitable for fillet weld, so it is necessary to study the special scanning device for fillet weld in station. [Methods] Firstly, 17 kinds of test blocks were designed based on defect analysis and base modeling. According to the specifications and models of the fillet weld head, branch pipe and base of the test block, two kinds of scanning devices were developed through 3D modeling according to the detection of the main pipe side and the detection of the branch pipe side by PAUT. Secondly, CIVA software was used for simulation analysis, considering the influence of different pipe diameters and wall thicknesses on simulation results, and suitable scanning devices were selected. The assembly and debugging were carried out in the laboratory, and the performance test of the special scanning device was carried out in a typical station. [Results] The results were verified by ultrasonic phased array technology of fillet welds at Xixia sub-transmission gas station of the West Third Line, Turpan Station of the West Fourth Line and Gudun Station. The defect response was significant. In addition to the low sensitivity of transverse crack defects, PAUT detected defects such as non-fusion, non-penetration and porosity of 17 fillet welds. The ultrasonic phased array scanning bracket is equipped to adapt to fillet welds of different pipe diameters in the station and effectively detect defects. PAUT can effectively detect internal defects, and by discovering potential internal defects, serious safety accidents and environmental pollution caused by defects can be avoided. Therefore, timely detection and repair of defects in fillet welds of tube seat through PAUT detection can effectively prevent leakage accidents and reduce huge economic losses caused by maintenance and shutdown. [Conclusion] The design and development of a special scanning device for phased array ultrasonic inspection of fillet weld of tube seat, real-time control and recording of the scanning position of phased array inspection, solved the influence of position deviation in the previous detection of fillet weld of tube seat on the test results, and formulated the non-destructive testing process plan of fillet weld of tube seat, realized the digital detection and storage of internal defects and near-surface defects of fillet weld of tube seat. (13 Figure, 3 Tables, 8 References)

Key words Seat fillet weld; Phased array ultrasonic inspection(PAUT); Simulation analysis; Special scanning device; Defect response amplitude; Internal defect

近几年管座角焊缝失效泄漏事件时有发生，西气东输管道公司从2015年发生多起站场支管与主管连接角焊缝泄漏事件，经分析发现泄漏角焊缝大多存在超标缺陷，缺陷类型包括裂纹、未熔合等。2018年某天然气站场双金属温度计套管与主管道间角焊缝处渗漏，经检验分析认为：温度计套管与工艺管线间角焊缝处存在气孔和夹渣缺陷，在工作内压及振动载荷共同作用下导致焊缝泄漏。2020年西气东输二线某放空管线根部与管座焊缝处泄漏，经分析认为焊缝焊趾外表面存在硬质点缺陷，硬质点缺陷邻近位置在外力及振动作用下萌生疲劳裂纹，裂纹自焊趾沿管道厚度方向扩展直至泄漏，且该焊缝在建设期未做内部缺陷检验。

目前管座角焊缝检测常用的方法包括磁粉（MT）、渗透（PT）、手工超声（UT）和射线（RT）检测。其中磁粉检测仅能对角焊缝表面及近表面缺陷进行检测，渗透检测仅适用于表面开口缺陷的检测。这两种方法对于管座角焊缝内部及根部缺陷均无法有效检出。超声和射线检测技术应用也有一定的局限性。首先，管座角焊缝的结构特殊、形状复杂、大多数情况下仅能从外表面对其实施检测，运用斜探头在管座角焊缝外表面进行检测时，焊缝上有些区域声束扫查不到，或仅能从一个方向扫查，手动常规超声检测容易误判或漏检重要缺陷；其次，由于焊缝结构形状复杂，超声波回波信号干扰因素多，缺陷信号难以识别；再次，由于检测面是曲面，特别是当主管及支管尺寸较小时，检测面曲率大，超声波探头及楔块与工件耦合困难，超声声束难以有效进入工件；最后，射线检测不能对缺陷的高度进行量化外，受角焊缝结构的限制，只能从有限的几个方向进行透照，焊缝根部和熔合面上的缺陷容易漏检。同时，受站场或储气库内设备复杂，检测空间受限，存在无法进行射线检测的区域。

国内管座角焊缝的检测主要是在焊接完成后及运行期间的定期检验中实施，定期检验一般为三年或更长时间检验一次。定期检验多为抽检，按照一定的规则抽取一定数量的焊缝进行检测，检测方法多为磁粉或渗透检测，每次检测前均需要清除角焊缝及其附近的油漆，将检测部位打磨干净，存在检测效率低、速度慢，焊缝质量100%检验成本高的缺点。如何能在运行期间，在不对角焊缝表面油漆清理的条件下对角焊缝进行快速全面的检测，也是生产实践中亟需解决的一个问题。

因此，需要探索开发适用于管座角焊缝缺陷的检测技术，提高缺陷的检出率，确保角焊缝内在质量，提升设备运行的安全性和可靠性，同时满足智能管道建设、无损检测结果数字化的需要。目前管座角焊缝相控阵超声检测的相关研究较少，尤其是机械扫查装置不适合管座角焊缝，需对站场管座角焊缝专用扫查装置进行研究。

1 扫查方案研究

采用超声相控阵聚焦检测方式用二次波对安放式管座角焊缝进行检测。平面线性相控阵为超声相控阵中典型阵列,将其各阵元近似为点声源,通过特定延时法则使各个点声源的声场进行叠加,从而实现声束聚焦和偏转。装有斜楔块的平面线阵声束偏转聚焦示意图(图1),建立直角坐标系,以第1个阵中心的坐标点 $O(x_0, z_0)$ 为参考点,x 轴与楔块底面平行,且其正向向右,而 z 轴正向向下垂直于楔块底面,假设声束聚焦于焦点 $Q(x_q, z_q)$,聚焦深度为 H。假设探头阵元数目为 N,阵元中心间距为 P,第 m 个阵元的坐标点为 $M(x_m, z_m)$,其中 m 为整数,且 $1 \leq m \leq N$,第 m 个阵元声束在楔块和工件两种介质中传播路径分别为 l_m、$L_m(m=1, 2, \cdots, N)$,第 m 个阵元声束到达楔块和工件的界面时所在交点为 $B_m(xb_m, zb_m)$,第 m 个阵元声束到达楔块和工件的界面时入射角为 $\alpha_m(m=1, 2, \cdots, N)$,其在工件中的折射角为 $\beta_m(m=1, 2, \cdots, N)$,第一个阵元中心到楔块底部距离为 H_0。另外,楔块倾斜角为 θ,其声速为 c_w,而工件的声速为 c_s。

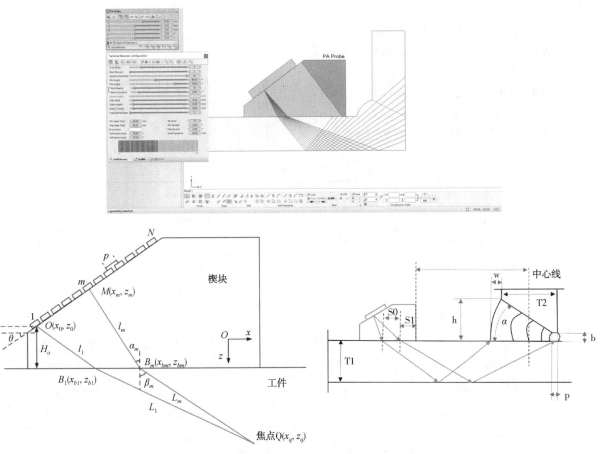

图1 PA扫查角焊缝示意图

楔块和工件检测面接触进行角焊缝检测时,第 m 个阵元的声束到焦点的时间为:

$$T_m = \frac{\sqrt{(x_{bm}-x_m)^2+(z_{bm}-z_m)^2}}{c_w} + \frac{\sqrt{(x_q-x_{bm})^2+(z_q-z_{bm})^2}}{c_s} \quad (1)$$

假设一次激发 N 个阵元,由式(1)可以求得每个阵元的声束到达焦点的时间,从而确定出所有阵元的声时最大值 T_{max},因而可计算得到第 m 个阵元的延迟时间为:

$$\Delta T_m + T_{max} = T_m \quad (2)$$

根据斯涅耳定律可得:

$$\frac{\sin\alpha_m}{C_w} = \frac{\sin\beta_m}{C_s} \quad (3)$$

由几何关系可知:

$$\sin\alpha_m = \frac{x_{bm} - x_m}{l_m}, \quad \sin\beta_m = \frac{x_q - x_{bm}}{L_m} \quad (4)$$

$$l_m = \sqrt{(x_{bm} - x_m)^2 + (z_{bm} - z_m)^2}$$
$$L_m = \sqrt{(x_q - x_{bm})^2 + (z_q - z_{bm})^2} \quad (5)$$

已知 $x_m = x_0 + (m-1)p\cos\theta$，$z_m = z_0 - (m-1)p\sin\theta$，$z_{bm} = z_0 - (m-1)p\sin\theta + H_0$，$H = z_q - z_{bm}$，将它们与式(4)和式(5)代入式(3)，并且 x_q 已知，则可以求得 x_{bm}，然后将已知参数代入式(1)中求得每个阵元的声束到焦点的时间，进而求出各个阵元的延时。

若使声波覆盖整个被测焊口，PAUT 采用扇形扫描，声束 l_1 最小角度设为 α_1，声束 l_m 最大角度设为 α_m。超声波从探头发射出来，沿着楔块到与钢交界面处，从交界面折射入钢管中，声束 L_1 应能至少覆盖焊缝最左侧，声束 L_m 至少能覆盖到焊缝根部最右侧，方能保证检测过程覆盖整个被检区域。

声束 L_1 与声束 L_m 之间的距离应大于 $T_2 + W$：
$$T_2 + W \leq S_0 + 2T_1\sin\beta_m - 2T_1\sin\beta_1 \quad (6)$$
计算探头前沿距离管座中心线的距离：

$$S \geq 2T_1\sin\beta_m - S1 - \frac{T_2}{2} \quad (7)$$

其中，S_0 和 S_1 还与楔块角度和晶片的选择相关。在设置 S 时，可以通过调整晶片数量和扫查角度来改变探头前沿到中心线的距离，使扫查区域覆盖整个焊缝。首先设置探头前沿距接管中心线 1.5mm 的位置，如果不能完全覆盖焊缝区域，仿真过程中可以调整此数值。

设计扫描方式为扇形扫描，使用二次波进行检测。步进方向为在主管及支管座上沿周向 0°~360°对焊缝处进行扫查，在不同管径、壁厚、探头下形成最佳扫查方案，确保可以检测出所有缺陷。

2 专用扫查装置设计

通过前期调研，选取西三线西峡分输压气站、西四线吐鲁番站和了墩站三个站场作为研究对象，三个站场含 17 种管座角焊缝，主管、支管和管座规格型号如表 1 所示。

表 1 三个站场管座的钢管规格

编号	特征	主管规格/mm	支管规格/mm	支管座规格/mm
1	碳钢/管座对接角焊缝	$\Phi1219\times22$	$\Phi168.3\times8.8$	DN1200×150T=38
2	碳钢/管座对接角焊缝	$\Phi1219\times22$	$\Phi114.3\times6.3$	DN1200×100T=26
3	碳钢/管对接角焊缝	$\Phi1219\times22$	$\Phi60.3\times5$	DN1200×50T=20
4	碳钢/管座对接角焊缝	$\Phi1016\times21$	$\Phi168.3\times8.8$	DN1000×150T=38
5	碳钢/管座对接角焊缝	$\Phi914\times23.6$	$\Phi168.3\times8.8$	DN900×150T=38
6	碳钢/管座对接角焊缝	$\Phi914\times23.6$	$\Phi60.3\times5$	DN900×50T=20
7	碳钢/管座对接角焊缝	$\Phi813\times21$	$\Phi114.3\times6.3$	DN800×100T=26
8	碳钢管对接角焊缝	$\Phi813\times21$	$\Phi60.3\times5$	DN800×50T=20
9	碳钢管对接角焊缝	$\Phi711\times20$	$\Phi114.3\times6.3$	DN700×100T=26
10	碳钢管对接角焊缝	$\Phi711\times20$	$\Phi60.3\times5$	DN700×50T=20
11	碳钢管对接角焊缝	$\Phi610\times20$	$\Phi114.3\times6.3$	DN600×100T=26
12	碳钢管对接角焊缝	$\Phi610\times20$	$\Phi60.3\times5$	DN600×50T=20
13	碳钢/低合金钢管对接角焊缝	$\Phi406\times14.2$	$\Phi114.3\times6.3$	DN400×100T=26
14	碳钢/低合金钢管对接角焊缝	$\Phi406\times14.2$	$\Phi60.3\times5$	DN400×50T=20
15	碳钢/低合金钢管对接角焊缝	$\Phi273\times10$	$\Phi60.3\times5$	DN250×50T=20
16	碳钢管对接角焊缝	$\Phi219.1\times12.5$	$\Phi60.3\times5$	DN200×50T=20
17	碳钢管对接角焊缝	$\Phi168.3\times8.8$	$\Phi60.3\times5$	DN150×50T=20

其中，主管规格在 DN150mm（Φ168mm）~ DN1200mm（Φ1219mm）之间，支管有三种类型 DN50mm（Φ60.3mm×5mm）、DN100mm（Φ114.3mm×6.3mm）、DN150mm（Φ168mm×8.8mm）。

管座角焊缝 PAUT 检测试验分为在主管侧检测和在支管座侧检测。

2.1 在主管侧检测

由表 1 可知，主管管径跨度较高，因此设计为小管径扫查器和大管径扫查器两套装置。探头在主管侧检测的运行轨迹（图 2）。

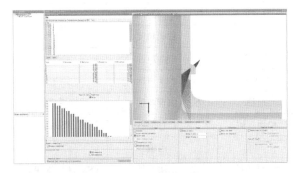

图 2 主管上检测轨迹

扫查器支架设计方案分为两种，一种是爬行装置直接架在管座上检测，一种是架在支管上检测。考虑到管座平台长度过低、尺寸过小可能影响到检测稳定性，因此选择将卡紧装置安装在支管上。根据主管扫查装置机械装配总设计以及零部件设计图，进行 3D 建模，主管扫查装置 3D 建模（图 3）。

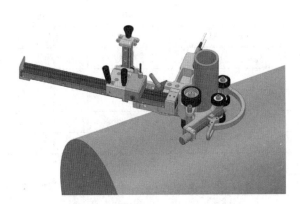

图 3 主管扫查装置 3D 建模

2.2 在支管侧检测

探头在支管座上，可以类比于对接焊缝 PAUT 检测，扫查轨迹为圆周（图 4），因此扫查装置设计也可参照对接焊缝扫查装置。

同样将爬行装置安装在支管上，探头在支

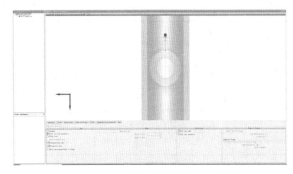

图 4 探头在支管座检测轨迹示意图

管座转动一圈。由于支管座只有三种管径，因此设计一套装置（手动扫查器）即可。支管手动扫查器零部件设计包括圆弧框架、探头滑轨安装块、导轨安装座、移动探头支臂设计、编码器滚轮、编码器固定支架设计、探头固定摆臂等。根据支管扫查装置机械装配总设计以及零部件设计图，进行 3D 建模，支管扫查装置 3D 建模（图 5）。

图 5 支管扫查装置 3D 建模

3 PAUT 整体试验

对研发的专用扫查装置进行性能检测，以西三线西峡分输压气站、西四线吐鲁番站和了墩站的 17 种管座角焊缝作为测试对象。

首先确定扫查方案，其次在实验室进行扫查装置的测试，最后在施工现场进行相控阵超声检测技术的现场验证。

3.1 扫查方案

本文采用 CIVA 软件对声束覆盖进行模拟仿真分析，PAUT 采用扇形扫查，扫查角度 40°~70°，采用的探头型号为 5MHz-16，PAUT 扫查方案详见表 2。

表2 扫查方案

试块规格	详见表1	焊接方法	GTAW↑+SMAW↑
坡口型式	角焊缝	坡口角度/(°)	35~50
坡口钝边/mm	1.6±0.8	对口间隙/mm	2.5~4.0
角焊缝焊脚差/mm	≤3.0	角焊缝凸度/凹度/mm	≤1.5
检测条件及标准			
检测时机	焊后	温度差	±15℃
表面状况	打磨平整	检测比例	100%	合格级别	Ⅱ
相控阵设备及工艺参数			
设备型号	ZETEC	楔块型号	定制	探头频率	5MHz
角度步进	1°	使用晶片	1-16	激发晶片数	16
扫查速度	40mm/s	聚焦深度	2.2倍壁厚	检测角度	40°~70°
波型模式	横波	PA1声速	实测值	PA2声速	实测值
波型模式	纵波	PA3声速	实测值	PA4声速	实测值
扫查器类型	自动扫查器	扫描类型	扇扫	扫查类型	纵向垂直扫查
检测区域	角焊缝	扫查面	单面双侧	扫查覆盖	≥50mm
扫查分辨率	1mm	表面补偿	3dB	步进偏移	±25mm
记录方式	编码器	耦合剂	水/防冻液	试块型号	CSK-ⅠA/CSK-ⅡA-1
扫查灵敏度	Φ2×35 横孔满屏高度的80%增益值+2dB		表面补偿	+3dB

1. 扫查说明：探头距离焊缝中心线的位置以工艺要求步进值为准，误差距离±1mm。
2. 耦合通道底面回波峰值信号调整满屏高度80%，增加6-10dB作为耦合监视通道灵敏度。

3.2 实验室测试

根据确定的扫查方案，首先将选定的探头安装在扫查装置上，进行扫查装置的组装工作，随后进行了扫查装置的现场调试，确定扫查角度、扫查速度等参数，最后利用实验室的17种管座角焊缝试块，进行扫查装置的性能测试(表3)。

表3 模拟试块检测结果

缺陷编号	缺陷位置/(°)	缺陷位置/mm	长度/mm	深度/mm	幅值/%	缺陷类型
D1	0	0	8	5	>100	点状
D2	135.9	269	8	38	>100	未焊透
D3	176.8	350	8	19.4	72.1	裂纹
D4	203	402	38	27	>100	线状
D5	234.4	464	9	13.3	65.3	点状
D6	270.2	535	28	14.1	85.8	裂纹
D7	291.9	578	16	28.6	>100	裂纹
D8	46.9	93	36	30.5	>100	线状
D9	84.9	168	19	28.9	>100	裂纹
D10	132.8	263	16	26	>100	线状
D11	221.2	438	9	38	>100	未焊透
D12	265.2	525	12	14.3	>100	线状
D13	0	0	14	8.2	97.4	裂纹
D14	64.7	83	10	26	>100	未焊透
D15	133.4	171	12	5.1	44.2	线状

续表

缺陷编号	缺陷位置/(°)	缺陷位置/mm	长度/mm	深度/mm	幅值/%	缺陷类型
D16	58.8	79	18	12.8	>100	线状
D17	119.9	161.5	13	24.3	>100	未焊透

实验室检测结果显示，除了横向裂纹信号低，PAUT检出了17种试块的未熔合、根部未焊透、气孔、裂纹、夹渣等缺陷。PAUT的检测效率相对较高，在被检工件尺寸和形貌确定的前提下，能够在较短的时间内完成对大量部件或结构的检测，且实验中模拟试块缺陷100%检出，因此PAUT工艺、检测方法具备实用性和可推广性。

3.3 现场验证

在实验室完成对相控阵超声检测(PAUT)的分步研制以及整体试验后，选取西三线西峡分输压气站、西四线吐鲁番站和了墩站的现场，对17种管座角焊缝相控阵超声检测技术进行了现场验证。

PAUT检出了17种管座角焊缝的点状缺陷、线状缺陷等，部分检测结果如表4所示。

表4 三座站场部分检测结果

序号	焊口编号	主管外径/mm	支管尺寸/mm	缺陷数量	缺陷类型
1	XQD4T01-SP003-NG010301-20-S-S-SCZC03	1219	168.3	1	点状缺陷
2	XQD4T01-SP003-NG010301-21-S-S-SCZC03	1219	168.3	1	点状缺陷
3	XQD4T01-SP003-NG010302-71-S-S-SCZC03	1016	60.3	0	—
4	XQD4T01-SP001-UT010312-18-S-S-SCZC01	219.1	88.9	0	—
5	XQD4T01-SP001-NG010302-68-S-S-SCZC01	1016	168.3	0	—
6	XQD4T01-NG010302-66-S-S-SCZC07	1016	219.1	1	线状缺陷

现场验证包含17种不同管座的角焊缝接头类型，对缺陷产生原因进行现场分析。通过分析得出，检出高度0.5mm、长5mm的缺陷，PAUT不仅能检测出缺陷，还可以准确地确定缺陷信息（如位置、长度、深度等），因此PAUT检测具备检测准确性和可靠性。

4 结论

（1）国内首次研制成功站场管座角焊缝专用扫查装置，填补了站场无损检测领域的空白。本文分别研究了主管扫查器和支管扫查器，能够满足主管侧和支管座侧双侧检测。主管侧设计了双轴扫查装置，能够适用于不用弧度的马鞍形检测面，满足现场需求的不同管径、不同壁厚的PAUT检测。支管扫查器能够适用不同壁厚的检测需求。

（2）该成果在西三线西峡分输压气站、西四线吐鲁番站和了墩站开展了超声相控阵技术的验证，实现了超声相控阵扫查支架配装对站内不同管径管座角焊缝的适应及缺陷的有效检出。

（3）站场管座角焊缝相控阵超声检测专用扫查装置的设计及开发，实时控制并记录相控阵检测扫查的位置，解决了以往管座角焊缝检测中的位置偏差对检测结果的影响。

参 考 文 献

[1] 齐高君,王耀礼,丁成海,等.小径管管座角焊缝相控阵超声检测工艺[J].无损检测,2019,41(10):44-49.

[2] 李守彬,夏中杰,孔晨光,等.相控阵超声检测技术在核电厂不等厚对接环焊缝检测中的应用[J].压力容器,2020,37(10):64-69.

[3] 于达,龙华明,孙亚娟,等.小直径管管座角焊缝相控阵检测探究[J].焊管,2015,38(8):16-19.

[4] 张义磊.相控阵技术在插入式管座角焊缝检测中各类缺陷的识别探究[J].化工装备技术,2021,42(1):45-48.

[5] 吴家喜,张子健,张小龙,等.承压设备插入式接管角焊缝超声相控阵检测工艺[J].无损检测,2020,42(3):43-49.

[6] 陈乐,胥杨,邵晗烽,等.基于正交设计的管座焊缝相控阵检测参数寻优方法研究[J].锅炉技术,2021,52(增刊1):20-24.

[7] 钱盛杰,黄焕东,沈成业,等.插入式管座角焊缝缺陷的柔性相控阵检测技术研究[J].机械强度,2022,44(4):813-818.

[8] 梁国安, 姚叶子, 郑凯, 等. 基于超声相控阵的角焊缝缺陷信号重构方法研究[J]. 计算机测量与控制, 2022, 30(3): 222-228.

[9] 曹燕亮, 云维锐. 超声相控阵多聚焦技术在奥氏体不锈钢检测中的应用[J]. 无损探伤, 2022, 46(6): 43-45.

[10] 张昊, 陈世利, 贾乐成. 基于超声相控线阵的缺陷全聚焦三维成像[J]. 电子测量与仪器学报, 2016, 30(7): 992-999.

[11] XIE Q, TAO J H, WANG Y Q, GENG J H, CHENG S Y, LÜ FC. Use of ultrasonic array method for positioning multiple partial discharge sources in transformer oil[J]. Review of Scientific Instruments, 2014, 85(8): 084705.

[12] SCHICKERT M. Three-dimensional ultrasonic imaging of concrete elements using different SAFT data acquisition and processing schemes[J]. AIP Conference Proceedings, 2015, 1650(1): 104-113.

[13] 赵湘阳, 曹学文, 曹恒广, 等. 超声回波法海底管道外涂层检测技术[J]. 中国石油大学学报(自然科学版), 2022, 46(5): 162-169.

[14] 李玉坤, 于文广, 李玉星, 等. 超声临界折射纵波测量应力的温度影响[J]. 中国石油大学学报(自然科学版), 2021, 45(2): 134-140.

[15] 叶晓同, 赵鹏, 郑珂, 等. 基于超声相控阵的水下船体表面成像方法研究[J]. 计量学报, 2020, 41(1): 79-84.

[16] 孙明明, 方宏远, 赵海盛, 等. 不规则缺陷管道失效压力影响因素及评价方法[J]. 中国石油大学学报(自然科学版), 2022, 46(4): 152-159.

[17] 寿乐勇. 天然气长输管道内腐蚀原因分析及控制措施[J]. 中国石油和化工标准与质量, 2020, 40(16): 31-32.

[18] 靳世久, 杨晓霞, 陈世利, 等. 超声相控阵检测技术的发展及应用[J]. 电子测量与仪器学报, 2014, 28(9): 925-934.

[19] DRINKWATER B W, WILCOX P D. Ultrasonic arrays for nondestructive evaluation: a review[J]. NDT & E International, 2006, 39(7): 525-541.

[20] 谢飞, 李佳航, 王新强, 等. 天然气管道 CO_2 腐蚀机理及预测模型研究进展[J]. 天然气工业, 2021, 41(10): 109-118.

[21] 江雁山, 邓进, 李喆, 等. 长输管道内壁腐蚀缺陷的超声波衍射时差法检测[J]. 无损检测, 2021, 43(7): 57-59, 78.

[22] 盛沙, 戴波, 谢祖荣. 管道超声内检测三维成像技术研究[J]. 北京石油化工学院学报, 2012, 20(1): 1-5.

基于 IMU 数据的管道力学响应分析

余晓峰[1]　黄忠宏[1,2]　陈翠翠[3]　吕志阳[2]　李云涛[2]

[1. 国家管网集团工程技术创新有限公司；2. 中国石油大学(北京)安全与海洋工程学院；
3. 国家管网集团西部管道有限责任公司]

摘　要　【目的】随着惯性测绘单元(IMU)在内检测中的应用，产生了大量的管道中心线数据，如何根据这些数据评估管道受到的载荷与变形，成为管道应力分析的重要问题。【方法】为了解管道的变形和应力应变状态，采用非线性土弹簧模型，提出了基于 IMU 数据的管道应力分析方法。针对某沉降管道，基于 IMU 数据绘制管道的走向，计算管道沉降量，并建立沉降作用下埋地管道有限元模型。利用三次 B 样条计算方法，对管道沉降数据进行处理，获得管道沉降插值数据，将其作为位移载荷施加至模型中。【结果】研究结果表明，基于某条管道的应力监测数据，对基于 IMU 数据的管道应力分析方法进行了验证，管道截面的应力监测最大值与数值计算最大值位置基本一致、大小相近，应力监测结果与有限元计算结果的平均相对误差为 14.5%，表明有限元模型及插值方法具有较好的准确性；该沉降管道最大沉降值约为 0.66m，管道影响范围约为 58m，轴向应变范围为 -0.11% ~ 0.15%，轴向应力范围为 -137MPa ~ 375MPa，最大等效应力为 376MPa；根据基于应力的评估准则，最大等效应力未超出 0.9 倍的屈服强度，管道处于安全的运行状态；管道发生沉降时，存在三处应力峰值，分别位于沉降区中部、沉降区与非沉降区交界处，其中最危险位置为沉降区中部。【结论】该研究可为沉降区域管道的安全运行及维抢修提供一定的技术依据。

关键词　IMU；土弹簧模型；沉降；三次 B 样条；应力分析

Pipeline mechanical response analysis based on the IMU data

Yu Xiaofeng[1]　Huang Zhonghong[1,2]　Chen Cuicui[3]　Lv Zhiyang[2]　Li Yuntao[2]

(1. PipeChina Engineering Technology Innovation Co., Ltd.;
2. College of Safety and Ocean Engineering, China University of Petroleum(Beijing);
3. PipeChina West Pipeline Co., Ltd.)

Abstract　[Objective] With the application of IMU(Inertial Measurement Unit) in internal detection, a large amount of pipeline centerline data are generated. How to evaluate the load and deformation of the pipeline according to these data has become an important problem for pipeline stress analysis. [Methods] In order to understand the deformation and stress-strain state of the pipeline, an analysis method of pipeline stress based on IMU data is proposed by using nonlinear soil spring model. For a settlement pipeline, the direction of the pipeline is drawn based on IMU data, the settlement amount of the pipeline is calculated, and the finite element model of the buried pipeline under the action of settlement is established. The cubic B-spline calculation method is used to process the pipeline settlement data, and obtain the pipeline settlement interpolation data, which is applied to the model as displacement load. [Results] The research results show that, based on the stress monitoring data of a pipeline, the analysis method of pipeline stress based on IMU data is verified. The position of the maximum stress monitoring value of the pipeline cross-section is basically the same as that of the numerical maximum value, and the average relative error between the stress monitoring result and the finite element calculation result is 14.5%. The results show that the finite element model and interpolation method have good accuracy. The maximum settlement value of the settlement pipeline is about 0.66m, the influence range of the pipeline is about 58m, the axial strain range is -0.11% ~ 0.15%, the axial stress range is -137MPa ~ 375MPa, and the maximum equivalent stress is 376MPa. According to the stress-based evaluation criteria, the maximum equivalent stress does not exceed 0.9 times of the yield strength, and the pipeline is in a safe operating state. There are three stress peaks when the pipeline is settling, which are located in the middle of the settlement zone and the junction between the settlement zone and non-settlement area, among which the most dangerous position is the middle of the settlement zone. [Conclusion] This study can provide some technical basis for the safe operation, maintenance and emergency repair of the pipelines in the settlement zone.

Key words IMU; soil spring model; settlement; the cubic B-spline; stress analysis

1 引言

随着经济的发展，对油气资源的需求不断加大，越来越多的管道穿越复杂的自然地理环境，导致管道失效事故屡有发生，对社会经济发展及人员安全造成较大的后果。为确保管道的安全运行，诸多学者对地质灾害作用下的管道力学进行了分析。帅健等人采用解析法、数值法求解管道的应力应变，判断管道的运行状态。石磊等人通过现场测量获得管道沉降数据，对其进行谐波拟合，开展沉降管道受力分析。张玉等人针对塌陷和沉降作用，开展管道力学试验研究，提出了土压力计算方法。

对地质灾害频发重点区域埋地管道采取监测手段已成为管道管理部门的重要管理内容，但埋地管道输送距离长，难以掌握管道全线的变形状况。随着 IMU 的发展，管道全线变形检测得以实现。于希宁等人开展了中心线测绘检测试验，融合卫星及里程轮快速精准地定位管道缺陷。刘燊等人基于机器学习理论，对冻土区管道的变形特征进行智能识别。

目前，大多数学者主要是基于 IMU 数据开展地质灾害作用下管道的变形识别及应变计算，对于管道的应力状态分析考虑不充分，IMU 数据利用效率仍有待提高。同时，针对管道变形问题，现场工作人员无法快速判断管道运行状态。因此，本文提出了基于 IMU 数据的管道力学分析方法，采用三次 B 样条计算方法对基于 IMU 数据计算得到的管道沉降数据进行插值处理，建立应力有限元分析模型，可有效判断管道的应力状态，为管道的维抢修提供一定的理论支撑。

2 三次 B 样条计算方法

三次 B 样条法能够使曲线在各数据点上更加平滑，且具有计算简便、精度高、较好的收敛性等优势。

假设 $V_i(i=0,1,\cdots,n)$（其中 $n\geq 3$）为多边形的 $n+1$ 个特征顶点，则由其定义的三次 B 样条曲线的数学表达式为：

$$P(u)=\sum_{i=0}^{n}B_{i,3}(u)V_i,\ u_3\leq u\leq u_{n+1} \quad (1)$$

式中，$B_{i,3}(u)$ 是三次 B 样条的基函数。

三次 B 样条曲线的矩阵表达式可表达为：

$$P_i(u)=\begin{bmatrix}u^3 & u^2 & u & 1\end{bmatrix}\begin{bmatrix}b_{11} & b_{12} & b_{13} & b_{14}\\ b_{21} & b_{22} & b_{23} & b_{24}\\ b_{31} & b_{32} & b_{33} & b_{34}\\ b_{41} & b_{42} & b_{43} & b_{44}\end{bmatrix}\begin{bmatrix}V_i\\ V_{i+1}\\ V_{i+2}\\ V_{i+3}\end{bmatrix} \quad (2)$$

式中，u 为参数；$u\in[0,1]$，i 为曲线段序号，$i=0,1,\cdots,n-3$。

3 有限元模型建立

3.1 管道模型

管材的本构关系使用 Ramberg-Osgood 模型，其表达式为：

$$\varepsilon=\frac{\sigma_s}{E}\left[\frac{\sigma}{\sigma_s}+\alpha\left(\frac{\sigma}{\sigma_s}\right)^N\right] \quad (3)$$

式中，σ 为管道应力，MPa；σ_s 为管材的屈服应力，MPa，取 549MPa；α 为硬化指数；N 为幂硬化指数，取 15.6。

管道的建模采用管单元模拟，其材料参数如表 1 所示。

表 1 管道材料参数

管径/mm	弹性模量/GPa	内压/MPa	泊松比
660	200	6.3	0.3

3.2 管土相互作用模型

极限抗力与屈服位移是管土相互作用模型中存在的两个重要影响因素，管土相互作用对管道的影响可认为是土壤阻力与位移对管道的影响。根据国家标准 GB/T 50470—2017《油气输送管道线路工程抗震技术规范》的相关规定，可分别用三个弹簧模拟管轴方向的土摩擦力、水平横向及垂直方向的土压力，如图1所示，p_u、f_u、q_u和q_{u1}分别表示水平横向、轴向、垂直三个方向的极限抗力，x_u、z_u、y_u和y_{u1}表示三个方向的屈服位移。

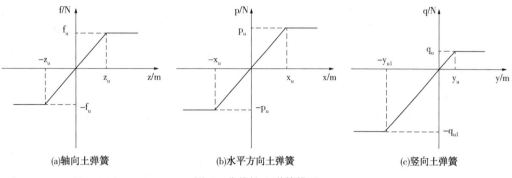

图1 非线性土弹簧模型

管道沿线主要为一般黏性土，其物理力学性质见表2，黏聚力取19kPa。基于以上数据，可计算得到土弹簧参数，其结果如表3所示。

表2 土壤力学参数

有效容重/(kN·m^{-3})	内摩擦角/(°)	黏聚力/kPa
18.56	22	19

表3 土弹簧参数

方向	极限抗力/(kN/m)	屈服位移/m
轴向	51.030	0.010
侧向	149.583	0.074
竖直向上	79.864	0.153
竖直向下	375.714	0.132

约束管道两端的位移，将插值得到的位移载荷施加至管道上，以模拟管道沉降。

4 准确性验证

对某条输气管道进行应力监测，将监测点的有限元应力结果与监测数据进行对比，验证基于IMU数据的管道力学分析方法的准确性。

基于IMU数据，结合其他资料，可知在监测点周围不存在弯管或弯头，即均为直管段。根据IMU数据得到管道的竖直及水平走向，如图2所示，水平方向发生最大位移约为0.05m，竖直方向发生最大位移约为0.22m。

由于各焊缝之间距离较大，缺少焊缝之间的位移数据，因此利用三次B样条曲线对位移数据进行插值处理，得到管道的位移，补充焊缝之间缺少的位移数据，将插值前后的竖直位移数据进行对比，较为符合管道变形实际情况，如图3所示。

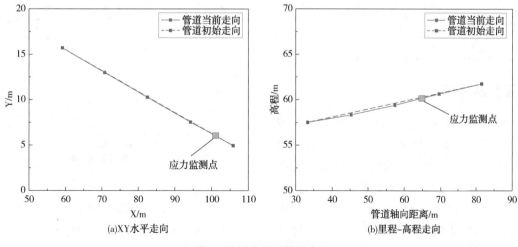

图2 监测点所在管段走向

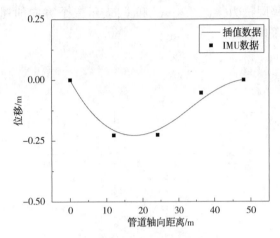

图3 管道竖直位移数据插值前后对比

建立该段管道的有限元模型，将管道位移施加至有限元模型中，得到管道的应力状态，其中监测点截面的应力云图如图4(b)所示，管道截面最大轴向拉应力为20.3MPa，最大轴向压应力为-24.5MPa。

针对应力监测点，其截面监测结果与有限元计算结果如图4所示，由图可知，二者截面的最大轴向拉应力与最大轴向压应力位置基本一致，分别位于1点钟、7点钟方向附近。其最大拉应力与最大压应力对比情况如表4所示，监测结果与IMU检测数据计算得到的结果相近，二者的平均误差为14.5%，说明基于IMU检测数据计算得到的管道应力结果与应力监测结果基本保持一致，同时也验证了本文方法的可行性与准确性。

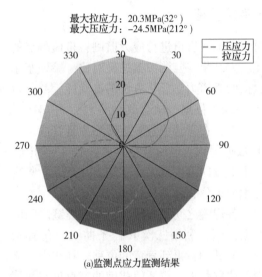

(a)监测点应力监测结果

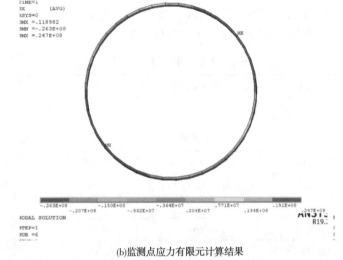

(b)监测点应力有限元计算结果

图4 监测点应力对比

表4 监测点最大拉压应力对比

监测点应力		监测结果/MPa	有限元结果/MPa	平均误差
监测点	最大拉应力	20.3	24.7	14.5%
	最大压应力	-24.5	-26.3	

5 工程案例分析

经IMU检测发现，管线中存在弯曲应变较大的管段。为进一步了解该管段的变形状况，对其进行位移分析。结合该管段的设计资料、施工资料及内检测数据等相关资料，排除该管段存在弯管的可能性，即管节均为直管段。基于IMU数据得到管道的走向，可知该管段发生沉降变形，最大沉降位移约0.66m，变形影响范围为58m，如图5所示。基于上述方法，采用三次B样条曲线对管道沉降位移数据进行插值处理，得到管道的连续沉降数据。

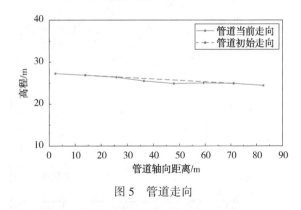

图5 管道走向

将连续沉降数据施加至有限元模型中，对管道发生沉降时的应力状态进行分析。由图6可知，最大轴向拉应力为375MPa，最大轴向压应力为138MPa，最大等效应力为376MPa，均位于管道沉降区中部，此时管道受到的应力主要以轴向为主。这是由于管道发生沉降时，沉降区中部受到的弯曲与拉伸变形最为严重，产生轴心应力与弯曲应力。此外，在沉降区边缘附近存在较大的轴向应力，约为310MPa，这是由于管道发生沉降时，该处主要受管道自身的重力、土壤对管道的轴力，由此产生的拉力引起的。

在沉降区内，管顶受压，管底受拉。除变形区中部外，在管道沉降区与非沉降区交界处均出现应力峰值。管道最大轴向应力与最大等效应力基本处于同一位置，约处于沉降区中部，该处为管段危险位置。管道材质为X70，根据管材拉伸试验可知，管道的最低屈服强度为549MPa，此时管道的最大等效应力小于$0.9\sigma_s$，管道依旧处于安全的运行状态。

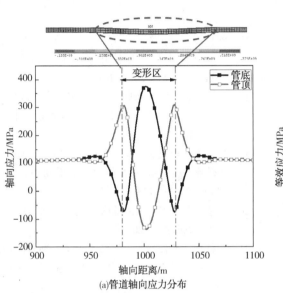
(a)管道轴向应力分布

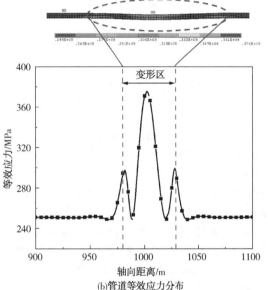
(b)管道等效应力分布

图6 管道应力分布

沉降区管道轴向应变分布如图7所示，与应力分布相似，在管道沉降区中部、沉降区与非沉降区交界处均存在应变峰值，最大轴向应变出现在沉降区中部。管道上表面为压应变，下表面为拉应变，管道沉降区边缘上表面为拉应变，下表面为压应变，这主要是由于管道沉降时受到的弯矩分布不均匀造成的。

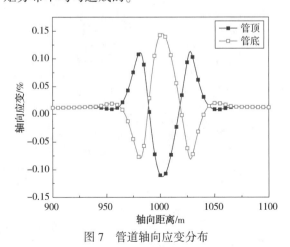

图7 管道轴向应变分布

6 结论

基于IMU数据，采用三次B样条计算方法，对管道沉降数据进行插值处理，建立基于土弹簧模型的沉降管道有限元分析模型，得到以下主要结论：

（1）基于应力监测数据，对基于IMU数据的管道力学分析方法进行了验证，有限元结果与监测结果的平均相对误差为14.5%，验证了该方法的合理性和有效性，该方法能够满足工程的实际应用。

（2）以某条沉降管道为研究对象，对其进行力学分析。当管道发生沉降时，在沉降区中部、沉降区与非沉降区交界处等位置存在应力应变峰值，其中危险位置为沉降区中部，最大等效应力为376MPa，未超出$0.9\sigma_s$，该管道仍处于安全运行状态。

参 考 文 献

[1] 帅健, 王晓霖, 左尚志. 地质灾害作用下管道的破坏行为与防护对策[J]. 焊管, 2008, (05): 9-15+93.

[2] 滕振超,周亚东,池林林,等.基于单元生死技术的埋地管道塌陷过程力学分析[J].中国安全科学学报,2024,34(06):73-81.

[3] 狄彦,帅健,王晓霖,等.油气管道事故原因分析及分类方法研究[J].中国安全科学学报,2013,23(07):109-115.

[4] 王晓霖,帅健.洪水中漂浮管道的应力分析[J].工程力学,2011,28(02):212-216.

[5] 陈翠翠,黄忠宏,帅健,等.纵向滑坡条件下顺坡敷设管道力学响应分析[J].安全与环境学报,2022,22(06):3034-3040.

[6] 石磊,叶远锡,单克,等.基于谐波沉降的管道力学分析[J].中国安全生产科学技术,2020,16(09):122-126.

[7] 张玉,梁昊,林亮,等.不同沉降方式下埋地管道力学响应试验研究[J].岩土力学,2023,44(06):1645-1656.

[8] 林睿南,罗敏,汪波,等.天然气站场沉降监测与报警系统设计[J].油气储运,2023,42(06):653-660.

[9] 陈朋超.油气管网安全状态监测传感系统构建与创新发展[J].油气储运,2023,42(09):998-1008.

[10] 张银辉,帅健,张航,等.1种基于云服务平台的滑坡管道状态远程实时监测系统[J].中国安全生产科学技术,2020,16(02):124-129.

[11] 刘啸奔,刘燊,季蓓蕾,等.基于IMU数据的管道弯曲变形段智能识别方法[J].油气储运,2021,40(11):1228-1235.

[12] 王琳,马林杰,徐建,等.基于深度学习的油气管道变形管段识别方法[J].石油机械,2023,51(11):11-19.

[13] 于希宁,彭鑫,冯丽丽,等.基于惯性的组合导航管道中心线测绘及缺陷定位[J].化学工程与备,2021(05):215-216.

[14] 刘燊,刘啸奔,李睿,等.基于机器学习的冻土区融沉变形管段识别方法[J].石油机械,2022,50(03):106-114.

[15] 李睿,蔡茂林,董鹏,等.地震区油气管道的应变与位移检测技术[J].油气储运,2019,38(01):40-44.

[16] 赵晓明,李睿,陈朋超,等.中俄东线天然气管道弯曲变形识别与评价[J].油气储运,2020,39(07):763-768.

[17] 何仁洋,刘艳贺,王海涛,等.基于IMU数据的应变解析技术及其应用[J].科学技术与工程,2024,24(07):2683-2689.

[18] 霍晓彤,王宇,刘点玉,等.基于中心线坐标的管道弯曲应变计算研究[J].中国安全生产科学技术,2021,17(S1):15-20.

[19] 吴硕琳,李亚娟,邓重阳.基于PIA的非均匀三次B样条曲线Hermite插值[J].计算机报,2023,46(11):2463-2475.

[20] 王宇.基于IMU数据的管道力学分析[D].中国石油大学(北京),2023.

[21] 胡行华,蔡俊迎.一类Caputo-Fabrizio型分数阶微分方程的三次B样条方法[J].应用数学和力学,2023,44(06):744-756.

[22] Shuai Y, Shuai J, Xu K. Probabilistic analysis of corroded pipelines based on a new failure pressure model[J]. Engineering Failure Analysis, 2017, 81.

[23] GB/T 50470—2017,油气输送管道线路工程抗震技术规范[S].

LNG 接收站罐容及其配套外输管道规模确定方法研究

候宇坤　明　亮　吴国其　陈　彬　于润泽

(国家管网集团工程技术创新有限公司)

摘　要　LNG 储罐和外输管道规模是影响工程投资和运行费用的关键因素之一，目前地方政府、LNG 接收站运营企业以及管道运营企业难以就合理确定 LNG 接收站及其配套外输管道规模达成共识，导致项目进度缓慢，因此迫切需要建立一套合理确定 LNG 接收站及其配套外输管道规模的确定方法。基于全国及各地区天然气供气格局，分析在季度调峰、基荷消费、应急保供、市场储备以及政策要求等方面的天然气需求，结合调研及历史统计数据，确定不同类型 LNG 接收站的市场空间及其规模，根据细分市场空间研究 LNG 接收站外输管道规模、气化设施能力以及储罐外输罐容、应急保供罐容、市场储备罐容和不可连续作业储备罐容的相互关系，建立一套合理的 LNG 接收站罐容及配套外输管道规模确定方法。通过选取国内某 LNG 接收站作为案例进行分析，通过明确市场结构，可以更为准确、科学地反映各行业实际用气需求，合理确定 LNG 接收站罐容和配套外输管道建设规模。研究结果为优化 LNG 接收站及其配套外输管道建设提供指导，从而规避大量的投资和运行费用的浪费，加快项目决策速度和建设进程。

关键词　LNG 接收站；外输管道；规模；市场空间

A Method for Determining the Tank Capacity of LNGTerminal and the Scale of Pipeline

Hou Yukun　Ming Liang　Wu Guoqi　Chen Bin　Yu Runze

(Pipechina Engineering Technology Innovation Co. LTD)

Abstract　The capacity of LNG storage tanks and the scale of export pipelines is one of the key factors affecting engineering investment and operating costs. At present, it is difficult for local governments, LNG terminal company, and pipeline company to reach a consensus on the reasonable determination of the capacity of LNG terminal and the scale of pipelines, resulting in slow project progress. Therefore, it is urgent to establish a set of methods for determining the reasonable scale of LNG receiving stations and their supporting export pipelines. Based on the national and regional natural gas supply situation, this study analyzes the natural gas demand in terms of quarterly peak shaving, base load consumption, emergency supply, market reserves, and policy requirements. Combining research and historical statistical data, the market space and scale of different types of LNG terminals are determined. Based on segmented market space, the study investigates the scale of export pipelines, gasification facility capacity, and the interrelationships between tank export tank capacity, emergency supply tank capacity, market reserve tank capacity, and non continuous operation reserve tank capacity. A reasonable method for determining LNG terminals tank capacity and export pipeline scale is established. By selecting a domestic LNG terminals as a case study for analysis, clarifying the market structure can more accurately and scientifically reflect the actual gas demand of various industries, and reasonably determine the tank capacity of LNG terminals and the construction scale of export pipelines. The research results provide guidance for optimizing the construction of LNG terminals and their supporting export pipelines, thereby avoiding the waste of a large amount of investment and operating costs, accelerating project decision-making speed and construction process.

Key words　LNG terminal; Pipeline; Scale; Market space

1 引言

我国是天然气消费大国,高速增长的天然气需求要求管网基础设施及储气调峰能力的适度超前部署。LNG 接收站是我国重要的储气调峰设施,同时 LNG 接收站所在的沿海地区也为我国天然气的市场中心,陆上气源与进口 LNG 海陆互济,可以有效保证市场中心的用气安全和灵活调配。而 LNG 储罐又是 LNG 接收站最重要的设施之一,储罐的设置直接影响 LNG 接收站的接收能力,以及投资和运行维护费用。冷绪林等综合考虑 LNG 运输船的运输及装卸方案、天然气的外输方案、LNG 接收站最大连续不可作业天数或应急储备天数,同时结合用户用气的不均匀性确定 LNG 接收站的储存能力。宋鹏飞等对国内外设计公司常用的静态计算方法进行了梳理,并结合案例分析对不同类型的 LNG 接收站推荐了适用的罐容计算方法。黄洁馨等对调峰保供型 LNG 接收站和普通型 LNG 接收站分别使用静态和动态罐容计算法进行了有效罐容的计算并做了简要对比。

LNG 外输管道是 LNG 接收站与国家天然气干线管网连接,实现资源疏散和调峰保供的重要通道。国家管网集团成立以来,我国天然气产供储销体系建设进入了新时期。但受近年出现的几次全国较大范围天然气供应紧张的局面影响,政府部门要求 LNG 外输管道与接收站气化能力匹配建设。对于 LNG 接收站投资主体来说,其对以价格为导向的竞争市场期望值较高,由于外输管道的建设不占用接收站投资,也要求外输管道与接收站气化能力匹配建设。根据国家管网集团工程技术创新公司预测,到 2030 年我国 LNG 接收站负荷率将降至 40%,这也将导致 LNG 外输管道负荷常年保持较低水平,盲目的要求外输管道规模与接收站气化能力匹配对于管道投资主体和运行企业来说既浪费了投资和运行费用,也增加了设备设施的碳排放量。为避免以上情况的发生,本文根据运营模式将国内的 LNG 接收站进行分类,确定其市场空间后核定外输管道规模,继而确定 LNG 罐容。

2 LNG 接收站罐容及外输管道规模确定方法

2.1 LNG 接收站的分类

国内的 LNG 接收站根据运营模式的不同可分为开放型 LNG 接收站、自营型 LNG 接收站和自营兼储备型 LNG 接收站。

开放型 LNG 接收站是指不参与资源的采购和销售,向所有满足条件的企业开放的 LNG 接收站,接收站只提供储存、液态转运及气态加工服务,覆盖基荷市场和调峰市场,并具备储备及区域应急保供功能。

自营型 LNG 接收站是指参与 LNG 资源的采购和销售的 LNG 接收站,接收站运营的主要目的是服务自我资源的接收及自我市场的供应,提供储存、液态转运及气态加工服务,仅覆盖基荷市场和调峰市场,不具备储备及区域应急保供功能。

自营兼储备型 LNG 接收站是指不仅覆盖基荷市场和调峰市场,还兼备一定水平的应急和储备功能的 LNG 接收站。

2.2 市场空间的确定

2.2.1 调峰市场需求

根据全国储气库和 LNG 接收站逐年调峰分析,并结合区域内储气库、LNG 接收站和外输管道实际情况以及价格竞争性分析,综合确定本 LNG 接收站的调峰气量。

2.2.2 基荷市场需求

统筹考虑全国一张网和区域供气格局,并结合区域内所有资源(气田气、管道气、LNG 接收站、储气库等)供应情况和储运设施能力,参考已经签订的托运商协议,综合确定本 LNG 接收站的市场空间。市场空间的天然气年需求总量除调峰需求外的部分为基荷市场需求量。

2.2.3 应急保供需求

区域内储运设施计划维修、事故抢修期间供需失衡时触发应急保供,期间由国家或地方政府相关部门统一指挥并总体协调调度,统筹考虑区域的应急保供任务气量,通过对区域内的储气库、LNG 接收站等所有储气设施统筹进行应急保供任务分配从而确定本 LNG 接收站的应急保供气量。

应急保供需求量的确定可考虑区域内最重要的一个主供气源失效,此条件下需要本项目承担的保供量。

应急保供市场必须保证不可中断用户需求,条件具备的情况下可考虑满足其他用户用气需求。

2.2.4 市场储备需求

市场储备需求主要考虑淡季采购低价LNG资源，在市场消费旺季向下游销售的需求。

2.2.5 政策储备需求

根据国家相关要求，管道企业和供气企业承担季节调峰/月调峰责任和应急保供责任，城镇燃气企业承担所供应市场的小时调峰供气责任，地方政府承担本区域一定时间的应急保供责任。

供气企业、管道企业、地方政府、城市燃气等通过协议购买储气服务，LNG接收站根据协议储备量配置储罐。

2.2.6 市场空间的确定

由此，开放型LNG接收站和自营兼储备型LNG接收站包含全部五种市场空间，自营型LNG接收站则仅包含调峰市场需求、基荷市场需求以及应急保供需求。其中开放型LNG接收站的市场储备需求宜根据区域市场情况按照比例配置储罐。

2.3 外输管道设计输量的确定

2.3.1 外输管道设计日输气能力核算

开放型LNG接收站应按以下公式核算：

$$Q_G = Q_1 + Q_2 + Q_3 \times Y \quad (1)$$

自营型LNG接收站应按以下公式核算：

$$Q_G = Q_1 + Q_2 \quad (2)$$

自营兼储备型LNG接收站应按以下公式核算：

$$Q_G = Q_1 + Q_2 + Q_3 \times Y \quad (3)$$

式中，Q_G 为外输管道设计日输量，Nm^3/d；Q_1 为均日基荷气量，Nm^3/d；Q_2 为高月均日日调峰量，Nm^3/d；Q_3 为应急日输气量（应急增输部分），Nm^3/d；Y 为不可中断比例。

2.3.2 外输管道预留能力分析

预留能力分析时应当充分发挥全国天然气一张网的调配优势，外输管道调峰或应急保供能力不足部分由区域内管道气、储气库、其他LNG接收站等储气设施协同供应。若仍无法满足，则应结合泊位、码头能力和路由稀缺性，综合确定外输管道预留能力和预留方案。预留方案宜根据技术经济比选结果，采用预留压气站场地、扩大管径等方式。

2.4 储罐罐容的确定

LNG罐容受卸船情况、市场需求以及公司运营模式等影响，LNG总罐容基于以下5项进行考虑：

2.4.1 卸船所需的LNG容积

卸船所需的容积考虑1艘LNG船的容积及卸船期间的天然气外输量。

$$V_1 = V_b \times i \times n - \left(\frac{Q_G}{24} \times \frac{\rho_g}{\rho_l} + \frac{Q_L}{24}\right) \times \frac{V_b \times i \times n}{v} \quad (4)$$

式中，V_1 为卸船所需的LNG容积，m^3；V_b 为船容，m^3；i 为有效卸载率；n 为码头泊位数量；ρ_g 为气态密度，kg/m^3；ρ_l 为液态密度，kg/m^3；Q_L 为LNG最大槽车外输量，m^3/d；v 为卸船速率，m^3/h。

2.4.2 码头最大连续不可作业天数储备需求

LNG储罐需满足码头最大连续不可作业D_1内的外输需求。

$$V_2 = \left(Q_G \times \frac{\rho_g}{\rho_l} + Q_L\right) \times D_1 \quad (5)$$

式中，D_1 为码头最大连续不可作业天数，d；

2.4.3 应急保供罐容

承担应急保供任务的LNG接收站，若区域内的应急保供天数D_2大于码头最大连续不可作业的天数，还应考虑应急保供罐容。

$$V_3 = Q_3 \times \frac{\rho_g}{\rho_l} \times (D_2 - D_1) \quad (6)$$

式中，D_2 为应急保供天数，d；

2.4.4 市场储备罐容

市场储备量q_m应综合考虑供气企业用气高峰月份天然气销售量和储存成本，则市场储备罐容为：

$$V_4 = q_m \times \frac{\rho_g}{\rho_l} \quad (7)$$

式中，q_m 为市场储备量，Nm^3/d；

2.4.5 政策储备罐容

政策储备罐容由两部分组成，一部分是在本项目预定的需按政策具备一定储气能力的企业或地方政府的储备罐容V_5，另一部分则是LNG接收站所属企业按照政策配备一定储气罐容V'_5。

2.4.6 储罐罐容的确定

（1）开放型LNG接收站

开放型LNG接收站总罐容为卸船所需的LNG容积、码头最大连续不可作业天数储备需求罐容、应急保供罐容和储备罐容之和，即 $V = \max(V_1 + V_2 + V_3 + V_4, V'_5) + V_5$。

(2) 自营型LNG接收站

自营型LNG接收站总罐容为卸船所需的LNG容积、码头最大连续不可作业天数储备需求罐容之和，即 $V=V_1+V_2$。

(3) 自营兼储备型LNG接收站

自营兼储备型LNG接收站总罐容为卸船所需的LNG容积、码头最大连续不可作业天数储备需求罐容、应急保供罐容和储备罐容之和，即 $V=\max(V_1+V_2+V_3+V_4, V'_5)+V_5$。

3 案例分析

选取国内某LNG接收站作为案例进行分析。该LNG接收站新建1个可靠泊 $8\sim21.7\times10^4 m^3$ LNG船的码头及配套设施，设计年通过能力 $602\times10^4 t$。贫液LNG的液相和气相（标况）的密度分别为422.66和 $0.67 kg/m^3$，体积比631；富液LNG的液相和气相（标况）的密度分别为468.79和 $0.77 kg/m^3$，体积比609。按照分类，该LNG接收站为自营兼储备型LNG接收站。

3.1 外输管道设计输量的确定

LNG接收站的定位是为已签订意向协议的用户提供基荷用气 Q_1 和调峰用气 Q_2 以及为接收站所在市（A市）地区和所在省份的省会（B市）提供应急保供用气 Q_3，本项目达产后各部分用气需求量构成见表1。

表1 LNG外输管道用户用气需求量构成表

序号	项目	Q_1/ $(10^4 Nm^3/d)$	Q_2/ $(10^4 Nm^3/d)$	Q_3/ $(10^4 Nm^3/d)$
1	定向用户	651	502	—
2	A市	—	—	639
3	B市	—	—	463
4	合计	651	502	1102

根据调研，A市和B市不可中断用户比例为35%，故根据式2-3，本LNG外输管道的设计日输量 Q_C 为 $1539\times10^4 Nm^3/d$。

3.2 储罐罐容的确定

3.2.1 卸船所需的LNG容积

LNG接收站建设1座可靠泊 $8\sim21.7\times10^4 m^3$ LNG船的码头，计算时选取 $21.7\times10^4 m^3$ 作为基准，LNG船有效卸载率为98%，卸船速率为 $1.2\times10^4 m^3/h$。选取富液LNG的密度作为基准，由式4得知，本LNG接收站 V_1 为 $19.87\times10^4 m^3$。

3.2.2 码头最大连续不可作业天数储备需求

参考历史气象资料，本LNG接收站最长不可作业天数为5~7天，选取7天作为最大连续不可作业天数，根据式5可得 V_2 为 $7.96\times10^4 m^3$。

3.2.3 应急保供罐容

根据协议，本区域内管道每年可以安排两次维检修，期间管道可以减供甚至停供最长7天；天然气管道在运行过程中难免发生事故造成供气中断，正常情况下3天恢复供气，特殊条件下考虑7天完成抢修。

由于计划性检修和事故应急抢修具备区域性特征，大面积发生的可能性不大，因此按计划性维检修和事故抢修不会同时发生考虑，故应急天数 D_2 取7天。

应急天数与码头最大连续不可作业天数相等，故不另外考虑应急保供罐容。

3.2.4 市场储备罐容

本LNG接收站地处北方，每年12月至次年2月为用气高峰，根据不均匀性分析确定在此期间天然气销售量约为 $8\times10^8 Nm^3$，综合考虑LNG的储存成本，本接收站的市场储备罐容按照 $4\times10^8 Nm^3$ 考虑，V_4 约合 $65.68\times10^4 m^3$。

3.2.5 政策储备罐容

LNG接收站所属企业天然气年合同销售量为 $21\times10^8 Nm^3$，根据国家相关部委政策要求，供气企业应当拥有不低于其年合同销售量10%的储气能力，故本LNG接收站应具有 $2.1\times10^8 Nm^3$ 的储备罐容，V'_5 约合 $34.48\times10^4 m^3$。

3.2.6 储罐罐容的确定

综上，本LNG接收站总罐容为 $93.5\times10^4 m^3$，即需建设5座 $20\times10^4 m^3$ LNG储罐，实际有效罐容为 $100\times10^4 m^3$。

4 结论

本文建立了一套LNG接收站罐容及其外输管道规模确定的综合方法，研究结果为优化LNG接收站及其配套外输管道建设提供指导。将LNG接收站分类确定其市场空间后得到外输管道的规模，继而确定总罐容。将外输管道的规模作为LNG接收站规模确定的输入条件，基于市场的规模确定方法可以有效规避造成的LNG接收站和外输管道实际负荷率均处于较低水平的问题，从而节省了投资和降低了后期的运维费用。

参 考 文 献

[1] SHI Xiaoxing. Optimization on Peak Shaving of LNG Receiving Terminal[D]. China University of Petroleum, 2019.

[2] ZHOU Shouwei, ZHU Junlong, SHAN Tongwen, et al. Development status and outlook of natural gas and LNG industry in China[J]. China Offshore Oil and Gas, 2022, 34(1): 1-8.

[3] SU Kehua, LI Wei, LIU Jianxun, et al. The peak shaving practice of foreign LNG terminals and its enlightenments[J]. International Petroleum Economics, 2020, 28(12): 34-44+53.

[4] CHEN Zhenghui, SONG Mingguo, FAN Hui. Research on the operation mode and trend of LNG regasification terminal under the operation of PipeChina[J] International Petroleum Economics, 2022, 30(01): 77-84.

[5] GAO Jie, SUN Chunliang. Overview of construction and operation of LNG receiving terminals in China[J]. China Oil & Gas, 2022, 29(05): 42-47.

[6] WANG Xiaoxiao. Analysis on the construction of LNG receiving terminal[J]. Chemical Engineering Design Communications, 2023, 49(06): 147-149.

[7] HOU Jun. Comparative analysis of calculation methods for tank capacity of LNG terminals[J]. Technology Supervision in Petroleum Industry, 2018, 34(07): 38-39+42.

[8] ZHAO Sisi, ZHANG Ying, XU Shuangshuang, et al. Construction of main technical and economic indicators of LNG terminal[J]. Construction Economy, 2023, 44(05): 99-104.

[9] LENG Xulin, WANG Yan, JIAN Chaoming, et al. Calculation method for tank capacity of LNG Terminals[J]. Oil & Gas Storage and Transportation, 2007, (09): 17-18+63+13.

[10] SONG Pengfei, CHEN Feng, HOU Jianguo, et al. Design and calculation of storage tank capacity and quantity of LNG terminal[J]. Oil & Gas Storage and Transportation, 2015, 34(03): 316-318+339.

[11] HUANG Jiexin, HAN Yinshan, LIU Fang, et al. Calculation of tank capacity of peak-shaving and supply-guaranteedand ordinary LNG receiving stations[J]. Gas & Heat, 2021, 41(05): 23-25+29+45-46.

[12] FU Zihang, SHAN Tongwen, YANG Yuxia, et al. Interoperability of LNG terminals and gas pipeline networks[J]. Natural Gas Industry B, 2021, 8(1): 48-56.

[13] HU Yunyue. Research on integrated operation mode of station and line[J]. Chemical Engineering Management, 2022, (27): 39-42.

[14] WANG Liang, CAI Llin, JIAO Zhongliang, et al. Innovation and exploration of natural gas infrastructure business model based on "one network in whole country"—Research on the operation mechanism of natural gas pipeline network in China III[J]. International Petroleum Economics, 2022, 30(01): 67-76.

[15] CHEN Xianlei, WANG Manqi, WANG Bin, et al. Energy consumption reduction and sustainable development for oil & gas transport and storage engineering[J]. Energies, 2023, 16(4).

[16] ZHAO Wenzhong, SUN Lili, LI Fengqi, et al. Analysis on investment and cost characteristics of LNG receiving stations[J]. Technology & Economics in Petrochemicals, 2023, 39(05): 5-9.

[17] TIAN Meng, ZHENG Chengli. Insight of the Construction of the carbon neutrality LNG terminals[J]. Chemical Engineering Management, 2024, (01): 165-168.

[18] LIU Jun. Research on the anti-transport market of shanghai LNG reception terminal[J]. Shanghai Gas, 2018, (06): 1-4+9.

[19] ZHANG Hongliang, WEN Haifeng, LIU Liantao, et al. Business model of gas storage service in coastal LNG terminals[J]. International Petroleum Economics, 2022, 30(04): 92-97.

[20] XU Jianmin, LI Xiao, YANG Zhen. Reform of LNG sales in traditional oil and gas enterprises after pipechina establishment: an example from sinopec[J]. Natural Gas Technology and Economy, 2021, 15(03): 54-61.

[21] WEN Xizhi, XU Wenping. Construction and development of the second tier LNG terminal station project[J]. Petroleum and New Energy, 2022, 34(02): 29-37.

[22] KONG Linghai, JIANG Lu. Discussion on a new-type storage management mode of large-scale LNG terminal[J]. Natural Gas Technology and Economy, 2019, 13(03): 80-84.

[23] ZHOU Shuhui, LIANG Yan, WANG Zhanli. Fair opening-up and outlook of china's LNG terminal[J]. Petroleum and New Energy, 2022, 34(03): 1-10.

国家管网集团新建油气管道与航道、港口相遇相交跨行业系统性协作管理创新与探索

王丽[1] 李维[2] 刘垒[3]

(1. 国家管网集团工程技术创新有限公司；2. 国家石油天然气管网集团有限公司工程部；
3. 交通运输部水运科学研究院)

摘要 为了更好地保障国家能源安全，促进油气管道行业高质量发展，国家管网集团积极落实国家能源部署，构建横跨东西、纵贯南北、覆盖全国、联通海外的天然气"全国一张网"，持续完善西北、东北、西南、海上四大战略通道，形成"四大战略通道+五纵五横"的干线管网格局，截止2023年底运营在役油气干线管道里程约10.4万公里。随着油气干线管网加速建设的同时，有关单位也在积极布局全国港口航道的发展规划，加快推进航道、港口建设，导致油气管道建设与规划升级航道相交的情况及管道与港口、锚地等配套设施的间距不足500米的情况日益增多。以上两种与航道、港口等相遇相交的管道穿越工程，需充分论证穿越管道安全及对航道和港口等设施的影响，并通过相关管理部门的审查和批复后方能开工建设，工程涉及多个相关企业和管理部门，管理流程复杂、内容繁多、时序性强，具体工作涉及管道、航道、海事、水利等多行业交叉、协调难度大，完成以上开工前置手续办理的周期通常超过一年。为了提升管道与航道、港口及其配套设施相遇相交管理工作效率，加快推进管道工程建设，保障穿越管道的安全，开展管理模式创新与探索。在西气东输三线中段项目、川气东送二线天然气管道工程川渝鄂段，国家管网集团通过建立多部门协调沟通机制全面识别风险因素、多部门全方位风险管理、跨行业协同合作、优化管理流程、积极推动系统性解决问题，形成一套跨行业系统性协作管理模式(简称 CISC 管理模式)，将有效缩短项目开工前置手续办理时间，首次实现油气管道项目管理与关联方跨行业系统性协同合作，对后续油气干线管道工程与公路、铁路、电力等线形工程相遇相交协调管理提供参考。

关键词 新建油气管道；航道、港口；相遇相交；跨行业系统性协作

The Pipe China's Innovation and Exploration of Cross-Industry Systematic Collaborative Management for New Oil and Gas Pipelines Meeting and Intersecting with Waterway and Ports

Wang Li[1] Li Wei[2] Liu Lei[3]

(1. PipeChina Engineering Technology Innovation Co. Ltd;
2. China Oil & Gas Pipeline Network Corporation Engineering Department; 3. China Waterborne Transport Research Institute)

Abstract In order to better safeguard national energy security and promote high-quality development of the oil and gas pipeline industry, China Oil & Gas Pipeline Network Corporation actively implements the national energy deployment, builds a national gas pipeline network that spans east and west, runs north and south, covers the whole country and connects overseas, and continuously improve the four strategic channels of northwest, northeast, southwest, and offshore, and form a grid of five major trunk pipelines running horizontally and vertically, with a total of approximately 104,000 kilometers of operational oil and gas trunk pipelines by the end of 2023.

As the oil and gas trunk pipeline network is being accelerated, relevantdepartments are also actively laying out the development plan of national port and waterway, and accelerating the construction of waterways and ports, resulting in an increasing number of situations where oil and gas pipeline construction intersects with upgraded waterways and situations where the distance between pipelines and ports, anchorages, etc. is less than 500 meters. The above two kinds of pipeline crossing projects, which meet and intersect with waterways and ports, need to fully demonstrate the safety of the pipeline and the impact on waterways and ports and other facilities, and can only start construction after the review and approval of relevant management departments. The project involves multiple relevant enterprises and management departments, with a complex management process, numerous contents, and strong time sequence. The specific work involves coordination among multiple industries, including pipelines, waterways, maritime, and hydraulic engineering, with a high degree of difficulty. The time required to complete the preliminary procedures before construction typically exceeds one year. In order to improve the efficiency of the management of the meeting and intersection of pipelines and waterways, ports and their supporting facilities, accelerate the construction of pipeline projects, ensure the safety of crossing pipelines, and carry out management model innovation and exploration. In the middle of the Third West-East Gas Pipeline Project and the Second Sichuan-East Pipeline Project (Sichuan, Chongqing, Hubei section), the PipeChina has formed a set of Cross-industry systematic collaborative management model by the establishment of a multi-departmental coordination and communication mechanism, comprehensively identifies risk factors, multi-departmental all-round risk management, cross-industry collaboration, optimizes management processes, and actively promotes systematic problem solving. Form a set of Cross industry systematic collaborative management model (CISC management model for short), which will effectively shorten the time required for project start preliminary procedures. For the first time, the oil and gas pipeline project management with related parties have realized cross industry systematic cooperation, which provides a reference for the coordination and management for subsequent oil and gas trunk pipeline projects with road, railway and linear projects such as road, railway and power line projects.

Key words new oil and gas pipeline; Waterways and ports; Meet and intersect; Cross-industry systematic collaboration

为了更好地保障国家能源安全，促进油气管道行业高质量发展，国家管网集团积极落实国家能源部署，构建横跨东西、纵贯南北、覆盖全国、联通海外的天然气"全国一张网"，持续完善西北、东北、西南、海上四大战略通道，形成"四大战略通道+五纵五横"的干线管网格局，截止2023年底运营在役油气干线管道里程约10.4万公里。随着油气干线管网加速建设的同时，航道、港口建设也在积极推进，布局全国港口航道发展规划。同为全国线性工程的油气管道建设与航道不可避免会出现相遇相交。

随着川二西、川二东、西三中、百色-文山、中俄东线等项目的建设，油气管道建设与规划航道相交的情况及管道与港口、码头、锚地等配套设施的间距不足500米的情况日益增多。油气管道与航道相交区域犹如能源输送网络与交通运输通道的"关键节点"，直接关系管道和航道两大系统的安全与发展。因此以上两种管道与航道、港口等相遇相交的建设工程，都要开展穿越管道安全以及建设对航道和港口等设施的影响论证，并通过相关管理部门的审查和批复后方能开工建设，工程涉及管道、航道、海事、水利等多行业交叉、协调管理复杂，完成以上开工前置手续办理的周期通常超过一年，对部分工程的进度产生了制约。

为了提升管道与航道、港口及其配套设施相遇相交管理工作，亟需开展管理模式探索与创新，加快推进管道工程建设，保障管道和航道的建设和运行安全，促进油气运输与航道运输协同发展，为行业间的可持续发展筑牢根基。

1 新建管道与航道相遇相交管理难点原因分析

探索跨行业系统性协作管理模式创新，旨在全方位、深层次剖析油气管道建设与航道、港口、码头等相遇相交管理难点，寻求跨行业管理新路径，提升管理效能。通过法律法规研究、工程案例分析、实地调研等多种途经分析管理困难的原因，探索管理模式创新，聚焦跨行业协同联动，提升管理效率，保障多行业的关键节点安全。

1.1 管道建设和航道通航安全存在多种相互影响风险因素

油气管道与存在升级建设的航道航道相交,管道穿越轴线与码头、港口、锚地、停泊区不足500m时,管道与航道、码头、港口、锚地、停泊区之间存在多种相互影响的风险因素。

新建油气管道穿越现状等级未达到航道发展规划技术等级的航道,后续航道升级整治工程的航道加深加宽建设,可能对江底管道产生安全影响。管道穿越中心线上下游500m的既有或规划码头、港口、锚地、停泊区,后续规划港口、码头建设、船舶抛锚都可能对穿越管道产生影响。反之,当后续航道、码头、港口建设对已建管道可能造成安全影响时,其建设将受到制约。

油气管道安全和航道通航安全,直接关系社会稳定和经济发展,至关重要。如何保证管道与航道的建设和运行安全,是油气管道建设与航道、港口、码头等设施相遇相交管理的首要目标和主要任务。对相互影响风险因素的评估和消减是管理工作的重点。

1.2 所需支持性文件多,环环相扣

新建油气管道与航道、港口、码头等交叉,涉及多个管理部门和运营单位,需要针对性开展相互影响论证、安全防护方案设计,签署安全防护协议、法律适用性说明,各成果间环环相扣。管道建设中需要充分识别所需支持性文件,如前期遗漏一个支持性文件,后续补充,将延长。

（1）需开展多项针对性的相互影响论证,论证管道建设和运行与航道、港口、码头的建设和运行的安全影响,确定需采取的安全防护措施。

（2）管道企业与相关单位签订安全防护协议,对后续双方建设和运行达成一致意见,对后续建设工作进行约定。

（3）管道穿越设计方案对安全风险因素提出对策,提出管道安全防护方案,保证管道建设、运行的安全。

（4）在保证管道安全的前提下,对于后续航道升级建设和码头港口建设、船舶抛锚,进行法律适用性说明。

1.3 涉及管理部门多,相互影响大

油气管道与航道、港口、码头等相遇相交的管理涉及多个部门,各部门职责不同,要求也不同。审批流程中,管道项目审批涉及能源、规划、环保、水利、安全等部门,航道工程审批为交通、海事等部门。各部门审批要求、侧重点不同,能源部门聚焦管道输送能力、安全性,交通部门关注航道通航安全保障。

对于管道与航道相遇相交的工程,各部门审批意见可能出现需要重新优化方案的情况,这种反复对工期造成较大影响,严重降低管理效率。

2 系统性跨行业管理模式的创新

深入分析导致管理困难的成因,建立多部门协调机制、多方位风险管理,跨行业协作,从源头解决问题,形成一套跨行业系统性协作管理模式（简称CISC管理模式）,将有效缩短项目开工前置手续办理时间。首次建立油气管道建设项目管理与关联方跨行业系统性协同合作模式,将对后续油气管道工程与公路、铁路、电力等线形工程相遇相交协调管理提供参考。

2.1 建立多部门协调沟通机制,全面识别风险因素

针对新建油气管道与航道、港口、码头等交叉,涉及多个管理部门和运营单位的情况,建设单位在项目前期组织设计单位、评价单位识别所涉及的航道、海事等管理部门和运营单位,建立沟通渠道和定期沟通机制,及时与航道、海事、水利、港口、码头等相关部门和单位沟通,全面梳理风险影响因素,针对性开展相互影响论证。

管理创新聚焦于多方协同联动,着力打造管道建设、航道管理、海事、水利、环保等多主体建立沟通协调机制,实现多行业交叉区域信息互通,提前规划管道运行管理原则,提升管理效率与应急响应速度,保障交叉区域运行流畅。

2.2 设计管理、项目管理、行业主管部门进行全方位风险管理

通过设计管理、项目管理、行业主管部门管理,实现不同层次、不同角度全方位的风险管理,全面保证管道和航道安全。

设计管理,针对相互影响论证识别的风险影响因素进行风险管理,结合评价意见、各部门审批意见,形成管道安全防护方案,降低工程风险。项目管理,通过协调落实航道、水利、规划、环保等多方意见,征集相关企业意见,降低

管理风险。行业主管部门对主管领域的航道和管道建设、运营安全风险进行把控,通过专家审查进行风险管理。

2.3 跨行业协同合作、优化管理流程、积极推动系统性解决问题

(1)组织跨行业协作、系统性解决管理难点的新模式:国家管网集团发挥行业领军企业的作用,基于丰富的油气管道建设管理经验和对油气管道与航道相遇相交管理难点的深刻认识,主动开展跨行业合作,与航道管理部门、研究单位积极沟通,推进以规范性文件的形式,规范管道建设与航道相遇相交行业间管理的要求和流程,构建新建油气管道与航道相遇相交管理新模式,从而提升管理效率,系统性解决该管理问题。

(2)优化管理流程,内部管理无缝衔接:国家管网集团构建一套完善的管理沟通机制,实现油气管道建设、生产等部门之间的无缝衔接。建立专项评价管理流程、专项评价管理指南等体系文件,明确各部门职责和工作内容,确保从项目规划到落地执行的每个环节工作流程清晰、分工明确、协同组织顺畅、管理高效。

(3)通过跨行业联合技术攻关,实现交叉领域技术提升:通过跨行业联合攻关,综合管道、航道、海事行业科研机构和高校的技术优势,联合攻克跨行业交叉领域的技术难题,实现多行业整体技术提升,促进行业间资源共享与协同发展,营造良好的行业生态。

3 总结

本次油气管道建设与航道相遇相交管理模式的创新,对目前存在的问题原因进行了深入探究,通过建立项目沟通协调机制,开展了全方位风险管理,实现了首次跨行业的协同合作,对今后新建油气管道与航道、港口、码头相遇相交管理要求和程序达成共识,并形成了一套跨行业系统性协作管理模式,从源头系统性解决了管理的痛点。

通过此次系统性跨行业管理模式的创新,打破了油气管道行业和交通水运行业之间的技术和管理壁垒,实现跨行业的对话和沟通,促进行业间和谐发展,将明显提升今后管道建设与航道、港口、码头相遇相交的管理协同效率。也为今后管道建设与公路、铁路等线性工程的相交管理提供了新的模式。

参 考 文 献

[1] 池洪建,我国油气长输管道项目建设管理模式探讨[J].国际石油经济,2012,11.

[2] 王光,韩春梅,郭伟.长距离油气输运管道外部安全风险辨识与评估[J].管道保护,2018,7.

[3] 张上,冯伟,吴桐,田伟力.城市轨道交通穿越输气管道安全评估方法和管理措施[J].管道保护,2021,3.

[4] 罗强贤,唐杰,廖东波,第三方建设项目与油气管道交叉安全审查要点[J].管道保护,2012,3.

[5] 油气管道完整性管理规范[S].Q/CR 9247—2016.

竞合关系在海外油气管道建设 EPC 项目界面管理中的应用

杨 洋

(国家管网集团工程技术创新有限公司)

摘 要 随着世界经济增速渐缓，国际油气行业投资市场低迷，为了建设更多"小而美"的项目、精品项目，现有国际市场油气管道建设不得不追求精细化管理方能走出困局。海外油气管道建设 EPC 项目管理具有界面复杂、信息传递频繁的特点。项目建设的相关方为了共同的目标将异质性资源聚集起来，也为了各自的利益在项目开展过程中存在着隐性的竞争关系，具有显著的竞合关系特征。深入探究竞合关系对海外油气管道建设 EPC 项目界面管理的影响，可以有效降低界面位势，充分激发项目效益。本文以实际工程为例构建了竞合关系作用于海外油气管道建设 EPC 项目界面管理的概念模型，深入分析了竞合关系对海外油气管道建设 EPC 项目界面管理的影响，详细阐述了项目建设各阶段项目执行策略选择。研究表明，在该海外油气管道建设的不同阶段，总包和业主之间一直保持着强竞争强伙伴的关系；总包和设计之间是强伙伴弱竞争的关系；对应不同的供应商类别，供应商与总包之间呈现不同的竞合关系类型，对于单一来源采购或者关键设备材料的供应商，总包和供应商之间呈现出强伙伴弱竞争关系，对于非单一来源采购的供应商尤其是散材供应商总包和供应商之间呈现出强竞争弱伙伴的关系；根据施工难易度不同，对施工难度大的工程，总包和施工分包呈现出强伙伴强竞争关系，对施工难度不高的工程，总包和施工分包呈现出弱伙伴强竞争的关系；总包和金融机构间是强伙伴弱竞争的关系；总包和政府之间是强伙伴弱竞争关系；总包和监理之间是强竞争弱伙伴关系。该研究成果有助于油气管道 EPC 承包商重视竞争关系在项目界面管理上的应用。

关键词 竞合关系；海外油气管道 EPC 项目；界面管理；项目执行策略

Practice of coopetition relationship on interface management of international oil and gas pipeline EPC project

Yang Yang

(Pipe China Engineering Technology Innovation Co., Ltd.)

Abstract With the global economic growth slowing down, the international oil and gas industry investment market remains slump, and more "small yet smart" projects corner the market share further. In order to get out of the current dilemma, the existing oil and gas pipeline project in the international market has to pursue delicacy management. Complex interface and frequent information transmission feature prominently in the overseas oil and gas pipeline EPC project management. For the common goal the related parties of project construction gather heterogeneous resources as partners, but for respective interests they are in competition with one another for the project finite resources. This is the notable feather of coopetition relationship. An in-depth study of the impact of the coopetition relationship on the interface management of international oil and gas pipeline EPC projects can effectively reduce the interface potential and fully maximize the project performance. This paper takes the actual project as an example to build a conceptual model of the coopetition relationship application on the interface management of overseas oil and gas pipeline EPC project, analyzes in-depth the impact of the coopetition relationship on the interface management of overseas oil and gas pipeline EPC project, and elaborates the project execution strategy selection in each stage of project execution. The research shows that in the different project execution stages of the overseas oil and gas pipeline project, the relationship between the general contractor and the project owner always maintains strong competition and strong partnership. The relationship between the general contractor and the engineering subcontractor is strong partnership and weak competition. Corresponding to different procurement types, the relationships between general contractor and vendor differ significantly. One between the general contractor and the project vendors is strong partnership and weak competition when it is single source purchase or the critical equipment and material supplier. The other between the general contractor and the project vendors is strong competition and weak partnership when it is not single source purchase especially the bulk material supplier. Corresponding to the construction difficulty, the relationships between general contractor and construction subcontractor also polarize. One between the general contractor and the construction subcontractors is strong partnership and strong competition when the construction section is in high risk or construction technique is difficult or it is the single source purchase construction subcontractor. The other between the general contractor and the construction subcontractors is weak partnership and strong competition when the construction section or major is easy such as civil of architecture construction subcontractor. The relationship between general contractor and the financial institutions is strong partnership and weak competition. The relationship between general contractor and the local authorities is strong partnership and weak competition. The relationship between general contractor and the project PMC is strong competition and weak partnership. This research result is helpful for EPC contractors in the oil and gas pipeline construction area to pay attention to the application of competition relationship in project interface management to reduce interface potential among project related parties to maximize the project performance.

Key words Coopetition relationship; international oil and gas pipeline EPC project; interface management; project execution strategy

在大流行病的后继影响下，随着大量出海建设的在建工程的延期、待建项目的搁置，投资市场低迷，承包方也极尽所能地探索更有效的项目系统管理方法。海外油气管道建设EPC项目管理由于投资巨大、建设周期长、项目参与方众多，造就了其界面复杂、信息传递频繁的特点。界面管理已经逐渐成为各类EPC项目合同控制管理的重点方向。

EPC模式是海外油气管道建设的主流承包模式，发承包双方不再以施工图为合作起点，而是根据发包人要求实行定制化的项目建设，根据FEED(Front End Engineering Design)由承包商来负责实施油气管道建设项目的整体设计、采购、施工、试运行、培训及运行移交。与传统的建设项目模式不同，业主(即建设方)的界面管理位置前移，管理界面减少，风险降低，有利于缩短项目建设工期，早日回收投资。而总包负责项目设计、采购、施工等各项目相关方的统一管理，资源配置整合，资金投入增大，管理界面增多，管理难度加大，风险增加，利益回报更高。区别于传统的委托代理关系，杜亚灵等认为在EPC项目建设模式下建设方与承包方之间是既有竞争有有合作的竞合关系。项目建设的相关方为了共同的目标将异质性资源聚集起来，也为了各自的利益在项目开展过程中存在着隐性的竞争关系，具有显著的竞合关系特征。现有的研究多数集中在建立问卷、量表的形式收集竞合关系在EPC项目管理中的应用，鲜少应用于海外油气管道建设EPC项目。

因此，有必要结合实际建设项目研究竞合关系对海外油气管道建设EPC项目界面管理的影响，可以有效降低界面位势，充分激发项目效益。

1 海外油气管道EPC项目界面管理主要内容

海外油气管道EPC项目界面依附于项目管理组织结构和工程合同关系建立。一旦项目启动，组织机构成立或是签订工程合同后，项目的参与方为了实现各自的项目目标将与合作方形成交流的界面。例如物资采购合同签订后，供应商需要和设计方进行往复的技术澄清，澄清的过程是通过设计和采购之间的交互界面完成。

界面管理是将不同的管理界面进行整合，在不同的项目建设阶段，最大程度地让相关方完成信息交互，同时积累项目知识，完成后续的资源调配。

界面管理需要统一信息口径,避免重复或有误的信息传输,降低项目各方的界面位势,从而提高整体项目参与方的效益。这就要求建立高效的界面管理方法,同时配备专业的管理人员才能完成项目整体管理。界面管理本质上是项目资源的协调、调配。界面管理的目标是在规定的工期、预算内利用有限资源完成项目的全过程建设。在项目开始执行后将搭建整体的EPC项目执行计划,根据目标完工时间倒排工期,确定每一项工作的预计开始时间和预计完成时间。

在设计阶段,设计单位在技术层面关注是否能根据FEED完成满足业主需求的相应设计,其获得的有效信息的衰减较少,设计的界面位势较高,从进度层面,需要尽快进行设计文件的升版关闭,完成既定的项目效益。采购团队需要在这一阶段根据设计进度跟踪技术规格书等请购文件,对接业主的准入供应商名单中的满足技术要求的供应商并建立联系,完成供应商准入以及技术澄清等工作,同时调研项目所在国市场,对施工措施材料的市场资源情况进行分析。施工团队需要根据采购物资的到场计划完成施工部署,进行人员的前期招聘,签证手续办理,施工沿线的踏勘,在业主的协助下完成临时征地手续的办理,完成营地建设选商及临时营地的建设等工作。可以看出在设计阶段,管理界面较为复杂,不光是设计和总包/业主沟通频繁,采购和施工在这一阶段界面前移,和设计有大量的信息交互,项目的前期准备工作在这一阶段完成。

在采购阶段,采购团队根据设计出具的技术规格书等请购文件进行物资的采购工作,设计单位需要需要参与与供应商的技术澄清工作。采购工作主要分为长周期、短周期和散材的采购,对于不同原产地的设备物资尤其是进口货物,采购团队需要提前了解设备物资所到港口的政府海关部门的清关政策,提前调研港口周边设施,布置转运站,与运输公司签订货运协议,设置现场仓库。在货物到达现场后,需要协调业主、监理进行现场检验。同时与施工团队协调,在这一阶段完成施工设备、措施材料的运输工作。施工团队将完成人员的许可办理等进场施工准备工作。

在施工阶段,施工团队将按照设计图纸施工,对施工过程中的变更,设计和采购团队根据实际与监理/业主对接处理。

EPC项目界面管理界面示意如图1所示。

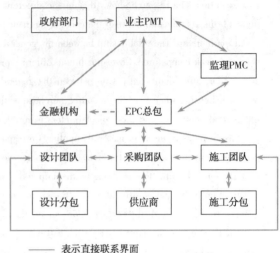

—— 表示直接联系界面
---- 表示辅助

图1 EPC项目界面示意图

2 海外油气管道EPC项目界面管理的方法

从界面管理的工作内容可以看出,界面管理的工作重点是加强信息的交互,而难点在于协作的相关方由于技能水平不同从界面获取的有效项目信息不同,从而无法达到高水平的业务处理。这就需要界面管理者从中协调,提高低界面位势一方的信息获取能力。

无论是界面协调、资源调配,界面管理对人员的工程经验,计划编制水平,工程控制水平要求都较高。可以采取以下措施满足人员的需求。

(1)对于在海外执行的建设项目,为了更好地和当地居民政府沟通,同时也为了满足所在国的属地化政策,需要招聘有大量工程协调经验的工程师扩充界面管理队伍。

(2)组织定期培训,在停工期组织有经验的项目管理人员或邀请管理专家进行经验分享,将界面管理常态化,分享项目知识,让项目界面管理经验在项目团队中流动起来。

界面管理的进程和项目WBS(Work Breakdown Structure,工作分解结构)、项目计划是同步的。跟踪项目CP(Critical Path,关键路径)实施情况,解决制约项目进行的矛盾冲突是界面管理的核心。

3 竞合关系与界面管理

竞合关系是项目协作方之间既有竞争又有合

作的关系。学者通常认为伙伴关系在 EPC 项目的执行中尤为重要。但是即便是同一项目，项目各参与方都有着不同的利益诉求，因此竞争关系也广泛存在。如果孤立地认为各参建方是纯伙伴关系，可能使承包商投入过量的资源来达成协作方的利益诉求，却最终造成承包商的亏损。杜亚灵等的研究也证实了这一观点。

当干系方为伙伴关系时，界面管理中信息资源的交互阻力小，双方都有强烈的意愿促成合作，界面管理的难度低。例如总包和单一来源供应商之间的关系，总包在安装设备下订单之后，需要供应商按期交货，从而避免现场的人员窝工设备闲置；供应商安排生产后倾向于尽快交付产品从而回收生产成本；因此，总包和供应商在设备制造运输的过程中建立紧密的伙伴关系，反复确认技术细节，定期更新生产状态。总包需要保证工期，供应商需要保证资金，双方没有可见的竞争关系。而面临非单一来源的物资采购尤其是散材采购时，这种情况又截然相反了。由于可供货源多，技术难度低，总包能在技术满足的前提下，倾向价格更低、供货周期更短的供应商。货值低，供应商可替代性强，双方呈现出低合作高竞争的关系，这时界面管理的难度增大，需要进行更多的界面协调，以此保证供应商对生产信息的及时反馈从而督促货物的按时交付。

竞合关系作于与界面管理的机理模型如图 2。

图 2　竞合关系作用于海外油气管道建设
EPC 项目界面管理的概念模型

4　工程应用

项目 X 是我国在西非某国家承揽的第一个油气管道建设项目，该项目横跨该国东西，建成后将承担该国的主要天然气输送任务。但是该项目执行难度大，主要原因如下：

（1）项目执行期间，该国货币贬值加剧，业主没有稳定的现金流支撑项目执行。

（2）建设方要求以联合体的形式承揽该 EPC 项目，联合体的界面复杂。

（3）业主供应商准入困难，对于工程物资的标准极高，监理强势，供应商替换的可能性低。

（4）设计水平不足，业主指定的项目所在国的设计公司进行管道设计，由于技术水平落后，管理难度大，给项目执行带了巨大的工期风险。

在该项目的执行过程中，总包采取了不同的关系模式和各干系方进行协作。

（1）业主

总包和业主之间保持强竞争强伙伴的竞合平衡关系。总包根据业主资金的投入情况，对项目监管的力度以及投入的监理资源适时调整执行策略，保证了总包的回款稳定，项目建设平稳，为合同的顺利履行打下了基础。

（2）设计

业主指定了设计公司，这也就造就了其资源垄断的地位，为了更好地执行项目，总包调动技术资源支持属地设计团队，保障了项目关键信息的及时传达，同时派遣了专业的设计审核人员为属地设计公司的成果把关，最终在总包和设计的密切协作下按照计划完成了约定的设计工作。

（3）施工

根据施工难易度不同，对施工难度大的定向钻穿越工程，由于其风险大、成本高，总包采取了强竞争强伙伴的竞合平衡关系。而对于施工难度低的线路扫线工作，总包采取了弱伙伴强竞争的竞争主导关系。

（4）其他干系方

总包和金融机构间采取了强伙伴弱竞争的关系；和政府之间采取强伙伴弱竞争关系；和监理之间采取强竞争弱伙伴关系。

在不同竞合关系策略的执行下，该项目平稳执行，不仅为总包创造了效益，也获得了业主和当地媒体的一致好评。

5　结论

本文以实际工程为例构建了竞合关系作用于海外油气管道建设 EPC 项目界面管理的概念模型，深入分析了竞合关系对海外油气管道建设 EPC 项目界面管理的影响，详细阐述了项目建设各阶段项目执行策略选择。该研究成果有助于油气管道 EPC 承包商重视竞争关系在项目界面管理上的应用。

通过分析海外油气管道建设 EPC 项目执行

风险，结合目前的研究趋势，未来研究将在界面管理的基础上详细探究竞合关系在不同国际合同文本下的应用，丰富竞合关系在 EPC 项目管理中的实际应用内涵。

参 考 文 献

[1] 杜亚灵，王孝宇. 竞合关系对 EPC 项目价值增值的影响研究[J/OL]. 系统工程理论与实践，1-16 [2024-12-23]. http://kns.cnki.net/kcms/detail/11.2267.N.20240702.1718.002.html.

[2] 孙洪昕，尤日淳，唐文哲. 国际工程 HSE 管理和项目绩效影响因素分析[J]. 清华大学学报(自然科学版)，2022，62(02)：230-241.

[3] 王腾飞，王运宏，沈文欣，等. 基于伙伴关系的国际 EPC 项目风险管理[J]. 清华大学学报(自然科学版)，2022，62(02)：242-249.

[4] 唐文哲，雷振，孙洪昕，等. 制度性差异对国际工程项目履约的影响[J]. 同济大学学报(自然科学版)，2017，45(10)：1569-1576.

[5] 唐文哲，雷振，王姝力，等. 国际工程 EPC 项目采购集成管理[J]. 清华大学学报(自然科学版)，2017，57(08)：838-844.

[6] 沈文欣，唐文哲，张清振，等. 基于伙伴关系的国际 EPC 项目接口管理[J]. 清华大学学报(自然科学版)，2017，57(06)：644-650.

[7] 王腾飞，唐文哲，漆大山，等. 基于伙伴关系的国际 EPC 水电项目设计管理[J]. 清华大学学报(自然科学版)，2016，56(04)：360-364+372.

[8] 林正航，强茂山，袁尚南. 大型国际工程承包商核心能力模型评价[J]. 清华大学学报(自然科学版)，2015，55(12)：1309-1314+1323.

[9] 宋莹琪，郝生跃，穆文奇，等. EPC 项目知识链组织间知识转移界面协同机制研究[J]. 情报科学，2024，42(04)：63-68+78.

[10] 谢晖. 基于界面管理的创新团队复杂系统运行机制研究[D]. 昆明理工大学，2015.

[11] 杨继柱. 基于 PDCA 循环法的石化项目设计阶段进度管理研究[J]. 化工管理，2018，(25)：175-179.

[12] 张文锦，张飞涟. 基于生态位理论大型建筑企业走出去战略选择[J]. 铁道科学与工程学报，2024，21(07)：2907-2916.

[13] 朱欣. FIDIC 在海外 DB 招标中的属地化应用实践与思考[J]. 铁道工程学报，2023，40(06)：104-108.

[14] 唐文哲，王腾飞，孙洪昕，等. 国际 EPC 水电项目设计激励机理[J]. 清华大学学报(自然科学版)，2016，56(04)：354-359.

[15] 王颖，马亮，白居，等. 基于神经网络的大型国际工程财务风险控制评价[J]. 同济大学学报(自然科学版)，2015，43(07)：1104-1110.

[16] 赵育梅，吕廷杰. 国际工程建设项目风险识别与对策[J]. 中国通信，2009，6(02)：31-34.

[17] 侯渡舟，林飞腾. 国际工程承包风险管理超文本组织的架构[J]. 西安建筑科技大学学报(自然科学版)，2003，(04)：379-382.

[18] 林飞腾，侯渡舟. 国际工程承包中的项目风险模糊层次分析[J]. 西安建筑科技大学学报(自然科学版)，2003，(01)：63-66.

Feasibility and challenges of digital twin technology development in the field of oil and gas pipelines

Jiang Yuyou[1]   Kou Bo[1]   Chen Ruolei[1]   Jiang Haibin[1]   Zhu Yanxiang[1]
Wang Can[2]   Cheng Yutao[1]   Wang Bo[1]

(1. Pipechina Yunnan Company; 2. Pipechina Guangxi Company)

Abstract Driven by the wave of Industry 4.0, digital twin technology, with its distinctive allure and immense potential, is gradually emerging as a hot topic in global scientific research and industrial circles. Ranging from construction to healthcare and extending to the automotive industry, its applications have permeated various sectors, infusing new impetus into innovative developments across multiple fields. In the oil and gas pipeline industry, where pipeline systems are increasingly complex, and operational requirements are continually increasing. How to effectively regulate and optimize becomes an urgent problem to be solved. Against this backdrop, the establishment of the digital twin system has become a pressing necessity for the pipeline industry. This paper first outlines the latest development trends of digital twin technology and deeply explores the application potential of this technology in seven major areas, including pipeline operation optimization, energy consumption optimization, geological disaster monitoring, and early warning, metering remote diagnosis and visualization, equipment reliability analysis and maintenance planning, leakage risk prediction, and life cycle management. However, despite the broad application prospects of this technology in the pipeline industry, there are still many challenges in implementing and applying the technology, including data perception, data transmission, data integration and management. To overcome these difficulties, this paper elaborates and analyzes the challenges and countermeasures faced in the construction of pipeline digital twin systems around the three key links, aiming to provide useful reference and guidance for the digital transformation of oil and gas pipeline systems. Through in-depth research and technological optimization, it can be anticipated that digital twin technology will play an even more prominent role in the realm of oil and gas pipelines. This technology will offer robust support for digital transformation thereby facilitating the pipeline industry's pursuit of more efficient, safer, and sustainable development.

Keywords Digital twin technology   Application potential   Operational optimization   Full lifecycle management   Challenges and countermeasures

1 Introduction

With the wave of the Fourth Industrial Revolution sweeping across the globe, digitization and intelligence have become the core engines driving industrial progress. Against the backdrop of this grand era, the traditional petroleum industry is facing profound changes in global energy transformation and the market environment full of unlimited opportunities. In the continuous iteration and development process of cutting-edge technologies such as information and communication technology, artificial intelligence (AI), data analysis technology, and industrial Internet of Things, digital transformation has become a universal consensus in the industry to achieve high-quality development.

In response to this trend, major domestic oil and gas companies have developed and expedited the implementation of digital transformation strategies. Within this process, oil and gas pipelines—vital conduits for energy transmission—are progressively transitioning towards digitization, informatization, and intelligence to meet the challenges of the 21st century regarding competitiveness, production efficiency, and sustainable development. This transition aims to realize an ambitious vision of cost reduction, enhanced efficiency, manageable risks, and maximized competitive advantages. Through digital transformation initiatives, pipeline enterprises can achieve more efficient resource management, optimize production processes, and lower operational costs.

The emergence of digital twin technology has created new opportunities for the management and operation of oil and gas pipelines. This innovative technology generates a virtual model of the pipeline that accurately reflects its real-time operational status. Utilizing this digital information enables simulation analysis and intelligent decision-making to forecast future operational trends within the pipeline system, thereby facilitating comprehensive lifecycle management of the infrastructure. The implementation of this technology not only enhancesthe operational level of pipelines but also reduces production and maintenance costs significantly. Most importantly, it plays a crucial role in mitigating management risks and enhancing safety performance across pipeline operations. Consequently, the adoption of digital twin technology offers novel perspectives and methodologies for effectively managing and maintaining pipeline systems.

Despite the substantial potential of digital twin technology, the oil and gas pipeline industry—characterized asa non-digital native traditional industrial sector—continues to encounter numerous difficulties and challenges during its digital transformation journey. A primary obstacle in practical applications is data integration. The inadequate integration of digital infrastructure within the pipeline sector, coupled with variations in equipment density, complicates data transmission processes. Furthermore, the absence of standardized data protocols has led to fragmented information systems and significant barriers to information flow, resulting in pronounced instances of 'information silos'. These challenges have resulted in ineffective utilization of data resources, thereby constraining the effective application of digital twin technology within the pipeline industry.

In response to the above issues, thispaper will delve into the feasibility and challenges of developing digital twin technology in the field of oil and gas pipelines, and analyze its specific applications in operational optimization, risk prediction and evaluation, equipment reliability analysis, and full lifecycle management. At the same time, feasible countermeasures will be proposed to address the challenges to ensure the safe and efficient operation of pipelines, providing reference and inspiration for the digital transformation of the pipeline industry.

2 Overview of Digital Twin Technology

The concept of digital twins was first proposed in a speech by Professor Grieves in 2003. Its initial definition referred to physical objects as digital representations of CAD models, but many scholars and technicians have recognized its potential and development prospects. Since then, the concept of the digital twin has frequently appeared in literature and has been continuously deepened and expanded. Scholars have made multiple improvements to the definition of digital twins, gradually applying it to multiple research fields. Glaessgen et al. described it as a multi-physics, multi-scale simulation of an already built vehicle, which utilizes the best physical model and sensor measurement data to reflect the entire lifecycle of the entity. Rosen et al. believe that digital twins are a realistic model of the current state of virtual processes, which can simulate the interactive behavior between virtual processes and the real environment. Li et al. further expanded this concept, stating that the digital twin is a model that integrates objects, models, and data in virtual space. This model is completely synchronized with real-world entities and is a perfect mapping of real-world entities.

Overall, digital twin technology is an integration based on physical models, sensor data updates, historical records, and real-time data. It accurately maps the simulation process of multiple physical quantities, scales, and probabilities into a virtual model to achieve full lifecycle simulation and prediction of physical entities. Its core lies in building a virtual model that is highly consistent with the physical entity, collecting and analyzing data to update and optimize the virtual model in real-time, thereby achieving accurate prediction and decision support for the entity. The initial practical implementation of this technology was in the aerospace industry, but with the continuous development of technology, it is increasingly being used in fields such as construction, healthcare, industrial production, meteorology, and automobiles. In the future, this technology can also be applied to the construction of smart cities. In the field of oil and gas pipelines, the application of digital twin technology means that real-time monitoring of pipeline operation status, prediction of potential leakage risks, optimization of scheduling operations, and other purposes can be achieved, thereby improving the overall efficiency and reliability of pipeline management (Figure 1).

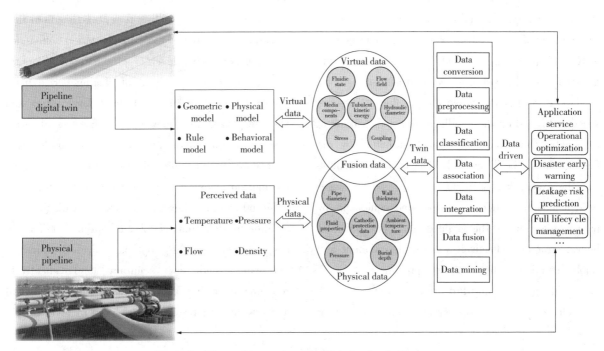

Figure 1 The working mode of the digital twin pipeline system

3 The potential application of digital twins in the pipeline field

As early as 2003, when China Petroleum began construction of the Jining pipeline, the oil and gas industry proactively proposed the grand goal of pipeline digitization. By fully utilizing advanced technological means such as SCADA and GIS systems, the pipeline industry has gradually built a comprehensive digital design system and full lifecycle database, laying a solid foundation for digital transformation. Subsequently, the concept of smart pipeline construction was further proposed, focusing on real-time monitoring, integrity management, optimized operation, and safety improvement of pipelines, aiming to achieve core functions such as "twin rehearsal, dynamic interaction, autonomous decision-making, and virtual real symbiosis" of pipeline digital twins.

3.1 Operation optimization

The rapid advancement of pipeline construction has resulted in a more dynamic and variable operational landscape for oil and gas pipelines, thereby significantly increasing the complexity and difficulty of pipeline regulation and operation optimization. In this context, the demand for digital twin technology is becoming increasingly urgent, which is expected to provide strong support for the safe, efficient, and stable operation of pipeline networks.

The pipeline digital twin can collect and analyze real-time operational production data through various sensors installed along the pipeline, thereby achieving the goal of flexibly and accurately grasping the pipeline operation status. This technology can dynamically simulate different operating scenarios based on intelligent algorithms and numerical simulations, predict the performance of pipelines under different working conditions, provide decision support for dispatchers, and greatly avoid the current phenomenon of relying on personnel technical experience. Taking the mixed oil treatment of finished oil products as an example, this technology can accurately simulate and analyze the amount of mixed oil between different oil products, while combining intelligent sensors to monitor oil indicators in real-time and dynamically optimize download strategies. During mixed oil cutting, real-time monitoring and simulation analysis of oil parameters can effectively improve the level of mixed oil cutting and greatly reduce the cost of mixed oil treatment.

More importantly, when the pipeline entities are scheduled based on the intelligent decision-making of the digital twin, the information transmitted by the digital twin can be dynamically changed in combination with the expert database data entered, and no longer need manual operation of the pipeline equipment by human intervention. This reduces the possibility of human error and effectively ensures the efficient and stable operation of the oil and gas pipeline.

3.2 Energy consumption optimization

The pipeline network system can fully utilize digital twin technology to construct process simulation models and closely integrate them with actual regulation, aiming to deeply explore the energy-saving potential of pipeline transportation and further reduce the energy consumption level of pipeline operation. This system is based on the hydraulic and thermal characteristics of oil and gas pipelines, and can dynamically optimize and adjust the operation strategy, pipeline flow, and medium flow direction, improving the transportation efficiency and energy-saving level of production and operation. For example, in terms of pigging period management, traditional methods have large errors and slow responses. In contrast, pipeline digital twins employ machine learning techniques to predict wax deposition within the pipeline while combining mechanism models to analyze variations in transportation efficiency, thereby establishing a pigging calculation model. This model can optimize efficiency and determine the optimal cycle—ultimately improving prediction accuracy and effectively reducing operating energy consumption during oil and gas pipeline transport. As shown in Figure 2.

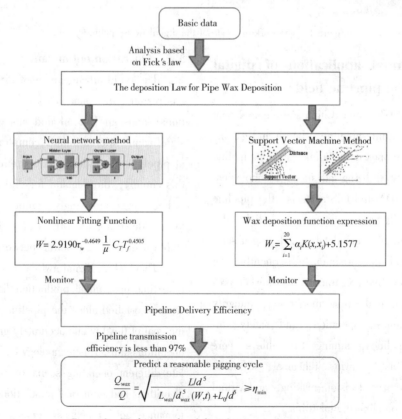

Figure 2 Working mode of digital twin pipeline system

In addition, thepipeline digital twin will conduct dynamic simulation analysis when the pipeline transportation volume changes, and obtain the optimal operating mode under the current working conditions. Through real-time interaction between this information and the pipeline system, a dynamic collaborative working mechanism between pipeline entities and digital twins is formed, thereby achieving the effect of reducing energy consumption and saving costs.

3.3 Monitoring and early warning of geological disasters

Oil and gas pipelines often cross complex geological environments. For example, the "China-Myanmar Oil and Gas Pipeline," one of China's four major energy strategic corridors, is a typical mountainous pipeline with a large drop, characterized by complex geological conditions along the route and frequent geological disasters. Therefore, some sections of the pipeline face the risk of exposure and water conservation collapse caused by geological disasters such as earthquakes, landslides, mudslides, and floods.

To achieve rapid perception and early warning of geological disasters, a highly integrated virtual model of pipeline digital twins can be used to construct a three-dimensional geometric model of the pipeline and surrounding geology using GIS systems and sensors to collect information on pipeline process parameters and surrounding environmental parameters. Numerical simulation analysis can be conducted using simulation software, enabling prediction of the occurrence rate of geological disasters. By real-time interaction of this information with the pipeline entity, timely response and implementation of emergency measures are carried out to provide sufficient guarantee for the safe operation of the pipeline.

3.4 Remote Diagnosis and Visualization of Metrology

Traditional oil and gas pipeline metering equipment requires professional personnel to conduct regular on-site inspections and calibration maintenance to ensure the accuracy of metering handover. In contrast, by building a digital twin system, remote fault diagnosis and calibration can be achieved, greatly reducing the frequency of on-site inspections and labor costs at grassroots stations. By collecting real-time data from measuring equipment and its corresponding pipeline segment, digital twin technology can determine the operating status of the equipment, predict possible faults, and remotely calibrate and adjust promptly. This method can effectively avoid the problem of delayed calibration and maintenance caused by a lack of allowed operating time for on-site personnel, greatly improving work efficiency.

By building a digital twin system for oil and gas stations, real-time transmission and visualization of grassroots station data (such as pipeline flow, tank level height, oil density, temperature, etc.) can be achieved through data transmission and integration. All levels and professionals can easily access the data, effectively realizing the efficient utilization of production data.

3.5 Equipment Reliability Analysis and Maintenance Planning

The digital twin technology for pipelines enables the real-time monitoring of operational status for both pipelines and associated equipment. By conducting a thorough analysis of historical operational data, alongside maintenance records and fault information, it is possible to identify periodic and seasonal failure patterns in equipment and facilities. During normal pipeline operations, digital twins facilitate real-time surveillance based on identified failure patterns, allowing for online assessment of equipment reliability and predictions regarding remaining useful life. The system autonomously determines the necessity for preventive maintenance or spare parts replacement, thereby significantly enhancing the overall operational stability of the pipeline.

In addition, the successful establishment and operation of the pipeline digital twin system can effectively adjust equipment maintenance plans. It transforms the traditional work plan formulation method based on personnel experience into a scientific work plan formulation method based on equipment reliability and failure laws, significantly improving the safety and stability of pipeline operation.

3.6 Leakage Risk Prediction

During long-term operation, pipelines are inevitably affected by factors such as erosion, corrosion, third-party construction, equipment failures, and natural disasters, leading to leaks. According to the "Analysis Report on Petrochemical Accidents at Home and Abroad", accidents caused by leaks account for 43.6% of the total number of accidents in the field of oil and gas storage and transportation, highlighting the seriousness of the leakage problem. Therefore, it is particularly important to use digital twin technology to predict the probability and risk of pipeline leaks.

From Figure 3, it can be seen that by constructing the digital twin of pipelines and utilizing the production data support network, seamless connection and real-time interaction with real production operation data can be achieved. The system deeply integrates various monitoring mechanisms such as fiber optic vibration, negative pressure waves, and infrasound waves, supplemented by advanced self-learning algorithms and model optimization mechanisms. This data is transmitted in real-time with the digital twin to continuously improve monitoring accuracy and warning response speed.

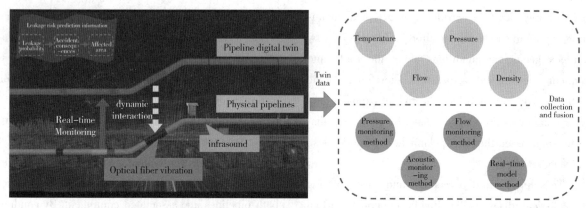

Figure 3 The leakage risk prediction function of digital twin pipelines

The digital twin of pipelines processes and analyzes historical data and operational status, and uses machine learning and other methods to accurately predict pipeline corrosion trends and cycles, thereby clarifying the probability of leakage in areas prone to corrosion. At the same time, digital twin technology can also simulate and evaluate the types of accidents that may be caused by pipeline leaks and the range of areas where serious consequences may occur. Operators can make scientific judgments based on the results and develop effective emergency measures to reduce pipeline failure and economic losses, providing strong guarantees for the safe and stable operation of the oil and gas storage and transportation industry.

3.7 Whole life cycle management

Digital twin technology can cover the entire lifecycle data of pipelines, including the design, operation, and maintenance stages. By integrating various types of data and establishing a data information database suitable for oil and gas pipelines, the operation data, maintenance records, spare parts, and other information on equipment and facilities are integrated and displayed. Based on the collected information, a unified method, data standard, and theoretical system are formed to improve the integrity management level of the pipeline network system. At the same time, digital twins can also receive real-time monitoring data from various sensors on the operation status of pipelines and equipment, and dynamically update and reconstruct virtual models based on these data, achieving in-depth comprehensive application analysis of the data. This ensures real-time interaction and dynamic consistency between the virtual model and the pipeline entity, thereby ensuring the effectiveness of digital twin prediction results and decision-making solutions.

In addition, thepipeline digital twin can provide decision support for each stage of the pipeline, achieving comprehensive control over the operation status of the pipeline. Therefore, utilizing digital twin technology for full lifecycle management of pipelines can optimize pipeline operation and maintenance strategies, thereby extending the service life of pipelines and providing a solid foundation for the sustainable development of the oil and gas industry.

4 Challenges and Countermeasures of Digital Twin Pipeline

Since the oil and gas pipeline industry is a non-digital native traditional industry, the comprehensive implementation and application of digital twin technology will inevitably face many challenges. To achieve the digital transformation of pipeline systems and fully leverage the potential of digital twin technology, it is necessary to conduct in-depth research and adopt corresponding solutions to overcome these challenges.

4.1 Data Perception

The successful construction and operation of the pipeline digital twin system are closely related to the real-time data collected by sensors. However, the maturity of the perception and interaction technology for current oil and gas pipelines is still insufficient. There is a lack of unified data standards, and the data is scattered and faces huge information barriers. In addition, the density and synergy effect of the pipeline industry in digital infrastructure, intelligent perception, and intelligent monitoring have not reached the ideal state, making it difficult for the types and quantities of data collected to meet the needs of big data analysis and decision-making. To solve this problem, it is necessary to increase technological research and development efforts, enhance the sensing capability of on-site monitoring equipment, improve the accuracy and reliability of sensing technology, and optimize sensor layout to ensure the comprehensiveness and effectiveness of data collection.

4.2 Data Transmission

The core of digital twin systems lies in data transmission. Although transmission technologies such as fiber optic, satellite, and industrial Internet of Things have been deployed, effective integration has not yet been achieved. In addition, unexpected events such as cable breakage accidents may lead to the interruption of instant communication in oil and gas pipelines, making it impossible for equipment and facilities to receive remote commands, thereby causing abnormal shutdown accidents in pipelines. Therefore, it is necessary to deepen research on transmission technology, promote the integration of multiple transmission methods, ensure the stability and reliability of data transmission, and optimize the efficient operation of pipeline digital twins.

4.3 Data Integration and Management

Data integration and management is one of the huge challenges for the successful implementation and application of digital twin technology in the pipeline industry. The information of pipeline systems is extremely complex, with diverse data sources and huge information barriers, which makes it difficult to interconnect data information and leads to a serious phenomenon of "information silos". To solve this problem, it is necessary to establish unified data management rules to achieve data sharing and interoperability. At the same time, it is necessary to strengthen the research and application of data integration technology, integrate data into a unified management platform, and achieve centralized management and analysis of data.

5 Conclusion

Digital twin is not a single technology, but a technology system that integrates multiple technologies such as simulation, the Internet of Things, and big data. It has the characteristics of high simulation, data-driven, and real-time interactivity. Applying this technology to the field of oil and gas pipelines will inevitably bring significant application prospects, such as operational optimization, disaster monitoring and early warning, leakage risk prediction, and full lifecycle management. This will make the pipeline more digital and intelligent. However, the development of pipeline digital twin systems is still in its infancy and needs to overcome challenges in data perception, data transmission, data integration and management. And the formation of a unified standard and theoretical system still requires continuous exploration and improvement. But by gradually digging and optimizing, it is believed that digital twin technology will play a greater role in the oil and gas pipeline industry and provide more powerful support for the digital transformation of oil and gas pipelines.

References

[1] Cozmiuc D, Petrisor I. Industrie 4.0 by Siemens: Steps Made Today[J]. Journal of Cases on Information Technology, 2018, 20(2): 30-48.

[2] Cimino C, Negri E, Fumagalli L. Review of digital twin applications in manufacturing [J]. Computers in Industry, 2019, 113: 103130.

[3] Ma J, Wu Z, Hou H, et al. Prospect of the application of digital twin technology in oil and gas pipelines, 2022 [C]. SPIE, 2022.

[4] Ma X, Qi Q, Tao F. A Digital Twin-Based Environment-Adaptive Assignment Method for Human-Robot Collaboration[J]. Journal of Manufacturing Science and Engineering, 2024, 146(3): 31004.

[5] Glaessgen E, Stargel D. The Digital Twin Paradigm for Future NASA and U.S. Air Force Vehicles [J]. 53rd AIAA/ASME/ASCE/AHS/ASC Structures, Structural Dynamics and Materials Conference20th AIAA/ASME/AHS Adaptive Structures Conference14th AIAA, 2012.

[6] Rosen R, von Wichert G, Lo G, et al. About The Importance of Autonomy and Digital Twins for the Future of Manufacturing [J]. IFAC-PapersOnLine, 2015, 48(3): 567-572.

[7] Li B, Wang X, Wang J. Digital Twin and its application feasibility to intelligent pipeline networks[J]. Oil & Gas Storage and Transportation, 2018, 37(10): 1081-1087.

[8] Hribernik K, Cabri G, Mandreoli F, et al. Autonomous, context-aware, adaptive Digital Twins—State of the art and roadmap[J]. Computers in Industry, 2021, 133: 103508.

[9] Ma X, Qi Q, Tao F. An ontology-based data-model coupling approach for digital twin [J]. Robotics and Computer-Integrated Manufacturing, 2024, 86: 102649.

[10] Tao F, Zhang M, Cheng J, et al. Digital twin workshop: a new paradigm for future workshop[J]. Computer Integrated Manufacturing Systems, 2017, 23(01): 1-9.

[11] Koo B, Chang S, Kwon H. Digital twin for natural gas infrastructure operation and management via streaming dynamic mode decomposition with control[J]. Energy, 2023, 274: 127317.

[12] Yu B, Feng C, Cong R, et al. Application of Digital Twin in Intelligent Construction of China Myanmar Crude Oil Pipeline [J]. Petroleum Engineering Construction, 2022, 48(04): 1-6.

[13] Analysis report on petroleum and chemical accidents at home and abroad[R]. SINOPEC Safety Engineering Institute, 2017.

[14] Liang J, Ma L, Liang S, et al. Data-driven digital twin method for leak detection in natural gas pipelines [J]. Computers and Electrical Engineering, 2023, 110: 108833.

[15] Li B; Gai J; Xue X. The Digital Twin of Oil and Gas Pipeline System[J]. IFAC-PapersOnLine, 2020, 53(5): 710-714.

[16] Xue X; Li B; Gai J. Asset management of oil and gas pipeline system Based on Digital Twin[J]. IFAC-PapersOnLine, 2020, 53(5): 715-719.

油气管道完整性管理智能化实施方案研究

薛鲁宁[1,2]　杨秦敏[1]　田明亮[2]　陈钻[2]　赵俊丞[2]

(1. 浙江大学；2. 国家管网集团浙江省天然气管网有限公司)

摘　要　针对传统完整性管理数据利用率不高、对管道本体及周边环境状态的评价全面性不够的问题，基于智慧系统的"感知、传输、认知、应用"闭环控制逻辑，提出了油气管道完整性管理的智能化实现方案。该方案将管道完整性管理的六步循环整合成为四步工作，分别为数据采集与清洗、状态评价、维修维护与效能评价。数据采集和清洗需要将建设期移交数据、感知数据和人工录入数据进行清洗入库，以便后续计算模型抓取计算。状态评价将传统的高后果区识别、风险评价和完整性评价进行整合，通过融合运用大数据、机理模型、人工智能等方法，完成对管体和周边环境的状态评价，为维修维护工作提供依据。维修维护过程则运用大模型技术，实现对标准、体系文件、历史维修方案等知识的智能提取，自动形成更为科学的维修维护方案。通过工作流程的整合，将数据进行融合利用，对管体和周边环境状态的评价更为全面准确，维修维护方案更为科学合理。

关键词　管道；智能化；完整性；大数据；大模型

Research on the Intelligent Implementation Plan of Oil and Gas Pipeline Integrity Management

Xue Luning[1,2]　Yang Qinmin[1]　Tian Mingliang[2]　Chen Zuan[2]　Zhao Juncheng[2]

(1. Zhejiang University; 2. PipeChina Group Zhejiang Natural Gas Pipeline Network Co., Ltd)

Abstract　A smart implementation scheme for oil and gas pipeline integrity management is proposed based on the closed-loop control logic of "perception, transmission, cognition, and application" of intelligent systems to address the problems of low data utilization and insufficient comprehensive evaluation of pipeline and surrounding environment status in traditional integrity management. This plan integrates the six step cycle of pipeline integrity management into four steps, namely data collection and cleaning, status evaluation, maintenance and efficiency evaluation. Data collection and cleaning require the cleaning and storage of handover data of construction, perception data, and manually entered data, in order to facilitate subsequent model capture and calculation. State evaluation integrates traditional high consequence area identification, risk assessment, and integrity assessment. Methods such as big data, mechanism models, and artificial intelligence to will be used to complete the state evaluation of the pipe body and surrounding environment, to provide basis for maintenance work. The maintenance process utilizes big model technology to intelligently extract knowledge such as standards, system documents, and historical maintenance plans, automatically forming a more scientific maintenance plan. Through the integration of workflow, data is utilized integrally to provide a more comprehensive and accurate evaluation of the status of the pipeline and surrounding environment, resulting in a more scientific and reasonable maintenance plan.

Key words　pipeline; intelligent; integrity; big data; large model

2000 年以来，中国将完整性管理引入管道行业，结合中国管道实际情况，通过引进、消化、吸收、再创新，已形成规范化做法，并于2015 年发布了国家标准 GB 32167《油气输送管道完整性管理规范》，指导管道运营单位开展完整性管理。

根据 GB 32167 规定，管道完整性管理包括数据采集与整合、高后果区识别、风险评价、完整性评价、风险消减与维修维护、效能评价等六个步骤，为系统实行管道完整性管理提供了规范化流程。国外在管道完整性管理方面普遍采用强制性技术法规和自愿性标准相结合的标准化管理模式。美国机械工程师协会（ASME）发布的ASME B31.8S《管道系统完整性规范》和美国石油学会（API）发布的标准 API 1160—2019《危险液体管道的完整性管理》是国际上实施管道完整性管理的重要参考标准。

目前国内外执行的管道完整性管理程序，虽然为管道完整性业务提供了规范性做法，但各个环节的划分，客观导致了对各类数据的分散利用，数据利用率不高，缺少对各类数据的融合分析，对管道本体及周边环境状态的评价全面性不够，对指导管道建设和维修维护业务的科学性有待进一步提升。随着智慧管网的建设稳步推进，管道完整性管理也将迎来智能化变革。

1 油气管道完整性管理基本做法

美国机械工程师协会（ASME）发布的《管道系统完整性规范》（ASME B31.8S）认为管道完整性管理是从好的设计、选材和管道施工开始的，将完整性管理程序分为识别管道的潜在威胁、数据搜集和整合、风险评价、完整性评价、采取响应措施、数据的整合更新、风险再评价等7个阶段，并形成闭环（见图1）。

管道的潜在威胁分为与时间相关的威胁、固有威胁和随机或与时间无关的威胁三大类。在数据收集与整合阶段，B31.8S 强调对不同的威胁进行评估时所需的数据也不一样，并针对不同的威胁因素给出了所需收集的数据列表。例如，对于外腐蚀因素，至少要搜集以下数据：

- 运行时长
- 防腐层类型
- 防腐层状态
- 阴极保护良好时间
- 阴极保护有问题的时间
- 土壤特性
- 管道检测报告
- 泄漏历史
- 壁厚
- 管径
- 运行压力水平
- 以往水压测试信息

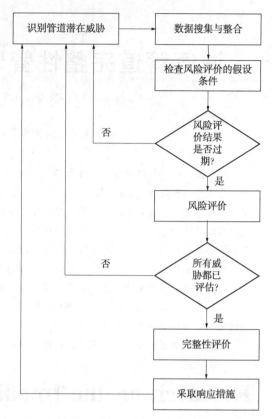

图 1　ASME B31.8S 中管道完整性管理程序

对于数据整合，首先需要确认一个参照系统，以使得不同数据源的数据能够整合并准确关联到管道的位置。例如，内检测的数据参照的是里程轮数据，而风险评价大多参照里程桩信息。

在美国石油学会（API）发布的标准 API 1160—2019《危险液体管道的完整性管理》中，完整性管理工作基于 PDCA 的理念开展，详细分为收集数据并识别完整性威胁、识别对关键地区的潜在影响、风险评价并分级、建立并修订评估计划、开展管道检测、试验和测试、整合完整性评价数据、收集项目实施数据、检查变更管理措施、整合项目实施和变更管理数据、检查施工方、行业和政府要求、评估完整性项目实施情况、管道完整性评价、管道修复、实施风险预防和削减措施、计算管道再评价间隔、开展完整性项目提升行动等阶段。

GB 32167—2015 规定了管道完整性管理是持续循环的过程，包括数据采集与整合、高后果区识别、风险评价、完整性评价、风险削减与维修维护、效能评价等6个环节。目前已成为我国管道运营企业实施完整性管理的通用做法。文献[4]根据管道建设、运行、废弃处置等各阶段管理与技术的特点，建立了管道全生命周期

"564"完整性管理工作流程体系(图2)，使管道完整性管理的理念扩展到了管道建设期和废弃处置阶段。

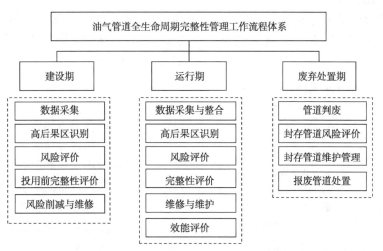

图2　油气管道全生命周期完整性管理体系

综合分析国内外相关指导文件，可以看出管道完整性管理基本都是遵从 PDCA 思想，根据管道完整性管理的主要业务活动分步骤开展。

2　油气管道完整性管理智能化实现

对于智能化系统，国内外基本按照"状态感知-数据传输-分析计算-决策应用"的总体技术路线进行建设。这为管道完整性管理的智能化框架提供了参考。

状态感知是利用利用各种智能传感终端实现数据采集、通信协议适配、虚拟交换、安全防护、边缘数据存储等功能，达到对管体和周边环境的全面感知。

数据传输是利用移动网络、卫星通信和无线网络等多种通信手段，在云-端之间提供高可靠、低延迟的数据传输，实现整个管网的全面互联。

分析计算是利用时空大数据与云计算架构，构建数据接入和智能分析的数据管理与计算平台，实现多源异构数据的存储、集成、融合、分析与管理，为智能业务应用提供数据与模型支撑。

决策应用是根据计算层的输出结果，提供智能解决方案，辅助管理决策。

随着近年来人工智能技术的飞速发展和配套算力资源的极大提升，参照智能系统建设思路，现有管道完整性管理的六步循环中的高后果区识别、风险评价和完整性评价已具备整合的条件，整合为管道及周边环境状态评价，输出结果包含了高后果区管段、各管段风险等级、管道适用性等。整合后的管道完整性管理程序可以按照图3所示开展。

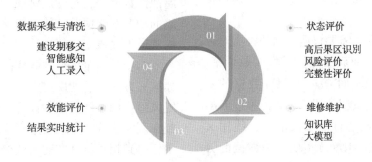

图3　智能化管道完整性管理步骤

其中数据采集与清洗对应了智能系统的状态感知和数据传输，状态评价对应分析计算，维修维护则对应决策应用，效能评价则是 PDCA 循环所需的必要步骤。

2.1　数据采集与清洗

以后续开展高后果区评价、风险评价和完

整性评价和维修维护工作的数据需求为依据，开展数据采集工作。根据数据获取的方式，可以将数据分为建设期移交数据、感知数据和人工录入数据，随着智能化水平的提高，智能感知获取的数据越来越多。不同方式获取的典型数据见表1。

表1 管道完整性管理典型数据与获取方式

获取方式	典型数据
建设期移交	中心线数据
	阴极保护基础数据
	管道设施基础数据
	第三方设施基础数据
	试压数据
	沿线地理和社会环境数据
感知数据	管道本体检测数据
	应力应变检测数据
	管道腐蚀监测数据
	光纤振动监测数据
	视频监控数据
	地灾监控数据
	正射影像数据
	气象数据
人工录入数据	土壤属性数据
	水工保护数据
	维修维护记录
	人工巡线数据
	失效记录

以上三种数据来源，并不是相互独立的，同一个数据项可能有多个数据来源，比如管道设施数据中的环焊缝数据，既来源于建设期移交，同时也来源于管道本体检测。在利用数据开展分析计算时，应做对比验证，总体上优先采用人工录入的数据，其次是感知的数据，最后是建设期移交的数据。

数据采集过程中，需要根据数据的种类对数据进行清洗入库，以便后续计算模型抓取计算。可使用关系型数据库（如Oracle）保证数据的安全、可靠；使用非关系型数据库（如Redis、Cassandra）保证数据存储的性能和灵活性，使用文件系统（如Hadoop的HDFS）实现相关数据的分布式存储。通过智慧管网建设实践，目前初步探索出了构建基于Postgres数据库集群的结构化数据库集群、构建基于MongoDB的非结构化数据库集群、基于OpenIO的对象存储系统支持文件对象存储，保证大数据存储、读取、检索、分析

的一致性、高效性和安全性。

对于数据的传输，一般选择消息队列遥测传输（message queuing telemetry transport，MQTT）为物联网网关与平台双向通信系统的消息传输协议，实现数据的传输与数据模型的同步更新。

2.2 状态评价

该过程是区分传统完整性管理的关键步骤，在传统的完整性管理程序中，首先要识别高后果区，为针对性的开展风险评价提供依据，风险评价目前主要采用评分法开展定性评价，对于高后果管段则可采用定量风险评价，对失效后果进行定量评价。风险评价之后，根据管道内外检测信息，开展管道适用性评价，并根据完整性相关标准给出管道缺陷的维修维护建议。按照这三个步骤开展完整性管理，往往容易造成对各类数据的分散利用，造成部分工作量的重复和评价结果的不精确。随着数据库技术的发展和算力资源的丰富，多源数据存储和多模型关联计算成为可能，因此管道完整性管理程序的三个识别评价步骤可以整合为状态评价，包括管体状态和周边环境状态评价。

主要评价思路是采用机理模型和大数据模型相结合的方式开展，机理模型主要用于由确定性事件导致管道本体发生破坏概率的计算，比如内压和外力导致的含缺陷管段的失效概率，此类机理模型主要计算思路是结构可靠性的极限状态方程（见式1），也是目前完整性管理中"完整性评价"步骤的主要工作。

$$Z = R - S = g(R, S) \tag{1}$$

对于管道内外检测间隔期缺陷的发展状况，则需要融合阴极保护数据、管输介质数据、土壤腐蚀性数据等，对管道腐蚀速率进行预测得到缺陷的动态变化情况，从而能够对内外检测间隔期的管道失效概率进行计算。

对于非确定性事件导致的管道失效概率，例如第三方施工、地质灾害，目前主要通过合理设置风险指标，开展定性评价，将定性评价结果以概率调整系数的方式对该管段的失效概率进行调整，得到相对更真实的失效概率。现在则可以通过对历史事件开展特征分析，提取致灾因子，开展数据关联规则分析，通过融合卫星、无人机等监测手段获取的数字高程模型（DEM）、降雨量、植被、土壤特性等多维数据，得到评价区域地质灾害发生的概率。对于地灾事件发生后管道破破

坏的概率，则通过管土耦合机理模型开展评价。对于第三方施工破坏，可以利用分布式光纤传感、无人机核查、现场视频监控等多种监测手段确定是否有第三方施工事件发生，目前通过多种监测手段对第三施工事件识别的准确率已到达97%以上。第三方施工对管道造成破坏的概率则可通过对历史破坏事件的因素进行提提取，获取现场施工类型、机械类型、敷设区域地质条件等多维数据，开展大数据分析，得到管道失效概率。

对于管道失效后的后果，可以通过无人机正射影像、卫星图形、手机信号热点图等数据，获取管道周围的人员、设施等数据，利用现有池火、喷射火、蒸气云爆炸、热辐射等模型开展定量评价。

本框架的状态评价，包含了传统完整性管理中的高后果区识别、风险评价和完整性评价，主要利用机理模型和大数据相结合的方式开展。在数据采集情况理想的条件下，通过融合多种智能感知手段，可以实现管体及周边环境状态的短周期评价。随着边缘算力的增强和算法的优化，对于威胁事件基本可以实现实时预警。该步骤可以输出管体和周边的多种状态信息，包括管体的可靠性、管段风险等级、缺陷状态、防腐状态、沿线地质灾害情况、第三方施工状态等等，为开展维修维护工作提供全面支持。整个评价过程可以用图4所示。

2.3 维修维护和效能评价

维修维护和效能评价，工作内容和传统完整性管理一致，维修维护同样要根据风险评价的结果，提出日常管理与巡护、缺陷修复、第三方损坏风险控制、自然与地质灾害控制、腐蚀风险控制等维护和维修措施，主要区别在于这些措施的提出方式。可以通过建立管道维修维护知识库，知识库包括各类标准、体系文件、处置流程、历史类似时间的处置方案等，利用大模型技术，对于预警的高风险点自动推送维修维护方案，使维修维护方案比传统方式更为科学。效能评价则可以根据平台中的统计数据自动给出结论。

3 结论

本文基于智慧系统的"感知、传输、认知、应用"闭环控制逻辑，提出了油气管道完整性管理的智能化实现方案，该方案将管道完整性管理传统的六步循环整合成为四步工作，能够解决传统完整性管理中分析计算工作分散、对数据利用效率不高的问题。整合后的完整性管理工作，综合运用智能感知、大数据、深度学习、大模型等人工智能技术，可以实现多源数据融合计算分析，根据实际业务需要，输出管道缺陷状态、腐蚀状态、管体可靠性、周边环境风险、管道风险等级分布等，评价结果更为全面准确。利用大模型技术，实现对标准、体系文件、历史维修方案等知识的智能提取，形成更为科学的维修维护方案。

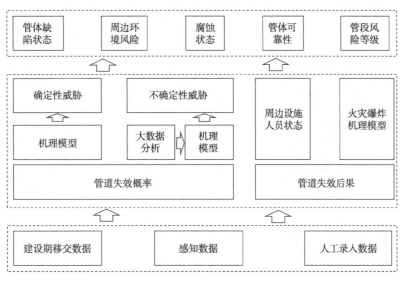

图4 状态评价总体框架

参 考 文 献

[1] GB 32167—2015,油气输送管道完整性管理规范[S].北京:中国标准出版社.

[2] ASME B31.8S-2018, Managing System Integrity of Gas Pipelines, ASME Code for Pressure Piping, B31 Supplement to ASME B31.8[S]. New York: The American Society of Mechanical Engineers, 2018.

[3] API Recommended Practice 1160, Managing System Integrity for Hazardous Liquid Pipelines[S]. Washington, DC: American Petroleum Institute.

[4] 陈朋超,冯文兴,燕冰川.油气管道全生命周期完整性管理体系的构建[J].油气储运,2020,39(01):40-47.

[5] 张晓华,刘道伟,李柏青,等.智能电力物联网功能架构体系设计及创新模式探讨[J].电网技术,2022,46(05):1633-1640.

[6] 陈朋超.油气管网安全状态监测传感系统构建与创新发展[J].油气储运,2023,42(09):998-1008.

[7] 陈文艺,梁宁宁,杨辉.基于MQTT的物联网网关双向通信系统设计[J].传感器与微系统.2022,41(08):100-103.

[8] 张华兵,程五一,周利剑,等.管道公司管道风险评价实践[J].油气储运.2012,31(02):96-98,168.

[9] 王新,刘建平,张强,等.油气管道定量风险评价技术发展现状及对策[J].油气储运.2020,39(11):1238-1243.

[10] 罗富绪.结构可靠性评估在管道风险管理中的应用[J].油气储运.2001(09):6-7,51-52.

[11] 刘鹏,李玉星,张宇.典型地质灾害下埋地管道的应力计算[J].油气储运,2021,40(02):157-165.

[12] 马云宾,董红军,孙万磊,等.新冠肺炎疫情下管道线路智能感知技术的思考与探索[J].油气储运,2020,39(04):389-394,424.

[13] 杨茜,赵红涛,麻克君,等.面向铁路遥感影像的自动化处理技术研究与应用[J].电脑知识与技术,2024,20(31):112-115.

[14] 税碧垣.智慧管网的基本概念与总体建设思路[J]油气储运,2020,39(12):1321-1330.

Optimization of water blending in unheated gathering based on low-temperature flow characteristics of heavy oil with high water contents

Lv Yuling[1]　Hu Weili[1]　Shi jiakai[1]　Ma lihui[2]　Song kunlin[1]　Wang jiaxin[1]

[1. College of Pipeline and Civil Engineering, China University of Petroleum(East China);
2. Technical Inspection Center, Sionpec Shengli Oilfield Company]

Abstract　Heavy oil production involves a high water-cut period, and a hot water blending process is used to reduce the viscosity and drag, despite problems such as significant energy consumption. In this context, unheated gathering technology was introduced; however, in this process, the crude oil easily aggregates and blocks the pipeline. To ensure the safety of the unheated gathering process for high water-cut heavy oil, the gathering and transportation system of the X block was investigated by halting the water blending of oil wells, in order to explore the oil-water flow characteristics and variation law of the pipeline pressure under the different wall sticking stages of heavy oil. The temperature at which the pressure decreases abruptly with dropping temperature is defined as the temperature of adhesion and deposition; this was used as the critical temperature of the heavy oil unheated gathering, and a mathematical model for the temperature of adhesion and deposition was established. The least squares method was used to calculate the total heat transfer coefficients of the pipeline during operation and an optimization method was established for the water blending parameters based on the energy conservation law and model of the temperature of adhesion and deposition. Furthermore, as interface calculation program was developed for water blending optimization, which could calculate the optimal watering volume of heavy oil with a high water-cut in the oilfield block under different working conditions. A three-day gathering and transportation experiment was conducted according to the optimized water blending amount. The results revealed that the wellhead back pressure maintained stable fluctuations, thereby verifying the reliability of the model of the temperature of adhesion deposition and the optimization method for water blending parameters.

Key words　heavy oil with high water-cut; unheated gathering; flow characteristics; temperature of adhesion and deposition; water blending optimization

1　Introduction

As a special type of crude oil, heavy oil is rich in polar macromolecules such as colloids and asphaltenes, which exhibit high viscosity, high density, and poor fluidity. Therefore, appropriate viscosity and drag reduction methods must be adopted during gathering and processing, including heating, dilution, water blending, and emulsion viscosity reduction transportation methods. Among these methods, the water blending transportation viscosity reduction method has been widely used owing to its wide application range, sufficient sewage resources, simple processing, and low cost. During the middle and late stages of oilfield development, the water content of the produced fluid increases. However, maintaining the gathering temperatures 3–5℃ above the pour point of crude oil inevitably results in significant heat loss and does not reflect the actual high water-cut period of an oilfield. A high water-cut provides favorable conditions for unheated gathering. Since the 1970s, Shengli and Daqing oilfields have been used for experimental studies on the unheated gathering and transportation technology of mixed oil-gas-water transportation; such studies have contributed toward effectively reducing the energy consumption of the gathering and transportation system, affording considerable economic benefits.

However, the multiphase flow of oil, gas, and water in high water-cut pipelines exhibits new flow characteristics. Especially when the gathering temperature is excessively low, gelled crude oil is accumulated and deposited on the wall; this results in a decrease in the effective flow area of the pipeline, an increase in the wellhead back pressure, and a deterioration in the flow situation. Therefore, understanding the variations in the hydraulic and thermal conditions during the unheated gathering of high water-cut crude oil is considerably important to achieve safe production and reduce energy consumption.

The temperature of the gelled crude oil deposited on the pipeline wall is repeatable and regular, and the temperature is defined as the wall sticking occurrence temperature (WSOT), which is related to the water content, flow rate, and crude oil composition. Zheng used the pressure drop change in an experimental pipeline to calculate the deposition thickness on its wall and established a function prediction model describing the WSOT. Based on the aforementioned model, Li investigated the rheological properties of crude oil at low temperatures, recognizing that, below the pour point, crude oil exhibits the characteristics of a pseudoplastic fluid, such as shear dilution. Accordingly, a rheological coefficient and a consistency index were introduced to improve Zheng's model. Based on an analysis of the diffusion pressure and adhesion mechanism, Zhang observed that, when the shear force acts against adhesion, the kinetics generated by the thermal motion and adhesion force are balanced, leading to the critical state of oil droplet adhesion; in addition, a theoretical model of the WSOT was deduced. Lv applied the Derjaguin – Landau – Verwey – Overbeek (DLVO) theory to subdivide the adhesion force into repulsive and attractive forces between oil droplets and proposed an adhesion energy model that better described the minimum oil gathering and transportation temperature. In general, scholars have conducted in-depth research on the WSOT during unheated gathering and transportation, forming an empirical model that is convenient for field use and a theoretical model that explains the wall adhesion mechanism; this is considerably important for promoting unheated gathering and transportation technology.

Scholars have also discussed the optimization of watering and investigated the gathering radius in high water-cut periods. Wang analyzed the hydraulic and thermal conditions of the ring oil gathering process with different diameters, to obtain the end temperature of the return pipeline. The optimal water mixing parameters were obtained by adjusting the water mixing temperature and observing the operation of the system both inside and outside the station. Liang considered the critical condition where the inlet oil temperature of a single well is equal to the pour point and simulated the thermal characteristics of the pipeline to determine the gathering and transportation radius of a single well. Yang used the black oil model in PIPEPHASE to calculate the gathering and transportation radius of heavy oil blocks, under the constraint of a pipeline pressure drop of less than 0.2MPa. Considering the high moisture content in the later stage of production, Gao established a thermodynamic and hydraulic model for multiphase mixed transportation and corrected the model according to the actual operation parameters of a pipeline. A single-factor sensitivity analysis method was used to plot the gathering radius map, which offered technical support for low-temperature gathering and transportation. According to big data analyses, Zhu established a mathematical model of the optimal oil gathering temperature of different types of pipelines for various liquid production, water contents, and pipeline length changes. Based on this model, an intelligent control scheme for oilfields was proposed based on the boundary temperature.

To summarize, many studies have discussed the WSOT and water blending optimization during high water – cut periods; however, limited efforts have been devoted toward using the low-temperature flow characteristics of heavy oil gathering pipelines in high water-cut periods and the WSOT model to guide field water blending optimization. Therefore, in this study, we refined the field dual-pipe water blending gathering system of heavy oil, explored the flow characteristics of oil-water gathering pipelines and the temperature and pressure changes of crude oil under different sticking stages during high water periods, and established a model for the temperature of adhesion and deposition via dimensional analyses of unheated gathering and transportation. Subsequently, the least squares method was used to calculate the total heat transfer coefficient of the pipeline during production. Based on the energy conservation law and the model of the temperature of adhesion and deposition, we herein propose an optimization method for water blending parameters.

2 Experimental

2.1 Equipment

The heavy oil production in Block X of the M Oilfield entails a dual-pipe water mixing process. To study the flow characteristics of the pipe and the blocking characteristics during the process of unheated gathering, the water mixing system was optimized and adjusted. A field experimental device was designed and installed to form an oil-water flow experimental system, as shown in Figure 1, based on the existing facilities and processes from the wellhead to the metering room.

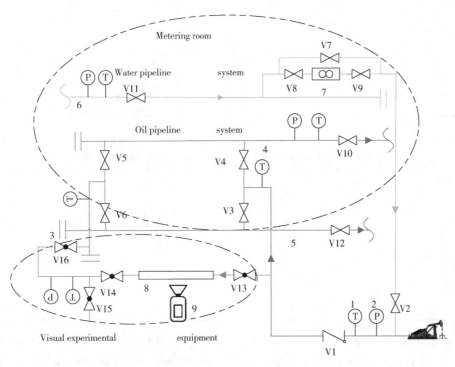

Figure 1 Experimental pipeline system process for Well 1

1—Remote temperature meter; 2—Remote pressure temperature meter; 3—Plug; 4, 5—Oil return manifold; 6—Water mixing manifold; 7—Mass flowmeter; 8—Transparent glass tube; 9—GoPro motion camera; V1—Nonreturn valve C; V2-V12—Gate valve; V13-V16—Ball valve

Figure 2 Physical map of visualization device

This experimental procedure features the following characteristics:

(1) The temperature, pressure, and other data during the experiment were recorded using an active wireless remote transmission instrument and transmitted to the network cloud in real time.

(2) The experimental water blending adjustment relied on the original water blending system. The water volume was adjusted using a precision needle valve and displayed in real time on a wireless electromagnetic flowmeter. The data were remotely transmitted to the network cloud at intervals of 5 min.

(3) The visualization device was composed of a wireless pressure sensor, wireless temperature sensor, pressure resistance (maximum pressure of 2MPa), and heat-resistant glass. The inner diameter of the glass tube was equal to that of the oil gathering pipeline. A GoPro high-speed camera was placed in front of the visualization device to observe and photograph the flow pattern of the oil-water two-phase flow and oil sticking to the wall.

2.2 Experimental materials

Six wells with high water-cuts were selected from Block X for the unheated gathering experiments. The wellhead temperature, pipeline length, water content, liquid production, and other well parameters of the six test wells are listed in Table 1, and the corresponding physical parameters of the dehydrated crude oil are listed in Table 2. The six test wells were high water-cut wells with a low gas-oil ratio. According to previous research, in this case, the gas phase has little effect on the flow state of the oil-water two-phase system. Therefore, this experiment essentially focuses on an oil-water two-phase flow system.

Table 1 Test well parameters

Well number	Liquid production/ (m^3/d)	Wellhead temperature/℃	Comprehensive water/%	Length of pipeline/m	Diameter/ (mm×mm)	Soil temperature/℃
W1	9.6	44.1	92.80	550	89×9	7.0
W2	11.6	42.0	90.02	201	89×9	6.5
W3	11.1	40.1	93.30	223	89×9	6.8
W4	12.1	51.2	87.71	416	89×9	7.1
W5	9.1	46.1	92.33	462	89×9	7.2
W6	12.0	43.5	95.60	550	89×9	7.0

Table 2 Physical parameters of the crude oil in the test wells

Well number	Viscosity-temperature equation	Density/(kg/m^3)	Abnormal point/℃
W1	32.034~0.0675T	987.2	78.5
W2	35.229~0.0768T	982.6	79.9
W3	33.992~0.0729T	981.0	77.7
W4	23.558~0.0451T	970.6	79.4
W5	34.931~0.0768T	983.2	76.3
W6	32.034~0.0675T	987.2	78.5

2.3 Experimental procedures

Because the properties of the products of the six wells were similar, as were the experimental steps and phenomena, well W1 is considered for the following discussion.

(1) The visualization device was installed at the end of the gathering pipeline of the corresponding oil well, V5 was opened, and the oil-water mixture was returned to the oil-collecting manifold in the metering room through the visualization device. Furthermore, V3 and V4 were closed, and the original pipe section was used as a bypass pipe.

(2) The oil-water mixture was run in the experimental pipeline to check the smoothness and airtightness of the pipeline.

(3) V8 were turned off to stop the flow of mixing water; thus, the well flow product was transported to the metering room at its own temperature and flow rate.

During the experiment, the state of the oil-water two-phase flow in the pipeline and the deposition state of the oil sticking to the wall was visualized by GoPro camera, the pressure and temperature data of the test point on the experimental pipeline were recorded and the pressure and temperature of the wellhead were monitored. The temperature was recorded when the thickness of the adhesive layer increased rapidly and the oil-water two-phase flow decelerated or even stopped. In the later stage of the experiment, the wellhead pressure was emphasized. Once the wellhead pressure exceeded the limit pressure (850kPa), V8 was opened to remixed with water to ensure safe production.

(4) Valves 3, 6, 7, and 15 remained fully closed, whereas the other valves were fully open.

3 Results and discussion

3.1 Flow characteristics without watering

The hot water blending process is an important measure for realizing the safe gathering and transportation of heavy oil. After water blending was halted, the temperature at the end of the oil return pipeline decreased, and the pressure difference between the beginning and end of the oil gathering pipeline fluctuated and increased. Thereafter, the wellhead pressure exceeded the limit pressure of the oil well, as shown in Figure 3.

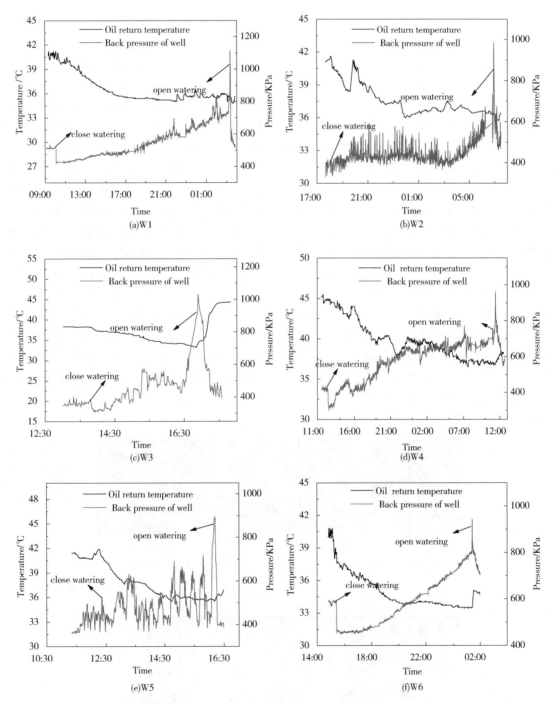

Figure 3 Pressure and temperature change after closing water supply

The trends of the return temperature and wellhead back pressure of the six oil wells after shutting down water blending were roughly the same. The change in the wellhead back pressure can be divided into the following stages:

(1) After shutting down water mixing, the wellhead back pressure decreased temporarily owing to the decrease in the resistance loss, such as the friction along the flow caused by the decrease in liquid volume.

(2) With a decrease in the pipe flow temperature, the wellhead back pressure increased and fluctuated. This can be explained by the rheological properties of the heavy oil. On the one hand, the viscosity of heavy oil is relatively high at room temperature. When the temperature decreased, the interaction between polar molecules such as asphaltene increased significantly, resulting in a further increase in viscosity. When the temperature decreased below the abnormal point, heavy oil exhibited Bingham fluid characteristics and a certain yield value. The yield stress forms a yield zone in the oil layer, causing a flow core without a velocity gradient, and the flow velocity in the oil layer decreases significantly. On the other hand, the rate of increase in the momentum diffusivity of heavy oil with a decrease in temperature is considerably larger than that of the water phase; this leads to faster dissipation of the momentum of the oil layer than that due to the water phase and eventually results in the adhesion and deposition of certain oil phases at the wall. The oil slips at a lower speed across a larger cross-section, whereas the water phase slips under the opposite conditions. A stable initial adhesion region was formed near the wall, and the oil phase aggregation accelerated the growth of the cohesive region, as shown in Figure 4. Therefore, the increase in viscosity leads to an increase in the resistance loss, and the formation of the yield zone and the non-isothermal properties of oil and water lead to a decrease in the cross-sectional area of the pipe flow. The combined effect of these two phenomena causes an increase in the wellhead back pressure.

(3) As the temperature continued to decrease, the oil-water interface changed its position. During this period, the fast-flowing water phase sheared off the fresh condensate, which was not strongly combined with the cohesive layer, resulting in a temporary recovery of the cross-sectional area of the pipe flow and a decrease in the wellhead back pressure. The falling condensate continued to cool during the flow process as the viscosity increased, and it easily adhered to the condensate cohesive region at the rear end of the detachment point. The adhesion became stronger at this time and finally caused the flow cross-sectional area to decrease again. Eventually, local deposition occurred and spread across the entire pipeline until the back pressure of the wellhead reached the limit value. This entire process is illustrated in Figure 5.

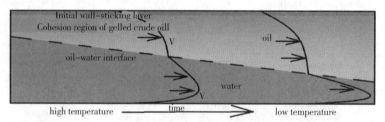

Figure 4 Oil-water interface at the same location at different times

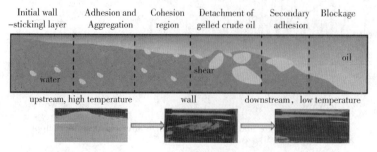

Figure 5 State of adhesion layer at different positions at the same time

As the temperature decreased, the pressure drop in the pipeline fluctuated and increased gradually. After reaching a critical point, the rate of increase grew significantly. Figure 6 shows the variation in pressure with respect to temperature for Well 2, where the pressure difference refers to the pressure between the wellhead and end of the return pipeline. Combined with the visualization of the oil-water flow characteristics and the stick-wall situation of the pipeline, the stick-wall thickness of crude oil increased very slowly above the critical temperature, and higher temperatures were associated with thinner adhesive layers. The oil-water two-phase flow pattern changed from an oil-in-water dispersed flow in a water-dominated region to an annular flow with a continuous water phase as the ring and a water-in-oil emulsion as the core. In addition, the pressure drop fluctuated and was relatively flat. When the temperature was close to the critical temperature, a small amount of crude oil adhered to the pipe wall to form an initial viscous layer. The flow pattern was also converted to an intermittent flow of water and water-in-oil emulsion, and large oil blocks appeared intermittently as the pressure drop of the pipe fluctuated violently. After cooling below the critical point, crude oil filled the flow section and moved slowly, a situation where the pipeline tends to become blocked. This temperature is termed as the temperature at which adhesion and deposition occur. Through single-factor influence analyses, this value was determined to be positively correlated with factors such as the liquid volume and water content and negatively correlated with factors such as the crude oil viscosity and abnormal point.

The temperature change under the condition of unheated gathering and transportation was the main reason for the change in the flow state of the oil-water two-phase flow. As the temperature decreased, the interaction force between oil droplets increased. When the shearing action of the water flow can no longer overcome the viscous force between the oil layers, the adhesion and aggregation of crude oil will occur, leading to the blockage of the pipeline. Therefore, if the endpoint temperature of

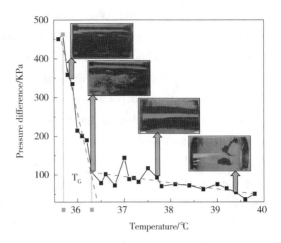

Figure 6 Variation law of pressure difference in Well 2 with a decrease in temperature

the unheated gathering line is maintained above the critical temperature at which fouling occurs, smooth operation for the unheated gathering of high water-cut crude oil can be maintained for a long time.

3.2 Temperature of adhesion and deposition

Zheng used waxy crude oil to explore the characteristics of oil sticking to walls during unheated gathering and transportation, via loop experiments in a laboratory. It was found that crude oil is suspended in water in the form of gelled oil during the process of unheated gathering, and that the wall sticking of the condensate is a result of the interaction between the adhesion of the condensate particles and the shear stripping of the pipe flow. Finally, the following empirical formulas were obtained:

$$T_G = T_n - a\varphi^m \tau^n \tag{1}$$

where, T_G is the WSOT, ℃; T_n is the pour point, ℃; φ is the volumetric water content; τ is the average shear stress, Pa; and a, m, and n are nondimensional parameters.

The oil-water two-phase flow pattern for this calculation model is an oil-in-water emulsion, and the rheological properties of waxy crude oil at low temperatures are considered; therefore, the calculation model needs to incorporate the experimental results of this study. The wax content of heavy oil is low but the contents of polar macromolecular structures such as colloids and asphaltenes are high. The decrease in temperature is associated with a large increase in viscosity. In addition, the colloids and asphaltenes form a supramolecular structure due to the intermolecular forces, although the structural solidification of crude oil does not result in a loss in its fluidity. Therefore, the pour point is of little significance for the unheated gathering and transportation of heavy oil, as compared with that in the case of waxy crude oil; however, the abnormal point is a turning point of the rheological property changes, which are of great significance for the continuous sliding of heavy oil, enabling continuous transport at temperatures exceeding the abnormal point. Considering the rheological properties of heavy oil and the experimental results, the correlation formula for the temperature of adhesion and deposition under the flow condition of a heavy oil pipeline with a high water-cut can be expressed as follows:

$$T_G = f(T_A, \varphi, Q_L, \rho_o, \rho_w, D, \mu_w, \mu_o) \quad (2)$$

where, T_G is the temperature of adhesion and deposition, ℃; T_A is the abnormal point, ℃; Q_L is the liquid production, m³/d; ρ_o and ρ_w represent the oil and water phase densities, respectively, kg/m³; D is the inside diameter, m; and μ_o and μ_w represent the oil and water phase viscosities, respectively, mPa·s.

Because the oil flow pattern is stratified, calculating the shear stress of the pipeline flow medium based on existing data is difficult. Therefore, considering the factors influencing the temperature of adhesion and deposition and drawing on Equation(2), the temperature of adhesion and deposition can be determined using dimensional analyses and the Π theorem. The symbols, units, and dimensions of the eight physical quantities in Equation(2) are listed in Table 3.

Table 3 Experimental physical quantities and dimensions

Variable	Symbol	Unit	Dimension
Temperature of adhesion and deposition	T_G	℃	θ
Abnormal point	T_A	℃	θ
Liquid production	Q_L	m³/s	$L^3 T^{-1}$
Density	P	kg/m³	ML^{-3}

续表

Variable	Symbol	Unit	Dimension
Viscosity	M	mPa·s	$M L^{-1} T^{-1}$
Inside diameter	D	m	L
Water content	φ		

According to the Π theorem, there are four dimensionless Π numbers, and length (L), mass (M), time (T), and temperature (θ) are selected as the basic dimensions to establish the dimensional matrix, as shown in Table 4.

Table 4 Physical dimension matrix

Dimension \ Variable	T_G	T_A	Q_L	ρ	M	D	φ
L	0	0	3	-3	-1	1	0
T	0	0	-1	0	-1	0	0
M	0	0	0	1	1	0	0
θ	1	1	0	0	0	0	0

The following expressions are obtained after solving and analyzing the homogeneous linear equation of the dimensional matrix using the null function in MATLAB(MathWorks, Natick, MA, USA):

$$\pi = \frac{T_G}{T_A} \quad (3)$$

$$\pi_1 = \frac{Q_L \rho}{\mu D} \quad (4)$$

Substituting Equations(3) and(4) into Equation(2), we obtain the following dimensionless equation:

$$f\left(\frac{T_G}{T_A}, \frac{Q_L \rho}{\mu D}, \varphi\right) = 0 \quad (5)$$

This equation is expressed in the power form of the basic dimensions, and the power of each basic dimension is guaranteed to be equal. The power function form is expressed as follows:

$$T_G = k \left(\frac{Q_L \rho}{\mu D}\right)^a \varphi^b T_A \quad (6)$$

where ρ is the mixed density of the mixed media, $\rho = (1-\varphi)\rho_o + \varphi \rho_w$, kg/m³, and μ is the viscosity coefficient of the mixture, mPa·s.

The expression $\frac{Q_L \rho}{\mu D}$ is similar to that of the Reynolds number, which characterizes the ratio of inertial forces to viscous forces acting on a fluid; therefore, Equation(6) has a certain theoretical significance.

The viscosity coefficient of the oil-water mixture is an important factor that affects the temperature of adhesion and deposition. The oil-water mixture obtained from oil wells often fail to form stable emulsions when oil field development enters the middle and late stages. In addition, the flow patterns captured via the visualization pipeline during the process of unheated gathering were mostly oil-water laminar flows. Therefore, traditional emulsion viscosity models suitable for low moisture contents and stable homogeneous phases, such as those of Einstein, Vand, Taylor, and Richardson, are not suitable for this unstable mixture.

Arirachakaran proposed a mixed model for oil-water stratified flows using the volume fraction weighting method, as shown in Equation(7); this model was used to predict the pressure gradient of a pipe flow. The results show that the predicted values of the model were in good agreement with the experimental values.

$$\mu_{mix} = \mu_o \varphi_o + \mu_w \varphi_w \quad (7)$$

where μ_{mix} is the viscosity of the mixture; φ_o is the oil content, vol.%; and φ_w is the water content, vol.%.

Based on the work of Arirachakaran and Richardson, Luo considered two extreme flow patterns in a high water-cut oil-water mixed flow, namely completely stratified flow and completely dispersed flow. The apparent viscosity in any other mixed state can be determined by these two most extreme flow regimes. The weighted average of the extreme mixing state is described in Equation(8), and the average relative deviation between the apparent viscosity calculated with this model and the actual apparent viscosity is 5.3%.

$$\mu_{mix} = (1-k_m)\mu_{sep} + k_m \mu_{dis}$$
$$= (1-k_m)(\mu_o \varphi_o + \mu_w \varphi_w) + k_m \mu_w e^{\varphi_o} \quad (8)$$

where K_m is the coefficient of the mixing state, μ_{sep} is the oil-water mixed viscosity of a fully stratified flow, and μ_{dis} is the oil-water mixed viscosity of a completely dispersed flow.

Based on the results of the aforementioned studies, the actual flow patterns of oil and water in the case of unheated gathering were combined, and Equation(9) was used to calculate the temperature of adhesion and deposition.

$$T_G = kT_A \left(\frac{Q_L((1-\varphi)\rho_o + \varphi \rho_w)}{D((1-\varphi)\mu_o + \varphi \mu_w)} \right)^a \varphi^b \quad (9)$$

According to the experimental results, Equation(9) is solved using multivariate nonlinear regression, and the coefficients $a = -0.081$, $b = -0.1247$, and $k = 0.2699$.

According to Table 5, the maximum error between the experimental value and the calculated value was 1.84℃, indicating that the model reflects the adhesion and deposition temperature of heavy oil well.

Table 5 Error analysis of empirical formulas for T_G

Well number	Experimental measurement/℃	Calculated value/℃	Error/℃
1	35.60	33.76	1.84
2	36.30	37.09	0.79
3	33.80	33.66	0.14
4	37.50	37.11	0.39
5	35.70	35.13	0.57
6	33.50	33.51	0.01

3.3 Water blending optimization

As the temperature limit of unheated gathering, the temperature of adhesion and deposition is an important index for determining whether an oil well can facilitate the unheated gathering process and optimize water blending. In addition, the operation and management of the unheated gathering process are affected by many factors, such as the oil production temperature, liquid production, water content, pipeline length, diameter, and ambient temperature. Therefore, based on a study of the unheated gathering flow characteristics and the temperature of adhesion and deposition, the influence of the abovementioned factors was analyzed, a feasibility judgment criterion was established, and a water mixing optimization method was developed.

According to the first law of thermodynamics and Joule-Thomson's law, the temperature drop formula for a multiphase flow pipe can be derived as shown in Equation (10).

$$T_Z = (T_0+b)+(T_R-T_0-b)\exp(-aL) -J\frac{x_{wg}c_{pg}}{c_m}\left(\frac{p_R-p_Z}{aL}\right)[1-\exp(-aL)] \quad (10)$$

$$a = \frac{K\pi D_0}{Gc_p} \quad (11)$$

$$b = \frac{ig(1-x_{wg})}{ac_m} \quad (12)$$

$$i = \frac{p_R-p_Z}{g\rho_l L} \quad (13)$$

$$c_m = x_{wg}c_{pg}+(1-x_{wg})c_{pl} \quad (14)$$

$$J = \frac{x_{wg}c_{pg}}{c_p}J_g + \frac{x_{wl}c_{pl}}{c_p}J_l \quad (15)$$

where T_Z is the endpoint temperature of the oil gathering pipeline, ℃; T_0 is the soil temperature of the pipeline depth, ℃; T_R is the temperature of the wellhead, ℃; L is the pipe length, m; J is the Thomson coefficient, ℃/Pa; D_0 is the diameter, m; G is the mass flow, kg/s; ρ_l is the fluid density; x_{wg} is the mass gas quality; x_{wl} is the liquid holdup; i is the hydraulic gradient; J_g and J_l are the gas phase and liquid phase Thomson coefficients, respectively; c_{pg} is the gas phase specific heat capacity, J/(kg·℃); c_{pl} is the liquid phase specific heat capacity, J/((kg·℃); c_m is the specific heat capacity of the gas-liquid mixture, J/(kg·℃); P_R is the starting pressure of the pipeline, MPa; P_z is the ending pressure of the pipeline, MPa; K is the total heat transfer coefficient for pipes, W/(m²·℃); and a and b are nondimensional parameters.

Considering that the oil gathering pipeline is short and that the gas content is low, Equation (10) can be simplified to realize the temperature drop calculation method and obtain Equation (16), namely the Sukhov formula.

$$T_Z = T_0+(T_R-T_0)\exp(-aL) \quad (16)$$

The total heat transfer coefficient is an important parameter affecting the calculation of the decrease in the pipeline temperature. According to this definition, the calculation method is shown in Equation (17).

$$K = \frac{1}{\frac{1}{\alpha_1}+\sum\frac{\delta_f}{\lambda_f}+\frac{1}{\alpha_2}} \quad (17)$$

where α_1 is the heat release coefficient between the crude oil and inner wall of the pipeline, W/(m²·℃); α_2 is the heat release coefficient between the pipe wall and surrounding soil, W/(m²·℃); δ is the thickness of layer f, m; and λ_f is the corresponding f layer thermal conductivity, W/(m²·℃).

According to the defined calculation, the K value is affected by many factors such as the soil moisture and liquid flow characteristics; hence, obtaining an accurate value is difficult. Therefore, the temperature decrease, pressure drop, and other data measured during the water mixing operation of the six pipelines can be used to inversely calculate the K value, as shown in Equation (18). The least squares method makes full use of the actual operating parameters of the pipeline, eliminates the influence of various random factors to a certain extent, and ensures the reliability of the inversely calculated K value; thus, this method was used to inversely calculate the total heat transfer coefficient, K.

$$K = \frac{Gc_m}{\pi D_0 L}\ln\frac{T_R-T_0}{T_Z-T_0} \quad (18)$$

where c_m is the specific heat capacity of the oil-water mixture, J/(kg·℃);

The basic principle of using the least squares method to inversely calculate the K value is to determine the appropriate K value to minimize the sum of squares of the deviation between the incoming temperature calculated according to the K value and the actual measured value. Therefore, the constraint equation is constructed, as shown in Equation (19).

$$\min F(K) = \sum_{i=1}^{n}[T_{Zi}(K)-T_Z^{(i)}]^2 \quad (19)$$

where $n=6$, expressed as the oil well number; $I=1-6$; $T_{Zi}(K)$ is the calculated temperature value at the end of the pipeline; and $T_Z^{(i)}$ is the monitoring value.

To reduce the complexity of the calculation, the lsqnonlin function in MATLAB was used to optimize the solution; the corresponding results are shown in Figure 7.

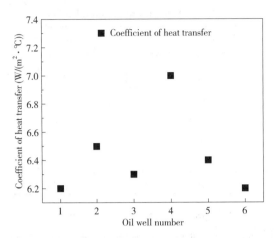

Figure 7 Total heat transfer coefficient for different oil wells

Using the law of conservation of energy, the starting temperature of the gathering pipeline under the condition of water mixing was obtained, as shown in Equation(20).

$$c_W G_w(T_w - T_R) = c_o G_J(1-\varphi)(T_R - T_J) + c_W G_J \varphi(T_R - T_J) \quad (20)$$

where G_w is the quality of the blending water, kg/s; T_w is the temperature of the blending water, ℃; T_R is the starting temperature of the pipeline, ℃; G_J is the liquid production, kg/s; φ is the water-cut for the wellhead fluid, vol.%; and T_J is the temperature of the produced fluid for the wellhead, ℃.

Based on the total heat transfer coefficient of the pipeline and the temperature of adhesion and deposition, a water blending optimization programming calculation was conducted to determine if the unheated gathering and transportation technology was suitable for the oil well and perform the corresponding adjustments in the water blending amount. A flowchart of the optimization module is presented in Figure 8.

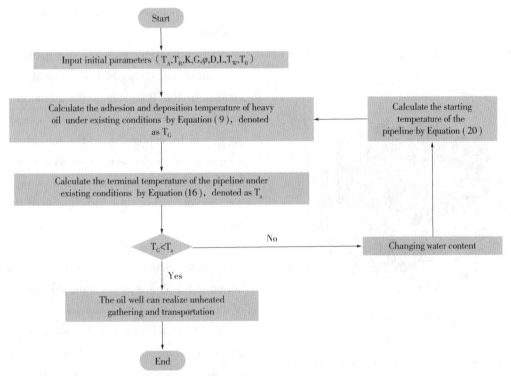

Figure 8 Flowchart of water blending optimization calculation

The above flowchart was achieved using MATLAB, and Figure 9 shows the calculation results for Well 1.

The calculation results of water blending in each well are listed in Table 6.

After the optimization of water blending, the water mixing amount was significantly reduced to only 73.4% of the original value; this can help reduce the energy consumption associated with the oilfield.

To verify the accuracy of the optimization method for water mixing and the model of the temperature of adhesion and deposition, a three-day gathering and transportation experiment was conducted for six oil wells, according to the optimized water mixing amount; the changes in the wellhead back pressure were observed, as shown in Figure 10.

Table 6 Comparison of terminal temperature of gathering pipeline before and after optimization

Well number	Watering temperature/ ℃	Amount of water before optimization/(m³/d)	Amount of water after optimization/(m³/d)	Watering difference/ (m³/d)
1	48.0	36.00	21.60	14.40
2	48.0	19.20	10.32	8.88
3	47.0	14.40	8.40	6.00
4	57.0	19.20	11.52	7.68
5	49.0	26.40	17.52	8.88
6	48.0	24.00	18.72	5.28

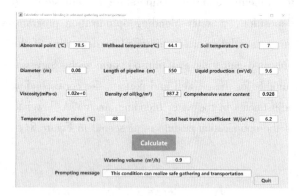

Figure 9 Calculation results of water blending in Well 1

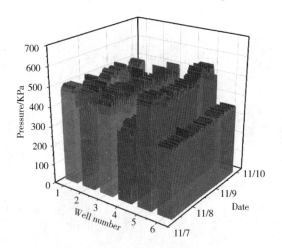

Figure 10 Wellhead pressure variation of optimized water blending

The wellhead back pressure of each well exhibited stable fluctuations after water blending, and the wells that exhibited a rapid increase remained below the limit pressure value (850kPa). This indicated that the oil wells could maintain stable operation under this water blending amount.

4 Conclusions

In this study, a visual experimental system was designed for a field-gathering pipeline. By shutting off oil well water blending, the multiphase flow characteristics of oil and water in the unheated gathering process and the variations in the pressure drop of the pipeline were investigated. A model for the temperature of adhesion and deposition suitable for high water-cut heavy oil was established, and a water blending optimization method was proposed. The main conclusions of this work can be summarized as follows:

(1) Afterwater blending was halted, the back pressure of the oil wells underwent three stages of change. In the first stage, the liquid flow in the pipeline was reduced, resulting in a temporary reduction in back pressure. In the second stage, with a decrease in temperature, the heavy oil exhibited Bingham fluid characteristics and formed a yield zone without a flow core, which caused the flow velocity of the oil layer to decrease. In addition, the momentum of the oil phase dissipated faster than that of the water phase, increasing the viscosity and adhesion of oil phases at the wall. The combined effect of these two caused the wellhead backpressure to increase. In the third stage, the water phase with an accelerated flow rate removed the fresh condensate that was not strongly combined, but the separated condensate easily adhered again, resulting in pipeline deposition and a rapid increase in the back pressure.

(2) During the unheated gathering of heavy oil, a decrease in temperature caused the oil-water two-phase flow to change from an oil-in-water dispersed flow to an annular flow with a continuous water phase as the ring and a water-in-oil emulsion as the core. Subsequently, intermittent flow of the

water and water-in-oil emulsion occurred, with intermittent large oil blocks. Finally, the oil phase was filled with a flow section, and the flow was stopped.

(3) For heavy oils with high water contents, the critical point of the sudden increase in the pressure drop during temperature reduction is the temperature of adhesion and deposition; this value is used as the temperature limit to guide the operation of unheated oil gathering. Using the dimensional analysis method to establish a calculation model for the temperature of adhesion and deposition.

(4) The applicability of the unheated oil gathering process is jointly affected by many factors, such as the temperature limit, wellhead oil outlet temperature, liquid production, water content, pipeline length, diameter, and ambient temperature. Using the least squares method to calculate the total heat transfer coefficient during the production and operation of the pipeline. Based on the energy conservation law and the model of the temperature of adhesion and deposition, we proposed an optimization scheme for the water blending parameters and established a computer program for optimizing the water blending amount of heavy oils with high water contents. The interface program can be used to calculate the optimal watering volume of heavy oils with a high water-cut by using different parameters.

(5) According to the optimized water blending amount, a three-day gathering and transportation experiment was conducted, and the wellhead back pressure exhibited stable fluctuations. This verified the reliability of the model for the temperature of adhesion and deposition and the water blending optimization method.

References

[1] Hasan S W, Ghannam M T, Esmail N. Heavy crude oil viscosity reduction and rheology for pipeline transportation[J]. Fuel, 2010, 89(5): 1095-1100.

[2] YIN J. Research on the transportation of heavy oil mixed with water[D]. Xi'an: Xi'an Shiyou University, 2014.

[3] Liu H, Meng S, Zhao Z, et al. Best Practice of Increasing the Income and Reducing the Expenditure, Lowering Cost and Improving Benefit on the Brown Oilfields[A]. OnePetro, 2017.

[4] Liu Y, Pan C, Jin H. Research Progress on rheological properties of polymer-bearing produced fluid[J]. International Journal of Science, 2017: 4(5): 104-108.

[5] Hoffmann R, Amundsen L. Single-Phase Wax Deposition Experiments[J]. Energy & Fuels, American Chemical Society, 2010, 24(2): 1069-1080.

[6] Gan Y, Cheng Q, Chu S, et al. Molecular Dynamics Simulation of Waxy Crude Oil Multiphase System Depositing and Sticking on Pipeline Inner Walls and the Micro Influence Mechanism of Surface Physical – Chemical Characteristics[J]. Energy & Fuels, 2021, 35(5): 4012-4028.

[7] Zheng H M, Huang Q Y, Wang C H. Wall sticking of high water – cut crude oil transported at temperatures below the pour point[J]. Journal of Geophysics and Engineering, 2015, 12(6): 1008-1014.

[8] LI J. Study on Flow Characteristics of Oil-Water Two-Phase Flow and Process Application during Non-heating Oil Gathering[D]. Qingdao: China University of Petroleum(East China), 2020.

[9] Zhang Y, Huang Q, Cui Y, et al. Estimating wall sticking occurrence temperature based on adhesion force theory[J]. Journal of Petroleum Science and Engineering, 2020, 187: 106778.

[10] Lyu Y, Huang Q, Zhang F, et al. Study on adhesion of heavy oil/brine/substrate system under shear flow condition[J]. Journal of Petroleum Science and Engineering, 2022, 208: 109225.

[11] MAO Q J. Study on Temperature Limit and Energy Saving Technology of Circular Gathering and Transferring Process in Oil Field[D]. Daqing: Daqing Petroleum Institute, 2009.

[12] Liang Y R, Xue H B, Zhang S Q. Boundary condition for oil transportation of single well in Yan Chang oilfield[J]. Oil Production Engineering. 2013(4): 4.

[13] Yang Z R, Research on Gathering and Transportation Critical Conditions of High-Water-Cut Thick Oil in LiaoHe Oilfield[D], China University of Petroleum, Beijing, 2016.

[14] Wei G, Qinglin C, Zuonan H, et al. Research and application on the gathering and transportation radius chart of multi-phase pipelines in high water cut stage[J]. Case Studies in Thermal Engineering, 2020, 21: 100654.

[15] Zhu C, Liu X, Xu Y, et al. Determination of boundary temperature and intelligent control scheme for

heavy oil field gathering and transportation system[J]. Journal of Pipeline Science and Engineering, 2021, 1(4), 407-418.

[16] Rastogi E, Saxena N, Roy A, et al. Narrowband Internet of Things: A Comprehensive Study [J]. Computer Networks, 2020, 173: 107209.

[17] Lv Y L, Tan H, Li J, et al. Flow characteristics of high water fraction crude oil during the non-heating gathering and transportation[J]. China Petroleum Processing & Petrochemical Technology, 2021, 23(1): 88.

[18] Souas F, Safri A, Benmounah A. A review on the rheology of heavy crude oil for pipeline transportation [J]. Petroleum Research, 2021, 6(2): 116-136.

[19] Huang Z Y, Michael, et al. Wax deposition modeling of oil/water stratified channel flow[J]. American Institute of Chemical Engineers, 2010, 57(4): 841-851.

[20] Lyu Y, Huang Q, Li R, et al. Effect of temperature on wall sticking of heavy oil in low-temperature transportation[J]. Journal of Petroleum Science and Engineering, 2021, 206: 108944.

[21] Tan J, Luo P, Vahaji S, et al. Experimental investigation on phase inversion point and flow characteristics of heavy crude oil-water flow[J]. Applied Thermal Engineering, 2020, 180: 115777.

[22] LIU X Y. The Limit Confirming and Hydraulic/Thermodynamic Calculation Method Research for Oil-gas-water Mixing Transportation Safe in Pipeline during Oil Producing With Supper High Water Cut[D]. Daqing: Daqing Petroleum Institute, 2005.

[23] Xu P, He L, Yang D, et al. Blocking characteristics of high water-cut crude oil in low-temperature gathering and transportation pipeline[J]. Chemical Engineering Research and Design, 2021, 173: 224-233.

[24] Liu B, Long J. The mechanism of how asphaltene affects the viscosity of Tahe heavy oil[J]. SCIENTIA SINICA Chimica, 2018, 48(4): 434-441.

[25] CHEN Y J. The Physical and Chemical Property Study of Viscous Crude Oil of Shanjiasi Oilfield in Shengli [D]. Zhejiang University, 2002.

[26] Reddy G M, Reddy V D. Theoretical Investigations on Dimensional Analysis of Ball Bearing Parameters by Using Buckingham Pi-Theorem [J]. Procedia Engineering, 2014, 97: 1305-1311.

[27] Reeves M T, Billam T P, Anderson B P, et al. Identifying a superfluid Reynolds number via dynamical similarity[J]. 2014.

[28] Li A, Zhu M, Hao P, et al. Wall sticking inhibition of high water cut crude oil (below pour point) by underwater superoleophobic PA-FC modification [J]. Colloids and Surfaces A: Physicochemical and Engineering Aspects, 2020, 607: 125427.

[29] S.A, D.O K, S.M M, et al. An Analysis of Oil/Water Flow Phenomena in Horizontal Pipes[C]. 1989.

[30] Luo H, Wen J, Lv C, et al. Modeling of viscosity of unstable crude oil-water mixture by characterization of energy consumption and crude oil physical properties [J]. Journal of Petroleum Science and Engineering, 2022, 212: 110222.

[31] Wang H, Xu Y, Shi B, et al. Optimization and intelligent control for operation parameters of multiphase mixture transportation pipeline in oilfield: A case study[J]. Journal of Pipeline Science and Engineering, 2021, 1(4): 367-378.

[32] Huang Z N. Study on Thermal and Hydraulic Characteristics and Calculation Method of Gathering Radius at Normal Temperature of Multiphase Pipeline in Oil Field [D]. Northeast Petroleum University, 2020.

[33] Li W, Huang Q, Wang W, et al. Estimating the wax breakingforce and wax removal efficiency of cup pig using orthogonal cutting and slip-line field theory [J]. Fuel, 2019, 236: 1529-1539.

[34] Wilson K C, Thomas A D. Analytic Model of Laminar-Turbulent Transition for Bingham Plastics[J]. The Canadian Journal of Chemical Engineering, 2006.

[35] Xiong P, Qiu Z, Lu Q, et al. Simultaneous estimation of fluid temperature and convective heat transfer coefficient by sequential function specification method [J]. Progress in Nuclear Energy, 2021, 131: 103588.

阀芯结构对蝶阀内部空化演化及抑制的影响

张 光[1,2] 吴 璇[1] 张浩钿[1] 林 哲[1,2] 朱祖超[1,2]

(1. 浙江理工大学全省复杂流动与流体工程装备重点实验室;
2. 浙江理工大学流体传输系统技术国家地方联合工程实验室)

摘 要 蝶阀是流体输送系统的重要元件,凭借其结构简单,操作简单,启闭迅速等优点被广泛运用于航空航天、生物医疗、能源、石化等领域。在系统运行过程中,当液体局部压力低于饱和蒸汽压时会发生空化现象,这种现象是不可避免的并且会对阀芯和管路造成严重的损坏。本研究搭建可视化实验平台,结合实验进行数值模拟,探究轴径大小对蝶阀内部空化的影响。通过数值模拟获得不同轴径下流场中空化体积变化及周期性空化演化,蝶板表面压力、涡、剪切力、壁面熵产分布等信息,并对其进行了详细的观察和分析。研究发现,蝶阀内的空化具有明显的周期性。随着轴径比的增大,蝶阀上表面低于饱和蒸汽压的区域越多,空化总体积也越大。蝶板表面的最大壁面剪切应力和壁面熵产均出现在阀轴位置、蝶板头部和尾部。此外,随着轴径比的增大,壁面剪切应力和壁面熵产均增大。

关键词 蝶阀;空化;壁面剪切应力;熵产;空化抑制

Effects of core structure on cavitation evolution and inhibition in the butterfly valve

Zhang Guang[1,2] Wu Xuan[1] Zhang Haotian[1] Lin Zhe[1,2] Zhu Zuchao[1,2]

(1. Zhejiang Key Laboratory of Multiflow and Fluid Machinery, Zhejiang Sci-Tech University;
2. National-Provincial Joint Engineering Laboratory for Fluid Transmission System Technology, Zhejiang Sci-Tech University)

Abstract Butterfly valve is a common component of fluid conveying system for controlling flow conditions. It is widely used in aerospace, biomedical, energy, petrochemical and other fields due to its simple structure, simple operation and fast opening and closing. In the process of system operation, when the local pressure of the liquid is lower than the saturated vapor pressure, the inevitable cavitation phenomenon will occur. This phenomenon will cause serious damage to the valve core and pipeline. In this study, a visualization experimental platform was built, and numerical simulation was carried out in combination with the experiment to explore the influence of shaft diameter on the internal cavitation of butterfly valve. The change of cavitation volume and periodic cavitation evolution were obtained and discussed in detail under different shaft diameter ratios. Distributions of vortex, wall shear stress and entropy generation induced by wall shear stress on the surface of butterfly plate were obtained and analyzed as well. It is found that the cavitation in the butterfly valve has obvious periodicity. With the increase of shaft diameter ratio, the areas of pressure filed lower than saturated vapor pressure on the upper surface of butterfly valve are lager and the cavitation volume is also lager. The maximum wall shear stress and entropy generation induced by wall shear stress on butterfly plate surface appear at the position of valve shaft, the head and end of butterfly plate. In addition, the wall shear stress and entropy generation induced by wall shear stress increase with the increase of shaft diameter ratio.

Key words Butterfly valve; Cavitation; Wall shear stress; Entropy generation; Cavitation inhibition

阀门是流体输送系统的控制元件,承担起截止、调节、分流等重要任务。阀门广泛的应用于工业生产和输运,它的存在就像是工业的"血管"。阀门种类繁多,可以根据不同的要求和工作环境选择适用的阀门,如用于截断或接通管路介质流的截止阀、球阀、蝶阀,用于组织介质倒流的止回阀，还有用于改变管路中介质流动方向的调节阀、节流阀等。由于阀门的广泛使用及其在系统中的重要性，阀门的可靠性及寿命对工业稳定生产和降低生产成本有着重要意义。在种类丰富的阀门中，蝶阀因其结构简单、操作简单、启闭迅速等优点被广泛应用于生活和生产中，因此市场对蝶阀的要求越来越高，耐腐蚀、高温高压、抗空化的要求也越来越高。

空化是一种会对流体机械造成持续破坏的现象，会对阀门性能和寿命造成严重威胁。空化是局部流速过高导致压力低于饱和蒸汽压，液体发生相变汽化产生气泡的现象。这些气泡在流向高压区时会发生溃灭。溃灭的过程伴随着能量的释放，从而造成阀门的损坏。研究阀门的抑制空化，提高阀门的使用寿命，保证输送系统的稳定运行具有很大的应用空间。

空化产生的空泡持续时间短，很难被捕捉。随着科学技术的快速发展，学者们有了更好的测量技术和计算模型来研究空化的动态演化过程。Reuter 等利用高速摄像机和阴影成像技术进行了激光诱导空化的可视化实验，清晰地观察到了单个空泡溃灭过程，发现只有当非轴对称能量自聚焦起作用时，才会对金属表面造成损伤。Meng 等提出并开发了可视化实验系统，发现以油为工作介质时，攻角为-7°的 NACA0012 水翼在 0Mpa 工作压力条件下速度大于 10m/s 会产生空化现象，且水翼的临界空化数低于水。Han 等建立了基于 PIV 双头激光器和传统工业相机的纳秒级分辨率的摄影系统，详细研究了激光诱导初始空化和溃灭的过程，揭示了空化的非对称演化规律。Zhang 等利用相位分辨光学探针和高速相机研究了片状/云状空化内部复杂的物理和水气混合特性，提出了一个经验气泡尺寸分布，并讨论了云空化的相应统计量。Banks 等利用红外（IR）成像和平面激光诱导荧光（PLIF）技术观测了连续波激光诱导空化泡周围的温度场。他们观察到，在成核位置显热显著降低，气泡坍塌后温度降低，液体的初始的热体积收缩。Soyama 等使用高速 X 射线成像来观察文丘里管内产生的空化现象。他们发现文丘里管内的空化不是球形气泡，而是由成角度的气泡组成，并评估了气泡表面的切向速度。

空化现象在流体机械中十分常见，会造成空蚀、振动、噪声等诸多不利影响。Shi 等通过数值模拟量化了表面中心处的压力峰值、压力持续时间和压力脉冲对仿生表面造成的损伤，并讨论了四类表面条件对空化、水射流和冲击压力载荷动力学行为的影响。Firly 等采用流固耦合方法研究了空泡溃灭对金属和聚合物的损伤机制。研究发现，在金属中，冲击载荷主要以压缩和剪切弹性波的形式传递。它会对聚合物产生过度的塑性变形，即所谓的空化坑。空化坑的体积和深度与聚合物的屈服强度有很强的相关性。Zou 等发现颗粒的存在改变了气泡周围液体介质的稠度，导致气泡溃灭时产生水击冲击波和内爆冲击波。Khavari 等基于实验研究了液体物理性质对空泡溃灭过程中产生的冲击波特性的影响。在低黏度液体中，表面张力和惯性力决定了空化的动力学，而在高黏度液体如甘油中，黏性力将主导表面张力和动量。Li 等利用可渗透声学模拟方法研究了 NACA66 水翼周围云空化引起的噪声。在不同空化条件下，偶级噪声分量所占比例随空化数的降低而增加。

对于流体机械而言，空化的破坏是巨大且不可避免的，很多学者对空化的抑制进行了研究。Li 等基于仿生学原理设计了一种周期性射流方案，用于抑制流体机械在高速流动条件下的空化。Yang 等通过结合两种常见的空化抑制仿生结构，开发了一种新的水翼空化被动控制方法。两种结构控制效果的叠加和互补不仅抑制了耐空蚀，而且增强了流场的稳定性。Qiu 等通过压力传感器和高速摄像机观察到微涡流发生器诱导产生微涡流并促进空化稳定性，带有微涡流发生器的水翼对空化溃灭有抑制作用。Timoshevskiy 等利用高速成像、时间分辨 LIF 可视化、二维 PIV 技术和水声压力测量，研究了利用连续切向液体喷射通过二维水翼表面展向槽产生的壁射流主动控制空化的新方法。Sezen 等研究了粗糙度对螺旋桨水力性能的影响，发现粗糙的叶片避免降低了叶尖涡流内部的轴向和切向速度，压力增加，从而抑制了叶尖涡流空化。Sun 等首次利用计算流体力学方法研究了空化发生单元（CGU）结构对典型 ARHCR 性能的影响，得到了不同形状、直径、相互作用距离、高度和倾角下轴的空化量和所需扭矩。之后，又首次采用基于简化流场策略的方法研究了 CGU 布置方式对典型 ARHCR 性能的影响，得到了空化效率最高的布置位置。

目前，针对阀门空化抑制的研究较少。本研

究拟搭建实验平台，对空化进行可视化和分析。通过实验结果验证了数值模拟的准确性，并通过数值研究获得了更详细的流场信息，如空化分布、涡分布、壁面切应力分布以及壁面切应力诱导的熵分布等，获得了阀芯结构对抑制空化的影响，并进行了详细的讨论。

1 数值计算方法

1.1 控制方程

在本实验中，由于水的比热容较大，可以假设介质保持恒温，不考虑液相和气相之间的传热和可压缩性。控制方程主要包括连续性方程和动量守恒方程。

混合物的连续性方程为：

$$\frac{\partial}{\partial t}(\rho_m) + \nabla \cdot (\rho_m \vec{v}_m) = 0 \tag{1}$$

式中，$\vec{v}_m$ 为质量平均速度（m/s）；ρ_m 为混合物密度（kg/m³）。

混合物的动量方程可以通过求和所有相位的单个动量方程。表达式如下：

$$\frac{\partial}{\partial t}(\rho_m \vec{v}_m) + \nabla \cdot (\rho_m \vec{v}_m \vec{v}_m) = -\nabla p + \nabla \cdot [\mu_m(\nabla \vec{v}_m + \vec{v}_m^T) + \rho_m \vec{g} + \vec{F} - \nabla \cdot (\sum_{k=1}^{n} \alpha_k \rho_k \vec{v}_{dr,k} \vec{v}_{dr,k})] \tag{2}$$

式中，n 为相数；F 为体积力；μ_m 为混合黏度表达式为 $\mu_m = \sum_{k=1}^{n} \alpha_k \mu_k$，$\vec{v}_{dr,k}$ 为第二相滑移速度。

1.2 空化模型

气液两相在流动中的转化模型可以使用空化模型来描述，空化模型可分为状态方程模型和输运方程模型。在输运方程模型中，基于 Ralyleigh-Plesset 方程的空化模型是目前应用最广泛的，它可以更好的描述空化初生和发展时空泡体积变化。Ralyleigh-Plesset 方程描述了一个空泡在内外差作用下体积膨胀或溃灭速度，其表达式为：

$$R_B \frac{d^2 R_B}{dt^2} + \frac{3}{2} \cdot \left(\frac{dR_B}{dt}\right)^2 = \left(\frac{P_b - P}{\rho_l}\right) - \frac{4v_l}{R_B}\frac{dR_B}{dt} - \frac{2\sigma}{\rho_l R_B} \tag{3}$$

式中，R_B 为空泡半径，m；v_l 为液体运动黏度，m²/s；σ 为液体表面张力系数；P_b 为气泡表面压力，Pa。

本实验选择使用的 Schnerr-Sauer 空化模型方程基本形式由 Ralyleigh-Plesset 方程推导而来，它认为许多球形液泡液体组成气液混合物，气相和液相掺混在一起，两相之间的质量传输使用近似的传输方程来模拟，具有模型中不存在经验系数的特点，气液相方程表达式分别为：

$$\frac{\partial}{\partial t}(\alpha \rho_v) + \nabla \cdot (\alpha \rho_v \vec{V}) = R \tag{4}$$

$$\frac{\partial}{\partial t}((1-\alpha)\rho_l) + \nabla \cdot ((1-\alpha)\rho_v \vec{V}) = -R \tag{5}$$

式中，R 为传质速率，可得混合物密度 ρ 和气相体积分数 α，Schnerr-Sauer 模型将气相体积分数与单位体积的液体中的气泡数量及半径进行联系：

$$\rho = \alpha \rho_v + (1-\alpha)\rho_l \tag{6}$$

$$\alpha = \frac{n \frac{4}{3}\pi R_B^3}{1 + n \frac{4}{3}\pi R_B^3} \tag{7}$$

得到 Schnerr-Sauer 模型的最终形式如下：

$$R_e = \frac{\rho_v \rho_l}{\rho} \alpha(1-\alpha)\frac{3}{R_B}\sqrt{\frac{2(P_v - P)}{3\rho_l}}, (P_v \geq P) \tag{8}$$

$$R_c = \frac{\rho_v \rho_l}{\rho} \alpha(1-\alpha)\frac{3}{R_B}\sqrt{\frac{2(P - P_v)}{3\rho_l}}, (P_v \leq P) \tag{9}$$

1.3 湍流模型

湍流的数值模拟方法主要有直接数值模拟（DNS）、雷诺时均（RANS）和大涡模拟（LES）。其中 LES 方法是一种介于 DNS 和 RANS 方法之间的一种方法，它可以克服 DNS 和 RANS 方法各自的不足，在准确计算流场脉动信息的同时还能大大降低计算量，能较好地模拟流场脉动的细节特性。LES 方法通过滤波函数将湍流的瞬时运动分解成大尺度和小尺度涡两部分，对大尺度涡通过控制方程直接进行数值求解，小尺度涡部分使用亚格子应力尺度模型（SGS）进行模拟。使用滤波函数得到的不可压缩流动的控制方程为：

$$\nabla \cdot (\tilde{v}) = 0 \tag{10}$$

$$\frac{\partial}{\partial t}(\tilde{\rho v}) + \nabla \cdot (\widetilde{\rho v \times v}) = -\nabla \cdot \tilde{p}I + \nabla \cdot (T + T_t) + f_b \tag{11}$$

式中，∇ 为梯度算子；ρ 为密度（kg/m³）；$\tilde{v}$ 和 $\tilde{p}$ 分别为过滤后的速度（m/s）和压力（pa）；I

为等同张量；T 为应力张量；f_b 为体积力；T_t 为湍流应力张量，其表达式为：

$$T_t = 2\mu_t S - \frac{2}{3}(\mu_t \nabla \cdot \tilde{v} + \rho k)I \qquad (12)$$

式中，S 为应变率张量；k 为亚格子湍动能；μ_t 为亚格子湍流黏度。

2 计算域与可行性验证

2.1 计算域及边界条件

蝶阀模型(图1)和蝶阀流道的模型(图2)的实验管道内径尺寸为10mm×10mm，数值模拟与实验条件保持一致。为了使阀门上下游流动的充分发展，将上游进口设置在蝶阀来流方向距离5L处，将出口设置在蝶板下游距离10L处。无量纲数轴径比R定义为蝶阀轴的长度与直径的比值，其表达式如下：

$$R = \frac{2r_2}{D} \qquad (13)$$

在 ANSYS ICEM 中对蝶阀流道进行结构网格划分(图3)。对壁面处进行网格加密。把数值模拟出入口条件设置为压力出入口，根据实验得到的进出口数据设置压力大小，采用无滑移壁面条件。为了更好地分析瞬态数值模拟，采用 PISO 算法来进行压力速度耦合。根据 CFL 数设置时间步长能保证数值模拟的计算精度，时间步长设置为 2×10^{-5}s。

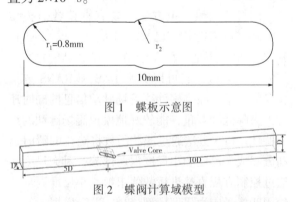

图1 蝶板示意图

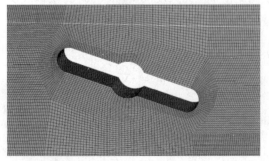

图2 蝶阀计算域模型

图3 计算网格

2.2 计算域及边界条件

空化实验平台的实物图(图4)和实验平台的结构图(图5)如图所示。实验装置主要由压力控制系统、数据采集系统和图像采集系统组成。

图4 空化可视化实验平台

压力控制系统包括储水罐、管道、安全阀和截止阀。储水罐容量为120L，可承受3.0MPa的压力。储水罐顶部设有安全阀，以保证实验中不会因施加的压力过大而产生安全问题。左侧的储水罐控制进水压力，并与空气压缩机相连。两个储水罐通过10mm×10mm的可视化管道连接。可视化段的上游和下游分别装有球阀，用于控制实验管道的入口和出口。数据采集系统由高精度的压力和温度传感器组成。在储水罐中安装了温度传感器和加热装置，对实验温度进行控制。高精度压力传感器安装在可视化部分的左右两侧。压力传感器将信号传输至信号采集系统，信号采集系统以100Hz的采集频率记录实验过程中的压力变化数据。图像采集系统由高速摄像机和由高透明有机玻璃制成的蝶阀可视化管道组成。可视化段内部设有可调节开度的蝶阀。通过有机玻璃，可以清晰地看到管道内部的空化现象。在可视化段的正前方设置高速摄像机进行图像采集。

实验前，打开位于蝶阀上游的球阀，关闭下游球阀，将左侧储水罐注满水，打开下游球阀1至2秒，使蝶阀被完全浸没。由于空化现象产生气泡，类似于残余空气，这个操作可以保证实验不受装置内残余空气的干扰。排气完成后，打开空气压缩机，观察进出口压力传感器的读数。当入口压力传感器读数满足实验要求时，迅速关闭空气压缩机，打开蝶阀下游的球阀，开始实验。水迅速通过阀门，当进出口压差足够大时，空化发生。可视化段正面布置帧数为2200fps的高速摄像机进行图像采集，侧面采用无频闪 LED 光源，保证拍摄环境光照充足。将高速摄像机和信号采集器与计算机相连，记录和存储实验现象和进出口的压力变化曲线。

2.3 可行性验证

数值结果与实验结果的比较如图 6 所示。通过轴径比为 0.2、开度为 80% 的蝶阀进行验证，采用与实验测试相同的边界条件。通过分析一个典型周期的空化演变，可以得出数值空化分布与试验结果吻合较好。因此，该数值方法可用于准确预测蝶阀空化的发展过程。

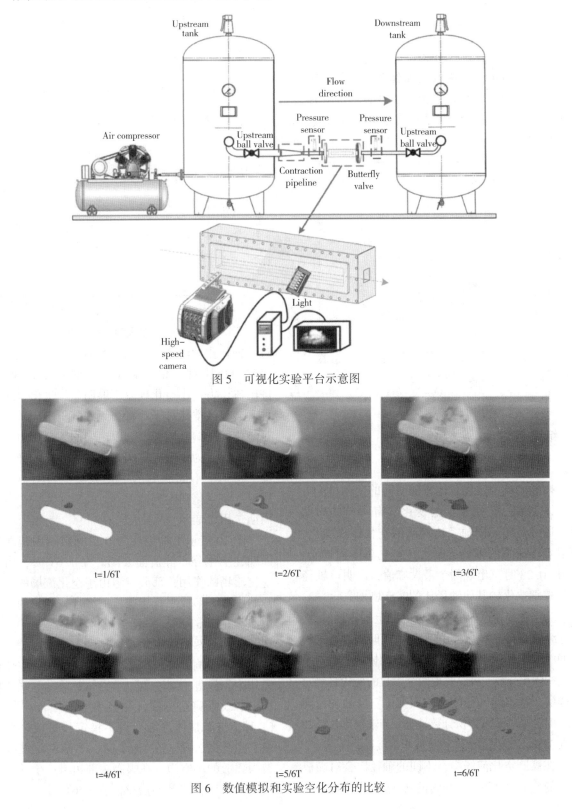

图 5 可视化实验平台示意图

图 6 数值模拟和实验空化分布的比较

2.4 网格无关性

为保证网格数量不影响计算结果的准确性，采用 5 组不同的网格进行网格无关性验证。蝶阀流量系数是表征蝶阀工作性能的主要参数之一，通过计算蝶阀的流量系数 K_v 验证了网格无关性。蝶阀的流量系数计算如下所示：

$$K_v = 10Q\sqrt{\frac{\rho}{\Delta P}} \quad (14)$$

式中，ΔP 为压降；ρ 为液体密度（kg/m^3）。

如图 7 所示，网格数增加到 384 万网格时，流量系数开始趋于稳定，但为了计算的准确性，又不浪费计算资源，选择 468 万这套网格来进行数值模拟研究。

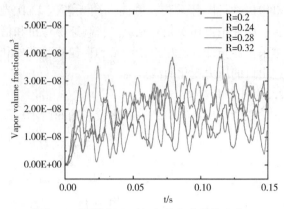

图 8 不同轴径比下气相体积分数的变化

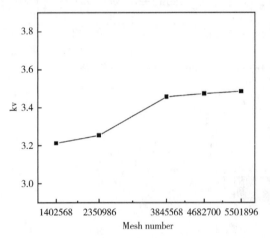

图 7 不同网格数下的流量系数

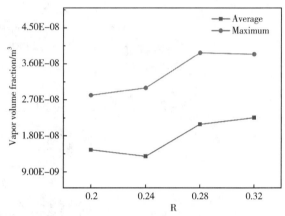

图 9 不同轴径下最大和平均气相体积分数

3 结果与讨论

3.1 空化分布

蝶阀流场中空化总体积随时间的变化（图 8）从整体上看，空化体积的变化具有明显的周期性。通过对比不同轴径比蝶阀的空化体积变化，可以发现蝶阀的轴径比越大，空化总体积越大。当 $R=0.28$ 时，出现两个最大峰值，说明该轴径比的蝶阀空化较其他轴径比的蝶阀剧烈。

为了更好地反映出空泡在时间上的累计特性，对连续的 0.15s 内的空泡体积分数随时间进行平均（图 9）。对比不同轴径下的最大气相体积分数与平均气相体积分数，可以发现两个参数的变化趋势是一样的，基本随轴径的增大而变大。当轴径比为 0.24 时，平均蒸汽体积分数最小。当轴径比为 0.28 和 0.32 时，最大蒸汽体积分数非常接近。

在蝶阀空化流场中，不同的轴径比会对蝶板附近的空化流场产生不同的影响。当开度较大时，轴径比对流场的影响更为明显。为了更好地探究轴径比对空化的影响，选取开度为 80%。进出口压力过小空化现象不够明显，进出口压力过大会导致空化过于严重，无法分析不同轴径比对空化的影响。因此，选取适中的入口压力 4bar 作为工况。为了研究蝶阀内部空化演化的详细过程，得到了不同时刻的空化分布，如图 10-13 所示。将空化初生到溃灭的时间定义为一个空化周期，用 T 表示。

图 10 展示了轴径比为 0.2 的蝶阀内部典型周期性空化分布等值面云图。在 1/6T 时，由于空化流场的作用，部分空泡由于空化流场的作用开始脱落，形成平行于蝶板前缘且较均匀的一排空泡。在 2/6T 时，空化从蝶板中心线处开始分裂，分裂出两个小空化。这是由于管道内壁受到黏性力的影响，蝶板中心线处流速最快，空化首先从中心线处开始分裂。在 3/6T 时，空泡从蝶板中心线处完全分裂，并与小空泡融合。在 4/6T 时，空泡整体逐渐变得不稳定，蝶板尾部边缘空泡开始脱落。在 5/6T 时，下游空化继续分裂为离散的小空泡。在蝶阀头部边缘产生一个新的空化周期，并开始脱落。在 6/6T 时，在该周

期内下游的空泡完全溃灭,而头部边缘新周期的空化形态与 3/6T 相似,在中心线两侧各形成一个较大空泡。

轴径比为 0.24 时的空化分布(图 11)与轴径比为 0.2 相比,可以明显发现轴径比的增大对空化向下游的发展起到了一定的阻滞作用。在 1/6T 时,由于阀轴的阻挡作用,空泡在阀轴处积聚形成大气泡。在 2/6T 时,头部边缘产生新的空化,并与 1/6T 的空化融合,阀轴不再能够阻挡体积较大的空泡,空化泡开始往蝶阀后端发展,并呈对称分布。在 3/6T 时,空化进一步发展,体积变大,空化整体上仍呈对称分布。由于中心线处速度较快,空化在此处比左、右两侧靠近壁面处移动得更远,呈凸起状,中心有一大片无空化区域。从 4/6T 到 5/6T,空泡发展到蝶阀后端。在蝶板后半部分右侧形成较大空泡,在蝶阀左侧和尾部形成许多离散的小空泡,呈溃灭趋势。当 6/6T 时,该周期空泡完全溃灭。

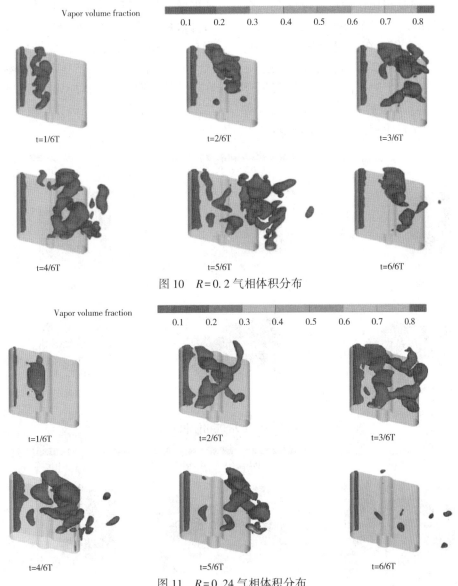

图 10　$R=0.2$ 气相体积分布

图 11　$R=0.24$ 气相体积分布

当 $R=0.28$ 时,一个周期内的空化分布如图 12 所示。在 1/6T 时,蝶板头部边缘分离出一排整齐的大空泡。在 2/6T 时,阀轴的阻滞效果比 $R=0.24$ 时更加明显。几乎所有的空化都滞留在阀轴的前方区域,只有少数空化穿过阀轴。在 3/6T 时由于湍流作用,空泡并没有停留在轴处,而是跨过阀轴向下游发展。在 4/6T 时,由于流速较快,轴处的空化明显比近壁面处的空化更加靠后,形成了光滑的弧形空泡。在轴处,两个小空泡被滞留在蝶板的前半部分。在 5/6T 时,蝶板头部边缘出现片状空化,产生一个表面光滑的空穴,空穴是不稳定的,存在一些空隙,并且在空泡的尾部不稳定区域发生脱落现象,形成明显的云状空化。由于蝶阀尾部存在回射流,回射流向前运动时,会对阀板上方的空化产生剪切应力,使得空化结构不稳定,出现空空泡断裂的现象。上一时刻观察到的阀轴后端的空化开始溃灭,6/6T时已完全溃灭,空穴也消失,下一周期的空化产生,一个条状空化位于轴处,还有一个较稳定的大空泡位于阀板后半部分。

当 $R = 0.32$ 时,一个周期内的空化分布(图13)与较小的轴径相比,特点十分明显,在1/6T时,蝶板头部产生多个条形空化,并在轴处被滞留。蝶板后方仍存在前期未溃灭的空泡。与其他工况相比,$R = 0.32$ 时的空化具有明显更长的演化距离。在2/6T时,滞留的空泡聚集合并成一个大的空泡,蝶板后方上一个周期的空泡开始坍缩变形。3/6T时刻,蝶阀头部开始产生

新的条状空泡与聚集在轴处的气泡融合,由于空泡的大量产生,轴已无法阻挡所有空泡,空泡开始跨过轴,轴的阻挡作用和空泡的运动相互作用,使得空泡形态十分不稳定。到了4/6T时刻,相互作用力暂时达到一个平衡,蝶阀头部开始产生稳定的片状空化,生成稳定且表面光滑的气泡,将整个阀板前半部分完全包住。由于轴径比为0.32时的阀轴对回射流的阻隔作用更强,相比于轴径比为0.28时发生剪切的位置略有提前,回射流更早与空泡完成剪切,使得空泡呈条状向后大尺度脱落。5/6T时刻,片空化消失,轴处仍有一整条空泡被滞留,而轴后的空泡开始溃灭。6/6T时,前缘又产生新的条状空泡,被挡在轴处的空泡开始整体向后脱落,分离出一些小空泡,而尾部的空泡已经完全溃灭。

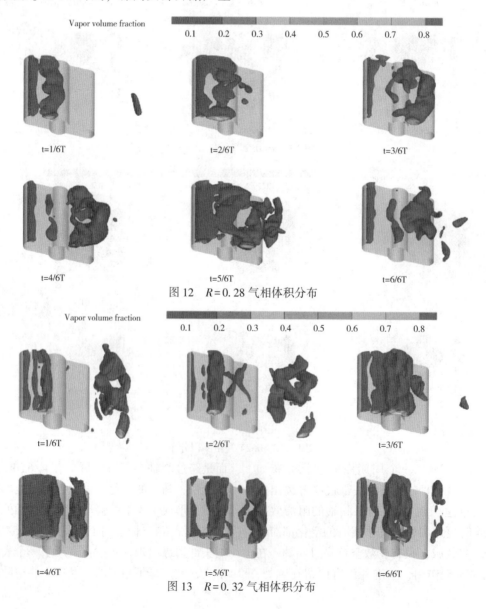

图12 $R = 0.28$ 气相体积分布

图13 $R = 0.32$ 气相体积分布

当 $R=0.2$ 时,阀轴对空化的发展影响不大。随着轴径比的增大,阀轴对空化演化的阻滞作用加强。阀板前半部分的空化体积占比更高,更多的空化积聚在阀轴前方区域,在此处会出现更为稳定的片状空化。

气相体积越大,并不意味着空化对蝶板的影响越大。空化会产生巨大的能量波动,对设备表面造成破坏。因此,空化与蝶板的接触面积越小,对蝶板的破坏就越小。为了探究空化对蝶板的实际作用,定义空化与蝶板的接触面积为 SC,蝶板面积为 SA,SC/SA 为空化与蝶板接触面积占蝶板总面积的百分比。不同轴径比下不同时刻空化与蝶板接触面积的比值如图 14 所示。从空化与蝶板接触面积的数值来看,当轴径比为 0.2、0.24 和 0.28 时,接触面积随着轴径的增大而增大。$R=0.32$ 时的平均气相体积最大,但接触面积最小。虽然产生了较多的空化,但由于轴径的阻挡作用,使得再入射流的位置提前。强回射流在轴径前半部分,导致形成较稳定的片状空化,空化不附着在蝶板上。从空化与蝶板接触面积的变化来看,当轴径比为 0.2、0.24 和 0.32 时,接触面积在一个周期内先增大后减小,轴径比为 0.28 时接触面积则不同,经历了两次增大和减小。轴径越大,阻滞作用越强。0.2 和 0.24 均在轴径处接触较为轻微,因此在周期中间 3/6T 时接触面积最大。当 $R=0.28$ 时,轴径足够大,足以阻挡大部分空化积聚在轴径处。部分空化尚未成功越过轴径,新的空化就已在此处累积。空化需要积累到足够的程度才能跨过轴。因此,在周期中期 3/6T 时,空化在转轴处积聚足够多并跨过轴,向下游发展,此时接触面积最小。此后,新的空化开始在转轴处累积,接触面积再次增大,进入新一轮的循环。因为 R=0.32 时轴径大到可以产生较强的再入流,空化无法在轴处聚集,故并没有经历两次先增大后减小的过程,接触面积的变化与 0.2 和 0.24 时类似。

3.2 涡分布

涡的强度和数量与空化的演化呈正相关,空化和涡在时间和结构上具有很好的一致性。探究涡的分布可以更直观地观察空化对流场的影响,为此引入了能够反映涡三维结构的 Q 准则。图 15 是 $R=0.2$ 时一个典型空化周期的涡分布。通过与气相体积分布进行对比,发现空化的位置与涡有很好的相关性,空化的边缘位置与涡结构

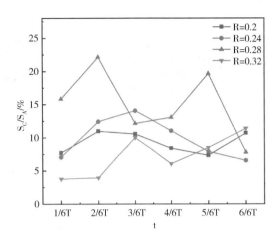

图 14 空化与蝶板接触面积的比值

的分布基本一致。1/6T 至 3/6T 时,空化在初始阶段,大部分空化分布在阀板表面,因此蝶板表面存在小尺度的涡。蝶板下游的涡开始减小,反映了上一周期空泡的溃灭。4/6T 到 6/6T 时刻,随着空泡的持续发展和脱落溃灭,蝶板下游开始形成大量的涡,下游流场的湍流程度增加。

图 16 和图 17 分别为 $R=0.24$ 和 $R=0.28$ 时的涡分布。对比 $R=0.2$ 可以发现,随着直径的增大,阻滞作用变强,轴前半部分流速减小,流场紊乱程度降低。蝶板前半部分表面的小尺度涡变少,蝶板下游的涡分布比重随着直径的增大明显后移。由于轴径的增大,阀轴所在流道截面面积减小,速度增大,导致下游更加不稳定,并且随着直径的增大,影响范围变大,因此在下游产生了更多的涡。这种现象在 $R=0.28$ 时尤为明显,下游涡带相对粗壮。这些现象表明直径越大,对下游流场的扰动作用越强。

图 18 展示了 $R=0.32$ 时的涡分布。此时,在蝶板前半部分和下游均分布有大量的涡,其形状基本与阀轴平行,这与气相体积分布相对应。由于轴直径较大,蝶板前半部分的回射流较强,空泡暂时无法跨过轴,且不断受到轴的扰动,加强了该区域的湍流程度。蝶阀下游的涡分布与小轴径情况相比,$R=0.32$ 时,涡较多且粗壮,但形状如蝶板前半部分的涡一样,呈与轴平行的条状。

3.3 蝶板表面压力分布

为了定量分析空化对蝶板的影响,在蝶板上监测了五条线上的压力(图 19),得到了不同轴径下蝶板表面的压力分布。为了更好地区分低压区,小于饱和蒸气压的区域被标记为红色。由于空化是由于液体中的局部压力低于饱和蒸汽压引起的,因此可以认为低于饱和蒸汽压的标红部分即为空化发生的区域。

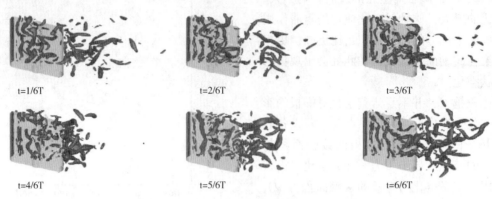

图 15　$R=0.2$ 一个空化演化周期内的涡分布

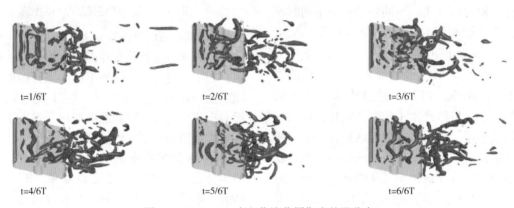

图 16　$R=0.24$ 一个空化演化周期内的涡分布

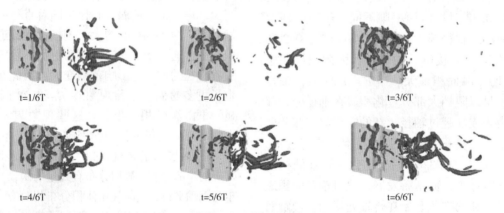

图 17　$R=0.28$ 一个空化演化周期内的涡分布

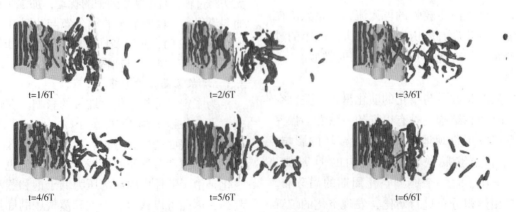

图 18　$R=0.32$ 一个空化演化周期内的涡分布

阀芯结构对蝶阀内部空化演化及抑制的影响

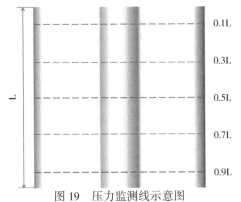

图 19 压力监测线示意图

$R=0.2$ 时蝶板表面的压力分布(图 20)可见空化主要产生在蝶阀的头部边缘。到 2/6 至 4/6T 空化发展阶段,阀轴部分有 1-3 条监测线低于饱和蒸气压以下。在 6/6T 空化溃灭阶段,也有两条监测线为红色,与气相体积分布结合分析,可知这是下一个周期的空化正在发展变大。

$R=0.24$ 时蝶板表面的压力分布(图 21)与 $R=0.2$ 类似,空化在蝶板头部边缘产生。2/6 至 4/6T,监测线都有低于饱和蒸气压的区域。在 2/6T 时,阀轴的 5 条监测线都标红,可知此工况下空化产生更多。与 $R=0.2$ 相比,$R=0.24$ 时压力波动更加明显,流场更不稳定。

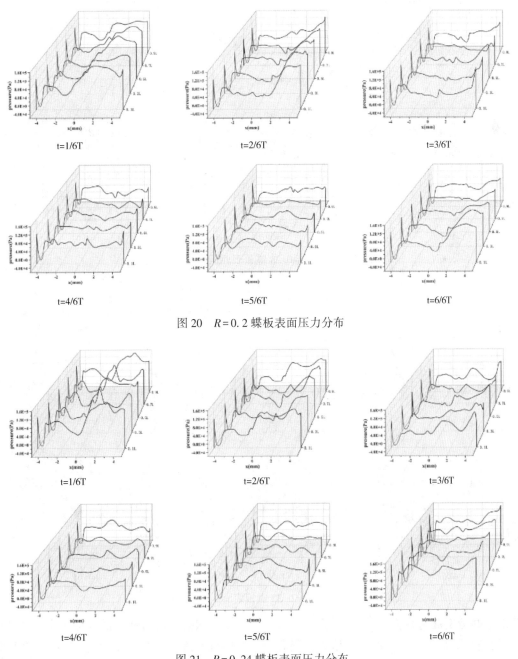

图 20 $R=0.2$ 蝶板表面压力分布

图 21 $R=0.24$ 蝶板表面压力分布

$R=0.28$ 时蝶板表面的压力分布（图 22）对比前两种工况，可以明显发现更多的区域在饱和蒸气压以下。从 1/6T 到 5/6T，均存在低于饱和蒸汽压的区域，在 2/6T 和 5/6T 时空化的产生尤为严重。当 $t=4/6T$ 时，蝶阀尾部也开始产生空化。

图 23 展示了 $R=0.32$ 时蝶板表面的压力分布。$R=0.32$ 时空泡的演变更具有同步性，5 条监测线在不同时刻的变化都非常相似，且同一时间的 5 条线相差不大，可以对应 $R=0.32$ 时的气相体积分布特征，1/6 至 4/6T，都有小部分区域低于饱和蒸气压。在 5/6T 和 6/6T 时，阀轴部分全部低于饱和蒸汽压。值得注意的是，在 5/6T 时，蝶板尾部的大部分区域也开始产生空化。这与涡分布对应较好，证明了该工况下下游流场受影响较大，紊乱程度较高。

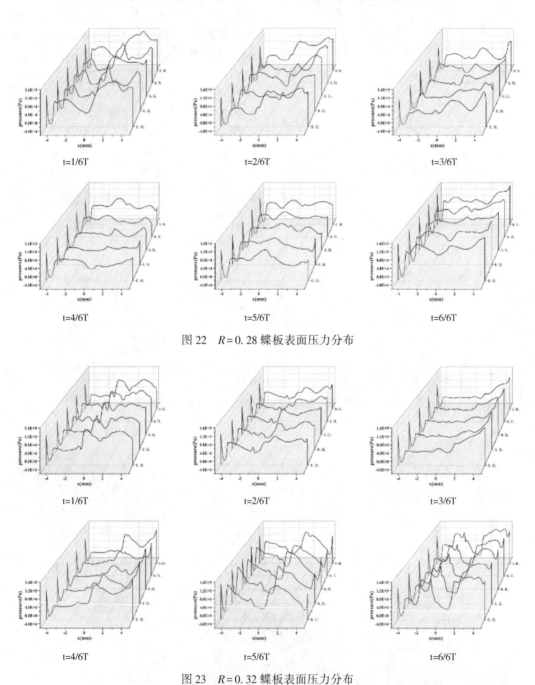

图 22 $R=0.28$ 蝶板表面压力分布

图 23 $R=0.32$ 蝶板表面压力分布

3.4 壁面剪切力引起的熵产

蝶阀的空化流场中空泡的产生和溃灭伴随着不可逆的能量损失。熵的概念通常用来衡量流体在某一状态下所含能量的高低。由壁面切应力引起的近壁流动的熵产称为壁面熵产（EPWS），其表达式为：

$$EPWS = \frac{\tau_w \nu}{T} \tag{15}$$

式中，τ_w 为壁面上的剪切应力，Pa；ν 为近壁面处的流体速度，m/s。

图 24 – 27 分别为 $R = 0.2$，$R = 0.24$，$R = 0.28$ 和 $R = 0.32$ 时蝶板表面剪切力引起的熵产分布图。由于壁面熵产与壁面剪切应力有关，因此壁面熵产等值线与壁面剪切应力等值线对应较好，与空化区域吻合度也较高。熵产值较高的区域在蝶板头部和空化区域末端附近，是空化的起始和溃灭位置，证明空化会给流场带来剧烈的能量变化，而能量的损失主要集中在蝶板头部边缘。对比四张熵产分布图可以发现，随着轴径比的增大，壁面熵产极值增大，且多出现在轴处和蝶板尾端。表明轴径对流体的干扰作用增强，下游回射流对蝶板下游流场的扰动作用增强，在这两个位置处能量损失增加。

4 结论

本文采用数值模拟的方法研究了蝶阀不同轴径比下的空化特性。获得并详细讨论了不同轴径比下的空化体积和演化过程，以及蝶板上的压力、涡分布和壁面剪切应力引起的熵产分布。主要结论如下所示。

（1）随着轴径比的增大，蝶板表面压力低于饱和蒸气压的区域越多，空化总体积也越大。在 $R = 0.2 \sim 0.28$ 时，空化与蝶板的接触面积随轴径比的增大而增大，但当轴径比达到 $R = 0.32$ 时，由于较大的轴径导致强烈的回射流在蝶板头部形成片状空化，使得接触面积最小。

（2）在 $R = 0.2 \sim 0.28$ 时，蝶板前半部分的小尺度涡分布随着轴径比的增大而减少。在 $R = 0.32$ 时，蝶板前半部分的涡分布最多，其形状呈与轴平行的条状。随着轴径比的增大，蝶板下游的流场受到的扰动距离更长，湍流程度增大。

（3）空化是导致蝶板表面产生较大壁面剪切应力和熵产的主要因素。壁面切应力的分布与蝶板表面的熵产具有高度的对应关系。壁面剪切应力和熵产较大的区域主要分布在蝶板头部和空化区末端附近。随着轴径比的增大，壁面剪切应力和熵产的极值变大，出现在轴、蝶板头部和尾部。

该研究可为优化蝶阀的抗空化结构提供理论支撑，对延长蝶阀的使用寿命也具有重要意义。

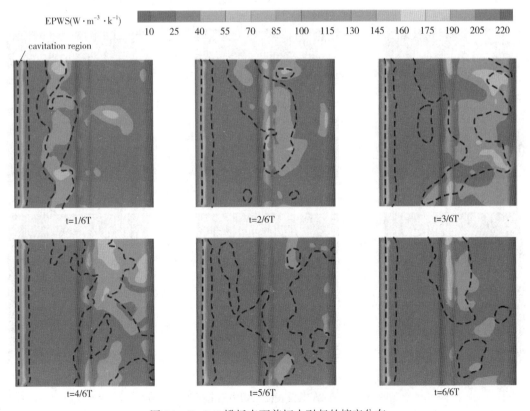

图 24　$R = 0.2$ 蝶板表面剪切力引起的熵产分布

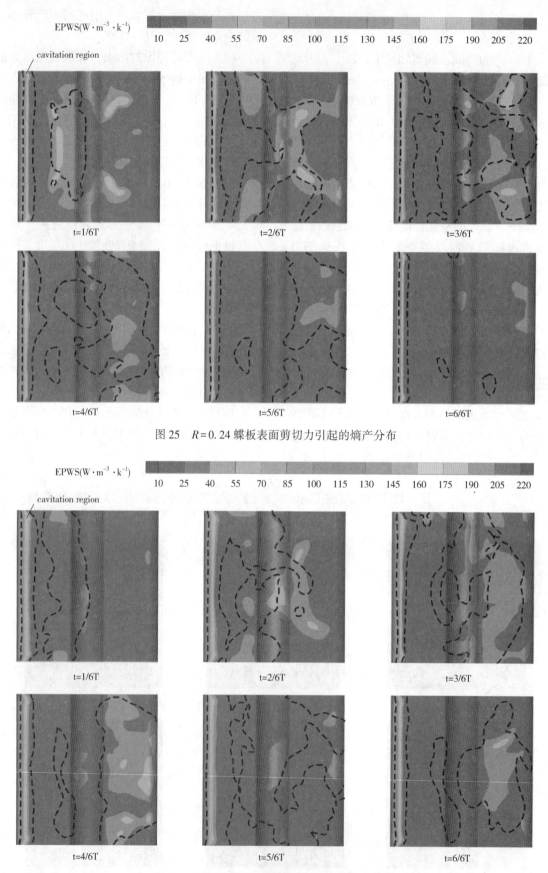

图 25 $R=0.24$ 蝶板表面剪切力引起的熵产分布

图 26 $R=0.28$ 蝶板表面剪切力引起的熵产分布

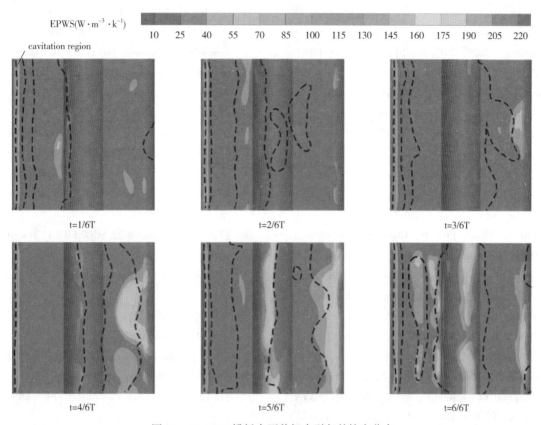

图27 $R=0.32$ 蝶板表面剪切力引起的熵产分布

参 考 文 献

[1] REUTER F, DEITER C, OHL C D. Cavitation erosion by shockwave self-focusing of a single bubble[J]. Ultrasonics Sonochemistry, 2022, 90, 106131.

[2] GUO M, LIU C, LIU S Q, ZHANG J H, KE Z F, YAN Q, et al. On the transient cavitation characteristics of viscous fluids around a hydrofoil[J]. Ocean Engineering, 2023.

[3] HAN D X, YUAN R, JIANG X K, GENG S Y, ZHONG Q, ZHANG YF, et al. Nanosecond resolution photography system for laser-induced cavitation based on PIV dual-head laser and industrial camera[J]. Ultrasonics Sonochemistry, 2021, 78: 105733.

[4] ZHANG H, LIU Y Q, WANG B L, CHENG X S. Phase-resolved characteristics of bubbles in cloud cavitation shedding cycles [J]. Ocean Engineering, 2022, 256, 111529.

[5] BANKS D, ROBLES V, ZHANG B, DEVIA-CRUZ L F, CAMACHO-LOPEZ S, AGUILAR G. Planar laser induced flu-orescence for temperature measurement of optical ther-mocavitation[J]. Experimental Thermal and Fluid Science, 2019, 103, 385-393.

[6] SOYAMA H, LIANG X Y, YASHIRO W., KAJIWARA K, ASIMAKOPOULOPU E M, Bellucci V, et al. Revealing the origins of vortex cavitation in a Venturi tube by high-speed X-ray imaging[J]. Ultrasonics Sonochemistry, 2023, 101, 106715.

[7] SHI H B, ZHANG H, GENG L L, QU S, WANG X K, Nikrityuk P A. Dynamic behaviors of cavitation bubbles near bio-mimetic surfaces: A numerical study [J]. Ocean Engineering, 2024, 292, 116628.

[8] FIRLY R, INABA K, TRIAWAN F, KISHIMOTO K, NAKAMOTO H. Numerical study of impact phenomena due to cavita-tion bubble collapse on metals and polymers[J]. European Journal of Mechanics-B/Fluids, 2023, 101, 257-272.

[9] ZOU L T, LUO J, XU W L, ZHAI Y W, Li J, QU T, FU G H. Experimental study on influence of particle shape on shockwave from collapse of cavitation bubble [J]. Ultrasonics Sonochemistry, 2023, 101, 106693.

[10] KHAVARI M, PRIYADARSHI A, MORTON J, PORFYRAKIS K, Pericleous K, ESKIN D, TZANAKIS I. Cavitation-induced shock wave behaviour in different liquids[J]. Ultrasonics Sonochemistry, 2023, 94.

[11] LI Z J, WANG W, JI X, WU X Y, WANG X F. Re-

vealing insights into hydrodynamic noise induced by different cavitating flows around a hydrofoil[J]. Ocean engineering, 2024, 291 (Jan. 1): 116431. 1 - 116431. 21.

[12] LI J, YAN H, WANG F. Suppression of hydrofoil unsteady cavitation by periodic jets based on fish gill respiration[J]. Ocean Engineering, 2024, 293116584-.

[13] YANG Q, LI D Y, XIAO T L, CHANG H, FU X L, WANG H J. Control mechanisms of different bionic structures for hydrofoil cavitation. [J]. Ultrasonics sonochemistry, 2023, 102106745-106745.

[14] QIU N, XU P, ZHU H, GONG Y F, CHE B X, ZHOU W J. Effect of micro vortex generators on cavitation collapse and pressure pulsation: An experimental investigation[J]. Ocean Engineering, 2023, 288.

[15] TIMOSHEVSKIY M V, ZAPRYAGAEV I I, PERVUNIN K S, MALTSEV L I, MARKOVICH D M, HANJALIC K. Manipulating Cavitation by a Wall Jet: Experiments on a 2D Hydrofoil [J]. International Journal of Multiphase Flow, 2017: S0301932217304068.

[16] SAVAS S, DOGANCAN U, OSMAN T, MEHMET A. Influence of roughness on propeller performance with a view to mitigating tip vortex cavitation[J]. Ocean Engineering, 2021, 239.

[17] SUN X, YOU W B, XUAN X X, JI L, XU X T, WANG G C, et al. Effect of the cavitation generation unit structure on the performance of an advanced hydrodynamic cavitation reactor for process intensifications [J]. Chemical Engineering Journal, 2021, 412.

[18] SUN X, XIA G J, YOU W B, JIA X Q, MANICKAM S, YANG T, et al. Effect of the arrangement of cavitation generation unit on the performance of an advanced rotational hydrodynamic cavitation reactor [J]. Ultrasonics Sonochemistry, 2023.

[19] ZHANG G, ZHANG H T, WU Z Y, WU X, Kim H D, Lin Z. Experimental studies of cavitation evolution through a butterfly valve at different regulation conditions[J]. Experiments in Fluids: Experimental Methods and Their Applications to Fluid Flow, 2024, 65(1).

[20] YUAN W X, SAUER J, SCHNERR G. Modeling and computation of unsteady cavitation flows in injection nozzles[J]. Mecanique & Industries, 2001, 2(5), 383-394.

[21] SCHNERR G, SAUER J. Physical and numerical modeling of unsteady cavitation dynamics[C]. Proceedings of the 4th International Conference on Multiphase 2001.

[22] FLUENT Theory Guide. Canonsburg, PA: ANSYS Inc, 2011.

[23] ZHANG G, WANG W W, ZHANG H T, KIM H D, LIM Z. Characteristics of Cavitation Evolution through a Butterfly Valve under Transient Regulation[J]. Physics of Fluids, 2023, 35, 025113.

[24] ZHANG G, WANG W W, WU Z Y, CHEN D S, KIM H D, LIN Z. Effect of the opening degree on evolution of cry-ogenic cavitation through a butterfly valve [J]. Energy, 2023, 283, 128543.

Integrated process optimization of natural gas pretreatment and pressurized liquefaction based on low temperature packed bed decarburization

Fu Juntao[1] Liu Jinhua[2] Li Zihe[1] Zhu Jianlu[1]

[1. College of Pipeline and Civil Engineering, China University of Petroleum(East China);
2. China Petrochemical Sales Co., Ltd. Hebei Shijiazhuang Petroleum Branch]

Abstract PLNG technology refers to the liquefaction, storage, and transportation of natural gas under pressurized conditions. Increasing the liquefaction pressure raises the condensation temperature, reduces refrigeration energy consumption, and enhances the solubility of CO_2 in LNG, thereby simplifying the liquefaction process and improving the economy of LNG. [Purpose] To fully utilize the cold energy of LNG, an integrated process of pretreatment and pressurized liquefaction is proposed. The low-temperature packed bed method is a decarbonization method for gas-solid separation through low-temperature phase transitions, demonstrating strong adaptability and high safety. [Method] Therefore, the low-temperature packed bed method is chosen as the decarburization method of the integrated process of pressurized liquefaction. In this paper, an integrated process model for natural gas pretreatment and pressurized liquefaction, utilizing low-temperature packed bed decarburization. A genetic algorithm is used to optimize specific power consumption, and the performance of the optimized process is thoroughly analyzed. [Results] The results indicate that the specific power consumption of the process is 8.74% lower than that prior to optimization, while efficiency has increased by 8.73% compared with that before optimization. The exergy loss of the equipment has been reduced by 10.883% relative to the previous configuration, demonstrating a significant improvement in process performance. Furthermore, the process exhibits sensitivity to increases in CO_2 content in the feed gas and shows good adaptability to slight increases in system pressure, minor decreases in flow rate, and increases in C_2H_6 content in the feed gas. [Conclusion] (1) A steady-state model of integrated process of natural gas pretreatment and pressurized liquefaction based on low-temperature packed bed decarburization is established. (2) After optimization by genetic algorithm, the process performance of the integrated process model of low-temperature packed bed decarburization and pressurized liquefaction is significantly improved, and the energy consumption of the process is reduced while the efficiency is improved. (3) The process has a certain scope of application and has a relatively high sensitivity to CO_2 content.

Keyword Pressurized liquefied natural gas, Low temperature packed bed, GA

1 Introduction

PLNG (Pressured Liquid Natural Gas) technology is an innovative liquefaction technology that can significantly lower the construction costs of LNG modules and enhance the economy of LNG applications in marginal and small gas fields. PLNG technology refers to the liquefaction of natural gas under pressure conditions (1.0MPa ~ 2.0MPa), and storage and transportation at the corresponding pressure. The increase in liquefaction pressure raises the liquefaction temperature. This, in turn, reduces the energy consumption required for refrigeration and minimizes the necessary equipment. Additionally, higher temperature increases the solubility of CO_2 in LNG, which can simplify or even eliminate the pretreatment device. In summary, the reduction in energy consumption, equipment quantity, and floor space makes PLNG more cost-effective than conventional liquefaction technologies.

For the FLNG project using PLNG technology, the researchers presented an advanced conceptual design of the PLNG supply chain, encompassing the stages of marine production, transportation, and utilization, while estimating the life cycle cost of the chain. However, the design of the PLNG process remains in a simplified stage, and several scholars have investigated the design and optimization of the PLNG liquefaction process. Xiong et al. initially developed a gas expansion refrigeration pressurized liquefaction process and optimized it from two perspectives: process parameters and refrigerant components. They concluded that pure CH_4 exhibited the lowest energy consumption when employed as a refrigerant. Consequently, the incorporation of methane refrigerant can further decrease the energy consumption of the nitrogen expansion refrigeration pressurized liquefaction process. Subsequently, Aspen HYSYS software was utilized to simulate three common liquefaction processes that integrated PLNG technology, namely cascade, single mixed refrigerant(SMR), and single expander natural gas liquefaction processes under pressure. A genetic algorithm(GA) optimization was employed to minimize power consumption. Compared to conventional processes, the power consumption and heat transfer area of the three processes were significantly reduced, theoretically confirming the advantages of the PLNG process. Lee et al. added a diethanolamine (DEA) aqueous solution deacidification process to the nitrogen expansion refrigeration liquefaction process, analyzed the impact of CO_2 concentration in the feed gas on the deacidification gas device, and estimated that the life cycle cost (LCC) of the deacidification gas and liquefaction process of the PLNG process was significantly lower than that of LNG.

To fully utilize the cold energy of LNG and further improve the economic efficiency of PLNG process, a novel integration of the pretreatment process and pressurized liquefaction process is proposed. This integrated approach establishes a cohesive system for both pretreatment and pressurized liquefaction. In the current decarbonization processes, the low-temperature separation method is characterized by its environmentally friendly nature, low carbon footprint, and suitability for high pressure environment. Due to the simple equipment, no tower and deep removal degree, it can well adapt to the complex sea conditions. The low-temperature separation method employs various removal processes tailored to different phase transition types. Xiong et al. combined the CO_2 low-temperature removal process based on gas-solid separation with the expansion-type pressurized liquefied natural gas process and the cascade-type pressurized liquefied natural gas process, and analyzed and optimized them.

In summary, several studies have been conducted on the new liquefaction process integrating pretreatment and pressurized liquefaction. In this paper, based on the principle of gas-solid separation, an integrated process model of low-temperature packed bed decarburization and nitrogen expansion liquefaction under pressure is established, and the steady-state model is optimized. The parameter adaptability of the optimized model is examined.

2 Model establishment

2.1 Theoretical basis

In this paper, Aspen HYSYS software is used to simulate the integrated process of low temperature packed bed pretreatment and pressure liquefaction. Peng-Robinson state equation(PR EOS) is widely used in the simulation of natural gas liquefaction process as a calculation model with high calculation accuracy for phase equilibrium of natural gas system. Therefore, PR EOS is selected for calculation in this paper.

Compared with the conventional liquefaction process, one of the significant advantages of the pressurized liquefaction process is the increase of CO_2 solubility in PLNG, which reduces the purification requirements of CO_2 in pretreatment. Therefore, the CO_2 purification index is one of the key parameters of the process for the establishment of the pressurized liquefaction process model. In order to ensure the safe operation of the process, the solubility of CO_2 in the pressurized liquefaction temperature zone is used as the basis for determining the purification index. For the solubility of CO_2 in LNG, the experimental measurement is carried out before, and the theoretical model is established and optimized based on the experimental data and the data in references. The phase equilibrium calculation model of CO_2 solid solubility is briefly introduced below.

The calculation model of CO_2 solid solubility phase equilibrium used in this paper is based on the principle of liquid-solid phase equilibrium and the principle of equal fugacity. The fugacity of the fluid phase is calculated by the PR EOS, and its standard form is expressed as:

$$p = \frac{RT}{v-b_m} - \frac{a_m}{(v+(1+\sqrt{2})b_m)(v+(1-\sqrt{2})b_m)} \quad (1)$$

Where p is pressure, Pa; r is the molar gas constant, J/(mol·K); v is molar volume, L/mol; a_m and b_m are the mixing parameters given by the van der Waals mixing rule, which are related to the relationship between the mole fraction of component i in the mixture x_i and the pure component parameters a and b.

According to the principle of equal fugacity, the fugacity coefficient of component i in liquid phase is equal to that of solid phase in liquid-solid phase equilibrium, as follows:

$$x_i^L \cdot \varphi_i^L = \varphi_{pure,i}^S \quad (2)$$

Where x_i^L is the molar fraction of component i in the liquid phase; φ_i^L is the liquid phase fugacity coefficient of component i; $\varphi_{pure,i}^S$ is the fugacity coefficient of pure component i solid phase.

For PR EOS, the fugacity coefficient φ_i^L of component i in the liquid phase of the mixture is calculated as follows:

$$\ln\varphi_i^L = (Z-1)\frac{b_i}{b_m} - \ln(Z-B) - \frac{A}{2\sqrt{2}B}\left(\frac{2}{a}\sum_{j=1}^n x_j a_{ij} - \frac{b_i}{b_m}\right)$$
$$\ln\left(\frac{Z+(1+\sqrt{2})b_m}{Z+(1-\sqrt{2})b_m}\right) \quad (3)$$

Where:

$$A = \frac{ap}{R^2T^2}, \quad B = \frac{b_m p}{RT} \quad (4)$$

For the solid phase fugacity coefficient, the solid phase fugacity model proposed by Prausnitz et al. is used, as shown in Equation(5). The equation represents the Gibbs free energy change when the component i changes from the liquid phase to the solid phase under the system temperature and pressure. Among them, the first two terms after the liquid fugacity coefficient of pure component i represent the change caused by the difference between the melting temperature and the system temperature T, which are related to the melting enthalpy and the isobaric heat capacity, respectively. The third item after the liquid fugacity coefficient of pure component i represents the influence of the difference between the system pressure and the three-phase point pressure (generally atmospheric pressure) on the melting volume. The parameter values used in Equation(5) are shown in Table 1.

$$\ln\varphi_{pure,i}^S = \ln\varphi_{pure,i}^L - \frac{\Delta H_{f,i}}{RT_{m,i}}\left[\frac{T_{m,i}}{T}-1\right] + \frac{\Delta c_{p,i}^{L-S}}{R}\left[\frac{T_{m,i}}{T}-1+\ln\left(\frac{T}{T_{m,i}}\right)\right] - \frac{\Delta v_i^{L\to s}(p-p_m)}{RT} \quad (5)$$

In the formula: $\varphi_{pure,i}^L$ is the fugacity coefficient of the liquid phase of pure component i; $\Delta H_{f,i}$ is the corresponding molar enthalpy of melting, J·mol^{-1}; $T_{m,i}$ is the melting temperature of component i, K; $\Delta c_{p,i}^{L\to S}$ is the change value of isobaric specific heat capacity when pure component i changes from pure liquid phase to pure solid phase; $\Delta v_i^{L\to S}$ is the volume change of component i from liquid phase to solid phase, cm^3·mol^{-1}; p_m is the reference pressure, here is the atmospheric pressure of 0.101325MPa.

Table 1 The parameters of methane and carbon dioxide used in formula(5)

Medium	T_c/K	p_c/MPa	ω	T_m/K	ΔH_f/(J·mol^{-1})	Δc_p/(J·mol^{-1}·K^{-1})	ΔV/(cm^3·mol^{-1})
Methane	190.56	4.641	0.0115	90.67	9 284	4.641	90.67
Carbon dioxide	304.10	7.370	0.23894	216.58	9 019	7.370	216.58

The critical temperature T_c, the critical pressure p_c, the eccentricity factor ω and the binary interaction coefficient k_{ij} are all parameters used in the fugacity calculation, which will not be repeated here. For the binary interaction coefficient k_{ij}, a temperature-related correlation based on experimental data optimized by genetic algorithmis used, as shown in Equation(6).

$$k_{ij} = 0.0599 + \frac{9.2847}{T} - \frac{36.134}{T^2} \quad (6)$$

The comparison between the optimized model calculation results and the experimental data is shown in Fig. 1, and the calculation results are in good agreement with most of the experimental data.

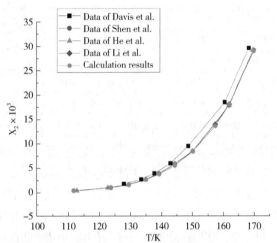

Fig. 1 Solubility of CO_2 solid in $CH_4 + CO_2$ binary system (mole fraction)

Therefore, it is reliable to determine the CO_2 purification index under certain temperature and pressure conditions according to the established CO_2 solid solubility phase equilibrium calculation model.

2.2 Model establishment

The low-temperature packed bed method is used to decarbonize the raw gas, and the integrated process flow model based on low-temperature packed bed decarburization and natural gas liquefaction is established as shown in Figure 2. Low temperature packed bed is a method to remove CO_2 solid particles from feed gas based on the principle of low temperature phase change separation. The liquid nitrogen is used to cool the packed bed, so that the CO_2 in the feed gas passes through the packed bed to form a solid and adsorbs on the packed bed to achieve the removal of CO_2. When the packed bed is covered by CO_2, the temperature of the packed bed is increased, and the CO_2 solid particles adsorbed on the packed bed are sublimated to realize the recovery of CO_2. Therefore, two packed beds are set up to operate alternately, one of which cools and captures CO_2 in natural gas, and the other recovers CO_2 condensed in the packed bed and operates alternately. In the HYSYS simulation, the heat exchanger and solid separator are used to replace the low temperature packed bed.

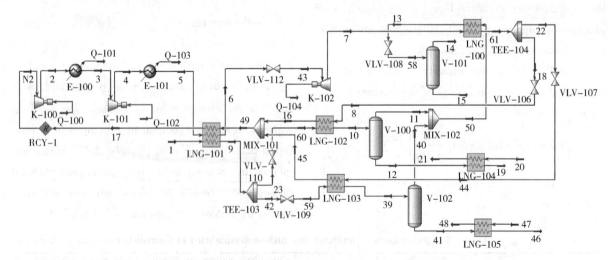

Fig. 2 Integrated process of decarbonization and LNG based on low temperature packed bed

For the pressurized liquefaction process, pressure is the key parameter to determine the liquefaction temperature and purification index. While ensuring that the increase of liquefaction pressure has obvious energy-saving effect, it is also necessary to avoid the high cost of LNG storage and transportation caused by excessive liquefaction pressure. Therefore, the liquefaction pressure is selected as 1.5MPa, and the corresponding working temperature is −114.9℃. According to the calculation results of the CO_2

solubility prediction model, Figure 1 shows that the solubility of CO_2 is about 1.4% at this temperature, so the CO_2 purification index is determined to be 1% on the basis of considering the safety margin.

In summary, an integrated process model of pretreatment and pressurized liquefaction based on low-temperature packed bed decarburization with liquefaction pressure of 1.5MPa and CO_2 purification index of 1% was established. In this simulation, the composition of CO_2 in the feed gas is set to 10%, the CH_4 content is 89.45%, the N_2 content is 0.1%, and the C_2H_6 content is 0.45%. The product in this boundary area is LNG. The inlet pressure of the feed gas is 1500kPa, the temperature is 20℃, and the standard flow rate is 2000Nm^3·d^{-1}. In this paper, Aspen HYSYS software is used to simulate the steady state of the process, and Peng-Robinson (P-R) state equation is used to calculate. The key parameters in the process are shown in Table 2. In addition, the following assumptions are made for the process:

(1) The outlet pressure of the expander is the same as the inlet pressure of the compressor, and both are higher than the atmospheric pressure;

(2) The storage temperature of LNG is 1500kPa, -114.9℃;

(3) The adiabatic efficiency of pump and compressor is 75%, and the minimum temperature difference of heat exchanger is 3℃.

Table 2 Key node parameters

Node name	Temperature/ ℃	Pressure/ kPa	Molar flow rate/ (kgmole·h^{-1})
1	20	1500	3.72
9	-45.76	1500	3.72
11	-110	1500	3.00
13	-114.5	1500	3.00
N_2	27.05	220	30
2	141.2	534.8	30
4	152.1	1300	30
6	-73	1300	30
7	-136.5	220	30
8	-114.4	220	30
16	-95	220	30

3 Model establishment and optimization method

3.1 Process optimization method

At present, genetic algorithm is widely used at home and abroad to seek the optimization results or close to the optimization results. Genetic algorithm is an efficient global optimization search algorithm. Based on natural selection and genetic theory, it combines the survival rules of the fittest in biological evolution and the random information exchange mechanism of chromosomes within the population. The genetic algorithm generates a set of random initial populations, and uses replication, crossover, mutation and other methods to obtain better target fitness function values. According to the evolutionary principles of survival of the fittest and survival of the fittest, the population is continuously improved to obtain better solutions, and finally the optimal solution satisfying the constraints is obtained. In this paper, Aspen HYSYS is used to simulate the natural gas liquefaction process, and the genetic algorithm is optimized by MATLAB. The calculation block diagram is shown in Fig. 3. Firstly, the data of 'nitrogen flow rate, nitrogen inlet pressure and nitrogen secondary compression outlet pressure' are received from MATLAB and brought into HYSYS liquefaction process. HYSYS is used for simulation calculation to verify whether the simulated liquefaction process converges or not. If the convergence reads data from HYSYS to MATLAB in turn to check whether the temperature difference of the heat exchanger is greater than 3 K, a set of objective functions is output. If the conditions are not met, the penalty is imposed, and other parameters are selected. After crossover, mutation and reproduction, a new population is formed, and then it is substituted into HYSYS to recalculate until the termination conditions are met, and the optimal process parameters are output.

3.2 Details of optimization

For the parameters to be optimized, within a certain range, the genetic algorithm is used to optimize the search of multiple parameters at the same time, and the minimum objective function value is obtained. The parameters of the genetic algorithm are set as shown in Table 3.

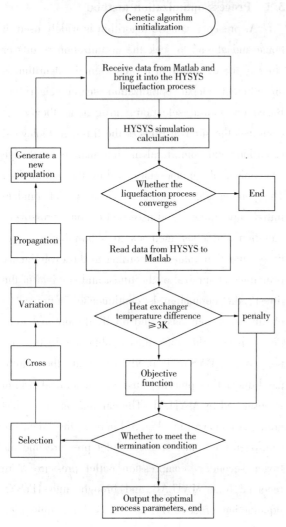

Fig. 3 Genetic algorithm calculation block diagram

Table 3 Setting parameters

Parameters	Value
Population size	200
Maximum generations	200
Cross probability	0.8
Mutation probability	0.1

(1) Optimized parameters

For the simulated low-temperature packed bed decarburization and nitrogen expansion refrigeration LNG integrated process, the main reason affecting the power consumption of the system is mainly the key parameters in the feed gas system and the nitrogen refrigeration cycle. The inlet pressure of the feed gas will affect the heat transfer effect of the feed gas and nitrogen, so it is necessary to optimize the pressure of the feed gas after the compressor. The pressure of the feed gas after throttling, that is, the storage pressure, will affect the liquefaction rate after throttling, so it is necessary to optimize the storage pressure of the feed gas. However, in order to ensure that the standard of CO_2 removal is reached, and the feed gas system is mainly used as the dependent variable, the pressure is set to a fixed value.

The flow rate of nitrogen will affect the cooling capacity of the system and affect the heat transfer effect of the heat exchanger, so it is necessary to optimize this parameter. The pressure after secondary compression of nitrogen will affect the power of the system, so the outlet pressure of nitrogen needs to be optimized. The temperature of nitrogen out of the heat exchanger LNG-101, that is, the temperature entering the expander, will affect the temperature difference of the heat exchanger of the system and the decisive factor for the improvement of the subsequent cooling capacity. Therefore, it is necessary to optimize the temperature of nitrogen entering the expander. The outlet pressure of the expander will affect the outlet temperature, so the pressure needs to be optimized. The temperature of nitrogen out of the low temperature packed bed will affect the heat transfer characteristics of the low temperature packed bed. The upper and lower limits of the parameters to be optimized are shown in Table 4.

Table 4 Upper and lower limits of parameters to be optimized

Parameters	Lower limit	Upper limit
Q Nitrogen flow rate/(kgmole/h)	20	80
Stream 6 temperature/℃	-100	-50
N Nitrogen secondary pressurization/kPa	400	2000
P Expander outlet pressure/kPa	101	300
Stream 16 temperature/℃	-110	-80

(1) Objective function

Nitrogen expansion refrigeration liquefaction process is a high energy consumption liquefaction process, so it is very important to optimize the energy consumption of the system. This section will still use genetic algorithm to optimize the process system, with power consumption as the objective function, power consumption is the ratio of nitrogen system power consumption to LNG product flow. For the integrated process of low temperature packed bed decarburization and nitrogen expansion refrigeration liquefaction, the power consumption w can be expressed as follows:

$$w=\frac{W_{total}}{Q_{LNG}}=\frac{W_1+W_2}{Q_{LNG}} \quad (7)$$

Where w is power consumption, $kW \cdot h \cdot kg^{-1}$; W_{total} is total power consumption, kW; Q_{LNG} is mass flow rate of LNG, $kg \cdot h^{-1}$; W_1 and W_2 represent the power of two nitrogen compressors, kW, respectively.

(2) Constraint conditions and penalty function

The optimization of the steady-state process will be used to guide the design of the actual device. Therefore, while minimizing the power consumption of the natural gas liquefaction process, the feasibility of the heat exchanger device needs to be considered. When the temperature difference of the heat exchanger in the optimized process is too small, the heat exchanger area required for the actual heat exchanger device will be too large, and temperature crossover will occur easily. Therefore, the heat transfer temperature difference of the heat exchanger in the liquefaction process is taken as the constraint condition, as shown in formulas (8) to (11):

$$\Delta T_{LNG-100} \geqslant 3 \quad (8)$$
$$\Delta T_{LNG-101} \geqslant 3 \quad (9)$$
$$\Delta T_{LNG-102} \geqslant 3 \quad (10)$$
$$\Delta T_{LNG-103} \geqslant 3 \quad (11)$$

Where ΔT represents the minimum temperature difference of the heat exchanger, ℃; the subscripts LNG-100, LNG-101, LNG-102, and LNG-103 represent each heat exchanger.

The design of the interstage pressure as shown in (12)

$$P=\sqrt{P_{in}P_{out}} \quad (12)$$

Where P is the interstage pressure, kPa; Pin is the inlet pressure of the first-stage compressor, kPa; Pout is the outlet pressure of the secondary compressor, kPa.

When any of the formulas is not satisfied, we use Eq. (13) to punish.

$$z=w(1+\exp(\max(a))) \quad (13)$$

Which

$$a=[3-\Delta T_{LNG-100}, 3-\Delta T_{LNG-101},$$
$$3-\Delta T_{LNG-102}, 3-\Delta T_{LNG-103}] \quad (14)$$

4 Results and analysis

4.1 Optimization results and analysis

The genetic algorithm used in this paper performs global search optimization on the upper and lower limits of the optimization parameters to minimize the power consumption of the system. The parameters of the low-temperature packed bed and the nitrogen expansion refrigeration liquefaction process optimized by the algorithm are shown in Table 5, which shows the values of the optimized parameters before and after optimization.

Table 5 Optimization parameters Before and after optimization

Parameters	Before optimization	After optimization
Q Nitrogen flow rate/ (kgmole/h)	30.00	32.92
Stream 6 temperature/℃	−73.00	−80.48
N Nitrogen secondary pressurization/kPa	1300.00	1030.33
P Expander outlet pressure/kPa	220.00	230.73
Stream 16 temperature/℃	−95.00	−100.32

In the process of optimization, the natural gas system is not optimized as an independent variable, but in order to ensure the liquefaction rate of natural gas, a fixed natural gas outlet temperature is set in the final heat exchanger LNG-102; in order to reduce the power consumption, it is generally to reduce the nitrogen flow rate or increase the inlet pressure, that is, the outlet pressure of the expander, while ensuring sufficient cooling capacity. From the table, it can be found that the optimized molar flow rate in the nitrogen system is 32.92 kgmole $\cdot$ h^{-1}, which is higher than the previous 30 kgmole $\cdot$ h^{-1}, and the flow rate is increased. In the optimization result, the nitrogen flow rate is increased by 9.73% compared with that before optimization; the outlet pressure of the nitrogen two-stage compressor is 1030.33kPa, which is 20.74% lower than the previous 1300kPa. For this reason, the optimized parameters in this paper adopt the method of increasing the nitrogen flow rate and reducing the pressure after the secondary compression of nitrogen. The outlet pressure of the expander increases from 220kPa to 230.73kPa, and the pressure changes little in this process, but the increase of the outlet pressure of the expander improves the power consumption of the first-stage compressor. The temperature of logistics 6, that is, the temperature of entering the nitrogen expander from the heat exchanger LNG-101, is reduced from -73.00℃ to -80.48℃, and the optimized temperature is reduced by 10.25%. This is due to the reduction of the pressure after the two-stage compression of nitrogen. However, it is still necessary to provide sufficient cooling capacity for natural gas to reduce the temperature entering the expander, so that the expanded temperature can meet the liquefaction demand. Compared with the outlet temperature of the nitrogen outlet heat exchanger LNG-103, that is, the temperature of the nitrogen outlet low-temperature packed bed increases from -95℃ to -100.32℃, and the difference before and after optimization is not significant.

(1) Optimization results

The genetic algorithm optimizes the power consumption iteration number curve as shown in Fig. 4. It can be found from the figure that the power consumption is in the initial stage.

The decline is relatively rapid, and the curve changes smoothly in the subsequent process. When iterating to 122 generations, the optimal result of power consumption is obtained. In the optimization process, the power consumption is reduced from 0.8074kW·h·kg^{-1} before optimization to 0.7368kW·h·kg^{-1}, which is reduced by 8.74%. The power of the nitrogen primary compressor was reduced from 28.02kW to 25.78kW. The power of the nitrogen secondary compressor was reduced from 28.76kW to 26.04kW。

(2) Performance analysis

The minimum temperature difference of the four heat exchangers before and after optimization is shown in Table 5. The heat exchanger LNG-100 is mainly used to rewarm the system. In the process of optimization algorithm, it does not affect the overall specific power consumption, so the temperature difference before and after optimization is unchanged. The minimum heat transfer temperature difference of the other three groups of heat exchangers is 3.032℃, 3.478℃ and 3.013℃, respectively, which is 60.99%, 20.54% and 12.97% lower than that before optimization. The total heat flux of the optimized heat exchanger is 19.61% higher than that before optimization. On the one hand, it is due to the increase of nitrogen flow rate after optimization, and on the other hand, it is due to the increase of heat transfer temperature difference of the heat exchanger. Figure 5 shows the temperature curve of the heat exchanger before and after optimization. The logarithmic mean temperature difference (LMTD) of the heat exchanger LNG-101 and LNG-103 is reduced from 13.48℃, 9.064℃ and 9.978℃ to 7.579℃, 7.545℃ and 7.288℃, respectively. The temperature difference between the cold side and the hot side of the two heat exchangers is reduced. The logarithmic average heat transfer temperature difference of the heat exchanger LNG-102 increases from 9.978℃ to 12.56℃, an increase of about 25.88%. The temperature difference between the cold side and the hot side of the heat exchanger increases, because the inlet temperature of the expander decreases in order to reduce the specific power consumption, and the temperature

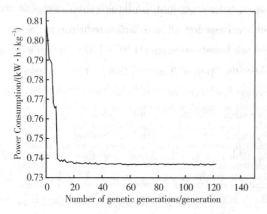

Fig. 4 Iteration diagram of genetic algorithm

of the feed gas entering the low-temperature packed bed increases, so the logarithmic average heat transfer temperature difference of the heat exchanger LNG-102 becomes larger.

Table 6 Optimizing the temperature difference between front and rear heat exchangers

Parameters	Before optimization	After optimization	Relative error
LNG-100 Minimum heat transfer temperature difference/℃	5.000	5.000	0.00%
LNG-101 Minimum heat transfer temperature difference/℃	7.773	3.032	−60.99%
LNG-102 Minimum heat transfer temperature difference/℃	4.377	3.478	−20.54%
LNG-103 Minimum heat transfer temperature difference/℃	3.462	3.013	−12.97%
Heat flow/(kJ·h^{-1})	142271.39	170174.31	19.61%

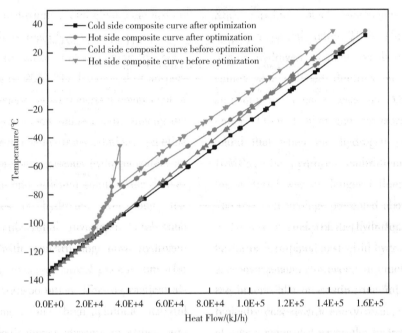

Fig. 5 Heat transfer curve before and after optimization

(1) Analysis

The analysis method is one of the important methods for evaluating and analyzing the comprehensive utilization of energy in chemical processes. One of the main purposes of exergy analysis is to reveal the location, type and quantity of exergy loss in the system, so as to reduce these losses and maximize the efficiency of the system, as shown in equations (16) to (18).

$$e = h - h_0 - T_0(s - s_0) \quad (16)$$

$$W = W_1 + W_2 \quad (17)$$

$$\eta = \frac{me}{W} \quad (18)$$

Where e is the mass effective energy, kJ·kg^{-1}; h is the enthalpy under operating conditions, kJ·kg^{-1}; h_0 is the enthalpy under standard conditions, kJ·kg^{-1}; T_0 is the temperature under the standard condition, K; s is the entropy under operating conditions, kJ·kg^{-1}·K^{-1}; s_0 is the entropy under the standard condition, kJ·kg^{-1}·K^{-1}; η is exergy efficiency, %, m is mass flow rate, kg·h^{-1}.

The optimized exergy efficiency is 21.64%, which is 8.73% higher than that before optimization (19.75%). At the same time, the exergy loss of different equipment is also discussed. The exergy loss of the main equipment before and after the base optimization is shown in Fig. 6, and the error result is shown in Table 7.

After genetic algorithm optimization, the total loss of the equipment is 32.572kW, which is 10.883% lower than the 36.55kW before optimization.

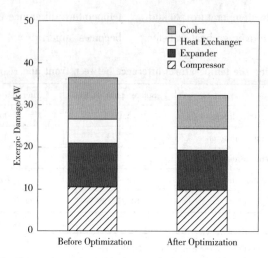

Fig. 6 Comparison of exergic damage before and after optimization

Table 7 Exergic damage results

Equipment name	Exergic damage before optimization/kW	Exergic damage after optimization/kW	Error
Compressor	10.440	9.879	5.374%
Expander	10.850	9.742	10.212%
Heat exchanger	5.563	4.885	12.188%
Cooler	9.697	8.066	16.820%
Total	36.55	32.572	10.883%

4.2 Process adaptability analysis

The natural gas flow rate in this model is 2000Nm$^3 \cdot$d^{-1}, but in the actual mining process, the composition, pressure and flow rate of raw gas are constantly changing. Moreover, when the current gas field exploitation is completed, it is necessary to transfer this device to other gas fields for exploitation. The parameters of the new gas field are not necessarily consistent with the original mining parameters, so the process flow needs to have certain adaptability to the raw gas parameters. Therefore, in this paper, three different gas sources with different parameters are selected to analyze the adaptability of the integrated process of decarburization and liquefaction in low temperature packed bed. Three kinds of natural gas with different parameters are shown in the table. The first and second only change the composition of natural gas. The first feed gas is to increase the volume fraction of CO_2 in the system, and the temperature, pressure and flow rate of natural gas are still in the original state. The second feed gas is to change the content of methane and ethane, other conditions remain unchanged. The third feed gas is to explore the adaptability to the working conditions of the gradual reduction of the mining flow, so the flow of natural gas is reduced from 2000Nm$^3 \cdot$d^{-1} to 1700Nm$^3 \cdot$d^{-1}, and the composition, pressure and flow of natural gas are still in the original state. The fourth is to change the pressure of natural gas. At the same time, in order to ensure the decarburization effect, the pressure is adjusted to 1600kPa, and the temperature, composition and flow rate of natural gas are still in the original state. The specific parameter changes are shown in Table 8.

Table 8 Natural gas with different parameters

Parameters	Original state	1	2	3	4
Natural gas temperature/℃	20	20	20	20	20
Natural gas flow/(Nm$^3 \cdot$d^{-1})	2000	2000	2000	1700	2000
Natural gas pressure/kPa	1500	1500	1500	1500	1600

续表

Parameters	Original state	1	2	3	4
CH_4 Mole fraction/%	89.45	84.45	87.45	89.45	89.45
C_2H_6 Mole fraction/%	0.45	0.45	2.45	0.45	0.45
CO_2 Mole fraction/%	10	15	10	10	10
N_2 Mole fraction/%	0.1	0.1	0.1	0.1	0.1

The most important equipment in the natural gas liquefaction process is the LNG heat exchanger, and the low-temperature packed bed used this time is also simulated by the heat exchanger in the HYSYS software. Therefore, the applicability of various working conditions is analyzed according to some key parameters of LNG heat exchanger. After the process parameters change, the heat transfer temperature difference during the rewarming process of CO_2 is fixed at 5℃, so the parameter changes of the other three heat exchangers are mainly analyzed. Aspen EDR software is used to simulate the parameters of different working conditions and different heat exchangers. Table 9 shows the size of each heat exchanger. It can be seen from the table that the size of the heat exchanger LNG-1 will exceed the size of the original design under condition 1. The heat exchanger LNG-102 meets all the conditions. The design size of LNG-103 heat exchanger cannot meet the demand in the second and fourth conditions.

Table 9 Design results of heat exchanger dimensions

Heat exchanger name	Condition	Width/m	Height/m	Length/m	Whether to meet the original heat exchanger size
LNG-101	Original	1.45	1.67	20.70	
	1	1.67	1.77	30.43	NO
	2	1.41	1.67	17.53	YES
	3	1.30	1.58	11.58	YES
	4	1.39	1.67	17.07	YES
LNG-102	Original	1.19	0.91	5.47	
	1	1.19	0.91	4.51	YES
	2	1.18	0.71	1.93	YES
	3	1.11	0.91	3.79	YES
	4	1.19	0.71	1.49	YES
LNG-103	Original	1.20	0.98	4.48	
	1	1.19	0.88	3.65	YES
	2	1.19	0.91	4.75	NO
	3	1.16	0.88	3.23	YES
	4	1.19	0.91	5.58	NO

The results of the parameters under each working condition are shown in table 10, and the parameters of the three heat exchangers are mainly analyzed. In the first case, the content of CO_2 increases and the content of CH_4 decreases. The heat transfer area required by the heat exchanger is 117.2 m^2, and the total heat transfer rate UA is 33887.48 kJ·℃$^{-1}$·h^{-1}, which is significantly larger than the heat transfer area required by the heat exchanger and the total heat transfer rate of the heat exchanger under the original working condition, indicating that the heat transfer required in this case is relatively large, and the optimized parameter setting is not applicable. This is because increasing the content of CO_2 in the case of constant volume flow, the overall mass flow rate increases, and the heat transfer required in the LNG-101 heat exchanger increases. At the same time, the specific power consumption is related to the mass flow rate of the feed gas, so the specific power consumption of the process is reduced. In addition, the minimum heat transfer temperature difference of LNG - 101 is 1.03℃, which is less than the specified minimum value of 3℃, but the minimum temperature difference of the other two heat exchangers is slightly larger than the original working condition, which can meet the minimum heat transfer temperature difference condition.

The second case is to increase the content of C_2H_6 and reduce the content of CH_4. Although the length of the heat exchanger LNG-103 in the initial state does not meet the demand, the width and height can meet the demand. It can be seen from Table 10 that the heat exchange area required by the heat exchanger is 52.8m² smaller than the original 69.2m², and the total heat transfer rate UA is also smaller than the initial working condition, indicating that the heat transfer required in this case is smaller than the initial working condition. The reason for this change is that the increase of C_2H_6 content increases the overall mass flow rate, so that the specific power consumption of the process is less than the initial condition. In addition, the minimum temperature difference of the three heat exchangers is greater than the specified 3℃, which can meet the design requirements.

The third case is to reduce the feed gas flow. At this time, the size of the three heat exchangers designed meets the requirements of the initial situation, and it can be seen from Table 10 that the heat transfer area required by the heat exchanger and the total heat transfer rate UA of the heat exchanger are smaller than the original working condition. It shows that the heat transfer required in this case is smaller than the initial value, and the optimized pa-rameters are suitable for this case. The minimum heat transfer temperature difference is greater than the design specification of 3℃, meet the design requirements. However, the specific power consumption is greater than the initial specific power consumption at this time. Because the flow rate decreases, the parameters of the nitrogen refrigeration expansion system do not change, that is, the overall power consumption does not change. Therefore, the cooling capacity provided is relatively excessive, which not only increases the specific power consumption, but also increases the minimum heat transfer temperature difference of each heat exchanger.

The fourth case is to increase the pressure of the feed gas. Although the length of the heat exchanger LNG-103 in the initial state does not meet the demand, the width and height can meet the demand. It can be seen from Table 10 that the heat transfer area and the total heat transfer rate required by the heat exchanger are less than the initial working condition, indicating that the heat transfer amount required in this case is smaller than the initial value, and the optimized parameters are suitable for this case. The minimum temperature difference of the three heat exchangers is greater than the specified 3℃, which meets the design requirements, and the specific power consumption does not change much at this time.

In general, the low-temperature packed bed decarburization and nitrogen expansion refrigeration liquefaction process have different applicability to different working conditions. The process is more sensitive to the increase of CO_2 content in the feed gas, and the system pressure is slightly increased, the flow rate is slightly reduced, and the C_2H_6 content in the feed gas is increased.

Table 10 Process adaptability analysis results

Parameters	Original state	1	2	3	4
TotalUA/(kJ·℃⁻¹·h⁻¹)	20967.25	33887.48	15963.86	10702.39	15608.51
Heat exchange area/m²	69.2	117.2	52.8	30.17	51.1
LNG-101 Minimum heat transfer temperature difference/℃	3.032	1.03	4.249	10.02	4.526

续表

Parameters	Original state	1	2	3	4
LNG-102 Minimum heat transfer temperature difference/℃	3.478	4.346	13.14	6.24	16.549
LNG-103 Minimum heat transfer temperature difference/℃	3.013	4.701	9.246	10.23	6.179
Power consumption/(kW·h·kg^{-1})	0.7368	0.6883	0.7245	0.8568	0.7349

5 Conclusion

In this paper, the integrated process of low temperature packed bed decarburization and nitrogen expansion liquefied natural gas is established. The steady - state model is optimized by genetic algorithm, and the performance of the optimized process is analyzed. The conclusions are as follows:

(1) In the optimization process, the power consumption is reduced from 0.8074kW·h/kg to 0.7368kW·h/kg, which is reduced by 8.74%. The power of the nitrogen first-stage compressor was reduced from 28.02kW to 25.78kW. The power of the nitrogen secondary compressor was reduced from 28.76kW to 26.04kW.

(2) The minimum heat transfer temperature difference of the three groups of heat exchangers is 3.032℃, 3.478℃ and 3.013℃, respectively, which is 60.99%, 20.54% and 12.97% lower than that before optimization. The logarithmic mean heat transfer temperature difference (LMTD) of LNG-101 and LNG-103 heat exchangers decreased from 13.48℃, 9.064℃ and 9.978℃ to 7.579℃, 7.545℃ and 7.288℃, respectively, but the heat flux increased by 19.61%. After optimization, the exergy efficiency is 21.64%, which is 8.73% higher than 19.75% before optimization. The total exergy loss of the equipment is 32.572kW, which is 10.883% lower than the 36.55kW before optimization.

(3) The low temperature packed bed decarbonization and nitrogen expansion refrigeration liquefaction process have different applicability to different working conditions. The process is sensitive to the increase of CO_2 content in the feed gas, and has good adaptability to the slight increase of system pressure, the slight decrease of flow rate and the increase of C_2H_6 content in the feed gas.

Author introduction

Fu Juntao, male, born in 1987, is a doctoral student. In 2012, he graduated from the oil and gas storage and transportation engineering major of China University of Petroleum. The current research direction is liquefied natural gas technology. Address: China University of Petroleum (East China), No. 66 Changjiang West Road, Huangdao District, Qingdao City, Shandong Province, 266580. Telephone: 15963006487; email: 570062601 @ qq.com

Acknowledgment

This work acknowledges funding from the National Natural Science Foundation of China under grant no. U21B2085.

References

[1] Papka, S.D., Gentry, M.C., Leger, A.T., et al. 2005. Pressurized LNG: A New Technology For Gas Commercialization. Presented at the The Fifteenth International Offshore and Polar Engineering Conference, p. ISOPE-I-05-015.

[2] Fairchild, D.P., Smith, P.P., Biery, N.E., et al. 2005. Pressurized LNG: Prototype Container Fabrication. Presented at the The Fifteenth International Offshore and Polar Engineering Conference, p. ISOPE-I-05-017.

[3] Bowen, R.R., Gentry, M.C., Nelson, E.D., et al. 2005. Pressurized Liquefied Natural Gas(PLNG): A New Gas Transportation Technology.

[4] Nelson, E.D., Papka, S.D., Bowen, R.R., et al. 2005. Pressurized LNG: A Paradigm Shift in Gas Transportation. Presented at the SPE Middle East Oil and Gas Show and Conference, p. SPE - 93633 - MS. https://doi.org/10.2118/93633-MS.

[5] Heerden, F., Putter, A., 2021. Alternative Modes of Natural Gas Transport.

[6] Lee, S.H., Seo, Y.K., Chang, D.J., 2018. Techno-economic Analysis of Acid Gas Removal and Liquefaction

for Pressurized LNG. IOP Conf. Ser.: Mater. Sci. Eng. 358, 012066. https://doi.org/10.1088/1757-899X/358/1/012066.

[7] Xiong, XJ., Lin, W S., Gu, A Z., 2013. Simulation and optimal design of a natural gas pressurized liquefaction process with gas expansion refrigeration. Natural Gas Industry 33(06): 97-101.

[8] Xiong, X., Lin, W., Gu, A., 2016. Design and optimization of offshore natural gas liquefaction processes adopting PLNG (pressurized liquefied natural gas) technology. Journal of Natural Gas Science and Engineering 30, 379-387.

[9] Lee, S., Seo, Y., Lee, J., et al. 2016. Economic evaluation of pressurized LNG supply chain. Journal of Natural Gas Science and Engineering 33, 405-418. https://doi.org/10.1016/j.jngse.2016.05.039.

[10] Zhu, J L., Liu, J H., Li, Z H., et al. 2024. Research status and development trends of FPLNG pretreatment and liquefaction integrated process. Low-Carbon Chemistry and Chemical Engineering 49(05): 112-122.

[11] Babar, M., Bustam, M. A., Maulud, A. S., et al. 2019. Optimization of cryogenic carbon dioxide capture from natural gas. Materialwissenschaft Werkst 50, 248-253. https://doi.org/10.1002/mawe.201800202.

[12] Dashliborun, A. M., Larachi, F., Taghavi, S. M., 2019. Gas-liquid mass-transfer behavior of packed-bed scrubbers for floating/offshore CO_2 capture. Chemical Engineering Journal 377, 119236. https://doi.org/10.1016/j.cej.2018.06.025.

[13] Xiong, X., Lin, W., Gu, A., 2015. Integration of CO_2 cryogenic removal with a natural gas pressurized liquefaction process using gas expansion refrigeration. Energy 93, 1-9. https://doi.org/10.1016/j.energy.2015.09.022.

[14] Lin, W., Xiong, X., Gu, A., 2018. Optimization and thermodynamic analysis of a cascade PLNG (pressurized liquefied natural gas) process with CO_2 cryogenic removal. Energy 161, 870-877. https://doi.org/10.1016/j.energy.2018.07.051.

[15] Faramarzi, S., Nainiyan, S. M. M., et al. 2021. A novel hydrogen liquefaction process based on LNG cold energy and mixed refrigerant cycle. International Journal of Refrigeration 131, 263-274. https://doi.org/10.1016/j.ijrefrig.2021.07.022.

[16] Morosuk, T., Tesch, S., Hiemann, A., et al. 2015. Evaluation of the PRICO liquefaction process using exergy-based methods. Journal of Natural Gas Science and Engineering 27, 23-31. https://doi.org/10.1016/j.jngse.2015.02.007.

[17] Shen, T., Gao, T., Lin, W., et al. 2012. Determination of CO_2 Solubility in Saturated Liquid $CH_4 + N_2$ and $CH_4 + C_2H_6$ Mixtures above Atmospheric Pressure. J. Chem. Eng. Data 57, 2296-2303. https://doi.org/10.1021/je3002859.

[18] He, T., Lin, W., Du, Z., 2022. Measurement and Theoretical Calculation of CO_2 Solubility Data in Liquid $CH_4 + C_2H_6$ Mixtures at Cryogenic Temperatures. J. Chem. Eng. Data 67, 3222-3233. https://doi.org/10.1021/acs.jced.2c00237.

[19] Davis, J. A., Rodewald, N., Kurata, F., 1962. Solid-liquid-vapor phase behavior of the methane-carbon dioxide system. AIChE Journal 8, 537-539. https://doi.org/10.1002/aic.690080423.

[20] Li, Z H., Zhu, J L., Miao, Q., et al. 2024. Experimental determination and theoretical calculation for CO_2 liquid-solid phase equilibrium in PLNG. Oil and Gas Storage and Transportation 43(09): 1039-1047.

[21] ZareNezhad, B., Eggeman, T., 2006. Application of Peng-Rabinson equation of state for CO_2 freezing prediction of hydrocarbon mixtures at cryogenic conditions of gas plants. Cryogenics 46, 840-845. https://doi.org/10.1016/j.cryogenics.2006.07.010.

[22] De Hemptinne, J.-C., 2005. Benzene crystallization risks in the LIQUEFIN liquefied natural gas process. Process Safety Progress 24, 203-212. https://doi.org/10.1002/prs.10084.

[23] Huang, T., Wang, W., Yuan, Y., et al. 2021. Optimization of high-temperature proton exchange membrane fuel cell flow channel based on genetic algorithm. Energy Reports 7, 1374-1384. https://doi.org/10.1016/j.egyr.2021.02.062.

[24] Qin, Y., Li, Z., Ding, J., et al. 2023. Automatic optimization model of transmission line based on GIS and genetic algorithm. Array 17, 100266. https://doi.org/10.1016/j.array.2022.100266.

Numerical simulation study on pressure reduction characteristics of dense phase CO_2 pipeline leakage

Yin Buze[1]　Huang Weihe[2]　Li Yuxing[1]　Hu Qihui[1]　Yu Xinran[1]　Zhao Xuefeng[3]　Meng Lan[3]

[1. College of Pipeline and Civil Engineering, China University of Petroleum(East China)//Shandong Key Laboratory of oil & Gas Storage and Transportation Safety; 2. China National Petroleum Corporation; 3. PetroChina Daqing Oilfield Co]

Abstract　Pipeline transport plays an important role in Carbon Capture Utilization and Storage(CCUS) processes. As operational duration extends, pipelines may experience leaks due to corrosion, perforation, or damage from external sources. Upon a leak, the pipeline experiences a pressure drop, expansion, and a pronounced phase change, which maintain a high pressure at the leakage. This high pressure not only accelerates crack propagation but also cause suffocation threat to surrounding personnel or organisms. Therefore, it is imperative to develop a numerical simulation methodology for CO_2 pipeline leaks. A numerical simulation method for CO_2 leakage is developed in this work, which adopts the S-W(Span-Wagner) state equation to establish a CO_2 physical property data table to accurately describe CO_2 physical properties. The model is verified through experimental results of pressure and decompression waves. At the same time, the state characteristics of CO_2 inside the pipeline during the leakage process and the transient characteristics at the leakage port was analyzed. The calculation results indicate that there are five states of CO_2 in the pipeline during the leakage process: initial state, single-phase expansion state, saturation equilibrium state, gas-liquid state, and gaseous state. The mass flow rate of the outlet rapidly increases after leakage. As the time increases, the gas phase generated by phase transition increases, the mixing density decreases, and the mass flow rate of leakage decreases accordingly. By adjusting the evaporation coefficient, the numerical simulation method can effectively and accurately calculate the changes of pressure through CO_2 pipeline leakage.

Keywords　CCUS; CO_2 pipeline; Leakage; Decompression wave; Numerical simulation

1　Introduction

In recent years, the emergence of CCUS has alleviated global climate issues and is considered the most effective way to achieve CO_2 emissions reduction. Pipeline transportation is considered the most flexible and economical way to transport CO_2 over long distances on an industrial scale. As the running time increases, CO_2 pipelines may leak or even break due to various reasons such as corrosion, perforation, or third-party damage. CO_2 has strong Joule Thomson effect and phase transition characteristics. After pipeline leakage or fracture occurs, the leakage outlet will produce strong throttling effect and phase transition will produce low temperature and high-pressure level. Low temperature can cause brittle transformation of pipes, reducing their toughness. Higher pressure level can provide continuous energy supply for crack propagation, both of which have adverse effects on the toughness and crack prevention of pipelines. So the prediction and simulation of transient characteristics after CO_2 pipeline leakage are very important for the consequences and safety evaluation of CO_2 pipeline leakage.

To reveal the mechanism of CO_2 leakage and provide a data basis for theoretical research, many scholars have conducted experimental studies. The most representative experimental study is Botros et al., which conducted CO_2 decompression wave experiments with different initial pressures, initial temperatures, and impurity contents. And the predictive model of decompression waves was established based on the assumption of isentropic, and the experimental data was verified by combining the PR (Peng-Robinson) and GERG-2008 equations. The experimental and computational results show that the GERG-2008 state equation is more accurate than the PR equation in both the high-pressure stage and the decompression wave platform. Cosham et al. conducted CO_2 decompression wave experiments with different concentrations of non-polar impurities such as H_2, N_2, O_2, CH_4, etc. By comparing the experimental results with the prediction of the one-dimensional isentropic homogeneous flow DECOM model, it can be seen that the decompression wave results predicted by model are relatively conservative compared to the experimental results. Log et al. conducted full-scale release experiments of dense phase CO_2 at different temperature and pressure, which observed the phenomenon of CO_2 overheating. A homogeneous flash evaporation model was established by considering the principle of delayed nucleation of bubbles and validated with experimental data. The model is applicable to all leakage situations under constant parameters, without the need to adjust model parameters according to operating conditions like non-equilibrium relaxation models. Guo et al. conducted a series of pure CO_2 leakage experiments with different diameters of hole in horizontal pipelines, studying the jet characteristics in the near field and the concentration and temperature diffusion in the far field after CO_2 leakage. They analyzed the leakage mass flow characteristics and the safety distance law of CO_2 diffusion. Cao et al. designed a multi-point temperature sensor and discovered temperature stratification in different CO_2 release experiments, revealing the phase transition law and heat transfer mechanism inside large diameter CO_2 pipelines when small holes leak.

Compared to experimental research, numerical calculations are more comprehensive and cost-effective, surely, which need to be continuously improved to ultimately obtain experimental data for verification. Wareing et al used the PR equation of state to describe the physical properties of CO_2 and simulated the field distribution of physical quantities such as jet structure and temperature after CO_2 small hole leakage. The overall accuracy of the calculation results needs to be improved because the PR equation has a large calculation error for the physical properties of CO_2, especially for liquid CO_2. Elshahomi et al. used the GERG-2008 equation of state to calculate the physical properties of CO_2 containing impurity, and created a property table using the UDRGM model implanted in Fluent software. However, the model assumed a homogenous-equilibrium, resulting in higher plateau pressure calculation of the decompression wave compared to experimental data. Liu et al. introduced the Mixture multiphase flow model and Lee phase transition model based on this model, using UDRGM model to describe liquid CO_2 and UDF model to describe gas CO_2, respectively. By adjusting the coefficients in the phase transition model, the calculated results were consistent with the experimental data. Xiao, Zhang, and Wang conducted numerical simulation analysis on the pressure reduction characteristics of CO_2 leakage under different initial states, pipe flow rates, and gas-liquid volume fractions at equilibrium. These works have further deepened the understanding of the impact of different parameters on the CO_2 pipelines leakage, but there is a lack of mechanistic analysis on the phase change of CO_2 inside the pipeline and the transient characteristics of the leakage port.

In this paper, the S-W equation of state is used to calculate the physical properties of pure CO_2 and prepares a two-dimensional property table of different CO_2 properties in advance regarding temperature and pressure. UDRGM and UDF models are implanted into Fluent software. Simultaneously combining the Mixture multiphase flow model and Lee phase transition model to describe the phase transition behavior of CO_2. The accuracy of the model was verified using experimental data from Botros et al., and the phase change law and transient characteristics of the leakage port during the high-pressure CO_2 depressurization process were analyzed.

2 Model Framework

2.1 Physical properties of CO_2

The PR state equation is often used in CO_2 engineering calculations, which has a simple structure and high computational efficiency and is widely used. However, due to its simple cubic structure, there is a significant error in the calculation results of CO_2 physical properties under high pressure. Therefore, this article describes the physical properties of pure CO_2 using the S-W state equation, which has a higher recognition for pure CO_2. The S-W state equation is based on the Helmholtz free energy theory, expressed in terms of temperature and pressure, and fitted based on a large amount of experimental data. When the pressure is less than 30MPa and the temperature is less than 500K, the calculation error of the S-W state equation regarding physical properties such as density, specific heat, and sound velocity is less than ±1.5%.

Both gas and liquid CO_2 have strong compressibility, and their thermodynamic and kinetic properties are functions of temperature and pressure. To accurately describe the physical properties of CO_2 and calculate the phase transition and decompression process of CO_2, both gas and liquid CO_2 need to be set as compressible phases in the simulation. This requires to apply User Defined Real Gas Model (UDRGM) and User Defined Functions (UDF) to define liquid and gaseous CO_2, respectively. UDRGM needs to define 12 parameters of liquid CO_2, including density ρ, specific enthalpy h, specific entropy s, speed of sound w, specific heat c_p, molecular weight M, dynamic viscosity μ, thermal conductivity λ, partial derivative of ρ w.r.t. T ($\partial\rho/\partial T$), partial derivative of ρ w.r.t. P ($\partial\rho/\partial P$), partial derivative of h w.r.t. P ($\partial h/\partial P$), and partial derivative of h w.r.t. T ($\partial h/\partial T$) and UDF needs to define 7 parameters of gaseous CO_2, including density ρ, specific enthalpy h, speed of sound w, specific heat c_p, molecular weight M, dynamic viscosity μ, thermal conductivity λ.

Calculate the above physical properties into a two-dimensional data table within a certain temperature and pressure range at a certain interval, as shown in the figure 1. Compared to directly compiling the S-W state equation, the efficiency can be greatly improved by calling the database. For states where temperature and pressure are not at the node, linear or bilinear interpolation is used to calculate the physical properties of CO_2. This paper will select a temperature range of 190K-373K, a pressure range of 0.005MPa-20MPa, a temperature step size of 0.1K, and a pressure step size of 0.1MPa. After verification, the error between the physical properties of different temperatures and pressures calculated through interpolation and direct use of S-W is less than 0.1%. These parameters are not fixed, and readers can make appropriate adjustments according to their own calculation range and accuracy requirements.

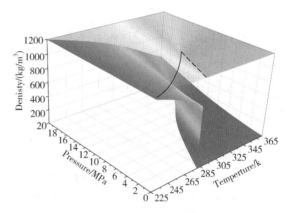

Figure 1 Three-dimensional plot of pure CO_2 density as a function of temperature and pressure

2.2 Mixture model

A non-uniform multiphase flow model considering slip velocity is used to calculate the decompression process of gas-liquid two-phase CO_2. The continuity equation of CO_2 gas-liquid mixture is shown in formula (1):

$$\frac{\partial}{\partial t}(\rho_m) + \nabla \cdot (\rho_m \vec{v}_m) = 0 \qquad (1)$$

Where t is time, s; $\rho_m = \alpha_l \rho_l + \alpha_v \rho_v$ is mixture density, ρ_l and ρ_v the density of liquid and gas CO_2 are respectively, kg/m³; $\vec{v}_m = \dfrac{\alpha_l \rho_l \vec{v}_l + \alpha_v \rho_v \vec{v}_v}{\rho_m}$ is the mass-average velocity, $\vec{v}_l$ and $\vec{v}_v$ are the velocity of liquid and gas, m/s; α_l and α_v are the volume fraction of liquid and gas CO_2, respectively.

The momentum equilibrium equation of CO_2 gas-liquid mixture is shown in equation (2):

$$\frac{\partial}{\partial t}(\rho_m \vec{v}_m) + \nabla(\rho_m \vec{v}_m \vec{v}_m) =$$
$$-\nabla P + \nabla \cdot [\mu_m(\vec{v}_m + \nabla \vec{v}_m^T)] \quad (2)$$
$$+ \rho_m g + \vec{F} + \nabla \cdot (\sum_{k=1}^{n} \alpha_k \rho_k \vec{v}_{dr,k} \vec{v}_{dr,k})$$

Where n is the number of phases; P is pressure, Pa; $\mu_m = \alpha_l \mu_l + \alpha_v \mu_v$ is hybrid viscosity, μ_l and μ_v are viscosity of liquid and gas, Pa·s; g is the gravity, m/s²; $\vec{F}$ is volumetric force, N; $\vec{v}_{dr,k} = \vec{v}_k - \vec{v}_m$ is the slip velocity of the k-th phase, $\vec{v}_k$ is the velocity of the k-th phase, m/s;

The energy balance equation of CO_2 gas-liquid mixture is shown in equation (3):

$$\frac{\partial}{\partial t}\sum_{k=1}^{n}(\alpha_k \rho_k E_k) + \nabla \cdot \sum_{k=1}^{n}(\alpha_k \vec{v}_k(\rho_k E_k + p))$$
$$= \nabla \cdot (k_e \nabla T) + S_E \quad (3)$$

Where k_e is thermal conductivity, W/(m·K); S_E is the energy source term, J/s; $E_k = h_k - \frac{p}{\rho_k} + \frac{v_k^2}{2}$ is the total energy of the k-th phase, J/kg; h_k is the apparent enthalpy of the k-th phase, J/kg.

In Mixture model, in order to calculate the volume fraction of the secondary phase, the continuity equation of the secondary phase is introduced, as shown in formula (4):

$$\frac{\partial}{\partial t}(\alpha_k \rho_k) + \nabla(\alpha_k \rho_k \vec{v}_k) = -\nabla \cdot (\alpha_k \rho_k \vec{v}_{dr,k}) + \dot{m}_{lv} + \dot{m}_{vl} \quad (4)$$

Where $\dot{m}_{lv}$ and $\dot{m}_{vl}$ is mass source terms for evaporation and condensation, respectively, kg/s.

2.3 Phase transition model

The Lee model is widely used in phase transitions and evaporation processes. Evaporation occurs when the liquid temperature is above the saturation temperature, and condensation occurs when the gas temperature is below the saturation temperature. The process is represented by equations (5) and (6):

If $T < T_{sat}$ (vaporization):
$$\dot{m}_{lv} = C_{lv} \alpha_l \rho_l \frac{T_{sat} - T}{T_{sat}} \quad (5)$$

If $T > T_{sat}$ (condensation):
$$\dot{m}_{vl} = C_{vl} \alpha_v \rho_v \frac{T_{sat} - T}{T_{sat}} \quad (6)$$

Where C_{lv} and C_{vl} is the evaporation factor and condensation factor, s⁻¹. In simulation, it is necessary to continuously adjust its size to ensure that the calculated results match the experimental data. They can be understood as the reciprocal of relaxation time, with larger values indicating faster evaporation or condensation.

Unlike the phase transition driven by heat exchange, CO_2 leakage is a process of phase equilibrium instability caused by rapid pressure decompression, and pressure changes are the main driving force for CO_2 phase transition. So the phase transition conditions for CO_2 should be that when the pressure is below the saturation pressure, the liquid vaporizes, and when the pressure is above the saturation pressure, the gas liquefies. Change the temperature in equations (5) and (6) to pressure, as shown in equations (7) and (8):

If $P < P_{sat}$ (vaporization):
$$\dot{m}_{lv} = C_{lv} \alpha_l \rho_l \frac{P_{sat} - P}{P_{sat}} \quad (7)$$

If $P < P_{sat}$ (condensation):
$$\dot{m}_{vl} = C_{vl} \alpha_v \rho_v \frac{P_{sat} - P}{P_{sat}} \quad (8)$$

The energy term S_E in formula (3) is the enthalpy difference between the gas-liquid phase during the CO_2 phase transition process, as shown in formula (9):

$$S_E = (\dot{m}_{lv} + \dot{m}_{vl}) \cdot (h_v - h_l) \quad (9)$$

Where h_v and h_l are gas and liquid CO_2 and sensible enthalpy, respectively, in J/kg.

2.4 Computational domain and boundary conditions

To verify the accuracy of the model, this study used experimental data from Botros et al.. The experimental setup consisted of a shock tube with a length of 42m and an inner diameter of 38.1mm, as shown in Figure 2. To save computation cost, the difference between the horizontal directions on the same section of the pipeline is ignored and simplified into a 2D model. Due to the very short leakage time, the influence of gravity is ignored and the center of the pipe is used as the symmetrical boundary. The left side of the pipeline is set as the atmospheric pressure outlet, with no slip and adiabatic conditions on the pipe wall, and the right side of the pipeline is the closed end. ICEM was used to partition the computational domain into grids with quantities of 18000, 40000, 70000, and 100000, respectively. Grid independence tests were conducted using pressure data at the same position under the same initial conditions, as shown in Figure 3. Finally, a 7W grid was used for subsequent research simulations.

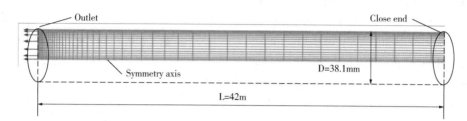

Figure 2　Schematic of Botros' test model

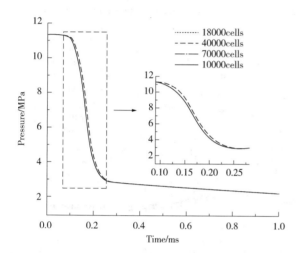

Figure 3　Grid independence verification

2.5 Numerical solution method

Due to the limitations of the Mixture model, the solver can only choose pressure-based solvers. Elshahomi et al. and Liu et al. both used the Realisable k-ε turbulence model for simulation, and the calculated results were in good agreement with experimental data. Therefore, Realisable k-ε turbulence model is used in this paper. The coupling format for pressure and velocity is PISO, and the spatial discretization format for pressure is PRESSO!. The spatial discretization format of the remaining variables is selected as QUICK format, and the time format is selected as first-order implicit. Residual of the energy equation is $1e^{-8}$, and other equations is set as $1e^{-4}$, with a time step of $5e^{-6}$. The simulated flowchart is shown in Figure 4.

3　Model Validation

The pressure and decompression wave data at

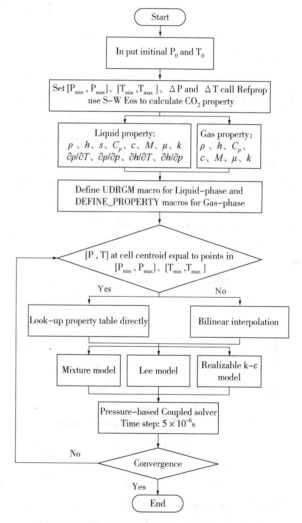

Figure 4　Schematic of simulation process

PT1A (0.0924m) of Botros experiment Test #32A (Initial pressure $P_0 = 11.27$MPa and Initial temperature $T_0 = 281.89$K) are selected for verification. Due to the adjustable empirical values of the evaporation and condensation coefficients C_{lv} and C_{vl} in phase transition model, it is necessary to first determine the two coefficients before validation. The default value is 0.1, so the condensation coefficient is kept at 0.1, and the evaporation coefficients are selected as 0.1, 1, 10, 15, 50, 100, and 1000 for calculation. The pressure and pressure decompression wave experimental data at PT1A position are compared, and the calculation results are shown in the figure 5-6. The plateau values of pressure and decompression waves at point PT1A are closest to the experimental values when the evaporation coefficient is equal to 15. When the evaporation coefficient is equal to 1000, the calculation result is the same as the homogeneous mode of Gu calculation result under the isentropic condition in. Then, the evaporation coefficient was fixed at 15, and the condensation coefficient was changed to 0.01, 0.1, 1, 10, and 100. The pressure and pressure decompression wave experimental data at PT1A were also used for verification, and the results are shown in the figure7-8. The sensitivity of the condensation coefficient to calculation results is very small, and the changing of condensation coefficient has almost no effect on results, so condensation coefficient is selected as 0.1.

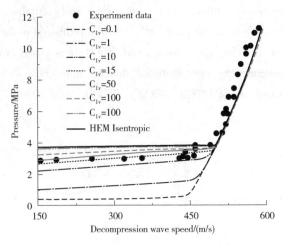

Figure 6 Comparison of Botros' Test #32A decompression wave data and simulated values under different evaporation coefficients

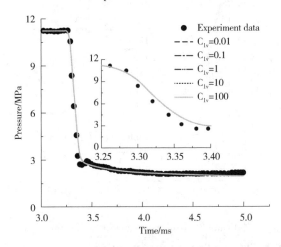

Figure 7 Comparison of Botros' Test #32A pressure data and simulated values under different condensation coefficients

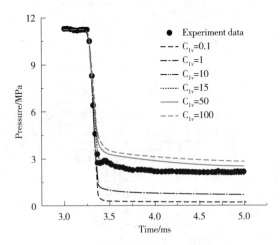

Figure 5 Comparison of Botros' Test #32A pressure data and simulated values under different evaporation coefficients

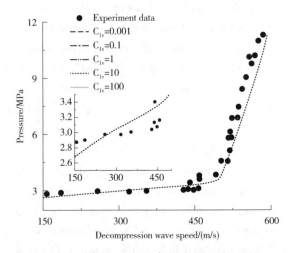

Figure 8 Comparison of Botros' Test #32A decompression wave data and simulated values under different condensation coefficients

4 Results and discussion

4.1 Characteristics of CO_2 state inside the pipeline during leakage process

The figure 9 shows decompression path at different positions in the Botros experiment. It can be seen that the closer to the leakage, the more severe the overheating of CO_2. Munkejord(2020) observed the same phenomenon in experiment, which is consistent with the simulation results obtained in this paper. When the distance from leakage port is greater than 3.1m, there is no overheating phenomenon. Without generating dry ice, there are five states of CO_2 in the pipeline after leakage: 1 initial state, 2 single-phase expansion state, 3 saturated equilibrium state, 4 gas-liquid two-phase state, and 5 gas state. The figures 10 and 11 show the schematic of the CO_2 decompression path and the decompression wave, respectively. The specific decompression process can be described as follows: CO_2 leaks in initial state, and expansion wave is generated at the leakage port and propagates towards the end of the pipeline, reducing the CO_2 pressure inside the pipeline. Before the pressure drops to saturation pressure, CO_2 belongs to a two-phase expansion state, where flow rate increases, sound velocity and decompression wave velocity decreases. When single-phase CO_2 continues to depressurize and expand to saturation pressure, the CO_2 near the leakage port overheats due to the rapid decompression rate and directly becomes a two-phase state of gas-liquid phases. The CO_2 far away from the leakage port maintains a dynamically stable saturation state due to the replenishment of high-pressure CO_2 behind it, which lead to a high-pressure platform in decompression wave curve. As the pressure further decreases, the overheated CO_2 near the leakage port gradually evaporates completely, and superheat gradually disappears. The decompression path returns to the phase envelope. The saturated CO_2 far away from the leakage port gradually evaporates along the phase envelope. When the pressure decreases to a certain extent, all liquid phase CO_2 evaporates completely, and the low temperature generated by the phase change is less than the heat absorption of the pipeline, making it difficult to maintain gas-liquid phase equilibrium. Eventually, only gas CO_2 leaks and the temperature rises slightly. At this point, the flow of gas CO_2 reaches stagnation,

where the velocity and speed of sound are equal, causing the decompression wave velocity to decrease to 0 before the pressure reaches 0.

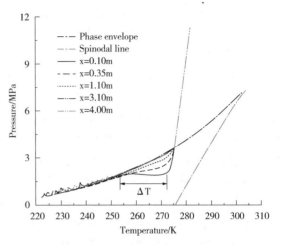

Figure 9 Decompression path at different positions in Botros experiment

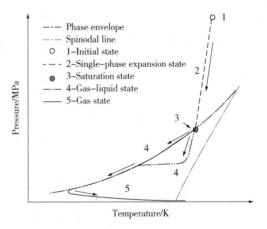

Figure 10 Schematic of CO_2 decompression path

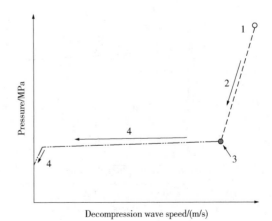

Figure 11 Schematic of decompression wave curve

4.2 Transient characteristics of leakage

At the moment of leakage, a strong evaporation wave is generated at the leakage port, causing phase change and a large amount of gas CO_2. Due to the pressure at the leakage port being atmospheric pressure, the gas CO_2 generated by phase change expansion cannot make pressure increase. According to energy conservation, it will inevitably make velocity at the leakage port increase, as shown in the figure12. From the Figure 13, it can be seen that mass flow rate at the leakage port rapidly increases to 32kg/s at 1ms. But as amount of gas CO_2 generated by phase transition increases, the mixing density of CO_2 at the leakage port decreases. The calculation formula for mass flow rate is: $M = \rho v A$. At the beginning of leakage, velocity is zero, so mass flow rate during initial period of leakage is mainly determined by the change of velocity. As the speed change toward stability, the mass flow rate is more affected by density.

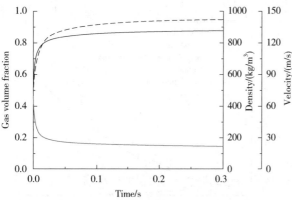

Figure 12 Density, velocity, and gas volume fraction curves at the leakage port

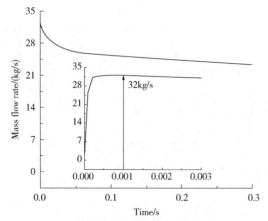

Figure 13 Mass flow rate curve at leakage port

5 Conclusion

The S-W state equation is used in this paper to accurately describe the properties of CO_2 and create a property table, through UDRGM and UDF model implanted into FLUENT. In addition, this simulation method combines the Mixture multiphase flow model and the Lee phase transition model to successfully consider the phenomenon of delayed phase transition caused by CO_2 overheating. The feasibility and accuracy of this method are verified by comparing with experimental data from Botros. And the following conclusions were drawn:

(1) The evaporation coefficient in phase change model has an impact on results. A lower evaporation coefficient will cause the pressure platform of decompression wave to be lower, while a higher evaporation coefficient will cause the results to be higher. When the evaporation coefficient is $15s^{-1}$, the calculated results are in good agreement with the experimental results. When the evaporation coefficient reaches $1000\ s^{-1}$, the calculated results are consistent with the results of the isentropic model. The condensation coefficient has almost no effect on results;

(2) During leakage, CO_2 in the pipeline will undergo five states: initial state, single-phase expansion state, saturation state, gas-liquid two-phase state, and gas state;

(3) About 1ms after leakage, the mass flow rate rapidly increases to a peak of 32kg/s, before which the mass flow rate is mainly determined by changes of velocity. Afterwards, as the changes of velocity tend to steady, the mass flow rate is more affected by density.

Acknowledgements

This work was supported by theChina University of Petroleum (East China) Independent Innovation Research Program Project(21CX06036A)

References

[1] Allen M, Dube O P, Solecki W, et al. Special report: Global warming of 1.5℃. Intergovernmental Panel on Climate Change(IPCC), 2018, 27: 677.

[2] Huang W, Li Y, Chen P. China's CO_2 pipeline development strategy under carbon neutrality. Natural Gas Industry B, 2023, 10(5): 502-510.

[3] Chen L, Hu Y, Yang K, et al. Experimental study of decompression wave characteristics and steel pipe applicability of supercritical CO_2 pipeline. Gas Science and Engineering, 2024, 121: 205203.

[4] Wang C, Li Y, Teng L, et al. Experimental study on dispersion behavior during the leakage of high pressure CO_2 pipelines. Experimental Thermal and Fluid Science, 2019, 105: 77-84.

[5] Botros K K, Geerligs J, Rothwell B, et al. Measurements of decompression wave speed in pure carbon dioxide and comparison with predictions by equation of state. Journal of Pressure Vessel Technology, 2016, 138(3): 031302.

[6] Botros K K, Geerligs J, Rothwell B, et al. Measurements of decompression wave speed in binary mixtures of carbon dioxide and impurities. Journal of Pressure Vessel Technology, 2017, 139(2): 021301.

[7] Cosham A, Jones D G, Armstrong K, et al. The decompression behaviour of carbon dioxide in the dense phase. International Pipeline Conference. American Society of Mechanical Engineers, 2012, 45141: 447-464.

[8] Log A M, Hammer M, Munkejord S T. A flashing flow model for the rapid depressurization of CO_2 in a pipe accounting for bubble nucleation and growth. International Journal of Multiphase Flow, 2024, 171: 104666.

[9] Log A M, Hammer M, Deng H, et al. Depressurization of CO_2 in a pipe: Effect of initial state on non-equilibrium two-phase flow. International Journal of Multiphase Flow, 2024, 170: 104624.

[10] Guo X, Yan X, Zheng Y, et al. Under-expanded jets and dispersion in high pressure CO_2 releases from an industrial scale pipeline. Energy, 2017, 119: 53-66.

[11] Guo X, Yan X, Yu J, et al. Under-expanded jets and dispersion in supercritical CO_2 releases from a large-scale pipeline. Applied Energy, 2016, 183: 1279-1291.

[12] Cao Q, Yan X, Liu S, et al. Temperature and phase evolution and density distribution in cross section and sound evolution during the release of dense CO_2 from a large-scale pipeline. International Journal of Greenhouse Gas Control, 2020, 96: 103011.

[13] Cao Q, Yan X, Guo X, et al. Temperature evolution and heat transfer during the release of CO_2 from a large-scale pipeline. International Journal of Greenhouse Gas Control, 2018, 74: 40-48.

[14] Wareing C J, Fairweather M, Falle S A E G, et al. Validation of a model of gas and dense phase CO_2 jet releases for carbon capture and storage application. International Journal of Greenhouse Gas Control, 2014, 20: 254-271.

[15] Elshahomi A, Lu C, Michal G, et al. Decompression wave speed in CO_2 mixtures: CFD modelling with the GERG-2008 equation of state. Applied Energy, 2015, 140: 20-32.

[16] Liu B, Liu X, Lu C, et al. A CFD decompression model for CO_2 mixture and the influence of non-equilibrium phase transition. Applied Energy, 2018, 227: 516-524.

[17] Xiao C, Lu Z, Yan L, et al. Transient behaviour of liquid CO_2 decompression: CFD modelling and effects of initial state parameters. International Journal of Greenhouse Gas Control, 2020, 101: 103154.

[18] Zhang Z X, Lu Z, Wang J Q, et al. Effects of initial flow velocity on decompression behaviours of GLE CO_2 upstream and downstream the pipeline. International Journal of Greenhouse Gas Control, 2022, 118: 103690.

[19] Wang J, Xiao C, Wu S, et al. Numerical study of the transient behaviors of vapour-liquid equilibrium state CO_2 decompression. Applied Thermal Engineering, 2024, 236: 121861.

[20] Peng D Y, Robinson D B. A new two-constant equation of state. Industrial & Engineering Chemistry Fundamentals, 1976, 15(1): 59-64.

[21] Span R, Wagner W. A new equation of state for carbon dioxide covering the fluid region from the triple-point temperature to 1100 K at pressures up to 800MPa. Journal of physical and chemical reference data, 1996, 25(6): 1509-1596.

[22] Andresen T, Skaugen G. Lookup Tables Based on Gibb's Free Energy for Quick and Accurate Calculation of Thermodynamic Properties for CO_2. 22nd International Congress of Refrigeration: Refrigeration creates the future. 2007.

[23] Lee W H. A pressure iteration scheme for two-phase flow modeling. Multiphase transport fundamentals, reactor safety, applications, 1980, 1: 407-431.

[24] SHUAIWEI GU, YUXING LI, LIN TENG, et al. A new model for predicting the decompression behavior of CO_2 mixtures in various phases. Transactions of The Institution of Chemical Engineers. Process Safety and Environmental Protection, Part B, 2018, 120237-247.

[25] Munkejord S T, Austegard A, Deng H, et al. Depressurization of CO_2 in a pipe: High-resolution pressure and temperature data and comparison with model predictions. Energy, 2020: 118560.

Effect of Epoxy ResinCoating on Hydrogen Embrittlement Behavior of X80 Pipeline Steel in Hydrogen Blended Methane Environment: Mechanism of Coating Delamination in Aggravating Hydrogen Embrittlement

Zhu Yuanchen[1,2]　Yang Hongwei[2,3]　Zhu Yun[4]　Liu Fang[2,3]　Chen Lei[1,2]　Liu Gang[1,2]

[1. Key Laboratory of Oil & Gas Storage and Transportation Safety, China university of Petroleum(East China);
2. CNOOC Key Laboratory of Liquefied Natural Gas and Low-Carbon Technology;
3. CNOOC Gas & Power Group, Research & Development Center;
4. Nanjing Jinling Petrochemical Alkyl Benzene Plant]

Abstract　Considering the substantial investment and time required for building dedicated hydrogen transportation pipeline, the hydrogen blended natural gas pipeline transportation has emerged as the most economically feasible method for large-scale hydrogen transportation. Currently, the main challenge hindering the development of hydrogen blended natural gas pipeline transportation is the hydrogen embrittlement damage of pipeline steel. To increase the hydrogen blended ratio in the transported mixed gas, it is crucial to address the impact of hydrogen on the integrity of pipeline steel. Although extensive research has been conducted on the hydrogen embrittlement behavior of pipeline steel in hydrogen blended methane environment, the influence of epoxy resin coating, applied as inner friction-reducing layer in practical pipeline engineering, on the hydrogen embrittlement behavior of pipeline steel remains unclear. In order to investigate the effect of epoxy resin coating on the hydrogen embrittlement behavior of pipeline steel during hydrogen blended natural gas transportation, constant strain rate tensile tests were performed on X80 pipeline steel specimens under different strain rates. The result showed that when the epoxy resin coating was fully adhered to the steel surface, it significantly mitigated the hydrogen embrittlement damage to the X80 pipeline steel. However, when the epoxy resin coating delaminated, not only was its protective effect against hydrogen embrittlement compromised, but also the hydrogen embrittlement damage to the X80 pipeline steel increased in the hydrogen blended methane environment. The delamination of epoxy resin coating allowed more hydrogen molecules to penetrate to the steel surface, thereby raising the risk of hydrogen embrittlement. To further understand the underlying mechanism of this phenomenon, molecular dynamics simulation was carried out. The simulation aimed to elucidate the aggravating effect of coating delamination on the hydrogen embrittlement behavior of X80 pipeline steel in hydrogen blended methane environment. The simulation result showed that when the epoxy resin coating was completely adhered to the iron surface, compared with the bare iron surface without the epoxy resin coating, although the existence of the epoxy resin coating prevented the diffusion of methane to the iron surface to play a competitive adsorption role, it greatly reduced the relative concentration of hydrogen on the iron surface, and thus reduced the number of hydrogen molecules dissociated and adsorbed into the iron. However, when the epoxy resin coating delaminated, the physical barrier effect of the epoxy resin coating on the dissociation and adsorption of hydrogen molecules on the iron surface disappeared, while its diffusion hindering effect on methane molecules still existed. The combination of two negative effects resulted in more severe hydrogen embrittlement damage to coated pipeline steel than bare steel. The findings of this work highlighted the delamination of epoxy resin coating in hydrogen blended natural gas pipeline represents substantial threat to the hydrogen embrittlement safety of pipeline steel. While epoxy resin coating provided significant protection when fully adhered to steel surface, delamination severely damaged the protective function and aggravated hydrogen embrittlement damage to pipeline steel. Therefore, to ensure the safe operation of hydrogen blended natural gas pipeline, it is essential to maintain stable adhesion of the internal friction-reducing epoxy resin coating in practical engineering application.

Keywords Hydrogen embrittlement; Hydrogen blended natural gas pipeline; Epoxy resin coating; X80 pipeline steel; Coating delamination

1 Introduction

Among the renewable energy sources discovered to date, hydrogen energy has emerged as the most promising alternative to fossil fuels due to its high calorific value, non-toxicity, and clean combustion without pollution. It is regarded as playing a pivotal role in the energy transition and achieving the net-zero emission target by 2050. The large-scale transportation of hydrogen energy is an important component of realizing the hydrogen energy value chain. Among the methods of hydrogen energy transportation, pipeline transport of high-pressure gaseous hydrogen offers an economically viable and efficient solution for delivering large volumes of hydrogen over long distance. Given the substantial investment and time costs associated with constructing new hydrogen pipelines, utilizing existing natural gas pipelines for hydrogen blended transportation offers a practical alternative, avoiding the high initial capital expenditure on dedicated hydrogen pipelines and further enhancing economic benefits.

The issue of hydrogen embrittlement in high-strength steel under hydrogen blended methane environment has been extensively studied. Research has shown that the addition of CH_4 to hydrogen environment can mitigate the hydrogen embrittlement of steel. First-principles calculations of the density of states indicate that methane adsorption on the surface of X80 pipeline steel can influence the dissociative adsorption of hydrogen molecules. The competitive adsorption behavior between CH_4 and H_2 molecules on the X80 pipeline steel surface reduces the catalytic activity of the iron surface, thereby inhibiting the dissociation of hydrogen molecules into hydrogen atoms. This process ultimately leads to a lower surface concentration of H_2 on X80 pipeline steel. The experimental observations are consistent with the theoretical studies. For instance, Frandsen et al. and San et al. observed that CH_4 inhibited hydrogen-accelerated fatigue crack growth. Wang et al., through slow strain rate tensile (SSRT) tests, evaluated the hydrogen embrittlement sensitivity of X80 pipeline steel specimens under varying hydrogen partial ratios. Their study revealed that higher hydrogen content resulted in lower strain values and reduced ductility at fracture. Moro et al. conducted tensile tests on X80 pipeline steel under different hydrogen pressure conditions and found that when the hydrogen pressure exceeded 10MPa, the hydrogen embrittlement sensitivity of the material was no longer influenced by pressure. The research findings presented above provide a comprehensive summary of the hydrogen embrittlement behavior of bare pipeline steel in hydrogen blended methane environment. However, in practical applications, the inner walls of existing natural gas pipelines are typically coated with epoxy resin, and the influence of epoxy resin coating on the hydrogen embrittlement behavior of pipeline steel under hydrogen blended methane environment remains unclear.

Epoxy resin coating is the most commercially valuable liquid coatings in the natural gas pipeline industry. These coatings can be conveniently applied to the inner surfaces of pipeline via air spraying or brushing, serving as friction-reducing layer to minimize friction between the pipeline surface and transported fluid, thereby lowering the energy consumption of gas transportation. Additionally, polymer coating has been studied as hydrogen barrier to mitigate hydrogen embrittlement in pipelines. However, due to the microporous voids formed by the inefficient packing of polymer chains, hydrogen molecules, with their small molecular size, may permeate through polymer coatings more readily under external environmental effect. This phenomenon renders the hydrogen embrittlement resistance of epoxy resin coating unclear. Therefore, although the hydrogen embrittlement behavior of bare steel in hydrogen blended methane environment has been extensively studied, further research is needed to better connect the practical engineering context of epoxy resin coating. The influence of epoxy resin coatings on the hydrogen embrittlement behavior of pipeline steel under hydrogen blended methane environment remains to be further investigated.

This work characterizes the hydrogen embrittlement behavior of X80 pipeline steel specimen coated with epoxy resin under hydrogen blended methane environment through constant strain rate tensile test. By comparing the stress-strain curve, fracture elongation, and section shrinkage between bare steel specimen and epoxy-coated specimen, the influence of the epoxy resin coating on the hydrogen embrittlement behavior of X80 pipeline steel is investigated. Additionally, molecular dynamic simulation is employed to elucidate the mechanism underlying the observed experimental phenomena.

2 Experimental

2.1 Material

The specimens used in this research were sourced from X80 pipeline steel base material. The chemical composition of the steel was determined using the SPECTRO MAX X07 direct reading spectrometer, and the results are presented in Table 1.

Table 1 Chemical composition of X80 pipeline steel experimental material (mass fraction%)

Elements	Content/%	Standard/%	Elements	Content/%	Standard/%
Fe	96.7	—	Cr	0.24	≤0.50
Mn	1.70	≤1.85	Mo	0.24	≤0.50
C	0.09	≤0.12	Si	0.31	≤0.45
P	0.004	≤0.025	Cu	0.19	≤0.50
Bi	0.005	—	Al	0.036	—
Nb	0.11	—	Sb	0.011	—
Ti	0.016	—	Co	0.008	—
Ni	0.26	≤1.00	Se	0.002	—
V	0.005	—	As	0.006	—
Pb	0.005	—	Ta	0.021	—

The epoxy resin, experimental reagents, and gases employed in the experiments were obtained from well-known manufacturers, both domestic and international. Detailed specifications and supplier information are listed in Table 2.

Table 2 Experimental materials

Name	Molecular Formula	Specification	Manufacturer
Hydrogen	H_2	≥99.99%	Qingdao Xinkeyuan Technology Co., Ltd.
Methane	CH_4	≥99.999%	Qingdao Xinkeyuan Technology Co., Ltd.
Nitrogen	N_2	≥99.999%	Qingdao Xinkeyuan Technology Co., Ltd.
Epoxy Resin E44	$C_{21}H_{24}O_4$	≥85%	Nantong Xingchen Synthetic Materials Co., Ltd.
Benzyl Alcohol	C_7H_8O	AR, ≥99%	Shanghai Aladdin Biochemical Technology Co., Ltd.
Polyetheramine D230	$C_{12}N_2O_3H_{28}$		Shanghai Aladdin Biochemical Technology Co., Ltd.
Distilled Water	H_2O	Conductivity<0.1us/cm	South China High-Tech Co., Ltd.
Anhydrous Ethanol	C_2H_6O	AR	Tianjin Fuyu Fine Chemical Co., Ltd.

2.2 Experimental environment

The experiments were conducted at a total gas pressure of 10MPa, with the temperature controlled at 25 ± 1℃. To replicate the gas environment of hydrogen-blended natural gas pipelines, the hydrogen partial pressures were set at 3, 4, and 5MPa, with methane(CH_4) constituting the remainder of the gas mixture. In this study, the minimum hydrogen blended ratio was 30%, exceeding the typical proportions found in current hydrogen-natural gas pipeline systems, where hydrogen content is generally below 20%.

2.3 Experimental equipment

The primary experimental equipment required for the preparation and application of epoxy resin coatings on pipeline steel specimens, as well as for conducting constant strain rate tensile tests indifferent methane-hydrogen environments, is provided in Table 3.

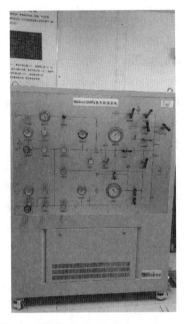

Fig. 1 15MPa hydrogen booster system.

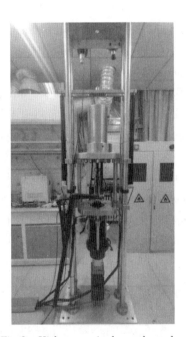

Fig. 2 High pressurized gas phase slow strain rate tensile test machine.

Table 3 Main instruments and equipment

Equipment	Model	Manufacturer
UV-Visible Spectrophotometer	SRECTRO MAX X07	SPECTRO Analytical Instruments GmbH (Germany)
Constant Temperature Magnetic Stirrer	81-2	Shanghai Sile Instruments Co., Ltd.
Ultrasonic Processor	FS-900N	Shanghai Shenxi Ultrasonic Equipment Co., Ltd.
Vacuum Drying Oven	DZF-6021	Shanghai Qixin Scientific Instrument Co., Ltd.
Blast Drying Oven	DHG-9023A	Shanghai Qixin Scientific Instrument Co., Ltd.
Water Bath Ultrasonic Cleaner	YS0203	Shenzhen Yuyi Technology Co., Ltd.
High Pressurized Gas Phase Slow Strain Rate Tensile Test Machine	YYF-50	Shanghai Bairoe Testing Instrument Co., Ltd.

2.4 Experimental Method

The experimental operation should follow the requirements of ASTM G142. The specific test procedure is as follows:

(1) Smooth tensile specimens were machined from X80 pipeline steel base material, with details shown in Figure 3. The gauge sections were polished sequentially with 400[#], 800[#], 1200[#], and 2000[#] sandpapers to remove rust, and inspected under a magnifying glass to ensure defect-free surfaces. The

polished specimens were then ultrasonically cleaned in anhydrous ethanol, dried with cold air, and stored in a vacuum drying oven for subsequent coating.

(2) This procedure applies exclusively to specimens covered with epoxy resin coatings. Before coating preparation, epoxy resin E44 was preheated to 80℃ to improve its fluidity. In the first step, the epoxy resin was mixed with 20% (by mass) benzyl alcohol and stirred at 2000 r/min for 30 minutes. In the second step, polyether amine D230 was added as a curing agent at a mass ratio of 10 : 2.68 to the epoxy resin, and the mixture was stirred again at 2000 r/min for 30 minutes. In the third step, the mixture was evacuated at room temperature for 15 minutes under 0.1MPa vacuum to remove any microbubbles. Finally, the mixture was applied to the X80 pipeline steel specimens, forming a polymer coating with a thickness of 150±20 microns by rotating the specimen during curing.

(3) The bottom of the reactor was wiped with an alcohol-dampened cotton ball to remove impurities. The tensile specimens were securely fixed in the autoclave, and the autoclave was sealed with all hexagonal screws tightened to ensure air-tightness. To minimize the influence of residual oxygen, the autoclave was evacuated until the internal pressure dropped below 100 Pa. It was then purged with 0.5MPa of N_2 gas to reduce the oxygen content to less than 0.5 vol%, after which the remaining N_2 was replaced with 0.5MPa of H_2 gas to establish the hydrogen environment.

(4) H_2 and CH_4 were introduced into the autoclave sequentially according to the required ratios. The specimens were then exposed to the mixed gas environment for 24 hours to facilitate hydrogen pre-charging and ensure stable hydrogen content within the specimens. After pre-charging, tensile tests were conducted in the hydrogen environment at the constant strain rate specified by the experimental requirement until the specimens fractured.

(5) Following fracture, the gas mixture was discharged from the autoclave, and the specimens were carefully removed. The fracture size was measured, and section shrinkage and hydrogen embrittlement index were calculated.

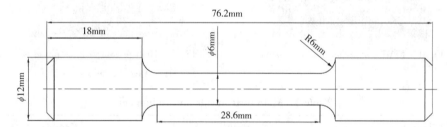

Figure. 3 Smooth tensile specimen size.

2.5 Calculation Method

The molecular dynamics Forcite module in Materials Studio software was used to simulate and calculate the adsorption behavior of a mixed gas of methane and hydrogen on different surface states of pipeline steel.

(1) Due to the lower surface energy of Fe (110) surface compared to other Fe high index crystal planes, Fe(110) surface is selected to simulate the metal surface that is most easily exposed during the rolling process of pipeline steel. After structural optimization, the thickness of the crystal cell was extended to 7 layers to represent the surface of pipeline steel (110), and the crystal plane was extended to a 20×20 supercell structure, and a periodic structural model was established.

(2) In the construction of bisphenol A-type epoxy resin polymer chains, 10 epoxy groups are incorporated into the polymer backbone. Subsequently, 20 epoxy resin polymer chains are generated to form amorphous unit cells, which are used as epoxy resin coatings. The structure is relaxed after the addition of each monomer to ensure proper integration and stability of the polymer chains. The amorphous unit cell structure undergoes both annealing and geometric optimization processes, with an annealing temperature

range from 300 K to 500 K, in order to eliminate local stresses within the unit cell. In the initial geometry of CH_4/H_2 on the iron surface with a delaminated epoxy resin coating, a distance of 20 Å was maintained between the epoxy resin coating and the Fe (110) crystal plane.

(3) Under a hydrogen-enriched methane environment with a hydrogen partial pressure of 3MPa and a methane partial pressure of 7MPa, the compressibility factor of CH_4 was 0.88, while that of H_2 was 0.83. Using Formula (1), the calculated number of CH_4 molecules under these conditions was 751, and the number of H_2 molecules was 341.

$$pV = ZnRT \quad (1)$$

where p represents the pressure of gas component (Pa), V represents the volume of gas (m^3), Z represents the compressibility factor of gas, n represents the amount of substance of gas (mol), R represents the ideal gas constant, T represents the thermodynamic temperature of gas (K).

(4) Geometric optimization was performed on three adsorption models, with COMPASS III selected as the charge force field and Ewald used as the calculation method for the Coulomb interaction between atoms; The calculation method for van der Waals forces interaction is Atom based; The calculation accuracy is selected as Fine. Then, the NVT ensemble was selected for molecular dynamics simulation of the geometrically optimized adsorption model, fixing the positions of Fe atoms and epoxy resin coating, and only analyzing the motion of the two gases. The total simulation time is 2000ps, the simulation time step is 1fs, and the simulation temperature is 298K. The velocity of gas molecules is set to random, the temperature controller is Andersen, and the calculation accuracy is selected as Fine.

2.6 Evaluation parameters

The plastic deformation behavior of X80 pipeline steel under varying hydrogen partial pressure conditions was evaluated using elongation (δ) and section shrinkage (ψ). These parameters were calculated using the following formulas:

$$\delta = \frac{l_1 - l}{l} \times 100\% \quad (2)$$

$$\psi = \frac{A - A_1}{A} \times 100\% \quad (3)$$

where l and l_1 represent the initial and fractured lengths of the specimen's gauge section, respectively, while A and A_1 denote the cross-sectional areas of the specimen before and after testing.

To quantify the material's susceptibility to hydrogen embrittlement, the hydrogen embrittlement index (F_H) was employed. This index is calculated using the formula:

$$F_H = (Z_{air} - Z_{h_2})/Z_{air} \quad (4)$$

where Z represents the plasticity parameter (either elongation or section shrinkage) obtained from specimens tested in air (Z_{air}) or in a hydrogen environment (Z_{h_2}).

In this study, section shrinkage (ψ) was used to determine the hydrogen embrittlement index.

3 Result

3.1 Effect of Strain Rate on the Hydrogen Embrittlement Behavior of X80 Pipeline Steel

Tensile tests were conducted on X80 bare steel specimens at strain rates of $1 \times 10^{-3} s^{-1}$ and $1 \times 10^{-5} s^{-1}$, respectively, in hydrogen blended methane environment with total pressure of 10MPa. The partial pressures of hydrogen were 0, 3, 4 and 5MPa, respectively. From Figure 4, it can be seen that the stress-strain curves of the specimens under different gas environments basically overlap in the elastic deformation and yield stages, indicating that hydrogen partial pressure does not have a significant impact on the mechanical properties of the specimens in the elastic deformation and yield stages. Meanwhile, under both strain rates, X80 pipeline steel exhibits a trend of lower plasticity with higher hydrogen partial pressure. When the hydrogen partial pressure was 0MPa, the strain value at fracture was the largest and the plasticity was the best. With the increase of hydrogen partial pressure, the strain value of at fracture decreased gradually, and the plasticity decreased.

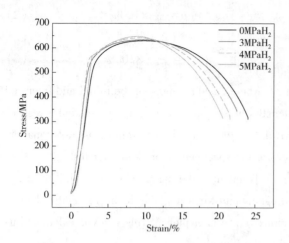

Figure. 4 Stress-strain curves of X80 pipeline steel obtained from constant strain rate tensile tests conducted at strain rate of $1\times10^{-5} s^{-1}$ under different hydrogen partial pressures.

As shown in Figure 5, the stress-strain curves of X80 pipeline steel obtained from constant strain rate tensile tests conducted at two different strain rates under different hydrogen partial pressures nearly overlapped, and the elongation atfracture was very close, indicating that the change in strain rate did not significantly affect the characterization of hydrogen embrittlement susceptibility in pipeline steel. Under three hydrogen partial pressure conditions, the specimens tested at the slower strain rate exhibited lower elongation at fracture, indicating poorer plasticity. This reduction in plasticity may be attributed to the longer exposure time to the hydrogen blended environment under the slower strain rate, which could be beneficial for hydrogen diffusion.

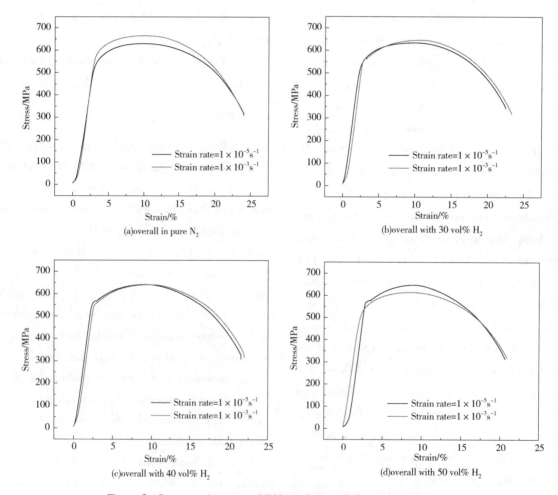

Figure. 5 Stress-strain curves of X80 pipeline steel obtained from constant strain rate tensile tests conducted at different strain rates in different hydrogen partial pressures

3.2 Effect of Epoxy Resin Coating on Hydrogen Embrittlement Behavior of X80 Pipeline Steel

Firstly, the effect of epoxy resin coating on hydrogen embrittlement behavior of X80 pipeline steel was tested under strain rates of $1\times10^{-3} s^{-1}$. As shown in Fig. 6, the elongation of X80 pipeline steel specimens with epoxy resin coatings progressively decreased with increasing hydrogen partial pressure. Under the protection of the epoxy resin coating, the elongation at fracture for coated specimens was consistently higher across various hydrogen partial pressure environments.

As shown in Fig. 7, the elongation and section shrinkage of the epoxy resin-coated specimens were greater than those of the bare steel specimens under all tested hydrogen partial pressure environments.

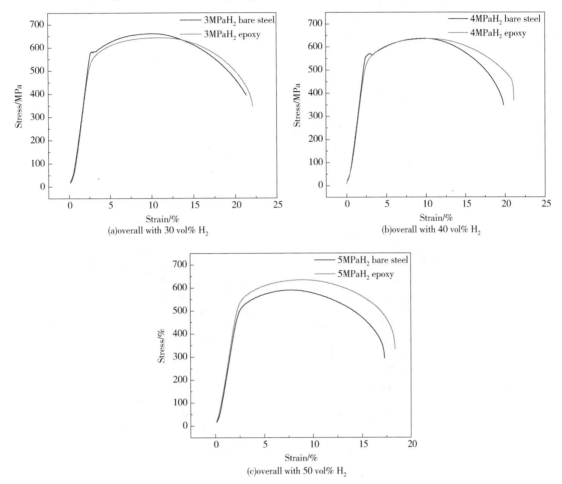

Figure. 6 Stress-strain curves of bare and epoxy resin coated X80 pipeline steel obtained from constant strain rate tensile tests conducted at strain rate of $1\times10^{-3}\,\mathrm{s}^{-1}$ under different hydrogen partial pressures

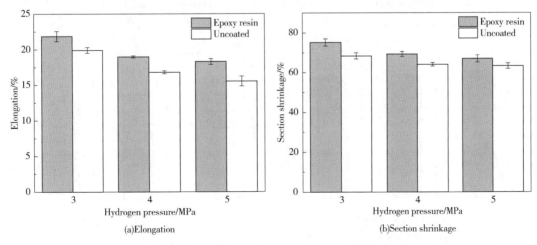

Figure. 7 Elongations and section shrinkages of bare and epoxy resin coated X80 pipeline steel obtained from constant strain rate tensile tests conducted at strain rate of $1\times10^{-3}\,\mathrm{s}^{-1}$ under different hydrogen partial pressures

Subsequently, the strain rate was reduced to $1\times10^{-5}\,\text{s}^{-1}$ while keeping all other experimental parameters and setup unchanged to evaluate the effect of the epoxy resin coating on the hydrogen embrittlement behavior of X80 pipeline steel. From Figure 8, it can be seen that the elongation at fracture of X80 pipeline steel specimens coated with epoxy resin also decreased with the increase of hydrogen partial pressure. However, the elongation at fracture of the epoxy-coated specimens was consistently lower than that of the bare steel specimens.

As shown in Figure 9, specimens coated with epoxy resin exhibited lower elongation and section shrinkage after fracture under all tested hydrogen partial pressure environments.

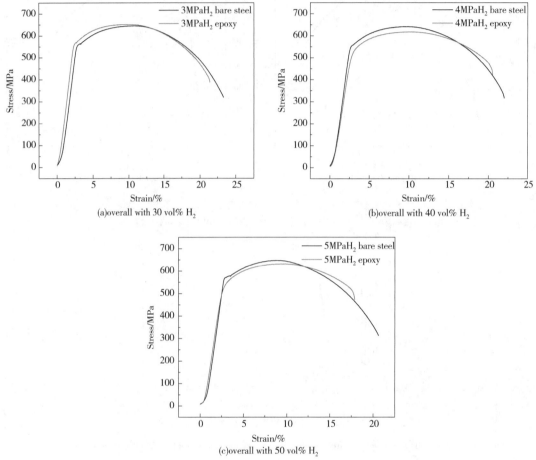

Figure. 8 Stress-strain curves of bare and epoxy resin coated X80 pipeline steel obtained from constant strain rate tensile tests conducted at strain rate of $1\times10^{-5}\,\text{s}^{-1}$ under different hydrogen partial pressures.

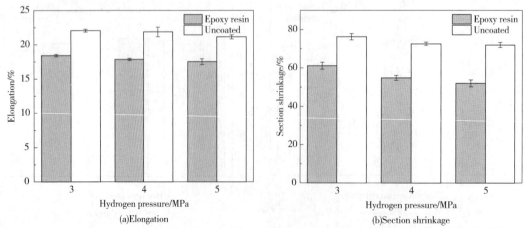

Figure. 9 Elongations and section shrinkages of bare and epoxy resin coated X80 pipeline steel obtained from constant strain rate tensile tests conducted at strain rate of $1\times10^{-5}\,\text{s}^{-1}$ under different hydrogen partial pressures.

4 Discussion

4.1 Influence of epoxy resin coatings on the hydrogen embrittlementbehavior of X80 pipeline steel in hydrogen blended methane environment

Figures 10 and 11, respectively, compare the hydrogen embrittlement indices of epoxy resin coated and bare steel specimens under two different strain rates in the test environment. It can be inferred from figures that, under all tested conditions, the hydrogen embrittlement index of the specimen consistently increases with rising hydrogen partial pressure. However, the effect of the epoxy resin coating on the hydrogen embrittlement behavior of X80 pipeline steel varies depending on the strain rate.

At a higher strain rate ($1\times10^{-3}s^{-1}$), the epoxy resin-coated specimens exhibit lower hydrogen embrittlement indices across all hydrogen partial pressures, indicating that the coating effectively protects X80 pipeline steel and mitigates hydrogen embrittlement damage. In contrast, at a lower strain rate ($1\times10^{-5}s^{-1}$), the epoxy resin coated specimens show higher hydrogen embrittlement indices under all hydrogen partial pressures. This suggests that, under these conditions, the presence of the epoxy resin coating not only fails to provide protection but also exacerbates the hydrogen-induced damage to pipeline steel in the same hydrogen blended methane environment.

To further investigate the mechanism by which the epoxy resin exacerbates hydrogen embrittlement damage in pipeline steel under relatively low strain rate, the tensile testing process was closely examined. As shown in Fig. 12, it was observed that during the necking stage, the deformation of the pipeline steel specimen resulted in reduction of cross-sectional area, inducing curvature change on the specimen surface. However, the tensile stress acting on both sides of the coating stretched it, disrupting its adhesion to the steel surface. As a result, the epoxy resin coating delaminated from the steel surface.

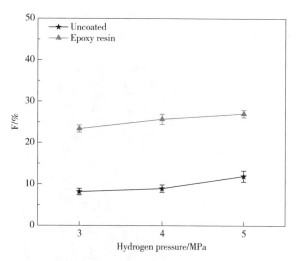

Fig. 10 Hydrogen embrittlement indices of bare and epoxy resin coated X80 pipeline steel obtained from constant strain rate tensile tests conducted at strain rate of $1\times10^{-5}s^{-1}$ under different hydrogen partial pressures.

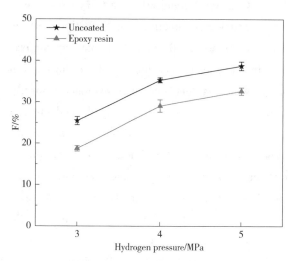

Fig. 11 Hydrogen embrittlement indices of bare and epoxy resin coated X80 pipeline steel obtained from constant strain rate tensile tests conducted at strain rate of $1\times10^{-3}s^{-1}$ under different hydrogen partial pressures.

Based on the observed phenomenon of coating detachment, it is reasonable to infer that the loss of protective efficacy of the epoxy resin coating at relatively low strain rates is due to the extended duration during which the pipeline steel specimen remained in delaminated state of the epoxy resin coating throughout the testing process. However, the mechanism underlying the more severe hydrogen embrittlement behavior of delaminated coated specimens compared to bare steel under the same gaseous environment remains unclear. Further investigation is necessary to elucidate the mechanism by which the delamination of epoxy coating exacerbates hydrogen embrittlement in X80 pipeline steel under hydrogen blended methane condition.

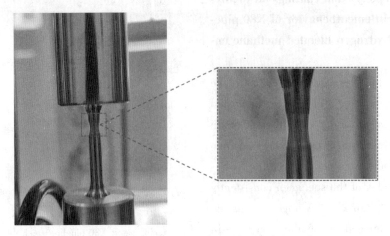

Figure. 12　Delamination of epoxy resin coating during tensile process.

4.2　Aggravation Mechanism of Epoxy Resin Coating Delamination on Hydrogen Embrittlement Behavior of X80 Pipeline Steel in Hydrogen Blended Methane Environment

In order to elucidate the aggravation mechanism of epoxy coating delamination on hydrogen embrittlement behavior of X80 in hydrogen blended methane environment, the molecular dynamics software Materials Studio was used to simulate the distribution of methane and hydrogen molecules on the iron surface. The initial geometries of CH_4/H_2 on the bare iron surface (Fig. 13), the iron surface coated with epoxy resin (Fig. 14) and the iron surface with delaminated epoxy resin coating (Fig. 15) were respectively constructed.

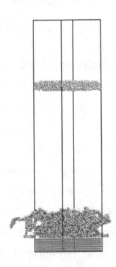

Fig. 14　Adsorption model of CH_4/H_2 on the iron surface coated epoxy resin.

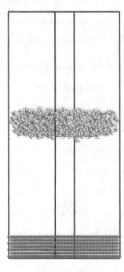

Fig. 13　Adsorption model of CH_4/H_2 on the bare iron surface.

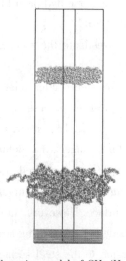

Fig. 15　Adsorption model of CH_4/H_2 on the iron surface with delaminated epoxy resin coating.

All three systems reached an energetically stable equilibrium state before 1000 ps. Therefore, the relative concentration of CH_4 and H_2 gas molecules in the 0–40 Å range of 200 frames from 1000ps to 2000ps were statistically analyzed to investigate the effect of epoxy resin on the distribution of CH_4 and H_2 molecules on iron surface. The statistical results are shown in Figure 16.

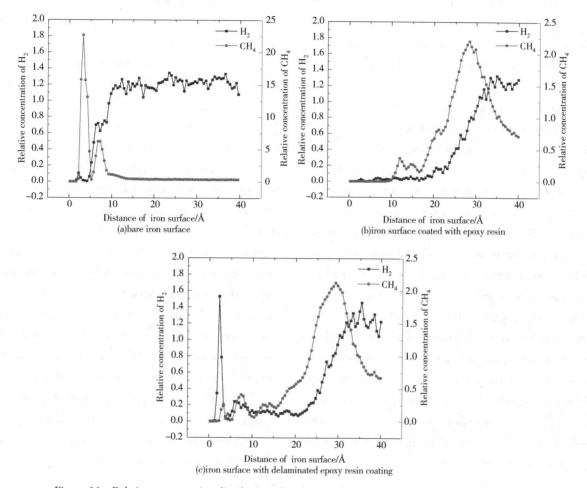

Figure. 16 Relative concentration distribution of methane and hydrogen in different ending geometries

As shown in Figure 16(a), it can be observed that the peak relative concentration of CH_4 occurs at a distance of 4.8 Å from the iron surface. Simultaneously, due to the aggregation of methane molecules, the relative concentration of hydrogen molecules within 5 Å of the iron surface is maintained below 0.1. This indicates that in hydrogen blended methane environment, the competitive adsorption between the two gases inhibits the accumulation of hydrogen molecules near the iron surface, thereby making their dissociative adsorption more difficult. This further validates that the competitive adsorption of methane and hydrogen on the bare pipeline steel surface plays a positive role in mitigating hydrogen embrittlement in pipeline steel.

As shown in Fig. 16(b), when the iron surface is covered with epoxy resin, the relative concentration of both gas molecules on the surface are significantly reduced. This indicates that the epoxy resin impedes the approach of gas molecules to the iron surface, making it more difficult for both gases to diffuse to the surface. Among them, methane molecules, due to their larger molecular size, face greater difficulty diffusing to the epoxy resin–covered iron surface compared to hydrogen molecules, resulting in a lower relative surface concentration. Meanwhile, the relative concentration of hydrogen molecules near the iron surface decreases, as the epoxy resin coating occupies adsorption sites on the surface. Therefore, although the epoxy resin reduces the relative concentration of methane, thereby diminishing the competitive adsorption between the two gases, the combined effect of the coating's obstruction and the weakened competitive adsorption ultimately leads to lower hydrogen molecule relative concentration near the iron surface compared to bare steel which provides better resistance to hydrogen embrittlement under the same hydrogen blended methane environment.

As shown in Fig. 16(c), when the epoxy resin coating on the iron surface is in delaminated state, the relative concentration of hydrogen molecules near the iron surface is significantly higher than that of CH_4 molecules. This phenomenon indicates that the delaminated epoxy resin coating acts as a selective barrier, allowing hydrogen molecules to permeate through the coating more rapidly. Consequently, the relative concentration of methane on the iron surface is reduced, severely weakening the competitive adsorption effect on the iron surface. Simultaneously, the delamination of the coating eliminates its ability to occupy adsorption sites and hinder the dissociative adsorption of hydrogen molecules on the iron surface. This explains the aggravation mechanism of epoxy coating delamination on hydrogen embrittlement behavior of X80 pipeline steel in hydrogen blended methane environment.

5 Conclusion

This work investigated the influence of epoxy resin coating on the hydrogen embrittlement behavior of X80 pipeline steel in hydrogen blended methane environment. The experimental results indicate that the effect of the coating on hydrogen embrittlement varies significantly with strain rate. At higher strain rate ($1\times10^{-3} s^{-1}$), the epoxy resin coating effectively mitigated hydrogen embrittlement, reducing the embrittlement indices across all tested hydrogen partial pressures. However, at lower strain rate ($1 \times 10^{-5} s^{-1}$), the coating not only failed to protect the steel from hydrogen embrittlement but also exacerbated the hydrogen embrittlement behavior. The observed delamination of the epoxy resin coating from the steel surface during tensile testing was identified as the key factor contributing to the aggravated hydrogen embrittlement at lower strain rate.

Molecular dynamics simulation further confirmed that when the epoxy resin coating remained intact on the iron surface, it provided effective protection by hindering the adsorption of hydrogen molecules onto the iron surface. Although the coating partially suppressed the competitive adsorption between hydrogen and methane, the overall effect was more beneficial than that of the bare steel surface. However, when the epoxy resin coating delaminated from the surface, it not only lost its protective function but also obstructed the competitive adsorption between two gases. In this delaminated state, the epoxy resin coating made it more difficult for methane molecules, which diffuse more slowly than hydrogen molecules, to approach the iron surface, leading to an accumulation of hydrogen molecules on the iron surface that was more severe than that observed on the bare steel surface. This elucidates aggravation mechanism of epoxy coating delamination on hydrogen embrittlement behavior of X80 pipeline steel in hydrogen blended methane environment.

Therefore, in the practical application of hydrogen blended methane transportation in pipelines, attention must be paid to the factors that could lead to coating delamination on the inner surface of pipelines. Future research should focus on enhancing the adhesion properties of epoxy resin coating and improving the coating's selectivity for methane gas diffusion through modification. This would help ensure that the coating provides effective protection during adsorption while minimizing the hindrance of competitive adsorption when delaminated.

6 Declaration of competing interest

The authors declare that they have no known competing financial interests or personal relationships that could have appeared to influence the work reported in this paper.

7 Acknowledgement

This work is supported by Open Fund Project of CNOOC Key Laboratory of Liquefied Natural Gas and Low-Carbon Technology.

References

[1] M. Momirlan and T. N. Veziroglu, 'Current status of hydrogen energy', *Renewable and Sustainable Energy Reviews*, vol. 6, no. 1, pp. 141-179, Jan. 2002.

[2] G. Nicoletti, N. Arcuri, G. Nicoletti, and R. Bruno, 'A technical and environmental comparison between hydrogen and some fossil fuels', *Energy Conversion and Management*, vol. 89, pp. 205-213, Jan. 2015.

[3] A. Sgobbi, W. Nijs, R. De Miglio, A. Chiodi, M. Gargiulo, and C. Thiel, 'How far away is hydrogen? Its role in the medium and long-term decarbonisation of the European energy system', *International Journal of Hydrogen Energy*, vol. 41, no. 1, pp. 19-35, Jan. 2016.

[4] J. O. Abe, A. P. I. Popoola, E. Ajenifuja, and O. M. Popoola, 'Hydrogen energy, economy and storage: Review and recommendation', *International Journal of Hydrogen Energy*, vol. 44, no. 29, pp. 15072-15086, Jun. 2019.

[5] A. D. Korberg et al., 'On the feasibility of direct hydrogen utilisation in a fossil-free Europe', *International Journal of Hydrogen Energy*, vol. 48, no. 8, pp. 2877-2891, Jan. 2023.

[6] M. E. Demir and I. Dincer, 'Cost assessment and evaluation of various hydrogen delivery scenarios', *International Journal of Hydrogen Energy*, vol. 43, no. 22, pp. 10420-10430, May 2018.

[7] D. Hardie, E. A. Charles, and A. H. Lopez, 'Hydrogen embrittlement of high strength pipeline steels', *Corrosion Science*, vol. 48, no. 12, pp. 4378-4385, Dec. 2006.

[8] N. E. Nanninga, Y. S. Levy, E. S. Drexler, R. T. Condon, A. E. Stevenson, and A. J. Slifka, 'Comparison of hydrogen embrittlement in three pipeline steels in high pressure gaseous hydrogen environments', *Corrosion Science*, vol. 59, pp. 1-9, Jun. 2012.

[9] J. Shang et al., 'Effects of stress concentration on the mechanical properties of X70 in high-pressure hydrogen-containing gas mixtures', *International Journal of Hydrogen Energy*, vol. 45, no. 52, pp. 28204-28215, Oct. 2020.

[10] S. Cerniauskas, A. Jose Chavez Junco, T. Grube, M. Robinius, and D. Stolten, 'Options of natural gas pipeline reassignment for hydrogen: Cost assessment for a Germany case study', *International Journal of Hydrogen Energy*, vol. 45, no. 21, pp. 12095-12107, Apr. 2020.

[11] J. Ogden, A. M. Jaffe, D. Scheitrum, Z. McDonald, and M. Miller, 'Natural gas as a bridge to hydrogen transportation fuel: Insights from the literature', *Energy Policy*, vol. 115, pp. 317-329, Apr. 2018.

[12] Y. Sun, Y. Ren, and Y. F. Cheng, 'Dissociative adsorption of hydrogen and methane molecules at high-angle grain boundaries of pipeline steel studied by density functional theory modeling', *International Journal of Hydrogen Energy*, vol. 47, no. 97, pp. 41069-41086, Dec. 2022.

[13] J. D. Frandsen and H. L. Marcus, 'Environmentally assisted fatigue crack propagation in steel', *Metall Trans A*, vol. 8, no. 2, pp. 265-272, Feb. 1977.

[14] 'Technical Reference for Hydrogen Compatibility of Materials', Hydrogen Materials Technical Database. Accessed: Dec. 25, 2024. [Online]. Available: https://www.sandia.gov/matlstechref/.

[15] 'Study on the lowest CO content to inhibit hydrogen embrittlement of hydrogen-blended gas transmission pipeline | Article Information | J-GLOBAL'. Accessed: Dec. 26, 2024. [Online]. Available: https://jglobal.jst.go.jp/en/detail?JGLOBAL_ID=202202229248076438.

[16] I. Moro, L. Briottet, P. Lemoine, E. Andrieu, C. Blanc, and G. Odemer, 'Hydrogen embrittlement susceptibility of a high strength steel X80', *Materials Science and Engineering: A*, vol. 527, no. 27, pp. 7252-7260, Oct. 2010.

[17] S. Pradhan, P. Pandey, S. Mohanty, and S. K. Nayak, 'Insight on the Chemistry of Epoxy and Its Curing for Coating Applications: A Detailed Investigation and Future Perspectives', *Polymer-Plastics Technology and Engineering*, vol. 55, no. 8, pp. 862-877, May 2016.

[18] M. Qian, A. Mcintosh Soutar, X. H. Tan, X. T. Zeng, and S. L. Wijesinghe, 'Two-part epoxy-siloxane hybrid corrosion protection coatings for carbon steel', *Thin Solid Films*, vol. 517, no. 17, pp. 5237-5242, Jul. 2009.

[19] Y. Lei, E. Hosseini, L. Liu, C. A. Scholes, and S. E. Kentish, 'Internal polymeric coating materials for preventing pipeline hydrogen embrittlement and a theoretical model of hydrogen diffusion through coated steel', *International Journal of Hydrogen Energy*, vol. 47, no. 73, pp. 31409-31419, Aug. 2022.

[20] Y. Lei, L. Liu, C. A. Scholes, and S. E. Kentish, 'Crosslinked PVA based polymer coatings with shear-thinning behaviour and ultralow hydrogen permeability to prevent hydrogen embrittlement', *International Journal of Hydrogen Energy*, vol. 54, pp. 947-

954, Feb. 2024.

[21] M. Wetegrove et al., 'Preventing Hydrogen Embrittlement: The Role of Barrier Coatings for the Hydrogen Economy', *Hydrogen*, vol. 4, no. 2, Art. no. 2, Jun. 2023.

[22] S. K. Dwivedi and M. Vishwakarma, 'Hydrogen embrittlement in different materials: A review', *International Journal of Hydrogen Energy*, vol. 43, no. 46, pp. 21603–21616, Nov. 2018.

[23] H. Li, R. Niu, W. Li, H. Lu, J. Cairney, and Y. -S. Chen, 'Hydrogen in pipeline steels: Recent advances in characterization and embrittlement mitigation', *Journal of Natural Gas Science and Engineering*, vol. 105, p. 104709, Sep. 2022.

[24] X. Li, X. Ma, J. Zhang, E. Akiyama, Y. Wang, and X. Song, 'Review of Hydrogen Embrittlement in Metals: Hydrogen Diffusion, Hydrogen Characterization, Hydrogen Embrittlement Mechanism and Prevention', *Acta Metall. Sin. (Engl. Lett.)*, vol. 33, no. 6, pp. 759–773, Jun. 2020.

[25] Y. -Q. Zhu et al., 'Advances in reducing hydrogen effect of pipeline steels on hydrogen-blended natural gas transportation: A systematic review of mitigation strategies', *Renewable and Sustainable Energy Reviews*, vol. 189, p. 113950, Jan. 2024.

[26] 'Prevention of Hydrogen Embrittlement in Steels'. Accessed: Dec. 26, 2024. [Online]. Available: https://www.jstage.jst.go.jp/article/isijinternational/56/1/56_ISIJINT-2015-430/_article.

[27] M. A. Otmi, F. Willmore, and J. Sampath, 'Structure, Dynamics, and Hydrogen Transport in Amorphous Polymers: An Analysis of the Interplay between Free Volume Element Distribution and Local Segmental Dynamics from Molecular Dynamics Simulations', *Macromolecules*, vol. 56, no. 22, pp. 9042–9053, Nov. 2023.

[28] 'G142 Standard Test Method for Determination of Susceptibility of Metals to Embrittlement in Hydrogen Containing Environments at High Pressure, High Temperature, or Both'. Accessed: Dec. 26, 2024. [Online]. Available: https://www.astm.org/g0142-98r16.html.

[29] J. Rehrl, K. Mraczek, A. Pichler, and E. Werner, 'Mechanical properties and fracture behavior of hydrogen charged AHSS/UHSS grades at high-and low strain rate tests', *Materials Science and Engineering: A*, vol. 590, pp. 360–367, Jan. 2014.

A Synergistic Approach to BOG Generation Calculation Using Horizontal Feature Fitting and Vertical Time Series Prediction

Wang Yuqian[1] Liu Gang[1] Wang Bo[2] Zhang Chao[3] Chen Lei[1]

[1. College of Pipeline and Civil Engineering, China University of Petroleum(East China);
2. PetroChina Kunlun Gas Co., LTD.; 3. CNOOC Supplies & Equipment Center]

Abstract During the storage process of Liquefied Natural Gas(LNG) in the tanks of the receiving station, a portion of LNG evaporates into the gas phase, referred to as Boil Off Gas(BOG). Accurate calculation of BOG generation is crucial for the safety management of the tanks at the receiving station and for the economic efficiency of the enterprise. Traditional calculation methods, which rely on empirical formulas and numerical simulations, exhibit significant disadvantages, including large calculation errors, high time costs, and poor generalization in extrapolation. To address these issues, we analyzed the historical multivariate time-series data of LNG tank systems from both horizontal and vertical perspectives, employing machine learning and deep learning methodologies. First, from the perspective of Horizontal Feature Fitting(HFF), we calculated BOG generation at the same moment using various sensor data and evaluated the performance of six classical machine learning algorithms under different data distribution scenarios. Second, from the perspective of Vertical Time Series Prediction(VTSP), we utilized sensor data from several historical moments of the LNG tank system to predict future sensor data. Moreover, to comprehensively account for the spatial dimensional dependencies of sensors with distinct physical significance within actual industrial systems, we constructed the Spatial Domain Encoder(SDE) architecture and compared its computational performance with three traditional time series methods, LSTM, GRU, and Transformer, over a variety of time spans. Finally, we conducted extensive comparative experiments on a real dataset sourced from a receiving station along the southeast coast of China. The experimental results indicated that under the HFF perspective, the three algorithms—Decision Tree, GBDT, and XGBoost—achieved excellent fitting performance when the data distribution was known. For instance, the evaluation metrics of Mean Squared Error(MSE), Mean Absolute Percentage Error(MAPE), Mean Absolute Error(MAE), and R-squared(R^2) for XGBoost reached 0.00028, 0.00798, 0.01304, and 0.99707, respectively. However, the six machine learning algorithms demonstrated considerable deviations when confronted with unknown data distributions. Taking XGBoost as an example, the evaluation metrics of MSE, MAPE, and MAE increased to 0.04009, 0.09911, and 0.18279, respectively, while R^2 declined to 0.22749, which indicates that the HFF perspective is not suitable for dealing with the situation of unknown data distribution. The VTSP perspective overcame the limitations associated with the HFF perspective. In the one-step prediction mode, the four evaluation metrics of MSE, MAPE, MAE, and R^2 of the SDE reached 0.00025, 0.00674, 0.01202, and 0.99446, respectively. However, in the three-step prediction mode, the three evaluation metrics of SDE—MSE, MAPE, and MAE—increased to 0.00054, 0.01063, and 0.01856, respectively, while R^2 is reduced to 0.98455. Although the prediction performance of the SDE improves compared to the three traditional time-series methods, its performance slightly decreases as the prediction time span increases. In summary, HFF and VTSP each have their own advantages and disadvantages. Based on the specific needs of the enterprise and the distribution of historical data in the industrial system, the synergistic approach that combines the two perspectives of HFF and VTSP enables highly accurate and generalized calculations of BOG generation in the LNG tank system.

Keywords Machine learning; Deep learning; BOG generation calculation

1 Introduction

LNG produces a certain amount of Boil-Off Gas (BOG) during storage. Accurate and reasonable predictions of BOG generation facilitate the optimization of BOG compressor operations, thereby ensuring the safe operation and economic benefits of LNG receiving stations.

In recent years, numerous studies have investigated the calculation of BOG generation in LNG tanks. Lee et al. analyzed the finite elements of fuel tanks through thermal conductivity modeling, performed a numerical analysis based on boundary conditions and empirical correlation derivation, and obtained a prediction model for BOG generation with an error rate of less than 10%. Jeon et al. considered the phase change of LNG in the tanks and the heat transfer from the adiabatic system to accurately estimate the BOG values required for designing cryogenic liquid containment cargo systems, and they performed high-fidelity multiphysics CFD simulations. Saleem et al. conducted a comprehensive dynamic CFD simulation of onshore LNG tanks using a multiphase flow model while reliably quantifying the BOG generation. Khan et al. discussed the effect of liquid level on the Boil-off Rate (BOR) and BOG, attributing the heat loss in the tanks to the liquid phase and the increased temperature of the gas phase. They ultimately predicted the BOG through numerical simulation methods as well. However, empirical equation-based research methods are subject to contingencies, resulting in significant final prediction errors. Numerical simulation-based research methods tend to be computationally intensive and time-consuming, and they face limitations when applied to scenarios with complex physical patterns. Additionally, the generalization of extrapolation is weak in both approaches. Therefore, there is a need for a BOG generation prediction model that offers high accuracy, rapid response times, and strong extrapolation generalization capabilities.

With the continuous advancement of data-driven methods, such as machine learning and deep learning, along with the ongoing improvement of computer hardware resources, the application of data-driven techniques to mine the vast historical data in actual factories extracts sufficient and effective patterns to develop highly robust computational models. Singh et al. interpreted the model using Shapley's additive method to extract the critical input features, thereby reducing the model's complexity. They also employed various machine learning algorithms to model LNG historical data and concluded that the Random Forest algorithm helped plant operators make decisions quickly. However, most machine learning algorithms can only achieve better performance on test sets with data distributions similar to the train set, a premise that essentially limits their application in actual factories. Additionally, the historical data of LNG tank systems consists of data from multiple specific moments in sequential order in the temporal dimension and includes sensor data collected from different functions or locations in the spatial dimension, which can be referred to as multivariate time-series data. Traditional time-series prediction methods focus on modeling the dependence of multivariate time-series data in the temporal dimension, without adequately considering the relationships between sensors with different physical meanings in real industrial systems. This oversight leads to challenges in overcoming the bottleneck in the model's computational performance. Therefore, based on the fitting of BOG generation using machine learning algorithms, it is necessary to consider the dependence of industrial historical data in the spatial dimension within the VTSP perspective and simultaneously build an architecture for the synergistic analysis of both dimensions.

To address the aforementioned problems, we established the Synergistic Approach to BOG generation calculation using Horizontal Feature Fitting and Vertical Time Series Prediction (SA - HFF - VTSP). First, we employed the LOF algorithm to detect anomalies in the historical multivariate time-series data of the LNG tank system while performing data preprocessing operations such as normalization, smoothing, and differencing. Second, we examined the impact of data distribution on the HFF perspective and developed the Spatial Domain Encoder (SDE) architecture to model the dependencies of multivariate time series data in the spatial dimension within the VTSP perspective. Finally, we conducted numerous experiments using a real dataset from a receiving station along the southeast coast of China to verify the effectiveness of the synergistic approach.

The remainder of this paper is organized as follows: Section II describes the methodology, Section III evaluates the method, and Section IV summarizes the entire paper.

2 Conceptual framework

A portion of the BOG generated in the LNG tanks accumulates above the tanks, while another portion is pumped out by the compressors (n_{dis}). The amount of BOG that accumulates on top of the tanks is relatively small compared to the BOG pumped out by the compressors. Therefore, only the BOG removed by the compressor is predicted. Additionally, assuming minor fluctuations in compressor outlet pressure (p_{dis}) and temperature (T_{dis}), the expression for the BOG volume pumped by the compressor concerning the time derivative is as follows:

$$\frac{d(n_{dis})}{dt} = \frac{p_{dis} Q_{dis}}{zRT_{dis}} \quad (1)$$

In Eq. (1), R represents the molar gas constant, and z denotes the compression factor, the exact value of which is determined based on the actual gas's temperature, pressure, and chemical composition. Therefore, we consider $\frac{p_{dis} Q_{dis}}{T_{dis}}$ as the final prediction targets. The overall BOG generation can be obtained from the product of the prediction targets and $\frac{1}{zR}$ in the actual plant, integrated over time. In the actual receiving station we studied, three tanks—A, B, and C—are connected to the same BOG compressor, as illustrated schematically in Fig. 1. Additionally, the characterization statistics collected from the on-site sensor measurement points are presented in Table 1.

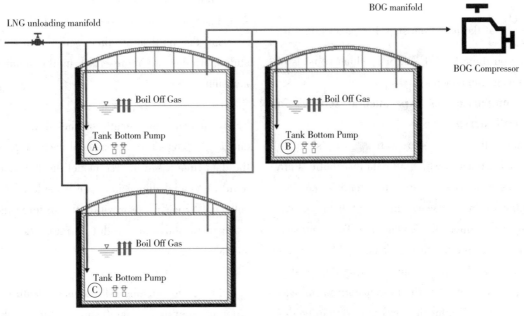

Figure 1 Schematic diagram of the LNG tank system.

Table 1 Comparison table of model input parameters and physical significance.

Input Parameters	Physical Significance
P_{pipe}	Unloading arm pressure
$T_{w\_(A,B,C)}$	Side temperatures of tanks A, B, and C
$T_{b\_(A,B,C)}$	Bottom temperature of tanks A, B, and C
$T_{t\_(A,B,C)}$	Top temperatures of tanks A, B, and C
$T_{x\_(A,B,C)}$	Tanks A, B, and C discharge temperatures
$L_{\_(A,B,C)}$	Tank A, B, and C levels
$P_{\_(A,B,C)}$	Tank pressure in tanks A, B, and C
$Q_{\_(A,B,C)}$	Tanks A, B, and C Bottom Pump Output Flow Rate
p_{dis}	BOG Compressor Outlet Pressure
Q_{dis}	BOG Compressor Outlet Flow
T_{dis}	BOG compressor outlet temperature

In addition to the volume to be predicted $\frac{p_{dis}Q_{dis}}{T_{dis}}$, 22 sensor features are identified, as shown in Table 1. We address the problem of BOG generation prediction for LNG tanks using two perspectives: HFF and VTSP. In the HFF perspective, we simultaneously use all 22 features to fit $\frac{p_{dis}Q_{dis}}{T_{dis}}$. In the VTSP perspective, we use the 22 features and $\frac{p_{dis}Q_{dis}}{T_{dis}}$ at the current time to predict the 22 features and $\frac{p_{dis}Q_{dis}}{T_{dis}}$ at the next time step. In this section, we elaborate on the SA-HFF-VTSP, which consists of the following components:

2.1 Comparison of experimental settings in HFF perspective

Under the HFF perspective, we simulate working conditions in which all data distributions are known, as well as scenarios where only some data distributions are known, utilizing both global sampling and sequential segmentation. This approach allows us to explore the fitting effects of different machine learning algorithms under varying data distributions, let the total length of the historical multivariate time series data for the LNG tank system be denoted as N, as outlined below:

(1) Global sampling

First, N is uniformly divided into subsets. Second, the first 70% of data from each subset is selected and combined to form the train set. Finally, the last 30% of data from each subset is selected and combined to create the test set. Based on this division, both the train set and the test set contain similar working conditions.

(2) Sequential division

First, the initial 70% of the historical multivariate time series data for the LNG tank system is designated as the train set. Second, the final 30% of the historical multivariate time series data is designated as the test set. Based on this division, the test set includes working conditions that do not appear in the train set.

2.2 SDE architecture construction

Transformer is currently effective in the sequence-to-sequence (seq2seq) task, where its Encoder component encodes the information in the input sequence, and the Decoder component generates the target sequence using the encoded information. In the multivariate time-series data prediction task, it is sufficient to model the hidden information within the input sequences without the need for further decoding. Therefore, we eliminate the Decoder structure in the Transformer to develop the SDE architecture, as illustrated in Fig. 2:

2.2.1 Spatial position vector embedding

Let the multivariate time-series data within the sliding window be represented in the temporal dimension as X:

$$X = \{x_1, x_2, \cdots, x_w\} \quad (2)$$

w denotes the length of the sliding window, and $x_{i,i \in (1,w)}$ is an n-dimensional vector representing the data from n sensors. To model the dependency of X in the spatial dimension, we considered the data collected by each sensor within the sliding window as its features and denoted the n sensors as S in the spatial dimension:

$$S = X^T = \{s_1, s_2, \cdots, s_n\} \quad (3)$$

First, to distinguish sensor features with different physical significance, the embedding vector PE is introduced:

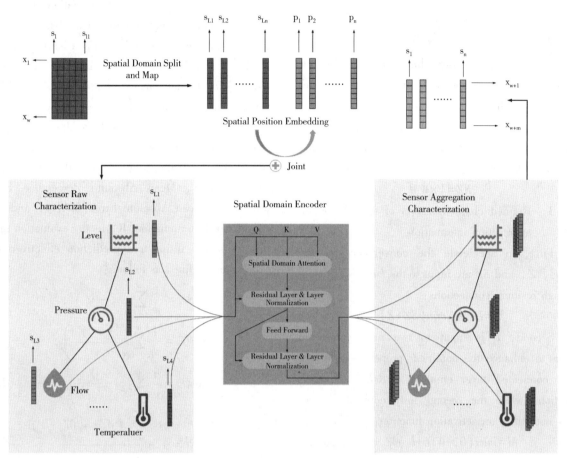

Figure 2 SDE architecture diagram.

$$PE = \{p_1, p_2, \cdots, p_n\} \quad (4)$$

$p_{i, i \in (1,n)}$ is an h-dimensional vector, and to map the features of each sensor within the sliding window to higher dimensions for a more comprehensive representation and to accomplish the dimensional matching with the positional embedding vectors, a dimensional mapping of S is performed to obtain S_L:

$$S_L = \{s_{L1}, s_{L2}, \cdots, s_{Ln}\} \quad (5)$$

$s_{Li, i \in (1,n)}$ is an h-dimensional vector, and the final input Z of Encoder is denoted as follows:

$$Z = PE + S_L \quad (6)$$

2.2.2 Spatial domain attention mechanism

We modeled the dependencies between different sensor features based on the Attention mechanism. First, a linear transformation of Z was performed to obtain Q, K, and V:

$$\begin{aligned} Q &= w_q Z + b_q, \\ K &= w_k Z + b_k, \\ V &= w_v Z + b_v, \end{aligned} \quad (7)$$

since Q, K, and V are obtained from Z by linear transformation, allowing Q, K, and V to represent Z to a certain extent. Noting that the row vector of the Q matrix is denoted as row_i, and the column vector of the K^T matrix is denoted as col_j, row_i and col_j represent the features of a particular sensor within a sliding window. The essence of the Attention mechanism lies in constructing the Attention weight matrix, which encapsulates the dependency coefficients between the features of a specific sensor and those of other sensors. The dot product of row_i and col_j reflects the feature dependency between the $it?$ sensor and the other sensors, thus enabling the matrix product of Q and K^T to construct the Attention weight matrix WM in the spatial domain:

$$WM = Q \times K^T \quad (8)$$

Second, to ensure that the new weight matrix maintains the original order of magnitude after aggregating different sensor features, further weight mapping is performed on all elements of each row vector in WM:

$$qk_i = \frac{e^{qk_i}}{\sum_{i=1}^{dim(col)} e^{qk_i}} \quad (9)$$

$qk_{i, i \in (1, dim(col))}$ denotes the element of the row vector in WM, and $dim(col)$ represents the number of columns of WM.

Finally, the process of information aggregation among different sensors in the spatial domain is expressed as follows:

$$SZ = V \times WM \quad (10)$$

2.2.3 Refined characterization of spatial domain aggregation information

First, to accelerate the convergence of the model, Z, and SZ are input to the *LayerNorm* layer through residual connections:

$$L = LayerNorm(SZ + Z) \quad (11)$$

Second, L is input to two fully connected layers to enhance the nonlinear properties. The first fully connected layer employs a *ReLU* activation function, while the second fully connected layer does not utilize an activation function:

$$M = max(0, W_1 L + b_1) W_2 + b_2 \quad (12)$$

Finally, L and M are fed into the LayerNorm layer again as a residual connection to obtain the final modeling result:

$$Out = LayerNorm(L + M) \quad (13)$$

In addition, the *Layer Norm* is computed as follows:

$$y = \frac{x - E[x]}{\sqrt{var[x] + \varepsilon}} * \gamma + \beta \quad (14)$$

$E[x]$ and $var[x]$ denote the mean and variance of the features of a particular sensor within a sliding window, respectively. ε represents a small constant to avoid division by zero, while γ and β denote the scaling and offset parameters, respectively, used to adjust the distribution of the normalized data.

3 Experiments

In this section, we first described the experimental setup and evaluation metrics. Second, we outlined the dataset. Next, we detailed the data preprocessing operations. Finally, we presented comparison experiments for the HFF perspective and the VTSP perspective, respectively.

3.1 Experimental Setup and Evaluation Metric

We implemented our method in Python 3.8.10 and PyTorch version 2.0.1, using a learning rate of 1×10^{-4}, a total of 100 training rounds, and a step size of 3 for the data smoothing operation. The sliding windows used in the time-series prediction model had lengths of 2, 4, 6, and 8 for the 1-step, 2-step, 3-step, and 4-step prediction angles, respectively. In addition, we employed Mean Square Error (MSE), Mean Absolute Percentage Error (MAPE), Mean Absolute Error (MAE), and Coefficient of Determination (R^2) as the evaluation metrics for model fitting and prediction effectiveness, which were defined as follows:

$$MSE = \frac{1}{w} \sum_{i=1}^{w} (y_i - \widehat{y_i})^2,$$

$$MAPE = \frac{1}{w} \sum_{i=1}^{w} \left(\frac{|y_i - \widehat{y_i}|}{|y_i|} \right),$$

$$MAE = \frac{1}{w} \sum_{i=1}^{w} |y_i - \widehat{y_i}|, \quad (15)$$

$$R^2 = 1 - \frac{\sum_{i=1}^{w} (y_i - \widehat{y_i})^2}{\sum_{i=1}^{w} (y_i - \widehat{y_i})^2}$$

where y_i represents the sample value, $\widehat{y_i}$ denotes the predicted value, and w indicates the sample size. According to Eq. 15, lower values of MSE, MAPE, and MAE signify better model fitting, while a higher value of R^2 indicates more robust model performance. In practical application scenarios, a minor error contributes to more accurate operations and enables timely response processing when the data exceeds the rated threshold.

3.2 Datasets

The dataset used in this paper comes from the tanks of an LNG receiving station on the southeast coast, which is a 160,000-cubic-foot full-capacity concrete-roofed tank, and the receiving station currently consists of three tanks. The period used for the train set is from 14:00 on December 6, 2020, to 07:00 on December 10, 2020, with a total of 2304 sets of actual measurement point data. The period used for the test set is from 12:00 on December 18, 2020, to 07:00 on December 19, 2022, with 802 sets of actual measurement point data.

3.3 Data Preprocessing

3.3.1 Outlier Detection

In actual working conditions, the environment in which the sensor is situated is influenced by temperature, humidity, vibration, and other factors. Consequently, it is necessary to remove the outliers from the original data before modeling with the data collected by the sensor. We selected the LOF algorithm to detect the outliers in each original data column. The trend curve of BOG generation after outlier detection is shown in Fig 3.

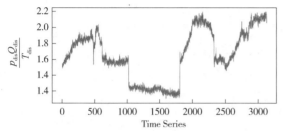

Figure 3 Trend curve of BOG generation after LOF removal of outliers.

3.3.2 Smooth Operation

Due to measurement errors, instrument failures, and other abnormal factors, the fluctuation of the trend curve of BOG generation after processing in Section 3.3.1 remains significant, making it difficult to observe the model fitting and testing effects. To reduce the influence of noise, ensure data continuity and stability, and more clearly represent the data trend while maximizing the retention of the local characteristics of the original data, we utilized the Weighted Moving Average (WMA) data smoothing method to smooth the data in both the train set and the test set every three steps, allowing for more accurate analysis and facilitating more reliable predictions and decisions. The trend curves of BOG generation in the train and test sets after processing are depicted in Fig. 4, where blue represents the smoothed values and green represents the original values.

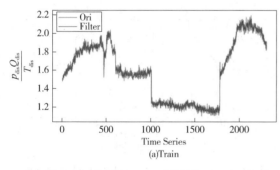
(a)Train

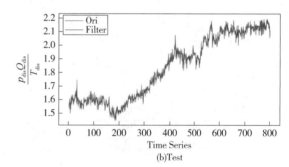
(b)Test

Figure 4 Trend curves of BOG generation in the train set and test set after the data smoothing process

3.4 Experiment Results

3.4.1 Performance analysis of different data distributions in HFF perspective

To explore the influence of different data distributions on the effect of fitting computation, we analyzed the fitting performance of six classical machine learning algorithms: SVR, AdaBoost, KNN, DecisionTree, GBDT, and XGBoost, from the perspectives of global sampling and sequential division. The curves of the predicted versus actual values of the six machine learning algorithms under the global sampling perspective are illustrated in Fig. 5.

It was observed that the SVR model exhibited the poorest fitting effect. Although AdaBoost could effectively capture the overall trend, it also demonstrated weak local fitting performance. The fitting results of KNN contained a greater number of outliers. In contrast, the Decision Tree, GBDT, and XGBoost algorithms achieved superior fitting performance. Additionally, a comparison of the evaluation metrics for the six machine learning algorithms under the global sampling perspective is presented in Table 2.

Table 2 Evaluation metrics of six machine learning algorithms under the global sampling perspective

	MSE	MAPE	MAE	R^2
SVR	0.01303	0.05676	0.08995	0.86535
AdaBoost	0.00119	0.01621	0.02653	0.98772
KNN	0.00081	0.01106	0.01832	0.99161
DecisionTree	0.00038	0.00942	0.01539	0.99602
GBDT	0.00028	0.00808	0.01317	0.99706
XGBoost	0.00028	0.00798	0.01304	0.99707

It was noted that the R^2 values of KNN, Decision Tree, GBDT, and XGBoost all reached 0.99 under the global sampling perspective, with the XGBoost model demonstrating the most superior performance when evaluated by combining the four metrics. The comparison curves of predicted versus actual values under the sequential division perspective are illustrated in Fig. 6.

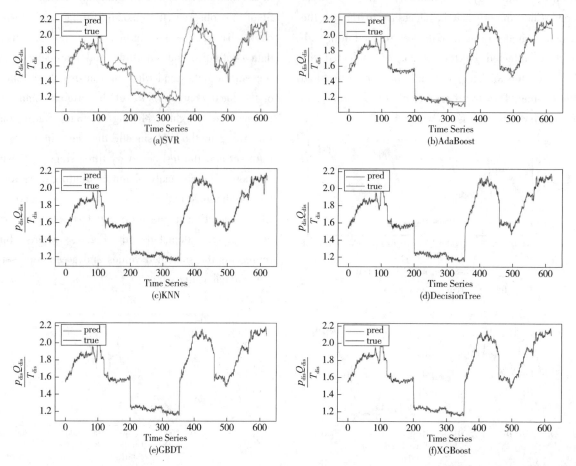

Figure 5 Curves of predicted versus true values for six machine learning algorithms with the global sampling perspective

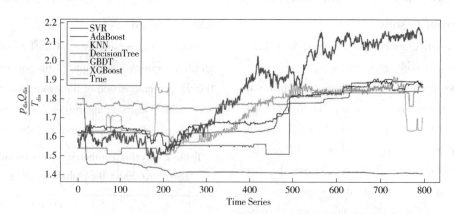

Figure 6 Curves of predicted versus true values for six machine learning algorithms with the sequential division perspective

The red curve in Fig. 6 represented the true value, and it was evident that the remaining six curves failed to capture the trend of the true value. None of the six machine learning algorithms demonstrated improved performance when addressing unknown data distributions. Additionally, the comparison of the evaluation metrics for the six machine learning algorithms in terms of sequential division is presented in Table 3.

Table 3 Evaluation metrics of six machine learning algorithms under the sequential division perspective

Model	MSE	MAPE	MAE	R^2
SVR	4.06964	0.70282	1.28355	-77.4130
AdaBoost	0.03452	0.07857	0.15330	0.33490
KNN	0.04044	0.08446	0.16179	0.22075
DecisionTree	0.07766	0.11459	0.22597	-0.49625
GBDT	0.02284	0.07035	0.13351	0.56001
XGBoost	0.04009	0.09911	0.18279	0.22749

None of the six machine learning algorithms achieved satisfactory results from the sequential division perspective, further validating that these algorithms were unsuitable for addressing scenarios with unknown data distributions.

3.4.2 Performance analysis of multiple models at different time spans for VTSP perspective

To validate the effectiveness of the SDE architecture, we conducted comparative experiments on the prediction performance of four models: LSTM, GRU, Transformer and SDE, across four time spans: 1-step, 2-step, 3-step, and 4-step. The comparison of the four evaluated metrics for these models across the specified time spans is presented in Table 4.

Table 4 Comparison of evaluation metrics of the four models under four time spans

Time Step	Model	MSE	MAPE	MAE	R^2
1	LSTM	0.00025	0.00662	0.01178	0.99466
1	GRU	0.00032	0.00790	0.01398	0.99285
1	Transformer	0.00026	0.00695	0.01239	0.99419
1	SDE	0.00025	0.00674	0.01202	0.99446
2	LSTM	0.00082	0.01257	0.02170	0.97607
2	GRU	0.00133	0.01484	0.02519	0.96141
2	Transformer	0.00057	0.01067	0.01853	0.98354
2	SDE	0.00048	0.00988	0.01715	0.98614
3	LSTM	0.00058	0.01099	0.01899	0.98344
3	GRU	0.00071	0.01213	0.02111	0.97959
3	Transformer	0.00117	0.01601	0.02765	0.96638
3	SDE	0.00054	0.01063	0.01860	0.98455
4	LSTM	0.00191	0.01981	0.03188	0.82828
4	GRU	0.00220	0.02268	0.03631	0.80285
4	Transformer	0.00260	0.02756	0.04505	0.76716
4	SDE	0.00084	0.01405	0.02273	0.92496

In the 1-step angle, SDE compared to LSTM, GRU, and Transformer, MSE decreased by 0%, 21.9%, and 3.8%, MAPE decreased by -1.8%, 14.7%, and 3.0%, MAE decreased by -2.0%, 14.0%, and 3.0%, and R^2 increased by -0.02%, 0.2%, and 0.02%, respectively.

In the 2-step angle, SDE compared to LSTM, GRU, and Transformer, MSE decreased by 41.5%, 63.9%, and 15.8%, MAPE decreased by 21.4%, 33.4%, and 7.4%, MAE decreased by 21.0%, 31.9%, and 7.4%, and R^2 increased by 1.0%, 2.6%, and 0.3%, respectively.

In the 3-step angle, SDE compared to LSTM, GRU, and Transformer, MSE decreased by 6.9%, 23.9%, and 53.8%, MAPE decreased by 3.3%, 12.4%, and 33.6%, MAE decreased by 2.1%, 11.9%, and 32.7%, and R^2 increased by 0.1%, 0.5%, and 1.9%, respectively.

In the 4-step angle, SDE compared to LSTM, GRU, and Transformer, MSE decreased by 56.0%, 61.8%, and 67.7%, MAPE decreased by 29.1%, 38.1%, and 49.0%, MAE decreased by 28.7%, 37.4%, and 49.5%, and R^2 increased by 11.7%, 15.2%, and 20.6%, respectively.

The comparison of the aforementioned evaluation metrics indicates that SDE demonstrates an advantage over the three traditional time-series models across all four prediction spans. Furthermore, when the prediction span is increased to 4, the prediction performance of LSTM, GRU, and Transformer decreases significantly, while the R^2 of SDE remains above 0.9, thereby verifying the effectiveness of SDE. Finally, we presented the results of the SDE on BOG generation for the 1-step, 2-step, 3-step, and 4-step scenarios, as illustrated in Fig. 7:

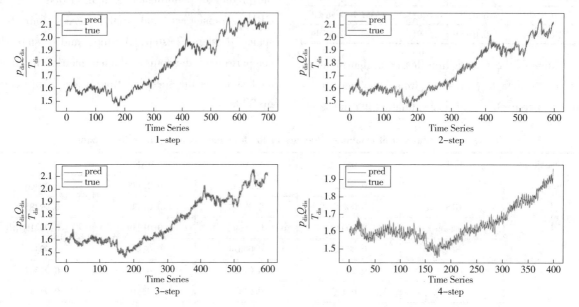

Fig. 7 Predicted versus true values of SDE in 1-step, 2-step, 3-step, and 4-step modes.

3.4.3 Synergistic Approach of HFF and VTSP

Combining the experimental analysis results from sections 3.4.1 and 3.4.2, the HFF perspective remains unaffected by the time span; however, its performance is inadequate in handling new data distributions. The VTSP perspective addresses the limitations of the HFF perspective in managing new data distributions, but its performance degrades when predicting multiple time steps into the future. Therefore, when applied in real industrial scenarios, a synergistic approach can be conducted based on the two perspectives.

In cases where the dataset is substantial and encompasses various working conditions for predicting BOG generation for one step, the R^2 values for both HFF and VTSP prediction results reach as high as 0.99, consequently, the prediction values from both perspectives are reliable, allowing for the selection of either perspective or the average of the two as the final prediction value. When predicting BOG generation over two or three steps, the R^2 for SDE slightly decreases but remains notably high at 0.98, consequently, a combination of HFF and VTSP can be chosen, with a somewhat higher percentage assigned to the HFF perspective, or the HFF perspective can be selected alone. It is advisable to select only the HFF perspective when predictions exceed three steps.

In cases of limited data or uncertainty regarding whether the historical data encompasses a sufficiently large number of working conditions, the HFF perspective should be discarded, and the VTSP perspective should be selected, ensuring that the number of time steps in the prediction does not exceed four.

4 Conclusions

To obtain a computational model of BOG generation with high accuracy, fast computation speed, and strong generalization, we established the Synergistic Approach of Horizontal Feature Fitting and Vertical Time Series Prediction (SA - HFF - VTSP). First, we preprocessed the historical data of the LNG tank system. Second, we analyzed the effect of data distribution on the HFF perspective. Next, we established the SDE architecture and concluded that it was superior to the three traditional time series prediction models. Finally, we analyzed the advantages and disadvantages of the HFF and SDE perspectives and provided suggestions for utilizing their synergistic approaches.

However, in this paper, BOG generation is analyzed solely from a data-driven mode, and the developed model lacks interpretability. Additionally, the computational performance of SDE in the 4-step perspective demonstrates significant degradation. Therefore, future work should focus on improving the interpretability of the data-driven model and exploring methods for calculating BOG generation that are less affected by the time span.

5 Declarations

This work was supported by the National Key Research and Development Program project 'New Algorithms for Optimizing Complex Oil and Gas Pipeline Networks through the Integration of Machine Learning and Mixed Integer Programming' (Project No. 80915861).

Reference

[1] Bouabidi Z, Almomani F, Al-musleh E I, et al. Study on boil-off gas(BOG) minimization and recovery strategies from actual baseload LNG export terminal: towards sustainable LNG chains[J]. Energies, 2021, 14(12): 3478.

[2] Shin M W, Shin D, Choi S H, et al. Optimization of the operation of boil-off gas compressors at a liquified natural gas gasification plant[J]. Industrial & Engineering Chemistry Research, 2007, 46(20): 6540-6545.

[3] Lee D H, Cha S J, Kim J D, et al. Practical prediction of the boil-off rate of independent-type storage tanks[J]. Journal of Marine Science and Engineering, 2021, 9(1): 36.

[4] Jeon G M, Park J C, Choi S. Multiphase-thermal simulation on BOG/BOR estimation due to phase change in cryogenic liquid storage tanks[J]. Applied Thermal Engineering, 2021, 184: 116264.

[5] Saleem A, Farooq S, Karimi I A, et al. A CFD simulation study of boiling mechanism and BOG generation in a full-scale LNG storage tank[J]. Computers & Chemical Engineering, 2018, 115: 112-120.

[6] Khan M S, Qyyum M A, Ali W, et al. Energy saving through efficient BOG prediction and impact of static boil-off-rate in full containment-type LNG storage tank[J]. Energies, 2020, 13(21): 5578.

[7] Singh S P, Srinivasan R, Karimi I A. Data-driven modeling to predict the rate of Boil-off Gas(BOG) generation in an industrial LNG storage tank[M]//Computer Aided Chemical Engineering. Elsevier, 2023, 52: 1293-1299.

[8] Xu D, Wang Y, Peng P, et al. Real-time road traffic state prediction based on kernel-KNN[J]. Transportmetrica A: Transport Science, 2020, 16(1): 104-118.

[9] Liang W, Luo S, Zhao G, et al. Predicting hard rock pillar stability using GBDT, XGBoost, and LightGBM algorithms[J]. Mathematics, 2020, 8(5): 765.

[10] Fan L, Ji Y, Wu G. Research on temperature prediction model in greenhouse based on improved SVR[C]//Journal of Physics: Conference Series. IOP Publishing, 2021, 1802(4): 042001.

[11] Elmachtoub A N, Liang J C N, McNellis R. Decision trees for decision-making under the predict-then-optimize framework[C]//International conference on machine learning. PMLR, 2020: 2858-2867.

[12] Cavalli S, Amoretti M. CNN-based multivariate data analysis for bitcoin trend prediction[J]. Applied Soft Computing, 2021, 101: 107065.

[13] LI S, MA L, PAN S, et al. Research on prediction model of gas concentration based on RNN in coal mining face[J]. Coal Science and Technology, 2020, 48(1).

[14] Chen Y. Voltages prediction algorithm based on LSTM recurrent neural network[J]. Optik, 2020, 220: 164869.

[15] Han P, Wang W, Shi Q, et al. A combined online-learning model with K-means clustering and GRU neural networks for trajectory prediction[J]. Ad Hoc Networks, 2021, 117: 102476.

[16] Chandra A, Tünnermann L, Löfstedt T, et al. Transformer-based deep learning for predicting protein properties in the life sciences[J]. Elife, 2023, 12: e82819.

[17] Breunig M M, Kriegel H P, Ng R T, et al. LOF: identifying density-based local outliers[C]//Proceedings of the 2000 ACM SIGMOD international conference on Management of data. 2000: 93-104.

[18] Chang C C, Lin C J. LIBSVM: a library for support vector machines[J]. ACM transactions on intelligent systems and technology(TIST), 2011, 2(3): 1-27.

[19] Freund Y, Schapire R E. A desicion-theoretic generalization of on-line learning and an application to boosting[C]//European conference on computational learning theory. Berlin, Heidelberg: Springer Berlin Heidelberg, 1995: 23-37.

[20] Peterson L E. K-nearest neighbor[J]. Scholarpedia, 2009, 4(2): 1883.

[21] Breiman L. Random forests[J]. Machine learning, 2001, 45: 5-32.

[22] Friedman J H. Greedy function approximation: a gradient boosting machine[J]. Annals of statistics, 2001: 1189-1232.

[23] Chen T, Guestrin C. Xgboost: A scalable tree boosting system[C]//Proceedings of the 22nd acm sigkdd international conference on knowledge discovery and data mining. 2016: 785-794.

[24] Hochreiter S, Schmidhuber J. Long short-term memory[J]. Neural computation, 1997, 9(8): 1735-1780.

[25] Cho K, Van Merriënboer B, Gulcehre C, et al. Learning phrase representations using RNN encoder-decoder for statistical machine translation[J]. arxiv preprint arxiv: 1406.1078, 2014.

[26] Vaswani A, Shazeer N, Parmar N, et al. Attention is all you need[J]. Advances in neural information processing systems, 2017, 30.

基于 WebGIS 的天然气站场甲烷泄漏监测可视化平台实现

詹佳琪 刘 标 洪 啸 宋 成 易 欣 刘 翔

(国家管网集团浙江省天然气管网有限公司)

摘 要 激光甲烷监测设备因其高可靠性、低响应时间和覆盖范围广等特性在天然气站场泄漏监测中得到一定规模推广应用并取得阶段性成效。但对甲烷泄漏监测设备的操作和管理主要通过安装专业化软件实现,仅提供单一局部视角观察,无法直观感受设备的运动状态和站场监测情况全貌,对其大范围推广应用还存在一些困难。本文基于 WebGIS 技术设计并实现了一套天然气站场甲烷泄漏监测平台,能够将天然气站场三维可视化,直观展现甲烷泄漏监测设备实时运转状态,且只需浏览器即可支持平台登录并进行设备的操作管理,降低了部署使用门槛,为大范围推广使用甲烷泄漏监测设备提供了应用案例。

关键词 WebGIS;Cesium;甲烷泄漏监测;可视化平台

Implementation of a Visualization Platform for Monitoring Methane Leakage in Natural Gas Stations Based on WebGIS

Zhan Jiaqi Liu Biao Hong Xiao Song Cheng Yi Xin Liu Xiang

(PipeChina Zhejiang Pipeline Network Co., LTD.)

Abstract Laser methane monitoring equipment has been widely promoted and applied in natural gas station leakage monitoring due to its high reliability, low response time, and wide coverage, and has achieved phased results. However, the operation and management of methane leak monitoring equipment are mainly achieved through the installation of specialized software, which only provides a single local perspective observation and cannot intuitively feel the overall movement status of the equipment and the monitoring situation of the station. There are still some difficulties in promoting its large-scale application. This article designs and implements a natural gas station methane leakage monitoring platform based on WebGIS technology, which can visualize the natural gas station in 3D, intuitively display the real-time operation status of methane leakage monitoring equipment, and support platform login and device operation management with only a browser, reducing the deployment and use threshold and providing application cases for the widespread promotion and use of methane leakage monitoring equipment.

Key words WebGIS; Cesium; Methane leak monitoring; Visualization platform

天然气作为清洁能源,具有高效、经济、环保的特点,使其需求量快速增长,推动着中国天然气产业的发展。随着管道行业的快速建设和发展,日益增长的管道里程和天然气站场数量给天然气安全生产运维带来了一定考验。天然气站场仪表设备繁多受到多种因素的影响,存在可燃气体泄漏的风险,需配备灵敏可靠的监测设备,建立高效、全面、可靠的天然气泄漏监测系统用以及时、有效、精准地发现泄漏险情。

激光甲烷泄漏监测设备因其灵敏度高、抗感染性强和使用方便等特性,逐渐替代传统的催化燃烧式、点式红外吸收型可燃气体探测器。目前为了实现设备的长期运行监测和操作,通常配备专业化软件实现,这种方式需要每台设备都要进行一次部署调试,且产品功能较为单一、不易扩展、设备运行监控不直观,不利于大量推广。本文基于WebGIS技术设计并实现一套天然气站场甲烷泄漏监测平台,利用WebGIS的三维渲染和B/S架构能很好地解决以上问题,为天然气站场激光甲烷泄漏监测推广使用提供应用案例。

1 技术及原理

1.1 WebGIS技术

随着互联网技术与WebGIS(Web Geographic Information System,地理信息系统)技术的成熟与发展,GIS(Geographic Information System,地理信息系统)的适用范围逐渐变广。WebGIS技术是现代地理信息系统(GIS)与Web技术相结合的产物,它利用Web技术将地理信息数据进行整合、处理、存储,并通过网络向用户提供地理信息查询、检索、可视化、空间分析等功能。用户只需通过浏览器,即可轻松访问WebGIS平台,获取所需的地理信息。

WebGIS具有全球化、大众化、可扩展性和跨平台的特性。WebGIS使得全球范围内的用户都可以访问到GIS服务。借助互联网的发展和便捷,WebGIS技术提供给更多用户使用GIS的机会,并降低了门槛。用户可以使用通用浏览器进行浏览、查询,降低了终端用户的经济和技术负担,也在一定程度上扩大了GIS的潜在用户范围。WebGIS很容易跟Web中的其他信息服务进行无缝集成,可以建立灵活多变的GIS应用。WebGIS对任何计算机和操作系统都没有限制,只要能访问Internet,用户就可以访问和使用WebGIS。

WebGIS技术发展体现在以下几个方面:(1)从封闭的Web GIS网站到基于Web服务的架构,基于Web服务的WebGIS架构使得服务器可以发布Web服务,客户端可以调用和展现这些服务,从而实现了信息和功能的共享;(2)自上而下的信息流向转为信息的双向流动,基于众包模式的地理信息客户端与服务器之间可以实现信息的双向流动,使得地理信息更加实时和准确;(3)从本地到云端,云GIS解决了本地海量数据处理和频繁软件更新的问题,使得WebGIS更加高效和易用;(4)从有线到无线,随着智能手机和平板电脑等移动设备的普及,WebGIS也越来越趋向于移动化,用户可以随时随地访问和使用WebGIS平台;(5)从二维到三维和虚拟现实,随着计算机图形学和虚拟现实技术的发展,WebGIS在地图和数据的可视化方面也越来越注重提供逼真的三维展示效果。

1.2 激光甲烷原理

激光甲烷监测的原理是基于激光吸收光谱技术,一般利用半导体激光器的波长调谐特性,通过测量甲烷分子对特定波长激光的吸收强度来推算甲烷气体的浓度。这种可调谐半导体激光吸收光谱法简称TDLAS技术,是一种高灵敏度、高选择性的气体监测技术,基于朗伯-比尔定律(Lambert-Beer law)实现气体浓度的测量的方法。

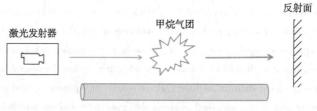

图1 激光甲烷监测原理

具体工作原理如图所示:(1)激光发射:激光甲烷监测仪通过激光发射器发射特定波长的激光束;(2)气体吸收:激光束穿过待测气体时,甲烷分子会吸收特定波长的激光,导致光强减弱;(3)光强测量:接收器接收经过气体吸收后的激光束,并将其转换为电信号;(4)信号处理:信号处理电路对电信号进行处理,通过比较吸收前后的光强差异,计算出甲烷气体的浓度。TDLAS技术具有以下几点优势:(1)高灵敏度:能够监测极低浓度的甲烷气体;(2)高选择性:通过选择特定的吸收谱线,避免其他气体的干扰;(3)实时监测:适合动态环境下的连续监测;(4)环境适应性强:能够在恶劣环境下稳定工作,如煤矿、石油化工等。由于TDLAS技术特点,激光甲烷监测设备已应用于煤矿安全和石油化工等领域中,监测煤矿井下的甲烷浓度以及天然气管道的泄漏情况,确保安全生产监测,保障燃气系统的安全运行。

2 平台设计

各天然气站场内的激光甲烷监测设备通过光纤与服务器连接，实现视频和通信讯号的传输，服务器承载平台软件的部署和站点计算机的访问请求，并将视频存储于指定位置。新接入的甲烷监测设备只需在平台中进行相应的信息配置即可实现入网，对应设备的告警信息和视频内容在设备上线后即可通过平台进行查看，通过调度中心的计算机及各门站计算机均可访问本平台，进行告警处理和设备管理等操作，还能进行设备交互，实现指令下单和自检等功能，如图2。

平台从实现角度可以分为以下 5 个层级：用户层、访问终端层、表现层、服务层和资源层。用户层主要是平台使用者，具体可分为管理员和普通用户。访问终端层是访问平台所支持的硬件设备，这里是 PC 端浏览器；表现层主要是前端 html、css 和 js 开发语言，Vue 前端框架和 Cesium 三维地球 js 库，实现用户可以看到的平台管理的界面、交互面板和三维窗口等内容；服务层包含服务通信、业务服务和后端服务部分，实现数据的通讯、三维模型数据调取、各业务所需数据接口提供和视频流推送等功能；资源层包括数据和三维模型数据存储等。

3 平台功能实现

基于以上架构并采用 Cesium 地图引擎作为可视化工具，开发了一套天然气站场激光甲烷泄漏监测平台，在具备了站场设备、告警信息、人员等信息管理的同时，实现了站场三维可视化，并增加了交互功能。平台能够实时直观地展现了激光甲烷云台运转状态，并进行交互式操作，为天然气站场提供了高效可靠的站场泄漏监测手段，提高站场安全运维水平。

3.1 三维可视化

前端进行三维可视化的步骤一般为：三维建模、模型数据处理和发布、三维可视化开发。首先将天然气站场进行三维建模，建模对象包括管道、仪表、阀门、其他设备、建筑以及环境。天然气站场常用的建模方式有倾斜摄影建模、激光扫描点云建模和人工建模，倾斜摄影三维建模可以快速获取实景三维模型，模型真实、建模速度快适合大场景室外建模，但对阀门、仪表等细节还原度低，且不能进行单体化是其弊端；激光扫描点云建模也具备高效快速的特点，且精度更高，但点云模型为离散点，也需要单体化或人工干预；人工建模是在数据处理初期就完成了设备单体化的目标，且模型数据在此阶段可以进行数据信息标记，更好地适配三维可视化的交互功能，但人工建模的成本较高，不适合大范围推广。本文采用激光扫描点云建模+人工建模的方式，通过激光扫描快速获取场景点云数据，扫描阶段完成了环境信息和贴图的获取，将大大提高人工建模效率，保证精度的同时降低建模成本。

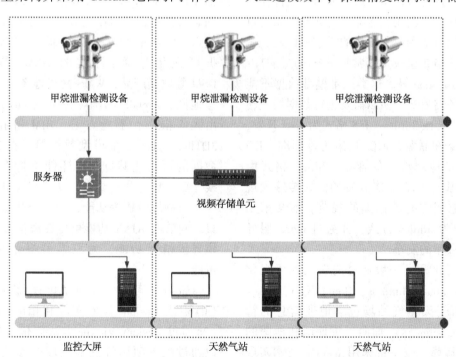

图 2　平台物理架构

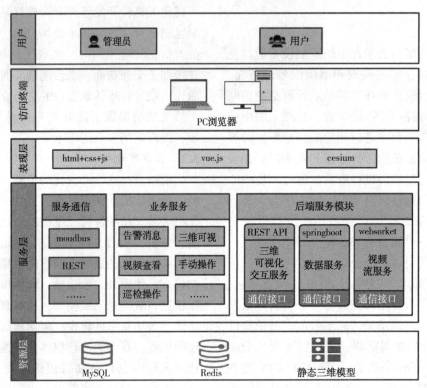

图 3 平台功能架构

表 1 三维建模常见方式优缺点比较

建模方式	优 点	缺 点
倾斜摄影三维建模	模型真实、速度快、自动化程度高	不能单体化、数据量大、不便于复用
激光点云扫描三维建模	模型精度高、速度快	需要一定的人工干预、不能直接进行单体化、不能复用
人工三维建模	模型精度高、可以复用	速度较慢、成本较高

模型数据处理和发布，是将数据进行格式转换后的模型数据通过数据库或者分布式文件存储系统进行保存，再利用服务器进行数据发布的过程。由于建模方式和业务场景的需求多样性，三维模型数据的格式也较多，常见的有 obj、glTF/glb、fbx、max、dae 等。这类三维模型数据不能直接通过服务器发布，所以需要进行数据格式转换，一般为 I3S 或 3dtiles。I3S 格式为 Esri 开发，由开放地理空间联盟（OGC）指定为该标准，I3S 用于任意大量地理数据的容器。3dtiles 格式由 Cesium 团队基于 glTF 二次开发的流式传输大规模异构 3D 地理空间数据集的规范。本文采用 fbx 格式转换为 3dtiles 方式，并通过 nginx 服务器进行三维数据的发布。

三维可视化开发一般通过前端三维可视化库，例如 Cesium、Three.js、Blend4Web、Babylon.js 等。由于天然气站场包含地理信息的特性，故采用 Cesium 这一面向三维地球和地图的 JavaScript 开源库。Cesium 采用 WebGL 技术实现硬件 3D 加速渲染，无需安装插件即可通过浏览器流畅展示 3D 场景和模型。Cesium 还具备数据动态可视化、图形在线渲染和跨平台跨浏览器的优点。Cesium 所支持的二维、三维地理信息数据可视化包括：标准的 DEM、DOM、DLG、DSM 等数据产品；WMS 网络地图服务、WFS 矢量地图、ArcGIS 等国际标准的 OGC 地图服务；由倾斜摄影、激光点云、BIM 等技术获取的 3DTiles、glTF 三维模型数据等。在支持二三维数据的可视化基础上，还能支持在线绘制点、线、面、体等几何要素，支持在线空间量测、路径、可视域分析等功能，为三维开发提供便捷工具。封装有 AJAX 功能可以在海量数据加载时提供异步请求功能，为三维可视化交互创造了可能。

Cesium 实现三维模型可视化的流程如下：（1）引入 Cesium js 库文件和相应的样式文件；（2）通过 Cesium.Viewer 方法创建三维地球，可以通过配置项调整三维地球的参数；（3）通过

viewer. scene. primitives. add 方法将三维模型添加到三维地球场景中；(4)对三维模型进行空间位置、显示效果调整。加载三维模型的方法如下：

//加载 3D Tiles 数据

var tileset = viewer. scene. primitives. add (new Cesium. Cesium^{3}DTileset({

url：'http：//localhost：9003/model/t0vT2M90i/tileset. json', //数据路径

}));

图 4 Cesium 实现三维模型加载

3.2 三维交互

三维交互是将用户在屏幕中的点击、绘制、指令下达等操作在场景中实现响应并进行动态三维可视化。由于用户操作是在屏幕中完成的所以前端直接获取到的是屏幕坐标系下的二维屏幕坐标，屏幕坐标系是以屏幕的左上角为原点，向右为 x 轴正方向，向下为 y 轴正方向。屏幕坐标系主要用于屏幕上的点定位，例如鼠标点击的位置、窗口的位置等。Cesium 是一款基于 GIS 系统的三维可视化引擎，三维可视化接口和方法主要以地心坐标系坐标作为输入，而实现激光甲烷云台三维交互是以云台为中心的运动计算方式，属于站心坐标系。所以要解决坐标转换问题。

用户操作指令一般有两种类型：绘制命令和按钮命令。绘制命令的流程是用户操作指令——屏幕坐标系坐标——地心坐标系坐标——站心坐标系坐标——运动计算——站心坐标系坐标——地心坐标系坐标——三维可视化；按钮命令的流程则是用户操作指令——实时数据请求——运动计算——站心坐标系坐标——地心坐标系坐标——三维可视化。通过坐标转换接口，交互过程即转换为以云台为中心的三维直角坐标空间运动计算，常用的三角函数和几何计算方法即可直接应用。基于此本文实现了云台按指定俯仰角或水平角运动、网格显示、告警点显示、告警热力图以及区域绘制等交互功能。再利用通信接口实现三维交互到硬件设备指令下达。

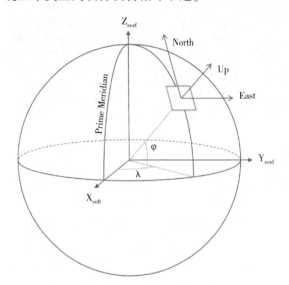

图 5 地心坐标系和站心坐标系的关系

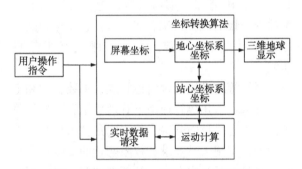

图 6 用户操作到三维可视化实现流程

站心坐标系(英文名称是 local Cartesian coordinates coordinate system)也叫作站点坐标系或东北坐标系 ENU，是以测站为原点的坐标系，主要是用于了解以观察者为中心的其他物体运动规律。地心地固坐标系(Earth-Centered, Earth-Fixed，简称 ECEF)简称地心坐标系，是一种以地心为原点的地固坐标系(也称地球坐标系)，是一种笛卡儿坐标系。同时为支持浏览器交互，还涉及为了便于云台运动计算并实现基于 cesium 引擎的可视化，需要开发坐标转换接口，解决坐标转换问题。

3.3 数据管理

该部分主要是针对人员登录信息、站点信息、设备信息和告警信息的管理。通过 vue + springboot 前后端分离技术将管理平台对数据的增删改查操作与数据库存储的数据进行联动。实现了告警管理、站点管理和系统管理三个模块。

告警管理模块包含设备告警和设备实时数据

图7 站心坐标系展示

两个功能：设备告警根据告警浓度分级策略将告警等级进行划分并差异化显示，同时根据该条告警是否处理将未处理告警显示在显著位置。

站点管理模块包含站点列表和设备管理两个功能：站点列表用于配置当前站点三维模型参数，包括三维模型存储路径及名称、三维模型地理坐标等；设备管理则用于配置云台摄像头参数，包括 IP 地址、端口、协议等，通过以上两个功能面板的配置，就能实现云台、站点和站点三维模型的三者绑定，保证三维交互中信息匹配。

系统管理模块包含角色管理、参数管理和定时任务管理：角色管理用于管理登录者账号密码等信息，通过权限分层，将不同角色用户配置不同功能，实现人员登录和权限配置的功能；参数管理则是将系统内允许进行调整的变量提供调整的功能窗口，例如实时数据获取的间隔时间、网格长度等；定时任务管理是用来配置设备定时检查功能的参数。

4 结论

针对目前天然气站场激光甲烷泄漏监测系统的不足设计并实现了一套基于 WebGIS 的天然气站场甲烷泄漏监测可视化平台，平台由用户层、访问终端层、表现层、服务层和资源层组成。相较于传统的天然气泄漏激光监测系统，从功能层面上：本平台在满足视频实时查看和指令下达功能的同时增加了三维可视化和人机交互的功能，使激光甲烷泄漏监测设备的使用更加灵活；从安装部署层面上：传统的天然气泄漏激光监测系统需对每台设备进行安装调试，必要时可能会安装相关插件和库文件，而本平台仅需浏览器即可直接运行较为方便；从维护升级层面上：传统的天然气泄漏激光监测系统为客户端程序，更新迭代需要重新部署安装并进行适配开发工作，而本平台采用 B/S 架构，只需更新服务端即可；从应用层面上：传统的天然气泄漏激光监测系统扫描周期较长，而本平台可以绘制重点区域进行扫描，更具有针对性。同时三维可视化功能可以更为直观地查看设备运行状态、告警发生点、当前扫描范围，能够及时调整设备、核查泄漏风险区。

虽然本平台相较于传统的天然气泄漏激光监测系统有诸多优点，但也存在不足：（1）平台的三维可视化部分依赖于三维模型的支持，虽然三维建模技术已较为成熟，但三维模型的建模和数据处理过程增加了平台开发成本，在实际推广中也是需要考虑的客观因素；（2）随着运行数据和功能的增多，可以结合人工智能和大数据等技术进行升级改造，例如利用人工智能技术开发智能决策模块，结合大数据技术进行数据挖掘和数据分析进而达到风险预测，使系统更加智能化，使用更加便捷。

参 考 文 献

[1] 马新杰，江星燕，魏文栋．我国天然气产业现状及未来展望[J]．生产力研究，2018，（04）：65-69+143+161．

[2] 何晓鹏，薛攀．天然气集输场站安全现状和管理对策分析[J]．山东工业技术，2017(4)：60．

[3] 严密，袁晓骏，管文涌．露天站场天然气泄漏激光监测系统的设计及应用[J]．油气储运，2021，40(6)：7．

[4] 刘欢，胡畔宁，魏莱．气云成像摄像机气体泄漏监测技术研究及应用[J]．天然气技术与经济，2019，13(1)：5．

[5] 孙旭．超声波气体泄漏检测仪在天然气分输站场的应用[J]．化工设计通讯，2018，44(11)：188．

[6] 刘广潆．激光式可燃气体探测器在天然气站场的应用[J]．石油化工自动化，2020，56(01)：83-85．

[7] 蔡亮．中美可燃气体探测报警系统应用标准差异研究[J]．天然气与石油，2016，34（06）：99-103+147．

[8] 时国明，周智，苏晔华，等．天然气站场扫描式激光气体监测系统的研制[J]．油气储运，2021．40(05)：554-560．

[9] 阚瑞峰，刘文清，张玉钧，等．基于可调谐激光吸收光谱的大气甲烷监测仪[J]．光学学报，2006(1)：67-70．

[10] 唐伟，龚建华，王桥，等．云台激光气体泄漏探测系统在天然气站场的应用[J]．石油与天然气化工，

2023, 52(1): 110-115.

[11] 马吉, 徐景德, 王理翔, 等. TDLAS 技术在天然气微量泄漏检测过程中的应用[J]. 华北科技学院学报, 2017, 14(3): 75-78.

[12] 杨翔宇. 基于 WebGL 的三维地形可视域分析算法研究[D]. 武汉: 华中师范大学, 2019.

[13] 朱丽萍, 李洪奇, 杜萌萌, 等. 基于 WebGL 的三维 WebGIS 场景实现[J]. 计算机工程与设计, 2014, 35(10): 3645-3650.

[14] 朱栩逸, 苗放. 基于 Cesium 的三维 WebGIS 研究及开发[J]. 科技创新导报, 2015, 34(009): 9-11+16.

[15] 张志荣, 孙鹏帅, 庞涛, 等. 激光吸收光谱技术在工业生产过程及安全预警标识性气体监测中的应用[J]. 光学精密工程, 2018, 26(08): 1925-1937.

[16] 聂伟, 阚瑞峰, 杨晨光, 等. 可调谐二极管激光吸收光谱技术的应用研究进展[J]. 中国激光, 2018, 45(09): 9-29.

[17] 朱荷欢, 武文, 孙玉婷. 三维建模不同技术方法的特点研究及应用思考[C]. 南京: 南京市国土资源信息中心 30 周年学术交流会, 2020: 28-31.

[18] 王逸凯, 徐萌, 罗建松, 等. 基于 Cesium 的 WebGIS 倾斜三维平台的实现[J]. 测绘与空间地理信息, 2019, 42(04): 88-89.

[19] 刘振华, 何望君, 张福浩, 等. CALPUFF 模型与 WebGIS 集成开发应急地理信息系统[J]. 地理信息世界, 2021, 28(05): 79-83+123.

[20] 吴昌政. 基于前后端分离技术的 web 开发框架设计[D]. 南京: 南京邮电大学, 2020.

水下气液分离与增压技术及装备研究进展

罗小明　李蔚迪　许欣怡

[中国石油大学(华东)储运与建筑工程学院]

摘　要　【目的】针对水下油气田储层低压、井口高背压等复杂开采条件,以及水合物易于堵塞管道、油品高黏性等流体特性导致的开采效率低、能耗高等难题,业界正积极探索水下环境中气液分离与增压输送技术的高效方案。此举旨在借助技术创新,克服开采过程中的技术瓶颈,有效提升油气采收率,同时降低生产成本、提高开采效率,为海洋油气资源的可持续开发利用提供坚实的技术支撑。【方法】聚焦于水下气液分离与增压技术及装备的最新研究成果,深度剖析七个已投产的国外先进水下工程项目及三个处于概念设计阶段的项目案例,全面揭示了这些工程的设备结构特征、关键技术参数以及在实际应用中面临的挑战和应用优势,为后续的技术研发与应用实践提供了宝贵的经验与启示。【结果】这些先进的水下系统通过优化分离流程、高效运用增压技术等创新手段,不仅显著提升了油气开采效率、降低了能耗,还增强了作业过程的安全性,为海洋油气资源的可持续开发奠定了坚实的基础。【结论】展望未来,随着深海油气资源的持续开发,水下分离器将朝着更大长径比、更高耐压等级的方向发展,以适应深海高外压、低温等极端作业条件。因此,必须不断推进相关技术的创新与优化,以满足日益严格的深海开采标准与市场需求,为海洋油气资源的可持续开发贡献更多力量。

关键词　水下分离；水下增压输送；流动保障；提高采收率；大长径比

Research Progress on Technologies and Equipment for Underwater Gas-Liquid Separation and Pressurization

Luo Xiaoming　Li Weidi　Xu Xinyi

[College of Pipeline and Civil Engineering, China University of Petroleum(East China)]

Abstract　[Objective] In response to the complex extraction conditions such as low reservoir pressure in underwater oil and gas fields and high back pressure at the wellhead, as well as challenges like pipeline blockages caused by hydrates and the high viscosity of oil products leading to low extraction efficiency and high energy consumption, the industry is actively exploring efficient solutions for gas-liquid separation and pressurized transportation technology in underwater environments. This initiative aims to leverage technological innovation to overcome technical bottlenecks during the extraction process, effectively enhance oil and gas recovery rates, reduce production costs, and improve extraction efficiency, thereby providing solid technical support for the sustainable development and utilization of marine oil and gas resources. [Methods] Focusing on the latest research achievements in underwater gas-liquid separation and pressurization technology and equipment, this analysis deeply examines seven operational advanced underwater engineering projects abroad, as well as three projects currently in the conceptual design stage. It comprehensively reveals the equipment structural characteristics, key technical parameters, and the challenges and advantages faced in practical applications of these projects. This provides valuable experience and insights for subsequent technological research and development, as well as practical application. [Results] These advanced underwater systems, through innovative approaches such as optimizing the separation process and efficiently utilizing pressurization technology, not only significantly enhance oil and gas extraction efficiency and reduce energy consumption but also improve the safety of operational processes. This lays a solid foundation for the sustainable development of marine oil and gas resources.

[Conclusion] Looking ahead, with the ongoing development of deep-sea oil and gas resources, underwater separators will evolve towards larger length-to-diameter ratios and higher pressure ratings to withstand the extreme operating conditions of deep-sea environments, such as high external pressure and low temperatures. Therefore, continuous innovation and optimization of related technologies must be pursued to meet the increasingly stringent standards for deep-sea extraction and market demands, contributing further to the sustainable development of marine oil and gas resources.

Key words underwater separation; underwater pressurized transportation; flow assurance; improve recovery rate; large aspect ratio

在全球能源结构转型及深海油气资源开发日益受到重视的背景下，水下生产系统作为深水油气田开发的重要装备，其技术水平的提升不仅关乎能源供应的安全与稳定，更是实现能源结构优化和绿色可持续发展的关键。水下生产系统由水下井口、水下采油树、水下管汇和水下控制系统等核心组件构成，通过高度集成的作业流程，显著降低了深海油气开发成本，缩短了项目建造周期，并展现出卓越的抗自然灾害能力，已成为深水油气开采领域不可逆转的发展趋势。

水下井口采出液经水下采油树初步调控后，通过水下管汇汇集并在复杂的管道与立管网络中输送至海上浮式生产平台（FPSO）进行后续处理。然而，直接输送未经处理的气液混合物不仅降低了运输效率，增加了能耗，还可能导致管道腐蚀，影响系统整体寿命。因此，通过在海底直接对气液混合物进行处理，并对分离后的液相和气相分别进行增压输送，可以显著提升油气开采的效率，同时有效降低开采的成本。

水下分离与增压的目标在于降低井口背压并进一步提高油气采收率。分离后的液相通过增压泵增压，确保长距离输送过程中的稳定性与效率；而气相则利用压缩机增压，以满足后续处理或输送的压力需求。这一技术策略不仅优化了深海油气资源的开采流程，还显著提升了整个水下生产系统的运行效率与经济效益。

近年来，随着深海油气勘探与开发的深入推进，水下生产系统朝着更高程度的集成化、智能化和高效化方向发展。水下分离与增压技术作为连接深海油气井与海底管线的核心环节，其性能优化与技术创新直接关系到整个水下生产系统的效能。因此，深入研究水下气液分离与增压技术的最新进展，分析当前技术面临的瓶颈与挑战，展望未来的发展方向与应用前景，对于推动我国深海油气开采技术的进步及提升水下生产系统整体效能具有深远的理论与实践意义。

本文旨在全面综述水下气液分离与增压技术及装备的最新研究成果，剖析技术发展的难点与机遇，为我国深海油气资源的高效、绿色开发提供坚实的技术支持与科学决策参考。

1 水下气液分离与增压技术基本原理

水下气液两相分离技术的关键在于利用气液两相的物性差异，包括密度、惯性和表面张力等，通过采用变速、变向或旋涡形成等手段调控气液混合物在分离设备内的流动状态，促使气液两相产生相对运动，从而实现有效分离。

水下气液分离的基础方法包括重力分离与离心分离。重力分离基于液体和气体的密度差异，通过静置使两相自然分层。这一方法适用于低流量和较大颗粒的分离场景，具有操作简单、成本低廉的优势。然而，在复杂的气液混合物或高流量情况下，重力分离的效率受到限制。

离心分离的原理在于利用旋转产生的离心力，结合物质间沉降系数或密度差异，使气液混合物中的各相沿不同半径方向迅速分离。该方法特别适用于处理含有细小颗粒、气泡或高流量、高黏度的复杂气液混合物，能够适应不同流量和黏度的流体，适用性广。但对于极细微的颗粒或气泡，可能无法完全分离；此外，离心分离设备的初期投资成本相对较高，且需要定期维护和保养。

增压技术在水下气液分离后各相的管输过程中起着至关重要的作用，其基本原理是利用外部动力源提供能量，便于将分离后的气/液相高效输送至平台，方便后续油气处理流程。电潜泵和压缩机是常用的增压设备，电潜泵通过电动机驱动叶轮旋转以产生压力差输送液体，而压缩机则通过压缩气体体积提升其压力。这些设备能够精确控制增压大小和流量，以适应不同的开采条件和分离需求。增压技术有助于克服管道输送过程中的阻力，提升了输送效率，同时增强了系统的稳定性与可靠性。

综上所述，水下气液分离与增压技术的基本

原理涵盖重力分离、离心分离及增压设备的运用,这些技术共同构成了水下生产系统不可或缺的一部分,为实现高效、安全的深海油气开采提供了强有力的支持。随着技术的不断进步与创新,水下气液分离和增压技术将在未来深海油气资源开发中发挥更加重要的作用。

2 水下气液分离与增压技术及装备现状

2.1 Marimbá油田VASPS系统

2001年,Petrobras公司在巴西里约热内卢Campos盆地Marimbá油田部署了垂直环空分离与泵送系统(VASPS),完成水下安装与初步测试后,因电潜泵机械故障,直至2004年才正式投产。由于该油田储层压力偏低,Petrobras公司在距离P-8平台一公里处(作业水深395米)部署了VASPS系统,该系统实现了高效的气液分离作业,其中气体通过自然举升方式上升,而液体则借助电潜泵进行增压,随后输送至P-8平台。

VASPS垂直环空分离与泵送系统由一系列关键组件构成,包括临时导向基座(TGB)、流动基座(Flowbase)、顶部构件、分离器和电潜泵(图1)。该系统采用柱状螺旋通道旋流分离技术与沉箱设计,沉箱直径为0.762米,高达72米。分离器由三个同心套管组成:外层承压套管,由六节组成,总长72米,每节12米,直径0.66米;中间层为螺旋导叶管,由六节构成,直径0.324米,其外壁焊接有螺旋导叶,与承压套管内壁紧密接触,形成螺旋通道;内层为液体排出管,直径在0.203~0.254米之间,与螺旋导叶管之间形成气体环空通道。

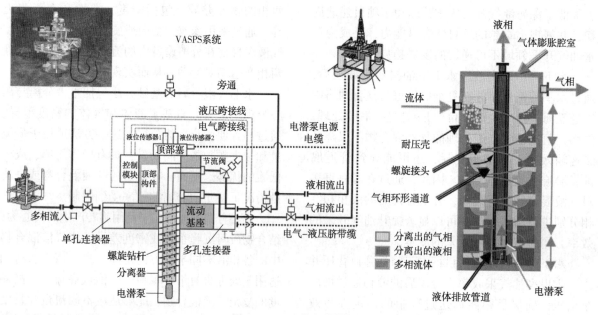

图1 VASPS分离器结构

分离器的液相流量设定为1500m³/d,气相流量为190000Sm³/d,设计压力达到20.68MPa,分离操作压力维持在0.8~1.2MPa。分离过程的温度范围为40~70℃,而海底温度为5℃。原油的物理特性为29° API,38℃时黏度为14.3mPa·s,60℃时降至7.6mPa·s,年含砂量控制在1m³。电潜泵配置参数为流量1900m³/d,增压能力为7MPa,达到平台时压力为0.7MPa,供电条件为1375V/60Hz,额定功率为150kW,泵的转速由变频器精确调控。

VASPS系统分离机制阐述如下:气液两相流体经承压套管上部导入外层与中间套管间螺旋通道,利用离心力高效分离气体。初步分离后,液体流至底部脱气,由电潜泵增压通过内层套管举升至平台。同时,气体经螺旋管壁孔道流入气体环空通道,上升至膨胀腔,最终由气相出口排出并举升。

分离器液位由承压套管顶部的双雷达液位计(主备冗余)监测,确保数据准确可靠。通过调节电潜泵转速和节流阀开度,液位控制在最优范围内±1.5米。控制策略需满足最优分离效率、避免电潜泵干运行及防止液相窜入气相通道,以保持系统稳定。由于VSAPS系统具备高效气液分离性能,能够将液相含气率降至最低,因此可配备单相泵及计量系统,简化流程并提升整体效率。

VASPS系统有效实现了海底气液分离和液相增压,降低井口背压,从而提高低储层压力油田的采收率。该系统利用成熟的钻井技术垂直部署海底假井,降低了实施难度。沉箱设计的分离器提供了充足空间,确保高效分离,并支持单相泵增压,简化了整个流程。此外,该系统与井口结构紧密结合,充分利用现有采油技术,便于实施和推广。

2.2 Perdido油田沉箱分离与增压系统

壳牌公司于2009年在美国弗里波特以南约320公里处启动了墨西哥湾Perdido油田的沉箱分离与增压系统(SBS),该油田是全球首个应用深水气液分离与举升系统进行开发的超深水油田。为应对低储层压力和高达13.79MPa的井口背压挑战,该油田采用了海底沉箱分离与增压系统结合Spar平台的开发策略。沉箱分离与增压系统部署在水深达到2450米的墨西哥湾,通过海底沉箱分离器实现气液的有效分离。气体通过预张力立管的环空区域自然举升至Spar平台,而液体则经过电潜泵增压后,通过预张力立管的内管输送至Spar平台。

Perdido工程的沉箱分离与增压系统由一个嵌入海床的沉箱构成(图2),该沉箱采用无缝钢管焊接工艺制造,外径0.941米,内径0.889米,总长度达到100.584米。沉箱顶部配置了气液柱状旋流分离模块(GLCC),以实现气液的高效分离。底部腔室中安装了电潜泵,以提供必要的增压功能。整个系统遵循模块化设计理念,确保各个模块能够独立调至地面进行维护和检修。

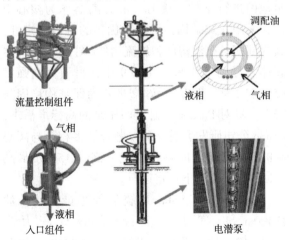

图2 Perdido沉箱分离与增压系统结构

Perdido沉箱分离与增压系统的处理能力为油量3975m³/d、气量1560000m³/d。该系统的设计压力为31.03MPa,原油重度范围在17~40°API之间,气液比为62.4~463.7。系统的分离指标要求液相中的含气率低于15%。电潜泵具有3975m³/d的流量处理能力,能够提供15.17MPa的增压效果,其运行功率为1.2MW。

沉箱分离器的液位检测采用三个竖直布置的井下压力传感器,其中底部两个用于测量液相密度。通过综合液相密度和顶部与底部传感器之间的压差数据,可以计算当前液位。为实现精确控制,系统采用具有反馈机制的PID控制回路,动态调节电潜泵转速,以确保液位维持在预设范围内。

Perdido系统能够实现海底气液分离与液相增压,显著降低井口背压并提升采收率,有效拓展低储层压力油田的开发潜力。沉箱分离器总长度超过100米,能适应广泛的气液相表观流速条件,对复杂流动变化有良好的适应性。电潜泵完全浸没于液体中,自然冷却马达。沉箱的直径和容积经过精密计算,能维持液位稳定,保障系统平稳运行。电潜泵周围设置的闭合挡板有效引导液体冷却马达,优化压降与传热性能。沉箱底部采用流线型设计,防止砂粒沉积,并配备过滤装置以阻挡大型碎片进入电潜泵,进一步提升系统可靠性与耐用性。

2.3 BC-10油田沉箱分离与增压系统

Shell公司于2009年启动了位于巴西维多利亚市东南海域的BC-10油田海底沉箱分离与增压系统。该油田面临的主要技术挑战是储层压力偏低及油品高黏度特性,因此,项目采用将海底沉箱分离与增压系统与FPSO相结合的开发策略。BC-10油田部署了四套由FMC提供的海底沉箱分离与增压系统,水深为1780米。系统通过海底沉箱分离器对井口产出物进行气液分离,气体通过直径0.1524米的管道自然升举至FPSO,液体则经过电潜泵增压,通过直径0.254米的管道输送至FPSO。

沉箱分离与增压系统包括一个直径为0.889米、全长106.68米的沉箱,采用高强度无缝钢管焊接而成,嵌入海床中。沉箱顶部集成了气液柱状旋流分离模块(GLCC),底部腔室则配备电潜泵,以实现液体介质的增压(图3)。该系统设计考虑到可维护性,整个装置能够提升至地面进行必要的检修和回收。

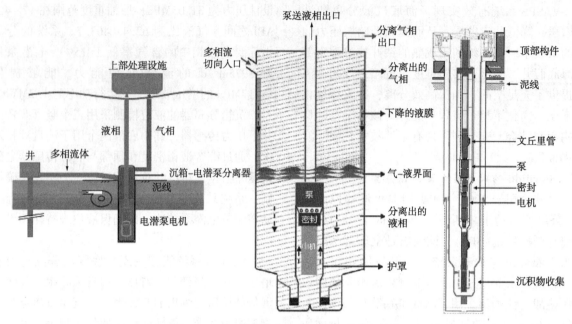

图 3　BC-10 沉箱分离与增压系统流程与结构

BC-10 油田沉箱分离与增压系统的处理能力为油量 4770m³/d、气量 99000m³/d。系统设计压力为 34.47MPa，适用于处理 16~44°API 的原油，能有效应对气液比高达 48.8 的工况。该系统的分离性能指标严格，确保液相含气率低于 15%。配套的电潜泵具备 4770m³/d 的流量处理能力，能够提供 12.41MPa 的增压效果，且其功率输出为 1.2MW，以满足系统高效运行的需求。

沉箱分离器的液位检测采用三个竖直布置的井下压力传感器，其中底部两个用于测量液相密度。通过关联计算液相密度及顶部与底部之间的压差，确定分离器的液位。

BC-10 系统的海底气液分离技术结合液相增压机制，能够有效降低井口背压，提升油田采收率，降低水合物和段塞流风险，实现低储层压力油田的经济高效开发。系统采用沉箱结构分离器，提供宽敞的分离空间以确保优良分离效果，同时便于直接集成电潜泵进行增压作业。系统在海底假井中垂直安装，充分利用了现有成熟钻井技术，降低了实施难度。分离结构与井口结构的巧妙结合，进一步促进了该技术的实际应用与推广。

2.4　Pazflor 油田 SSPS 系统

Total 公司于 2011 年在西非安哥拉几内亚湾的 Pazflor 油田投产了水下气液分离与增压系统（SSPS），使该油田成为全球首个采用整体式深水分离与增压系统开发的油田。该油田储量中有三分之二为 17~22°API 的重质原油，且最低储层压力仅为 20MPa。因此，项目在距离安哥拉 150 公里、水深 800 米的海域部署了海底气液分离系统，结合 FPSO 开展开发。该系统实现了井口产物的海底气液分离，分离后的气体自然举升至 FPSO，液体则通过混合泵增压后输送至 FPSO。在 FPSO 上进一步进行油水分离，分离出的水回注至注水井。

Pazflor 工程采用了三个来自 FMC 公司的海底气液分离模块（SSU），每个模块集成了一个立式重力分离器和两个混合泵，总重约 800 吨。该系统采用模块化设计，以确保各模块可独立提升至地面进行维护。重力气液分离器作为核心组件，采用立式结构，展现出优异的抗砂稳定性，其内部配备布液构件和旋流构件（图 4）。分离器壳体直径为 3.5 米，高约 9 米，材质为高强度 P500QL2 钢，壁厚达 96 毫米，确保结构稳固与安全。针对 Pazflor 项目对增压泵的高标准需求，Framo 公司研发了混合泵，融合了多相泵与离心泵的技术优势，能够在含气率高达 15% 及最高 40% 的极端工况下稳定运行。

Pazflor 油田 SSPS 系统的处理能力为：油 17490m³/d、气 100 万 m³/d。设计压力为 34.47MPa，实际操作压力维持在 2.3MPa，以确保系统远离水合物形成区。该系统分离指标要求液相含气率控制在 15% 以下。配备的混合泵性能卓越，流量可达 350m³/h，能够有效处理黏度

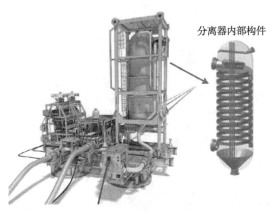

图 4 Pazflor 油田 SSPS 系统分离与增压单元

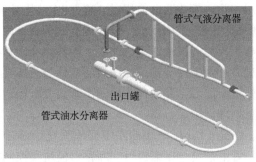

图 5 Marlim 油田 SSAO 系统管式分离模块

达 250mPa·s 的流体,并实现 10.5MPa 的增压效果。混合泵的吸入压力为 2.3MPa,并能在启动阶段应对高达 4500mPa·s 的最大黏度流体。

Pazflor 油田 SSPS 系统通过海底分离与增压技术有效降低了井口背压,使原油产量在较长时间内保持稳定,避免了产量下降。即使在系统停产期间,分离器分离出的气体和液体均能维持在远离水合物形成区的安全范围内,且上游多相管线仅需采用常规降压手段即可有效防止水合物形成,无需额外注入抑制剂。所有液体管线均以倾斜方式连接至 FPSO,确保停产时产物能够借助重力自然流回分离器,极大地方便了系统的再启动操作。在冷启动阶段,该系统能够在一个合理且低于水合物形成区压力的低压范围内稳定运行。

2.5 Marlim 油田 SSAO 系统的管式气液分离器

Petrobras 在巴西里约热内卢的 Campos 盆地部署的三相海底分离系统(SSAO),自 2011 年投产以来,成为全球首个深海重油-水分离及净化水回注系统。经过 20 年的开采,该区域储层压力持续下降,含水率显著上升,同时面临砂对生产系统潜在损害的风险。为应对这些挑战,Petrobras 在距离生产井 341 米、注水井 2100 米、水深 870 米的海域安装了海底分离与水回注系统,实现油气水砂四相的高效分离。在此过程中,油气水砂混合物首先通过多相除砂器去除大部分砂粒,随后在管式分离模块中进行气液分离与油水分离。分离后的砂、气、油及部分水通过一条总长 2.4 公里的混输管道(涵盖立管与管线)输送至 P-37 FPSO 进行处理,而净化后的水则通过注水泵加压,回注至注水井。

SSAO 系统集成了管式分离模块、砂处理模块和水力旋流模块,整体尺寸为长 29 米、宽 10.8 米、高 8.4 米,总重 392 吨。管式分离模块的核心组件包括管式气液分离器、管式油水分离器及出口罐(图 5)。其中,管式气液分离器是该系统的关键设备,位于多相除砂器下游,由一个主管道和五根(部分测试设施配置为六根)垂直管道组成,这些管道与共同旁通管线连接。液体作为重相通过较低水平管道向下流动,而气体则通过垂直管道上升,随后汇集并通过上部管道输送至气体出口。该垂直管道可部分充满液体,为气液入口流量波动(如段塞流)提供备用容积,从而增强系统的段塞容量。由管式气液分离器分离出的气体直接进入出口罐的碳氢化合物收集管,并与准备输送至上部的油重新混合。

管式气液分离器只能分离出游离气体,将气体体积分数降低至 30% 以下,但对油流中的溶解气体无法实现分离。采出液中残留的溶解气体将继续与从管式气液分离器液体出口流出的流体混合,并进入管式油水分离器进行重力分离。系统中的液锁装置有效阻止游离气体进入管式油水分离器。

由于深水环境中的极高内部和外部压力,必须采用非常规的分离器设计,以实现可接受的气液和油水分离性能,同时满足模块的尺寸和重量限制。在此背景下,应用于 Marlim 油田的 SSAO 水下紧凑型分离系统是一种新型设计。该系统的管式分离模块相比传统重力式分离器具有更小的直径,从而缩短了液滴的移动距离,减少了在较大水深下所需的壁厚，并缩短了停留时间。这一改进促进了更加紧凑的设计，满足了未来超深水应用的需求。

2.6 Ormen Lange 气田水下压缩系统

2012年，Shell公司在Ormen Lange气田成功部署了一套水下压缩系统，是挪威最大的工业项目之一。由于Ormen Lange气田采用天然储层压力进行衰竭式开采，储层压力逐渐下降，因此公司在距离挪威西北海岸约100公里、水深900米的地点安装了该系统，以实现高效的水下气液分离。天然气被输送至压缩机，而液体和固体则通过离心泵进行传输。加压后的气体与液体在出口处重新混合，并通过共用输出管线安全输送至陆上处理设施。

Ormen Lange 气田的水下压缩系统由基础系统、管汇系统、压缩机组、断路器模块和控制模块等多个关键组件构成，整体尺寸为 70×54×14 米，总重达 6500 吨。该系统采用模块化设计，确保每个可独立回收的工艺与动力模块干重控制在 250 吨以内。井流进入分离器后实现气液分离，天然气被导向 12.5MW 的水下压缩机，而烃类液体、水及固体则通过 400kW 的水下离心泵模块输送。加压后的气体与液体在出口处重新融合，并通过共用输出管线顺利输送至陆上处理设施。

Ormen Lange 气田水下压缩系统的分离器模块（图6）集成了水下分离器、机械连接器、电动隔离阀和一系列工艺仪表（包括液位控制仪表）。分离器设计为立式容器，内径约 3 米，总高度（T/T）大约 7 米，依靠重力原理实现高效气液分离，同时具备气体除雾和段塞流捕集功能。该模块配备了四个液位传感器，包括两个简单差压型传感器和两个核子型传感器。传感器协同工作，确保压缩机组中的液体泵变速操作对液位实现精确的控制。

压缩机组由四台压缩机以并联方式配置，入口压力为 8MPa，出口压力为 14MPa，整体生产能力约为 60MSm³/d。每台压缩机的设计轴功率为 12.5MW，以满足相应的负荷要求；每台泵的设计负荷设定为 400kW，以确保系统的稳定运行。

Ormen Lange 气田水下压缩系统通过将水下压缩机更靠近井口及立管或回接管道的上游位置，有效提高了采收率，同时降低井口背压，使

图6 Ormen Lange 气田水下压缩系统分离器模块

吸入压力相较于顶部或陆上部署时进一步减小。从资本支出角度来看，水下压缩站总重量为 6500 吨，远低于平台的 25000 吨，体现了其成本效益。由于该系统实行无人操作并直接安装于海底，原料使用量显著减少，从而将操作费用降至最低水平。从 HSE 角度看，水下压缩减少了平台所需直升机飞行次数，提高了作业安全性，同时方便对末端生产进行优化，进一步提升了综合效益。

2.7 Åsgard 水下压缩系统

Aker Solutions 公司于 2015 年在 Åsgard 油田成功部署了全球首个水下压缩站。该油田面临的主要挑战包括储层压力下降导致的管线流量减小和液相在管道内积聚，这进一步引发了天然气流速降低、液体输送效率低下及液体积聚造成的浪涌波问题。此外，流体量超出了平台的液体处理能力，同时伴随较高的水合物风险。为应对这些挑战，Aker Solutions 在距离 Åsgard B 平台 40 公里、水深 260 米的位置安装了水下气体压缩系统。该系统实现了高效的水下气液分离，将压缩后的气体与液体重新混合，并通过管道输送至 Åsgard B 平台。

Åsgard 水下压缩系统由入口冷却器、洗涤器、压缩机、泵和出口冷却器等核心模块构成，整体尺寸为 74×45×26 米，总重量达 5100 吨。该系统采用模块化设计，其中两个安装于水下的压缩机组共有 11 个可回收模块。工作流程中，井流首先经过入口冷却器进行冷却，然后进入洗涤器以有效分离气液两相。气体部分通过离心压缩机增压并再次冷却，而由水、MEG（单乙二醇）及冷凝液组成的液体部分则通过离心泵增

压。最终,增压后的气体与液体在系统中合并,并通过管道输送至Åsgard B平台。

洗涤器模块是Åsgard水下压缩系统的关键组件(图7),重210吨,尺寸为8×8×12米,设计中选用了约3米的最佳容器直径以优化性能。该模块集分离与洗涤功能于一体,主要由分离容器、管道、阀门、接头及水下专用模块构成。分离容器内部配备了精密内件,包括进气装置、轴向除雾旋流器、防涡器、冲砂系统和液位控制传感器。基于重力原理,该模块实现气液分离,有效去除气体中的液体和砂粒,为压缩机提供更优操作条件。设计采用垂直布局以提高气液两相分离效率,并具备段塞捕集功能。洗涤器的液位通过核子源与检测器进行精确测量,并由泵系统智能控制。

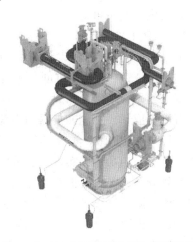

图7 Åsgard水下压缩系统洗涤器模块

压缩机模块重289吨,整体尺寸为11×9×10米。该模块集成了一台电机和一台七级直列径向离心压缩机。每台压缩机的电机额定功率为11.5MW,容量高达14000Am/h,运行速度可达到7200rpm,并实现2.6的增压比(输出压力与输入压力之比,$P_r = P_{out}/P_{in}$)。电机与压缩机采用刚性联接,确保高效动力传输。压缩机单元的速度与扭矩由先进的变速驱动(VSD)系统精确控制。

Åsgard水下压缩系统通过将压缩机部署于靠近储层的位置,最大限度地提升了油井的压降效率、产率和储层总体采收率,从而加速生产进程、增加前期经济回报,并有效降低资本与运营支出,实现经济高效的开发目标。此外,水下压缩机展现出更高的能效,显著降低了出油管的压降,延长了回接长度,并减小了所需管道尺寸。

该系统还减少了与立管中多相流相关的操作难题,增强了流动保障策略的灵活性,有效避免了液体积聚现象。水下压缩系统取代了顶部人员配置的需求,带来了显著的HSE效益,且在水深和外伸距离增加时愈发明显。同时,该系统还有助于减少碳足迹,对环境保护具有积极意义。

3 水下气液分离与增压技术及装备发展趋势

近年来,随着海洋油气资源开发向深海及超深海领域的深入推进,水下气液两相分离技术面临着更为复杂的挑战。在此背景下,大长径比成为水下气液分离器设计的重要趋势。大长径比分离器不仅具备优异的分离效果,还拥有更大的处理能力,通过优化分离器的结构设计和材料选择,可以进一步提升分离效率和稳定性,有效应对深海环境中的高外压条件。同时,增压技术能便捷地将分离后的气/液相输送至平台,从而为后续的油气处理流程提供便利。展望未来,随着深海油气资源开发需求的不断增长,大长径比分离器与增压技术的结合将成为水下气液分离技术发展的重要方向。

3.1 Saipem多管分离器

在深水油田开发过程中,实施海底气液分离及增压技术是有效降低井口背压、显著提升采油效率的关键策略。传统的大直径重力式分离器在深水环境中受到巨大静压的影响,导致其壁厚必须大幅增加,这无疑提高了制造成本并延长了生产周期。为此,Saipem公司创新设计了一种多管式大长径比分离器(图8),与传统压力容器相比,该设计显著降低了壁厚,通过多个细长管道替代传统压力容器,不仅能够降低成本、简化制造流程,缩短制造时间,还实现了现场组装的可行性,进一步拓宽了其在深水油田的应用范围。

Saipem多管分离器通过将传统单一且庞大的压力容器创新性地分割为多个更小容器单元,成功实现了壁厚的显著缩减,此分离器依托标准化的管道制造工艺流程,并融入了便捷的现场组装技术,使得其在实际应用中更加灵活高效。其气液分离机制基于重力原理,每根分离管均设计为类似于常规重力分离器。为了提升分离效率,特别设计了入口布置,并配备可选内件以进一步优化性能。工艺流程如下:多相流体从入口进入,随后垂直穿越中心线,进而被均匀分配至各

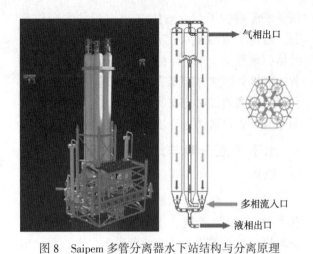

图8 Saipem 多管分离器水下站结构与分离原理

管道中,这些管道具备处理段塞流的独特能力,并能实现对气液的有效分离。在组件的顶部与底部,管道合并形成单一气体和液体出口。气体通过一条或两条流线自然流向地面,而液体则由液体泵增压后汇入单一流线。

Saipem 多管分离器集成了多个液位传感器,精确测量各管道内液位,并通过调节液泵的转速实现液位控制。由于液相与气相分别共用单一出口,系统设计确保所有管道内液位保持一致。此外,分配与回收系统采用对称设计。管道可采用不同类型的传感器技术(包括雷达传感器、核子传感器和压力传感器)以测量液位。整个分离器的设计融入了自动排水的几何构造,以预防砂沉积现象的发生。

对于 Saipem 多管分离器有以下两个设计案例(图9),给出的工作参数及尺寸数据可供参考。

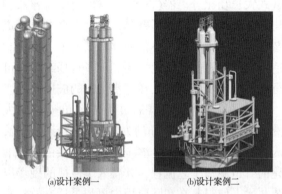

图9 Saipem 多管分离器设计案例

设计案例一分离器设计适用于800米水深环境,分离压力设定为3MPa,段塞体积为$12m^3$。多管分离器结构包含六个管道,每个管道直径为1.07m,高度为15m,壁厚25mm,并通过加强筋进行了优化设计。为每根管道都使用22毫米的 GSPU 进行隔热,以避免任何冷点,并确保在停机期间有最短的冷却时间。相应的水下生产系统配置了两个30m长的电潜泵,这些泵焊接在基桩上的两个沉箱中。水下管汇则集成了所有驱动阀,可回收至地面以便维护。整个系统重量约为370吨,可用深水安装船进行吊装。

设计案例二专注于处理较小流量但更大水深的场景。适用于3000米水深条件,设计压力为69MPa,操作压力为3MPa,段塞体积为$4m^3$。该分离器结构包含四根直径为0.762m、长度为11m的管道。为应对高压环境,所需壁厚达到约77mm。尽管这些管道并不属于标准管道类别,且在制造上采用了非传统的设计理念,但它们依然具备实际应用的可行性。相比之下,如果采用单个压力容器则不具备可行性。

3.2 SEPPUMP 水下分离器

SEPPUMP 水下分离器的设计理念源于北海 Highlander 项目,SEPPUMP 的创新亮点在于,它摒弃了传统的垂直电潜泵布局,转而采用水平配置,并直接集成于段塞流捕集器的内部。这一变革性设计使得 SEPPUMP 不仅具备段塞捕集的能力,还兼具了泵送功能,实现基于重力原理的气液两相高效分离,极大地提升了分离效率。

SEPPUMP 水下分离器的管道直径设计为0.6米,系统总长度为24.4米,以确保足够空间有效消除段塞流。分离后的液体通过约22米长的水平电潜泵泵送至系统外部。为实现理想的分离效果,SEPPUMP 采用双层管道设计:顶层用于气体流动,底层负责液体流动。交叉分支结构允许两层之间进行必要的流体交换(图10)。整个 SEPPUMP 系统的总重量约为160吨。

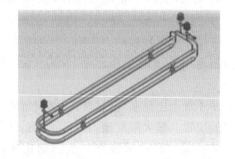

图10 SEPPUMP 水下分离器基础配置

3.3 WAVy 水下分离器

WAVy 水下分离器是一种气-液-固三相管式分离设备,类似于 SEPPUMP,其采用水平布置并基于重力原理实现气液两相分离。多相井流通过位于左上方的阀门进入分离器，随后流经多个向下和向上分支形成的"W-A-V"几何构型（图11）。在此过程中，密度较大的液体沉降到底部，而气体上升至顶部，最终各相流体被引导至右侧的独立流线中。为了应对砂粒在WAVy分离器"V"形底部可能积聚并造成流动阻碍的问题，该分离器特别设计并配备了冲洗管线和冲洗泵。这些设施能够有效清除掉底部积累的砂粒，确保分离器的顺畅运行。

WAVy水下分离器专为深水环境设计，能够承受高达68.95MPa的压力，整体重量约45吨。其主要支管采用直径0.61米、壁厚0.03米的管道，严格遵循DNV标准规范。交叉流动管的直径为0.3048米，冲洗管线的直径为0.1524米。WAVy的长度仅为SEPPUMP的一半，占地面积紧凑，约为12.19×4.57米。该分离器采用模块化设计理念，多个WAVy分离器能够根据实际需求灵活组合，以串联或并联方式适应不同油井间的流速及气油比（GOR）变化。

图11 WAVy水下分离器基础配置

4 结论

水下气液两相分离与增压输送技术的发展代表了深海油气开采领域的一次重要进展。这项技术成功克服了低储层压力、高井口背压等一系列技术障碍，从而有效满足了降本增效的迫切要求。

本文主要研究了水下气液分离与增压技术的最新成果，并详细探讨了国外先进系统所遇到的挑战、结构特点、关键参数及其显著优势。通过对分离流程的优化和输送效率的提升，这些先进系统为复杂多变的海洋环境提供了可靠而高效的解决方案，显著增强了油气开采的效率和安全性。

展望未来，水下分离工程将更加倾向于采用大长径比的设计策略，以适应高外压和高流量等极端水下作业条件。水下气液两相分离与增压技术将聚焦于精准匹配不同水下环境下的分离需求，优化分离与增压工艺选择，以实现高效、环保的物质分离与输送。该领域的研究将致力于开发能够适应深海高压、低温等极端条件的分离技术，同时，针对不同组分的物理化学特性，探索更为精准和高效的分离工艺。

参 考 文 献

［1］杨兆铭，陈建磊，何利民，等．水下旋流气液分离器研究进展［J］．油气储运，2019，38（08）：856-862.

［2］YANG L L, CHEN X D, HUANG C Y, LIU S, NING B, WANG K. A review of gas-liquid separation technologies: Separation mechanism, application scope, research status, and development prospects［J］. Chemical Engineering Research and Design, 2024, 201: 257-274.

［3］陈家庆，王强强，肖建洪，等．高含水油井采出液预分水技术发展现状与展望［J］．石油学报，2020，41（11）：1434-1444.

［4］LI Y, QIN G L, XIONG Z Y, FAN L. Gas-liquid separation performance of a micro axial flow cyclone separator［J］. Chemical Engineering Science, 2022, 249: 117234.

［5］QIU Z, ZHOU L, BAI L, EL-EMAM M A, AGARWAL R. Empirical and numerical advancements in gas-liquid separation technology: A review［J］. Geoenergy Science and Engineering, 2024, 233: 212577.

［6］TAVARES P L, FURTADO S F L, DINELLI D K, SANTOS L O S, LAMEIRA O D, JUNIOR E B L, et al. Campos basin revitalization: from small subsea tiebacks to complete redevelopment［C］. Houston: Offshore Technology Conference, 2024: D011S002R001.

［7］PEIXOTO G A, RIBEIRO G A S, BARROS P R A, MEIRA M A, BARBOSA T M. VASPS prototype in Marimba field-workover and re-start［C］. Rio de Janeiro: SPE Latin America and Caribbean Petroleum Engineering Conference, 2005: SPE-95039-MS.

［8］CAETANO E F, DO VALE O R, TORRES F R, SILVA JR A. Field experience with multiphase boosting systems at Campos Basin, Brazil［C］. Houston: Offshore Technology Conference, 2005: OTC-17475-MS.

［9］BENETTI M, VILLA M. Field tests on VASPS separation and pumping system［C］. Houston: Offshore Technology Conference, 1997: OTC-8449-MS.

［10］DO VALE O R, RIBEIRO G S, DO BENETTI M, DO VILLA M. VASPS pre-subsea phase development: high viscosity field tests and outlook for subsea prototype［C］. Houston: Offshore Technology Conference, 1998: OTC-8864-MS.

[11] DO VALE O R, GARCIA J E, VILLA M. VASPS installation and operation at Campos Basin[C]. Houston: Offshore Technology Conference, 2002: OTC-14003-MS.

[12] FIGUEIREDO M W, KUCHPIL C, CAETANO E F. Application of subsea processing and boosting in Campos Basin[C]. Houston: Offshore Technology Conference, 2006: OTC-18198-MS.

[13] LABES A, MYHRE T. Subsea hydrogen long duration energy storage[C]. Houston: Offshore Technology Conference, 2024: D011S002R004.

[14] ENTRESS J H, PRIDDEN D L, BAKER A C. The current state of development of the VASPS subsea separation and pumping system[C]. Houston: Offshore Technology Conference, 1991: OTC-6768-MS.

[15] JAHNSEN O F, STORVIK M. Subsea processing & boosting in a global perspective[C]. Ravenna: Offshore Mediterranean Conference and Exhibition, 2011: OMC-2011-143.

[16] LITTELL H S, JESSUP J W, SCHOPPA W W, SEAY M R, COULON T D. Perdido startup: flow assurance and subsea artificial lift performance[C]. Houston: Offshore Technology Conference, 2011: OTC-21716-MS.

[17] GRUEHAGEN H, LIM D. Subsea separation and boosting—an overview of ongoing projects[C]. Jakarta: Asia Pacific Oil and Gas Conference and Exhibition, 2009: SPE-123159-MS.

[18] JU G T, LITELL H S, COOK T B, DUPRE M H, CLAUSING K M, SHUMILAK E E, et al. Perdido development: subsea and flowline systems[C]. Houston: Offshore Technology Conference, 2010: OTC-20882-MS.

[19] GILYARD D, BROOKBANK E B. The development of subsea boosting capabilities for deepwater Perdido and BC-10 assets[C]. Florence: SPE Annual Technical Conference and Exhibition, 2010: SPE-134393-MS.

[20] BYBEE K. Perdido startup: flow assurance and subsea artificial Lift[J]. Journal of Petroleum Technology, 2011, 63(07): 90-93.

[21] ROJAS M, MERLINO A, LINEY D, OBST L, KOTTEMAN M, HORTON A, et al. Qualification and deployment of the highest power ESP in the world[C]. Virtual and The Woodlands: SPE Gulf Coast Section Electric Submersible Pumps Symposium, 2021: D021S003R002.

[22] SNYDER D, TOWNSLEY B. SS: Perdido development project: world[C]. Houston: Offshore Technology Conference, 2010: OTC-20887-MS.

[23] DEUEL C L, CHIN Y D, HARRIS J, GERMANESE V J, SEUNSOM N. Field validation and learning of the Parque das Conchas(BC-10) subsea processing system and flow assurance design[C]. Houston: Offshore Technology Conference, 2011: OTC-21611-MS.

[24] FREIRE DE CARVALHO T P, OLIJNIK L, BRODERICK R J, LABES A. The contribution from an operator's global subsea hardware standardization program to 10K subsea hardware and controls on Parque das Conchas(BC-10)[C]. Rio de Janeiro: Offshore Technology Conference Brasil, 2015: D011S002R002.

[25] ATAKAN Z, CHIN Y D, LANG P P, IYER S. Design and operability considerations of the gas flowline at Parque das Conchas(BC-10) Ostra field[C]. Houston: Offshore Technology Conference, 2010: OTC-20528-MS.

[26] SLEIGHT N C, OLIVEIRA N. BC-10-optimizing subsea production[C]. Rio de Janeiro: Offshore Technology Conference, 2015: OTC-26200-MS.

[27] OKEREKE N U, OGAZI I A, UMOFIA A, ABILI N, OHIA N P, EKWUEME S T. Adopting subsea separation technologies for deepwater West-Africa: a review study[C]. Lagos: SPE Nigeria Annual International Conference and Exhibition, 2021: D021S003R006.

[28] STINGL K H, PAARDEKAM A H M. Parque das Conchas(BC-10)-an ultra-deepwater heavy oil development offshore Brazil[C]. Houston: Offshore Technology Conference, 2010: OTC-20537-MS.

[29] IYER S, LANG P, SCHOPPA W, CHIN Y D, LEITKO A. Subsea processing at Parque das Conchas (BC-10): taking flow assurance to the next level[C]. Houston: Offshore Technology Conference, 2010: OTC-20451-MS.

[30] PARSHALL J. Brazil Parque das Conchas project sets subsea separation, pumping milestone[J]. Journal of Petroleum Technology, 2009, 61(09): 38-42.

[31] BARRIOS L, DEBACKER I, RIVERA R, BASILIO M, LINEY D. Brazil deepwater BC-10 ESP operation without downhole ESP gauges[C]. Virtual and The Woodlands: SPE Gulf Coast Section Electric Submersible Pumps Symposium, 2021: D041S009R002.

[32] BIBET P J, SANTOS R, LUCAS A, VUNZA D. Pazflor subsea separation, ten years after[C]. Virtual and Houston: Offshore Technology Conference, 2021: D021S026R003.

[33] ERIKSEN S, MCLERNON H, MOHR C. Pazflor SSPS project: testing and qualification of novel tech-

nology: a key to success[C]. Houston: Offshore Technology Conference, 2012: OTC-23178-MS.

[34] BON L. The Pazflor adventure[C]. Houston: Offshore Technology Conference, 2012: OTC-23173-MS.

[35] LIDDLE D. Subsea technology: a reflection on global challenges and solutions[C]. Houston: Offshore Technology Conference, 2012: OTC-23092-MS.

[36] DEOCLECIOA L H P, DE OLIVEIRA S S, CELESTE W C, CHAVES G L D, MENEGUELO A P. Subsea processing as a tool for cost reduction of deepwater projects[J]. Research, Society and Development, 2020, 9(1): e29911493-e29911493.

[37] CONSTANT A, RAMAT L. Start-up of a giant[C]. Houston: Offshore Technology Conference, 2012: OTC-23176-MS.

[38] BON L. Pazflor, a world technology first in deep offshore development[C]. Aberdeen: SPE Offshore Europe Conference and Exhibition, 2009: SPE-123787-MS.

[39] SANGWAI J, DANDEKAR A. Practical aspects of flow assurance in the petroleum industry[M]. Boca Raton: CRC Press, 2022.

[40] WILSON A. Startup of Pazflor giant offshore Angola tackles challenges from two reservoirs[J]. Journal of Petroleum Technology, 2012, 64(12): 107-111.

[41] EUPHEMIO M, OLIVEIRA R, NUNES G, CAPELA C, FERREIRA L. Subsea oil/water separation of heavy oil: overview of the main challenges for the Marlim field-Campos basin[C]. Houston: Offshore technology conference, 2007: OTC-18914-MS.

[42] DE OLIVEIRA D A, PEREIRA R D S, CAPELA MORAES C A, BARACHO V P, DE SOUZA R D S A, LOPES EUPHEMIO M L, et al. Comissioning and startup of subsea Marlim oil and water separation system[C]. Rio de Janeiro: Offshore Technology Conference Brasil, 2013: OTC-24533-MS.

[43] PEDROSO C A. Offshore mature fields in deepwaters: the final challenge[C]. Rio de Janeiro: Offshore Technology Conference Brasil, 2023: D021S020R003.

[44] PEREIRA R M, CAMPOS M C M M D, DE OLIVEIRA D A, DE SOUZA R D S A, FILHO M M C, Orlowski R, et al. SS: Marlim 3 phase subsea separation system: controls design incorporating dynamic simulation work[C]. Houston: Offshore Technology Conference, 2012: OTC-23564-MS.

[45] ORLOWSKI R, EIPHEMIO M L L, EIPHEMIO M L, ANDRADE C A, GUEDES F, TOSTA DA SILVA L C, et al. Marlim 3 phase subsea separation system-challenges and solutions for the subsea separation station to cope with process requirements[C]. Houston: Offshore Technology Conference, 2012: OTC-23552-MS.

[46] DUARTE D G, DE MELO A V, CARDOSO C A B R, VIANNA F, IRMANN-JACOBSEN T, MCCLIMANS O T, et al. Marlim 3 phase subsea separation system-challenges and innovative solutions for flow assurance and hydrate prevention strategy[C]. Houston: Offshore Technology Conference, 2012: OTC-23694-MS.

[47] CAPELA MORAES C A, DA SILVA F S, MARINS L P M, MONTEIRO A S, DE OLIVEIRA D A, PEREIRA R M, et al. Marlim 3 phase subsea separation system: subsea process design and technology qualification program[C]. Houston: Offshore Technology Conference, 2012: OTC-23417-MS.

[48] SAGATUN S I, GRAMME P, IIE G, HORGEN O J, RUUD T, STORVIK M. The pipe separator: simulations and experimental results[C]. Houston: Offshore Technology Conference, 2008: OTC-19389-MS.

[49] SKJEFSTAD H S, STANKO M. Subsea water separation: a state of the art review, future technologies and the development of a compact separator test facility[C]. Cannes: BHR International Conference on Multiphase Production Technology, 2017: BHR-2017-511.

[50] LABES A, MYHRE T. Subsea hydrogen long duration energy storage[C]. Houston: Offshore Technology Conference, 2024: D011S002R004.

[51] BERNT T, SMEDSRUD E. Ormen Lange subsea production system[C]. Houston: Offshore Technology Conference, 2007: OTC-18965-MS.

[52] LUPEAU A, BARRAS N, GODOE S, ALMQVIST J, VIELLIARD C. Subsea compression: sustainably unlocking subsea gas resources to any scale[C]. Kuala Lumpur: Offshore Technology Conference Asia, 2024: D031S024R003.

[53] BYBEE K. Ormen Lange subsea-compression-system pilot[J]. Journal of Petroleum Technology, 2009, 61(08): 48-49.

[54] BJERKREIM B, HARAM K O, POORTE E, SKOFTELAND H, ROKNE Ø, DIOP S, et al. Ormen Lange subsea compression pilot[C]. Houston: Offshore Technology Conference, 2007: OTC-18969-MS.

[55] ERIKSSON K G. Control system and condition monitoring for a subsea gas compressor pilot[C]. Newcastle: SUT Subsea Control and Data Acquisition (SCADA) Confer-

ence. SUT, 2010: SUT-SCADA-10-39.

[56] SKOFTELAND H, HILDITCH M, NORMANN T, ERIKSSON K G, NYBORG K, POSTIC M, et al. Ormen Lange subsea compression pilot-subsea compression station[C]. Houston: Offshore Technology Conference, 2009: OTC-20030-MS.

[57] LENDENMANN H, FLØISAND J O, VATLAND S. A new era: large subsea multiphase compressor-driven by subsea adjustable speed drive[C]. Virtual and Houston Offshore: Technology Conference, 2021: D021S026R005.

[58] LIMA F, STORSTENVIK A, NYBORG K. Subsea compression: a game changer[C]. Rio de Janeiro: Offshore Technology Conference Brasil, 2011: OTC-22411-MS.

[59] SETEKLEIV E, ANFRAY J, BOIREAU C, GYLLENHAMMAR E, KOLBU J. An evaluation of subsea gas scrubbing at extreme pressures[C]. Houston: Offshore Technology Conference, 2016: D031S033R005.

[60] HEDNE P E, AARVIK A, NORDSVEEN M, PETTERSEN B H, HAUGE L E. Åsgard subsea compression-technology overview and operational experience[C]. Cannes: BHR International Conference on Multiphase Production Technology, 2017: BHR-2017-243.

[61] HANSSEN B V. Technology focus: subsea systems (August 2020)[J]. Journal of Petroleum Technology, 2020, 72(08): 57-57.

[62] MICALI S. Åsgard project-another step versus the full subsea process plant[C]. Kuala Lumpur: Offshore Technology Conference Asia, 2014: OTC-25464-MS.

[63] MICALI S, ABELSSON C. Novel subsea boosting solutions to increase IOR[C]. Kuala Lumpur: Offshore Technology Conference Asia, 2016: D031S028R003.

[64] TIME N P, TORPE H. Subsea compression-Åsgard subsea commissioning, start-up and operational experiences[C]. Houston: Offshore Technology Conference, 2016: D021S024R006.

[65] IMS J. Modelling of Åsgard subsea gas compression station for condition monitoring purposes[J]. Project Thesis, Department of Chemical Engineering, Norwegian University of Science and Technology, 2017.

[66] MÆLAND D, BAKKEN L E. Wet gas hydrocarbon centrifugal compressor-performance test results and evaluation[C]. ASME International Mechanical Engineering Congress and Exposition. American Society of Mechanical Engineers, 2021, 85666: V010T10A005.

[67] 高原, 魏会东, 姜瑛, 等. 深水水下生产系统及工艺设备技术现状与发展趋势[J]. 中国海上油气, 2014, 26(04): 84-90.

[68] CUI Y, ZHANG M, WANG H, YI H, YANG M, HOU L, et al. Investigation of influence of high pressure on the design of deep-water horizontal separator and droplet evolution[J]. Processes, 2024, 12(12): 2619.

[69] DI SILVESTRO R, ABRAND S, MEVEL T, RIOU X, SHAIEK S. New way to use pipes for subsea separation in deepwater development and qualification of a novel gas/liquid separator[C]. Ravenna: Offshore Mediterranean Conference and Exhibition, 2009: OMC-2009-136.

[70] ABRAND S, SHAIEK S, ANRES S J, HALLOT R. Subsea gas-liquid and water-hydrocarbon separation: pipe solutions for deep and ultra deepwater[C]. Rio de Janeiro: Offshore Technology Conference Brasil, 2013: OTC-24359-MS.

[71] PRESCOTT N, MANTHA A, KUNDU T, SWENSON J. Subsea separation-advanced subsea processing with linear pipe separators[C]. Houston: Offshore Technology Conference. OTC, 2016: D031S033R006.

天然气超声速旋流分离技术现状与应用展望

边江[1]　王颖[1]　王泽润[1]　鞠淋[1]　邓涵玉[1]　张嘉伟[1]　曹学文[2]　曹恒广[2]

[1. 长江大学石油工程学院；2. 中国石油大学（华东）储运与建筑工程学院]

摘　要　天然气作为一种清洁、高效的能源，在全球能源结构中占据重要地位。然而，天然气在开采和运输过程中往往含有杂质和重烃成分，需要进行有效处理。超声速膨胀液化与旋流分离技术被认为是新一代气体分离处理技术，具有高效和清洁的优势。该技术通过超声速膨胀产生的冷却效果和旋流分离原理，有效去除天然气中的水蒸气、重烃和酸性气体，并可用于天然气液化。本文综述了超声速分离器的结构、工作原理、优化设计、数值模拟及实验研究进展，探讨了该技术在天然气处理中的应用，包括脱水、重烃去除和二氧化碳捕集。最后，强调了持续科学探索超声速分离器设计和应用的必要性，以满足行业不断变化的需求，并展望了该技术的未来发展方向，包括提高效率、拓展应用场景以及在碳捕获中的应用，研究成果可为天然气超声速旋流分离技术的研发和应用提供参考。

关键词　天然气；超声速；凝结相变；旋流分离

Current status and application prospects of natural gas supersonic separation technology

Bian Jiang[1]　Wang Ying[1]　Wang Zerun[1]　Ju Lin[1]
Deng Hanyu[1]　Zhang Jiawei[1]　Cao Xuewen[2]　Cao Hengguang[2]

[1. School of Petroleum Engineering, Yangtze University;
2. College of Pipeline and Civil Engineering, China University of Petroleum (East China)]

Abstract　Natural gas, as a clean and efficient energy source, plays an important role in the global energy structure. However, natural gas often contains impurities and heavy hydrocarbons during extraction and transportation, which require effective treatment. Supersonic expansion liquefaction and cyclone separation technology is considered a next-generation gas separation treatment method, offering advantages of high efficiency and cleanliness. This technology effectively removes water vapor, heavy hydrocarbons, and acidic gases from natural gas through the cooling effect generated by supersonic expansion and the principle of cyclone separation, and it can also be used for natural gas liquefaction. This paper reviews the structure, working principle, optimization design, numerical simulation, and experimental research progress of supersonic separators, and explores the applications of this technology in natural gas processing, including dehydration, heavy hydrocarbon removal, and carbon dioxide capture. Finally, it emphasizes the necessity of continuous scientific exploration in the design and application of supersonic separators to meet the evolving needs of the industry, and outlines the future development directions of the technology, including improving efficiency, expanding application scenarios, and its use in carbon capture. The research results can provide reference for the development and application of natural gas supersonic cyclone separation technology.

Key words　Natural Gas; Supersonic; Condensation; Swirling Separation

1　引言

天然气作为一种重要的清洁能源，在一次能源消费总量中的占比持续上升，在全球能源体系中占据重要比例，在解决能源短缺问题方面发挥了关键作用。然而，天然气在开采和运输过程中往往含有杂质和重烃成分。杂质的存在会降低天然气的热值，且水和重烃成分的凝结会提升管道的流动阻力，进而增加能源消耗，并有可能导致管道的阻塞。同时，酸性气体与水分的相互作用可能诱发腐蚀，进而损害管道和相关设备。因此，有效的天然气处理对于减少排放和确保天然气行业的可持续增长仍然至关重要。

超声速气体分离器是一种应用于天然气处理的新技术，其主要包括一个静置旋流器、一个拉瓦尔喷管和一个扩散器，如图1所示。静置旋流器产生具有巨大离心力的强漩涡流，含杂质天然气经超声速膨胀产生冷却效果，在喷管的扩张段形成低温低压环境，杂质气体在此环境下将冷凝成液滴。随后，液滴会在旋流器产生的强大离心力的作用下从天然气中被分离。扩散器产生的冲击波可将流速从超声速降至亚声速水平，从而提高压力能的利用率，但同时也会降低一定的压力。因此，这是一种静态装置，没有任何运动部件，具有很高的可靠性。纯物理的分离过程以及化学物质的零添加，使其成为了一种环境友好的天然气加工与处理技术。重要的是，该技术集膨胀制冷、气液分离、压力回收于一体，相比于传统工艺结构更加紧凑，为节省天然气处理装置的占地提供了可能。

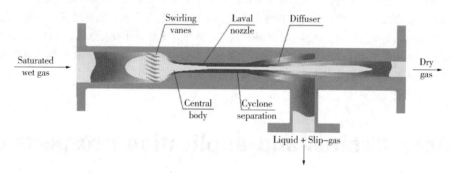

图1 超声速分离器结构示意图

在分离含有水蒸气和重烃的天然气时，超声速分离器的优化设计可确保杂质气体在超声速气流中发生相变，轻烃组分充当载气。因此，超声速分离技术已被用于去除天然气中的水蒸气和重烃。幸运的是，超声速分离技术还表现出了从天然气中去除二氧化碳（CO_2）与硫化氢（H_2S）的潜力。Sun等建立了超声速流动条件下CO_2自发冷凝的数学模型，得到了CO_2在天然气中高速膨胀时的自发冷凝特性。随后，Bian等研究分析了喷管收敛段线型和旋流对CO_2冷凝过程的影响。Chen等建立了考虑CH_4-CO_2混合物的喷管内CO_2冷凝流动的数值模型，并研究了操作参数和喷管结构对CO_2冷凝特性的影响。之后，Cao等考虑了CH_4-CO_2-H_2S三元体系，数值研究了喷管的冷却性能和旋流条件下的流动特性，证实了超声速分离技术同时去除CO_2和H_2S的可行性。最近，Wen等建立了非平衡冷凝流动模型，通过评估纯CO_2流的冷凝效果，进一步证实了利用超声速流动中的相变行为去除天然气中CO_2的可行性。

超声速分离技术可防止水合物问题，并且由于在分离装置中的停留时间短，因此无需抑制剂和再生系统，从而提供了一种环保设施。同时，与吸附、吸收、低温和膜等净化技术相比，超声速分离器在几个方面脱颖而出。与吸附和吸收方法不同，吸附和吸收方法通常需要大量的设施、复杂的系统，并且使用对环境有不利影响的化学品，超声速分离器提供了一种更精简、更经济的解决方案。此外，天然气液化在促进天然气运输和储存方面起着至关重要的作用。除了能够从天然气中分离CO_2和H_2S外，超声速装置还可以用于生产液化天然气。

这项新兴且高效的天然气处理技术不仅能够有效分离杂质气体、实现天然气液化，而且相较于其他净化技术，在成本效益、环境影响和操作灵活性方面展现出综合优势，突显了其在天然气预处理领域的重要前景。本文综述了天然气超声速膨胀液化与旋流分离技术的研究现状，旨在为天然气行业的研究人员和工程师提供参考，同时强调了对超声速分离器设计和应用进行持续科学探索的必要性，以满足行业不断变化的需求。

2 天然气中的杂质组分

天然气是指从储层中开采出的、由可燃性烃和非烃组成的混合气体，其主要成分是甲烷（CH_4），还包含少量乙烷（C_2H_6）、丙烷（C_3H_8）和丁烷（C_4H_{10}）等烃类组分（见表1）。此外，天然气中还可能含有CO_2、H_2S、氮气（N_2）和氦气

(He)等杂质。这些杂质不仅对天然气的性质和效用具有重要影响,还具有工业用途。例如,H_2S可用于生产硫磺、硫酸和硫铵,CO_2可用作制冷剂,He则是国防和核能工业的关键原料。

表1 天然气的组成和含量范围

编号	组成	含量 Vol. %
1	甲烷(CH_4)	>85
2	乙烷(C_2H_6)	3~8
3	丙烷(C_3H_8)	1~5
4	丁烷(C_4H_{10})	1~2
5	戊烷(C_5H_{12})	1~5
6	二氧化碳(CO_2)	1~2
7	硫化氢(H_2S)	1~2
8	氮气(N_2)	1~5
9	氦气(He)	<0.5

天然气的性能和效用显著受到其组分的影响。C_2H_6、C_3H_8和C_4H_{10}因其高热值,有助于提高天然气的整体燃烧性能。然而,重烃和其他杂质的存在可能降低天然气的热值,影响其高效燃烧,并在燃烧过程中生成有害副产品,从而影响设备运行效率。此外,CO_2和H_2S等酸性杂质不仅增加了上游处理设施的复杂性和运营成本,还在运输和使用中引发一系列问题。这些酸性气体会腐蚀管道设施,威胁基础设施的耐久性和使用寿命,尤其是在潮湿环境中更易形成腐蚀性水合物和管道堵塞,严重影响流动能力和传热效率。同时,颗粒物和重烃堆积在传热表面,会降低换热性能,影响系统整体效率。

天然气杂质的存在还对环境带来显著挑战。例如,CO_2作为一种温室气体,不仅对全球气候变化产生重要影响,还增加了天然气的碳足迹。而H_2S在燃烧中产生的硫氧化物会腐蚀设备并加剧空气污染。这些问题使得减少天然气中的酸性气体排放和去除其他杂质成为当前天然气处理中的重要任务。

为应对上述挑战,天然气通常需经过标准处理步骤,包括去除H_2S、脱水、调节烃露点和去除CO_2等。通过去除杂质,天然气的燃烧性能得以提高,能量转换更加高效,从而进一步增强其综合利用价值。此外,天然气净化过程对于保障设备安全、提高传热性能、控制污染排放及促进环境保护具有重要意义。

3 超声速分离技术

3.1 超声速分离器的结构

如图1所示,超声速分离器是一种紧凑的装置,在进气道安装了固定旋流叶片和拉瓦尔喷管。拉瓦尔喷管的制冷效果超过了膨胀器、涡流管和焦耳-汤姆逊(J-T)装置,如图2所示。在超声速分离器中使用拉瓦尔喷管可以实现绝热冷却,从而在分离阶段产生显著的低温,增强杂质的相变析出。与传统技术相比,超声速分离具有多项优势。其中一些主要优势包括(1)体积小:超声速分离器体积相对较小,与传统分离器相比,安装所需的空间更小。这种紧凑性使其更便于运输和处理。(2)成本效益高:与更大型、更复杂的分离系统相比,超声速分离器的安装费用通常更低。这种成本效益使其成为各种应用的有利选择。(3)运行成本更低:由于设计简化、能耗较低,超声速分离器的运行成本通常较低。这些设备能高效地分离天然气流中的重烃组分,从长远来看可节约成本。(4)环境友好:超声速分离器有效去除气流中的杂质和重金属,且由于停留时间短暂,防止水合物的形成,无需使用抑制剂,从而能够生产更清洁、更环保的天然气。

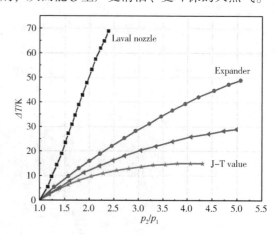

图2 不同装置的冷却性能

根据工作原理的不同,超声速分离器主要可分为两种典型结构。一种是旋流后置式超声速分离器。Twister BV 公司设计的 Twister I 超声速分离器是其中的代表,如图3所示,漩流装置安装在拉瓦尔喷管后面。另一种代表是旋流器前置式超声速分离器,例如 ENGO 公司设计的 3S 分离器,其中的漩流装置安装在拉瓦尔喷管的入口处,如图4所示。这两种分离器的组成元素基本相同,但旋流装置的安装位置不同。

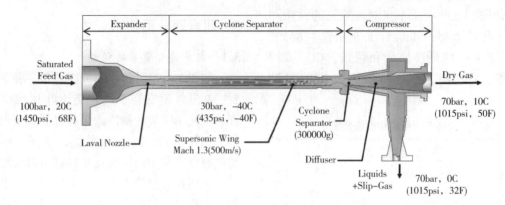

图 3　Twister I 旋流后置分离器

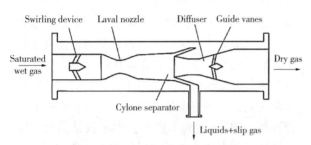

图 4　俄罗斯 ENGO 公司 3S 分离器

这两种结构设计的本质区别在于亚声速或超声速状态下的流动模式。在后置型分离器的拉瓦尔喷管中，流动是均匀的，没有明显的冲击波出现。但当气体流经旋流器时，在超声速状态下会发生轴向速度向切向速度的转换，在获得强大的旋流强度的同时，会在漩流段会出现明显的斜向冲击波，造成气体膨胀不充分、低温环境被破坏、液滴二次汽化等问题。而且控制旋流器和扩散器之间的冲击波，会限制分离器的压力恢复能力，增加能量损失。对于前置型分离器，轴向速度向切向速度的转换发生在亚音速条件下，旋流器后不会产生斜冲击波。同时，气体在拉瓦尔喷管的发散段以漩涡状态膨胀到超声速。漩涡和冷凝同时发生，可有效减少液滴再蒸发的负面影响，提高分离器的分离效率。

此外，旋流器的类型多样，包括轴向旋流器、切向旋流器等，如图 5 所示。不同类型的旋流器在结构和工作原理上略有差异，但都以实现高效旋流分离为目标。旋流器的关键设计参数包括入口直径、出口直径、旋流室长度、叶片角度等。入口直径影响天然气的进入流量和速度，出口直径决定了不同相态物质的排出路径，旋流室长度影响旋流的稳定性，叶片角度则直接影响旋流的强度。此外，旋流器的材质选择也很重要，需要考虑其耐腐蚀性、耐磨性等，以适应天然气中可能存在的杂质和腐蚀性成分。

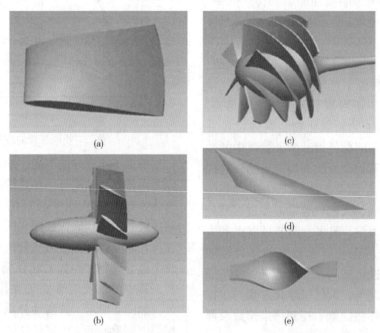

图 5　不同叶片旋转角度、形状的超声速分离器叶片结构

除了拉瓦尔喷管和旋流器外，该技术还涉及一些辅助部件。例如，在天然气入口处需要设置过滤器，以去除较大颗粒的杂质，防止其堵塞喷管和旋流器。同时，在系统中还需要配备温度、压力传感器等监测设备，用于实时监测天然气的状态参数，以便对系统进行优化控制。此外，为了保证系统的稳定运行，还需要有相应的稳压装置和流量调节装置。

3.2 喷管中的超声速膨胀与冷凝

拉瓦尔喷管是超声速膨胀液化过程中至关重要的核心部件，其独特之处在于通过绝热冷却显著降低温度，从而促使天然气中的杂质液化，这一特性是超声速分离器高效运行的基础。拉瓦尔喷管通常由收缩段和扩张段组成，通过将压力和温度转化为动能，将气流加速至超声速。流动特性可用马赫数（Ma）描述，马赫数是流体轴向速度（v）与声速（c）之比。如图6所示，在收敛段，流动开始时为亚音速（$Ma<1$），此时喷管截面逐渐减小。随着气流接近喷管喉部，马赫数达到1，此时喷管横截面积最小，称为最大收缩点。喉部之后，气流在扩张段迅速膨胀，导致喷管截面增大，流动转为超声速（$Ma>1$），并加速到极高速度。

拉瓦尔喷管内流动特性与凝结相变机理如图6所示。在已知温度、总压及杂质气体分压的条件下，含杂质的天然气进入拉瓦尔喷管。当混合气体流经喷管的渐缩段时，随着喷管横截面积逐渐缩小，Ma不断增加，气流在喉部达到Ma为1的临界状态，同时压力和温度显著降低，使可凝气体的过冷度不断增大。随后，混合气体进入渐扩段并进一步膨胀降温，达到超声速后，可凝气体因过饱和状态迅速发生凝结。凝结过程中释放的潜热会扰动流动场，使凝结点附近的轴线压力上升。与无凝结的膨胀流动相比，此时流场的压力分布出现显著变化；随着气体进一步膨胀，混合气体的压力和温度整体呈下降趋势。通过捕捉凝结过程中流动场轴线压力变化的位置和规律（即图中Δp的变化情况），可以直观反映出不同可凝气体在拉瓦尔喷管内的凝结特性。

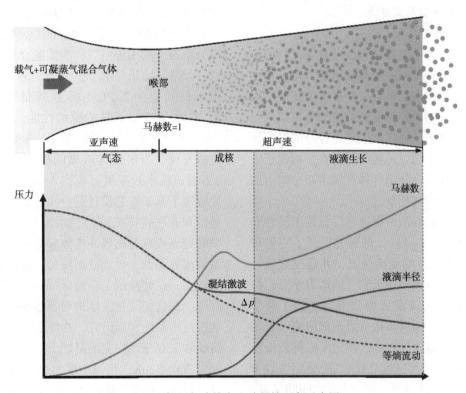

图6 拉瓦尔喷管中流动凝结现象示意图

喷管类型和几何形状的选择与优化在实现高效分离过程中至关重要。关键优化策略包括定制喷管以在出口处达到特定的马赫数、控制冲击波、调整扩张角、优化喉部尺寸、确保流量均匀分布，并考虑颗粒特性等。喷管的关键参数包括喉部直径、收缩角和扩张角。喉部直径直接决定气体的流量和流速；收缩段的形状影响气体的加速效率；扩张段则对维持超声速流动并防止冲击波的形成至关重要。此外，喷管的热力学特性会沿其轴向长度发生变化，因此喷管的位置在设计中是一个重要因素。由于每种喷管设计可能具有不同的几何形状，因此喷管的几何形状通常由一系列收缩-扩张角度和方程式决定。建立通用的喷管位置关系模型，有助于将其应用于不同几何形状的喷管设计。

拉瓦尔喷管根据流速分为三个区域：亚声速、声速和超声速部分。接下来是集液器和扩散器。当超声速流动通过扩散段时，会发生不可逆的正激波绝热跃迁。在这种亚稳态条件下，超声速流动转变为亚声速流动，导致熵、压力和温度的增加，但能量、动量和质量流量保持不变。为了防止由于再蒸发导致的分离损失，必须在激波上游收集凝结液。一旦激波形成，流体温度和压力恢复，并继续通过超声速分离器出口，然而这一过程会显著降低分离效率。因此，集液器的设计同样至关重要。此外，拉瓦尔喷管的设计与优化还需要考虑多种因素，包括入口和出口条件、流体特性、冲击波结构以及可靠性要求等。

3.3 超声速分离器研究进展

（1）数值模拟

合理的超声速分离器结构设计是确保其稳定和高效运作的关键。杨志毅系统阐述了超声速气体在不同组件中的流动特性，并提出了相应的设计理念与方法。刘恒伟对旋流后置型分离器进行了深入的理论剖析，并建立了超声速分离器的结构设计框架。他们通过适度简化装置，构建了一维数学模型，从而推导出气体在超声速分离器内部流动时各参数的解析解。

提升分离器的性能可以通过优化旋流特性和分离效率实现。Malyshkina 的研究设计了一种带有六片 45°倾斜叶片的旋流装置，并通过数值计算揭示了超声速分离器中天然气的动力学分布特性。Wen 等针对旋流前置型分离器进行了数学建模和深入分析，采用雷诺应力模型研究了旋流对天然气超声速流动的影响，得出了喉部旋流参数 S（切向速度与临界速度之比）与无量纲参数 M（有旋流与无旋流状态下流量之比）之间的关系。他们进一步通过对比 Boerner 等的实验数据，揭示了喷管膨胀制冷性能与旋流器旋流能力之间的制约机制。此外，通过离散粒子模型模拟了颗粒分离过程，结果显示大部分颗粒因离心作用聚集至分离器壁面或进入集液装置，少部分随气流逃逸。在此基础上，Wen 等优化了旋流叶片的设计参数，并在后续研究中探讨了三角翼在旋流后置型分离器中对流场特性和颗粒运动的影响。

在设计旋流后置型分离器时，激波干扰常被视为一大挑战，因为它可能破坏冷凝区域的稳定性，并导致已形成的液滴重新汽化。为了应对这一问题，Bian 等提出了一种优化方案，通过重新设计分离器的关键部件，例如优化 Laval 喷管的几何形状，具体包括减小其膨胀角、延长膨胀段，并用新型膨胀扩压段替代原有的膨胀结构和直管段。基于数值模拟的分析，这些设计改进显著增强了分离器的冷却效果和颗粒分离效率，同时有效缓解了激波对分离过程的负面作用。

（2）实验研究

针对先旋流后膨胀型分离器，2009 年马庆芬等开发了一种基于锥芯设计的超声速旋流分离装置，并详细阐述了旋流发生器的设计方法，同时优化了不同结构 Laval 喷管的几何参数，通过增加入口压力和相对湿度来改善分离。该装置因中心体采用锥芯设计，大大降低了加工难度。2014 年，文闯设计了一种带中心体的超声速旋流分离器，利用湿空气作为实验介质对脱水效果进行评估。结果显示，该分离器干气出口的水露点最低可达 -2.8℃，露点降低幅度最大为 34.9℃，表现出优异的脱水性能。2018 年，Niknam 等设计了一种实验装置，并通过低压脱水实验发现，当出口与入口的压力比由 0.8 降至 0.6 时，脱水效率可提高约 5%。2019 年，王荧光改进了基于锥芯设计的分离装置，采用了回流通道和齐平式排流结构（如图 7 所示），克服了喷管分流段容易出现冲击波和亚声速条件下出现漩涡流效率低，导致低温段短，冷却效果不理想的缺点。在此基础上，对导叶角度、回流管插入深度、回直径及排液结构等参数对脱水性能的影响进行了系统实验研究，在综合考虑良好的膨胀特性和离心分离性能的基础上，提出建议入口温度为 300~320K，出口角度为 50°~60°。

针对先膨胀后旋流型分离器，刘恒伟提出了设计的基本原则，并给出了预处理高压和低压天然气时喷管喉部尺寸的设计方法。根据天然气的组分、入口参数以及日处理量，即可获得所有部件的控制尺寸，完成超声速分离器的结构设计。Liu 等以湿空气为介质，通过室内实验对采用三种不同旋流器结构的分离器进行了性能测试(如图8所示)，发现流量、压损比、分离段尺寸、旋流结构以及激波是影响分离性能的关键因素。鲍玲玲等指出再循环超声速分离器能够有效分离出天然气中的烃类和液体，并通过循环提高分离效率。他们的实验结果表明，与现有装置相比，在相同压损比条件下，再循环分离装置能够实现更大的露点降，展现出显著的脱水效果。

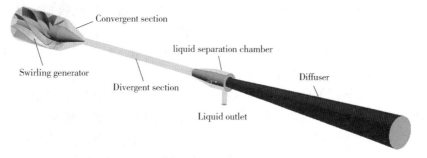

图7 带有分流锥的超声速装置结构图

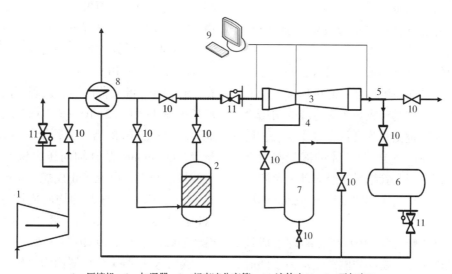

1—压缩机；2—加湿器；3—超声速分离管；4—液体出口；5—干气出口；
6—稳压罐；7—二次分离罐；8—换热器；9—计算机；10—截止阀；11—自力式调整阀

图8 室内超声速分离实验系统

4 超声速分离器内凝结机理研究

4.1 气体低温成核理论

在超声速喷管流动中，当气体通过喷管时，高速流动在过冷条件下诱导成核并克服能垒，随即发生相变。过饱和气体生成微小液滴，启动冷凝过程，使系统通过自发冷凝接近平衡状态。在超声速喷管流动中，自发成核显著依赖于过饱和度，因为过饱和度直接决定了气体的成核能力。

如图9所示，当气体流动接近饱和状态时，由于自由能垒的存在，蒸汽分子不会立即凝结，而是维持过饱和状态，直到系统达到临界过饱和条件，诱导成核并启动冷凝过程。更高的过饱和度会显著提高成核速率，而成核速率通常可通过吉布斯自由能 ΔG 进行表征：

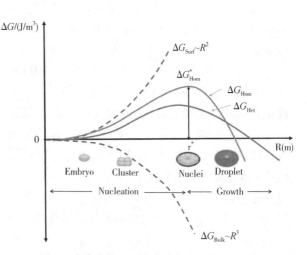

图9 液滴成核过程中吉布斯自由能的变化

$$\Delta G = 4\pi r^2 \sigma_r - \frac{4}{3}\pi r^3 \rho_L R T_G \ln(S) \quad (1)$$

式中，下标 L 代表液相；下标 G 代表气相；

r 为液滴半径；σ_r 为表面张力为；ρ_L 为液滴密度；T_G 为气体温度；S 为过饱和度。

通过计算液滴半径对应的最大自由能，可以确定形成液滴的临界半径。

$$r^* = \frac{2\sigma}{\rho_L RT_G \ln(S)} \quad (2)$$

式中，r^* 是液滴的临界半径。半径小于 r^* 的液滴会蒸发，而大于 r^* 的液滴会生长。利用所需半径可以估算出过冷蒸汽的凝结和成核速率。经典成核理论中成核率 J_{CNT} 可通过以下式计算：

$$J_{CNT} = \frac{\rho_G^2}{\rho_L}\sqrt{\frac{2\sigma_r}{\pi M^3}} \cdot \exp\left(-\frac{\Delta G^*}{KT_G}\right) \quad (3)$$

式中，ρ_G 为蒸汽密度；K 为玻尔兹曼常数；M 为分子质量。

在超声速气体分离技术中，揭示非平衡冷凝的复杂性，尤其是在纳米尺度下的机理，对深入理解成核过程至关重要。成核作为冷凝的初始阶段，对冷凝过程的量化和预测具有重要意义，这对于实现 H_2O、CO_2、CH_4 等气体的液化控制至关重要。然而，由于实验结果与经典理论预测之间存在显著差异，成核速率的定量测量仍面临巨大挑战。

近年来，分子动力学（MD）模拟为理解成核过程提供了微观视角，能够观测相变动力学随时间的演变。成核是一个随机过程，因此成核速率预测需要多个独立的 MD 模拟，以获得全面的见解。目前，MD 模拟的研究多集中在 H_2O 和 Lennard-Jones 流体上，对 CO_2、CH_4 成核的研究较少。

最近，Cao 等利用计算流体动力学（CFD）和 MD 模拟研究了 CO_2 在 CH_4 中的冷凝特性。CFD 模拟指出拉瓦尔喷管创造了促进 CO_2 冷凝的低温条件，通过控制入口温度和压力可以优化冷凝条件。成核阶段的 MD 模拟在分子尺度上揭示了 CO_2 气体在成核和生长途径中分子之间的复杂相互作用，潜热释放以及与周围分子能量相互作用影响的团簇稳定性。研究揭示了经典成核模型与 MD 模拟之间存在数量级的差异，强调了对经典理论进行修正的必要性。Guo 等研究了 CH_4 的均相和非均相冷凝特性，指出微量正己烷能显著提高 CH_4 的成核速率和液化率。预冷凝的正己烷团簇成为 CH_4 非均相成核的表面，显著降低了成核势垒。

4.2 气体超声速凝结研究

由于超声速膨胀与凝结过程的复杂性，一维模型在处理复杂流场问题时往往显得不足，因此需要采用二维或三维模拟方法。Matsuo 等基于粘性层效应构建了用于预测湿空气自发凝结行为的二维模型。White 和 Simpson 探讨了二维喷管边界层对凝结过程及流动参数的影响。Jassim 等则模拟了高压天然气在实际气体状态和喷管几何结构下的流动特性，研究表明，理想气体状态方程的应用会显著降低预测结果的准确性，而喷管几何参数的变化对激波位置的分布具有重要影响。王美利使用数值模拟方法研究了水蒸汽在超声速流动中的凝结过程及其对流场的作用。

Ghosh 等则针对超声速条件下的正构烷烃（C_iH_{2i+2}，$i = 7-10$）均质成核进行了实验研究。在 150~200K 温度范围内，发现不同烷烃的最大成核率在 $4×10^{15}-2×10^{18} cm^{-3} \cdot s^{-1}$ 之间。与经典成核理论计算值相比，实验结果高出 4.58 个数量级。Ogunronbi 结合轴向压力测量和小角度 X 射线散射技术，研究了正戊烷、正己烷和正庚烷在 109~168K 温度区间的均质成核行为。在喷管入口滞止压力为 30.1kPa、氩气作为载气时，不同烷烃的成核率为 $2.29×10^{16} ~ 5.48×10^{17} cm^{-3} \cdot s^{-1}$。

除了对超声速分离器内水蒸气、烷烃的凝结相变研究外，边江、孙文娟及 Chen 等还对天然气中 CO_2 和 H_2S 等酸性气体的超声速凝结特性进行了深入探讨。他们的研究表明，不同沸点的可凝组分在拉瓦尔喷管内的凝结行为存在显著差异。Deng 等针对 H_2O 与 CO_2 的超声速凝结特性进行了数值模拟，同一工作流体在亚临界和超临界条件下的液化率呈现显著差异；然而，在不同喷管入口条件下，不同工作流体的液化率尽管数值不同，但其变化趋势完全一致。这表明，喷管入口条件是决定液化率变化趋势的主要因素，而工作流体的类型并非关键。

5 超声速分离在天然气处理中的应用分析

5.1 天然气脱水

脱水在天然气处理与加工中起着至关重要的作用，因为它可以保护管道免受腐蚀并防止水合物的形成。天然气携带水蒸气的能力是有限的，取决于温度和压力。当管道内气相中的水分子开始凝结和聚集时，就会形成水合物，从而形成较大的颗粒。游离水形成的水合物有可能降低流量、输送效率，造成管道冲蚀，甚至堵塞管线。此外，气体中的水还会导致热值降低和管道腐蚀。水的露点温度会随着环境温度的降低而降低。

在天然气脱水技术中最主要的方法是吸收法和直接冷却法。Netusil 等比较了工业上三种广泛应用的方法：三甘醇的吸收法、固体干燥剂吸附法和冷凝法。从能源需求的角度来看，压力较高时推荐使用冷凝法。在可用压差不足的情况下，吸收法优于吸附法。其中分子筛吸附和三甘醇吸收是海上钻井平台天然气脱水的传统方法。然而，这些技术需要大量的基础设施投资，并产生大量的资本和运营成本。它们通常涉及运动部件，需要复杂的人工操作，带来安全问题，需要定期维护。此外，用作水合物抑制剂的传统化学添加剂可能对环境有害。相比之下超声速分离器方法在天然气脱水方面具有独特优势。

5.2 重烃组分去除

天然气中含有重烃，需要将其去除以增加其热容量、防止液化设备腐蚀并避免在液化过程中结晶。如果不能从天然气中分离出重烃，就会增加管道流量，并带来重大挑战，包括需要更大的管道直径、扩建工艺设施、增加电力需求以及成本大幅上升。

目前有多种方法可用于分离重烃，包括制冷过程、吸收过程、低温过程、表面吸收、膜分离和超声速分离器。在这些方法中，超声速分离器因其简单性、可靠性、安全性、较低的安装和操作成本、最小的压力降以及适用于沿海、海上和水下作业而受到欢迎。

5.3 二氧化碳捕集

在海上钻井平台中，化学吸收、膜吸收和物理吸收是常规从天然气中提取 CO_2 的主要方法。利用超声速分离器从高含碳的天然气中收集 CO_2 已经得到了广泛的研究。CO_2 的冷凝需要更低的温度，因此监测 CO_2 冻结对于避免超声速分离器堵塞至关重要。超声速分离器中 CO_2 的流动路径需要考虑冻结屏障的影响，以避免 CO_2 结晶，防止系统堵塞。Sun 等提出了高压下 CH_4-CO_2 进料中 CO_2 冷凝成核和液滴生长的模型。

Jiang 等最近采用了一种从天然气中分离 CO_2 的超声速分离器，研究了旋流结构对超声速流动特性的影响，获得了 CO_2 液滴和气体分离的机理。此外，他们还调查了入口压力、温度、CO_2 含量对冷凝参数影响。如图 10(a) 所示，液化效率 Y/Y_{in}（出口液态 CO_2 与入口气态 CO_2 的质量比）随着入口压力的增加而增加。当入口压力从 6MPa 增加到 9MPa 时，液化效率从 56.90% 增加到 79.97%，当入口压力低于 8MPa 时，液化效率随着入口压力的增加而显著增加。如图 10(b) 液化效率随入口温度的升高而降低。当入口温度从 273K 升至 288K 时，液化效率从 56.90% 降至 35.57%，而当入口温度低于 278K 时，液化效率随着入口温度的降低而显著增加。此外，入口 CO_2 含量对 CO_2 的液化效率没有显著影响，当入口 CO_2 分量从 20% 增加到 35% 时，液化效率保持在 57%～58% 之间。

5.4 天然气液化

天然气液化对于促进天然气的运输和储存至关重要，它通过将天然气转化为液化天然气（LNG）来实现。除了在分离天然气中的 CO_2 和 H_2S 方面的能力外，超声速分离器也可用于生成 LNG。图 11 展示了一个紧凑的天然气液化过程，天然气经过换热器预冷后，进入拉瓦尔喷管，在高速低温下液化。后续，气液混合物进入分离器，液体进入 LNG 储罐，而分离出的低温天然气与 LNG 储罐产生的蒸发气（BOG）混合后，进入换热器预冷入口天然气。该液化过程简单，通过利用 BOG 和低温气体的热量来提高能源效率，占地少，节能环保。此外，Bian 等还设计了一种充分利用冷能和分离能力的超声速旋流喷管，实现了天然气冷凝与旋流的耦合。在对二元天然气混合物进行模拟时发现，入口处温度的降低和压力的提高会导致冷凝区域向喷管入口方向偏移，同时对于 CH_4-C_2H_6 混合物，其成核速率峰值、液滴尺寸以及出口湿度均有所增加。温度的升高有助于增强喷管内部的切向速度，这有利于液滴与干燥天然气的有效分离。而进口压力的增加对切向速度的影响不大，其保持相对稳定。

因此，超声速分离器是一种以尽可能低的成本生产液化天然气的有效工具。分离器中的拉瓦尔喷管使流动变成超声速，导致温度急剧下降，最终发生冷凝。

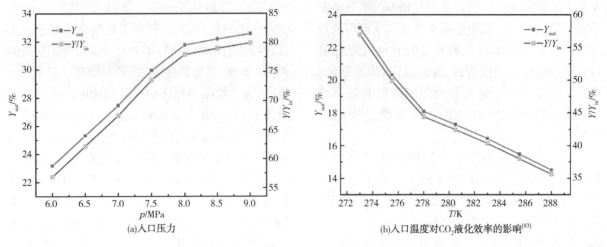

图 10 （a）入口压力、（b）入口温度对 CO_2 液化效率的影响

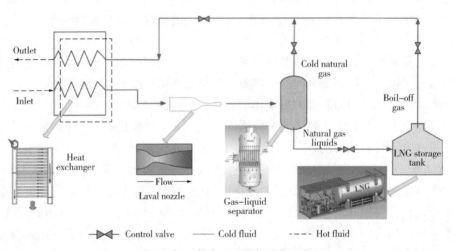

图 11 新型紧凑型天然气液化工艺

6 超声速分离技术发展的展望

天然气净化过程通常包括去除水、重烃、H_2S、CO_2 及汞等杂质。传统的天然气处理技术通常伴随着高资本和运营成本，且设施设计和运行受井口及天然气特性影响较大。这些传统方法（如脱水和脱碳系统）占地面积大、能耗高，且需要频繁的人工操作和定期维护。同时，使用水合物抑制剂等化学添加剂也带来了较大的环境风险。相比之下，超声速分离器作为一种新兴技术，结合了空气动力学、热力学和流体力学原理，为气体处理提供了创新的解决方案，具有显著的环境保护价值。该技术能高效去除天然气中的杂质，降低排放、改善空气质量，并最大限度减少环境污染。

随着超声速分离技术的不断发展，已有研究涉及其设计、功能和工业应用，特别是在天然气脱水方面的应用表现突出。该技术能够结合冷凝和分离过程，精确控制气体的水分和露点，改善天然气液化处理。然而，随着天然气消费需求的增加及环境挑战的日益复杂，仍需进一步优化技术流程，提高其效率和可操作性。

6.1 优化模型提高预测精度

为了提升超声速分离技术的应用效果，研究应聚焦于优化假设条件和参数估计方法，增强数值模拟的可靠性和精度。随着模拟技术的进步，MD 模拟已成为揭示超声速分离器中非平衡冷凝机理的重要工具。通过深入研究微观尺度的成核过程，可以全面理解气体冷凝行为。未来研究应结合最新成果，修正经典成核理论，优化天然气自发成核模型，并根据不同几何形状和运行条件调整成核速率及液滴生长方程，推动气体超声速冷凝理论，特别是在 CO_2 捕集领域的应用。

尽管已有大量研究，超声速分离器的流动特性仍需深入探索，以确保其具备更高的净化效率。建议在 CFD 技术应用中引入更精细的数值方法，并开发适用于更广泛温度和压力范围的热力学方程。此外，多组分模拟技术是研究不同组分浓度对流体流动及热力学特性影响的关键工具。未来的实验应聚焦于多组分气体的超声速凝结，通过对比不同混合物的凝结特性，定量表征多元烷烃混合物的凝结现象，进而建立多组分气体超声速凝结的理论模型。

6.2 优化设计提升分离性能

提高天然气超声速膨胀液化与旋流分离技术的效率是未来研究的一个重要方向，优化设备结构和工艺参数是实现这一目标的关键。旋流器和喷管结构是影响设备性能的关键因素，它们直接决定流体在设备内的速度分布和分离效果。开发新型拉瓦尔喷管和旋流器，以提升流动性能和分离效率，将有助于适应日益复杂的工业需求。

为了进一步提高效率，可以利用数值模拟等先进技术对整个过程进行深入分析，找出影响效率的关键因素并加以改进。此外，结合磁场辅助分离等新型辅助技术，可能为提升分离效率提供有效途径，弥补单一技术的局限性，进一步推动天然气处理系统性能提升。

在CO_2分离方面，仍面临高能耗和有限CO_2回收率的挑战。因此，设计时必须考虑几何形状、温度控制以及杂质对热力学性质的影响，以提高捕集效果、减少气体损失，并应对实际应用的复杂性。

6.3 适应复杂工况与技术集成

超声速分离器的工业集成至关重要，设备必须与现有的天然气处理系统兼容，涉及与压缩机、过滤器或热交换器等的集成。设计方案需精细规划，确保与过程控制和安全系统的无缝对接，从而保障整体系统的高效运行。将超声速膨胀液化与旋流分离技术与其他天然气处理技术集成，也是未来发展的趋势。例如，与膜分离技术、吸附分离技术结合，可以实现更高效、全面的天然气净化与液化。通过技术集成，能够弥补单一技术的不足，提升整体性能，为天然气产业发展提供更有力的技术支持。

随着天然气开采环境日益复杂，开发能够适应恶劣工况的超声速膨胀液化与旋流分离技术显得尤为必要。特别是在高H_2S或高CO_2含量等条件下的天然气处理技术，需要在设备材料选择、防腐措施和工艺调整等方面进行创新。此外，对于不同组成和状态的天然气，开发更具适应性的通用型技术，将有助于扩大技术的应用范围。因此，未来的研究可集中于工艺流程改进和技术集成，以进一步提升分离器的适用性。

7 结论

本文综述了天然气超声速膨胀液化与旋流分离技术的研究进展，重点探讨了该技术在天然气处理中的应用潜力。超声速分离技术因其高效、紧凑、低能耗且环保的特点，在天然气处理过程中展现出了显著的优势。通过超声速膨胀所带来的冷却效果和旋流分离原理，该技术能够高效去除天然气中的水蒸气、重烃以及酸性气体（如CO_2和H_2S），在提高天然气燃烧性能的同时，降低对环境的影响。此外，超声速分离技术还能在天然气液化过程中发挥重要作用，为天然气的高效运输和储存提供了可能。

尽管超声速分离器在实验研究和实际应用中已取得了显著进展，设备设计的进一步优化仍是未来研究的重点。特别是喷管形状、旋流器设计以及冷凝过程的结合优化，都是提升分离效率和能效的关键因素。当前的研究已证明数值模拟和实验研究在理解和改进技术性能方面的作用，未来应更多聚焦于设备结构设计和处理流程的优化，以适应多变的天然气组分和复杂工况。此外，如何提高预测模型的准确性，提升超声速分离器在不同操作条件下的稳定性，仍需深入探索。

参 考 文 献

[1] Cao HG, Cao XW, Wang ZX, et al. Advancing cryogenic carbon capture technology through understanding the CO_2 frosting crystallization mechanism[J]. Fuel, 2025, 384: 134006.

[2] Guo D, Cao XW, Ding GY, et al. Crystallization and nucleation mechanism of heavy hydrocarbons in natural gas[J]. Energy, 2022, 239: 122071.

[3] 边江, 曹学文, 孙文娟, 等. 气体超声速凝结与旋流分离研究进展[J]. 化工进展, 2021, 40(4): 1812-1826.

[4] 刘杨. 文23储气库超声速分离预脱水工艺技术研究[D]. 青岛: 中国石油大学(华东), 2017.

[5] 边江. 超声速分离器内CO_2相变凝结特性及结构优化研究[D]. 青岛: 中国石油大学(华东), 2017.

[6] Wen C, Cao XW, Yang Y, et al. Numerical simulation

of natural gas flows in diffusers for supersonic separators [J]. Energy, 2011, 37(1): 195-200.

[7] Haghighi M, Hawboldt A K, Abdi A M. Supersonic gas separators: Review of latest developments [J]. Journal of Natural Gas Science and Engineering, 2015, 27: 109-121.

[8] Medeiros L J, Arinelli O L, Teixeira M A, et al. Offshore Processing of CO_2-Rich Natural Gas with Supersonic Separator[M]. Switzerland: Springer Cham, 2019.

[9] Liu Y, Cao XW, Yang J, et al. Energy separation and condensation effects in pressure energy recovery process of natural gas supersonic dehydration[J]. Energy Conversion and Management, 2021, 245: 114557.

[10] Deng QH, Jiang Y, Hu ZF, et al. Condensation and expansion characteristics of water steam and carbon dioxide in a Laval nozzle[J]. Energy, 2019, 175: 694-703.

[11] Sun WJ, Cao XW, Yang W, et al. Numerical simulation of CO_2 condensation process from CH_4-CO_2 binary gas mixture in supersonic nozzles[J]. Separation and Purification Technology, 2017, 188: 238-249.

[12] Hou DY, Jiang WM, Zhao WX, et al. Effect of linetype of convergent section on supersonic condensation characteristics of CH_4-CO_2 mixture gas in Laval nozzle [J]. Chemical Engineering and Processing - Process Intensification, 2018, 133: 128-136.

[13] Chen JN, Jiang WM, Han CY, et al. Study on supersonic swirling condensation characteristics of CO_2 in Laval nozzle[J]. Journal of Natural Gas Science and Engineering, 2020, 84: 103672.

[14] Chen JN, Huang Z. Numerical study on carbon dioxide capture in flue gas by converging-diverging nozzle[J]. Fuel, 2022, 320: 123889.

[15] Cao XW, Guo D, Sun WJ, et al. Supersonic separation technology for carbon dioxide and hydrogen sulfide removal from natural gas[J]. Journal of Cleaner Production, 2021, 288: 125689.

[16] Bian J, Liu Y, Zhang XH, et al. Co-condensation and interaction mechanism of acidic gases in supersonic separator: A method for simultaneous removal of carbon dioxide and hydrogen sulfide from natural gas [J]. Separation and Purification Technology, 2023, 322: 124296.

[17] Wen C, Karvounis N, Walther J H, et al. An efficient approach to separate CO_2 using supersonic flows for carbon capture and storage [J]. Applied Energy, 2019, 238: 311-319.

[18] Cao XW Bian J. Supersonic separation technology for natural gas processing: A review[J]. Chemical Engineering and Processing-Process Intensification, 2019, 136: 138-151.

[19] Bian J, Jiang WM, Hou DY, et al. Condensation characteristics of CH_4-CO_2 mixture gas in a supersonic nozzle[J]. Powder Technology, 2018, 329: 1-11.

[20] Bian J, Cao XW, Yang W, et al. Supersonic liquefaction properties of natural gas in the Laval nozzle[J]. Energy, 2018, 159: 706-715.

[21] 张建, 孟庆华, 安文鹏, 等. 中国高含硫天然气集输与处理技术进展[J]. 油气储运, 2022, 41(6): 657-666.

[22] 王翀, 朱鑫鑫, 朱丽君. 天然气中硫化氢深度吸附剂的研究进展[J]. 石油化工, 2022, 51(11): 1354-1360.

[23] Teng L, Li XG, Lu SJ, et al. Computational fluid dynamics study of CO_2 dispersion with phase change of water following the release of supercritical CO_2 pipeline [J]. Process Safety and Environmental Protection, 2021, 154: 315-328.

[24] Yazdani S, Salimipour E, Moghaddam S M. A comparison between a natural gas power plant and a municipal solid waste incineration power plant based on an emergy analysis [J]. Journal of Cleaner Production, 2020, 274: 123158.

[25] George B V, Lara A O D, Luiz J M D, et al. Lowpressure supersonic separator with finishing adsorption: Higher exergy efficiency in air pre-purification for cryogenic fractionation [J]. Separation and Purification Technology, 2020, 248: 116969.

[26] Brouwer J M, Epsom H D. Twister supersonic gas conditioning for unmanned platforms and subsea gas processing [C]//SPE Offshore Europe Oil and Gas Exhibition and Conference. United Kingdom: 2003: SPE-83977-MS.

[27] 王协琴, 罗小米, 孙玉梅. 超音速分离器: 天然气脱水脱烃的新型高效设备[J]. 天然气技术, 2007 (05): 63-67.

[28] 陈佳男. 旋流器结构优化及CO_2超声速旋流凝结特性研究[D]. 青岛: 中国石油大学(华东), 2021.

[29] 安迪. 新型超音速气液分离器的结构设计和实验研究[D]. 西安: 西安石油大学, 2020.

[30] 许晶, 虞兰剑, 刘雪东. 基于叠加效应的拉瓦尔喷管结构参数优化设计分析[J]. 机械研究与应用, 2023, 36(04): 19-22.

[31] Zhang WW, Wang DA, Renganathan A, et al. Modeling and assessment of two-phase transonic steam flow with condensation through the convergent - divergent

nozzle[J]. Nuclear Engineering and Design, 2020, 364: 110632.

[32] 刘杨. 水/重烃组分超声速凝结机理研究[D]. 青岛: 中国石油大学(华东), 2024.

[33] Brigagão V G, Arinelli O D L, Medeiros D L J, et al. A new concept of air pre-purification unit for cryogenic separation: Low-pressure supersonic separator coupled to finishing adsorption[J]. Separation and Purification Technology, 2019, 215: 173-189.

[34] 杨志毅. 油气超音速旋流分离技术研究[D]. 成都: 西南石油学院, 2004.

[35] 刘恒伟. 超音速分离管的研发及其流动与传热传质特性的研究[D]. 北京: 北京工业大学, 2006.

[36] 刘恒伟, 刘中良, 张建, 等. 超声波旋流脱水装置及其内部流动的理论解[J]. 北京工业大学学报, 2006, 9: 829-831.

[37] Malyshkina M M. The structure of gasdynamic flow in a supersonic separator of natural gas[J]. High temperature, 2008, 46(1): 69-76.

[38] Wen C, Cao XW, Yang Y, et al. Swirling effects on the performance of supersonic separators for natural gas separation[J]. Chemical Engineering & Technology, 2011, 34(9): 1575-1580.

[39] Boerner C J, Sparrow E M, Scott C J. Compressible swirling flow through convergent-divergent nozzles[J]. Wärme-und Stoffübertragung, 1972, 5(2): 101-115.

[40] Wen C, Cao XW, Yang Y, et al. Supersonic swirling characteristics of natural gas in convergent-divergent nozzles[J]. Petroleum Science, 2011, 8(1): 114-119.

[41] Wen C, Cao XW, Yang Y, et al. Evaluation of natural gas dehydration in supersonic swirling separators applying the Discrete Particle Method[J]. Advanced Powder Technology, 2012, 23(2): 228-233.

[42] Yang Y, Li AQ, Wen C. Optimization of static vanes in a supersonic separator for gas purification[J]. Fuel Processing Technology, 2017, 156: 265-270.

[43] Wen C, Yang Y, Walther H J, et al. Effect of delta wing on the particle flow in a novel gas supersonic separator[J]. Powder Technology, 2016, 304: 261-267.

[44] Bian J, Jiang WM, Teng L, et al. Structure improvements and numerical simulation of supersonic separators[J]. Chemical Engineering and Processing: Process Intensification, 2016, 110: 214-219.

[45] 马庆芬. 旋转超音速凝结流动及应用技术研究[D]. 大连: 大连理工大学, 2009.

[46] Ma QF, Hu DP, He GH, et al. Performance of inner-core supersonic gas separation device with droplet enlargement method[J]. Chinese Journal of Chemical Engineering, 2009, 17(6): 925-933.

[47] 文闯. 湿天然气超声速旋流分离机理研究[D]. 青岛: 中国石油大学(华东), 2014.

[48] Niknam H P, Mortaheb H, Mokhtarani B. Dehydration of low-pressure gas using supersonic separation: Experimental investigation and CFD analysis[J]. Journal of Natural Gas Science and Engineering, 2018, 52: 202-214.

[49] Wang YG, Hu DP. Structure improvements and numerical simulation of supersonic separators with diversion cone for separation and purification[J]. RSC Advances, 2018, 8(19): 10228-10236.

[50] 王荧光. 循环超音速分离器的流体流动及实验性能研究[D]. 大连: 大连理工大学, 2019.

[51] Liu HW, Liu ZL, Feng YX, et al. Characteristics of a supersonic swirling dehydration system of natural gas[J]. Chinese Journal of Chemical Engineering, 2005, 13(1): 9-12.

[52] 鲍玲玲, 刘中良, 刘杰. 湿气再循环超音速分离管脱水性能的试验研究[J]. 石油机械, 2017, 45(7): 83-87.

[53] 刘兴伟, 刘中良, 鲍玲玲. 旋流后置型超音速分离管数值模拟与实验研究[J]. 低温工程, 2014, 1: 37-44.

[54] Kukushkin S A, Osipov A V. Kinetics of thin film nucleation from multi-component vapor[J]. Journal of Physics and Chemistry of Solids, 1995, 56(6): 831-838.

[55] Hosseini S A, Aghdasi M R, Lakzian E, et al. Multi-objective optimization of the effects of superheat degree and blade pitch on the wet steam parameters[J]. International Journal of Heat and Mass Transfer, 2023, 213: 124337.

[56] Noppel M, Vehkamäki H, Winkler P M, et al. Heterogeneous nucleation in multi-component vapor on a partially wettable charged conducting particle. I. Formulation of general equations: Electrical surface and line excess quantities[J]. The Journal of Chemical Physics, 2013, 139(13): 134107.

[57] Ebrahimi-Fizik A, Lakzian E, Hashemian A. Entropy generation analysis of wet-steam flow with variation of expansion rate using NURBS-based meshing technique[J]. International Journal of Heat and Mass Transfer, 2019, 139: 399-411.

[58] Taqieddin A, Allshouse M R, Alshawabkeh A N. Review—Mathematical formulations of electrochemically gas-evolving systems[J]. Journal of The Electrochemical Society, 2018, 165(13): E694-E711.

[59] Momeni A D, Samaneh M, Esmail L. Optimization variables of the injection of hot-steam into the non-equilibrium condensing flow using TOPSIS method[J]. International Communications in Heat and Mass Transfer, 2021, 129: 105674.

[60] Martin H, Jadran V, Martin B, et al. Homogeneous nucleation in supersaturated vapors of methane, ethane, and carbon dioxide predicted by brute force molecular dynamics[J]. The Journal of chemical physics, 2008, 128(16): 164510.

[61] Bian J, Guo D, Li YX, et al. Homogeneous nucleation and condensation mechanism of methane gas: A molecular simulation perspective[J]. Energy, 2022, 249: 123610.

[62] Dumitrescu L R, Smeulders D M J, Dam J A M, et al. Homogeneous nucleation of water in argon. Nucleation rate computation from molecular simulations of TIP4P and TIP4P/2005 water model[J]. Journal of Chemical Physics, 2017, 146(8): 084309.

[63] Halonen R. Atomistic insights into argon clusters and nucleation dynamics[J]. Journal of Aerosol Science, 2024, 181: 106406.

[64] Cao HG, Cao XW, Li H, et al. Nucleation and condensation characteristics of carbon dioxide in natural gas: A molecular simulation perspective[J]. Fuel, 2023, 342: 127761.

[65] Guo D, Cao XW, Zhang P, et al. Heterogeneous condensation mechanism of methane-hexane binary mixture[J]. Energy, 2022, 256: 124627.

[66] Matsuo S, Setoguchi T, Yu S, et al. Effect of non-equilibrium condensation of moist air on the boundary layer in a supersonic nozzle[J]. Journal of Thermal Science, 1997, 6(4): 260-272.

[67] Simpson D, White A. Viscous and unsteady flow calculations of condensing steam in nozzles[J]. International Journal of Heat and Fluid Flow, 2005, 26(1): 71-79.

[68] Jassim E, Abdi A M, Muzychka Y. Computational fluid dynamics study for flow of natural gas through high-pressure supersonic nozzles: Part 1. Real gas effects and shockwave [J]. Petroleum Science and Technology, 2008, 26(15): 1757-1772.

[69] 王美利. 高速膨胀流动中的非平衡凝结及其对流场影响的研究[D]. 合肥：中国科学技术大学, 2006.

[70] Ghosh D, Bergmann D, Schwering R, et al. Homogeneous nucleation of a homologous series of n-alkanes (C_iH_{2i+2}, $i=7-10$) in a supersonic nozzle[J]. The Journal of Chemical Physics, 2010, 132(2): 24307.

[71] Ogunronbi K E, Sepehri A, Chen B, et al. Vapor phase nucleation of the short-chain n-alkanes (n-pentane, n-hexane and n-heptane): Experiments and Monte Carlo simulations[J]. The Journal of Chemical Physics, 2018, 148(14): 144312.

[72] 孙文娟. 天然气超声速旋流脱酸气机理研究[D]. 青岛：中国石油大学(华东), 2018.

[73] Deng QH, Jiang Y, Hu ZF, et al. Condensation and expansion characteristics of water steam and carbon dioxide in a Laval nozzle[J]. Energy, 2019, 175: 694-703.

[74] Netusil M, Ditl P. Comparison of three methods for natural gas dehydration[J]. Journal of Natural Gas Chemistry, 2011, 20(5): 471-476.

[75] Liu Y, Mu HY, Lv XF, et al. Toward greener flow assurance: Review of experimental and computational methods in designing and screening kinetic hydrate inhibitors[J]. Energy & Fuels, 2024, 38(18): 17191-17223.

[76] Ashtiani S A, Haghnejat A, Sharif M, et al. Investigation on new innovation in natural gas dehydration based on supersonic nozzle technology [J]. Indian journal of science and technology, 2015, 8(S9): 450-450.

[77] 郭建，王刚，郑春来，等. 天然气液化中重烃和氮气脱除工艺优化研究[J]. 天然气化工(C1化学与化工), 2019, 44(5): 76-81.

[78] Teixeira M A, Arinelli O D L, Medeiros D L J, et al. Recovery of thermodynamic hydrate inhibitors methanol, ethanol and MEG with supersonic separators in offshore natural gas processing[J]. Journal of Natural Gas Science and Engineering, 2018, 52: 166-186.

[79] Chen B, Yang T, Xiao W, et al. Conceptual design of pyrolytic oil upgrading process enhanced by membrane-integrated hydrogen production system [J]. Processes, 2019, 7(5): 284-284.

[80] Arinelli O D L, Trotta F A T, Teixeira M A, et al. Offshore processing of CO_2 rich natural gas with supersonic separator versus conventional routes[J]. Journal of Natural Gas Science and Engineering, 2017, 46: 199-221.

[81] Sun WJ, Cao XW, Yang W, et al. Numerical simulation of CO_2 condensation process from CH_4-CO_2 binary gas mixture in supersonic nozzles[J]. Separation and Purification Technology, 2017, 188: 238-249.

[82] Jiang WM, Bian J, Wu A, et al. Investigation of supersonic separation mechanism of CO_2 in natural gas applying the Discrete Particle Method[J]. Chemical

Engineering and Processing – Process Intensification, 2018, 123: 272-279.

[83] Bian J, Jiang WM, Hou DY, et al. Condensation characteristics of CH_4-CO_2 mixture gas in a supersonic nozzle[J]. Powder Technology, 2018, 329: 1-11.

[84] Bian J, Cao XW, Yang W, et al. Supersonic liquefaction properties of natural gas in the Laval nozzle[J]. Energy, 2018, 159: 706-715.

[85] Bian J, Cao XW, Teng L, et al. Effects of inlet parameters on the supersonic condensation and swirling characteristics of binary natural gas mixture[J]. Energy, 2019, 188: 116082.

Comparison and selection of decarbonization process based on natural gas liquefaction under pressure

Li Zihe[1]　Liu Jinhua[2]　Zhu Jianlu[1]

[1. College of Pipeline and Civil Engineering, China University of Petroleum (East China);
2. China Petrochemical Sales Co., Ltd. Hebei Shijiazhuang Petroleum Branch]

Abstract　PLNG (Pressurized Liquefied Natural Gas) technology refers to the liquefaction of natural gas under pressure conditions (1.0MPa~2.0MPa) and storage and transportation under corresponding pressure. This technology can increase the solubility of CO_2 in LNG, reduce the purification standard required for liquefaction, simplify the pretreatment device in the liquefaction process, and further improve the economy of LNG. [Purpose] In order to give full play to the advantages of PLNG technology in reducing the CO_2 treatment index, a new integrated liquefaction process is proposed by combining the pretreatment process with the pressure liquefaction technology. Therefore, it is necessary to establish the corresponding integrated process and compare the decarburization process in combination with the actual production situation, so as to obtain a reasonable integrated scheme. [Method] In this paper, the process models of four pretreatment and pressurized liquefaction integrated processes based on semi-lean liquid decarbonization, amine decarbonization, CRS (Condensed Rotational Separation) technology and low-temperature packed bed decarbonization were established, and the steady-state simulation was carried out by HYSYS software. From the three aspects of floor space, process energy efficiency and safety, the results of process simulation are analyzed, the decarbonization methods are compared and selected, and the applicability of the process is analyzed. [Results] (1) The heat transfer area of the single circulating amine method is the largest, and the heat transfer area of the low temperature packed bed method is the smallest. The numbers of equipment for the semi-lean liquid deacidification method are the largest, and the number of CRS technology and low-temperature packed bed equipment is the least and there is no tower. (2) The CRS technology has the lowest total energy consumption and the lowest exergy loss, followed by the low-temperature packed bed method. (3) Compared with CRS technology, the separation temperature of low temperature packed bed is relatively fixed, so it has strong adaptability to natural gas source, and there is no tower, which is less affected by sea conditions. [Conclusions] (1) The traditional single-cycle amine method has high purification accuracy, but its required regeneration heat load is much higher than other processes, which is suitable for natural gas with low CO_2 content. (2) The semi-lean liquid decarburization method has high purification accuracy and is suitable for natural gas with high carbon content, but it covers a large area and has a large number of towers, which is not suitable for marginal gas fields and offshore natural gas generation. (3) The purification capacity of CRS decarbonization technology meets the needs of pressurized liquefaction process, with low energy consumption, small footprint and good safety. However, the gas-liquid separation method is sensitive to natural gas components, so it is suitable for natural gas liquefaction with stable gas source. (4) The low-temperature packed bed decarbonization method has low energy consumption and footprint. The gas-solid separation method has strong adaptability to natural gas components and is suitable for natural gas liquefaction of various gas sources. The conclusions of this paper can provide suggestions for the selection of decarburization methods for pressurized liquefaction process.

Keywords　Pressured liquid natural gas, decarburization process, low-temperature packed bed

Pressured Liquid Natural Gas (PLNG) technology is a new type of liquefaction process based on reducing the cost of liquefaction and refrigeration of FLNG (Floating Liquid Natural Gas) device. It means that natural gas is liquefied at high pressure (1.0~2.0MPa) and stored and transported under the corresponding pressure. The increase of liquefaction pressure makes the condensation temperature also increase significantly (about $-100 \sim -120\,^\circ\mathrm{C}$). The increase of natural gas condensation temperature not only reduces the power required for liquefaction, but also increases the solubility of CO_2 in LNG, making it possible to simplify or even cancel the pretreatment device. Therefore, the equipment required for refrigeration and pretreatment is reduced, and the area occupied by the device is also greatly reduced, which greatly enhances the overall reliability and adaptability of the FLNG device. In order to further improve the compactness of PLNG liquefaction device and improve the utilization rate of LNG cold energy, a new integrated process of pretreatment and pressurized liquefaction is proposed to realize the organic combination of low temperature impurity removal technology and pressurized liquefaction.

LNG liquefaction process mainly includes cascade liquefaction process, mixed refrigerant liquefaction process and expansion refrigeration liquefaction process. Based on the limitation of deck area and safety considerations, the FLNG device mainly uses the mixed refrigerant liquefaction process and the expansion refrigeration liquefactionprocess. According to statistics, in the common liquefaction process, the average relative energy consumption of single-stage nitrogen expansion is the highest, and the double mixed refrigerant (DMR) is the lowest. However, for offshore applications, the single mixed refrigerant (SMR) and nitrogen expansion process are considered to be the most promising options. Li et al. designed the FLNG mixed refrigerant liquefaction process, propane pre-cooled mixed refrigerant liquefaction process and nitrogen expansion refrigeration liquefaction process. The results show that although the nitrogen expansion refrigeration liquefaction process has high energy consumption and poor economy, its structure is simple and compact, the required deck area is small, and it is not sensitive to FLNG motion. It has better adaptability, higher safety and more convenient operation, and has obvious advantages in offshore applications. The study of Li et al. also showed that the propane pre-cooling double nitrogen expansion liquefaction process has better comprehensive performance and can adapt to harsh sea conditions. Therefore, for the pressurized liquefaction process, the gas expansion refrigeration liquefaction process is one of the main research directions. Xiong Xiaojun et al. first designed a gas expansion refrigeration with pressure liquefaction process, which has a significantly reduced floor space and energy consumption compared with the conventional process, and the process was optimized by two aspects of process parameters and refrigerant components. Subsequently, the cascade, single mixed refrigerant (SMR) and single expander natural gas liquefaction processes with PLNG technology are simulated. Compared with the conventional process, the three processes are significantly reduced in specific power consumption and heat transfer area, which theoretically confirms the advantages of the PLNG process. Subsequently, the CO_2 low-temperature removal process and the expansion-type pressurized liquefied natural gas process and combined, were analyzed and optimized. Lee et al. added the DEA aqueous solution deacidification process to the nitrogen expansion refrigeration liquefaction process, and analyzed the effect of CO_2 concentration in the feed gas on the deacidification gas device. It is considered that when the CO_2 concentration is 0.5mol% and the pressure is higher than 15 bar, the acid gas removal device is not required in the PLNG. However, in the above research, the comparison and selection of different decarburization processes are not involved. Therefore, this paper designs a nitrogen expansion refrigeration liquefaction process based on different decarburization processes, and compares the process performance.

1 Decarburization process comparison

The most mature CO_2 removal method is the chemical absorption method. However, the influence of complex sea conditions on the tower may cause the treatment effect to fail to meet the specification requirements. The ship motion and inclination caused by ocean swell will affect the hydrodynamics of the packed bed reactor. In addition, the large area is also one of the limitations of the chemical absorption method. In comparison, the membrane separation method has simple process equipment and system, less investment, small footprint, high reliability and flexible use, but its removal depth is limited, and it is more suitable for occasions with moderate removal depth requirements. In addition, the low temperature separation method has the characteristics of green environmental protection, low carbon footprint and suitable for high pressure environment. Because of the simple equipment, no tower and deep removal degree, it can well adapt to the complex sea conditions. The low-temperature separation method corresponds to different removal processes according to different phase transition types. Babar et al. designed and manufactured an experimental device composed of a low-temperature packed bed and an efficient control system to achieve CO_2 removal at low temperatures. Supersonic separation combines non-equilibrium condensation and swirl separation to achieve the separation of CO_2 droplets. At the same time, it is found that higher pressure makes the condensation process easier to achieve and can be applied to offshore natural gas treatment. Therefore, this paper chooses the decarbonization process based on different methods for comparison, which are single circulating amine method, semi-lean liquid deacidification, CRS technology decarbonization and low temperature packed bed decarbonization.

1.1 Introduction of common decarburization process

(1) Single circulating amine method

As shown in Fig. 1, natural gas (NG) first enters the bottom of the absorption tower (T-100), and flows from bottom to top in contact with the amine liquid (MDEA) entering the top of the tower to remove CO_2. The rich liquid absorbing CO_2 flows out from the bottom of the absorption tower, and the CO_2 is removed by throttling (VLV-100) depressurization flashing, then enters the heat exchanger (LNG-100) for heating, and then enters the regeneration tower (T-100) to regenerate into lean liquid (logistics 5). The lean liquid enters the heat exchanger for condensation, and is pressurized in turn with the added lean amine liquid, and is cooled by the air cooler (AC-100) into the absorption. This cycle.

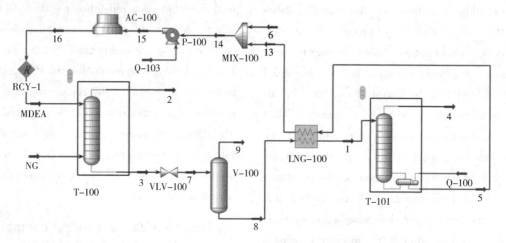

Fig. 1 Single cyclic amine method

(2) Deacidification of semi-lean solution

As shown in Fig. 2, the feed gas (stream 1) flows upward from the bottom of the absorption tower (T-100) while counter currently contacting with the activated MDEA solution downward from the top of the tower. This flow design enables CO_2 in the feed gas to be efficiently absorbed. The lean liquid (stream 3) at the top of the absorption tower and the semi-lean liquid (stream 2) in the middle play a key role in the absorption of CO_2. The rich liquid after absorbing CO_2 flows out from the bottom of the absorption tower. After decompression by throttling (VLV-100), it enters the flash tower (V-100) to further process and separate the hydrocarbon gas, which is generally transported to the fuel gas system or directly discharged. The rich liquid after flash evaporation enters the absorption tower (T-101) and contacts with the high temperature water vapor from the regeneration tower (T-102) and the high temperature semi-lean liquid gas from the gas-liquid separation tank (V-102). In this process, most of the CO_2 in the rich liquid is desorbed into the semi-lean liquid. Then, all the semi-lean liquid flows from the bottom of the absorption tower (T-101) to the gas-liquid separation tank (V-102) to achieve gas-liquid separation. After separation, part of the semi-lean liquid returns to the absorption tower and continues to be recovered. The other part of the semi-lean liquid enters the regeneration tower for heating and regeneration, and is completely regenerated into lean liquid under high temperature conditions. After heat exchange, cooling and pressurization, it is finally reused in the absorption tower (T-100).

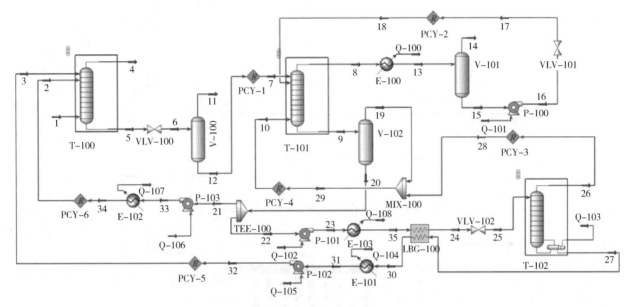

Fig. 2 Deacidification of semi-lean solution

(3) Decarbonization of CRS technology

The principle of condensed condensed rotational separation (CRS) technology is a gas-liquid separation method carried out by means of expansion cooling and liquefaction of rotational particle separator (RPS). In order to achieve higher removal accuracy, two-stage separation can be designed. As shown in Fig. 3, the feed gas (stream 7) is cooled to about -55 ~ -57℃ by a combination of isobaric heat exchange (E-100) and expansion (K-103). This temperature is 1~2℃ away from CO_2 freezing, otherwise there will be a possibility of solid formation blocking the pipeline. After the first PRS separation, high-purity CO_2 liquid (stream 11) and low-purity CO_2 vapor (stream 10) are obtained. The gas phase was compressed to 8MPa by compressor (C-104) and cooled to -57℃ by cooler (E-102). The pressure was then expanded to 1.5MPa by an expander (K-104), and the liquid phase was separated into the first PRS, and the gas phase was the obtained product.

(4) Low temperature packed bed decarburization

The low-temperature packed bed is a method for separating CO_2 by low-temperature phase transition. CO_2 is removed by cooling liquid nitrogen to form CO_2 solids and adsorbing solid particles. The model diagram is shown in Fig. 4, and the process diagram is shown in Fig. 5. The liquid nitrogen (LN_2) is first cooled in the low temperature packed bed 1. After the feed gas is introduced, the CO_2 is cooled for gas-solid separation, and the solid CO_2 is attached to the packing. The remaining feed gas flows out of the packed bed 1 and enters the next process. At the same time, the liquid nitrogen was introduced into the low temperature packed bed 2 for precooling. When the amount of CO_2 in the feed gas out of the low-temperature packed bed 1 exceeds 1%, the feed gas is introduced into the low-temperature packed bed 2 for further removal. At this time, the low temperature packed bed 1 should be rewarmed by CO_2, so that the solid CO_2 in the packing is heated to the gaseous state and then flows out. The same method is operated in the low temperature packed bed 2 to ensure the continuous decarburization of the feed gas.

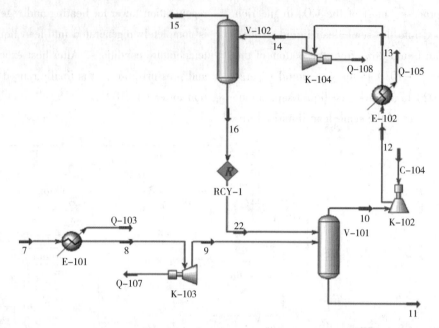

Fig. 3 Decarbonization of CRS technology

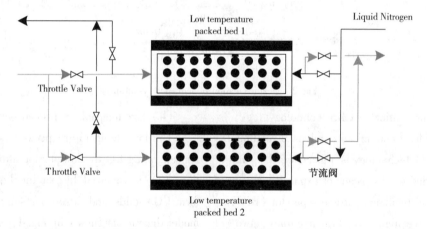

Fig. 4 Low temperature packed bed decarburization

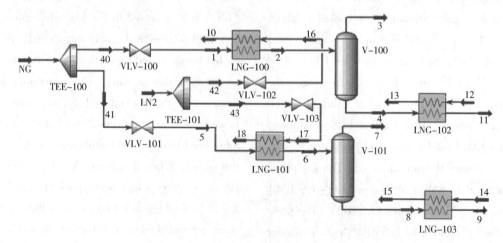

Fig. 5 Low temperature packed bed decarburization

1.2 Integrated process model of decarburization and pressurized liquefaction

(1) Semi-lean liquid decarbonization with liquefied natural gas

As shown in Fig. 6, the feed gas (stream 1) flows upward from the bottom of the absorption tower (T-100) and is in countercurrent contact with the activated MDEA solution at the top of the tower. This flow design efficiently absorbs CO_2 in the feed gas, in which the lean solution at the top of the absorption tower (stream 3) and the semi-lean solution in the middle (stream 2) play a key role in the absorption of carbon dioxide. The rich liquid after absorbing CO_2 flows out from the bottom of the absorption tower. After decompression by throttling (VLV-100), it enters the flash tower (V-100) to further process and separate the hydrocarbon gas, which is generally transported to the fuel gas system or directly discharged. The rich liquid after flash evaporation enters the absorption tower (T-101) and contacts with the high temperature water vapor from the regeneration tower (T-102) and the high temperature semi-lean liquid gas from the gas-liquid separation tank (V-102). In this process, most of the CO_2 in the rich liquid is desorbed into the semi-lean liquid. Then, all the semi-lean liquid flows from the bottom of the absorption tower (T-101) to the gas-liquid separation tank (V-102) to achieve gas-liquid separation. After separation, part of the semi-lean liquid returns to the absorption tower and continues to be recovered. The other part of the semi-lean liquid enters the regeneration tower for heating and regeneration, and is completely regenerated into lean liquid under high temperature conditions. After heat exchange, cooling and pressurization, it is finally reused in the absorption tower (T-100). The natural gas purified from the absorption tower is liquefied natural gas through the nitrogen expansion refrigeration liquefaction system.

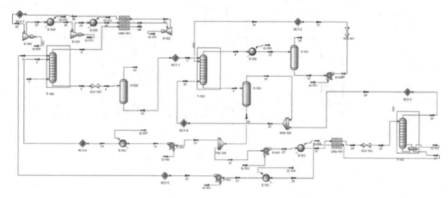

Fig. 6 Process flow chart of semi-lean liquid decarbonization with liquefied natural gas

(2) Single cycle amine deacidification with liquefied natural gas

As shown in Fig. 7, the natural gas enters the absorption tower to contact with the amine liquid to remove the CO_2, and the rich liquid absorbing the CO_2 enters the heat exchanger for heating, and then enters the regeneration tower to regenerate the lean liquid. The lean liquid enters the heat exchanger for condensation, and the supplementary lean amine liquid is pressurized in turn, and the air cooler is cooled to enter the absorption tower for circulation. Similarly, the natural gas purified in the absorption tower is liquefied natural gas through the nitrogen expansion refrigeration liquefaction system.

(3) Integration of decarbonization and LNG based on CRS technology

As shown in Fig. 8. The HYSYS model of natural gas pretreatment and pressurized liquefaction process is established based on condensation swirl separation technology (CRS) and expanded liquefied natural gas technology. The heat exchanger and expander are used to cool the natural gas rich in CO_2, which promotes the condensation phase change of CO_2 to produce droplets. The gas-liquid mixture mixed with a large number of CO_2 droplets enters the swirl particle separation device. In the microchannel of the swirl particle separation device, the centrifugal force generated by high-speed rotation is used to separate the condensate droplets. The liquid CO_2 is pressurized by the pump and then transported outside. After purification, the natural gas is pressurized by the compressor and then enters the heat exchanger for cooling and liquefaction.

(4) Integrated process of decarbonization and LNG based on low temperature packed bed

Based on the method of low temperature packed bed and mixed refrigerant liquefied natural gas technology, a new HYSYS model of integrated process of pretreatment and liquefaction of pressurized natural gas was established. This design uses two low-temperature beds, one of which cools and captures CO_2 in natural gas, and the other recovers CO_2 condensed in the packed bed, alternating operation; in the HYSYS simulation, the heat exchanger and solid separator were used to replace the low temperature packed bed. The process diagram is shown in Fig. 9, and the similar simulation diagram is shown in Fig. 10.

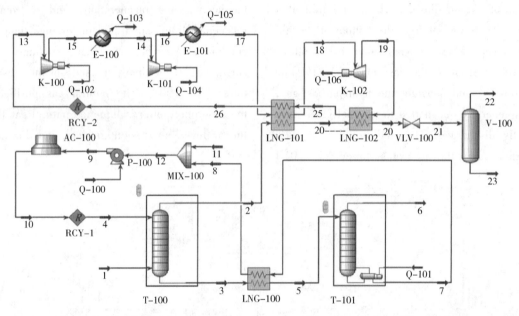

Fig. 7 Process flow chart of single cycle amine deacidification with liquefied natural gas

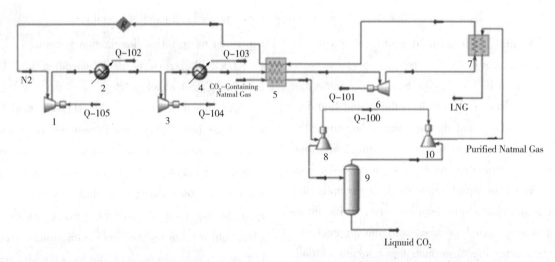

Fig. 8 Process flow chart of the integration of decarbonization and LNG based on CRS technology

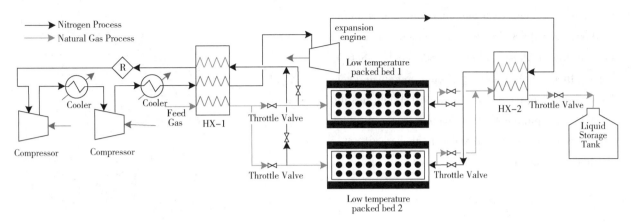

Fig. 9 Process flow chart of low-temperature packing bed decarburization and LNG integration

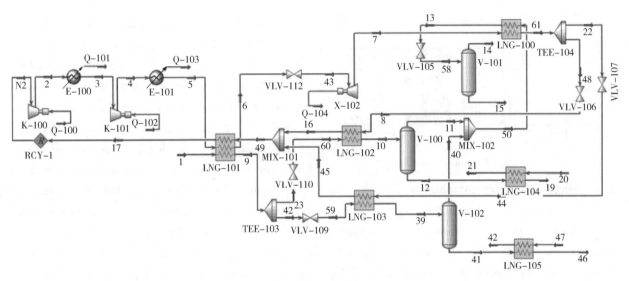

Fig. 10 Integrated process of decarbonization and LNG based on low temperature packed bed

2 Comparison of decarbonization process performance

In this paper, the performance of the above four integrated processes of pretreatment and pressurized liquefaction based on different decarbonization processes is compared from three aspects: occupied areas, energy efficiency and safety.

2.1 Occupied areas

(1) Comparison of heat transfer areas

Taking the traditional single-cycle amine deacidification and liquefied natural gas integration process as the comparison standard, the relative heat transfer area UA of the four forms of processes obtained is shown in Fig. 11. The heat transfer area of the single circulating amine process is the largest, which is 49.45%, 49.2% and 75.12% higher than that of the semi-lean liquid, CRS and low temperature packed bed processes, respectively.

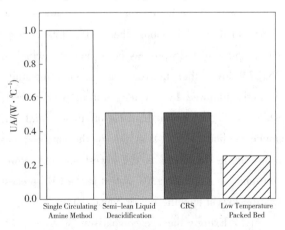

Fig. 11 Comparison of heat transfer areas

(2) Device quantity comparison

Taking the single cycle process as the comparison standard, the relative number of equipment of the four forms of processes is shown in Fig. 12. The semi-lean liquid process has the most equipment, which is 57% more than the single circulating amine method, and the CRS and low temperature packed bed are the smallest, which is reduced by 38.6%. Among them, only semi-lean liquid and single cycle have tower equipment, and semi-lean liquid has the most tower equipment. The complex sea conditions bring great technical challenges to the operation of the tower on the ship.

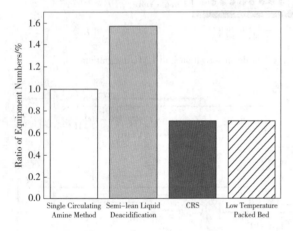

Fig. 12　Device quantity comparison

2.2　Energy efficiency comparison

(1) Power consumption comparison

Taking the single cycle process as the comparison standard, the specific power consumption is analyzed, including the total specific power consumption of the system, the specific power consumption of CO_2 and the specific power consumption of LNG. The specific power consumption of the four forms of processes is shown in Fig. 13. Among them, the total energy consumption of CRS process is the lowest, which is 76.27% lower than that of single circulating amine method, followed by low temperature packed bed, which is 68.53% lower than that of single circulating amine method. The CO_2 system in the single circulating amine method has the lowest specific power consumption, and the LNG system in the CRS process has the lowest specific power consumption.

(2) Exergic damage comparison

In order to compare the exergy efficiency of the integrated process of pretreatment and liquefaction, the above four liquefaction processes were analyzed for exergy loss, including compressor, expander, heat exchanger, cooler, fluid mixer, fluid mixer and pump exergy loss. The four forms of exergy loss are

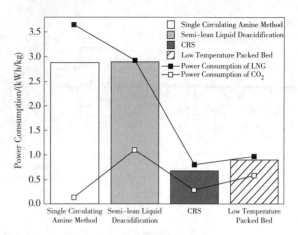

Fig. 13　Power consumption comparison

shown in Fig. 14. Among them, the exergy loss of semi-lean liquid is the largest, and the exergy loss of CRS is the smallest, mainly because the semi-lean liquid and single-cycle amine method are used in the absorption tower to remove CO_2. Some amine liquid will enter the liquefaction process with the purified natural gas, so that the nitrogen refrigerant will reduce the natural gas to the same temperature. The required flow rate increases, and the cold energy is used in the liquefaction process, resulting in a large cold loss ratio. In contrast, CRS technology and low-temperature packed bed are all low-temperature condensation methods, so some cold energy in liquefaction is used in decarburization, so that the cold energy utilization rate is higher; moreover, the heat loss of the heat exchanger is also mainly due to the energy consumption required for the regeneration of the amine solution. Therefore, the compressor, cooler, expander and heat exchanger of the first two methods will have a larger exergy loss than the other

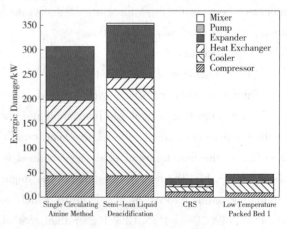

Fig. 14　Exergic damage comparison

two methods. Of course, this is also related to the number of equipment. Of course, the first two amine solution treatment methods also have pump loss, of course, this part of the loss is smaller.

(3) Heat loss comparison of CO_2

The CO_2 heat loss of the four processes obtained is shown in Fig. 15. Among them, semi-lean liquid has the largest heat loss and CRS has the smallest heat loss. This is mainly because the heat loss of CO_2 in the single-cycle amine method and semi-lean liquid is mainly composed of lean/rich liquid heat exchanger and reboiler, while the heat loss of CO_2 in CRS technology and low-temperature packed bed is mainly composed of heat exchanger, and the heat loss of CO_2 only accounts for the percentage of its removal, because the other part of the heat loss is used to liquefy natural gas, which is an effective utilization. Therefore, the heat loss of CO_2 removal by amine method is significantly higher than that of low temperature removal.

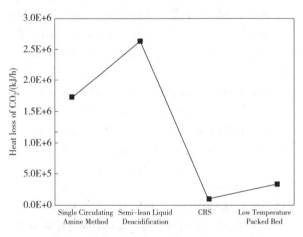

Fig. 15 Heat loss comparison

2.3 Safety comparison

The safety of the integrated process of pretreatment and liquefaction can be considered from the decarburization and liquefaction processes respectively. For the nitrogen expansion refrigeration liquefaction process, compared with the mixed refrigerant or cascade process may contain flammable and explosive substances, N_2 has stable chemical properties, non-combustible, non-explosive, non-toxic and less dangerous. Secondly, the process is simple. Compared with the mixed refrigerant and cascade process, it covers a small area and is suitable for offshore natural gas liquefaction operation. For the decarbonization process, although under the condition of pressure, the reduction of the removal depth requirement reduces the height of the absorption tower and the regeneration tower, the hull sloshing will still affect the flow and heat transfer of the refrigerant and natural gas in the liquefaction process system, thus affecting the operation stability of key equipment such as heat exchanger, tower and separator, including gas-liquid balance, logistics distribution and energy and mass transfer, and ultimately affecting the working performance of the liquefaction process system. The difference between the CRS technology and the low-temperature packed bed method is mainly based on the different phase change methods. The CRS technology is mainly through gas-liquid separation, while the low-temperature packed bed technology is mainly through gas-solid separation. The difficulty of gas-liquid separation is mainly the determination of separation temperature. In the simulation process, the temperature of gas-liquid separation is not easy to determine. If the temperature is too high, it will not be separated, and if it is too low, it will form a solid. The precipitation temperature of the solid is relatively fixed, which is convenient to determine the separation temperature, so the gas-solid separation will be more stable.

The main advantage of amine liquid separation is that the purification accuracy is high, but the purification accuracy required under pressurized liquefaction conditions is reduced. Compared with the amine liquid separation method which occupies a larger area and the process performance is easily affected by sea conditions, the low temperature packed bed method which occupies a small area and can handle high content of CO_2 has more advantages. Therefore, from the safety point of view, the integrated process of low temperature packed bed decarburization and nitrogen expansion refrigeration liquefaction is more suitable.

3 Conclusion

In this paper, four different natural gas decarburization processes including semi-lean liquid deacidification, amine deacidification, CRS technology and low temperature packed bed deacidification are selected and compared with the liquefaction process combined with the design of integrated process flow. The optimal process is obtained. The main conclusions are as follows:

(1) The process models of four kinds of pretreatment and pressurized liquefaction integrated processes based on semi-lean liquid decarburization, amine decarburization, CRS technology and low temperature packed bed decarburization are established, and the steady state simulation is carried out.

(2) By comparing the process performance of the four integrated processes, the heat transfer area of the low-temperature packed bed integration is the smallest, and the energy consumption of the CRS technology is the lowest, but in terms of comprehensive safety, the low-temperature packed bed method is the best.

(3) According to the process characteristics and process performance comparison of different decarburization processes, it can be seen that the single-cycle amine method has high purification accuracy, but its required regeneration heat load is much higher than other processes, which is suitable for natural gas with low CO_2 content. The semi-lean liquid decarbonization method has high purification accuracy and is suitable for natural gas with high carbon content, but it covers a large area and has a large number of towers, which is not suitable for marginal gas fields and offshore natural gas generation. The purification capacity of CRS decarbonization technology meets the needs of pressurized liquefaction process, with low energy consumption, small footprint and good safety. However, the gas-liquid separation method is sensitive to natural gas components, so it is suitable for natural gas liquefaction with stable gas source. The low-temperature packed bed decarbonization method has low energy consumption and floor space. The gas-solid separation method has strong adaptability to natural gas components and is suitable for natural gas liquefaction of various gas sources.

References

[1] SCOTT D P, MARK C G, ANN T L, RON R B, ERIC D N. Pressurized LNG: A New Technology for Gas Commercialization[C]//ISOPE-2005 conference; International offshore and polar engineering conference. 2005.

[2] JACKSON T S, SISAK W J, FARAH A M, et al. Pressurized LNG: Prototype Container Fabrication[J]. [2024-10-11].

[3] HEERDEN F V, PUTTER A, Alternative Modes of Natural Gas Transport. 2021.

[4] ERIC D N, SCOTT D P, BOWEN R R, GENTRY M C, ANN T L. Pressurized LNG: A Paradigm Shift in Gas Transportation. In: SPE Middle East Oil and Gas Show and Conference, Kingdom of Bahrain, 2005.

[5] BOWEN R R, GENTRY M C, NELSON E D, et al. Pressurized Liquefied Natural Gas (PLNG): A New Gas Transportation Technology., 2005.

[6] DU Q G, XIE B, XIE W H, ZHU X S, WANG J R, FENG J G. Preliminary Study on Development and Application of FLNG[J]. Oil Field Equipment, 2016, 45(08): 1-7.

[7] ZHU J L, LIU J H, LI Z H, LI Y X, WANG W C, LIU M E, LI E D, et al. Research status and development trends of FPLNG pretreatment and liquefaction integrated process [J]. Low - Carbon Chemistry and Chemical Engineering, 2024, 49(05): 112-122.

[8] HE T, KARIMI A I, JU Y. Review on the design and optimization of natural gas liquefaction processes for onshore and offshore applications[J]. Chemical Engineering Research and Design, 2018, 13289-114.

[9] XU Y J. On the Key Technologies of FLNG in South China Sea[J]. Ship & Ocean Engineering, 2017, 46(5): 148-152.

[10] HE T, KARIMI I A, JU Y. Review on the design and optimization of natural gas liquefaction processes for onshore and offshore applications[J]. Chemical Engineering Research and Design, 2018, 132: 89-114.

[11] LI Q, JU Y. Design and analysis of liquefaction process for offshore associated gas resources[J]. Applied Thermal Engineering, 2010, 30(16): 2518-2525.

[12] LI YX, ZHU J L, WANG W C. Key technologies of FLNG liquefaction process[J]. Oil & Gas Storage and Transportation, 2017, 36(2): 121-131.

[13] XIONG X J, LIN W S, GU A Z.. Simulation and optimal design of a natural gas pressurized liquefaction process with gas expansion refrigeration[J]. Natural Gas Industry, 2013, 33(06): 97-101.

[14] XIONG X J, LIN W S, GU A Z. Design and optimization of offshore natural gas liquefaction processes

adopting PLNG (pressurized liquefied natural gas) technology[J]. Journal of Natural Gas Science and Engineering, 2016, 30: 379-387.

[15] XIONG X J, LIN W S, GU A Z. Integration of CO_2 cryogenic removal with a natural gas pressurized liquefaction process using gas expansion refrigeration[J]. Energy, 2015, 931-9.

[16] LIN W S, XIONG X J, GU A Z. Optimization and thermodynamic analysis of a cascade PLNG (pressurized liquefied natural gas) process with CO 2 cryogenic removal[J]. Energy, 2018, 161: 870-877.

[17] LEE H S, SEO K Y, CHANG J D. Techno-economic Analysis of Acid Gas Removal and Liquefaction for Pressurized LNG[J]. IOP Conference Series: Materials Science and Engineering, 2018, 358(1).

[18] ROUSSANALY S, AASEN A, ANANTHARAMAN R, et al. Offshore power generation with carbon capture and storage to decarbonise mainland electricity and offshore oil and gas installations: A techno-economic analysis[J]. Applied Energy, 2019, 233: 478-494.

[19] DASHLIBORUN M A, LARACHI F, HAMIDIPOUR M. Cyclic operation strategies in inclined and moving packed beds—Potential marine applications for floating systems[J]. AIChE Journal, 2016, 62(11): 4157-4172.

[20] MOHAMMAD M, ISAIFAN J R, WELDU W Y, Rahman, M A. Progress on carbon dioxide capture, storage and utilisation[J]. International Journal of Global Warming, 2020, 20(2).

[21] BABAR M, BUSTAM M A, MAULUD A S, Ali A H. Optimization of cryogenic carbon dioxide capture from natural gas[J]. Materialwissenschaft und Werkstofftechnik, 2019, 50(3): 248-253.

[22] DASHLIBORUN A M, LARACHI F, TAGHAVI S M. Gas-liquid mass-transfer behavior of packed-bed scrubbers for floating/offshore CO_2 capture[J]. Chemical Engineering Journal, 2019, 377: 119236.

[23] BABAR M, BUSTAM M A, MAULUD A S, ABULHASSAN Ali, AHMAD M, SAMI U. Enhanced cryogenic packed bed with optimal CO_2 removal from natural gas: a joint computational and experimental approach[J]. Cryogenics, 2020, 105: 103010.

[24] DING H B, ZHANG Y, DONG Y Y, WEN C, YANG Y. High-pressure supersonic carbon dioxide (CO_2) separation benefiting carbon capture, utilisation and storage (CCUS) technology[J]. Applied Energy, 2023, 339: 120975.

[25] HONG Z P, YE C M, WU H, ZHANG P, DUAN C J, YUAN B. Research progress in CO_2 removal technology of natural gas[J]. Journal of Chemical Industry and Engineering (China), 2021, 72(12): 6030-6048.

A Review of Cavitation Problems of Cryogenic Fluids in Gathering Pipelines

Ao Di[1,2] Li Yuxing[1,2] Liu Cuiwei[1,2]

[1. College of Pipeline and Civil Engineering, China University of Petroleum (East China);
2. Shandong Provincial Key Laboratory of Oil and Gas Storage and Transportation Security,
China University of Petroleum (East China)]

Abstract As the market share of clean energy such as liquefied natural gas (LNG) and liquid hydrogen gradually increases, cryogenic fluid transportation pipeline system applications are becoming more and more common. In pipeline systems, a sudden change in cross-section leads to a sudden increase in the flow rate of cryogenic fluids and a sudden drop in local pressure. When the local pressure is lower than the saturation vapor pressure at that temperature, bubbles are generated and cavitation occurs. Cavitation damage caused by cavitation of cryogenic fluids is the main cause of leakage and even failure of pipeline transportation systems. A thorough grasp of the unsteady cavitation features and mechanisms in cryogenic fluids can improve the prediction of cavitation, prevent its occurrence, and ensure the safe, stable, and continuous operation of cryogenic fluid transport systems. Compared with water, the ratio of liquid-vapor phase density of cryogenic fluids is smaller, and the thermal conductivity is also smaller. However, the sensitivity of saturated vapor pressure to temperature increases significantly, and the cavitation mechanism is more complex. This article reviews the research progress of numerical calculation, flow field characterization, and cavitation inhibition methods of cavitation in recent years, so as to better understand the mechanism of cavitation in cryogenic fluids. The article focuses on the influence of thermodynamic effects on incipient low-temperature cavitation and finds that the thermal effects have an inhibitory effect on cavitation. It also provides ideas for the correction of the cavitation number. Secondly, the classification of cavitation models (semi-empirical models, cavitation models based on the Rayleigh-Plesset (R-P) equation, and cavitation models based on interfacial dynamics) is described in detail. A detailed review of their corrections at cryogenic conditions is also presented. In terms of cavitation flow field characterization, two aspects of cryogenic cavitation morphological features and unsteady state properties are analyzed. It is found that due to the thermal effect, the cavitation bubbles at low-temperature were not clear and were accompanied by many small bubbles, which are different from the cavitation bubbles at non-cryogenic. Moreover, the shape of the flow channel and the operating conditions play an important role in the formation of the bubble morphology. The interaction between the complex non-stationary properties of the flow field and the structural response is also an issue that needs to be addressed. Cavitation inhibition methods for different structures such as valves, impellers, and elbows under non-cryogenic conditions are also introduced. Cavitation is usually avoided by changing the cross-sectional shape, adding pressure-reducing devices, controlling the operating conditions, and optimizing the mechanical structure. This improves the safety and reliability of the pipeline system and provides a reference for cavitation inhibition under the low-temperature environment. Finally, the outlook and challenges of the current research on low-temperature fluid cavitation are further summarized, and it is proposed that the exploration of cryogenic cavitation fluid-solid coupling, the microscopic simulation of cavitation bubbles, and the strengthening of the dynamic measurement capability of the flow field are the important directions for the further research on cryogenic cavitation of pipelines in the future.

Keywords cryogenic fluids; thermal effect; cavitation number; cavitation inhibition

1 Introduction

As the market size of the pipeline transportation industry continues to expand, several issues continue to come to the fore. One of the most easily ignored and harmful is the liquid cavitation problem. In LNG pipelines, refined oil pipeline, crude oil pipelines, and water pipelines, when the pipe diameter is suddenly reduced due to the presence of valves and instruments, there is a sudden increase in flow rate and an instantaneous drop in pressure. When the pressure is lower than the saturated vapor pressure of the liquid at that time temperature, the liquid evaporation is caused by the explosive growth of micro-bubbles (known as gas nuclei) phenomenon, known as cavitation.

Cavitation occurs widely in the aerospace industry, wind tunnel system, chemical engineering, oil and gas gathering and transportation, etc. (as shown in Fig. 1). Cavitation was discovered on propeller blades as early as the second half of the 19th century. An early study of cavitation bubbles was carried out by Reynolds, who observed their creation in the flow and collapse in the downstream high-pressure region. In the early 20th century, with the development of steam engine technology, turbine steam engines replaced reciprocating steam engines. Severe cavitation damage occurred in the propellers of high-speed steam engine ships, and the cavitation phenomenon on the propeller blades attracted people's attention. Researchers conducted cavitation experiments on propellers. Ji et al. simulated unsteady cavitation turbulence around hydrofoils, which was used to predict cavitation damage with good results. In addition to impeller instruments, cavitation in structures such as orifice plates and venturi tubes has also been studied. Yan et al. analyzed the change of flow pattern during cavitation of an orifice plate by experiments and numerical simulations. Rudolf et al. obtained experimentally three cavitation states of the venturi: partial cavitation, complete cavitation, and supercavitation. Shi et al. investigated the effect of venturi with different shrinkage angles on cavitation performance and developed a semi-empirical model capable of predicting the cavitation phenomenon of liquid water in venturi. With the increasing understanding of cavitation phenomena, the cavitation mechanism of non-cryogenic fluids has been more thoroughly studied. In contrast, there is a relative lack of research on cavitation in cryogenic liquids.

The cavitation of cryogenic liquids is the problem of cavitation of fluids at temperatures ranging from those normally associated with ordinary liquid refrigerants (room temperature 298K) to those associated with liquid helium (4K). In late 2022, a leak was discovered at an elbow downstream of the orifice plate during an outgoing natural gas transfer at an LNG receiving terminal. Careful inspection revealed a perforation defect on the outer arc side of the elbow. Through the defect micro-morphology analysis and simulation results, the stress on the wall caused by bubble collapse in the failed elbow was greater than tens of MPa, higher than the design pressure of the pipeline. It indicated that the failure of this pipe section is mainly caused by cavitation collapse. Due to the extremely high flow rate after the orifice plate adjustment, it is very easy to cause the pipeline elbow to be in a severe operating environment with high pressure difference, triggering cavitation, and erosion phenomenon, leading to perforation failure of the elbow. Cavitation will cause damage to the pipeline triggering leakage or even pipe burst, affecting the normal operation of the system, and even causing major catastrophic engineering accidents.

Fig. 1　Cavitation occurrence scenario

Multi-stage centrifugal pumps are used for loading and unloading LNG in extremely cryogenic environments to strictly ensure the sealing and cryogenic preservation requirements of LNG during the LNG marine transportation process. However, continuous switching of loads to meet the required stability of the delivery can cause the pumping unit to deviate from its designed operation, which could potentially affect thesecure and stable operation of the pumping equipment. When the cryogenic fluid moves at high speed, sudden changes in velocity and pressure occur, and when the local pressure is lower than the saturated vapor pressure of the LNG, the fluid may undergo cavitation. Cavitation is an unavoidable and thorny problem for LNG cryogenic submerged pumps; In addition, under the condition of high pressure and high flow rate, the fluid usually accompanied by cavitation phenomenon occurs when flowing through the valve. Cavitation occurs frequently, resulting in the actual flow area of the valve flow path, the flow coefficient changes, thus causing the valve instability vibration, component life-shortening, and failure or even system failure. It can be seen that the oil and gas gathering pipeline is also the hardest hit by the cavitation phenomenon.

Research on the prediction and inhibition of cavitation incipient is the key to effectively guaranteeing the safe, stable, and continuous operation of cryogenic fluid transportation systems. This issue has also received extensive attention from scholars, and related research has increased exponentially in recent years. The key objects used for cryogenic cavitation research include venturi, valves, induced wheels, turbopumps, ogive, orifice plates, hydrofoils, pumps, and so on. Researchers have continued to focus on key aspects such as the evolution of cavitation patterns, similarity between cryogenic and non-cryogenic fluids, dynamic flow characteristics, and energy losses. Table 1 collects some representative articles on cavitation phenomena in cryogenic fluids.

Table. 1 Classification of cryogenic liquid cavitation studies

sections	fluids	Content of research	Research Methods	reference
venturi	Liquid hydrogen、liquid oxygen	Evolution of cavitation patterns	Experimentation	[11]
valve	liquid nitrogen	Similarity between cryogenic and non-cryogenic fluids	Experimentation and Simulation	[12]
	liquefied natural gas	Dynamic flow characteristics and energy loss	Simulation	[13]
induced wheel	liquid nitrogen	Differences in cavitation patterns between cryogenic and non-cryogenic fluids	Experimentation	[14]
turbopump	Liquid hydrogen	Cavitation and turbulent fields	Simulation	[15]
	liquid oxygen	Cavitation and thermal effects	Simulation	[16]
	liquid nitrogen	Cavitation and Vortices	Experimentation	[17]
ogive	Liquid hydrogen	Kinetic evolution and thermodynamic characterization of cavitation shedding	Simulation	[18]
		Cavitation and turbulence	Simulation	[19]
orifice plate	liquid nitrogen	Cavitation bubble dynamics of cryogenic and non-cryogenic fluids	Experimentation and Simulation	[20]
hydrofoil	liquid nitrogen	Cavitation model correction	Simulation	[21]
pump	liquid nitrogen、Liquid hydrogen	Cavitation and turbulence	Simulation	[22]

In summary, for the cavitation phenomenon, there is an urgent need to study the cavitation mechanism and damage-causing model ofcryogenic fluids under pressure perturbation. This can clarify the conditions of cavitation and monitor the oscillation characteristics of cavitation bubbles. Further, predict the consequences of collapse damage and develop cavitation inhibition methods. To provide a solid theoretical basis and application guarantee for the prevention and safety warning of pipeline damage caused by cryogenic liquid cavitation as well as leakage and pipe burst accidents. Therefore, this article to reduce the risk of cavitation in cryogenic fluid pipelines as a background, to the basic research of cryogenic media cavitation flow characteristics and other basic research as a traction, introduces the cavitation number formula for the judgment of cryogenic fluid cavitation. The current status of cavitation modeling research at home and abroad is investigated. The characterization method of the cavitation flow field of cryogenic fluid is highlighted. Finally, the methods of cavitation inhibition are reviewed. The summary of the study of cavitation in cryogenic fluids has important application value for safety, energy, environment, and other fields.

2 Judgment of cryogenic fluid cavitation incipient

Cavitation bubbles are generally generated when the local pressure is lower than the saturated vapor pressure corresponding to the current temperature of the liquid. In liquid pipelines due to valve operation, the presence of a flow meter causes a contraction of the pipeline cross-section and an increase in the flow rate, resulting in a decrease in pressure. The reason for the pressure decrease can be explained by Bernoulli's equation:

$$P_1 + \frac{1}{2}\rho v_1^2 = P_2 + \frac{1}{2}\rho v_2^2 \quad (1)$$

where ρ is the density of the fluid, kg/m^3; P_1, P_2 is the upstream and downstream pressure in the flow system, Pa; v_1, v_2 is the corresponding fluid velocity, m/s.

After the bubbles appear, as the fluid flows into the high-pressure region, the bubbles collapse causing a large amount of energy release to produce cavitation. However, when the fluid is cryogenic fluid, the ratio of the liquid phase to the gas phase density is much smaller than that of non-cryogenic fluid (Fig. 2a). To form cavitation bubbles of the same size as non-cryogenic fluid, cryogenic fluid requires more liquid vaporization. Thus, a large amount of latent heat of vaporization is absorbed, which reduces the temperature inside the vacuole. At the same time, cryogenic fluid is a heat-sensitive fluid, a small temperature change will cause a large change in saturation vapor pressure, as shown in Fig. 2b. Cryogenic fluid cavitation has significant thermodynamic effects, generates large interfacial mass and heat transfer as well as drastic changes in fluid properties, making the phase transition mechanism very different from that of water.

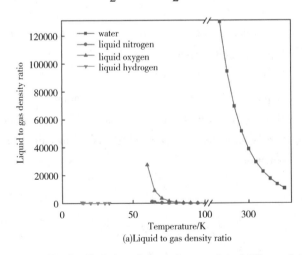
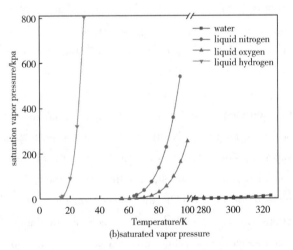

Fig. 2 Variation of physical properties of different fluids with temperature (data from REFPROP database)

When the flow velocity is certain and the pressure decreases (or the pressure is certain and the flow velocity increases), the critical state in which tiny cavities appear for the first time by chance in a very small region of the flow field is called cavitation incipient. The main factors affecting cavitation incipience are pressure and flow rate, so these two quantities are used to define the cavitation number σ. The cavitation number expression is given below:

$$\sigma = \frac{p_\infty - p_v}{\frac{1}{2}\rho v_\infty^2} \quad (2)$$

where p_∞ is the absolute pressure, Pa; v_∞ is the flow rate at a selected point in the flow system, m/s; p_v is the saturated vapor pressure at a given temperature, Pa; ρ is the liquid density, kg/m³. When $\sigma \gg 1$, the pressure margin is sufficient to provide the kinetic energy of the liquid, then cavitation does not occur; when $\sigma < 1$, cavitation occurs.

Salehi et al. also used the above equation to calculate the cavitation number of the spillway and categorized it into "no cavitation damage", "possible cavitation damage", "cavitation damage", "severe damage" and "major damage" levels as shown in Table 2.

Table. 2 Cavitation damage level

level	risk of cavitation damage	velocity/ (m/s)	Cavitation number range
1	no cavitation damage	$v \leq 5$	$\sigma > 1$
2	possible cavitation damage	$5 < v \leq 16$	$0.45 < \sigma \leq 1$
3	cavitation damage	$16 < v \leq 25$	$0.25 < \sigma \leq 0.45$
4	severe damage	$25 < v \leq 40$	$0.17 < \sigma \leq 0.25$
5	major damage	$v > 40$	$\sigma \leq 0.17$

In addition to the possibility of a detailed partitioning of the cavitation number interval, many other factors affect the cavitation number, such as liquid viscosity, surface tension, the number of gas nuclei contained in the liquid and geometric structure. Yan et al. found that the generation of cavitation does not necessarily require that the minimum static pressure at the constriction downstream of the orifice be equal to the saturated vapor pressure of the liquid. Cavitation occurs when this minimum pressure is close to the vapor pressure. The number of cavitation in the choked condition is a function related to the ratio of the orifice plate diameter (d) to the pipe diameter (D); Simpson et al. found through numerical simulation studies that the cavitation initiation point is related to the diffusion angle and the ratio of l (throat length) to d (diameter); Liu et al. proposed a new cavitation number expression (3) based on the amplitude of the dynamic pressure wave rather than on the velocity. This formula was verified to be valid by examples and can be used to study the phenomena of pressure fluctuation, leakage and corrosion in liquid pipelines, storage tanks and pressure vessels.

$$C_c = \frac{P_r - P_v}{P_r - P_c} = \frac{P_r - P_v}{\Delta p} = \frac{P_r - P_v}{\rho c \cdot \Delta v} = \frac{P_r - P_v}{\rho a \cdot \Delta s} = \frac{P_r - P_v}{\rho g \cdot \Delta s} / (a/g) \quad (3)$$

Where C_c is the cavitation number; P_r is the reference pressure, Pa; P_v is the liquid saturated vapor pressure, Pa; P_c has the leakage occurred after the leakage point of the pressure or flow cross-section changes in the location of the pressure, Pa; $\Delta p = P_r - P_c$; ρ is the density of the liquid, kg/m³; c is the speed of sound, m/s; a is the acceleration, m/s².

When the medium iscryogenic fluid, the cavitation phenomenon caused by the pressure perturbation will inevitably lead to local temperature changes in the liquid. Then its saturated vapor pressure also changes. To determine the temperature difference due to thermal effects, Frumana et al. used a thermally sensitive fluid for testing. A thermocouple was utilized to measure the temperature in the chamber and then the original fluid temperature difference was extrapolated. A model was proposed to calculate the temperature decrease by making certain assumptions:

$$\Delta T = T_{\text{ref}} - T_c = \frac{\rho_v L C_Q}{0.17 \rho c_p \sqrt{1+\sigma}} \text{Re}_l^{-0.1} \quad (4)$$

where ρ_v and ρ are the density of the gas and liquid phases, respectively, kg/m³; L is the latent heat of vaporization, kJ/kg; c_p is the Specific Heat Capacity, J/kg·K; σ is the cavitation number; Re_l and C_Q are the Reynolds number and volume flow coefficient, respectively; With the following equations:

$$\text{Re}_l = \frac{U_{\text{ref}} l}{v} \quad (5)$$

$$C_Q = \frac{Q}{U_{\text{ref}} \pi D l} = \frac{V_g}{U_{\text{ref}}} \quad (6)$$

where Q is the vapor flow rate, m³/s; U_{ref} is the free stream velocity, m/s; D is the model diameter, m; l is the cavity length, m; v is the viscosity of the liquid.

The method can be considered highly reliable despite its limitations. There is also a need to improve the accuracy of the cavity length measurement, as well as the measurement of the air (or non-condensable gas) flow rate in the ventilated cavity. Moore et al. have numerically investigated and obtained that the variation of the local saturated vapor pressure Δp due to heat transfer can be expressed as:

$$\Delta P = -\Delta T \frac{dP_{sat}}{dT} = -\frac{\alpha_v}{1-\alpha_v} \frac{\rho_v L}{\rho_l C_{pl}} G_{sat} \quad (7)$$

$$G_{sat} = \frac{dP_{sat}}{dT} \quad (8)$$

where P_{sat} is the saturated vapor pressure of the liquid phase as a function of temperature, pa; α_v is the volume fraction of the gas phase; G_{sat} is the rate of change of saturated vapor pressure with temperature, representing the temperature sensitivity.

Antoine's equation was frequently used to describe the saturated vapor pressure of a pure fluid, Where A, B, and C are constants depending on the type of fluid. The specific equation is as follows:

$$\lg P_{sat} = A - \frac{B}{T+C} \quad (9)$$

Then Eq. (7) can be changed to:

$$\Delta P = -\frac{\alpha_v}{1-\alpha_v} \frac{\rho_v L}{\rho_l C_{pl}} \frac{BP_{sat}}{(T+C)^2} \ln 10 \quad (10)$$

Considering the effect of thermal effects on saturated vapor pressure according to the above equation, the corrected saturated vapor pressure is shown below:

$$P_v = p_v(T_c) + \Delta P \quad (11)$$

where $p_v(T_c)$ is the saturated vapor pressure at T_c, pa; ΔP is the amount of change in saturated vapor pressure due to thermodynamic effects.

The above method allows us to determine the effect of thermal effects on the saturated vapor pressure, which in turn allows us to determine whether or not cavitation is occurring by using the cavitation number equation. A decrease in local temperature leads to a decrease in saturation vapor pressure and a subsequent increase in the cavitation number, inhibiting cavitation.

3 Modification of computational modeling of cryogenic fluid cavitation

The cavitation mechanism becomes complicated due to the phase change in the cavitation process. The selection of calculation methods and models is especially important. In recent years, therehave been two main development directions for cryogenic fluid cavitation simulation. One is to try to establish a model for describing the thermal effects of cavitation for cavitation prediction, and the other is to carry out cryogenic modification of the existing cavitation model (modification of the model coefficients, coupling the thermodynamic equations into the model), to optimize the model. The numerical calculation of cavitation flow mainly contains the boundary integral method, interface tracking method, and two-phase flow method. Their classification and drawbacks are shown in Fig. 3.

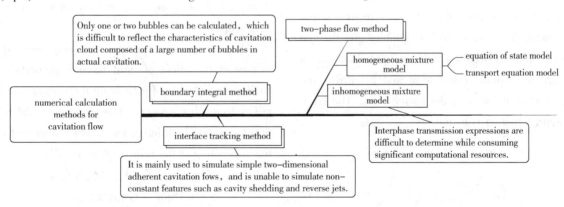

Fig. 3 Classification of numerical calculation methods for cavitation flow

Currently, the homogeneous mixture model is more widely used. Rahbarimanesh et al. investigated the effect of interfacial tension on the phase transition behavior of liquefied natural gas inside a Laval nozzle based on the homogeneous mixture model. The results showed that the presence of interfacial tension locally stabilizes the cavitation behavior by changing the phase transition mechanism. The key to implementing the homogeneous mixture model (HEM) is to calculate the variable density field of the mixture fluid during cavitation (due to the change in the volume content of the gas phase) thereby closing the set of equations. Typical approaches include equation of state modeling and transport equation modeling. Among them, the equation of state mainly relates pressure to thermodynamic properties such as density. For example, assuming that the fluid is positively pressurized, Pascarella et al. proposed the relation between the density of the mixture and the pressure in the flow field as:

$$\rho_m = \frac{\rho_l \rho_v}{2} + \frac{2}{\pi a_{min}^2}(P_l - P_v)\sin\left(\frac{p-p_v}{P_l-P_v}\frac{\pi}{2}\right) \quad (12)$$

Other researchers have also proposed various correlation equations on thisbasis. With the gradual deepening of the understanding of cavitation, researchers found that the equation of state approach could not well explain the convection and transport phenomena during cavitation flow because it did not consider the mass transport between the vapor and liquid phases. The transport equation model overcomes these shortcomings by characterizing the flow state of the liquid or vapor phase by a transport equation related to the volume fraction or mass fraction. The transport equation approach assumes a common pressure, velocity, and temperature between the phases and characterizes the transition between the vapor and liquid phases in terms of evaporation and condensation rates. The advantage of the transport equation model is that it can simulate the effects of inertial forces and cavitation zones during bubble growth and motion, but its biggest challenge is to establish an effective cavitation source term and to correct the empirical coefficients in the source term. The basic transport equation is as follows:

$$\frac{\partial}{\partial t}(\alpha_l) + \frac{\partial}{\partial x_j}(\alpha_l u_j) = \dot{m}^+ - \dot{m}^- \quad (13)$$

$$\frac{\partial}{\partial t}(f_v \rho_m) + \frac{\partial}{\partial x_j}(f_v \rho_m u_j) = \dot{m}^+ - \dot{m}^- \quad (14)$$

where α_l、f_v are the liquid volume fraction and vapor mass fraction, respectively. The superscripts + and – indicate condensation and evaporation, respectively. According to the proportion of the liquid-vapor mixture, the fluid properties can be determined as follows:

$$\varphi_m = \varphi_l \alpha_l + \varphi_v(1-\alpha_l) \quad (15)$$

Depending on how $\dot{m}^+$ and $\dot{m}^-$ are constructed, these models fall into three categories: semi-empirical models, Rayleigh-Plesset (R-P) equation-based cavitation models, and interfacial dynamics-based cavitation models.

(1) semi-empirical model

In the Merkle cavitation model, the rate of transformation from liquid to gas was assumed to be proportional to the pressure difference at a given location. The model uses specific time and velocity parameters to control how cavitation impacts the flow. The transport equations are shown as follows:

$$\begin{cases} \dot{m}^- = \dfrac{C_{vap}\rho_l \alpha_l \min[0, p-p_v]}{(0.5\rho_l U_\infty^2)\rho_v t_\infty}, & p \leqslant p_v \\ \dot{m}^+ = \dfrac{C_{cond}\max[0, p-p_v](1-\alpha_l)}{(0.5\rho_l U_\infty^2)t_\infty}, & p > p_v \end{cases} \quad (16)$$

where C_{vap} and C_{cond} are empirical coefficients; U_∞ is the reference velocity, m/s; t_∞ is the reference time, which is determined by the ratio of the characteristic length L and the reference velocity scale ($t_\infty = L/U_\infty$), s; Default constants for the Merkle cavitation model: $C_{vap} = 1.0$, $C_{cond} = 80.0$.

When the fluid is acryogenic fluid, the liquid-gas density is relatively small, which has a greater effect on the source term. The empirical coefficients at this point are significantly lower than for non-cryogenic fluid and the original empirical coefficients will no longer apply.

Esposito proposed a new semi-empirical model to predict the effect of fluid thermal properties on cavitation, capable of estimating the pressure at the orifice to reach cavitation. The semi-empirical relational equation is different for different ranges of initial subcooling at the orifice. It is as follows:

$$P_{min} = P_{sat}(T_{up})\Gamma_{sub} = P_{sat}(T_{up}) \\ [-0.18Rp^{0.87} + \exp(\Delta T_{sub})^{4.97}] \quad (17)$$

$$P_{min} = P_{sat}(T_{up})\Gamma_{sub} = P_{sat}(T_{up})[0.25(\Delta T_{sub})^{-0.27}] \quad (18)$$

Eqs. (17) and (18) correspond to $\Delta T_{sub} \geqslant 0.020(R^2 = 0.991)$ and $\Delta T_{sub} < 0.020(R^2 = 0.980)$. P_{min} is the minimum pressure at which the phase change occurs, pa; $P_{sat}(T_{up})$ is the saturated vapor pressure at the upstream pressure, pa; R^2 is the coefficient of determination; ΔT_{sub} is the degree of subcooling.

(2) cavitation model based on Rayleigh-Plesset (R-P) equation

While the semi-empirical model treats the gas phase as a whole, the cavitation model based on the Rayleigh-Plesset equation expresses the process of growth or rupture of individual bubbles in response to an internal and external pressure difference. Its expression is given by:

$$\rho_l\left(\frac{2}{3}\dot{R}^2+R\ddot{R}\right)=p_v-p+p_{g0}\left(\frac{R_0}{R}\right)^{3\gamma}-\frac{2\gamma}{R}-4\mu\frac{\dot{R}}{R} \quad (19)$$

where R is the radius of the bubble, m; R_0 is the initial radius of the bubble, m; $\dot{R}$ and $\ddot{R}$ are the first-order and second-order derivatives of the radius of the bubble for time, respectively; p_v is the pressure in the bubble, pa; p_{g0} is the partial pressure of the non-condensable gases in the bubble, pa; γ and μ are the coefficients of surface tension and viscosity of the liquid, respectively. Neglecting the quadratic term, surface tension, liquid viscosity, and non-condensable gases, then:

$$\frac{dR}{dt}=\pm\sqrt{\frac{2}{3}\frac{|p_v-p|}{\rho_l}} \quad (20)$$

$$m=\pm\rho_v\frac{d}{dt}\left(\frac{4}{3}\pi R^3\right)=\pm 4\pi\rho_v R^2\sqrt{\frac{2}{3}\frac{|p_v-p|}{\rho_l}} \quad (21)$$

On this basis, the cavitation model proposed by Schnerr and Sauer requires only the determination of the bubble number density:

$$\begin{cases}\dot{m}^+=\dfrac{3\rho_l\rho_v}{R_B\rho_m}\alpha_v(1-\alpha_v)\sqrt{\dfrac{2}{3}\dfrac{P_v-P}{\rho_l}}, & P<P_v \\ \dot{m}^-=\dfrac{3\rho_l\rho_v}{R_B\rho_m}\alpha_v(1-\alpha_v)\sqrt{\dfrac{2}{3}\dfrac{P-P_v}{\rho_l}}, & P>P_v\end{cases} \quad (22)$$

When the concentration of non-condensable gases in a liquid is low, cavitation becomes challenging. Hence, incorporating the non-condensable gas content as a parameter in the cavitation model is crucial. Direct predictions using this mass source term alone can be inaccurate due to the faster evaporation rate compared to condensation during cavitation. To address this issue, empirical constants must be introduced into the model. The final form of the mass transfer rate corrected by He et al. is:

$$\begin{cases}\dot{m}^+=C_e\dfrac{4.5nKT\rho_v}{\gamma}\sqrt{\dfrac{2}{3}\dfrac{P_v-P}{\rho_l}}(1-\alpha_v), & P<P_v \\ \dot{m}^-=C_d\dfrac{4.5nKT\rho_v}{\gamma}\sqrt{\dfrac{2}{3}\dfrac{P-P_v}{\rho_l}}\alpha_v, & P>P_v\end{cases} \quad (23)$$

The new cavitation model predicts cavitation flow over hydrofoils, cavitation flow in cusp holes, and unsteady cavitation flow in venturi, and the results are in good agreement with the experiments. The model could be further developed in future studies to incorporate thermal effects and predict cavitation-related corrosion. The cavitation model developed by Zwart et al. was much simpler:

$$\begin{cases}\dot{m}^+=C_e\dfrac{3\alpha_{nuc}(1-\alpha_v)\rho_v}{R_B}\sqrt{\dfrac{2}{3}\dfrac{P_v-P}{\rho_l}}, & P<P_v \\ \dot{m}^-=C_d\dfrac{3\alpha_v\rho_v}{R_B}\sqrt{\dfrac{2}{3}\dfrac{P-P_v}{\rho_l}}, & P>P_v\end{cases} \quad (24)$$

α_{nuc} is the volume fraction of the nucleation site; C_e and C_d are the empirical coefficients of evaporation and condensation, respectively.

To study cryogenic fluid cavitation in butterfly valves, Zhang et al. used numerical simulations to observe the fluid dynamics at different valve openings. They enhanced the traditional cavitation model by including thermal effects from the cryogenic medium and confirmed its effectiveness through experimental validation, leading to more precise predictions of cryogenic cavitation. The modified cavitation model is as follows:

$$\begin{cases}\dot{m}^+=C_e\dfrac{3\alpha_{nuc}(1-\alpha_v)\rho_v}{R_B}\sqrt{\dfrac{2}{3}\left|\dfrac{[P_v(T_c)-P]}{\rho_l}-\dfrac{\alpha_v}{1-\alpha_v}\dfrac{\rho_v LBP_v\ln 10}{\rho_l C_{pl}(T+C)^2}\right|}, & P<P_v \\ \dot{m}^-=C_e\dfrac{3\alpha_v\rho_v}{R_B}\sqrt{\dfrac{2}{3}\left|\dfrac{[P-P_v(T_c)]}{\rho_l}-\dfrac{\alpha_v}{1-\alpha_v}\dfrac{\rho_v LBP_v\ln 10}{\rho_l C_{pl}(T+C)^2}\right|}, & P>P_v\end{cases} \quad (25)$$

The cavitation model by Singhal et al. incorporated local turbulent kinetic energy, surface tension coefficients, and non-condensable gases:

$$\begin{cases} \dot{m}^+ = C_e \dfrac{\sqrt{k}\rho_l\rho_v}{\gamma R_B}\sqrt{\dfrac{2}{3}\dfrac{P_v-P}{\rho_l}}(1-f_v-f_g),\ P<P_v \\ \dot{m}^- = C_d \dfrac{3\alpha_v\rho_v}{R_B}\sqrt{\dfrac{2}{3}\dfrac{P-P_v}{\rho_l}}f_v,\ P>P_v \end{cases} \quad (26)$$

k is the turbulent kinetic energy, J; γ is the surface tension, mN/m; f_v is the vapor mass fraction; f_g is the mass fraction of non-condensable gas.

Pinho and Zhang investigated the low-temperature cavitation flow using a full cavitation model, which was found to have good robustness and predictive ability.

(3) cavitation model based on interfacial dynamics

Senocak and Shyy derived the mass transfer rates for gas-liquid phases and proposed an interfacial kinetic model. The objective was to overcome the constraints caused by empirical constants in the traditional semi-empirical cavitation model and the method based on the R-P equation. For a common fluid like water, a two-phase interfacial dynamics model characterized by a thin two-phase zone was proposed. Assuming the existence of a clear interface in the vapor-liquid mixing zone, the model's source terms were derived from mass and momentum balances at the cavity interface. The evaporation and condensation source terms were respectively:

$$\begin{cases} \dot{m}^+ = \dfrac{Max(0,\ p-p_v)(1-\alpha_l)}{(u_{v,n}-u_{I,n})^2(\rho_l-\rho_v)t_\infty} \\ \dot{m}^- = \dfrac{\rho_l Min(0,\ p-p_v)\alpha_l}{\rho_v(u_{v,n}-u_{I,n})^2(\rho_l-\rho_v)t_\infty} \end{cases} \quad (27)$$

$u_{I,n}$ is the interfacial velocity, m/s; $u_{v,n}$ is the centripetal velocity in the vapor phase, which can be expressed as:

$$u_{v,n} = \vec{u}\cdot\vec{n};\ \vec{n} = \dfrac{\nabla\alpha_l}{|\nabla\alpha_l|} \quad (28)$$

When the fluid was cryogenic, the gas-liquid interface was not clear as observed by experiments. Utturkar subsequently developed a model for the dynamics of the paste interface. Assuming the cavity interface divides the mixing zone from the liquid zone, the source term from mass and momentum balance conditions at the interface is given by the following equation:

$$\begin{cases} \dot{m}^+ = \dfrac{\rho_l Max(0,\ p-p_v)(1-\alpha_l)}{\rho_+(U_{m,n}-U_{I,n})^2(\rho_l-\rho_v)t_\infty} \\ \dot{m}^- = \dfrac{\rho_l Min(0,\ p-p_v)\alpha_l}{\rho_-(U_{m,n}-U_{I,n})^2(\rho_l-\rho_v)t_\infty} \\ \dfrac{\rho_l}{\rho_-} = \dfrac{\rho_l}{\rho_v}+\left(1.0-\dfrac{\rho_l}{\rho_v}\right)e^{-\dfrac{(1-\alpha_l)}{\beta}} \\ \dfrac{\rho_l}{\rho_+} = \dfrac{\rho_l}{\rho_m} \end{cases} \quad (29)$$

$U_{m,n}$ is the centripetal mixing velocity.

Although theoretically, the cavitation model based on interfacial dynamics eliminates the empirical coefficients. However, the selection of the time scale and how to determine the gas-liquid interface motion velocity are still difficult. The classification of existing cavitation models is summarized in Table 3.

Forcryogenic fluid cavitation, it is common for researchers to extend to numerical calculations of cryogenic fluid by correcting the cavitation source term with coefficients. Zhu et al. observed that the bubble number density (n) in the Schnerr-Sauer cavitation model has a considerable impact on cavity properties and cavitation source strength. For non-cryogenic fluids, this effect was significant when $n \geq 10^{13}$. The model achieves better correspondence with experimental data for liquid nitrogen and hydrogen in hydrofoil and ogive forms when n is 10^8. Li et al. proposed a new cavitation model by considering the vapor bubble growth rate and the vapor bubble number density, which was feasible for the cavitation of liquid nitrogen and hydrogen with the strongest thermodynamic effects, the organic fluids R114 and R245 with moderate thermodynamic effects, and the warm water with slight thermodynamic effects.

From the above studies, it can be seen that the cavitation model based on the transport equation is still able to predict the cryogenic cavitation characteristics well after calibration or correction. However, most of the above numerical studies of cryogenic fluid cavitation only consider the influence of thermodynamic effects on the saturated vapor pressure of cryogenic fluid, and do not investigate the mechanism

Table 3 Summarization of cavitation models

type	Cavitation model	$\dot{m}^+$	$\dot{m}^-$
semi-empirical model	Merkle et al.	$\dfrac{C_{cond}\max[0,\ p-p_v](1-\alpha_l)}{(0.5\rho_l U_\infty^2)t_\infty}$	$\dfrac{C_{vap}\rho_l\alpha_l\min[0,\ p-p_v]}{(0.5\rho_l U_\infty^2)\rho_v t_\infty}$
	Kunz et al.	$\dfrac{C_{prod}\rho_v(\alpha_l-\alpha_{ng})^2(1-\alpha_l-\alpha_{ng})}{t_\infty}$	$\dfrac{C_{dest}\rho_v\alpha_l\min[0,\ p-p_v]}{(0.5\rho_l U_\infty^2)t_\infty}$
Based on the Rayleigh-Plesset equation	Schneer and Sauer	$\dfrac{3\rho_l\rho_v}{R_B\rho_m}\alpha_v(1-\alpha_v)\sqrt{\dfrac{2}{3}\dfrac{P_v-P}{\rho_l}}$	$\dfrac{3\rho_l\rho_v}{R_B\rho_m}\alpha_v(1-\alpha_v)\sqrt{\dfrac{2}{3}\dfrac{P-P_v}{\rho_l}}$
	Singhal et al.	$C_d\dfrac{3\alpha_v\rho_v}{R_B}\sqrt{\dfrac{2}{3}\dfrac{P-P_v}{\rho_l}}f_v$	$C_e\dfrac{\sqrt{k}\rho_l\rho_v}{\gamma R_B}\sqrt{\dfrac{2}{3}\dfrac{P_v-P}{\rho_l}}(1-f_v-f_g)$
	Zhang et al.	$C_e\dfrac{3\alpha_{nuc}(1-\alpha_v)\rho_v}{R_B}\sqrt{\dfrac{2}{3}\left\vert\dfrac{[P_v(T_c)-P]}{\rho_l}-\dfrac{\alpha_v}{1-\alpha_v}\dfrac{\rho_v LBP_v\ln10}{\rho_l C_{pl}(T+C)^2}\right\vert}$	$C_e\dfrac{3\alpha_v\rho_v}{R_B}\sqrt{\dfrac{2}{3}\left\vert\dfrac{[P-P_v(T_c)]}{\rho_l}-\dfrac{\alpha_v}{1-\alpha_v}\dfrac{\rho_v LBP_v\ln10}{\rho_l C_{pl}(T+C)^2}\right\vert}$
	Zwart et al.	$C_d\dfrac{3\alpha_v\rho_v}{R_B}\sqrt{\dfrac{2}{3}\dfrac{P-P_v}{\rho_l}}$	$C_e\dfrac{3\alpha_{nuc}(1-\alpha_v)\rho_v}{R_B}\sqrt{\dfrac{2}{3}\dfrac{P_v-P}{\rho_l}}$
Based on interfacial dynamics	Senocak and Shyy	$\dfrac{C_{prod}\mathrm{Max}(0,\ p-p_v)(1-\alpha_l)}{(U_{v,n}-U_{I,n})^2(\rho_l-\rho_v)t_\infty}$	$\dfrac{C_{dest}\rho_l\mathrm{Min}(0,\ p-p_v)\alpha_l}{\rho_v(U_{v,n}-U_{I,n})^2(\rho_l-\rho_v)t_\infty}$
	Utturkar	$\dfrac{\rho_l\mathrm{Max}(0,\ p-p_v)(1-\alpha_l)}{\rho_+(U_{m,n}-U_{I,n})^2(\rho_l-\rho_v)t_\infty}$	$\dfrac{\rho_l\mathrm{Min}(0,\ p-p_v)\alpha_l}{\rho_-(U_{m,n}-U_{I,n})^2(\rho_l-\rho_v)t_\infty}$

of mass transfer between gas and liquid phases of cryogenic cavitation flow driven by temperature difference. And the effect of vacuole compressibility on cryogenic cavitation is also rarely considered, so the pressure wave transfer process cannot be correctly reflected. In the future, the influence of the thermodynamic effect and compressibility effect on cryogenic cavitation should be considered comprehensively to construct a full-flow field calculation model that can accurately predict the cavitation of cryogenic pipelines.

4 The method for characterizing the cavitation flow field of cryogenic fluids

Capturing the dynamic evolution of cavitation, from its inception, and development to collapse, is a transient process that is extremely challenging. Second, the cavitation process involves a two-phase flow, with temperature and pressure differences driving mass transfer between the gas and liquid phases. This dynamic gas-liquid bidirectional phase transition process is very complex. Therefore, accurately identifying and characterizing the complex cavitation flow field features is essential to understanding the cavitation mechanism of cryogenic fluids.

4.1 Morphological characteristics of cryogenic cavitation

Cavitation in liquid nitrogen, liquid hydrogen and water has been extensively studied and observed experimentally over the past decades. Ruggeri and Gelder, Hord, Stahl and Stepanoff found that the bubbles formed in water were clear and with a greater intensity of cavitation, whereas the cavitation regions in liquid nitrogen and liquid hydrogen showed a pasty structure accompanied with many small bubbles with relatively weak cavitation intensity. This difference was mainly due to the thermal effect in cryogenic fluid. Kikuta et al. further investigated the cavitation phenomenon in liquid nitrogen and cold water at different temperatures and found that the increase in cavity length was closely related to the extent of the thermodynamic effect. Giorgi et al. observed the transient growth process of cloud cavitation induced by the flow through a throat using high-speed video images. The bubbles formed in liquid nitrogen were found to exhibit a foamy appearance, which was more detailed than water under this ambient condition. The experimental results of Zhu et al. showed that the temperature decrease was linearly related to the vapor volume fraction. Chen et al. investigated the cavitation flow of liquid nitrogen in a convergent-divergent (C-D) nozzle by finding that under the same flow conditions, as the temperature increased, the vacuole area increases and the vacuoles become more numerous and viscous. Gao et al. established a set of liquid nitrogen cavitation bypassing flow experimental setup with visualized test sections, determined the length of the attached cavitation by the standard deviation method, and found that the cavitation length was closely related to the cavitation number. As the cavitation number decreased, the cavitation length increased sharply.

The formation of different cavitation patterns is significantly influenced by the shape of the flow channel and the operating conditions. Therefore, with the advancement of experimental devices and in-depth study of the relationship between flow parameters and cavitation length, methods to predict the different stages of cavitation occurrence are of particular importance. Cavitation length, as a key parameter to characterize different cavitation stages, is difficult to measure, but its relationship with flow parameters such as flow rate and pressure is of great significance in predicting cavitation behavior.

4.2 Cavitation unsteady state properties

Cavitation is a highly unstable process and the study of its cryogenic unsteady state properties is an important challenge. For example, in liquefied natural gas-based pumps, the complex cavitation behavior of cryogenic fluid results in a flow that exhibits a highly unstable vapor structure. The fluctuating properties of the cavitation zone are usually characterized by the St_c:

$$St_c = \frac{fc}{u_{th}} \quad (30)$$

where f is the main frequency of shedding or fluctuation in the cavitation zone, Hz; c is the length of the cavitation zone, m; u_{th} is the flow velocity in the throat, m/s. St_c Value can be used to distinguish between different types of cavitation phenomena, such as slice cavitation and cloudy cavitation. The value St_c is 0.04 ~ 0.08 for slice cavitation and 0.30 ~ 0.40 for cloudy cavitation.

Pressure ratio and thermal effects have a dominant role in bubble shedding during cavitation. The analysis of the transient cavitation flow of cryogenic liquids is crucial for an in-depth understanding of their non-stationary properties and flow field destabilization mechanisms. Ohira et al. studied cavitation patterns in saturated and supercooled liquid nitrogen, focusing on cavitation instability in C-D nozzles with varying throat diameters. By lowering the pressure, subcooled liquid nitrogen was achieved. They found that cooling the nitrogen to 76K shifted the flow from continuous to intermittent, with cavitation onset and dissipation occurring in milliseconds. To deeply investigate the dynamic cavitation characteristics in liquid nitrogen, the non-stationary flow of liquid nitrogen in a transparent venturi was experimentally investigated. As the pressure ratio increases, the photos showed two primary cavitation modes: quasi-steady and dynamic. Fourier transform analysis of the pressure wave was used to examine the cavitation shedding frequency. The findings indicated that both the maximum cavitation length and shedding frequency rise with the pressure ratio. When the pressure ratio exceeded 2.23, a kinetic pattern of periodic shedding of large-scale cavitation clouds appeared. Later, Zhu et al. further investigated the cavitation flow of liquid nitrogen in a venturi on a modified experimental setup, focusing on the influence of thermal effects on the cavitation kinetics. Bubble length and frequency were determined using image processing with the standard deviation method, and a dimensionless Strauhl number based on the length of cavitations was employed to characterize the shedding frequency properties of cloudy cavitation. Leroux et al. investigated the instability mechanism of the cavitation flow by combining experiments and numerical simulations, emphasizing the important role of the structural response in the development of the cavitation-induced instability.

In summary, the research on the unsteady properties of cavitation of cryogenic fluid has made remarkable progress in recent years but still faces many challenges. Future research should focus on solving the complex non-stationary properties of the flow field and its interaction with the structural response.

5 cavitation inhibition method for cryogenic fluid

Many scholars have worked to inhibit or mitigate the occurrence of cavitation in fluid mechanics, and many methods have been proposed so far.

In valves, cavitation can be avoided by changing the cross-sectional shape, adding pressure-relief devices, and optimizing the valve structure. Increasing the offset distance and bending radius of the throttle globe valve body can reduce the intensity of cavitation at the throttling place in the valve, thus reducing the damage to the downstream pipeline caused by the occurrence of cavitation. Cavitation inhibition methods often used for hydraulic machinery such as impellers are mainly through increasing the inlet pressure, improving the blade load distribution, and installing an inducer or multistage impeller in front of the main impeller. Imamura et al. inhibited the development of internal cavitation in the inducer wheel by installing a J-groove close to the gap of the top of the inducer wheel blade and improved the cavitation performance of the inducer wheel. Shimiya et al. stated that Full flow field cavitation can be inhibited by controlling the leakage vortex cavitation at the tip of the blade, and shallow grooves on the casing wall can reduce the occurrence area of rotating cavitation and asymmetric cavitation. Zhang et al. carried out a study on the cavitation suppression of hydrofoils by using numerical simulation and experimental methods, and analyzed the effects of different raised structures on the distribution of cavitation and shedding of hydrofoils. Meanwhile, it was found that setting the raised structure on the hydrofoil surface could effectively inhibit the cavitation flow. Kim et al. investigated the cavitation flow of a turbopump-induced wheel at different water temperatures. The temperature effect was quantified using a dimensionless thermal parameter, and two types of cavitation instabilities, rotating cavitation, and asymmetrically attached cavitation, were analyzed. It was found that increasing the water temperature or the dimensionless thermal parameter would make the critical cavitation number and the initial value of rotational cavitation decrease, thus reducing the intensity of rotational cavitation. Zhu et al. investigated cavitation in a centrifugal pump with a low specific speed. By placing the guide vanes near the suction surface of the impeller blade inlet, the flow in the impeller inlet region can be realized to control the flow, which can effectively inhibit the cavitation flow inside the centrifugal pump. Chang et al. established a simplified cavitation model based on the Rayleigh Plesset bubble equation and the Zwart cavitation model. The results showed that thermal effects can inhibit the occurrence and development of cavitation. Increasing the inlet velocity or decreasing the outlet pressure can enhance the cavitation process. The increase of turbulent viscosity ratio can enhance the thermal effect and reduce the temperature gradient in the cavity. Due to the long duration of centrifugal force, the cavitation process is enhanced with the decrease of the bend angle. A large number of results show that air entrainment can protect hydraulic structures and metal equipment from cavitation damage. That is, when the air concentration in the mixed phase reaches a certain level, the cavitation phenomenon disappears, but it has not been studied in cryogenic fluid machinery.

Qiao et al. explored the phenomenon of liquid-solid corrosion in elbows, and the local environmental deterioration caused by elbow disturbance was the main cause of downstream weld damage. To mitigate erosion-corrosion damage, the location of the downstream weld should be changed as much as possible and the welding quality should be ensured. Increasing the wall thickness of the elbow at the downstream outlet and the wall thickness of the downstream pipe will also help. The appearance and longitudinal sections of different elbows are shown in Fig. 4. Kadivar et al. experimentally investigated the effect of cavitation control on cavitation flow around a cylinder. Two types of fan corrugations machined on the surface of the cylinder were considered to control cavitation. Both horizontal and vertical fan corrugations positively affected the cavitation control around the cylinder as shown in Fig. 5. In most cases of cavitation, the horizontal scalloped corrugated sheet performed better in controlling the vibration and shedding frequency amplitude caused by cavitation. Therefore, corrugated sheets (especially horizontal scalloped corrugated sheets) can be used to control cavitation and fluctuations of cavity structures on different immersed bodies (e.g., hydrofoils and rudders in marine engineering). However, the limitation of passive control methods such as bellows may lie in the high angle of attack of the immersed body under cavitation. At high angles of attack, full-flow separation may occur, and thus this full-flow separation cannot be influenced by passive control methods.

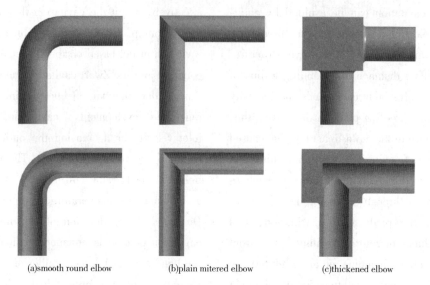

(a) smooth round elbow (b) plain mitered elbow (c) thickened elbow

Fig. 4 Appearance and longitudinal section of different elbows:
(a) smooth round elbow; (b) plain mitered elbow; (c) thickened elbow

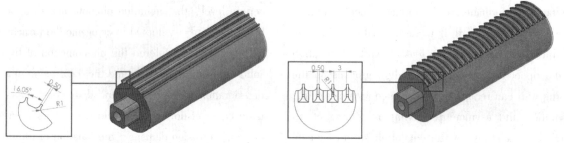

(a) Schematic diagram of a cylinder with vertical scalloped ribs   (b) Schematic diagram of a cylinder with horizontal scalloped ribs

Fig. 5 (a) Schematic diagram of a cylinder with vertical scalloped ribs,
(b) Schematic diagram of a cylinder with horizontal scalloped ribs

In summary, researchers have carried out a lot of research work on the internal cavitation flow inhibition of pumps, induced wheels, bends, and so on. By optimizing the design of their structures, the optimal structure for inhibiting cavitation flow was obtained. However, there are fewer studies on cryogenic cavitation inhibition methods. Moreover, most of the cavitation suppression methods are back-calculated using traditional optimization methods, without forming an intelligent optimization calculation model. The optimization content is longer and the period is longer, and the research work on the prediction model of cryogenic cavitation suppression effect also needs to be further carried out.

6 Perspectives and challenges

In the fields of chemical industry and energy, the problem of cryogenic fluid cavitation in the transportation system is unavoidable. Although scholars of cryogenic fluid transportation system process pumps, valves, pipelines, and other process components within the cryogenic fluid cavitation flow and cavitation inhibition using theoretical analysis, numerical simulation, and experimental research methods to carry out a lot of research, and made some progress, in the cryogenic fluid cavitation flow numerical computational modeling, did not take into account the full range of cryogenic fluid media properties, the actual working conditions, the flow characteristics of the cryogenic fluid and other factors on the impact of cavitation flow. At the same time, there is an urgent need to establish an effective numerical framework for gas-liquid-solid coupling in cryogenic cavitation, to accurately predict the degree of damage and its consequences. In addition, the multiphase flow problem in the existing cryogenic cavitation model should be further investigated, especially the behavior simulation under high pressure and extremely low-temperature conditions. To more accurately predict and analyze the potential damage of cavitation on materials; Current cryogenic cavitation models primarily address the vapor volume fraction distribution at a macroscopic level, but they do not capture the microscopic mechanisms of cavitation evolution. To better understand the microscopic mechanism of cavitation, multi-scale models based on molecular dynamics simulations and particle swarm optimization algorithms should be developed in the future. These methods can reveal the microscopic formation, growth, and collapse processes of cavitation bubbles.

In terms of cavitation inhibition research, most of the studies have been carried out on the cavitation of non-cryogenic fluid inside fluid machinery. Studies on cavitation inhibition strategies for cryogenic fluid transportation systems have not been reported. For cryogenic fluid cavitation full flow field calculation, the inhibition effect law of different structures on the internal cavitation flow of cryogenic fluid transportation system should be studied. The intelligent optimization calculation model should be formed to reduce the structure optimization cycle. Eventually, the best solution to inhibit the cavitation flow of the cryogenic fluid pipeline system should be obtained to provide support for the design development and engineering application of high-performance and high-reliability cryogenic fluid transportation systems. In addition, the presence of solid particles in the medium is also a problem that should not be ignored. The presence of multiple phases in the fluid affects the flow distribution and the formation of cavitation bubbles, which leads to different cavitation regions and different cavitation intensities, and therefore it is a challenge to study the cavitation inhibition of fluids containing multiphase flows.

7 conclusion

This article reviews several key issues involved in the analysis and prevention of damage tocryogenic fluid pipelines and leakage and pipe bursting accidents, and the relevant conclusions are as follows:

(1) As the saturation vapor pressure is greatly affected by temperature, the formula for determining the cavitation ofcryogenic fluid needs to be corrected for the saturation vapor pressure. In addition, experimental and theoretical studies show that the viscosity, density change, and mechanical structure of cryogenic fluid in the cryogenic environment have an important effect on the cavitation incipient.

(2) In terms of modeling thermal effects, the model should take into account the energy equation for the phase change effect of cavitation and consider the effect of temperature change on physical properties and vapor pressure. The cavitation model based on the transport equation is modified to reflect the influence of thermal effects on bubble growth and collapse.

(3) The characterization method of cryogenic fluid cavitation flow field based on vortex structure and entropy production analysis is constructed by considering the influence of the thermodynamic effect ofcryogenic fluid and the compressibility of the gas phase. An accurate description of the information of the cavitation flow field of cryogenic fluid can be realized.

(4) In the research of cavitation suppression, most of the research is carried out for the non-cryogenic cavitation inside the fluid machinery, and cavitation can be inhibited by structural optimization, and the suppression ofcryogenic fluid cavitation can be used as a reference.

References

[1] Li Q Y, Zong C Y, Liu F W, Xue T H, Zhang A,

Song X G. Numerical and experimental analysis of the cavitation characteristics of orifice plates under high-pressure conditions based on a modified cavitation model [J]. International Journal of Heat and Mass Transfer, 2023, 203: 123782. http://dx.doi.org/10.1016/j.ijheatmasstransfer.2022.123782

[2] Zhang J M, Yang Q. Cavitation Theory and Application: Study on Cavitation Scale Effect [M] Beijing: Science Publishing House, 2013.

[3] Weitendorf E. On the history of propeller cavitation and cavitation tunnels [C]. Pasadena: Proceedings of the 4th International Symposium on Cavitation, 2001.

[4] Ji B, Luo X, Wu Y, Peng X, Duan Y. Numerical analysis of unsteady cavitating turbulent flow and shedding horse-shoe vortex structure around a twisted hydrofoil [J]. International Journal of Multiphase Flow, 2013, 51: 33–43. https://doi.org/10.1016/j.ijmultiphaseflow.2012.11.008

[5] Yan Y, Thorpe R B. Flow regime transitions due to cavitation in the flow through an orifice [J]. International Journal of Multiphase Flow, 1990, 16: 1023–1045. https://doi.org/10.1016/0301-9322(90)90105-R

[6] Rudolf P H M, Griger M, Štefan D Characterization of the cavitating flow in converging-diverging nozzle based on experimental investigations [J]. The European Physical Journal Conferences, 2014, 67: 02101. http://dx.doi.org/10.1051/epjconf/20146702101

[7] Shi H, Li M, Nikrityuk P, Liu Q. Experimental and numerical study of cavitation flows in venturi tubes: From CFD to an empirical model [J]. Chemical Engineering Science, 2019, 207: 672–687. http://dx.doi.org/10.1016/j.ces.2019.07.004

[8] Ma X. Numerical study on cavitating flow of cryogenic fluids [D]. Harbin: Harbin Institute of Technology, 2013.

[9] Sun B, Yang X P, Wei Z Y, Zhen L, Yao P S, Xi L L. Piercing failure mechanism of pipeline elbows in LNG terminals and its influencing rules [J]. Oil & Gas Storage and Transportation, 2024, 43(4): 423–431. https://link_cnki_net.igw.upc.edu.cn/urlid/13.1093.TE.20231130.1717.002

[10] Besant W H. Hydrostatics and Hydrodynamics [M] Cambridae: Cambridge University Press, 1859.

[11] Zhu J K, Xie H J, Feng K S, Zhang X B, Si M Q. Unsteady cavitation characteristics of liquid nitrogen flows through venturi tube [J]. International Journal of Heat and Mass Transfer, 2017, 112: 544–552. http://dx.doi.org/10.1016/j.ijheatmasstransfer.2017.04.036

[12] Pinho J, Peveroni L, Vetrano M R, Buchlin J M, Steelant J, Strengnart M. Experimental and numerical study of a cryogenic valve using liquid nitrogen and water [J]. Aerospace Science and Technology, 2019, 93: 1–11. http://dx.doi.org/10.1016/j.ast.2019.105331

[13] Lin Z H, Li J Y, Jin Z J, Qian J Y. Fluid dynamic analysis of liquefied natural gas flow through a cryogenic ball valve in liquefied natural gas receiving stations [J]. Energy, 2021, 226: 120376. http://dx.doi.org/10.1016/j.energy.2021.120376

[14] Ito Y, Tsunoda A, Kurishita Y, Kitano S, Nagasaki T. Experimental Visualization of Cryogenic Backflow Vortex Cavitation with Thermodynamic Effects [J]. 2015, 32: 1–12. https://doi.org/10.2514/1.B35782

[15] Utturkar Y, Wu J, Wang G, Shyy W. Recent progress in modeling of cryogenic cavitation for liquid rocket propulsion [J]. Progress in Aerospace Sciences, 2005, 41: 558–608. https://doi.org/10.1016/j.paerosci.2005.10.002

[16] Xiang L, Tan Y, Chen H, Xu K. Numerical simulation of cryogenic cavitating flow in LRE oxygen turbopump inducer [J]. Cryogenics, 2022, 126: 103540. https://doi.org/10.1016/j.cryogenics.2022.103540

[17] Ito Y. The World's First Test Facility That Enables the Experimental Visualization of Cavitation on a Rotating Inducer in Both Cryogenic and Ordinary Fluids [J]. Journal of Fluids Engineering - Transactions of the Asme, 2021, 143: 1–14. http://dx.doi.org/10.1115/1.4051849

[18] Liang W, Chen T, Huang B, Wang G. Thermodynamic analysis of unsteady cavitation dynamics in liquid hydrogen [J]. International Journal of Heat and Mass Transfer, 2019, 142: 118470. https://doi.org/10.1016/j.ijheatmasstransfer.2019.118470

[19] Long X, Liu Q, Ji B, Lu Y. Numerical investigation of two typical cavitation shedding dynamics flow in liquid hydrogen with thermodynamic effects [J]. International Journal of Heat and Mass Transfer, 2017, 109: 879–893. https://doi.org/10.1016/j.ijheatmasstransfer.2017.02.063

[20] Grazia D M, Daniela B, Ficarella A. Analysis of thermal effects in a cavitating ofifice using rayleigh equation and experiments [C]. Brussels, Belgium: 17th International Conference on Nuclear Engineering, 2009. 763–774.

[21] Zhang S F, Li X J, Zhu Z C. Numerical simulation of

cryogenic cavitating flow by an extended transport-based cavitation model with thermal effects[J]. Cryogenics, 2018, 92: 98-104. https://doi.org/10.1016/j.cryogenics.2018.04.008

[22] Tseng C C, Shyy W. Modeling for isothermal and cryogenic cavitation[J]. International Journal of Heat and Mass Transfer, 2010, 53: 513-525. https://doi.org/10.1016/j.ijheatmasstransfer.2009.09.005

[23] Zhang X B, Qiu L M, Gao Y, Zhang X J. Computational fluid dynamic study on cavitation in liquid nitrogen[J]. Cryogenics, 2008, 48: 432-438. http://dx.doi.org/10.1016/j.cryogenics.2008.05.007

[24] Xia W H. Effects of water quality on cavitation priming [J]. Journal of Hydraulic Engineering, 1993, 11: 48-55.

[25] Salehi S, Mahmudi Moghadam A, Soori S. Effect of the pipe bend of the morning glory spillway on the cavitation number[J]. Flow Measurement and Instrumentation, 2023, 92: 102375. https://doi.org/10.1016/j.flowmeasinst.2023.102375

[26] Rajagopal K, Tripuraneni N. Bulk viscosity and cavitation in boost-invariant hydrodynamic expansion[J]. Journal of High Energy Physics, 2010, 2010: 18. https://doi.org/10.1007/JHEP03(2010)018

[27] Iwai Y, Li S. Cavitation erosion in waters having different surface tensions[J]. Wear, 2003, 254: 1-9. https://doi.org/10.1016/S0043-1648(02)00305-8

[28] Rodio M G, De Giorgi M G, Ficarella A. Influence of convective heat transfer modeling on the estimation of thermal effects in cryogenic cavitating flows[J]. International Journal of Heat and Mass Transfer, 2012, 55: 6538-6554. https://doi.org/10.1016/j.ijheatmasstransfer.2012.06.060

[29] Simpson A, Ranade V V. Modeling hydrodynamic cavitation in venturi: influence of venturi configuration on inception and extent of cavitation[J]. AIChE Journal, 2019, 65: 421-433. https://doi.org/10.1002/aic.16411

[30] Liu C, Li X, Li A, Cui Z, Chen L, Li Y. Cavitation onset caused by a dynamic pressure wave in liquid pipelines[J]. Ultrasonics Sonochemistry, 2020, 68: 105225. https://doi.org/10.1016/j.ultsonch.2020.105225

[31] Fruman D H, Reboud J L, Stutz B t. Estimation of thermal effects in cavitation of thermosensible liquids [J]. International Journal of Heat and Mass Transfer, 1999, 42: 3195-3204. https://doi.org/10.1016/S0017-9310(99)00005-8

[32] Tani N, Tsuda S, Yamanishi N, Yoshida Y. Development and validation of new cryogenic cavitation model for rocket turbopump inducer[J]. 2009, 63: 84260. https://hdl.handle.net/2027.42/84260

[33] Franc J P, Janson E, Morel P, Rebattet C, Riondet M. Visualizations of leading edge cavitation in an Inducer at different temperatures[C]. Shanghai: 4th International Symposium on Cavitation, 2001. 1-8.

[34] Brennen C E. Fundamentals of Multiphase Flow[M]. Fundamentals of Multiphase Flow, 2005.

[35] Deshpande M, Feng J, Merkle C L. Numerical Modeling of the Thermodynamic Effects of Cavitation[J]. Journal of Fluids Engineering, 1997, 119: 1-8. https://doi.org/10.1115/1.2819150

[36] Li J, Liu L, Feng Z. Numerical validation of the cavitation moddel and algorithm based on the liquid/vapor interface tracking method[J]. journal of engineering thermophysics, 2006, 27(2): 238-240.

[37] Singhal A K. Multi-Dimensional Simulation of Cavitating Flows Using a PDF Model for Phase Change[C]. Asme Fluids Engineering Division Summer Meeting, 1997. 3272.

[38] Rahbarimanesh S, Brinkerhoff J, Nejat A. Evaluation of the effects of interfacial tension forces in homogenous equilibrium mixture modeling of a cavitation flow of liquefied natural gas inside a Laval nozzle[J]. Chemical Engineering Research and Design, 2022, 188: 1029-1041. https://doi.org/10.1016/j.cherd.2022.10.037

[39] Pascarella C, Salvatore V, Ciucci A. Effects of speed of sound variation on unsteady cavitating flows by using a barotropic model[J]. 2022.

[40] Barre S, Rolland J, Boitel G, Goncalves E, Patella R F. Experiments and modeling of cavitating flows in venturi: attached sheet cavitation[J]. European Journal of Mechanics-B/Fluids, 2009, 28: 444-464. https://doi.org/10.1016/j.euromechflu.2008.09.001

[41] Goncalvès E, Patella R F. Numerical study of cavitating flows with thermodynamic effect[J]. Computers & Fluids, 2010, 39: 99-113. https://doi.org/10.1016/j.compfluid.2009.07.009

[42] Clerc S. Numerical Simulation of the Homogeneous Equilibrium Model for Two-Phase Flows[J]. Journal of Computational Physics, 2000, 161: 354-375. https://doi.org/10.1006/jcph.2000.6515

[43] Goncalves E, Rolland J, Challier G. Thermodynamic effect on a cavitation inducer in liquid hydrogen[C]. ASME Fluids Engineering Division summer conference, 2009.

[44] Sinibaldi E, Beux F, Salvetti M V. A numerical method for 3D barotropic flows in turbomachinery[J]. Flow Turbulence and Combustion, 2006, 76: 371-381. https://doi.org/10.1007/s10494-006-9025-7

[45] Barre S, Rolland J, Boitel G, Goncalves E, Patella R F. Experiments and modeling of cavitating flows in venturi: attached sheet cavitation[J]. European Journal of Mechanics B-Fluids, 2009, 28: 444-464. https://doi.org/10.1016/j.euromechflu.2008.09.001

[46] Utturkar Y, Thakur S, Shyy W. Computational Modeling of Thermodynamic Effects in Cryogenic Cavitation [D]. 2005.

[47] Ahuja V, Hosangadi A, Arunajatesan S. Simulations of cavitating flows using hybrid unstructured meshes [J]. Journal of Fluids Engineering-Transactions of the Asme, 2001, 123: 331-340. https://doi.org/10.1115/1.1362671

[48] Esposito C, Peveroni L, Gouriet J B, Steelant J, Vetrano M R. On the influence of thermal phenomena during cavitation through an orifice[J]. International Journal of Heat and Mass Transfer, 2021, 164: 120481. https://doi.org/10.1016/j.ijheatmasstransfer.2020.120481

[49] G. H. Schnerr and J. Sauer. Physical and Numerical Modeling of Unsteady Cavitation Dynamics [C]. in Fourth International Conference on Multiphase Flow, 2001.

[50] He J, Li C, Jia W, Qiu B, Yang F, Zhang C. A new cavitation model for simulating steady and unsteady cavitating flows[J]. Ocean Engineering, 2023, 273. https://doi.org/10.1016/j.oceaneng.2023.113925

[51] Zwart P, Gerber A G, Belamri T. A two-phase flow model for predicting cavitation dynamics [J]. Fifth International Conference on Multiphase Flow, 2004, 152.

[52] Zhang G, Wang W W, Wu Z Y, Chen D S, Kim H D, Lin Z. Effect of the opening degree on evolution of cryogenic cavitation through a butterfly valve [J]. Energy, 2023, 283: https://doi.org/10.1016/j.energy.2023.128543

[53] Singhal A K, Athavale M M, Li H, Jiang Y. Mathematical basis and validation of the full cavitation model [J]. Journal of Fluids Engineering: Transactions of the ASME, 2002, 124(3): 124. https://doi.org/10.1115/1.1486223

[54] Pinho J, Lema M, Rambaud P, Steelant J. Multiphase Investigation of Water Hammer Phenomenon Using the Full Cavitation Model[J]. Journal of Propulsion and Power, 2014, 30: 105-113. https://doi.org/10.2514/1.B34833

[55] Zhang X B, Qiu L M, Gao Y, Zhang X J. Computational fluid dynamic study on cavitation in liquid nitrogen[J]. Cryogenics, 2008, 48: 432-438. https://doi.org/10.1016/j.cryogenics.2008.05.007

[56] Senocak I, Shyy W. Interfacial dynamics-based modelling of turbulent cavitating flows, Part-2: Time-dependent computations[J]. 2004, 44: 997-1016. https://doi.org/10.1002/fld.693

[57] Senocak I, Shyy W. A Pressure-Based Method for Turbulent Cavitating Flow Computations[J]. Journal of Computational Physics, 2002, 176: 363-383. https://doi.org/10.1006/jcph.2002.6992

[58] Merkle C, Feng J, Buelow P. Computational modelling of the dynamics of sheet cavitation, 3rd Int[C]. Grenoble, France: Symposium on Cavitation, 1998.

[59] Kunz R F, Boger D A, Stinebring D R, Chyczewski T S, Lindau J W, Gibeling H J, et al. A preconditioned Navier-Stokes method for two-phase flows with application to cavitation prediction[J]. Computers & Fluids, 2000, 29: 849-875. https://doi.org/10.1016/S0045-7930(99)00039-0

[60] Schnerr G. Physical and Numerical Modeling of Unsteady Cavitation Dynamics[M]. 2001.

[61] Zhang X, Wu Z, Xiang S, Qiu L. Modeling cavitation flow of cryogenic fluids with thermodynamic phase-change theory[J]. Chinese Science Bulletin, 2013, 58: 567-574. https://doi.org/10.1007/s11434-012-5463-x

[62] Senocak I, Shyy W. Interfacial dynamics-based modelling of turbulent cavitating flows, Part-1: Model development and steady-state computations[J]. International Journal for Numerical Methods in Fluids, 2004, 44: 975-995. https://doi.org/10.1002/fld.692

[63] Zhu J, Chen Y, Zhao D, Zhang X. Extension of the Schnerr-Sauer model for cryogenic cavitation[J]. European Journal of Mechanics-B/Fluids, 2015, 52: 1-10. https://doi.org/10.1016/j.euromechflu.2015.01.008

[64] Li W, Yu Z, Kadam S. An improved cavitation model with thermodynamic effect and multiple cavitation regimes[J]. International Journal of Heat and Mass Transfer, 2023, 205: 123854. https://doi.org/10.1016/j.ijheatmasstransfer.2023.123854

[65] Ruggeri R S, Gelder T F. Cavitation and Effective Liquid Tension of Nitrogen in a Hydrodynamic Cryogenic

Tunnel [C]. Boston, MA: Advances in Cryogenic Engineering, 1964. 304-310.
[66] Hord J. Cavitation in liquid cryogens. 2: Hydrofoil [J]. 1973.
[67] Stahl H A, Stepanoff A J. Thermodynamic Aspects of Cavitation in Centrifugal Pumps [J]. Transactions of the American Society of Mechanical Engineers, 1956, 78: 1691-1693. https://doi.org/10.1115/1.4014152
[68] Kikuta K, Yoshida Y, Watanabe M, Hashimoto T, Nagaura K, Ohira K. Thermodynamic Effect on Cavitation Performances and Cavitation Instabilities in an Inducer [J]. Journal of Fluids Engineering, 2008, 130. https://doi.org/10.1115/1.2969426
[69] Chen T R, Chen H, Liang W D, Huang B, Xiang L. Experimental investigation of liquid nitrogen cavitating flows in converging-diverging nozzle with special emphasis on thermal transition [J]. International Journal of Heat and Mass Transfer, 2019, 132: 618 – 630. https://doi.org/10.1016/j.ijheatmasstransfer.2018.11.157
[70] Chen T R, Chen H, Liu W C, Huang B, Wang G Y. Unsteady characteristics of liquid nitrogen cavitating flows in different thermal cavitation mode [J]. Applied Thermal Engineering, 2019, 156: 63-76. https://doi.org/10.1016/j.applthermaleng.2019.04.024
[71] Gao X, Chen H, Zhu J K, Wang S H, Zhang X B. Visual Experimental Study on Liquid-Nitrogen Cavitating Flow on NACA 66 Hydrofoil [J]. Journal of Propulsion and Power, 2020, 36: 88-94. https://doi.org/10.2514/1.B37682
[72] Zhu J K, Wang S H, Zhang X B. Influences of thermal effects on cavitation dynamics in liquid nitrogen through venturi tube [J]. Physics of Fluids, 2020, 32: https://doi.org/10.1063/1.5132591
[73] Ohira K, Nakayama T, Nagai T. Cavitation flow instability of subcooled liquid nitrogen in converging-diverging nozzles [J]. Cryogenics, 2012, 52: 35 – 44. https://doi.org/10.1016/j.cryogenics.2011.11.001
[74] Leroux J-B, Coutier-Delgosha O, Astolfi J A. A joint experimental and numerical study of mechanisms associated to instability of partial cavitation on two-dimensional hydrofoil [J]. Physics of Fluids, 2005, 17. https://doi.org/10.1063/1.1865692
[75] Zhang Y, Liu X, Li B, Sun S, Peng J, Liu W, et al. Hydrodynamic characteristics and optimization design of a bio-inspired anti-erosion structure for a regulating valve core [J]. Flow Measurement Instrumentation, 2022, 85: 102173. https://doi.org/10.1016/j.flowmeasinst.2022.102173
[76] Xu Z F, Kong F Y, Zhang H L, Zhang K, Wang J Q, Qiu N. Research on Visualization of Inducer Cavitation of High-Speed Centrifugal Pump in Low Flow Conditions [J]. Journal of Marine Science and Engineering, 2021, 9. https://doi.org/10.3390/jmse9111240
[77] Imamura H, Kurokawa J, Matsui J, Kikuchi M. Suppression of Cavitating Flow in Inducer by J-Groove [C]. 2003. https://doi.org/10.1299/jsmemecjo.2003.2.0_35
[78] Shimiya N, Fujii A, Horiguchi H, Uchiumi M, Kurokawa J, Tsujimoto Y. Suppression of Cavitation Instabilities in an Inducer by J Groove [J]. Journal of Fluids Engineering, 2008, 130. https://doi.org/10.1115/1.2829582
[79] Zhang L, Chen M, Shao X. Inhibition of cloud cavitation on a flat hydrofoil through the placement of an obstacle [J]. Ocean Engineering, 2018, 155: 1-9. https://doi.org/10.1016/j.oceaneng.2018.01.068
[80] Kim J, Song S. Measurement of Temperature Effects on Cavitation in a Turbopump Inducer [J]. Journal of Fluids Engineering 2016, 138(1): 11304.
[81] Zhu B, Chen H X. Cavitating suppression of low specific speed centrifugal pump with gap drainage blades [J]. Journal of Hydrodynamics, Ser. B, 2012, 24: 729-736. https://doi.org/10.1016/S1001-6058(11)60297-7
[82] Chang H, Xie X, Zheng Y, Shu S. Numerical study on the cavitating flow in liquid hydrogen through elbow pipes with a simplified cavitation model [J]. International Journal of Hydrogen Energy, 2017, 42: 18325-18332. https://doi.org/10.1016/j.ijhydene.2017.04.132
[83] Qiao Q, Cheng G, Li Y, Wu W, Hu H, Huang H. Corrosion failure analyses of an elbow and an elbow-to-pipe weld in a natural gas gathering pipeline [J]. Engineering Failure Analysis, 2017, 82: 599 – 616. https://doi.org/10.1016/j.engfailanal.2017.04.016
[84] Kadivar E. Experimental and Numerical Investigations of Cavitation Control Using Cavitating-bubble Generators [J]. 2020. https://doi.org/10.17185/DUEPUBLICO/73171

基于状态空间的天然气管道系统模型预测主动控制

李 轩　陆洋帆　陈国龙　高 伟　伍梓文　金 凤　郭雨茜　张炜奇　宫 敬　温 凯

[中国石油大学(北京)机械与储运工程学院·城市油气输配技术北京市重点实验室·油气储运智能计量实验室]

摘 要　【目的】天然气瞬态流动偏微分方程的非凸性和强非线性使得目前的天然气管道瞬态控制方法存在计算精度与时间的两难问题，单依靠最优化理论无法满足如今智能化背景下天然气管道快速、精准调控的需求。【方法】本文建立了基于状态空间的天然气管道模型预测主动控制方法，以实现受限条件与用气波动下的天然气管道系统的压力实时控制。首先从基础的模型预测控制方法出发，经过矩阵变换将控制问题推导成了标准的二次规划形式，再考虑受约束条件的实际情况得到了嵌入状态变量约束、输入约束下的天然气管道模型预测控制算法。【结果】通过一个天然气管道系统进行了算例验证，结果表明模型预测控制方法能够实现出口压力的快速调节以及在出口流量波动的情况下将用户处分输压力控制在目标压力附近，此外对比了增加约束前后的控制策略和结果，证明本方法可以在有效考虑现实约束实现对下游压力的控制。【结论】基于状态空间的天然气管道系统模型预测主动控制方法可实现分钟级的实时控制，并且可以显式增加约束条件，为后续天然气管网的智能调控提供技术基础。

关键词　主动控制；天然气管道系统；状态空间；模型预测控制

Model predictive active control of natural gas pipeline system based on state space

Li Xuan　Lu Yangfan　Chen Guolong　Gao Wei　Wu Ziwen
Jin Feng　Guo Yuqian　Zhang Weiqi　Gong Jing　Wen Kai

(Beijing Key Laboratory of Urban Oil and Gas Distribution Technology/Intelligent Metering Laboratory of Oil and Gas Storage and Transportation, China University of Petroleum-Beijing)

Abstract　[Objective] The non-convexity and strong nonlinearity of the transient partial differential equation of natural gas flow make the current transient control method of natural gas pipeline have a dilemma between calculation accuracy and time. The optimization theory alone cannot meet the needs of fast and accurate regulation of natural gas pipelines under the current intelligent background. [Methods] This paper establishes a model predictive active control method for natural gas pipelines based on state space to achieve real-time pressure control of natural gas pipeline systems under constrained conditions and gas consumption fluctuations. First, starting from the basic model predictive control method, the control problem is derived into a standard quadratic programming form through matrix transformation. Then, considering the actual situation of the constraints, the model predictive control algorithm of natural gas pipelines under embedded state variable constraints and input constraints is obtained. [Results] A natural gas pipeline system was used for example verification. The results show that the model predictive control method can achieve rapid adjustment of the outlet pressure and control the user's outlet pressure near the target pressure when the outlet flow fluctuates. In addition, the control strategies and results before and after adding constraints are compared, proving that this method can effectively consider the actual constraints to achieve control of downstream pressure. [Conclusion] The model predictive active control method for natural gas pipeline systems based on state space can achieve real-time control at the minute level, and can explicitly add constraints, providing a technical basis for the subsequent intelligent regulation of natural gas pipeline networks.

Key words　Active control; Natural gas pipeline system; State space; Model predictive control

1 引言

由于天然气的高储能特性和快速发电能力，天然气管道系统对于缓解供需失衡至关重要。与此同时，中国天然气产业的市场化改革仍处于起步阶段。管网的独立运行要求运营商不仅要掌握业务流程，还要更加关注资源和用户。天然气管道系统的主动控制是过程与控制算法的深度集成，是管道运行管理从人为驱动到系统自驱、资源配置从局部优化到整体优化过渡的前提。在天然气管网系统的运行管理中，油气调控中心根据国家管网集团生产和销售计划，编制管道运行方案，通过季协调、月计划、周平衡、日指定对油气管网资源进行统一调配优化运行，保证天然气供应稳定和管道安全运行。日指定下很多分输站场都已经实现了自动分输功能，无需调度员亲力亲为，极大的节省了人力资源。然而，目前天然气管道系统上的自动控制仅限于分输站，对于更加常见的下游用气变化引起的压力波动情况，上游的压气分输站还没有具备主动的调节能力，需要调度员实时从SCADA系统中收集各种生产信息然后利用自身的专业知识和经验来做出调整，向站场设备发布指令。但是由于人类自身的限制无法同时操作整个系统的各个点位，很难实现整个生产过程的全局优化。因此，优化生产过程仍然是非常具有挑战性的，迫切需要采取一种更广泛、更系统的方法来进行天然气管道控制，将高级知识融入智能化工作。

天然气管网稳态运行控制的常见方法包括动态规划方法、粒子群算法、黄金分割法等等。然而，由于用户需求不断变化导致天然气管网的运行是不稳定的，稳态方法无法准确制定瞬态条件下的调度策略。因此，应考虑天然气管道的瞬态流动特性以解决控制问题。Mak TW提出了一种两阶段方法，对流动偏微分方程进行梯形空间和时间离散化。刘宇婷采用第二类和第三类特殊序列集合法(SOS)对流动方程线性化，而于肖雯采用中心差分方法将偏微分方程离散，并建立了改进粒子群方法提高了优化问题的求解速度，将压缩机的控制时步提高10min。但依然难以满足工业输送流程控制中的实时性。

高性能一直是工业控制应用的设计目标，二次型调节器(Linear Quadratic Regulator)、模型预测控制(Model Predictive Control, MPC)在现代最优控制理论中占有重要地位，由于其良好的鲁棒性，在天然气管道、输油管道、交通领域等控制过程都有一些应用。但是LQR由于无法显式添加约束而不能适用于复杂的天然气管道系统控制。MPC经常被应用于制定天然气网络的在线控制方案，模型预测控制的目标一般为经济成本最低，或尽快达到指定状态。Marques最早于1988年提出了天然气管网的模型预测控制方法，通过梯度搜索找到最优的控制方案，天然气网络的实时状态作为移动时域的初始状态被反馈给控制器，以实现天然气管网系统的在线控制。Gopalakrishnan提出了一个两阶段的MPC模型，在第一阶段用离线模型确定管道在周期终点时刻的分输压力，作为第二阶段在线MPC模型的约束。Chen等人基于神经网络模型应用MPC于枝状天然气管网控制下游压力跟踪设定值。Zhu等人以辨识的传递函数为基础应用MPC，提出通过统计方法开发时间常数公式。Aalto H在枝状天然气管网上应用了以经济性为目标的NMPC，但控制时步是小时级的。但以上方法存在由于模型复杂受限于计算速度过慢和模型的可解释性不足的矛盾，因此天然气管道瞬态运行控制问题的时间步长无法达到更小级别，这意味着当管道网络的运行状态波动频率较高时，控制方案会存在滞后性和较大误差。实时性要求优化问题的求解时间必须小于控制周期，故需要进一步探索快速的瞬态运行控制方法。

智能调控的本质在于基于外部需求变化，结合管网自身状态，自主完成控制任务，即"主动控制"。基于先前的工作中提出的线性状态空间模型，考虑天然气管网实际控制中存在的物理限制，本文建立了一种基于线性状态空间模型的天然气管网系统模型预测主动控制方法(MPAC)，从依赖管存平抑下游波动转变成对于控制的精度和灵活性要求更加高的从上游主动调整压力，实现了在下游流量持续波动的瞬态条件下对天然气管网系统的快速控制。

2 天然气管道模型预测主动控制方法

2.1 基本概念

模型预测控制(Model Predictive Control, MPC)是一种广泛应用于工业控制领域的高级控制策略。MPC的基本思想是利用当前时刻系统的状态及约束条件，对未来一段时间内的状态、输入变量进行预测，并求解出一组最优的控制输入序列。随后，只选取最优控制序列中的第一组结果，将其应用于系统中。在下一时刻，重复同样的操作，得到新的最优控制序列，直到系统达到期望状态(图1)。

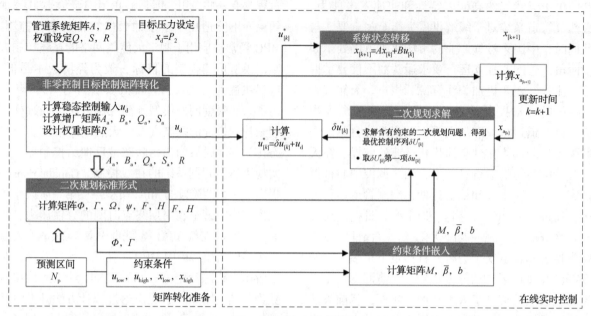

图 1 控制算法技术路线

MPC是一种滚动优化的控制方法。在每个采样时刻，MPC会通过求解一个有限时间内的最优化问题来计算最优控制序列，这一有限时间称为预测区间。考虑系统的不确定性、测量误差等因素，在实际控制中，只选取预测区间内最优控制序列中的第一项施加到系统中。MPC通常针对离散系统，因此预测区间通常指的是预测的离散步数。

下面通过一个单输入单状态系统的例子进行说明，其离散型状态空间方程为

$$x_{[k+1]} = f(x_{[k]}, u_{[k]}) \quad (1)$$

性能指标为

$$J = h(x_{[N]}, x_d) + \sum_{k=1}^{N_p-1} g(x_{[k]}, x_d, u_{[k]}) \quad (2)$$

其中预测区间定义为 N_p。

2.2 基础模型预测控制算法

取系统状态变量为 $x_{[k]} = [P_{1[k]}, P_{1[k-1]}, P_{1[k-2]}, M_{2[k]}, M_{2[k-1]}, M_{2[k-2]}, P_{2[k]}, P_{2[k-1]}, M_{1[k]}, M_{1[k-1]}]$，系统输入为 $u_{[k]} = \begin{pmatrix} u_1 \\ u_2 \end{pmatrix} = \begin{pmatrix} P_{1[k]} \\ M_{2[k]} \end{pmatrix}$，在本研究中将下游用气 $M_{2[k]}$ 作为系统干扰，系统输出为 $y_{[k]} = \begin{pmatrix} y_1 \\ y_2 \end{pmatrix} = \begin{pmatrix} P_{2[k]} \\ M_{1[k]} \end{pmatrix}$，可得状态空间形式如下：

$$x_{[k+1]} = Ax_{[k]} + Bu_{[k]} \quad (3)$$

其中 A, B 是管道的特征系数所组成的常数矩阵，决定了系统的输入输出特性。一般的二次型性能指标为：

$$J = \frac{1}{2}x_{[N_p]}^T S x_{[N_p]} + \frac{1}{2}\sum_{k=0}^{N_p-1}\left[x_{[k]}^T Q x_{[k]} + u_{[k]}^T R u_{[k]}\right] \quad (4)$$

$$S = \begin{bmatrix} s_1 & \cdots & 0 \\ \vdots & \ddots & \vdots \\ 0 & \cdots & s_n \end{bmatrix}, s_1, s_2, \cdots, s_n \geq 0$$

$$Q = \begin{bmatrix} q_1 & \cdots & 0 \\ \vdots & \ddots & \vdots \\ 0 & \cdots & q_n \end{bmatrix}, q_1, q_2, \cdots, q_n \geq 0 \quad (5)$$

$$R = \begin{bmatrix} r_1 & \cdots & 0 \\ \vdots & \ddots & \vdots \\ 0 & \cdots & r_p \end{bmatrix}, r_1, r_2, \cdots, r_n > 0$$

在模型预测控制中，N_p 代表预测区间，在 k 时刻可以基于当前状态推得 $k+N_p$ 时刻的系统状态：

$$x_{[k+1|k]} = Ax_{[k|k]} + Bu_{[k|k]} \quad (6)$$

$$\begin{aligned} x_{[k+2|k]} &= A(Ax_{[k|k]} + Bu_{[k|k]}) + Bu_{[k+1|k]} \\ &= A^2 x_{[k|k]} + ABu_{[k|k]} + Bu_{[k+1|k]} \end{aligned} \quad (7)$$

$$x_{[k+N_p|k]} = A^{N_p} x_{[k|k]} + A^{N_p-1} Bu_{[k|k]} + \cdots + ABu_{[k+N_p-2|k]} + Bu_{[k+N_p-1|k]} \quad (8)$$

为简化表达，定义

$$X_{[k]} \triangleq \begin{bmatrix} x_{[k+1|k]} \\ x_{[k+2|k]} \\ \vdots \\ x_{[k+N_p|k]} \end{bmatrix}_{(nN_p)\times 1} \quad (9)$$

$$U_{[k]} \triangleq \begin{bmatrix} u_{[k|k]} \\ u_{[k+1|k]} \\ \vdots \\ u_{[k+N_p-1|k]} \end{bmatrix}_{(nN_p)\times 1} \quad (10)$$

X_k 包含了 k 时刻预测的所有预测区间内的状态变量，得到紧凑的表达：

$$X_{[k]} = \Phi x_{[k|k]} + \Gamma U_{[k]} \quad (11)$$

其中

$$\Phi = \begin{bmatrix} A_{n\times n} \\ A^2 \\ \vdots \\ A^{N_p} \end{bmatrix}_{(nN_p)\times n} \quad (12)$$

$$\Gamma = \begin{bmatrix} B & 0 & \cdots & 0 \\ AB & B & \cdots & 0 \\ \vdots & \vdots & \ddots & \vdots \\ A^{N_p-1}B & A^{N_p-2}B & \cdots & B \end{bmatrix}_{(nN_p)\times(pN_p)} \quad (13)$$

为便于求解还需要将控制算法的二次性能指标转化为二次规划的标准形式：

$$J = \frac{1}{2} x_{[k+N_p|k]}^T S x_{[k+N_p|k]} + \frac{1}{2} \sum_{i=0}^{N_p-1} \left[x_{[k+i|k]}^T Q x_{[k+i|k]} + u_{[k+i|k]}^T R u_{[k+i|k]} \right] \quad (14)$$

代入 X_k，U_k

$$J = \frac{1}{2} x_{[k|k]}^T Q x_{[k|k]} + \frac{1}{2} X_{[k]}^T \begin{bmatrix} Q & \cdots & 0 \\ \vdots & Q & \vdots \\ 0 & \cdots & S \end{bmatrix} X_{[k]} + \frac{1}{2} U_{[k]}^T \begin{bmatrix} R & \cdots & 0 \\ \vdots & \ddots & \vdots \\ 0 & \cdots & R \end{bmatrix} U_{[k]} \quad (15)$$

简化得到

$$J = \frac{1}{2} x_{[k|k]}^T Q x_{[k|k]} + \frac{1}{2} X_{[k]}^T \Omega X_{[k]} + \frac{1}{2} U_{[k]}^T \Psi U_{[k]} \quad (16)$$

其中

$$\Omega \triangleq \begin{bmatrix} Q & \cdots & 0 \\ \vdots & Q & \vdots \\ 0 & \cdots & S \end{bmatrix}_{(nN_p)\times(nN_p)} \quad \Psi \triangleq \begin{bmatrix} R & \cdots & 0 \\ \vdots & \ddots & \vdots \\ 0 & \cdots & R \end{bmatrix}_{(pN_p)\times(pN_p)} \quad (17)$$

由于系统的预测状态 x_k 只和初始状态 $x_{[k|k]}$ 和输入序列 U_k 相关，因此 X_k 可以都用 U_k 来表示，经变换得

$$J = U_{[k]}^T F x_{[k|k]} + \frac{1}{2} U_{[k]}^T H U_{[k]} \quad (18)$$

其中

$$F \triangleq \Gamma^T \Omega \Phi \quad (19)$$
$$H \triangleq \Gamma^T \Omega \Gamma + \Psi \quad (20)$$

至此将控制问题转化成为了二次规划的标准形式，可以利用软件直接求解得到最优控制序列。

2.3 嵌入约束的模型预测控制算法

在实际天然气管道控制应用中，通常会遇到各种约束条件，例如压缩机出口压力限制、管道最大承压能力限制、下游分输点最大入口压力限制等等，为实现在控制策略的过程中满足诸多的现实约束，可将约束条件直接嵌入 MPC 控制算法中进行求解。

$$x_{[k+1]} = A x_{[k]} + B u_{[k]} \quad (21)$$

系统在 k 时刻做出预测时的二次型性能指标为

$$J = \frac{1}{2} x_{[k+N_p|k]}^T S x_{[k+N_p|k]} + \frac{1}{2} \sum_{i=0}^{N_p-1} \left[x_{[k+i|k]}^T Q x_{[k+i|k]} + u_{[k+i|k]}^T R u_{[k+i|k]} \right] \quad (22)$$

考虑系统的通用约束形式如下，可约束每个时刻的控制输入与状态变量，

$$\mathcal{M}_{[k+i]m\times n} x_{[k+i]n\times 1} + \mathcal{F}_{[k+i]m\times p} u_{[k+i]p\times 1} \leq \beta_{[k+i]m\times 1},$$
$$i = 0, 1, 2, \cdots, N_p - 1 \quad (23)$$

其中 $\mathcal{M}_{k+i}$ 是状态变量约束矩阵，$\mathcal{F}_{k+i}$ 是控制输入约束矩阵。

展开得

$$\begin{bmatrix} \mathcal{M}_{[k]} \\ 0 \\ \vdots \\ 0 \end{bmatrix} x_{[k|k]} + \begin{bmatrix} 0 & \cdots & \cdots & 0 \\ \mathcal{M}_{[k+1]} & 0 & \cdots & \vdots \\ 0 & \mathcal{M}_{[k+2]} & 0 & \vdots \\ \vdots & & \ddots & 0 \\ 0 & \cdots & 0 & \mathcal{M}_{N_p} \end{bmatrix} \begin{bmatrix} x_{[k+1|k]} \\ x_{[k+2|k]} \\ \vdots \\ x_{[k+N_p|k]} \end{bmatrix} + \begin{bmatrix} \mathcal{F}_{[k]} & 0 & \cdots & 0 \\ 0 & \mathcal{F}_{[k+1]} & \cdots & \vdots \\ \vdots & 0 & \ddots & 0 \\ 0 & \cdots & & \mathcal{F}_{[k+N_p-1]} \\ 0 & \cdots & \cdots & 0 \end{bmatrix} \begin{bmatrix} u_{[k|k]} \\ u_{[k+1|k]} \\ \vdots \\ u_{[k+N_p-1|k]} \end{bmatrix} \leq \begin{bmatrix} \beta_{[k]} \\ \beta_{[k+1]} \\ \vdots \\ \beta_{N_p} \end{bmatrix}$$

(24)

简化成以下形式

$$(\overline{\overline{\mathcal{M}}}\Gamma+\overline{\overline{\mathcal{F}}})U_{[k]} \leq \overline{\beta}-(\overline{\mathcal{M}}+\overline{\overline{\mathcal{M}}}\Phi)x_{[k|k]} \quad (25)$$

其中

$$\overline{\mathcal{M}} \triangleq \begin{bmatrix} \mathcal{M}_{[k]_{m\times n}} \\ 0_{m\times n} \\ \vdots \\ 0_{l\times n} \end{bmatrix}_{(mN_p+l)\times n} \quad (26)$$

$$\overline{\overline{\mathcal{M}}} \triangleq \begin{bmatrix} 0_{m\times n} & \cdots & \cdots & 0 \\ \mathcal{M}_{[k+1]_{m\times n}} & 0 & \cdots & \vdots \\ 0_{m\times n} & \mathcal{M}_{[k+2]_{m\times n}} & 0 & \vdots \\ \vdots & & \ddots & 0 \\ 0_{l\times n} & \cdots & 0 & \mathcal{M}_{N_{P_{l\times n}}} \end{bmatrix}_{(mN_p+l)\times (nN_p)} \quad (27)$$

$$\overline{\overline{\mathcal{F}}} \triangleq \begin{bmatrix} \mathcal{M}_{[k]_{m\times p}} & 0 & \cdots & 0 \\ 0_{m\times p} & \mathcal{F}_{[k+1]} & \cdots & \vdots \\ \vdots & 0 & \ddots & 0 \\ \vdots & \cdots & 0 & \mathcal{F}_{[k+N_p-1]} \\ 0_{l\times p} & \cdots & \cdots & 0_{l\times p} \end{bmatrix}_{(mN_p+l)\times (pN_p)} \quad (28)$$

$$\overline{\beta} \triangleq \begin{bmatrix} \beta_{[k]_{m\times 1}} \\ \beta_{[k+1]_{m\times 1}} \\ \vdots \\ \beta_{NP_{l\times 1}} \end{bmatrix}_{(mN_p+l)\times 1} \quad (29)$$

取 $M \triangleq \overline{\overline{\mathcal{M}}}\Gamma+\overline{\overline{\mathcal{F}}}$, $b \triangleq -(\overline{\mathcal{M}}+\overline{\overline{\mathcal{M}}}\Phi)$,得

$$MU_{[k]} \leq \overline{\beta}+bx_{[k|k]} \quad (30)$$

式(30)中只含有待求解的控制序列 U_k 以及 k 时刻的初始已知值 $x_{[k|k]}$,符合标准二次规划的约束形式。

3 案例分析

为了验证所提出模型预测控制方法的有效性,将其应用于某天然气管道系统(图2),经简化调整该系统由一个压气站、天然气管道及用户构成,管道长度为60km,管径为660mm,壁厚为6.4mm。下面通过静态和动态目标压力调节情景来说明算法可以有效地将下游压力控制到期望的目标值,再对比有无约束的下游流量波动恒压控制算例用于说明 MPC 算法的优势在于相比 LQR 可以显式的增加约束条件,允许控制过程符合诸多的物理限制,具备更高的落地实时应用价值和可能性,最后对本算法从预测区间和权重的角度进行了敏感性分析。

图2 天然气管道系统

3.1 目标压力调节控制

对于在当前下游流量不变,需要调高下游压力以提升管存量应对可能发生需求高峰的情况,在线 MPC 可以将下游压力 P2 调整至参考值附近并维持,而离线 MPC 无法将下游压力稳定在3.3MPa(图3)。由于离线 MPC 只在开始时刻计算一次控制输入,后续系统的控制输入只从该控制序列中取值(表1),而无法根据系统的状态变化调整其控制输入,因此难以在天然气管道的实时控制上发挥效果。MPC 的反馈特性体现在在线的修正上,在线控制过程中每个时步重新计算更新整个控制输入序列,但只取控制输入序列中的第一项,以此来适应这个系统的变化。

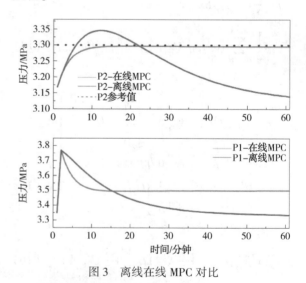

图3 离线在线 MPC 对比

在一天内用户可能存在多个用气高峰的时段,为保障下游用气以及管道平稳运行,调控人员需要多次调整下游压力。如图4可见,通过 LQR 和 MPC 算法都可以达到较好控制效果,平均控制偏差为 0.007MPa,最大控制偏差为0.027MPa,基本满足现场控制需求。

表 1　离线与在线 MPC 的输入取值

时间(分钟)		1	2	3	4	5	6	7	8	……	58	59	60
离线输入	U0	3.768	3.748	3.726	3.702	3.677	3.653	3.630	3.608	……	3.337	3.336	3.335
在线输入	U0	3.768	3.748	3.726	3.702	3.677	3.653	3.630	3.608	……	3.337	3.336	3.335
	U1		3.689	3.691	3.685	3.672	3.656	3.638	3.620	……	3.338	3.337	3.336
	U2			3.632	3.651	3.656	3.652	3.642	3.628	……	3.339	3.338	3.337
	U3				3.591	3.622	3.635	3.637	3.632	……	3.340	3.339	3.338
	U4					3.562	3.601	3.620	3.627	……	3.340	3.340	3.339
	U5						3.541	3.586	3.610	……	3.341	3.340	3.340
							……	……	……	……	……	……	……

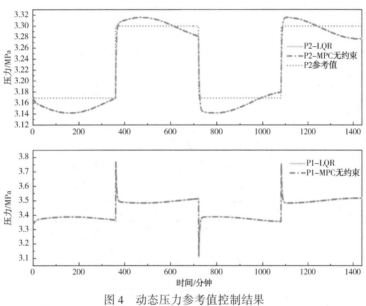

图 4　动态压力参考值控制结果

3.2　用气波动稳压控制

对于更加常见的一种情况是下游用户用气是频繁变换的，在此假设下游流量以某一个正弦函数形式波动(图 5)，如果不对管道系统施加控制，将会导致下游压力也随之波动。为此，对其分别应用 LQR 和 MPC 算法，控制目标是将下游用户压力 P2 维持在初始时刻 3.17MPa。

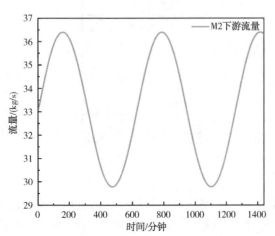

图 5　下游用气波动情况

如图 6 所示，可以发现经过控制后的下游压力 P2 波动幅值大大减小，平均偏差从 0.023MPa 降低至了 0.0035MPa，最大偏差从 0.038 降低至了 0.014MPa。另外，从本场景和前一节场景都发现 LQR 和 MPC 的控制输入和控制结果几乎是一致的，因为这两种方法求解的其实是同一个优化问题。LQR 使用贝尔曼最优化理论从末端向前递归求解，而 MPC 则将未来的控制序列作为求解项，预测未来的情况并使用二次规划进行求解。从形式上它们的区别在于 LQR 求解的是一个反馈矩阵，控制量需要用反馈矩阵乘以系统状态变量得到，而 MPC 直接求解得到最优控制序列。

由于现实情况中存在的诸多物理限制，例如压缩机的转速或者出口压力并非可以无限制提高的，因此在对天然气管道系统进行控制设计时需要考虑这些因素，MPC 区别于 LQR 最大的特点在于可以显式地添加约束条件，因此更加适合天然气管道系统是模型预测控制策略。

3.3 含约束下 MPC 控制

在本算例中下游流量同样以正弦形式波动，但是压缩机的出口压力存在限制，最高为 3.39MPa。如图 7 所示，可以看到在下游压力 P2 上升期间，由于没有设置压缩机出口压力下限，系统依然保持原有控制轨迹，而下游 P2 下降期间（70 分钟，730 分钟，1330 分钟附近）压缩机出口压力 P1 达到最高压力，下游压力 P2 呈现更大程度的压力偏离（图 7 橙线），证明了控制算法可以有效的被约束。

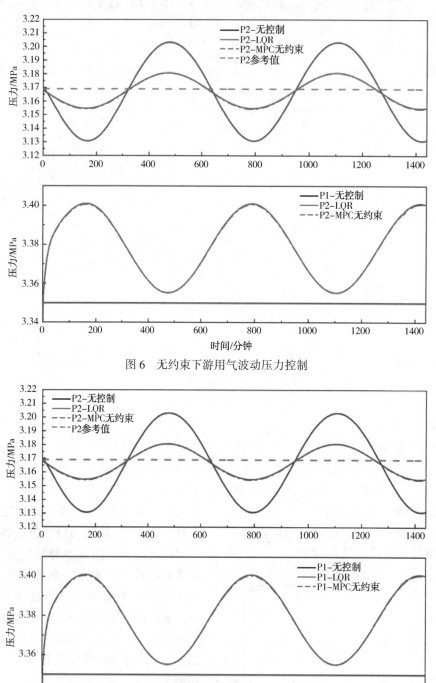

图 6 无约束下游用气波动压力控制

图 7 有约束的下游用气波动压力控制

3.4 敏感性分析

在 MPC 中预测区间是一个关键的设计参数，决定了控制算法需要计算未来多久时间区间内控制策略。预测区间越长可以提供更长远的系统预测，控制策略可以更好地响应系统的变化和扰动，有助于提升控制性能，同样，事物都是具备两面性的，由此需要付出的代价就是计算复杂度的增加，更长的预测周期就要求解更大规模的矩阵，导致计算时间也随之增加，可能无法用于实时控制。

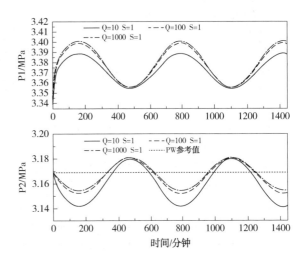

图9 不同权重下的控制结果

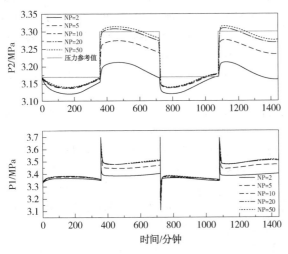

图8 不同预测区间控制结果（NP：预测区间）

由图8可见，随着预测区间的减小，下游压力P2逐渐偏离参考目标值，当预测区间为NP=5的时候（红线），系统的表现已不再理想，而当预测区间NP=2的时候，这意味着控制算法只考虑天然气管网系统在2分钟内的表现，而对于存在大时滞性的天然气管道，控制算法无法充分预测和适应系统的变化，导致控制性能显著下降。

另外，随着预测区间的增大，控制策略的计算时间也随之上升。当预测区间NP=50时，计算用时达到了108.509秒（表2），超过了1分钟，已经无法适用于当前控制时步下的实时控制了。

表2 不同预测区间的计算时间

预测区间/min	2	5	10	20	50
计算用时/s	5.467	6.686	7.804	15.780	108.509

以无约束下的MPC压力控制背景来直观地分析状态权重和输入权重对控制算法的影响。

因此在实际应用中，需要在计算复杂度和控制性能之间进行权衡，选择合适的预测区间以满足天然气管道系统的控制需求，同时也需要权衡控制成本和控制效果，选择合适的状态和输入权重。

4 结论

针对天然气管道系统实时控制研究计算实时性不足或者难以增加现实物理限制的问题，本研究建立了基于状态空间的天然气管道系统模型预测控制方法，并构建了目标压力调节跟踪、下游用户流量波动稳压控制以及物理受限情况下的压力控制共三个场景，通过数值案例应用得到如下结论：

（1）基于所提出的天然气管道状态空间模型，建立基于状态空间的天然气管道模型预测控制方法，经对比离线与在线MPC，可直观看到在线MPC能够有效考虑系统的时滞特性，快速控制下游压力使其达到目标参考值。

（2）在下游流量持续波动的时候本文所提出的控制方法同样能够将压力稳定在参考值附近，平均偏差为0.0035MPa，最大偏差为0.014MPa，符合现场控制需求。

（3）区别于LQR算法，MPC可以显式增加约束，通过将约束条件嵌入MPC以实现压缩机出口压力的限制，结果表明控制策略能够被设定约束有效约束。同时不同的预测区间以及状态输入权重会影响控制策略的计算时效性以及控制性能，需要妥善选取。

参 考 文 献

[1] Liu C, Shahidehpour M, Wang J. Coordinated scheduling of electricity and natural gas infrastructures with a transient model for natural gas flow[J]. Chaos, 2011, 21(2): 531.

[2] 梁永图, 邱睿, 涂仁福, 等. 中国油气管网运行关键技术及展望[J]. 石油科学通报, 2024(2).

[3] Lu Y, Wen K, Yang Y, et al. Intelligent process control ensures energy transmission safety in an ever more tumultuous world[J/OL]. The Innovation Energy, 2024, 1(3): 100044.

[4] 唐善华, 杨毅, 张麟, 等. 天然气管网智能调控初探[J]. 油气储运, 2021, 40(9): 7.

[5] 唐善华. 关于长输天然气管网运行高质量发展的几点思考[J]. 北京石油管理干部学院学报, 2019, 26(04): 16-20.

[6] 梁怿, 韩建强, 王磊. 远控自动分输系统在西气东输甘塘站的应用[J]. 油气储运, 2013, 32(02): 171-173.

[7] 赵国辉. 中俄东线天然气管道工程 SCADA 系统的设计与实现[J]. 油气储运, 2020, 39(04): 379-388.

[8] Peien D, Xinguo C, Xinze L, et al. Pipeline Network Steady-State Operation Optimization Based on Dynamic Programming and Improved Genetic Algorithm[J]. Oil & Gas Storage and Transportation, 2018, 37(3): 6.

[9] Ríos-Mercado R Z, Borraz-Sánchez C. Optimization problems in natural gas transportation systems: A state-of-the-art review[J]. Applied Energy, 2015, 147(jun. 1): 536-555.

[10] Wong P J, Larson R E. Optimization of natural-gas pipeline systems via dynamic programming[J]. IEEE Transactions on Automatic Control, 2003, 13(5): 475-481.

[11] F, Tabkhi, L, et al. Improving the performance of natural gas pipeline networks fuel consumption minimization problems[J]. Aiche Journal, 2010.

[12] Pratt K F, Wilson J G. Optimisation of the operation of gas transmission systems[J]. Transactions of the Institute of Measurement and Control, 1984, 6(4): 261-269.

[13] 周鹏, 魏琪. 天然气长输管网优化建模及求解研究进展[J]. 油气储运, 2023, 42(9): 968-977.

[14] Mak T W K, Van Hentenryck P, Zlotnik A, et al. Efficient Dynamic Compressor Optimization in Natural Gas Transmission Systems[C]//2016 American Control Conference (ACC). 2016.

[15] 刘恩斌, 匡建超, 吕留新, 等. 大型天然气管道稳态运行优化研究[J]. 西南石油大学学报(自然科学版), 2019.

[16] 于肖雯. 天然气管道系统离心压缩机出口压力优化控制方法研究[D]. 2017.

[17] Lu Y, Tan R, Gao W, et al. Active Control of Natural Gas Pipeline System Based on Box-Jenkins Method[J]. Available at SSRN 4995590.

[18] Priyanka E B, Maheswari C, Thangavel S. Remote monitoring and control of LQR-PI controller parameters for an oil pipeline transport system[J]. Proceedings of the Institution of Mechanical Engineers Part I Journal of Systems and Control Engineering, 2018, 233(8): 095965181880318.

[19] Dinh-Hoa, NGUYEN, Shinji, et al. Hierarchical Decentralized Controller Synthesis for Heterogeneous Multi-Agent Dynamical Systems by LQR[J]. Sice Journal of Control Measurement & System Integration, 2015.

[20] Marqués D, Morari M. On-line optimization of gas pipeline networks[J]. Automatica, 1988, 24(4): 455-469.

[21] Gopalakrishnan A, Biegler L T. Economic nonlinear model predictive control for periodic optimal operation of gas pipeline networks[J]. Computers & Chemical Engineering, 2013, 52: 90-99.

[22] Chen Q, Wu C, Zuo L, et al. Multi-objective transient peak shaving optimization of a gas pipeline system under demand uncertainty[J]. Computers & Chemical Engineering, 2021, 147(1): 107260.

[23] Liu Lorenz T. Zhang BingjianChen Qinglin K. Dynamic optimization of natural gas pipeline networks with demand and composition uncertainty[J]. Chemical Engineering Science, 2020, 215(1).

[24] Zhu G Y, Henson M A, Megan L. Dynamic modeling and linear model predictive control of gas pipeline networks[J]. Journal of Process Control, 2001, 11(2): 129-148.

[25] Aalto H. Model predictive control of natural gas pipeline systems - a case for constrained system identification[J]. IFAC-PapersOnLine, 2015, 48(30): 197-202.

甲醇对 X65 管线钢腐蚀性实验研究

陈更生[1] 姜子涛[1] 黄梓耕[1] 聂超飞[2] 刘罗茜[2] 刘世贸[1] 常滋茹[1] 刘冠一[3]

[1. 中国石油大学(北京)机械与储运工程学院；
2. 国家管网集团科学技术研究总院分公司；3. 中国石油大学(北京)安全与海洋工程学院]

摘 要 【目的】随着我国"双碳"目标的确立，甲醇作为清洁能源、储氢载体，在农药、医药、汽车、国防等行业应用广泛。伴随着用量的增加，用管道输送甲醇成为了最便捷、经济的方式。目前国内甲醇长输管道缺乏相关技术，对于改扩建或者新建甲醇管道，管道输送甲醇所带来的相关安全问题也亟需研究。【方法】利用高温高压反应釜开展 X65 管线钢在甲醇环境下的腐蚀浸泡实验，研究了含水率、温度和 pH 对管线钢试样腐蚀的影响规律，对管线钢试样腐蚀产物宏观形貌、微观形貌进行观察分析，腐蚀产物成分进行定性分析，确定了甲醇腐蚀反应机理。【结果】温度为 25 ℃，pH=7 时，随含水率增加管线钢试样均匀腐蚀速率呈线性增加，含水率在 10% 以下管线钢试样腐蚀速率小于 0.01 mm/a，为低腐蚀风险等级。含水率达到 30% 时，试样腐蚀速率较 10% 增加了约 4 倍，为中腐蚀风险等级，水的存在使得甲醇对管线钢试样的腐蚀性有极大增强作用。不同甲醇含水率下电阻率进测试，结果显示随含水率的升高溶液电阻率降低，即增加了溶液中的自由离子，腐蚀性增加；含水率为 5%，pH=7 时，温度从 25 ℃ 升高至 55 ℃，腐蚀速率从 0.005 mm/a 升高至 0.011 mm/a，此条件下温度对甲醇腐蚀性影响不大；含水率为 5%，温度为 25 ℃ 时，管线钢试样均匀腐蚀速率随 pH 增大而减小，pH 为 3.5 左右时有最大腐蚀速率，约 0.3 mm/a，碱性条件下腐蚀性不大。管线钢试样在甲醇中有点蚀发生，甲醇中含有水时点蚀速率快速增加，可达 5 mm/a。管线钢试样腐蚀产物主要成分为 Fe_2O_3。【结论】对于甲醇环境，实验条件下管线钢试样对含水率和 pH 较为敏感，因而严格控制甲醇中甲酸和水分能有效降低管道腐蚀，甲醇对管道腐蚀性研究可以为甲醇管道安全设计及运行提供依据和指导。

关键词 甲醇管道；腐蚀性 X65；管线钢；影响因素；腐蚀速率

Experimental study of methanol corrosion on X65 pipeline steel

Chen Gengsheng[1] Jiang Zitao[1] Huang Zigeng[1] Nie Chaofei[2]
Liu Luoqian[2] Liu Shimao[1] Chang Ziru[1] Liu Guanyi[3]

(1. School of Mechanical and Storage and Transportation Engineering, China University of Petroleum in Beijing;
2. PipeChina Institute of Science and Technology;
3. School of Safety and Marine Engineering, China University of Petroleum in Beijing)

Abstract [Objective] With the establishment of China's "dual-carbon" goal, methanol, as a clean energy and hydrogen storage carrier, is widely used in pesticide, pharmaceutical, automotive, defense and other industries. Along with the increase of usage, methanol transportation by pipeline has become the most convenient and economical way. At present, there is a lack of relevant technology for long-distance methanol pipelines in China, and the safety problems associated with the transportation of methanol by pipelines are also in urgent need of research for the expansion or construction of new methanol pipelines. [Methods] Using high temperature and high pressure reactor to carry out corrosion immersion experiments of X65 pipeline steel in methanol environment, studied the water content, temperature and pH on the corrosion of pipeline steel specimen corrosion law, observation and analysis of the corrosion products of pipeline steel specimen macroscopic morphology, microscopic morphology, corrosion product composition qualitative analysis, to determine the corrosion reaction mechanism of methanol. [Results] Temperature of 25℃, pH=7, with the increase of water content pipe steel specimen uniform corrosion rate increased linearly, the water content in the following 10% pipe steel specimen corrosion rate is less than 0.01mm/a, for low corrosion risk level. When the water content reaches 30%, the corrosion rate of the specimen increases by about 4 times compared with 10%, which is a medium corrosion risk level, and the presence of water makes the corrosion of methanol on the pipeline steel specimen greatly enhanced. Different methanol water content resistivity into the test, the results show that with the increase in water content solution resistivity decreases, that is, an increase in the free ions in the solution, corrosion increases; water content of 5%, pH=7, the temperature from 25℃ to 55℃, the corrosion rate from 0.005mm/a rise to 0.011mm/a, the temperature of the corrosive properties of methanol under this condition does not have much effect; water content of 5%, the temperature is 25℃, the pipe steel specimen corrosion rate increases from 10%, the corrosive effect is greatly enhanced. When the water content is 5% and the temperature is 25℃, the uniform corrosion rate of pipeline steel specimen decreases with the increase of pH, and there is a maximum corrosion rate of about 0.3mm/a when the pH is about 3.5, and there is not much corrosion under alkaline conditions. Pipeline steel specimens in methanol pitting corrosion occurs, methanol contains water pitting corrosion rate increases rapidly, up to 5mm/a. Pipeline steel specimens corrosion products are mainly composed of Fe_2O_3. [Conclusion] For the methanol environment, the pipeline steel specimens under the experimental conditions are more sensitive to the water content and pH, thus the strict control of formic acid and water in methanol can effectively reduce the corrosion of pipelines, and the study of corrosion of methanol on pipelines can provide a basis for the safe design and operation of methanol pipelines and guidance.

Key words Methanol Pipelines; Corrosivity; X65 pipeline steel; Influencing Factors; Corrosion rate

随着我国"双碳"目标的确立，使得氢储运技术的基础研究从多维度展开。其中，有机醇类，特别是甲醇为代表的循环储氢分子化学稳定，其氢能储放反应相关的催化和工程技术发展成熟，条件相对温和，因此成为备受关注的液态氢储存平台分子。随着液态阳光技术、太阳燃料合成甲醇工业的示范性应用，甲醇的生产和消费市场越来越大，甲醇管道有可能成为管道运输的重要发展方向。

目前在加拿大和中国有几条甲醇管道案例，运行压力低、口径小。国内外在甲醇输送管道方面都缺乏成熟经验，无论是新建还是将原有油气管道转换为甲醇管道，在腐蚀安全方面有待开展系统性研究。纯的甲醇为中性的，本身对于金属的腐蚀性不强。工业粗甲醇杂质含量较多，除了含有甲酸等有机杂质，含水率可达20%以上。工业上甲醇对金属设备和管道腐蚀出现过很多案例，某化工企业MTO装置气相甲醇管线发生腐蚀，经观察管道内壁出现区域性点腐蚀，严重区域已经穿透管壁，造成甲醇泄露。中国某甲基叔丁基醚/丁烯-1装置的甲醇回收系统E-201B管束在运行不到1年时间就存在大量腐蚀。国内外对于甲醇腐蚀进行了很多研究，许多学者研究了甲醇汽油的腐蚀性，结果表明工业上的甲醇汽油在实际应用中，腐蚀性明显要比汽油强很多，并且随着甲醇含量的增多而增强，金属构件表面的凹坑和氧化物层也越明显。另外，也有众多学者探究了各腐蚀因素对金属的影响规律，国外Shintani等探究了氯离子对甲醇/水体系的不锈钢腐蚀行为，表明氯离子会显著加剧甲醇环境下的腐蚀。加拿大Champion Technologies公司研究了氧浓度对甲醇腐蚀行为影响，结果表明氧气会加剧腐蚀速率与局部腐蚀失效风险。Morello等建立了甲醇浓度与腐蚀速率之间的量化关系，发现局部腐蚀失效风险随甲醇增加显著增大。国内贾伟艺等研究了甲醇汽油对金属的腐蚀，发现水分是造成金属腐蚀的重要因素，水含量的增加会加速金属腐蚀。张娟利等研究表明随着甲醇汽油中含水量增加，金属腐蚀速率先增大后减小，存在临界含水量，还探究了酸腐蚀对金属腐蚀度的贡献，加入碱性或弱碱性的添加剂将更有利于甲醇汽油中金属的防腐。针对这些研究，国内尚未开展甲醇对在役管材腐蚀性能的研究，甲醇对管材腐蚀规律、腐蚀机理和腐蚀控制还远远不够。

鉴于此，本工作针对甲醇环境，研究含水率、pH和温度变化对X65管线钢腐蚀速率、腐蚀形貌的影响，探讨甲醇对管线钢腐蚀影响机理，以期对输送的甲醇成分和工艺参数进行控制，保障管道运行安全。

1 实验

1.1 实验材料

本实验材料为 X65 管线钢，其化学成分(质量分数,%)为：C 0.04，Si 0.27，Mn 1.56，P 0.012，S 0.001，Mo 0.092，Cr 0.031，Ni 0.160，Al 0.019，Cu 0.003，V 0.03，Fe 余量。为了便于加工管材，且考虑到该反应釜挂片夹具尺寸大小，故将 X65 钢加工成尺寸为 40×10×3mm，孔径为 7mm 的试片。用砂纸逐级打磨试样表面至 1200 号，使之光滑并无明显划痕后，用丙酮超声波水浴清洗，以清除表面油脂，用无水乙醇冲洗并用冷风快速吹干，采用精度为 0.1mg 的电子天平称量并记录初始质量 m_1。实验介质为甲醇，甲醇组分及含量如下表。

1.2 实验装置

本实验利用高温高压反应釜开展腐蚀模拟实验，研究 X65 管线钢在甲醇环境下的腐蚀情况，实验装置原理图如图 1 所示，装置由反应容器、加热装置、冷却装置、搅拌装置等部分构成，通过调节反应釜内部的压力和温度，可以促进某些反应的进行或提高其反应速率。反应釜内部压力由外部加热和液体膨胀引起，而温度的控制则通过加热装置和温度传感器实现。

表 1 甲醇组分及性质

甲醇	甲醇	乙醛	甲醛	丙酮	水	pH	电阻率
甲醇	≥99.9%	≤0.001%	≤0.001%	≤0.001%	≤0.05%	7.2	160kΩ·cm

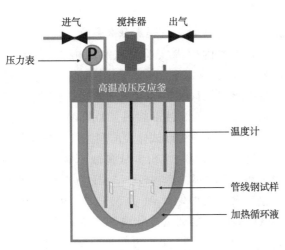

图 1 反应釜原理示意图

1.3 实验方法

每组实验设置四个平行样，实验开始前，将试样安装在聚四氟乙烯夹具上，将夹具安装好，加入调配好的甲醇，关闭反应釜。通入高纯氮气除去安装过程中反应釜内残留的空气，将反应釜调配至实验温度及压力，实验周期为 7×24h。

实验结束后，取出试样，使用尼康 Z-50 数码相机拍照记录其腐蚀样貌。用 Rigaku D/MAX-2600 型 X 射线衍射仪(XRD，Cu 靶，波长 0.15418nm，电压 40kV，电流 40mA)测定腐蚀产物的物相组成，用 LEXT OLS4000 观察微观腐蚀形貌和腐蚀纵向深度。在超声波清洗机中用清洗液(500mL 去离子水+500mL 浓盐酸+4~5g 六次甲基四胺)去除腐蚀产物，并利用丙酮溶液清洗，用去离子水清洗吹干。干燥后再次称量试样质量 m_2。采用失重法计算试样的腐蚀速率，计算公式如下：

$$V_{corr}=\frac{m_1-m_2}{\rho At} \quad (1)$$

式中，m_1 为实验前前试样质量；m_2 为去除腐蚀产物后试样质量；t 为实验周期；ρ 为试样密度；A 为试样表面积。

2 结果与讨论

2.1 甲醇含水率对管线钢腐蚀的影响

开展室内模拟实验，控制甲醇温度 25℃，压力 8MPa，pH 为 7，改变甲醇的含水率(0%~30%)，得到了不同含水率下管线钢试样的腐蚀形貌，如图 2 所示。通过观察可以发现随着甲醇含水率的增加，管线钢试样的腐蚀越来越严重，试样表面有不同程度的腐蚀斑点出现，含水率越高，斑点的密集程度越高，斑点及周围腐蚀产物呈黄褐色。纯甲醇浸泡试样酸洗前后表面无明显变化，表明纯甲醇腐蚀性很小。经酸洗后，观察到管线钢试样表面腐蚀为大小不一的腐蚀坑(呈现出暗黑色小点)，推测甲醇在含水条件下对金属造成的腐蚀类型为点蚀。

经计算得到了不同含水率下管线钢试样的均匀腐蚀速率，如图 3 所示。纯甲醇浸泡管线钢试样均匀腐蚀速率很低，可以忽略不计，此条件下基本无腐蚀，与形貌观察结果一致。随甲醇含水

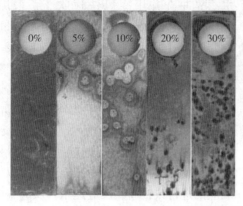
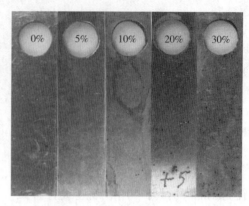

图 2　不同含水率管线钢试片形貌图，酸洗前（左）、酸洗后（右）

率增加，试样均匀腐蚀速率呈线性增加，含水率在10%以下时腐蚀速率小于0.01mm/a，根据GB/T 23258—2020《钢质管道内腐蚀控制规范》管道内腐蚀评价指标，该腐蚀属于低腐蚀风险等级。含水率30%时试样腐蚀速率比10%高4倍，约为0.027mm/a，属于中腐蚀风险等级。

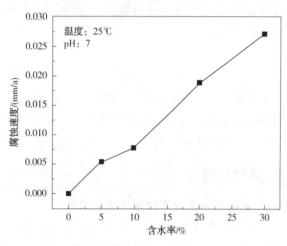

图 3　不同含水率管线钢试片腐蚀速率变化图

另外，在实验室测试了不同含水率甲醇的电阻率，发现甲醇含水率增大，电阻率降低，说明甲醇中的自由离子增多，溶液更容易电离，增强了甲醇的腐蚀性。由此说明，甲醇的含水率是影响管线钢试样腐蚀的重要因素。甲醇在实际管道运输时，含有甲酸等其他杂质，腐蚀会更严重，所以严格控制含水率在5%内保证管道在低腐蚀风险。

2.2　甲醇 pH 对管线钢腐蚀的影响

开展室内模拟实验，控制甲醇温度25℃，压力8MPa，含水率10%，利用甲酸和氢氧化钠改变甲醇的 pH（3.5~9.5），得到了不同 pH 管线钢试样表面的腐蚀形貌图，如图4所示。通过观察可以发现管线钢试样表面腐蚀产物呈黑色或黄褐色，pH 为 3.5~5.5 时腐蚀产物较多，pH 为 7 和 9.5 时腐蚀产物较少。pH 为 3.5、4.5 时腐蚀产物较为均匀，pH 为 5.5、7、9.5 时管线钢试样表面观察到斑点状腐蚀产物。经酸洗后，观察到 pH 为 3.5、4.5 时有部分金属遭受均匀腐蚀，pH 在 5.5、7、9.5 时管线钢试样表面腐蚀有大小不一的腐蚀坑。由此表明在酸性情况下可能反应更为剧烈，腐蚀已经由点蚀发展成为均匀腐蚀。

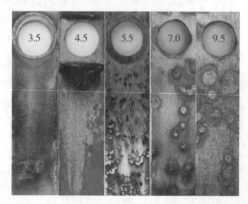
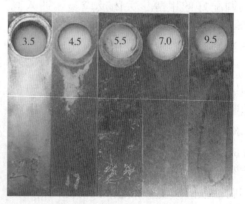

图 4　不同 pH 下管线钢试片腐蚀形貌图，酸洗前（左）、酸洗后（右）

经计算的到了不同pH管线钢试样的均匀腐蚀速率,如图5所示。均匀腐蚀速率随pH增大而减小。酸性条件下腐蚀速率较大,为高腐蚀风险,碱性条件下试样腐蚀速率较低,为低腐蚀风险,这与腐蚀形貌观察情况一致。pH为3.5左右时有最大腐蚀速率,最大腐蚀速率达0.3mm/a,酸性条件下溶液中的H^+浓度较大,易于管线钢发生析氢反应,腐蚀更严重。甲醇偏酸性时金属试片腐蚀更大,偏碱性时不容易发生腐蚀,在实际管道运输甲醇过程中,可控制甲醇在中性或者碱性条件。

经计算的到了不同温度管线钢试样的均匀腐蚀速率,如图7所示。不同温度管线钢试样腐蚀速率很小,平均腐蚀速率在0.005~0.011mm/a之间,这与腐蚀形貌观察一致,表明该实验条件下温度对管线钢试样的腐蚀影响并不大。实际管道运输甲醇过程中控制输送温度在常温即可。

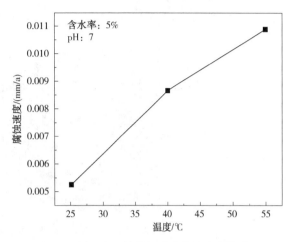

图7 不同温度下管线钢试片腐蚀速率变化图

2.4 腐蚀机理分析
2.4.1 局部腐蚀深度测试

通过对管线钢试样的腐蚀形貌的初步观察,发现有部分试样腐蚀呈现斑点状。因此,采用激光共聚焦进一步对微观形貌和腐蚀坑深度测试。通过激光共聚焦显微镜对纯甲醇、10%、20%、30%含水率甲醇浸泡后管线钢试样进行测试,所得腐蚀坑形貌如图8所示。

通过观察,在200微米标尺下,纯甲醇金属表面较为干净,随着甲醇含水率的增加,金属表面腐蚀坑的数量是增多的,腐蚀坑的大小也是增加的,表明腐蚀程度也是增加的,这与前述结果相同,这也证明在不同含水率下,金属会发生点蚀。

将所测最大腐蚀坑深度数据进行汇总,如表2。

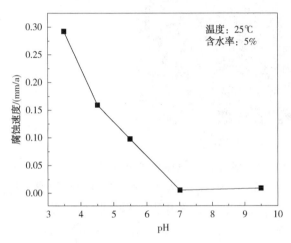

图5 不同pH下管线钢试片腐蚀速率变化图

2.3 甲醇温度对管线钢腐蚀的影响

开展室内模拟实验,控制压力8MPa、pH为7、含水率5%,改变甲醇浸泡温度(25℃~55℃),得到了不同温度管线钢试样表面的腐蚀形貌图,如图6所示。通过观察可以发现管线钢试样基本没有腐蚀产物,酸洗前后无明显变化,表明实验条件下管线钢试样腐蚀很小,温度对管线钢腐蚀的影响不大。

表2 不同含水率甲醇浸泡管线钢试样最大腐蚀坑深度

含水率	0%	10%	20%	30%
试样1	3.794	6.250	70.620	107.620
试样2	6.565	30.904	86.086	104.354
试样3	9.276	25.636	77.020	101.946
试样4	16.147	40.726	—	107.620

注:深度单位为μm。

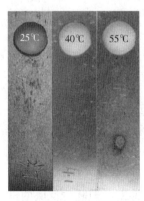 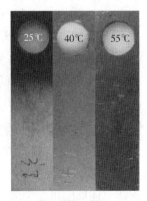

图6 不同温度下管线钢试样腐蚀形貌图,酸洗前(左)、酸洗后(右)

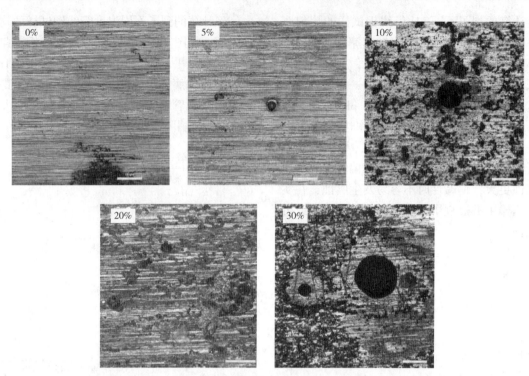

图 8 不同含水率甲醇浸泡后管线钢试样腐蚀坑形貌

根据最大腐蚀坑深度数据得到点蚀腐蚀速率随含水率的变化曲线，如图 9 所示。甲醇含水率对点蚀腐蚀速率影响很大，点蚀腐蚀速率随着含水率增加而增加，与均匀腐蚀速率的变化情况相同。利用点蚀因子来评价点蚀情况，点蚀因子为最大点蚀深度和均匀腐蚀深度的比值。当甲醇中含有一部分水时，点蚀速率快速增加，含水率 30% 可达 5mm/a，点蚀因子达到 224（一般土壤腐蚀约为 5~10），由此可见甲醇在有水条件下会对管线钢产生点蚀，点蚀的影响远大于均匀腐蚀，管道输送甲醇需要严格控制含水率在 5% 以内。

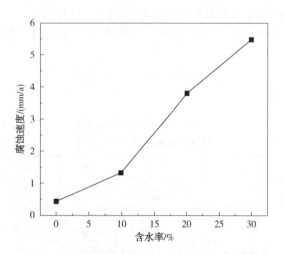

图 9 管线钢试样点蚀腐蚀速率随含水率变化

2.4.2 管线钢试样腐蚀产物分析

图 10 为在 pH 为 4.5、5.5 时管线钢腐蚀产物的 XRD 图谱，图中共出现了四处比较明显的特征峰。特征峰对应的物相为 Fe_2O_3 和 Fe。

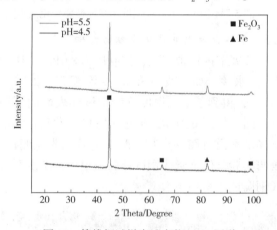

图 10 管线钢试样腐蚀产物 XRD 图谱

反应的方程式如下，甲醇中铁腐蚀后主要成分为 Fe 和 Fe_2O_3，腐蚀产物的生成过程为 Fe 和 H^+ 作用被氧化成为 Fe^{2+} 进入到溶液当中，阴极反应过程主要由 H^+ 的还原主导。除氧时间为 2h，并不完全，反应釜内会有残留溶解氧气，生成的 Fe^{2+} 被氧化成 Fe^{3+} 生成 $Fe(OH)_3$，酸性条件上述反应可能会受到抑制，但随着反应的进行，$Fe(OH)_3$ 还是会逐渐的生成。$Fe(OH)_3$ 极不稳定，会分解生成 Fe_2O_3，这也是试样表面呈黄褐色的原因。

$$2H^+ + 2e^- \longrightarrow H_2$$
$$Fe \longrightarrow 2e^- + Fe^{2+}$$
$$4Fe^{2+} + O_2 + 4H^+ \Longleftrightarrow 4Fe^{3+} + 2H_2O$$
$$2Fe(OH)_3 \Longleftrightarrow Fe_2O_3 + 3H_2O$$

3 结论

针对不同含水量、pH和温度条件下甲醇对X65管线钢腐蚀规律研究，分析腐蚀形貌和腐蚀产物得到如下结论。

（1）甲醇对金属材料的腐蚀主要受到温度、含水率和pH的影响，且在实验条件下温度的影响相对较小。通过一系列实验发现，随着温度的升高，管线钢试样的腐蚀速率呈现出逐步增加的趋势，在含水率为5%的条件下，金属的平均腐蚀速率在0.005~0.011mm/a之间。后续会增加实验组，探明不同条件下温度的影响规律是否一致。

（2）不同含水率对管线钢试样腐蚀速率的影响较为显著，随着含水率的增加，腐蚀速率显著上升。特别是在含水率达到30%时，腐蚀速率较10%时增加了约4倍，水的存在对甲醇的腐蚀性有极大增强作用。甲管输时应该严格控制含水量在5%以内。

（3）在pH方面，甲醇的腐蚀速率在pH为3.5左右时达到最大，为0.3mm/a，为高腐蚀风险。甲醇在偏酸性条件下腐蚀性更强，而在偏碱性条件下腐蚀性较低。甲醇管输时应严格控制甲酸含量以避免腐蚀，另外也可加入碱性缓蚀剂进一步抑制甲醇的腐蚀性。

（4）管线钢在甲醇环境下容易发生点蚀风险，且点蚀风险远高于均匀腐蚀风险，点蚀最大腐蚀速率达5mm/a。X65管线钢在甲醇下主要由甲酸、水共同作用加速了腐蚀，阴极反应过程主要由H^+的还原主导，最终生成黄褐色的腐蚀产物Fe_2O_3。

（5）本文在一定范围内研究了含水率、pH和温度对甲醇腐蚀性的影响规律，为控制内腐蚀风险，还需进一步细化参数，对甲醇进入管道的物性参数进行具体量化。

参 考 文 献

[1] 张丽平. 甲醇生产技术新进展[J]. 天然气化工, 2013, 38(01): 89-94.

[2] 孟翔宇, 陈铭韵, 顾阿伦, 等. "双碳"目标下中国氢能发展战略[J]. 天然气工业, 2022, 42(04): 156-179.

[3] TIERLING K S. Hydrogen in Storage and Transportation [J]. Offshore Technology Conference, 2024(35029).

[4] Andy L, GIUSEPPE B. Revolutionising energy storage: The Latest Breakthrough in liquid organic hydrogen carriers [J]. International Journal of Hydrogen Energy, 2024(63): 315-329.

[5] LI L L, JIANG Y F, ZHANG T H, et al. Size sensitivity of supported Ru catalysts for ammonia synthesis: From nanoparticles to subnanometric clusters and atomic clusters[J]. Chem, 2022, 8(3): 749-768.

[6] 舟丹. "液体阳光"是实现低碳能源的主要途径[J]. 中外能源, 2020, 25(07): 24.

[7] 许达, 刘启斌, 隋军, 等. 太阳能与甲醇热化学互补的分布式能源系统研究[J]. 工程热物理学报, 2013, 34(09): 1601-1605.

[8] 王集杰, 韩哲, 陈思宇, 等. 太阳燃料甲醇合成[J]. 化工进展, 2022, 41(03): 1309-1317.

[9] 我国第一条MTO甲醇长输管线正式投产[J]. 焊管, 2021, 44(06): 35.

[10] 黄鑫, 滕霖, 聂超飞, 等. 液氨/甲醇/成品油顺序输送技术研究进展[J]. 油气储运, 2023, 42(12): 1337-1351.

[11] 杨英, 沈显超. 甲醇长输管道可行性论证[J]. 当代化工研究, 2018, (09): 173-174.

[12] 姚硕, 卢俊文, 周璐璐, 等. 甲醇合成装置工艺管道腐蚀分析及选材原则[J]. 氮肥技术, 2023, 44(05): 41-44.

[13] 杨德忠, 张坤, 周明. 低温甲醇洗甲醇换热器腐蚀原因分析[J]. 化工管理, 2022(14): 118-121.

[14] 张大船. MTO装置的管道材料设计[J]. 石油化工腐蚀与防护, 2016, 33(03): 46-50.

[15] JIANG T, GONG Y, YANG Z. Failure analysis on abnormal damage and corrosion of methanol distillation column in MTBE production plant[J]. International Journal of Pressure Vessels and Piping, 2023, 205: 105011.

[16] 陶政灿. MTBE装置甲醇回收系统腐蚀原因及其对策[J]. 现代职业安全, 2024(07): 74-76.

[17] 滑海宁, 韩志歧, 李建锋, 等. 现行甲醇汽油腐蚀性评定方法及研究[J]. 交通节能与环保, 2014 (01): 29-33.

[18] 张志颖, 李慧明. 车用甲醇汽油的腐蚀性和溶胀性研究[J]. 材料导报, 2012, 26(19): 86-89.

[19] 滑海宁, 陈明星, 李阳阳. 甲醇汽油腐蚀性与挥发性的实验研究[J]. 贵州大学学报(自然科学版), 2013, 30(03): 65-68.

[20] SHINTANI D, FUKUTSUKA T, MATSUO Y, et al. Passivation Films Behavior Of Stainless Steel Under High Temperature And Pressure Anhydrous Methanol Solution Containing Chlorid[J]. CORROSION, 2008(08431): 1-9.

[21] 贾伟艺. 甲醇汽油对金属的腐蚀研究[J]. 现代化工, 2014, 34(06): 52-54.

[22] 张娟利, 黄勇, 张新庄, 等. 低比例甲醇汽油对金属的腐蚀研究[J]. 表面技术, 2016, 45(08): 34-39.

[23] 王海锋, 刘亮, 刘莹, 等. 湿气输送中甲醇对碳钢管道CO_2腐蚀的影响研究[J]. 石油化工腐蚀与防护, 2023, 40(06): 1-5.

[24] 聂超飞, 姜子涛, 刘罗茜, 等. 甲醇管道输送技术发展现状及挑战[J]. 油气储运, 2024, 43(02): 153-162.

[25] 周建成, 张永先. 粗甲醇对设备的腐蚀[J]. 山东工业技术, 2014(20): 20.

含蜡原油管道停输再启动初凝仿真研究

崔润麒　王　吉　苏　怀　李鸿英　张劲军

[中国石油大学(北京)机械与储运工程学院]

摘　要　原油管道停输再启动过程中管道可能出现压力上升、但流量不升反降的现象，即所谓"初凝"。初凝的发生是一个动态过程，管道的出力设备(泵)是管道运行的关键因素之一。在以往对于停输再启动问题的研究中，往往忽略泵特性对管道运行的影响。针对此问题，以某国内原油管道为例，考虑多个泵站密闭输送场景进行建模，对秋季和冬季工况分别设置停输40h后进行再启动模拟，对耦合泵特性和恒压启动两种不同的边界条件，进行对管道初凝分析的对比。耦合泵特性的模型计算结果表明，秋季环境下能够安全启输。在冬季环境下，再启输后出现流量下降，即发生初凝。但采用恒压启动边界条件模型，结果表明冬季停输后部分管段可以安全启动。结果表明，耦合泵特性的停输再启动模型更加贴近实际运行工况，而恒压力边界条件则变相提升了泵的工作能力，没有考虑流量变化所带来的启动压力变化。

关键词　管道；停输再启动；易凝高黏原油；初凝；泵特性

Initial Gelling for the Shutdown and Restart Processes of a Waxy Crude Oil Pipeline

Cui Runqi　Wang Ji　Su Huai　Li Hongying　Zhang Jinjun

(College of Mechanical and Transportation Engineering, China University of Petroleum in Beijing)

Abstract　During crude oil pipeline transportation, the pushing pressure may rise and the flow rate decreases instead of increasing during the restart of the pipeline. This is the mechanism of the initial gelling problem, which is a dynamic process. The pumps of a pipeline are one of the key factors that may decide whether initial gelling may occur. In the previous study on stopping and restart of crude oil pipelines, the influence of the pump on the pipeline was often ignored. In this study, a crude oil pipeline is simulated, considering several pump conditions. The simulation is carried out for autumn and winter pipeline restart operation. The transportation is stopped for 40 hours before restarting. Two boundary conditions, constant pressure inlet and dynamic inlet provided by pumps, are compared in predicting whether the pipeline initial gelling happens. With pump pushing boundary condition, the pipeline can restart safely in autumn environment, however the flow rate decreases soon after restart, which marks the occurrence of initial gelling problem for winter operation. With constant push pressure inlet, some pipes can restart safely after a 40-hour stopping in winter. The constant pushing pressure boundary condition improves the capacity of pumps without considering the change of pushing pressure caused by varying flow rate.

Key words　Pipeline; Shutdown and Restart; High-pour-point and viscous crude oils; Initial gelling; Pump Characteristics

原油管道在运行时不可避免的出现停输。停输过程中，原油的温度下降，黏度增加。如果原油的黏度过大，再启动过程中管道可能出现压力上升、流量下降的现象，即"初凝"现象。发生初凝的原理是：管内原油的黏度增加，管道流量下降，此时需要泵提供更大的压力才能推动原油流动。如果泵推动管内冷油移动速度过慢，随着原油温度进一步下降，其黏度继续增加。最终出现管道流量下降、压力上升的现象。若不能及时有效的推出冷油或者升高管内温度，最终将导致管道停流，发生凝管事故。发生初凝的根本原因是由于原油黏度的增加导致流动阻力的增大，出力设备提供的动力无法克服高黏度原油所带来的流动阻力。因此一些学者通过研究管道停输时原油的温度变化规律，分析原油的黏度变化。

李传宪等人通过构建停输温降模型，根据原油的析蜡量不同，提出一种划分管道停输温降阶段的方法，基于此方法进行原油管道停输温降的数值模拟，结果表明析蜡量最大的原油在停输时温降速率较低。唐渤使用FLUENT软件对鄯善-四堡段停输过程的温度场进行数值模拟，研究了环境温度、原油输送温度、保温层厚度、埋地深度等不同因素对停输时温降的影响。苏炳辉等人借助仿真软件，耦合原油物性数据，根据一年12个月停输时每小时温降的不同，给出各个月的最大停输时间。李浩建立停输过程的数学物理模型，并分析管道停输时原油温度降低规律，结果表明在停输时，管内存油的温度受到大气温度影响较大，大气温度越低，停输时原油的温降越大。刘晓燕等人认为自然对流是原油管道停输时原油最主要的传热方式，并根据原油管道停输时传热方式的不同，将停输时传热过程分为3个阶段，通过计算这三个阶段的瑞利数（自然对流相关的无量纲数），分析原油管道温度变化规律。陈思航等人构建管道径向温度场模型，通过数值模拟法对其进行求解。DONG等人构建了管道和土壤耦合的数学物理模型，研究了管道直径、环境温度和停输前输量等不同因素对停输时温降的影响。上述学者们的研究通过研究停输时原油管道的温度变化规律可以得到停输结束时的原油黏度分布。但是无法全面的反映管道启输时，驱动压力如何克服原油黏度所带来的流动阻力这一动态过程。一些学者基于管道原油的黏度分布，研究启输时管道内压力变化。

Luiz F. R. Dalla认为原油管道再启动时的最小启动压力为原油的屈服应力，结合原油的温度变化历史以及管道内部原油温度分布，采用实验方法测管道再启动的最小启动压力。Lomesh Tikariha等人研究出现胶凝的原油管道启动时的机理，发现如果胶凝原油内部的气囊数量增加可以减小胶凝原油的屈服应力。王东等人认为再启动过程的压降由四部分组成，其分别为被顶挤液压降、顶挤液压降、高程差压降以及惯性压降，再启动时所需的压力需要大于这四种阻力所带来的压降总和。尹晓云等人构建环形输送管道，采用恒定流量的方法对再启动过程压降变化进行研究，实验结果表明原油黏性增大导致再启动时压力较难恢复稳定。祝守丽等人通过缩比实验对再启动过程的压力变化进行研究，研究结果表明再启动时原油的温度越低，再启动时需要更大的压力才能正常启动。上述学者们的研究证明启输时管道压降变化规律可以确定管道最小启动压力，但是忽略了启输后土壤温度场仍然影响原油的温度变化。一些学者耦合了管道水力、热力过程，构建了原油管道停输再启动仿真模型。

蔡磊等人针对东临复线构建了耦合水力、热力过程的停输再启动模型，采用数值方法对其求解。采用恒定压力的再启动方式，利用管道末端的再启动流量衡量再启动的效果。结果表明，停输时间的增加会导致原油黏度增加，再启动后流量减小。宇波等人建立了冷热原油交替输送停输再启动过程的数学模型，通过设置不同的停输时间进行数值求解，通过管道的流量变化确定该管道的最大停输时间。同时，宇波等人建立了海底管道和热油管道大修期间的停输再启动模型。采用非结构化三角形网格、有限差分法分别对土壤、管道进行离散、求解，计算结果与现场数据吻合。

初凝是一个动态的过程，泵是管道的动力来源，其提供的压力、流量随着上下游的管道内的流动阻力进行调整，最终达到动态平衡。采用恒定压力作为入口条件进行计算时，相当于单独考虑某个管段的流动阻力的影响，而不是把管道作为完整的、封闭的系统进行流动阻力和动力耦合动态分析，无法给出初凝问题的合理判断。而已有的研究还没有完整的考虑这一影响因素。本文以多个泵站密闭输送场景进行建模，同时耦合土壤温度场，建立原油管道停输再启动模型。通过设置不同的环境温度进行停输再启动模拟，通过计算结果分析初凝发生的原因。之后对耦合泵特性和恒压启动两种不同的边界条件，进行对管道初凝分析的对比，验证耦合泵特性进行初凝分析的必要性。

1 理论模型

为了能够满足现场所需的计算需求，模型应该能够对三种运行状态进行仿真计算：1、稳态运行过程；2、停输降温过程；3、再启动过程。其中，稳态运行过程指的是停输之前管道的相对稳定运行状态。在停输过程开启之后，稳态运行过程的原油的温度以及土壤温度为停输温降过程的起始温度。停输降温过程就是获得规定停输时间内土壤和管内存油的温度下降水平。再启动过程是计算最复杂的过程，这个时候需要耦合泵和调节阀特性、再启动操作流程、管内流动传热过程、土壤非稳态传热过程计算得到管道内不同位置不同时刻的压力和流量变化结果。停输降温获得的土壤和管内存油的温度场作为再启动过程的起始状态进行后续计算。

1.1 水力模型

管内水力热力耦合计算主要针对的是稳态运行过程和再启动过程。采用连续性方程以及动量方程对这一过程进行描述。将管道流动简化成为一个一维过程，在差分求解过程中准确性较高同时求解速率较快。连续性方程如下所示：

$$\frac{\partial \rho}{\partial t} + \overline{w}\frac{\partial \rho}{\partial z} + \rho a^2 \frac{\partial \overline{w}}{\partial z} = 0 \quad (1)$$

式中，ρ 为原油密度，kg/m^3；t 为时间，s；z 为轴向坐标，m；$\overline{w}$ 为 z 方向的速度，m/s；a 为原油在管道内的声速，m/s。

对于动量方程需要考虑轴向的流速，同时仅考虑径向的剪切应力的作用。二维动量方程如下所示：

$$\frac{\partial p}{\partial z} + \rho\frac{\partial \overline{w}}{\partial t} - \frac{2\tau_w}{R} + \rho\overline{w}\frac{\partial \overline{w}}{\partial z} - \rho g\sin\alpha = 0 \quad (2)$$

式中，τ_w 为流体在管道内壁处所受到的切应力，Pa；R 为管道半径，m；g 为重力加速度，m^2/s；α 为管道与水平线的夹角，°。

1.2 热力模型

对于停输再启动过程的热力过程可以分为原油与管壁面及土壤的换热过程，上层土壤与空气的换热过程以及土壤的导热过程。

分析管内原油流动的热力过程，可知存在不同温度原油由于质量输运所带来的向下游的温度输运，主要的热量传递主要是原油与管壁面的传热。另外，还有原油内部摩擦所产生的热量。将这三者综合在一维能量方程如下所示：

$$\rho c_p\left(\frac{\partial \overline{T}}{\partial t} + \overline{w}\frac{\partial \overline{T}}{\partial z}\right) + \frac{1}{\pi R^2}\int_0^R (2\pi r\tau_{rz}\dot{\gamma})dr + \frac{2}{R}q_w = 0 \quad (3)$$

式中，c_p 为原油比热容，$J/(kg \cdot K)$；$\overline{T}$ 为管内原油的截面平均温度，℃；$\dot{\gamma}$ 为原油的剪切率，s^{-1}；q_w 为原油通过管壁面的热流密度，$W/(m^2 \cdot K)$。

为了能够加快计算速度，可以忽略沿着管长方向的土壤温度变化，进而将土壤温度计算约束在二维平面内，每一个二维平面对应管道的一个计算节点。事实上，沿着管长方向的土壤温度梯度相比于其他两个方向要小很多。方程如下所示：

$$\frac{\partial(\rho_i c_i T)}{\partial t} = \frac{\partial}{\partial x}\left(\lambda_i\frac{\partial T}{\partial x}\right) + \frac{\partial}{\partial y}\left(\lambda_i\frac{\partial T}{\partial y}\right) \quad (4)$$

式中，T 为土壤温度，℃；ρ_i 为土壤密度，kg/m^3；c_i 为土壤比热容，$J/(kg \cdot K)$；λ_i 为土壤导热系数，$W/(m \cdot ℃)$；x 为横坐标，m；y 为纵坐标，m。

在求解稳态运行状态的时候，土壤温度场认为已经达到了稳态，因此方程左边的非稳态项直接忽略掉；在计算停输过程和再启动过程中，则需要把非稳态项也考虑在内。

1.3 稳态运行过程物理模型

稳态运行现场可知的是每一个站点的进出站压力、流量、温度，还有一些阀室的温度、压力和地温。以可测参数进出站压力作为模型的入口条件，计算得到压力和温度分布。

稳态条件下流量是确定的，只需要计算在给定流量条件下的压力和温度分布即可，也就是需要稳态的动量方程和能量方程。稳态动量方程变化为：

$$\frac{\partial p}{\partial z} = \frac{4\tau_w}{D} \quad (5)$$

在稳态运行时，由于流量为恒定值仅摩擦阻力导致稳定的压力损失。

能量方程去掉非稳态项后变化为：

$$\rho c_p\left(\overline{w}\frac{\partial \overline{T}}{\partial z}\right) + \frac{1}{\pi R^2}\int_0^R (2\pi r\tau_{rz}\dot{\gamma})dr + \frac{2}{R}q_w = 0 \quad (6)$$

动量和能量方程可以沿着管道流动方向进行向前差分显式求解，每两个站点设定上游出站压力和流量（来自于现场数据），计算得到沿程的温度和压力。计算温度的时候需要耦合求解土壤稳态温度场获得管壁不同位置的热流密度，土壤稳态温度场求解方程在去掉非稳态项后变化为：

$$0 = \frac{\partial}{\partial x}\left(\lambda_i\frac{\partial T}{\partial x}\right) + \frac{\partial}{\partial y}\left(\lambda_i\frac{\partial T}{\partial y}\right) \quad (7)$$

1.4 停输温降过程物理模型

停输之后管道内流速降为零，管内压力也随

之下降，仅剩高程差所带来的重力水头作用。停输的温度起点即为停输前稳态运行的沿程管内温度和土壤温度，因此只需要求解传热方程即可获得停输温降。

停输后管道内原油的能量方程变为：

$$\rho c_p \left(\frac{\partial T}{\partial t}\right) = -\frac{2}{R_{in}} q_w \quad (8)$$

这里仅包含了原油温度降低过程和通过管壁向土壤传热的过程，这里的热流密度依然需要耦合土壤温度场求解获得，而土壤温度场求解就需要对非稳态过程进行逐时差分求解。求解得到的土壤和原油管道温度场将会为后续的再启动过程提供数据起点。

1.5 再启动过程物理模型

再启动过程需要同时求解连续性方程、动量方程、能量方程，并且逐时逐点耦合求解土壤温度场方程，计算的复杂度高。

连续性方程如下所示：

$$\frac{\partial p}{\partial t} + \overline{w}\frac{\partial p}{\partial z} + \rho a^2 \frac{\partial \overline{w}}{\partial z} = 0 \quad (9)$$

动量方程如下所示：

$$\frac{\partial p}{\partial z} + \rho \frac{\partial \overline{w}}{\partial t} - \frac{2\tau_w}{R} + \rho \overline{w}\frac{\partial \overline{w}}{\partial z} - \rho g \sin\theta = 0 \quad (10)$$

1.6 模型求解

采特征线法求解连续性方程和动量方程。为了求解特征线方程，需要初始条件和边界条件。对于初始条件，需要先计算初始条件支配区内的参数，即稳态工况内的压力分布和流量分配，作为瞬变过程计算的初值。对于再启动来说，其初始条件即为停输一段时间之后的管道内流量、压力分布。原油能量方程采用向前差分，显示求解。采用有限容积法对土壤导热模型进行离散，Jacobi 预条件方法进行求解。

1.7 边界条件

管道系统中的边界有内部边界和外界边界之分。一般把独立的管道起点、终点称为外部边界，管道系统中的串并联（或多管）接点、设备、中间站场等都称为内部边界。管道瞬变流动过程中的边界条件就是瞬变流动过程中，边界处压力、流量与时间的变化对应关系。建立于求解边界条件，就是使用辅助方程描述管道边界处的压力、流量随时间变化的某种函数关系，联立求解辅助方程和对应的特征方程的过程。通过求解边界特性传递给模拟的管道内部结点。

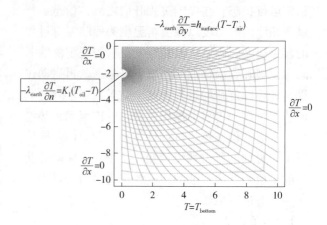

图 1　土壤温度场求解的网格和边界条件

土壤的对称边界（图中左边界）和垂直远边界（图中右边界）设置为绝热边界条件，不存在换热。土壤的下表面 10 米深度的位置设置为恒温边界，其土壤温度根据经验关联式计算得到。土壤上表面与空气接触，设置为与空气的自然对流换热，自然对流换热方程如下所示：

$$-\lambda_{earth}\frac{\partial T}{\partial y} = h_{surface}(T - T_{air}) \quad (11)$$

式中，$h_{surface}$ 为地表与空气的自然对流换热系数，W/(m²·℃)；T_{air} 为空气温度，℃。

土壤温度场求解的网格划分以及边界条件如图 1 所示。土壤与管外壁面接触的位置存在从油品传递过来的热量，用综合传热系数进行计算，并且应该与最贴近这一层的土壤传热量相当如下所示：

$$-\lambda_{earth}\frac{\partial T}{\partial n} = K_1(T_{oil} - T) \quad (12)$$

式中，n 为垂直管壁方向，由管道内部指向土壤侧；T_{oil} 为管内原油平均温度，℃。

2 设计参数

2.1 管道设计参数

管线的高程信息如图 2 所示。某原油输送管道设计长度为 691.2km。从管线上游输送过来的几种油经过掺混后由站点 1 输送至站点 6。其中站点 1-站点 6 管径 813mm。站点 1-站点 5 为泵站。此原油管道存在较为平缓的地势（站点 1-站点 3）同时还存在站间大落差的地势（站点 3-站点 6）。最大落差为站点 5 与站点 6 之间的落差，约为 753m。地形复杂的管道相对于平缓的地势，管道输送的安全性较差，因此选取该条管道能够更好的对初凝过程进行分析。

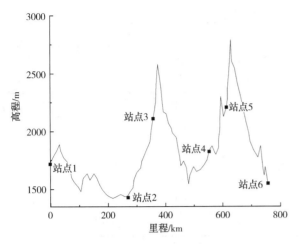

图 2 原油管道纵断面图

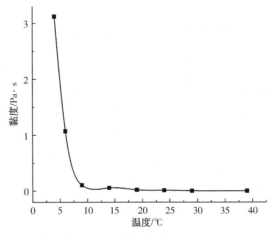

图 3 原油黏温曲线

2.2 油品物性参数

为了更好的验证模型以及分析发生"初凝"的原因，本文的不同工况的计算采用的为相同的原油。

表1为原油物性数据，为了更好的验证模型以及分析发生"初凝"的原因，本文的不同工况的计算采用的为相同的原油。

表 1 油品物性参数

温度/℃	稠度系数$K/(Pa \cdot s^n)$	流动行为指数n	$20s^{-1}$黏度/$(Pa \cdot s)$	$50s^{-1}$黏度/$(Pa \cdot s)$	$100s^{-1}$黏度/$(Pa \cdot s)$
4	3.120	1		3.120	
6	1.072	1		1.072	
7	0.205	1		0.205	
8	0.112	1		0.112	
9	0.104	1		0.104	
14	0.057	1		0.057	
19	0.0213	1		0.0213	
24	0.0828	1		0.0828	
29	0.00544	1		0.00544	
39	0.00304	1		0.00304	

此油品为某次注入管道的前所测的油品物性参数。该油品在 6℃ 以上时，黏度较低大约为 $0.205Pa \cdot s$ 以下，动力黏度较小。当温度低于 6℃ 时，原油的黏度突然恶化。原油温度为 7℃ 时，动力黏度为 $0.2Pa \cdot s$，为 6℃ 时原油黏度的 5 倍左右。因此，原油的黏度突变点为 7℃。在本次模拟分析时，不考虑流体性质的影响，仅考虑黏度的影响，将输入的原油作为牛顿流体处理（图3）。

2.3 泵参数

将泵耦合计算模型中，各个站点的泵特性曲线如表 2 所示。

表 2 泵特性曲线

站点名称	泵特性曲线（ΔP/MPa、$Q/(m^3 \cdot h)$）
站点 1	$\Delta P=-1.89 \times 10^{-7}Q^2+4.15$
站点 2	$\Delta P=-1.89 \times 10^{-7}Q^2+5.77$
站点 3	$\Delta P=-1.95 \times 10^{-7}Q^2+5.64$
站点 4	$\Delta P=-2.10 \times 10^{-7}Q^2+4.21$
站点 5	$\Delta P=-2.21 \times 10^{-7}Q^2+4.51$

3 模拟结果及分析

对所建立的管道的停输再启动过程进行分析，将天气温度设置为 10℃（秋季工况）、天气温度 -10℃（冬季工况）停输时间设置为 40 h。管道在夏天运行时，天气温度较高，较难发生"初凝"现象。秋季以及冬季天气温度较低，原油输送的安全性较低。因此选取上述两种季节的天气温度作为边界条件进行模拟仿真并对其结果进行分析。

3.1 稳态运行工况

在稳态运行工况，将各个站的出站压力、温度、表 1 的原油作为输入条件。流量为恒定的 $2000m^3/h$，通过方程(5)、(6)、(7)对管道的水力、热力方程以及土壤的热力过程进行差分求解，得到各个站的进站压力以及温度。

秋季、冬季两种运行工况的稳态运行过程的运行参数如表 3 所示。在冬季工况，由于环境温度较低，原油沿线温降较高，黏度较大；在秋季工况，环境温度较高，原油沿线温降较小，原油黏度低于冬季的原油黏度。因此冬季各个站间的压降明显高于秋季工况的各个站间的压降。

表3 稳态运行数据

站点信息	冬季运行				秋季运行			
	进站压力/MPa	出站压力/MPa	进站温度/℃	出站温度/℃	进站压力/MPa	出站压力/MPa	进站温度/℃	出站温度/℃
站点1	—	4.70	—	20.0	—	4.70	—	20.0
站点2	1.28	6.97	12.0	13.0	1.30	6.97	17.7	18.7
站点3	0.87	6.09	11.4	12.4	1.11	6.09	18.1	19.2
站点4	0.73	6.83	7.9	8.9	2.32	6.83	17.5	18.5
站点5	0.58	5.13	7.8	8.8	1.25	5.13	18.0	19.0
站点6	9.10	—	6.5	—	9.10	—	17.8	—

3.2 停输温降结果分析

表3中稳态运行过程的温度作为停输温降过程的温度起点，此时原油的流量为0。原油与周围土壤进行换热，管道上层土壤与空气进行换热计算，10m处土壤边界条件为恒定温度，土壤内部的导热方式为热传导。

图4反应了秋季、冬季停输40h沿线温降变化趋势。秋季的天气温度大约为10℃，停输40h时如图4(a)所示，大约降低0.5℃。由于冬天的天气温度较低，原油在停输过程中温降较大。停输40h如图4(b)所示，整体原油的温度大约降低4℃。土壤深层10m处地温在秋季以及冬季计算时保持一致均为25℃，在秋季运行工况时，原油在停输时温降较小，而冬季停输时温降较大。在相同的停输时间下，冬季的停输温降大于秋季的停输温降。

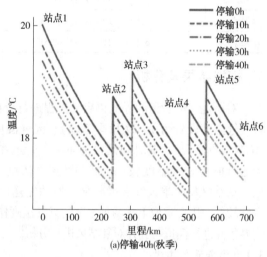
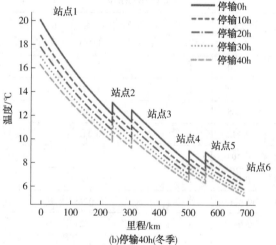

图4 冬季、秋季停输温降变化趋势

3.3 再启动过程分析

站点1开始逐渐启动各个站的泵，站点1注入新的原油温度稳定为16℃，每种工况再启动计算50h。采用常温不加热输送工艺，根据历史运行数据，原油过泵温度会升高大约0.7℃。

同时，由于站点6没有泵，站点5-站点6之间的管道为高落差管道，为了防止末站的流量、压力过高，因此设置末站为节流阀，释放一定的压力，节流阀设置截止压力为9.1MPa。

采用特征线法对再启动过程进行求解，同时耦合土壤温度场非稳态计算，将泵、节流阀等设备耦合到计算模型中。

3.3.1 计算结果分析

图5表示秋季、冬季工况的再启动流量、压力随时间的变化趋势。在冬季运行时，如图5(a)图5(b)所示，在管道再启动初始阶段，各个站的出站流量已经低于1000m³/h，仅有稳态运行时流量的1/2，且流量随时间的增加持续下降。冬季再启动时各个站的出站压力较高，除了站点1，其余各站的出站压力均有明显的上升趋

势。图5(c)和图5(d)为秋季工况的再启动流量、压力随时间的变化趋势。秋季各个站的再启动流量、压力很快恢复稳定，随着时间的增加各个站再启动流量、压力未出现明显的变化。

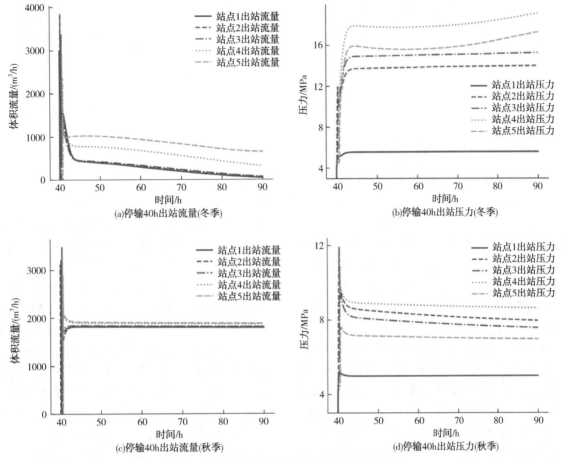

图5 冬季、秋季再启动压力流量变化趋势

3.3.2 初凝原因分析

为了更好分析启输过程中发生"初凝"现象的原因，图6为上述工况再启动过程的沿线温度变化。在秋季运行工况中，如图6(a)所示，秋季原油再启动后，虽然再启动计算时设置入口油温为16℃，但是管内的原油温度较高，管道再启动50h后全线原油的温度全部大于16℃。根据表1，秋季再启动50h之内，全线的原油黏度均小于0.042Pa·s。秋季工况管道再启动后，各个站的压力以及流量能够很快恢复稳定。

冬季停输40h后，管道再启动初始阶段，如图6(b)所示，管道下游650km(站点5-站点6)处的温度接近原油的原油黏度突变点，650km处至末站的原油的黏度均大于1Pa·s。泵提供的压力无法使高黏度原油的流量恢复至稳态运行时的流量。因此图5(a)中，站点4、站点5的出站流量在启输时流量仅有1000m³/h，而上游站点(站点1-站点3)在启输后5个小时之内，出站流量由1500m³/h下降至500m³/h左右。这是由于下游的原油黏度较大、流量较低，导致上游的原油流动受阻，流量下降。随着时间的增加，原油全线温度受到环境温度的影响持续下降。再启动50h时，400km处的原油温度接近原油黏度突变点，400km处上游原油的温度仍然高于原油的原油黏度突变点，上游站点的出站流量仍然持续降低。下游的高黏度原油造成了管道堵塞，造成了上游原油的流量降低。因此全线出现了流量下降、压力升高的初凝现象。

3.3.3 恒压启动边界计算

为了更好的体现耦合泵的必要性，选取上述管道的管道1、管道2，边界条件设置为压力入口、出口。管道的进出口温度选取管道启输50h后的这两条管道的入口温度作为输入条件。为了保障对比结果的准确性，选取管道1、管道2稳态运行时的数据开始进行稳态运行计算，设置停输时间为40h。设置两条管道的边界条件与稳态

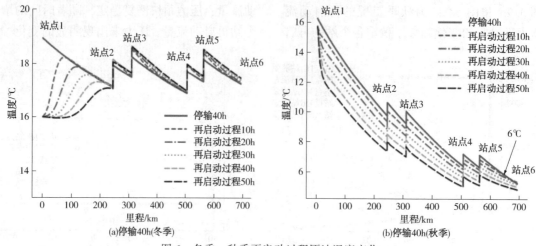

图6 冬季、秋季再启动过程原油温度变化

运行时的边界条件一致。选取这两条管道的原因是管道1的全线原油温度始终高于该原油的原油黏度突变点,管道中原油黏度始终低于 0.2Pa·s。管道2的全线温度略高于原油的黏度突变点,随着时间的增加,管道2下游温度逐渐低于原油的黏度突变点。

压力入口的计算方法是将上述泵特性曲线二次变量的系数无限接近于0,常数项设置为稳态运行时的入口压力。即在求解时,无论流量如何变化,压力始终保持为泵特性曲线的常数项的数值,实现恒定压力入口计算。压力出口的计算方法是将管道出口设置为调节阀,阀的截止压力为稳态运行时这两条管道的出口压力。

如图7所示,管道1与管道2经过停输40h后,管道全线温度与图4(a)保持一致。管道1、管道2的再启动计算的入口温度为16℃、8℃,这与全线计算时的再启动计算50h管道1、管道2的入口温度保持一致。

图8为管道独立计算停输40h后管道1、管道2再启动过程的流量压力变化趋势。如图8(a)管道1在管道启输后的流量始终保持不变,但是管道2的流量发生了下降的趋势,但是流量未低于稳态输量的一半。但是在耦合泵的全线计算结果中,管道1、管道2的入口流量持续下降,而且流量在管道启输50h后已经接近于0。管道1的全线温度始终未高于黏度突变点,所输送的油品的黏度较低,但是仍然受到下游的影响流量下降。这与管道1在独立计算时的流量变化的结果不同。对于管道2,由于入口温度较低,在再启动计算中,管道2的下游部分的温度较低,原油的黏度较大。若采用恒定压力作为边界条件,管道中的黏度较大时,才会发生初凝现象。采用恒定压力入口作为边界条件,入口压力不受到上下游流量变化的影响,对管道内原油持续提供动力使其流动。但是在实际运行时,全线密闭运行,整条管道作为一个整体,上下游管道的运行参数由于黏度的不同会相互影响。因此管道单独计算的结果较为保守,只能预测较为极端的情况。

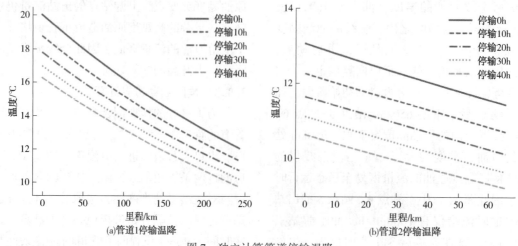

图7 独立计算管道停输温降

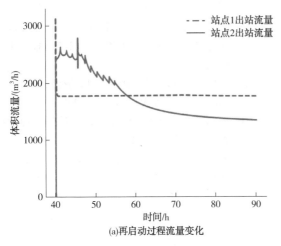
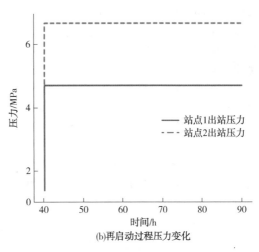

图 8　独立计算停输 40h 后再启动压力流量变化

4　结论

通过建立耦合泵特性作为边界条件的原油管道模型，与恒定压力边界条件的原油管道模型对比分析，得到以下结论。

（1）原油管道再启动后，若管道全线的温度高于原油的黏度突变点，原油可以安全启输。若管道下游的温度低于原油的黏度突变点，管道下游的黏度较大，即使上游的原油黏度较小，上游原油会受到下游较高的流动阻力的影响，流量降低。

（2）在冬季工况，采用耦合泵特性作为边界条件的模型计算时，黏度较低的上游管道受到下游高黏度原油的影响，上游管道发生流量下降，压力增高的初凝现象。将上游两条管道采用恒定压力作为边界条件的独立计算时，黏度较低的管段不会发生初凝现象，黏度较高管段会发生初凝现象。采用恒定压力作为边界条件研究初凝时，只有较为极端的情况才能预测初凝的发生，这相当于变相的高估了泵的能力。

参 考 文 献

［1］万军，郝建斌，王军防.鲁宁输油管道初凝应急处置与思考［J］.油气储运，2022，41（05）：575-582.

［2］井懿平，史建刚，段永军，齐文元，张衍礼.火三含蜡原油管道初凝现象分析［J］.油气储运，2001，20（10）：11-15.

［3］李传宪，纪冰，魏国庆.含蜡原油管道停输降温的数值计算［J］.石油化工高等学校学报，2019，32（3）：95-102.

［4］唐渤.鄯善—四堡站间原油管道停输温度场数值模拟［D］.成都：西南石油大学，2014.

［5］苏炳辉，邓子旋，邵游凯，王炳人.定靖复线停输再启动过程研究［J］.管道技术与设备，2016，（01）：8-11.

［6］李浩.热输含蜡原油管道安全停输时间与再启动工艺研究［D］.西安石油大学，2020.

［7］刘晓燕，周正，李从，范玉泽，张楠迪，等.Ra 数对停输管道原油相变特征的影响研究［J］.工程热物理学报，2024，45（04）：1055-1061.

［8］陈思杭，李其抚，闫锋，江璐鑫，雷连风，等.热油管道温度场模拟研究［J］.石油工程建设，2024，50（05）：19-24+49.

［9］DONG H, ZHAO J, ZHAO W Q, SI M L, LIU J Y. Study on the thermal characteristics of crude oil pipeline during its consecutive process from shutdown to restart［J］. Case Studies in Thermal Engineering. 2019, 100434. DOI：https://doi.org/10.1016/j.csite.2019.100434.

［10］DALLA L F R, SOARES E J, SIQUEIRA R N. Start-up of waxy crude oils in pipelines［J］. Journal of Non-Newtonian Fluid Mechanics, 2019, 263：61-68.

［11］TIKARIHA L, KUMAR L. Pressure propagation and flow restart in a pipeline filled with a gas pocket separated rheomalaxis elasto–viscoplastic waxy gel［J］. Journal of Non Newtonian Fluid Mechanics, 2021, 294：104582.

［12］王东，李群海，李静，代影春，张磊等.埋地含蜡原油管道停输再启动压力计算［J］.油气储运，2005，24（10）：21-25.

［13］尹晓云，付林浩，李佳忆，程思杰，敬加强等.稠油水环输送管道再启动压降特性分析［J］.化工进展，2023，42（11）：5669-5679.

［14］尹晓云，敬加强，孙杰，刘力华，蒲欢.水平管内黏稠油水环输送管道停输再启动特性［J］.石油机械，2022，50（04）：124-129.

[15] 祝守丽, 李长俊, 马志荣, 彭鹏, 田雨等. 含蜡原油停输再启动环道优化及模拟试验[J]. 油气储运, 2019, 38(04): 412-418.

[16] 蔡磊, 袁庆, 于红梅, 张春, 宇波等. 东临复线停输再启动研究[J]. 北京石油化工学院学报, 2018, 26(04): 29-33.

[17] 宇波, 徐诚, 张劲军. 冷热原油交替输送停输再启动研究[J]. 油气储运, 2009, 28(11): 4-16.

[18] 张争伟, 宇波, 孙长征, 窦丹. 海底管道油水两相混合输送的数值模拟[J]. 油气储运, 2009, 28(9): 13-15, 26.

[19] 宇波, 付在国, 李伟, 毛珊. 热油管道大修期间停输与再启动的数值模拟[J]. 科技通报, 2011, 27(06): 890-894.

[20] 袁庆. 含蜡原油管道停输再启动高效数值方法研究[D]. 中国石油大学(北京), 2021.

[21] 杨筱蘅, 张国忠. 输油管道设计与管理[M]. 东营: 石油大学出版社, 1994: 97-98.

[22] 蒲家宁. 管道水击分析与控制[M]. 北京: 机械工业出版社, 1991: 11-17.

A Mapless Path Generation and Control Method for UAV Pipeline Inspection

Ma Yinghan　Zhao Hong

(China University of Petroleum, Beijing)

Abstract　Objective: To Ensure energy security through regular inspection of offshore pipelines, this study introduces a mapless pipeline inspection path planning method using UAV, addressing real-time flexibility and robustness issues in dynamic environments. Methods: The method involves real-time depth image preprocessing, path generation, and motion control. Depth image preprocessing includes noise reduction and background removal. Path generation uses B spline curves driven by physical state, while the motion control method is based on variable structure sliding mode control. Results: The proposed depth image preprocessing method can perform background removal and path generation online, consume less computer resources, and does not accumulate errors as the pipeline length increases. The B spline curve driven by physical state is second-order continuous under the given cruising speed, ensuring the kinematic feasibility of the drone. The path generation method based on this curve has a low time-space complexity. The attitude controller designed based on the variable structure sliding mode control method is stable under appropriate parameter selection. Conclusion: The real-time preprocessing approach for depth images effectively filters noise and accurately removes the background. It features a rapid processing speed and can be extensively applied in the domain of computer vision. The B spline curve driven by physical state ensures that the tangent direction of the path aligns with the velocity direction of the current UAV, and is widely applicable to the path generation of unmanned vehicles, unmanned aerial vehicles, and industrial robots. For the sliding mode control method, the overshoot is within 1% and its response speed is faster than that of PID control method. Consequently, it constitutes a suitable control methodology under the working conditions of pipe inspection.

Keywords　Pipeline; UAV; Path Generation; Physical State; Sliding Mode Control

1 INTRODUCTION

Pipeline networks are critical infrastructure for the transportation of oil and naturalgas. Offshore overhead pipelines, exposed to environmental factors such as rock impacts and corrosion from humid air, require regular inspection to maintain their integrity. In recent years, unmanned aerial vehicles (UAVs) have emerged as an efficient and effective modality for inspecting offshore overhead pipelines. The autonomous navigation and control section in UAV pipeline inspection is one of the crucial technologies for achieving efficient and accurate inspection, encompassing three components: flight path planning, autonomous positioning, and flight control.

Flight path planning requires the planning of a flight route that aligns with the inspection requirements based on the actual alignment of the pipeline.

Quadrotor UAVs employ various path planning algorithms, including A*, grey wolf optimization (GWO), and dung beetle optimization (DBO). While these methods demonstrate considerable versatility in unfamiliar environments with well-defined endpoints, they typically rely on Euclidean Signed Distance Field (ESDF) modeling for path generation. In the context of pipeline inspection, the computational complexity of model reconstruction poses significant challenges UAV performance, consuming substantial computational resources and limiting operational duration. Gao et al. introduced EGO-Planner, a mapless path planning approach that achieved a 47% reduction in time consumption compared to traditional ESDF methods. However, EGO-Planner operates under the assumption of known endpoint coordinates and exhibits increasing error accumulation as the pipeline length increases. Pipeline inspection necessitates more precise tracking, wherein the UAV's target point dynamically shifts with the pipeline's trajectory. Furthermore, pipeline tracking becomes increasingly challenging in complex background environments.

During the pipeline inspection with UAVs, it is necessary to acquire the position information of themselves in real time. Commonly used autonomous positioning methods include GPS positioning technology applicable to open areas; visual SLAM technology that collects images through cameras to achieve simultaneous localization and mapping, which is suitable for environments with poor GPS signals; and LiDAR positioning technology that scans the surrounding environment by using LiDAR to achieve mapping, positioning, and obstacle avoidance. Nevertheless, with the escalation of environmental complexity, the computational volume of navigation and control algorithms rises accordingly, thereby imposing higher demands on the hardware performance of UAVs.

Flight control mainly encompasses attitude control, speed control, and altitude control. During the pipeline inspection process of UAVs, these three control aspects are interrelated and mutually influential. It is necessary to comprehensively apply various control methods and technologies, thereby enhancing the flight performance and reliability of UAVs in response to the pipeline inspection environment and task requirements.

To address these challenges, we have developed an advanced UAV system capable of real-time background filtering and pipeline tracking. Given the physical limitations of UAVs, specifically their inability to change speed instantaneously, inertia-induced overshooting is inevitable. While the electrical industry typically requires a safety distance of $L > 2R$ for UAVs, pipeline inspection necessitates a reevaluation of this parameter to ensure operational accuracy and safety.

In this study, we analyze overshoot characteristics by inducing sudden changes in the UAV's expected height along the z-axis. This analysis aims to establish a reference framework for determining appropriate safety distances for UAVs in the context of pipeline inspection. Meanwhile, we establish a kinematic model for the drone to track the pipeline along the z-axis based on the characteristics of the pipeline deformation. We then plan its trajectory based on this kinematic model and design its attitude controller based on the variable structure sliding mode control method.

2 METHODS

2.1 UAV State Acquisition

The UAV employed in this study, as illustrated in Figure 1, is a custom-built 5-inch quadrotor drone featuring a carbon fiber protective frame. The UAV is equipped with an Intel D435i binocular depth camera and an Intel NUC (Next Unit of Computing) for onboard processing. The onboard camera system simultaneously captures both RGB and RGB-D images, with RGB-D data utilized for path generation and tracking during pipeline inspection. The UAV's state is defined by its position in the x, y, and z axes, and its orientation described by yaw, pitch, and roll angles. The x, y, and z coordinates determine its spatial position, while the yaw, pitch, and roll angles specify its orientation. Utilizing both current and desired state information, the system generates an optimal flight path.

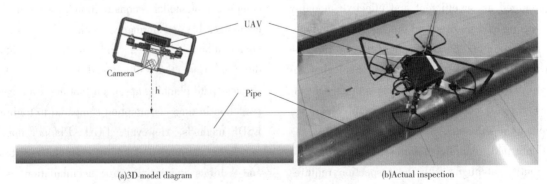

(a) 3D model diagram (b) Actual inspection

Figure 1 UAV pipeline inspection. (a) 3D model diagram. (b) Actual inspection

Duringthe flight, the desired values for pitch and roll relative to the yaw axis are calculated based on the UAV's velocity and acceleration vectors, ensuring that the desired direction aligns with the pipeline orientation. These vectors are derived from the onboard depth camera data, while the actual z-axis value (height) is measured concurrently, as illustrated in Figure 1. Within the depth camera's field of view, the distance between the UAV and the nearest object represents the actual height (h). The determination of the optimal flight height is based on two key considerations: detection accuracy and operational safety. In terms of accuracy, the depth camera exhibits optimal performance at closer ranges within its effective range (distance>120mm). From a safety perspective, it is crucial to account for non-linear changes in pipeline orientation within the UAV's field of view. Consequently, the path planning algorithm must carefully consider potential overshoots when adjusting the desired flight height. The minimum desired flight height is set as the sum of the maximum expected overshoot and the depth camera's minimum effective operational height.

2.2 Depth Image Preprocessing and Trajectory Generation

The UAV's trajectory generation strategy, illustrated in Figure 2, represents each grid cell as a pixel, with the UAV's current position located at the midpoint of the bottom edge. The raw depth image acquired by the camera contains significant noise, which is initially filtered using a Modified Rolling Average Median Filter (Modified-RAMF) to remove isolated noise points. Subsequently, the image undergoes preprocessing through a series of morphological operations, specifically erosion-dilation and dilation-erosion. Based on these preprocessing results, a mask is generated, from which pixel positions corresponding to pipeline corner points (depicted as red squares) are extracted. These corner points are subsequently utilized to determine the pipeline's midpoint (represented by yellow squares). The upper yellow square denotes the UAV's target point, while the dashed arrow connecting the current position to the target point indicates the desired direction of travel. Finally, the generated optimal path is represented by a solid arrow.

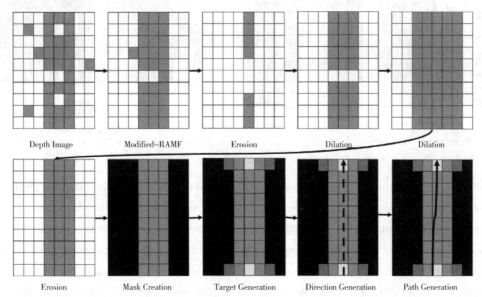

Figure 2 Depth image preprocessing and trajectory generation strategy

The mathematical expression for the improved adaptive median filter is as follows:

$$p_{i,j} = \begin{cases} p_{i,j}, & p_L < \sum |p_{i,j} - p_{i-1,j-1}| + |p_{i,j} - p_{i+1,j+1}| \\ \mathrm{MR}(i, j, N), & \text{otherwise} \end{cases}$$

$$\begin{cases} p_m = \mathrm{median}(p_{i-N,j-N}, p_{i-N,j-N+1}, \cdots, p_{i+N,j+N-1}, p_{i+N,j+N}) \\ \mathrm{MR}(i, j, N) = \begin{cases} p_m, & p_L < \sum |p_{i,j} - p_{i-1,j-1}| + |p_{i,j} - p_{i+1,j+1}| \\ \mathrm{MR}(i, j, N+1), & \text{otherwise} \end{cases} \end{cases} \quad (1)$$

Here, $p_{i,j}$ represents the pixel value at coordinates (i, j) in the image; p_L is the threshold obtained from the differences between the pixel values; p_m is the median pixel value of all pixels within the range from $p_{i-N,j-N}$ to $p_{i+N,j+N}$; N represents the size of the filter. If p_m falls within the threshold range, it will replace $p_{i,j}$. Then, the window will be moved to process the next pixel. If not, the filter size will be increased, and then the process will be repeated. Given that the image is a three-channel image, the calculation of differences between pixel values is performed separately for each channel. To reduce the computational load, only the absolute values of the differences between the diagonal pixels are calculated.

Let A be the original image, and B and C be the images probed by the structuring element. The formula for the opening and closing operations is as follows:

$$A' = A \circ B \cdot C = \{[(A-B) \oplus B] \oplus C\} - C \quad (2)$$

2.3 Deep Image Processing and Trajectory Generation Results

The real depth images captured by UAV can be processed through preprocessing methods to obtain noise-free and background-removed images. Figure 3 depicts the actual trajectory generation process of the UAV. The current direction of the UAV is indicated by a red arrow, while its target direction is represented by a white dashed line. The UAV's target point is denoted by a red marker, while the generated target path is represented by a white solid line.

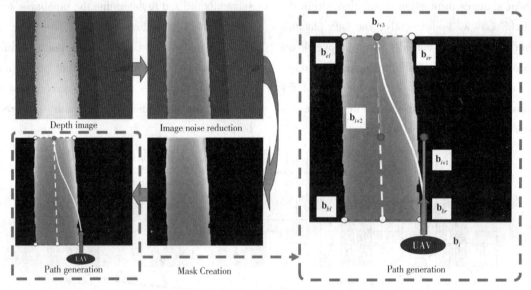

Figure 3 UAV Actual Trajectory Generation

The target point and direction are ascertained utilizing corner points (b_{bl}, b_{br}, b_{el}, b_{er}) extracted from the denoised RGB-Depth (RGB-D) image of the pipeline. The midpoint of the line connecting the upper corner points (the red dot b_{i+3}) serves as the target position. The target direction is established by the line connecting the midpoints of the lower and upper corner points (white dashed line).

To smooth the path and meet the feasibility of UAV kinematic and dynamic, B spline curve driven by physical state is proposed. After the initial and final states of the UAV are determined, a path is generated using the proposed B spline curve driven by physical state. Unlike B spline curve, B spline curve driven by physical states consider the initial and final velocity directions of unmanned aerial vehicles, as the speed of unmanned aerial vehicles cannot undergo sudden changes. Considering the physical characteristics of UAV, the acceleration and jerk of UAV should be minimized as much as possible Therefore, the B spline curve based on physical state driving proposed in this article is tangent to the initial velocity direction and the final velocity direction, respectively Thus ensuring the kinematic and dynamic feasibility of the UAV.

The mathematical expression of B spline curve

driven by physical state is as follows:

$$P_i(u) = (1-u)^3 \times b_i(t) + 3u(1-u)^2 \times b_{i+1}(t) \\ + 3u^2(1-u) \times b_{i+2}(t) + u^3 \times b_{i+3}(t) \quad (3)$$

Here, $i = 0, 1, \cdots, n-1$; $0 \leq u \leq 1$; $b_i(t)$, $b_{i+1}(t)$, $b_{i+2}(t)$, $b_{i+3}(t)$: four control points of B spline curve driven by physical state at time t; $b_{bl}(t)$, $b_{br}(t)$, $b_{el}(t)$, $b_{er}(t)$: corner points extracted from the denoised RGB-D image at time t.

Set $b_i(t)$: The current coordinates of UAV

b_{i+3}: The coordinates of the endpoint of the current segment of the UAV's path

$$b_{i+2}(t) = \frac{b_{el}(t) + b_{er}(t)}{2} + \frac{b_{bl}(t) + b_{br}(t)}{2}$$

$$b_{i+3}(t) = \frac{b_{el}(t) + b_{er}(t)}{2}$$

$$\vec{V}_{bi} = k[b_{i+1}(t) - b_i(t)] = \begin{bmatrix} V_{bix}(t) \\ V_{biy}(t) \end{bmatrix}$$

$$\vec{V}_{ei} = k[b_{i+3}(t) - b_{i+2}(t)] \\ = k\left[\frac{b_{el}(t) + b_{er}(t)}{2} - \frac{b_{bl}(t) + b_{br}(t)}{2}\right]$$

$$|b_{i+3}(t) - b_{i+2}(t)| = |b_{i+1}(t) - b_i(t)|$$

$$k = \left|\frac{2V}{b_{el}(t) + b_{er}(t) - b_{bl}(t) - b_{br}(t)}\right|$$

Here, $\vec{V}_{bi}$: initial velocity of the B spline curve segment; $\vec{V}_{ei}$: final velocity of the B spline curve segment;

We have:

$$|\vec{V}_{bi}| = |\vec{V}_{ei}|$$

At $u = 0$,

$$K = \frac{6|b_i(t) - 2b_{i+1}(t) + b_{i+2}(t)|}{\{9[b_i(t) - b_{i+1}(t)]^2 + 1\}^{\frac{3}{4}}}$$

Here, K: curvature of B spline curve driven by physical state.

Set the predetermined cruising speed to V, then $V^2 = \vec{V}_{bi}^2 = \vec{V}_{ei}^2$

$$\begin{bmatrix} a_{bin}(t) \\ a_{bit}(t) \end{bmatrix} = \begin{bmatrix} \dfrac{\vec{V}_{bi}^2}{R} \\ 0 \end{bmatrix}$$

$$= \begin{bmatrix} \dfrac{6|b_i - 2b_{i+1} + b_{i+2}|}{[9(b_i - b_{i+1})^2 + 1]^{\frac{3}{4}}} [k(b_{i+1} - b_i)]^2 \\ 0 \end{bmatrix} \quad (4)$$

$$= \begin{bmatrix} \dfrac{6|b_i - 2b_{i+1} + b_{i+2}|}{[9(b_i - b_{i+1})^2 + 1]^{\frac{3}{4}}} V^2 \\ 0 \end{bmatrix}$$

Here, $a_{bin}(t)$: normalized tangential acceleration of point $b_i(t)$ at time t; $a_{bit}(t)$: tangential acceleration of point $b_i(t)$ at time t.

Since $b_i(t)$, $b_{bl}(t)$, $b_{br}(t)$, $b_{el}(t)$, $b_{er}(t)$ are continuously changing coordinates, $b_{i+1}(t)$, $b_{i+2}(t)$ and $a_{bin}(t)$ are continuous.

In Figure 4, the yellow component represents the depth camera. The red arrow indicates the direction of the UAV's advance, which is parallel to the overall direction of the pipeline. The distance between the UAV and the pipeline in the forward direction is $Z_1(t)$; The vertical distance between the UAV and the pipeline is $Z_2(t)$; The distance from the UAV to the pipeline in the direction inclined backwards at a 45° angle to the vertical direction is $Z_3(t)$. Forward and upward are plus directions.

We have:

$$Z(t) = \min\left\{Z_1(t), Z_2(t), \frac{\sqrt{2} Z_3(t)}{2}\right\} \quad (5)$$

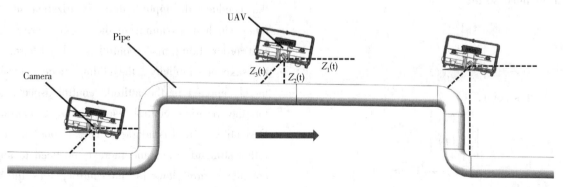

Figure 4 Z-direction planning for UAV

$$e_z(t) = Z(t) - l \tag{6}$$

$$\begin{cases} e_{z_1}(t) = Z_1(t) - l \\ e_{z_2}(t) = Z_2(t) - l \\ e_{z_3}(t) = \frac{\sqrt{2}}{2} Z_3(t) - l \end{cases}$$

UAV speed planning is as follows:

$$V_z(t) = \text{sgn}[Z_3(t) - Z_2(t)] \cdot \sqrt{\max\{V^2 - V_x^2(t) - V_y^2(t) - a, 0\}} \tag{7}$$

$$\begin{bmatrix} V_x(t) \\ V_y(t) \end{bmatrix} = \text{sgn}[e_{z_1}(t)] e^{-\left|\frac{1}{e_{z_1}(t)}\right|} k(\boldsymbol{b}_{i+1} - \boldsymbol{b}_i) \cdot \text{sgn}[e_{z_3}(t)] \frac{1}{ce^{z_2(t) - \frac{\sqrt{2}}{2} z_3(t)}} \tag{8}$$

Here:

$0 < a < V^2$, $1 < c$, $V_x(t)$, $V_y(t)$, $V_z(t)$: speed of UAV in x, y and z directions at time t.

The UAV tracks the pipeline in the Z direction, and there are three stages: the pipeline ascends, the pipeline is horizontal, and the pipeline descends.

When the pipe is horizontal,

$$\begin{cases} Z_1(t) \to +\infty \\ e_{z_2}(t) \to 0 \\ e_{z_3}(t) \to 0 \end{cases}$$

At this point,

$$\begin{cases} \begin{bmatrix} V_x(t) \\ V_y(t) \end{bmatrix} \to k(\boldsymbol{b}_{i+1} - \boldsymbol{b}_i) \cdot \frac{1}{c} \\ V_z(t) \to \sqrt{\max\{V^2 - V_x^2(t) - V_y^2(t) - a, 0\}} \end{cases}$$

Choose suitable values of a, c so that when the pipe orientation is horizontal,

$V_z(t) = 0$. And $V_x(t)$, $V_y(t)$ are continuously changing.

When the pipeline route changes from horizontal to an inclined stage,

$$\begin{cases} e_{z_1}(t) \to 0 \\ e_{z_2}(t) \uparrow \\ e_{z_3}(t) \uparrow \end{cases}$$

At this point,

$$\begin{cases} \begin{bmatrix} V_x(t) \\ V_y(t) \end{bmatrix} \to 0 \\ V_z(t) \to \sqrt{V^2 - a} \end{cases}$$

And $V_x(t)$, $V_y(t)$, $V_z(t)$ are continuously changing.

When the pipeline route changes from horizontal to a downward slope,

$$\begin{cases} Z_3(t) - Z_2(t) < 0 \\ e_{z_1}(t) \to +\infty \\ e_{z_3}(t) \to 0 \end{cases}$$

At this point,

$$\begin{cases} \begin{bmatrix} V_x(t) \\ V_y(t) \end{bmatrix} \to k(\boldsymbol{b}_{i+1} - \boldsymbol{b}_i) \cdot \frac{1}{ce^{z_2(t) - \frac{\sqrt{2}}{2} z_3(t)}} \\ V_z(t) \to -\sqrt{\max\{V^2 - V_x^2(t) - V_y^2(t) - a, 0\}} \end{cases}$$

Choose suitable values of a, c so that when the pipe trajectory decreases,

$V_x(t)$, $V_y(t)$ decrease, and $V_z(t)$ points downward.

And $V_x(t)$, $V_y(t)$, $V_z(t)$ are continuously changing.

2.4 UAV Control

The control block diagram for control of a quadrotor is shown in Figure 5. One can see in the figure that the inner loop is an attitude control loop. It is logical to infer that the dynamics of the inner loop must be faster than the dynamics of the outer loop.

To reduce control errors, thereby minimizing overshoot and reducing the possibility of danger, a Sliding Mode Control system is used. The Sliding Mode Control is a Non Linear Robust Control Technic which can be used to nullify the effect of disturbances and uncertainities. Sliding Mode Control is a type of Variable Structure Control (VSC). A high speed switching control is used to minimise the tracking error. The employed control is discontinuous. The control can be categorized into two sections, one that conducts the input/output linearization and the other which helps minimize the tracking error. The first-order sliding mode control is utilized here. For the sake of simplicity, the sliding mode control is merely applied in the attitude control aspect, and for the position control aspect, the Proportional Derivative Control is adopted. The altitude control is rather straightforward and there is no need to apply complex control laws for its control, thus the PD control is employed for this part.

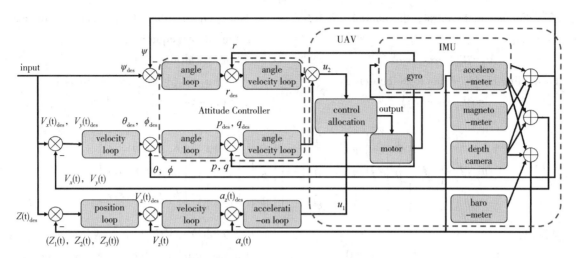

Figure 5 UAV Control Block Diagram

The dynamic equations of the Quadrotor are given in Equations 9 and 10. The Equation 8 concerns the dynamics of linear equation and the Equation 8 concerns the dynamics of angular motion.

$$m\begin{bmatrix} a_x(t) \\ a_y(t) \\ a_z(t) \end{bmatrix} = \begin{bmatrix} 0 \\ 0 \\ -mg \end{bmatrix} + R_b^e \begin{bmatrix} 0 \\ 0 \\ u_1 \end{bmatrix} \quad (9)$$

$$J\begin{bmatrix} \dot{p} \\ \dot{q} \\ \dot{r} \end{bmatrix} = u_2 - \begin{bmatrix} p \\ q \\ r \end{bmatrix} \times J\begin{bmatrix} p \\ q \\ r \end{bmatrix} \quad (10)$$

Where,

$$R_b^e = \begin{bmatrix} \cos\psi\cos\theta & \sin\phi\cos\psi\sin\theta - \cos\phi\sin\psi & \cos\phi\cos\psi\sin\theta + \sin\phi\sin\psi \\ \cos\theta\sin\psi & \sin\phi\sin\psi\sin\theta + \cos\phi\cos\psi & \sin\psi\sin\theta\cos\phi - \sin\phi\cos\psi \\ -\sin\theta & \sin\phi\cos\theta & \cos\phi\cos\theta \end{bmatrix}$$

R_b^e: Rotation matrix of any vector from the body coordinate system to the ground coordinate system.

We consider near hover configurations, so we can have the following approximation: $p \approx \dot{\phi}$, $q \approx \dot{\theta}$ and $r \approx \dot{\psi}$. The dynamics of angular motion is given in the following Equation 12:

$$J\dot{\omega} = u_2 - \dot{\omega} \times J\dot{\omega} \Rightarrow \ddot{\omega} = -J^{-1}\dot{\omega} \times J\dot{\omega} + J^{-1}u_2$$

(12)

We have:

$$\omega = \begin{bmatrix} \phi \\ \theta \\ \psi \end{bmatrix}, \quad \omega_c = \begin{bmatrix} \phi_c \\ \theta_c \\ \psi_c \end{bmatrix}$$

$$\begin{bmatrix} a_x(t) \\ a_y(t) \\ a_z(t) \end{bmatrix} = \begin{bmatrix} \dot{V}_x(t) \\ \dot{V}_y(t) \\ \dot{V}_z(t) \end{bmatrix}$$

$$\begin{bmatrix} \dot{\phi} \\ \dot{\theta} \\ \dot{\psi} \end{bmatrix} = \begin{bmatrix} 1 & \tan\theta\sin\phi & \tan\theta\cos\phi \\ 0 & \cos\phi & -\sin\phi \\ 0 & \dfrac{\sin\phi}{\cos\theta} & \dfrac{\cos\phi}{\cos\theta} \end{bmatrix} \begin{bmatrix} p \\ q \\ r \end{bmatrix} \quad (11)$$

$[p, q, r]$: body angular accelerations measured by the gyroscope; $[\phi, \theta, \psi]$: Roll, Pitch and Yaw angles; m: system mass; J: system moment of inertia; u_1: The thrust input; u_2: The moment input (3×1 vector) and

The error is defined as $e(t)$ as $e(t) = \omega - \omega_c$

From the Equation 12, it can be seen that the relative degree of the system is 2, so the First Order Sliding Mode Control variable can be defined as:

$$s = \dot{e} + \lambda e$$

Differentiating once,

$$\dot{s} = \ddot{e} + \lambda \dot{e}$$

$$\dot{s} = \ddot{\omega} - \ddot{\omega}_c + \lambda(\dot{\omega} - \dot{\omega}_c)$$

$$\dot{s} = -J^{-1}\dot{\omega} \times J\dot{\omega} + J^{-1}u_2 - \ddot{\omega}_c + \lambda(\dot{\omega} - \dot{\omega}_c) \quad (13)$$

$$\dot{s} = -J^{-1}\dot{\omega} \times J\dot{\omega} - \ddot{\omega}_c + \lambda(\dot{\omega} - \dot{\omega}_c) + J^{-1}u_2$$

$$\dot{s} = \alpha_{SM} + \beta_{SM}u_2$$

Where $\alpha_{SM} = -J^{-1}\dot{\omega} \times J\dot{\omega} - \ddot{\omega}_c + \lambda(\dot{\omega} - \dot{\omega}_c)$ and

$\beta_{SM} = J^{-1}$. For nominal control, the control law will be chosen as:

$$u_2 = \beta_{SM}^{-1}(-\alpha_{SM} + v)$$

For sliding mode, v has to be chosen as:

$$v = -K_s \cdot sign(s)$$

Here, K_s is the sliding mode gain matrix. Using the above equations, the control input for controlling the altitude is:

$$u_2 = \beta_{SM}^{-1}[-\alpha_{SM} - K \cdot sign(s)]$$

For the given problem at hand:

$$u_2 = J\left(-\left(-J^{-1}\begin{bmatrix}\dot{\phi}\\\dot{\theta}\\\dot{\psi}\end{bmatrix} \times J\begin{bmatrix}\dot{\phi}\\\dot{\theta}\\\dot{\psi}\end{bmatrix} - \begin{bmatrix}\ddot{\phi}_c\\\ddot{\theta}_c\\\ddot{\psi}_c\end{bmatrix}\right) + \lambda\left(\begin{bmatrix}\dot{\phi}\\\dot{\theta}\\\dot{\psi}\end{bmatrix} - \begin{bmatrix}\dot{\phi}_c\\\dot{\theta}_c\\\dot{\psi}_c\end{bmatrix}\right)\right) - \begin{bmatrix}K_s \cdot sign(s_1)\\K_s \cdot sign(s_2)\\K_s \cdot sign(s_3)\end{bmatrix} \quad (14)$$

Where,

$$\lambda = \begin{bmatrix}\lambda_1 & 0 & 0\\0 & \lambda_2 & 0\\0 & 0 & \lambda_3\end{bmatrix}, \quad K_s = \begin{bmatrix}K_1 & 0 & 0\\0 & K_2 & 0\\0 & 0 & K_3\end{bmatrix}$$

The proof of system stability is as follows:

Select the following Lyapu-nov function candidate

$$F(s) = \frac{s^2}{2} = \frac{[\dot{\omega} - \dot{\omega}_c + \lambda(\omega - \omega_c)]^2}{2} \geq 0 \quad (15)$$

$$\dot{F}(s) = s\dot{s} = [\dot{\omega} - \dot{\omega}_c + \lambda(\omega - \omega_c)]\{-J^{-1}\dot{\omega} \times J\dot{\omega} - \ddot{\omega}_c + \lambda(\dot{\omega} - \dot{\omega}_c) + J^{-1}\beta_{SM}^{-1}[-\alpha_{SM} - K_s \cdot sign(s)]\}$$

When $\dot{\omega} - \dot{\omega}_c + \lambda(\omega - \omega_c) \geq 0$, $-J^{-1}\dot{\omega} \times J\dot{\omega} - \ddot{\omega}_c + \lambda(\dot{\omega} - \dot{\omega}_c) + J^{-1}\beta_{SM}^{-1}[-\alpha_{SM} - K_s \cdot sign(s)] \leq 0$

When $\dot{\omega} - \dot{\omega}_c + \lambda(\omega - \omega_c) \leq 0$, $-J^{-1}\dot{\omega} \times J\dot{\omega} - \ddot{\omega}_c + \lambda(\dot{\omega} - \dot{\omega}_c) + J^{-1}\beta_{SM}^{-1}[-\alpha_{SM} - K_s \cdot sign(s)] \geq 0$

Choose suitable values for λ, K_s to ensure stability of the control system.

3 RESULTS AND DISCUSSION

As shown in Table 1, proposed method has a space complexity of O(1), as the storage space occupied does not increase with the increase of path distance. Meanwhile, due to the absence of complex loops in our proposed method, its time complexity is O(1). Therefore, compared with the traditional Dijkstra algorithm, A* algorithm, D* Lite algorithm, and Prim algorithm, proposed method has the smallest time and space complexities.

Table 1 The spatiotemporal complexity of different path planning methods

Model	Proposedmethod	Dijkstra	A*	D* Lite	Prim
Time complexity	O(1)	O(n·log(m))	O(n·log(m))	O(n²)	O(n²)
Spatial complexity	O(1)	O(n)	O(n)	O(n)	O(n)

To verify the control effect, a simulation model was established using MATLAB. The simulation path is a circular motion in the horizontal direction, because for the x and y directions, the pipeline is exposed within the field of view. TheB spline curve driven by physical state ensures the smoothness of the curve. At the same time, the circular motion process can simulate the continuous changes in the speed direction of the drone, making the simulation space small and easy to visualize. If the pipeline in the z-direction bends 90 degrees, the resulting undulations will cause a sudden change in the z-direction of the planned path. Therefore, the disturbance is a positional deviation of z=-8m, which is used to simulate significant changes in the position of the pipeline within the field of view, applied as a step input at time t=0s.

Proportional-Integral-Derivative (PID) control and variable structure sliding mode control are commonly used control methods for UAVs. PID control, known for its simplicity and computational efficiency, is widely used in UAV applications. Sliding-mode control, a nonlinear robust technique, effectively mitigates disturbances and uncertainties. The trajectory tracking results for the two different control methods are shown in Figure 6.

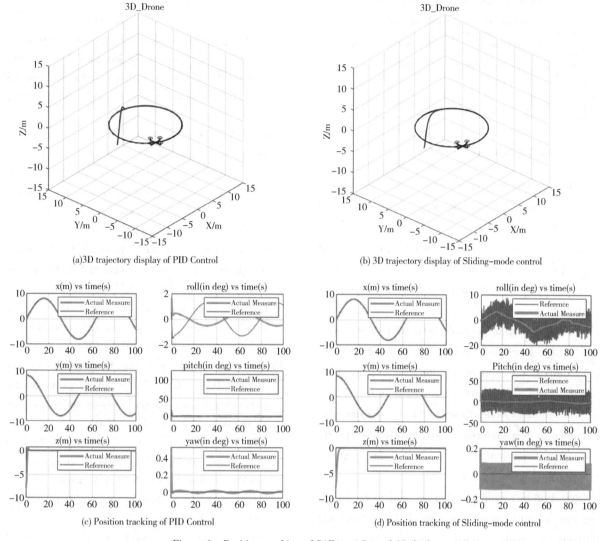

Figure 6 Position tracking of Different Control Methods

To optimize detection accuracy, the UAV must maintain close proximity to the pipeline while fulfilling detection requirements, necessitating the minimization of overshoot to reduce the required safety margin.

Figure 6(c) illustrates that under extreme conditions, the traditional PID control exhibited a position overshoot of 267mm (Parameters are listed in Table 2). Considering the UAV's outer sphere radius of 200mm, this overshoot necessitates a minimum safety distance of 467mm. However, detection accuracy diminishes significantly at flight heights exceeding 500mm, attributable to the reduced size of potential defects in the image (less than 1/32 of the frame). Considering the disturbances and fluctuations inherent in normal flight operations, this method proves inadequate for the task at hand.

Sliding-mode control, as illustrated in Figure 6(d), demonstrated superior performance with a minimal overshoot of 51mm, permitting a reduced minimum safety distance of 251mm (Parameters are listed in Table 3). This control methodology exhibited rapid response characteristics and non-divergent behavior, rendering it highly suitable for pipeline inspection tasks.

Based on these findings, we recommend a rated flight height of 450mm with a minimum safety distance of 400mm, which satisfies the requirements of at least two of the evaluated control methodologies.

Table 2 The control parameter of PID control method

Position controller		Attitude controller	
Kp(x) = 10	Kd(x) = 10	Kp(ϕ) = 625	Kd(φ) = 50
Kp(y) = 10	Kd(y) = 10	Kp(θ) = 625	Kd(θ) = 50
Kp(z) = 10	Kd(z) = 10	Kp(ψ) = 625	Kd(ψ) = 50

Table 3 The control parameter of
Sliding-mode control method

Position controller		Attitude controller	
Kp(x) = 10	Kd(x) = 10	λ_1 = 100	K_1 = 55
Kp(y) = 10	Kd(y) = 10	λ_2 = 100	K_2 = 55
Kp(z) = 10	Kd(z) = 10	λ_3 = 100	K_3 = 55

4 CONCLUSION

1. The proposed depth image-based path generation methodology demonstrates robust real-time path planning capabilities.

2. The proposed depth image preprocessing technique effectively mitigates background noise, significantly enhancing image quality and subsequent analysis.

3. The B spline curve driven by physical state fulfills the kinematic and dynamic constraints of UAV and possesses relatively low time-space complexity, with a time-space complexity of $O(1)$.

4. Sliding-mode control exhibits superior performance in overshoot reduction and response speed, rendering it particularly well-suited for complex pipeline inspection scenarios.

5. Future research endeavors will focus on integrating characteristics of overhead obstructions and obstacles to further optimize and refine path planning strategies for diverse inspection environments.

References

[1] Chang L, Fan S. Verification and evaluation methods for pipeline leak detection algorithms. InJournal of Physics: Conference Series 2024 Jun 1 (Vol. 2782, No. 1, p. 012008). IOP Publishing.

[2] Qian G, Niffenegger M, Zhou W, Li S. Effect of correlated input parameters on the failure probability of pipelines with corrosion defects by using FITNET FFS procedure. International Journal of Pressure Vessels and Pixi. 2013 May 1; 105: 19-27.

[3] Yu C, Yang Y, Cheng Y, Wang Z, Shi M, Yao Z. UAV-based pipeline inspection system withSwin Transformer for the EAST. Fusion engineering and design. 2022 Nov 1; 184: 113277.

[4] Zhou X, Wang Z, Ye H, Xu C, Gao F. Ego-planner: Anesdf-free gradient-based local planner for quadrotors. IEEE Robotics and Automation Letters. 2020 Dec 28; 6(2): 478-85.

[5] Joeaneke P C, Val O O, Olaniyi O O, Ogungbemi O S, Olisa A O, Akinola O I. Protecting autonomous UAVs from GPS spoofing and jamming: A comparative analysis of detection and mitigation techniques. Journal of Engineering Research and Reports. 2024, 26(10), 71-92.

[6] Chang Y, Cheng Y, Manzoor U, Murray J. A review of UAV autonomous navigation in GPS-denied environments. Robotics and Autonomous Systems. 2023, 104533.

[7] Macario Barros A, Michel M, Moline Y, Corre G, Carrel F. A comprehensive survey of visual slam algorithms. Robotics. 2022, 11(1), 24.

[8] Kazerouni I A, Fitzgerald L, Dooly G, Toal D. A survey of state-of-the-art on visual SLAM. Expert Systems with Applications. 2022, 205, 117734.

[9] Chen W, Shang G, Ji A, Zhou C, Wang X, Xu C, et al. An overview on visual slam: From tradition to semantic. Remote Sensing. 2022, 14(13), 3010.

[10] Li X, Liu C, Wang Z, Xie X, Li D, Xu L. Airborne LiDAR: state-of-the-art of system design, technology and application. Measurement Science and Technology. 2020, 32(3), 032002.

[11] Jia J, Guo K, Yu X, Guo L, Xie L. Agile flight control under multiple disturbances for quadrotor: Algorithms and evaluation. IEEE Transactions on Aerospace and Electronic Systems. 2022, 58(4), 3049-3062.

[12] Bu X, Lv M, Liu Z. Intelligent safety flight control with variable prescribed performance. IEEE Transactions on Aerospace and Electronic Systems. 2023.

[13] Yu X, Zhou X, Guo K, Jia J, Guo L, Zhang Y. Safety flight control for a quadrotor UAV using differential flatness and dual-loop observers. IEEE Transactions on Industrial Electronics. 2021, 69(12), 13326-13336.

[14] Zhang WL, Wang WY, Ji GY, Shi H, Pan J, Niu PJ, Su GQ. Research on the safe distance of UAV carrying manipulator for inspection and repair of transmission line. InE3S web of conferences 2020 (Vol. 218, p. 01031). EDP Sciences.

[15] Cao N, Liu Y. High-noise grayscale image denoising using an improved median filter for the adaptive selection of a threshold. Applied Sciences. 2024 Jan 11; 14(2): 635.

[16] Liu J, Xie J, Li B, Hu B. Regularized cubic B-spline collocation method with modified L-curve criterion for impact force identification. IEEE Access.

2020 Feb 14; 8: 36337-49.

[17] Dijkstra, E. W. A note on two problems inconnexion with graphs. In Edsger Wybe Dijkstra: his life, work, and legacy. 2022 (pp. 287-290).

[18] Tang G, Tang C, Claramunt C, Hu X, Zhou P. Geometric A-star algorithm: An improved A-star algorithm for AGV path planning in a port environment. IEEE access. 2021 9, 59196-59210.

[19] Ren Z, Rathinam S, Likhachev M, Choset H. Multi-objective path-based D * lite. IEEE Robotics and Automation Letters. 2022 7(2), 3318-3325.

[20] Mohan A, Leow W X, Hobor A. Functional correctness of C implementations of Dijkstra's, Kruskal's, and Prim's algorithms. In Computer Aided Verification: 33rd International Conference, CAV 2021, Virtual Event, July 20-23, 2021, Proceedings, Part II 33 (pp. 801-826). Springer International Publishing.

[21] Yazdannik S, Sanisales S, Tayefi M. A novel quadrotor carrying payload concept via PID with Feedforward terms. International Journal of Intelligent Unmanned Systems. 2024 May 14.

[22] Zhang C, Xu J, Niu Y. Event-triggered adaptive deep neural network sliding mode control design for unmanned aerial vehicle systems. Franklin Open. 2024 Jun 26: 100120.

Research on Obstacle Evasion and Joint Angle-Minimized Trajectory for Pipeline Grinding Robot based on PSO

Yan Zhouyu Zhao Hong

[China University of Petroleum (Beijing)]

Abstract Inner wall defects are primary cause of pipeline safety incidents. Manual grinding exhibit several limitations, including low precision, inefficiency, and potential safety hazards. In this paper, an innovative pipeline grinding robot is designed, which consists of one linear moving joint and four rotating joints arranged in a $p_z-r_z-r_z-r_z-r_y$ configuration. This pipeline grinding robot can efficiently grind the inner wall of the pipeline with a diameter of 700mm to 1000mm. Taking the robot trajectory between defect points as the research object, this paper establishes the cost function based on the joint angle, and uses the Particle Swarm Optimization (PSO) algorithm to plan the trajectory. Aiming at the narrow space inside the pipeline, this paper applies the equivalent volume method to calculate the maximum distance between the centreline and the grinding module, thereby enabling autonomous obstacle avoidance by introducing a penalty coefficient. In order to verify the effectiveness of the proposed method, the trajectory planning of 8 trajectory points inside a pipeline with a diameter of 780mm is carried out in Matlab. The experimental results illustrate that the proposed method effectively prevents rigid contact between the grinding module and the pipeline wall. Compared to the quintic non-uniform rational B-spline (NURBS) method, proposed method achieves a 13.8% reduction in total joint angle.

Keywords Inner wall defects; Pipeline grinding robot; Trajectory planning; Particle swarm optimization; Obstacle avoidance; Matlab

1 Introduction

At present, as the main carrier of energy transportation and living resources, the safety of pipeline operation has attracted much attention. The main cause of pipeline safety accidents is the inner wall defects of pipelines, as shown in Figure 1 (a). For the inner wall defects, the cost of replacing the pipeline is high, and the efficiency, quality and safety of manual grinding cannot be guaranteed. As a research hotspot in recent years, in-pipeline robot has gradually become a mature means of in-pipeline inspection and maintenance. Xu et al. designed a new type of wheeled pipeline grinding robot, which can move axially and grind circumferentially inside pipelines with a diameter of 550 ~ 714mm. Yu developed a modular in-pipeline robot specifically for welding seams inside nuclear power plant pipelines. Zhu et al. added a two-degree-of-freedom linear motion platform at the end of a six-degree-of-freedom industrial robot to achieve accurate grinding of pipeline welds. Zhang et al. proposed an adaptive crab-shaped grinding robot with excellent pass-ability and adaptability under different working environments. Most of the pipeline grinding robots focus on the grinding of welds, and there is little research on the grinding of various complex defects in pipeline.

Pipeline defects frequently manifest as defect clusters, where multiple independent defects occur within a specific segment of the pipeline, as illustrated in Figure 1 (b). Compared with the weld, the distribution and shape of defects are more complex, which requires more stringent obstacle avoidance ability and efficiency of grinding robots. In order to avoid rigid contact between obstacle areas and working components, Liu et al. used an improved RRT* algorithm to quickly search for the shortest expected path to effectively avoid obstacles. Zou et al. proposed a method combining A* algorithm and improved ant colony algorithm in configuration space to plan the expected trajectory. Based on the idea of task priority transformation, Yan et al. realized the trajectory planning of the robot arm with obstacle avoidance as the main and trajectory tracking as the auxiliary. Most existing studies primarily address simple obstacle avoidance in free space, with limited attention given to optimizing trajectories for minimal joint angles in confined spaces within pipelines.

To address the aforementioned challenges, a flexible five-degree-of-freedom grinding robot is designed in this paper, which can perform the grinding work efficiently in pipeline. Regarding the grinding trajectory, we employ particle swarm optimization (PSO) to plan the minimum joint angle trajectory, thereby achieving comprehensive obstacle avoidance in narrow pipeline spaces.

2 Structure design of the grinding robot

The objective of the in-pipeline grinding robot is to perform high-quality and flexible grinding on the inner wall of pipeline with diameters ranging from 700mm to 1000mm. The structural of the grinding robot designed in this paper is presented in Figure 1(a). Figure 1(b) shows the grinding state of the robot. Figure 1(c) depicts the walking and support module, which facilitates the robot's movement during non-grinding operations and ensures its fixation during grinding. The variable diameter structure adapts to varying pipeline diameters by adjusting the working space. Figure 1(d) shows the grinding module, which consists of one linear moving joint and four rotating joints arranged in a $p_z-r_z-r_z-r_z-r_y$ configuration. Both the walking and grinding modules are equipped with high-definition scanners for inspecting the inner wall of the pipeline and monitoring the grinding process. The grinding range can cover the pipeline wall within 400mm ahead of the robot. This design enables the robot to effectively detect and grind inner wall defects, ensuring the integrity and quality necessary for subsequent repair operations.

The comprehensive operational procedure of the robot is outlined as follows: (1) Adjust the dimensions of the walking module for the robot and apply the preload force. Subsequently, position the grinding module horizontally and place it at the pipeline origin. (2) The scanner, motor and driver are started. Utilize the scanner to inspect the inner wall of the pipeline. (3) The robot capture point cloud data of the inner wall and transmits information to the

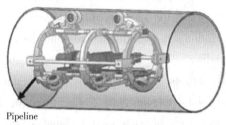
(a) The initial attitude of the robot

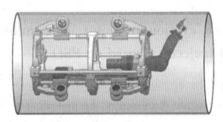
(b) attitude of the robot

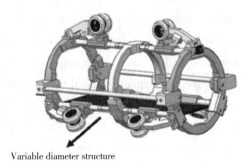
(c) Walking modules

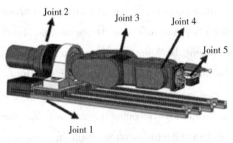
(d) Grinding module

Figure 1 Structure of grinding robot

controller. Based on the planned trajectory, the grinding module performs highly accurate and efficient grinding operations. (4) Upon completion of the task, the grinding module retracts into the robot and is positioned horizontally within it. The robot as a whole continues to move forward.

For the pipeline grinding robot, kinematic analysis of the grinding module is essential to ensure the safety and efficiency of the operational process. As illustrated in Figure 2, its structure is simplified using the Denavit–Hartenberg (D–H) coordinate system. Joint 1 exhibits linear motion along the positive direction of the Z-axis. Joints 2, 3, 4, and 5 are capable of rotational movement around their respective Z-axes. The D–H parameters for each connecting rod are detailed in Table 1.

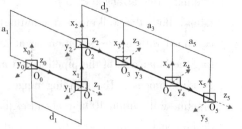 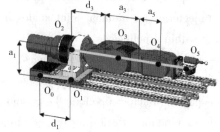

Figure 2 D-H coordinate system of grinding module

Table 1 Connecting rod parameter table

Link i	Joint angle $\theta_i/(°)$	Joint twist angle $\alpha_{i-1}/(°)$	Joint deviation distance a_{i-1}/mm	Joint distance d/mm	Variable
1	$\theta_1(0)$	$\alpha_0(0)$	$a_0(0)$	$d_1(860)$	—
2	θ_2	$\alpha_1(90)$	$a_1(104)$	$d_2(0)$	θ_2
3	θ_3	$\alpha_2(0)$	$a_2(0)$	$d_3(325)$	θ_3
4	θ_4	$\alpha_3(90)$	$a_3(390)$	$d_4(0)$	θ_4
5	θ_5	$\alpha_4(0)$	$a_4(0)$	$d_5(150)$	θ_5

3 Obstacle evasion and joint angle-minimized trajectory based on PSO

Through the structural and kinematic analysis of the grinding module discussed in the preceding section, we can ascertain the feasibility of deploying a grinding robot within pipeline. However, given that the operation of the grinding robot constitutes a precision machining process, it is imperative to ensure both the efficiency and safety of its operations. Continuous movement between grinding tracing points forms the foundation of the grinding module's functionality. When multiple defects are present in the workspace, efficiency becomes paramount for the robot, as random trajectories often result in numerous redundant actions. Moreover, the confined space within the pipeline, characterized by its cylindrical internal structure, imposes certain limitations on the movement of the grinding module. Unnecessary rigid contact between the robot and the pipeline wall can lead to operational failure. Consequently, this chapter aims to plan the grinding trajectory for the pipeline grinding robot when confronted with multiple defects, optimizing the trajectory to minimize joint angles while avoiding obstacle spaces.

3.1 Calculation of obstacle avoidance within the pipeline's internal space

When the grinding robot performs the grinding operation, the walking module and the Joint 1 of the grinding module are in a static state, so this paper only carries out obstacle avoidance calculation for the Joint 2 to Joint 5 and connecting rod. First, the center point of each joint is labelled $P_{i,j} = \{p_{i,j} = (x_{i,j}, y_{i,j}, z_{i,j}), i \in [2, 3, 4, 5], j = 0\}$, and each center point P_i is set to a sphere C of radius r as the equivalent volume of the real model of the joint. Secondly, the connecting rod between the center points is equally divided into 10 parts, where each new endpoint is denoated as $P_{i,j} = \{p_{i,j} = (x_{i,j}, y_{i,j}, z_{i,j}), i \in [2, 3, 4, 5], j \in [1, 2, 3, \cdots, 10]\}$, the calculation process Equation (1) shows:

$$P_{i,j} = P_i + \frac{j}{10} \cdot (P_{i+1} - P_i) \qquad (1)$$

Similarly, $P_{i,j}$ is set to a sphere C_0 of radius r_0

as the equivalent volume for the real model of the connecting rods. The joints and connecting rods are converted into a number of connected spheres with a certain volume. Finally, the working space is set to the inside of the pipeline with a radius of r_p, and the inner wall of the pipeline is set to the obstacle area. In order to accurately calculate the distance between each sphere and the pipeline wall during the movement of the grinding module, this paper adopts the calculation of the distance between each sphere and the Z-axis to determine, as shown in Figure 4. The calculation process and judgment criteria are shown in Equation (2) and Equation (3):

$$D_{i,j} = \sqrt{x_{i,j}^2 + y_{i,j}^2} \quad (2)$$

$$H = \begin{cases} 0° & \text{if } D_{i,j} + r < r_p \\ 100° & \text{if } D_{i,j} + r \geq r_p \end{cases} \quad (3)$$

In the XY plane of each node, if the sum of the distance D_i from the Z-axis and the radius r of the sphere is less than the pipeline's radius r_p, it is determined that the node does not make contact with the pipeline wall, and the additional angle of consumption of the current attitude is 0°. Conversely, if the sum of the distance D_i and the radius r is greater than or equal to the pipeline's radius r_p, it is concluded that the attitude is in contact with the pipeline wall, and the additional angle of consumption of the current attitude is 100°.

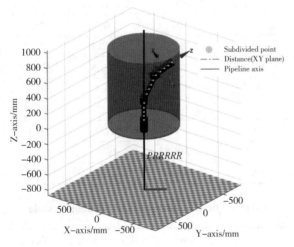

Figure 3　Obstacle avoidancecalculation

3.2　Joint angle-minimized planning based on PSO algorithm

When the grinding module solves the kinematics of tracing points, multiple inverse solutions corresponding to various attitudes may arise. Although obstacle avoidance planning can eliminate some attitudes that would contact the inner wall, random attitude selection is likely to result in redundant actions, thereby reducing the efficiency of the grinding module. This paper employs the PSO algorithm for trajectory planning across multiple tracing points to minimize the sum of joint angles.

Initially, the entire trajectory traversed by multiple points can be considered as the aggregate of trajectories between adjacent points. The trajectories between these points are planned using quintic non-uniform rational B-spline (NURBS). Subsequently, a cost function based on minimizing the angle of each joint is established, as presented in Equation (4):

$$Cost(Q) = \sum_{i=2}^{5} q = \sum_{i=2}^{5} \int_{t_0}^{t_f} |\dot{q}(t)| dt + H \quad (4)$$

Where, q is the sum of the absolute value of the angular change of a single joint between adjacent trajectory points. H is the obstacle avoidance detection label, which is equivalent to a penalty factor. Ultimately, the PSO algorithm is employed to plan the trajectory. Within this framework, particles are dispersed throughout a multidimensional search space. Each particle represents a potential solution in the problem and encapsulates the associated control parameters. The cost function serves as the fitness function, assessing the quality of each particle's solution.

The dynamic behaviour of particles within the search space is influenced by two pivotal factors: social experience and individual experience. Social experience encapsulates the exchange of information among particles, wherein the global optimal solution, denoted as g_{best}, signifies the best solution discovered thus far by the entire population and is disseminated across all particles. Individual experience pertains to the personal optimal solution for each particle, represented as p_{best}, which documents its own historical best position. Each particle subsequently updates its velocity and position in accordance with Equations (5) and (6):

$$v_i(t+1) = w \cdot v_i(t) + c_1 \cdot rand \cdot [p_{i,best} - x_i(t)]$$
$$+ c_2 \cdot rand \cdot [g_{best} - x_i(t)] \quad (5)$$
$$x_i(t+1) = x_i(t) + v_i(t+1) \quad (6)$$

Where $v_i(t)$ denotes the velocity of particle i at time t, and $x_i(t)$ represents the position of particle i at time t. $p_{i,best}$ signifies the individual optimal position for particle i, while g_{best} indicates the global optimal position for the entire group. w is the inertia weight, which governs the impact of the previous velocity on the current velocity. c_1 and c_2 are learning factors that influence the balance between exploration and exploitation. $rand$ is a random number within the interval [0, 1], reflecting the influence of both individual and social experience on the search process. At the same time of algorithm iteration, the fitness value of the optimal solution is calculated and recorded. The process terminates when the maximum number of iterations is reached. Ultimately, the global optimal solution set is outputted to generate joint angle – minimized trajectory. The workflow of the proposed method is illustrated in the Figure 4.

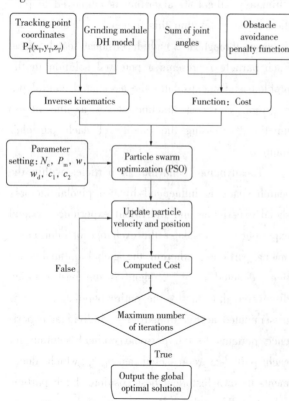

Figure 4　The flow of the proposed method

4　Trajectory planning and obstacle avoidance simulation experiment

4.1　Parameter Settings

In order to verify the accuracy of the proposed method, this paper uses Matlab to simulate the obstacle avoidance function and trajectory planning function of the grinding module. The simplified model of the grinding module is constructed according to the information of each joint and connecting rod in Table 1, and the working area is inside the pipeline with a radius of 390mm. Set 8 required tracing points, starting point and ending point are the end position of the Joint 5 when grinding the initial attitude of the module. The remaining 6 tracing points are the positions of defects in the inner wall of the pipeline. The coordinates of each point are shown in the Table 2, and the inverse solution of the corresponding grinding module is shown in the Table 3. Joint 2 has a movement range of [−180°, 180°]; Joint 3 has a movement range of [−120°, 120°]; Joint 4 has a movement range of [−110°, 110°]; and Joint 5 has a movement range of [−180°, 180°]. The equivalent sphere radius of joint and connecting rod is set to 60mm. The established working condition model is shown in the Figure 6.

Table 2　The coordinates of the tracing point

Tracing point	X-axis/mm	Y-axis/mm	Z-axis/mm
1	0	0	865
2	−293.9	−252.1	426.7
3	235.7	−306.5	396.2
4	367.5	123.1	541.6
5	−186.5	341.3	649.0
6	−208.1	327.7	509.0
7	−386.2	28.7	617.3
8	0	0	865

Table 3　Inverse kinematics of grinding module

	Joint 2/ (°)	Joint 3/ (°)	Joint 4/ (°)	Joint 5/ (°)
Tracing point 1	0.0	90.0	90.0	0
Tracing point 2	−139.4	36.5	−7.0	0
Tracing point 3	−52.4	32.5	−9.9	0

续表

	Joint 2/(°)	Joint 3/(°)	Joint 4/(°)	Joint 5/(°)
Tracing point 4	18.5	48.6	11.1	0
Tracing point 5	118.7	52.1	44.2	0
Tracing point 6	122.4	45.7	4.9	0
Tracing point 7	175.7	52.3	31.3	0
Tracing point 8	0.0	90.0	90.0	0

Before using the PSO algorithm to plan the trajectory, the parameters that affect the output results need to be set. The detailed parameters set in the simulation experiment are shown in the Table 4.

Table 4 Parameter setting of PSO algorithm

Descriptor	Parameter	Value
Number of iterations	N_c	100
Population size	P_m	50
Inertia weight	w	1
Inertia weight damping ratio	w_d	0.98
Personal learning coefficient	c_1	1.5
Global learning coefficient	c_2	1.5

4.2 Analysis of simulation results

The proposed method sequentially plans the 7 sub-trajectories connecting eight trajectory points. By employing Equation (4) as the cost function, the algorithm continuously iterates to find the trajectory that minimizes the sum of angles, as illustrated in the Figure 5. During the first 18 iterations, no improvement was observed, and the total angle remained approximately at 2400°. From the 19th to the 65th iteration, the algorithm progressively identified better solutions, updating the global optimum, and reducing the calculated minimum angle sum to approximately 1815°. Finally, from the 66th to the 100th iteration, the minimum angle sum stabilized at 1814.7°, indicating convergence to the optimal solution for the problem addressed.

The solution derived from the aforementioned process is utilized for simulation experiments. The simulation data for the 7 sub-trajectories is presented in the Figure 6. To more intuitively evaluate the

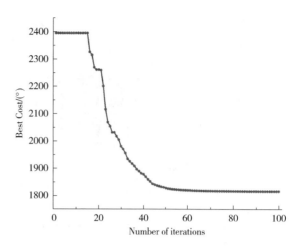

Figure 5 Iterative process of PSO algorithm

obstacle avoidance performance of the proposed method, the spatial extent occupied by each connecting rod and joint during the movement has been illustrated, as shown in the Figure 7. It is evident that there is no collision between the grinding module and the pipeline wall throughout the movement process. Moreover, the designed penalty factor effectively eliminates potential collision solutions, ensuring safe operation of the grinding module.

To further substantiate the advantages of the proposed method, this paper employs quintic NURBS for trajectory planning involving the same 8 tracing points. The changes in joint angles under B-spline trajectory and PSO trajectory are compared, as illustrated in the Figure 8. The sum of joint angles required for the B-spline trajectory and the PSO trajectory is 2104.4° and 1814.7°, respectively. This demonstrates that the proposed method can reduce the sum of joint angle movement by 13.8% while ensuring collision avoidance.

5 Conclusion

In this paper, a new type of pipeline grinding robot is designed, which can grind the inner wall of pipeline with diameter from 700mm to 1000mm with high precision and high efficiency. To enhance the operational efficiency when addressing multiple defects, this study develops a cost function based on the sum of joint angles and employs the PSO algorithm for trajectory planning. Furthermore, to prevent rigid contact between the pipeline wall and the

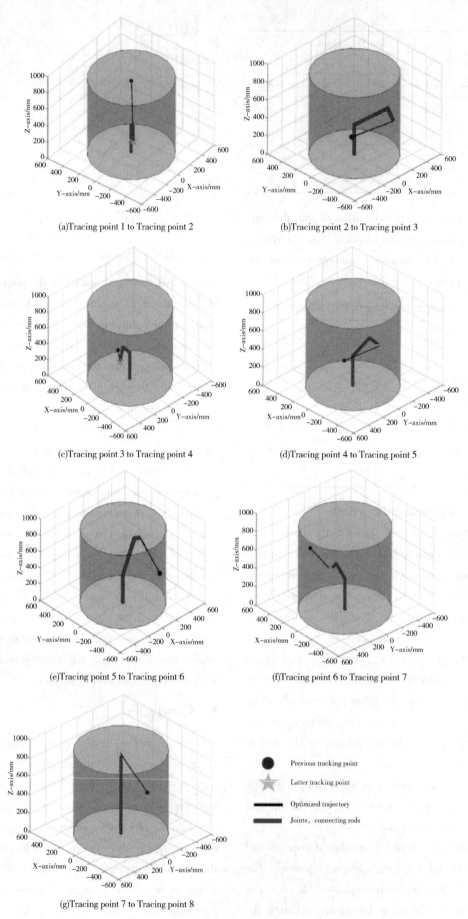

(a)Tracing point 1 to Tracing point 2

(b)Tracing point 2 to Tracing point 3

(c)Tracing point 3 to Tracing point 4

(d)Tracing point 4 to Tracing point 5

(e)Tracing point 5 to Tracing point 6

(f)Tracing point 6 to Tracing point 7

(g)Tracing point 7 to Tracing point 8

Figure 6　The trajectory planned by the proposed method

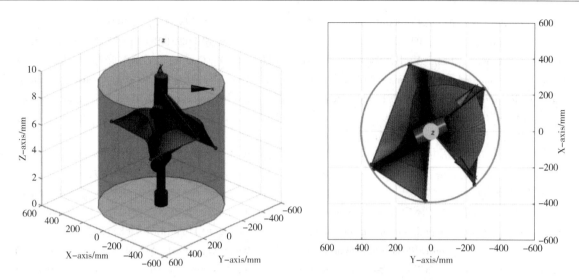

Figure 7 The obstacle avoidance effect of the proposed method

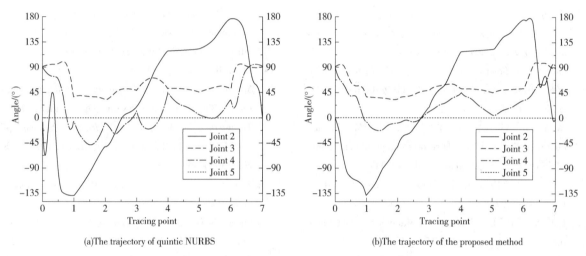

(a) The trajectory of quintic NURBS

(b) The trajectory of the proposed method

Figure 8 The angle of each joint changes.

robot, the equivalent volume method is utilized to simplify the robot model. A penalty factor is constructed by simplifying the distance between each sphere in the model and the pipeline's central axis, which is then integrated into the PSO algorithm. Simulation experiments are conductedby Matlab, selecting 8 defects as tracing points of a 780mm diameter pipeline. Compared to the quintic NURBS method, the results demonstrate that the proposed method can reduce the sum of joint angle by 13.8% while mitigating adverse effects caused by confined pipeline spaces. Future work will focus on optimizing the grinding sequence among multiple tracing points to further enhance the efficiency of the robot.

6 Funding source

This work is supported by National Key Research and Development Program of China (2024YFE0211500); the National Natural Science Foundation of China (Grant No. 51575528); the Science Foundation of China University of Petroleum, Beijing (No. 2462020XKJS01).

References

[1] Xu. H, Li. Y, Zhou. T, Lan. F, Zhang. L, An overview of the oil and gas pipeline safety in China, *Journal of Industrial Safety*. 1(2024) 100003, https://doi.org/10.1016/j.jinse.2024.100003.

[2] Yan. Z, Zhao. H, Miao. X, Gao. G, Design and Trajectory Optimization of a Large-Diameter Steel Pipe Grinding Robot, *Pipeline Systems Engineering and Practice*. 15(2024) 04024020, https://doi.org/10.1061/JPSEA2.PSENG-1581.

[3] Ahmed. Z, Amur. A. Y, Riadh. Z, Issam. B, Multi-objective design optimization of an in-pipe inspection

robot, *Franklin Open*. 6(2024) 100071, https://doi.org/10.1016/j.fraope.2024.100071.

[4] Xu. Z. L, Yang. J, Feng. Y. H, Shen. C. T, A wheel-type in-pipe robot for grinding weld beads, *Advances in Manufacturing*. 5(2017) 182-190, https://doi.org/10.1007/s40436-017-0174-9.

[5] Yu. P, Research and application of piping inside grinding robots in nuclear power plant, *Energy Procedia*. 127(2017) 54-59, http://dx.doi.org/10.1016/j.egypro.2017.08.066.

[6] Zhu. Y, He. X, Liu. Q, Guo. W, Semiclosed-loop motion control with robust weld bead tracking for a spiral seam weld beads grinding robot, *Robotics and Computer-integrated Manufacturing*. 73(2022) 102254, https://doi.org/10.1016/j.rcim.2021.102254.

[7] Zhang. L, Yang. Z, Ni. X, Cao. A, Design and development of adaptive crabshaped pipe grinding robot, 2023 *8th International Conference on Intelligent Informatics and Biomedical Sciences* (*ICIIBMS*). (2023) 166-168, https://doi.org/10.1109/ICIIBMS60103.2023.10347746.

[8] Wang. Z, Li. X, Liu. Y, Lv. Y, Li. M, Enhancing precision of defect 3D reconstruction in metal ultrasonic testing through point cloud completion, *Ultrasonics*. 142(2024) 107381, https://doi.org/10.1016/j.ultras.2024.107381.

[9] Liu. X, Cao. L, Simulation of Manipulator Path Planning Based on Improved RRT* Algorithm, 2022 *5th International Conference on Pattern Recognition and Artificial Intelligence* (*PRAI*). (2022) 1115-1120, https://doi.org/10.1109/PRAI55851.2022.9904064.

[10] Zou. J, Liu. C, Chen. L, Zhang. J, Multi-objective Operation Path Planning Method for Robotic Arm Based on Configuration Space, 2023 *5th International Symposium on Robotics & Intelligent Manufacturing Technology* (*ISRIMT*). (2023) 306-311, https://doi.org/10.1109/ISRIMT59937.2023.10428515.

[11] Yan. Y, Cui. H, Lin. B, Han. P, A Task-Priority-Transition-Based Obstacle Avoidance Strategy for Space Manipulator, 2020 *Chinese Automation Congress* (*CAC*), (2020) 3857-3862, https://doi.org/10.1109/CAC51589.2020.9326869.

[12] Wen. C, Wang. L, Ren. X, General particle swarm optimization algorithm, 2023 *IEEE 2nd International Conference on Electrical Engineering, Big Data and Algorithms* (*EEBDA*). (2023) 1204-1208, https://doi.org/10.1109/EEBDA56825.2023.10090725.

[13] Wang. H, Qi. H, Xu. M, Tang. Y, Yao. J, Yan. X, Research on the Relationship between Classic Denavit-Hartenberg and Modified Denavit-Hartenberg, 2014 *Seventh International Symposium on Computational Intelligence and Design*. (2014) 26-29, https://doi.org/10.1109/ISCID.2014.56.

[14] Chai. R, Tsourdos. A, Savvaris. A, Chai. S, Xia. Y, Solving Constrained Trajectory Planning Problems Using Biased Particle Swarm Optimization, *IEEE Transactions on Aerospace and Electronic Systems*. 57 (2021) 1685-1701, https://doi.org/10.1109/TAES.2021.3050645.

[15] Yan. Y, Cui. H, Lin. B, Han. P, A Task-Priority-Transition-Based Obstacle Avoidance Strategy for Space Manipulator, 2020 *Chinese Automation Congress* (*CAC*). (2020) 3857-3862, https://doi.org/10.1109/CAC51589.2020.9326869.

[16] Shi. H, Fang. H, Guo. L, Multi-objective optimal trajectory planning of manipulators based on quintic NURBS, 2016 *IEEE International Conference on Mechatronics and Automation*. (2016) 759-765, https://doi.org/10.1109/ICMA.2016.7558658.

光电+光热耦合相变蓄能供热系统在输油管道站场应用初探

王思杨[1]　张灿灿[2]　王新宇[2]

[1. 中国石油大学(北京)新能源与材料学院；2. 北京工业大学传热与能源利用北京市重点实验室]

摘　要　在油气管道输送中，电力和热力的消耗是管道输送过程中能耗组成的重要部分。尤其在输油站场上，通过燃气锅炉对原油进行加热，以及电伴热等保障原油顺利输送。随着"双碳"目标和发展战略新兴技术的提出，管道输送技术加速绿色转型。光伏和光热是太阳能的两种主要利用技术，相变储热是一种先进的储热技术，光伏+光热+相变储热耦合系统对原油输运技术绿色化转型具有重要意义。该系统在光照充足时，光伏系统所产电能可供给站场电力需求，同时可将多余电能转化为热能为管道供热，光热系统所产热能为相变储热系统输送热能，加热管道输运原油。当光照不足时，将相变储热系统中存储的热能加热原油输运系统，以满足管道输运的连续稳定运行。光伏+光热+相变储热耦合技术是一种供电供热储热一体化系统，可以满足24小时连续不间断的原油输运的热能和电力需求，有效减少原油输运站场的碳排放。本文综合分析了光伏+光热+相变储热耦合应用的技术路线，通过时空互补、能量协同耦合提升系统效能、验证技术路线可行性，探讨了该系统要素对光电+光热耦合相变蓄能供热系统在管道输送维温及日常使用上的应用前景。

关键词　太阳能光热；相变储能；太阳能光伏；原油输运

Preliminary Study on the Application of Photovoltaic and Photothermal Coupled Phase Change Energy Storage Heating System in Oil Pipeline Stations

Wang Siyang[1]　Zhang Cancan[2]　Wang Xinyu[2]

(1. College of New Energy and Materials, China University of Petroleum-Beijing;
2. Beijing Key Laboratory of Heat Transfer and Energy, Beijing University of Technology)

Abstract　In oil and gas pipeline transportation, electricity and heat consumption are significant components of energy consumption in pipeline transportation. This is particularly critical at oil transmission stations, where crude oil is heated by using gas boilers while ensuring smooth transportation through electric heating and other methods. With the introduction of the "dual carbon" goals and new strategic emerging technologies, pipeline transportation technology is accelerating its green transformation. Photovoltaic (PV) and photothermal are the two main utilization technologies of solar energy, and phase change thermal storage is an advanced thermal storage technology. The photovoltaic, photothermal, and phase change thermal storage coupled system has significant importance for the green transformation of crude oil transportation technology. When sunlight is abundant, the electricity generated by the PV system can meet the station's power demands, and can be converted into thermal energy for pipeline heating. The thermal energy generated by the photothermal system is delivered to the phase change thermal storage system to heat the pipelines for transporting crude oil. During periods of insufficient sunlight, the stored heat from phase change thermal storage system is supplied to the crude oil transport system, ensuring uninterrupted operation of pipeline transport. This coupling technology of PV, photothermal, and phase change thermal storage is an integrated system of power supply, heat supply and heat storage, which can effectively address the 24-hour heat and power needs of crude oil transportation while significantly reducing carbon emissions at transport stations. This paper provides a comprehensive analysis of the technical concepts underlying the of coupling applications of PV, photothermal, and phase change heat storage technologies. Through the system efficiency, technical routes, and solution of the collaborative coupling of time and heat complementarity, the application of system elements in the photoelectric and photothermal, coupling phase change energy storage heating system for maintaining transport temperatures and daily operational needs is discussed.

Key words Solar thermal; Phase change energy storage; Solar photovoltaic; Crude oil transport

原油输送是我国石油工业的重要部分，我国的原油具有高凝、高粘的特性，为保证管道安全顺利的运行，该类原油是通过加热进行输送。目前，输油站场普遍采用燃气锅炉进行加热，该方式会消耗大量化石能源。同时，输油站场日常传统的电力供应主要依赖于电网供电，这种模式不仅成本较高，而且不利于环保和能源的可持续发展。随着绿色能源技术的发展，特别是太阳能利用技术的进步，输油站场开始探索使用太阳能发电和电池储能系统来替代传统的电力供应方式。将太阳能利用技术应用于输油站场中，可以有效降低站场能耗，减少对传统化石能源依赖，降低温室气体排放。光热技术满足站场的热能需求，光电技术满足电能需求，实现能源的多元利用，提升整体能源利用效率，减少能源损耗。光电和光热技术可以就地取材获得能量，独立供电供热，降低能源供应中断风险，增强能源供应稳定性与可靠性，保障站场稳定运行。

中国石化在 30 座输油站安装光伏电站，一年省电费 81 万，这不仅体现了绿色能源的经济效益，还展示了其环保减排的优势。因此，引入太阳能发电和电池储能系统，不仅可以节约用电成本，还能减少碳排放，实现绿色发展。辽河油田设计了太阳能加热输送原油装置技术方案，并在辽河油田兴隆台采油厂建立了原油集输太阳能加热节能系统。原油最高温度达到 83℃，该系统投入使用以来，平均每天节约天然气消费量 30%。因此，可以通过太阳能集热作用使原油升温降黏减阻，提高原油的储运维温和加工处理能力，从而降低油气田能耗水平。盖忠睿等人设计搭建双级聚光集热槽式太阳能光热系统，其热效率可达到 27.35%，比单级槽式聚光集热系统提高约 0.9%~1.5%。刘鹏等人对带翅片的双层管槽式太阳能集热系统进行模拟研究，发现该系统集热管相比光滑圆管的综合传热性能的 1.16~1.78 倍，热损降低 42.4%~47.8%。刘魁华等以某厂房为例对光伏系统参数进行优化分析，并提出施工关键技术示范方案。Kolahi 等人基于数字孪生技术开发了一种光伏电站自主检测和管理系统，模拟了不同场景和配置下的光伏电站控制策略，给出最优控制策略。

目前油气管道企业站场、阀室供电主要为地方电力公司供电，少部分阀室建有太阳能发电装置、偏远地区建有发电机发电设施。目前仅有少部分油气站场应用太阳能用于生产、生活辅助电源，自建太阳能发电较少，并且在站场供电系统中占有的比重较低，主要还采用传统供电模式，新能源利用在油气站场中推广还处于初期阶段，创新改造空间还很大。太阳能是清洁能源，将太阳能应用与输油站场改造，可以有效降低站场对传统化石能源依赖，降低站场电费支出，缓解能源紧张，减少传统能源发电、供热带来的污染物和温室气体等排放，独立供电、供热系统可以有效保障站场内关键设备持续性运行，保障输油站场能源供应稳定，确保输油安全

1 技术思路和研究方法

1.1 光伏+光热+相变储热耦合应用技术

光伏+光热+相变储热耦合应用技术包括光伏发电制热系统、光热系统、相变蓄热器、水-原油换热器以及原油输运系统组成，如图 1 所示。太阳辐照强度充足时，阀门 1 关闭，阀门 2 打开，低温水经过光热系统加热，一部分热水分流至水-原油换热器中，直接加热原油并流回至低温水罐中继续循环；另一部分热水流经相变蓄热器，将热量储存以满足长时间热能需求，光伏发电系统中一部分电能用于场站其他系统供电，令一部分电能通过电加热器，将电能转化为热能，加热相变蓄热器出口的水，再流至水-原油换热器中加热原油，最终流回低温水罐继续循环。太阳辐照强度不足时，阀门 1 打开，阀门 2 关闭，低温水直接流至相变蓄热器中带走热能，再流至水-原油换热器中加热原油，最终回到低温水罐中继续循环。

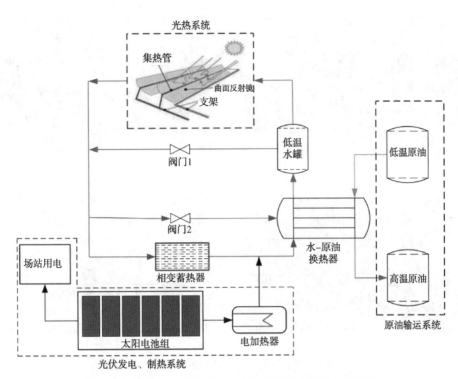

图 1 光伏发电+光热制热+相变储热耦合应用技术示意图

1.2 光伏发电、制热系统组成

光伏发电、制热系统由太阳电池组、电加热器、控制器、热交换装置以及系统辅助元件组成。系统工作原理是在一定辐照强度条件下，太阳电池将太阳能转化为直流电，一部分用于场站用电，也可储存在储能装置中，另一部分由电加热器加热导热介质，将电能转化为热能，完成系统发电与制热。

太阳电池组是由光伏板、充电控制器、逆变器和蓄电池共同组成。根据系统所需能量自动控制电能按需分配到电加热器和场站用电中，确保整体系统的安全平稳运行。太阳电池组安装前要充分考察安装场地，确保场地光照充足，合理设计太阳电池板的安装角度和朝向，确保其能最大程度接收阳光。电加热器是将电能转化为热能的设备。目前电阻式电加热器是操作最简便，适用范围最广的电加热器。

1.3 光热系统组成

光热系统主要包括集热系统、储热系统、换热系统、跟踪系统。集热系统按照镜场结构可分为塔式、碟式、线性菲涅尔式和槽式，如图2所示。槽式集热系统是目前应用最广泛，技术最成熟，商业化最成功的太阳能集热系统。槽式集热系统中集热管是该系统的关键部件，将反射镜聚集的太阳能辐射能转化为热能，通常采用真空集热管以达到更高的集热温度和更小的热损失，提高集热系统的总体热效率。集热镜场和镜场支撑结构是槽式集热系统的主要部件，反射镜大多采用超白玻璃，玻璃背面镀镜面反射膜，支撑结构固定反射镜，根据控制系统对太阳进行实时追踪，聚焦更多太阳能辐射能，提高系统光热效率。集热管中常见的传热流体为导热油和水，集热器将聚集的太阳光转化为热能，传热流体流经集热器吸收热量，完成太阳辐射能到热能的转化。大部分光热系统都需要配备储能系统，本技术通过相变蓄热系统进行储热，将相变蓄热器作为技术整体的蓄能系统，有效减少系统额外装置，降低建造成本。

1.4 相变蓄热器

相变蓄热是一种以相变储能材料为基础的储能技术。相变蓄热是利用储热工质在相变过程中，吸收或放出相变潜热的原理来进行能量储存的技术。固-固相变和固-液相变是实际中采用较多的相变类型。相变储热具有储放热温度波动小、蓄热密度大、化学稳定性好和安全性高等优势，尤其适用于热量供给不连续或供给与需求不协调的工况下。在该技术工作过程中，当系统有多余的热量输入时，相变材料吸收热量发生相变，将热量储存起来；当需要热量时，发生逆向相变，释放出储存的热量。在该系统中，相变蓄

(a)槽式

(b)塔式

(c)线性菲涅尔式

(d)碟式

图2 光热发电聚光集热系统类型示意图

热器在光照充足时,将光热系统与光伏系统所转化的热能储存起来;在光照不足时,将储存的热能用来加热冷水,使水-原油换热器连续换热,通过合理的控制系统,以达到24小时不间断的原油换热。

1.5 水-原油换热器

本技术中的水-原油换热器是利用高温水与低温原油进行热量交换,将原油温度升温达到流程规定的指标,以满足工艺条件的需要,同时该设备也是提高系统能源利用率的主要设备之一。由于原油中含有硫化物、盐分、环烷酸等腐蚀性物质,换热器需要采用耐腐蚀材料或使用防腐涂层,以减少腐蚀速率,延长换热器寿命。原油一般粘度较大,易结垢,流动复杂,因此换热器在设计、选型、计算方面需要进行特殊设计与计算。特别是在系统运行过程中,应该密切监测换热器的温度、压力、流量等参数,通过调节系统参数,防止出现温度变化波动大,影响换热效果。应定期检查维护,查看管束结垢、腐蚀情况,根据实际情况采用对应解决方法,如机械清洗或化学试剂清洗除垢,对腐蚀部位修复或更换部件。

2 可行性分析

2.1 耦合供热的技术可行性

光伏+光热+相变储热耦合应用技术的核心目的是为了对原油进行加热,因此确定原油的流量、原油所需温升和总换热量应是首要考虑的。总热负荷按公式(1)计算:

$$Q_T = q \cdot C_P \cdot \Delta T \quad (1)$$

式中,Q_T为总热负荷,kW;q为原油流量,kg/s;C_P为原油定压比热容,kJ/(kg·℃);ΔT为原油所需温升,℃。

槽式光热系统的镜场面积按公式(2)计算:

$$A_R = \frac{Q_T}{I_B \eta_{OP} \eta_{OT}} \quad (2)$$

式中,A_R为槽式光热系统镜场总面积,m^2;I_B为当地太阳辐射强度,kW/m^2;η_{OP}为系统光学效率;η_{OT}为系统光热转换效率。

槽式光热系统主要由集热管、抛物面反射镜、支架等部分组成。槽式聚光系统的反射镜为一块完整的镜面,其截面呈抛物线形,利用了抛物线的光学性质:与抛物线对称轴平行的光线射入抛物线后,其反射光线会汇聚于抛物线的焦点。集热管布置在抛物面的焦点上,使得平行照射到抛物面反射镜上的光线可以聚焦到集热管上。槽式集热器反射镜轮廓方程为:

$$f(y) = \frac{y^2}{4f} \quad (3)$$

式中,f为槽式集热器的焦距,m;

根据公式(4)、(5)、(6)、(7)、(8)、(9)即可计算槽式光热系统总长度。

$$\frac{w}{f} = -\frac{4}{\tan\psi_{r_rm}} + \sqrt{\frac{16}{\tan^2\psi_{r_rm}} + 16} \quad (4)$$

$$d = 2\sin r_r \frac{\theta_{sun}}{2} \quad (5)$$

$$r_r = \frac{w}{2\sin\psi_{r_rm}} \quad (6)$$

$$d = \frac{w\sin\frac{\theta_{sun}}{2}}{\sin\psi_{r_rm}} \quad (7)$$

$$S = \frac{w}{2}\sqrt{1+\frac{w^2}{16f^2}} + 2f\ln\left(\frac{w}{4f} + \sqrt{1+\frac{w^2}{16f^2}}\right) \quad (8)$$

$$L_T = \frac{A_R}{s} \quad (9)$$

式中，ψ_{r_rm} 为聚光器边缘角，（°）；w 为开口弦长，m；r_r 为聚光器半径，m；d 为接收器直径，m；θ_{sun} 为太阳张角，（°）；L_T 为槽式聚光镜场总长度，m。

太阳电池组是由太阳电池板、充电控制器、逆变器和蓄电池共同组成。其中太阳电池板是核心组件，根据系统所需热负荷以及太阳电池板的光电效率、光热效率以及当地太阳辐射强度可以计算出太阳电池组所需总面积，即：

$$A_D = \frac{Q_T}{I_B \eta_{OV} \eta_H} \quad (10)$$

式中，A_D 为太阳电池组所需总面积，m^2；η_{OV} 为太阳电池的光电转换效率；η_H 为电加热器的电热转换效率。

光伏+光热+相变储热耦合应用技术中，光伏发电系统和光热系统的设计功率与原油换热的总热负荷一致，应确保光伏发电系统与光热系统单独情况下都可以满足系统总热负荷。光伏系统的度电成本比光热系统较低，在太阳辐照强度充足时，光伏发电、制热系统作为主要热量来源与原油换热，此时光热系统应将大部分热能储存在相变蓄热器中，小部分热能用来与原油换热，以满足系统的连续性换热。相变蓄热器中相变材料的质量应按照公式（11）计算：

$$M = \frac{\int Q_T \cdot adt}{\Delta H_{tr}} \quad (11)$$

式中，a 为光热系统中用于蓄热的热量比例，随时间变化的函数；t 为系统蓄热时间，s；ΔH_{tr} 为相变材料的相变焓，kJ/kg。

光伏+光热+相变储热耦合应用技术通过系统设计可以满足原油输运管道24小时连续供热，充分利用太阳能和相变储能技术，实现光、热、电的时间互补、热量互补协同耦合，提高能源的综合利用效率。减少原油输运站场对传统化石能源的消耗，降低二氧化碳等污染物的排放，有利于降低油气田能耗水平，实现绿色可持续发展。

2.2 耦合供热的经济可行性

原有的输运场站通过直接加热炉燃烧天然气对原油进行加热，能耗极高。据统计，某原油输运场站2023年度消耗天然气268万方，耗电6773万度，累计二氧化碳排放量为8085吨。传统的输油站场热能补给技术，不仅耗能高，而且碳排放量也大。光伏发电+光热制热+相变储热耦合技术是新能源的综合利用技术，有望成为输油站场热能清洁替代技术及方案，降低站场的能耗以及碳排放量。

光伏+光热+相变储热耦合应用技术具有较高的经济可行性。光伏发电可满足场站部分能源需求，降低电网购电成本，也可以通过电加热器将产生的电能转化为热能，维持原油输运管道温度；光热系统可以充分聚集太阳能，将光能转化为热能，为原油输运管道供热，降低化石能源消耗，减少二氧化碳排放。光伏、光热系统的设备寿命较长，可以系统全自动化运行，维护简单，相变蓄热系统稳定可靠，安全性高。随着技术进步和市场发展，光伏、光热设备及相变蓄热材料成本的逐渐下降，系统的初始投资成本会降低。光伏+光热+相变储热耦合应用技术可以提升太阳能和相变材料在未来能源输配体系中的地位和作用，有助于国家管网集团实现绿色、低碳的发展目标，为构建清洁低碳安全高效的新型能源体系做出重要贡献。

3 结论

本文通过综合分析太阳能光热耦合相变储热提供热能和光电耦合电池提供电能的技术，通过时间互补、热量互补协同耦合的系统效能、技术路线和解决方案，探讨了该系统要素对光电+光热耦合相变蓄能供热系统在原油管道维温供热及站场日常使用供电。光伏+光热+相变储热耦合应用技术在输油场站的应用不仅具有经济性，还能提高能源供应的可靠性和环保性。通过合理的设计和控制，可以实现24小时的电伴热用电，节约用电成本，减少碳排放，实现绿色发展。输油站场的能源综合利用技术是实现"双碳"目标、推动能源结构优化和转型升级的重要手段，可为多能互补清洁供热的工程化应用提供一定参考，同时也符合国家能源安全新战略和能源革命的要求。未来，随着技术的进步和成本的降低，光伏、光热以及相变蓄热系统有望在更多的输油站场得到广泛应用。

参 考 文 献

［1］张玉龙. 原油储运中的安全风险及控制措施［J］. 石化技术，2017，24（11）：178.

［2］陈朋超. "双碳"愿景下管网多介质灵活输运与智能化高效利用发展战略思考［J］. 油气储运，2023，42（07）：721-730.

［3］吴洋洋. 太阳能协同储能原油维温系统能流输运研究［D］. 黑龙江：东北石油大学，2022.

［4］康萍. 油气田企业新能源产业合作模式研究［J］. 能源技术与管理，2024，49（05）：168-170.

［5］曲虎，陆诗建，杨佳朋，宋义伟，刘静. 双碳背景下油田清洁能源利用工艺可行性探讨［J］. 低碳化学与化工，2024，49（4）：79-89.

［6］盖忠睿，赵凯，杨天龙，饶琼，潘莹，金红光. 双级聚光集热的槽式太阳能热发电系统研究［J］. 西安交通大学学报，2024-12-25：1-9.

［7］刘鹏，马海文，巫云雷，陈林根. 带翅片双层管槽式太阳能梯级集热系统性能分析［J］. 工程热物理学报，2024，45（10）：3143-3150.

［8］刘魁华，张焱. 工业建筑屋顶分布式光伏电站施工关键技术［J］. 价值工程，2024，43（35）：107-109.

［9］KOLAHI M, ESMAILIFAR S M, SIZKOUHI A M M, AGHAEI M. Digital-PV: A digital twin-based platform for autonomous aerial monitoring of large-scale photovoltaic power plants［J］. Energy Conversion and Management, 2024, 321: 118963. DOI: https://doi.org/10.1016/j.enconman.2024.118963.

［10］赵悦婧. 三大油企三季度营收近五万亿元［N］. 中国电力报，2024-11-19（001）.

［11］仇中柱，倪行睿，朱群志，叶勇健，张涛，蔡靖雍. 基于分光谱利用的新型塔式热-光伏复合发电系统的发电量预测［J］. 太阳能学报，2024，45（9）：428-434.

［12］SADEGHI A, HASSANZADEH H, HARDING G T. A comparative study of oil sands preheating using electromagnetic waves, electrical heaters and steam circulation［J］. International Journal of Heat and Mass Transfer, 2017, 111908-111916. DOI: https://doi.org/10.1016/j.ijheatmasstransfer.2017.04.060.

［13］李石栋，叶家万，莫才颂，苏楚然. 太阳能热利用在稠油输送及开采中的应用现状［J］. 广东石油化工学院学报，2016，26（01）：48-51.

［14］王洪明，程勃. 相变蓄热型空气源热泵系统与太阳能互补供暖系统的优化研究［J］. 储能科学与技术，2024，13（06）：1980-1982.

［15］王国林. 套管式相变蓄热器蓄放热过程数值研究［J］. 能源工程，2024，44（02）：7-12.

［16］陈光. 原油稳定装置换热器运行节能探讨与实践［J］. 石油石化节能，2023，13（7）：44-48.

［17］孙涛，孙洪斌，王胜兵. 原油换热器换热效率提升技术研究［J］. 设备管理与维修，2023，174-176.

［18］付在国，高欢欢，张涛，朱群志. 太阳能加热原油储运系统热负荷匹配计算［J］. 油气储运，2019，38（11）：1306-1310.

［19］崔士波，刘玉岩. 太阳能光热结合吸收式热泵用于油田生产加热技术研究与设计［J］. 科技创新与应用，2017，7（24）：46-47.

［20］MA J, WANG C L, ZHOU Y, WANG R D. Optimized design of a linear Fresnel collector with a compound parabolic secondary reflector［J］. Renewable Energy, 2021, 171: 171141-171148. DOI: https://doi.org/10.1016/j.renene.2021.02.100

［21］孙秋分，段晓文，戴传瑞，梁坤，付玲，冯乔，等.【CCS/CCUS】"双碳"目标下石油公司发展CCUS-EGR的思考与建议［J］. 天然气与石油，2024，42（5）：28-32.

［22］王向宏，叶朝曦. 太阳能光热技术在油气田中的节能应用［J］. 石油石化节能，2013，3（07）：22-25.

国内外成品油管网公平开放机制研究

张澍原 薛 庆 王建良 马青天 刘 钰

[中国石油大学(北京)]

摘 要 管道输送因其安全、连续、低碳、低损耗等特点，一直以来都是国外成品油输送的首选运输方式，也是我国成品油运输的转型方向。在深入推进油气管网基础设施公平开放的背景下，我国成品油管网仍然存在设施互联性差等一系列瓶颈问题，制约了成品油行业高质量发展。如何协调市场主体责任与利益，指导我国成品油管网实现灵活调运和高效集输，是亟需解决的问题。为此，本文从典型国家和典型公司两个维度入手调研，详细比较了中外成品油管网所有权模式、公平开放监管制度、管输价格监管制度、管道公司运营实践等方面的异同，系统总结了国外成品油管网公平开放的理论基础和实践经验。对比发现，目前我国成品油管网存在多种油品的管理、接收、储存和混油处理困难，基础设施产权限制管输资源高效利用，市场监管不完善导致变票或无票销售油品行为频发，混油处理涉及经验资质、成本分摊、重复征税等诸多问题。为进一步促进成品油管道增输增效，完善我国成品油管网运营制度，本文建议以搭建全国统一大市场契机，打造新型管网设施商务合作模式，发展共享经济；搭建公共信息服务平台，实现成品油多式联运；优化"标签化"输送，实行批次输送模式；完善混油处理及定价制度，指导成品油运输市场健康发展。

关键词 成品油管网；公平开放；管道运输

Research on the Open-Access Mechanism of Products Pipeline Networks at Home and Abroad

Zhang Shuyuan Xue Qing Wang Jianliang Ma Qingtian Liu Yu

(China University of Petroleum Beijing)

Abstract Pipeline transportation has long been the preferred method for transporting refined oil products internationally due to its safety, continuity, low carbon, and low loss characteristics. It is also the direction of transformation for products transportation in China. In the context of actively promoting the open-access to oil and gas pipeline infrastructure, China's products pipeline network still faces a series of bottlenecks, such as poor facility interconnectivity, which hampers the high-quality development of the refined oil industry. Coordinating the responsibilities and interests of market participants and guiding China's products pipeline network to achieve flexible dispatching and efficient transportation are pressing issues to be addressed. So, this paper begins by studying typical countries and companies, comparing the ownership models, regulatory regimes, pricing regulation of transportation, and operational practices of pipeline companies domestically and internationally. And it systematically summarizes the theoretical foundations and practical experiences of open-access in foreign products pipelines. A comparison reveals that there are numerous issues with the products pipeline transportation in China, such as difficulties in the management, reception, storage, and blending of various oil products; limited efficient utilization of pipeline resources due to infrastructure capacity constraints; frequent occurrences of altered or undocumented oil sales due to inadequate market supervision; and challenges in blending processes including issues related to experience qualifications, cost sharing, and repeated taxation. To further promote increased throughput and efficiency in products pipelines and improve the operation system of China's products pipeline network, this paper proposes leveraging the opportunity to build a unified national market to establish a new business cooperation model for pipeline facilities and to foster a sharing economy. It also suggests constructing a public information service platform to enable multimodal transportation of refined oil products, optimizing "tagged" transportation through batch delivery modes, and improving blending and pricing systems to guide the healthy development of the products transportation market.

Key words products pipeline; open-access; Pipeline transport

在全球能源格局深刻变革与我国"双碳"目标的推动下，能源基础设施的高效利用与公平开放成为提升资源配置效率、保障能源安全的重要议题。成品油管网作为能源输送的关键环节，其公平开放不仅关系到市场竞争环境的优化，还对推动经济绿色转型具有重要意义。

目前，我国成品油管网在政策设计和实践操作上仍处于起步阶段，与国际成熟市场相比，在设施资源分配、价格监管和信息透明度等方面存在显著差距。尽管美国、欧盟等国家和地区早已构建了完善的管网开放体系，其经验为我国提供了重要参考，但因国情差异，直接照搬显然难以奏效。基于此，本文旨在系统分析国内外成品油管网公平开放的理论机制与实践经验，梳理国内成品油管网开放的现状与挑战，提出符合我国国情的政策建议，助力能源市场的高质量发展与绿色转型。

1 国际成品油管网公平开放的理论机制

1.1 美国

对于石油管道，美国国会于1906年通过的《Hepburn法令》修正了州际商业法，起因是公众认为标准石油公司通过滥用和歧视性利用管道来建立全国性的石油垄断。修正后的州际商业法规定所有的州际石油管道公司就像铁路、航空等运输商一样是公共运输商，禁止它们持有所运输油品的所有权，同时要接受商业委员会（ICC）对其运输费率的管辖，公平、无歧视地对待所有托运人。

1997年联邦能源管理委员会（FERC）接替了ICC的权限监管石油管道的运输费率，目前FERC在成品油管输领域的职责主要是监管州际石油管道公司的费率及相关规章制度的合理性，并确保管输容量的公平开放使用，但其权限不包括对石油管道输入/输出节点商业交易的监管。为进一步完善石油管输费率的管理，FERC先后发布了No.154-B《原始成本拟合模型（TOC）》、No.572《石油管道市场基准费率制定（终稿）》以及No.780《石油管道文件提交、价格指数和服务要求规定（终稿）》等法规，从而逐步优化美国石油管道的市场化运作模式。

1.1.1 公平准入

在美国成品油管网的公平开放机制中，管道公司根据联邦、州和地方政府的相关标准，制定油品准入条件，确保所运输的油品符合质量要求，包括黏度、密度、蒸气压、水含量和杂质含量等参数。只有符合这些标准的油品才能获得管道公司的运输服务，保证了公平性。

对于无储存设施的管道公司，要求用户在管道系统的上下游建设储存设施，确保油品的顺利接收与交付。同时，管道公司对储存设施的泵送速率和压力有严格要求，以保证运输过程的高效和安全。大多数管道公司还要求用户提供隔离液，并按照标准收费，隔离液会在运输终点等量返还，确保运输过程的平衡。在用户准入方面，新用户需要填写托运商注册信息表，提交相关资料，以审核企业资质、财务状况和信用评级，确保用户具备可靠的支付能力。对于已有用户，管道公司会定期评估其财务和信用状况，若不符合要求，则只在预付费用的前提下才能继续提供服务。这一机制保障了管道运输服务的公平性，同时维护了管道公司运营的安全与稳定。

1.1.2 托运商管理和管容分配

用户分类是拥塞管理的首要环节，对最终的管输调度计划有着重要影响。对于准入的第三方用户，通常根据基准期内是否存在历史运输量或运输量是否达到特定阈值，将其划分为长期托运商或新托运商，以美国管道企业BPL公司为例，其基准期为当前月份之前连续的12个日历月。当新托运商在一定时间内完成运输作业或运输量达到规定阈值后，可被重新定义为长期托运商。

在成品油管道市场化运营中，管道公司结合运输能力的紧缺程度和用户的需求差异，通常采用四种不同的运输能力分配方法：年度预订分配、紧急运输分配、市场化竞标分配及历史需求分配。年度预订分配是美国成品油管道运输能力分配的主要方式之一。用户可根据自身需求提前一年预订固定的运输能力，这种方式通常优先保障长期合同用户的运输稳定性。紧急运输分配是为应对短期市场需求波动而设计的一种机制。在出现突发运输需求时，管道运营商会将剩余运输能力分配给需要快速运输的用户。这种方式通常通过排队或先到先得的原则来分配运输能力，确保短期需求得到及时响应。市场化竞标是一种以市场机制为核心的运输能力分配方式，用户通过竞价的方式获取剩余的运输能力。运输能力根据出价高低优先分配给支付能力强的用户。这种机制通常用于高需求或管网运输能力紧张的情况下，帮助管道运营商最大化运力资源的经济效益。在某些特殊情况下，美国成品油管道市场也采用历史需求分配机制，即根据用户过去的使用记录分配运输能力。这种机制常用于用户基础稳定、需求规律性较强的市场环境。

1.1.3　价格监管

美国成品油管道公司主要通过指数型费率、协商型费率、市场型费率和服务成本型费率来制定管输费率。FERC 在 No. 561《根据1992 年能源政策法对石油管道规定修改（终稿）》中明确指出，石油管道公司可以在同一管输系统或不同路径上采用一种或多种费率制定方式。FERC 对管道运输费率实施监督，要求管道公司在设定价格时遵循"成本加成"原则，即运输费用应合理反映管道公司运营、维护和资本成本。

成品油管道公平开放的核心在于制定标准管输费率表。管道公司通常采用"点到点"计价模式，即明确规定各接收节点至交付节点的费率，并按照目的地实际交付的成品油量计算费用。用户必须在交付前支付管输费用，管道公司对此保留成品油的留置权和担保权。如果用户未在支付期限内结清全部费用，还需支付延期款项的利息。

1.1.4　信息披露

信息披露是确保市场透明、公平竞争的重要环节。管道公司必须定期公开与运输服务相关的关键信息，包括但不限于运输能力、管道状况、价格结构、服务条款以及可能的服务中断或调整等。管道公司通常会建立管道物流信息交流平台，同时美国还设有专门为管道公司和用户服务的物流信息交流中心，其中成立于1997 年的 Transport4 平台是美国规模最大、技术最先进的成品油管道物流信息交流中心。根据联邦能源监管委员会的规定，管道公司需在其官方网站或其他适当平台上发布运输费率、可用容量、运营计划和价格调整等信息，以便所有潜在用户了解并作出知情决策。此外，管道公司还需及时向监管机构报告有关业务运营的详细数据，包括运输量、收入、成本以及任何可能影响服务的重大变化。为了增加透明度，管道公司还应在用户请求时提供关于管道容量、运输安排和费用等方面的详细说明，确保各类用户能够平等地获取所需信息。

1.2　欧盟

欧洲的油品管道运营企业既有独立管道公司，又有一体化石油公司（关联管道公司），例如，大型一体化石油企业荷兰皇家壳牌公司拥有并运营着自己的管道网络，这些管道主要用于运输公司自产的石油和天然气。与此同时，独立的管道运营商，如荷兰的 Trans Europese Pipeline（简称 TENP）和德国的 OPAL Pipeline，则提供第三方服务，允许不同的生产商和供应商使用其管道系统进行油品运输。这种模式确保了市场的公平竞争，同时也提高了整个欧洲能源供应链的灵活性和效率。

欧盟成品油管网的开放机制建立在完善的法律框架之上。《第三能源法案》作为核心法律，为成品油管网开放提供了明确的规则和指导。法案要求各成员国实施油品所有权与管道运营权分离，确保管网企业的运营行为独立于所输油品的所有权，从而避免垄断企业利用管网优势排挤竞争者。同时，各成员国需设立独立的能源监管机构，对管网开放的实施进行监督，并对价格制定、运力分配等关键环节进行审核。欧盟委员会和法院体系负责确保法律的有效执行，针对不符合开放原则的行为进行处罚。这一体系使得成品油管网的开放具有法律保障，为市场化运作奠定了坚实基础。欧盟对于油品管道公司的监管还体现在工程授权监管、油品计量、经营税收及环境安全监管等方面。个别国家会对成品油管道执行相对严格的公平开放监管，如波兰对于液体燃料物流公司 OLPP 的所有权改革。

1.2.1　非歧视性准入

欧盟成品油管网的准入机制强调非歧视性，旨在降低市场准入门槛，为所有市场主体提供平等的使用机会。管道运营商需向所有符合条件的用户提供公平的运输服务，并公开技术规范、接入条件以及服务条款，以避免信息不对称问题。管网容量的分配采用公开透明的市场化规则，例如通过竞价机制或先到先得原则进行分配，保障资源的高效利用。在跨境成品油管道运输方面，欧盟建立了区域协调机制，为跨国用户提供便利，避免因国界分隔造成的资源利用效率低下。这种非歧视性准入模式不仅促进了市场竞争，也为中小企业提供了进入能源运输市场的机会。

1.2.2 价格监管

价格监管机制是保障成品油管网公平开放的关键环节。欧盟的价格体系以"成本回收+合理利润"为核心，确保运营商在维持正常运营的同时不会对用户施加过高的价格压力。成员国的独立监管机构负责价格的审批与调整，并要求运营商提前公布价格调整计划及其依据，增强用户对市场的信任度。欧盟广泛采用"入口-出口费率"模式，根据油品在管道网络中具体的进出节点进行收费，用户可以根据运输需求灵活选择路径，从而降低运输成本。这一机制实现了费用分配的公平性与效率的平衡，为资源优化提供了有力支持。例如连接意大利港口（的里雅斯特）与德国和奥地利的 TAL 成品油管道，入口费在成品油通过港口进入管道系统时支付，出口费在不同出口点（如慕尼黑或因斯布鲁克）卸载油品时支付，费率由管道运营商和区域市场协定，不与物理运输路径直接挂钩。

1.2.3 信息披露

完善的信息披露机制是欧盟成品油管网开放的显著特征。运营商通过电子平台实时更新管网运力、服务价格、维护计划等关键信息，为市场参与者提供透明的数据支持。用户可以根据这些信息合理安排运输需求，优化物流计划，从而提高整个市场的运行效率。欧盟还要求运营商每年发布服务质量报告，评估管网运行状况并提出改进建议。此外，用户咨询与投诉机制的设立，为解决用户与运营商之间的纠纷提供了有效渠道，进一步提升了市场的公信力。

1.2.4 可持续发展导向

在欧盟的能源绿色转型战略中，成品油管网的开放机制融入了可持续发展理念。管道运营商被鼓励采用多式联运模式，将管道运输与铁路和水运相结合，进一步减少碳排放并提高运输效率。同时，欧盟支持对管网进行现代化升级，例如引入智能化监控技术和高效输送设备，以减少能源消耗和运行损耗。政府通过税收优惠和补贴政策推动运营商加大绿色投资力度，为能源市场的低碳转型提供政策支持。这些措施不仅提升了管道运输的环境友好性，还增强了欧盟能源市场的竞争力。

1.3 英国

与欧盟一样，英国的油品管道也是独立管道公司与一体化石油公司（关联管道公司）共存的局面。英国陆上石油输送系统既包括 CLH PS、BPA（British Pipeline Agency）这类专业化管道公司，也有 ExxonMobil 的 ESSO Pipeline 管网系统、道达尔公司的 Fina-Line 管道系统等一体化公司经营管理的关联管道公司。政府对于油品管道公司的监管主要体现在工程授权监管、运营监管、油品计量、经营税收及环境安全监管等。

英国虽已脱离欧盟，但其成品油管网开放机制的法律基础与欧盟法案一脉相承。英国通过独立监管机构（如 Ofgem）对管网服务进行严格监管，确保价格合理性、服务公平性和资源分配效率。此外，英国还依赖市场化规则，如通过多方参与的合作框架管理成品油管网投资与运营，进一步提高了管网管理的透明性与效率。

欧盟规定统一的准入规则与技术规范，为用户提供公平的接入机会。跨境成品油管道的准入由区域协调机制保障，确保跨国用户能够顺利接入资源。英国在此基础上结合市场化机制，采用透明的竞标或拍卖方式分配剩余运力。特别是针对中小型炼油厂和分销商，英国通过公开的接入条件降低了市场进入壁垒，促进了市场多元化。

英国的价格机制在欧盟广泛采用的"入口-出口费率"模式基础上增加了动态调整的市场化元素，例如允许用户协商附加服务费用，如快速运输和特殊油品输送。英国对于价格信息公开的要求也较高，运营商需提前公布费用调整计划，为用户提供清晰的成本预期。

1.4 俄罗斯

俄罗斯的成品油管网由国家管控和大型能源企业运营的双重结构构成。俄罗斯输油管道运输公司（Transneft）是成品油管网的主要运营商，拥有全国大部分的成品油运输管道，负责成品油的国内分配和出口运输。在管网管理模式上，Transneft 采取"统一调度"机制，对国内和出口的成品油运输进行全面的计划与调配。管网的容量分配和运输计划需经过政府部门审批，确保能源供应的稳定性和安全性。这种高度集中的管理模式在提升管网运行效率的同时，也限制了市场主体的自主选择权，俄罗斯通过一系列政策改革，尝试在管网管理与市场开放之间寻找平衡。

1.4.1 第三方准入政策

俄罗斯在成品油管网的第三方准入方面，遵循非歧视原则，允许符合条件的市场主体使用管道运输成品油。Transneft 需按照政府核定的标准，向第三方提供管网接入服务。然而，由于管网容量有限，第三方准入在实际操作中常受到限制，特别是在出口管道上，优先权通常给予国有能源企业。

1.4.2 价格监管与定价机制

成品油管网的运输费用由联邦反垄断局（FAS）负责监管，实行政府定价和市场化定价相结合的机制。国内运输的费用主要由政府核定，而出口运输则允许一定程度的市场化调整，以适应国际市场的变化。这种双轨制的价格机制既保证了国内市场的稳定，也为出口创造了灵活的竞争环境。

1.4.3 信息公开与透明化

俄罗斯要求管网运营商定期公开管网容量、运输费用和使用情况等信息，为市场主体提供透明的参考数据。同时，Transneft 通过其在线平台发布运输计划和可用容量，方便企业提交运输申请。由于能源行业对国民经济和国家安全的特殊重要性，俄罗斯在推进管网开放时采取了循序渐进的策略，表现为准入政策的逐步放宽以及价格机制的逐步市场化，以确保在扩大市场准入的同时维护能源安全。

虽然俄罗斯通过政策试图扩大市场主体的参与，但国有企业在管网资源分配中的主导地位依然明显。Transneft 作为主要运营商，对管网的规划、分配和管理拥有高度控制权，民营企业和中小型市场主体在获取管网资源时面临较大竞争压力。同时，俄罗斯成品油管网的开放受到管网容量的限制，特别是在出口运输上，优质运输路径（如通往欧洲和亚洲的主要出口管道）经常处于高负荷状态，难以满足所有市场主体的需求（表1）。

表1 美国、欧盟、英国和俄罗斯油品管道运营机制对比表

要素	美国	欧盟	英国	俄罗斯
1）所有权形式	关联公司与独立管道公司并存	一体化公司、关联公司与独立管道公司并存	一体化公司与独立管道公司并存	国家管控与大型能源企业运营管理
2）公平开放	州际管道强制要求，各州存在差异	欧盟层面不强制要求，个别成员国严格开放	不强制要求	国家管道强制要求
3）服务价格监管	指数定价、市场定价、协商定价	指数定价、市场定价、协商定价	指数定价、市场定价、协商定价	政府定价、市场定价

2 国外成品油管道公司实践经验

2.1 美国成品油管道公司科洛尼尔

美国的科洛尼尔管道运输公司是美国最大的成品油管道运营商，承担了美国东海岸45%的燃油供应，运营着长达5500英里的成品油管道，每日输送量达到238万桶，包括精炼汽油、柴油和航空燃料。1964年12月建成了世界上第一条通过计算机控制管理并对外开放的成品油管道，这一技术使运输安排更为高效科学。

科洛尼尔的基础设施规划和制度设计相辅相成，构建了多输入站和分输站系统，优化管道利用率。其输入站分布于休斯敦等多个炼油厂附近，每小时输送能力达50320桶。为了提高输送效率，公司在管道中设置双线结构，分离输送轻质油与重质油，同时通过灵活的输入站协作安排，避免不必要的停输问题。管道共有37条支线，分输站既是干线分输点，又是支线的起点。亚特兰大、格林斯巴勒和多尔西分输站承担了大量分输任务，合理的站点布局显著提升了运输效率，减少了因混油造成的资源浪费。

科洛尼尔储油罐及站场设计较为巧妙，其运输的所有油品需先暂存于管道站场或中间储油点的储罐内，并不直接从炼油厂储罐进入管道，这便于从源头把控质量。为明确货源与货主关系，美国成品油管道公司选择采用"分储分输"的管理方式。分储分输要求首站、末站及注入、分输站场具备多种油品的储存能力，并拥有足够的储油罐数量和容量。站场的设计紧凑而有条理，在不同油品批次之间，采用了隔离液技术（如水或惰性液体）作为油品的物理隔离屏障，有些管道系统还采用了可膨胀的隔离膜，这种膜在油品流动过程中膨胀，紧密地隔离不同批次的油品，有效减少了混油风险。储油罐内部也有分隔板或独立罐区设计，不同批次的油品被分隔存储，同时还安装有液位传感器，实时监控油品的液位，避

免过度充填导致油品溢出或混合。

科洛尼尔的管容分配制度也极具代表性，其将托运商分为新托运商和长期托运商：新托运商只有在周期管输量超过 18750 桶时才有可能成为长期托运商。长期托运商的最终分配量取决于固定月计划与历史计算分配量（CCHA）中较小的数值，基于托运商在基准期（通常为前 12 个月）的历史运输比例计算得出。若固定月计划低于 CCHA，则产生的剩余管容将分配给计划高于 CCHA 的其他长期托运商。长期托运商按历史数据和固定计划分配，新托运商则通过平均分配、抽签或按需求比例分配剩余管容。此外，竞拍也常用于管容分配，遵循价高者得原则，优先满足高价需求，直到剩余管容分配完毕或需求满足。当多个托运商出价相同且需求超出剩余管容时，再按需求比例分配。

科洛尼尔高度重视运行程序的编制。其官网以 5 天为一个周期滚动更新下一年度的运输程序。运行程序编好后即发送给各站，作为调度指令看待。编制运行程序的计算精度高，运行程序可靠。对于公司的 37 条支线，则单独编制计划，列入科洛尼尔成品油管道的运行程序中，以便准时将需要的油品输送到分输站。科洛尼尔会提前测算和约定混油处理方案。在编制管道运行计划过程中，管道与委托商就已经计算出沿线卸出油量、到达终点站所余的纯油量和混油量，对应的混油处理方法也已经协商确定。

科洛尼尔采取批次输送制度，将客户的产品分为两类：第一类是"可替代批次"，指符合既定规格的一批石油产品，该批石油产品可以与符合相同规格的其他批次产品混合，也就是去标签化输送。第二类是"分离批次"（表2），指符合既定规格的一批石油产品，但不得掺混，其运输批次是分离的，也就是标签化输送。科洛尼尔运输的产品大多数是按照可替代批次运输。某个托运商的产品可以与管道内类似产品混合在一起同一批次输送，交付给托运商产品可能不一定是指定输入点的产品，但它满足该类产品的标准规格。可替代批次将在终端进行个性化调配。批次交付后，托运商在各自品牌下销售的燃料油中注入特殊添加剂，以使其产品呈现出特定的品牌特质，具有独一无二的品牌价值。分离批次运输占用储运资源更多，操作更复杂，需要专门的储罐容量，在运输产品前后均需清空储罐。但随着从墨西哥湾沿岸到东北部运输产品的需求增大，科洛尼尔于 2020 年开启了分离运输服务。鉴于分离批次并非主要运输类别，因此该类产品的限制相对较多。由于分离批次对管道容量和储罐容量有较大影响，只有指定的一些上载点和下载点才可提供分离批次输送服务，其次科洛尼尔对于单批次运输量有上下限规定（下限视情况为 1.5 万或 7.5 万桶，上限 10 万桶）。未经承运人事先批准，分离批次不得在两个或多个目的地之间分割。此外，分离批次运价也更高。分离运输的产品需要增收额外费用，所使用的隔离液也将视同产品，收取从始发地到目的地的管输费，以及不可退还的分离批次设施使用费（0.21 美元/桶）。

表 2　分离批次上下载点名单

分离批次起点	分离批次终点（1A 产品）	分离批次终点（其他产品）
Houston, Harris County, TX Hebert, Jefferson County, TX Lake Charles, Calcasieu Parish, LA Baton Rouge, E. Feliciana Parish, LA Baton Rouge-Dock Facility, W. Feliciana Parish, LA Collins, Covington County, MS Carteret, Middlesex County, NJ Port Reading, Middlesex County, NJ Sewaren, Middlesex County, NJ	Linden, Union County, NJ Linden-Buckeye, Union County, NJ Newark, Essex County, NJ Newark-Sun, Essex County, NJ Gulfport-Linden, Richmond County, NY Gulfport, Richmond County, NY Port Mobil - Linden, Richmond County, NY Philadelphia-Fort Mifflin/Philadelphia Airport, Delaware County, PA Philadelphia - Girard Point, Philadelphia County, PA Woodbury, Gloucester County, NJ Woodbury - Buckeye/Malvern, Gloucester County, NJ	Baton Rouge- East Baton Rouge Parish, LA Collins - CLP nomination code: TMO or TM2 tankage code, Covington County, MS Linden, Union County, NJ Linden- Buckeye, Union County, NJ Newark, Essex County, NJ Newark- Sun, Essex County, NJ Gulfport- Linden, Richmond County, NY Gulfport, Richmond County, NY Port Mobil- Linden, Richmond County, NY Philadelphia-Fort Mifflin/Philadelphia Airport, Delaware County, PA Philadelphia-Girard Point, Philadelphia County, PA Woodbury, Gloucester County, NJ Woodbury- Buckeye/Malvern, Gloucester County, NJ

对于分离批次，托运人需要先行提供产品信息、分析证书或其他质量数据，科洛尼尔会在三个工作日内回复是否接受该笔订单。如果科洛尼尔确认运输该批产品，则托运人需要先行支付0.21美元/桶的设施费用。支付费用后，科洛尼尔再根据产品信息，安排缓冲批次，根据历史流量分配管输量。之后托运人需要自行准备缓冲材料及产品，若管输量不足时还需根据历史转移制度向其他托运人购买。接着科洛尼尔便会着手清空并清理部分目的地的储罐。与此同时，托运人还需要关注下游线路运输的容量限额，超过的部分则需要通过产品转让服务（PTO）转让给其他托运人。在确定容量无误后，科洛尼尔便会将托运人的缓冲材料及产品从限定上载点运至限定目的地。

2.2 美国成品油管道公司巴克艾伙伴

作为美国最大的独立石油产品管道运营商之一，巴克艾伙伴（Buckeye Partners）的主要业务是将液态石油产品通过管道系统网络将这些产品运输到美国中西部和东北部的需求市场。该公司运输网络由134个液态石油产品码头组成，总储罐容量约为1.3亿桶，拥有超过5000英里的管道。

相比科洛尼尔，巴克艾的历史更为悠久。巴克艾在1886年至1940年期间建造了从芝加哥到布法罗的炼油厂原油干线网络，专注于原油采集。1940年至2000年间，巴克艾将部分闲置的原油管线改造为精炼产品运输管道，以满足客户日益增长的需求，并成为美国第一家管道主有限合伙企业（Master Limited Partnership）。在2000年至2010年期间，巴克艾通过收购多个码头资产，打造了美国最大的独立终端和管道网络之一。2011年至2019年，巴克艾进一步转型，收购了位于巴哈马的全球最大的海洋原油和精炼石油产品储存设施之一，提供覆盖全球的石油产品物流与混合服务，成为多元化的能源物流供应商，其管道、内陆码头和海运码头主要位于美国东海岸、中西部、东南部和墨西哥湾沿岸地区以及加勒比地区的主要石油物流枢纽。

巴克艾不同于科洛尼尔，该公司的业务领域广阔，除了运输业务外，还涉足炼化、储存和销售等环节。在产品线方面，公司不仅提供成品油，还包括原油在内的多种液态石油产品，用户和业务种类多、管道系统复杂。为应对大量不同品质和数量的产品，巴克艾在基础设施及制度安排上也有不少特点：

第一，巴克艾充分利用地理优势，以四大码头为核心，构建特色运输网。其中芝加哥综合设施（CCX）由五个相互连接的码头设施组成，总储存能力超过880万桶，拥有26条入站/出站管道连接。纽约港储存量超过1800万桶，分销到纽约、新泽西和宾夕法尼亚州的主要需求市场。南德克萨斯州由三个聚合和出口码头组成，提供超过1690万桶的储存量。加勒比海总储存能力超4000万桶，分布在加勒比海的三个主要码头地点（巴哈马、波多黎各和圣卢西亚）。巴克艾管道系统拥有超过5000英里的管道，每日输送超100万桶产品。该系统在连接主要炼油中心、东海岸海运进口和墨西哥湾沿岸管道流到美国中西部和东北部的主要成品油需求市场方面发挥着关键作用。

第二，巴克艾大力使用Synthesis系统（表3）。Synthesis是一个完整的订单到现金系统，有助于码头的商业运营，并与码头自动化系统和后台系统无缝集成。通过使用完全集成的管理信息系统，巴克艾每月结账时间从6天减少到2天，而吞吐率提高了25%。由于更好的规划能力以及订单整合能力，终端工作流程中的所有流程都得到了简化。不同于科洛尼尔，巴克艾伙伴以10天为一个周期。在不需要额外资源的前提下，巴克艾公司的终端数量增加了近100%，管道数量也显着增加，管理任务所花费的时间和资源减少了30%以上。

表3 Synthesis系统用于管道运营商物流管理部分功能

项目	功能
订单管理与规划	提交并管理管道提名的审批流程；根据实际容量分析运输和存储请求；管理通过采购和销售实现的产品交接
调度与预测	集成行业领先的PipelineScheduler®应用程序，用于模拟运输流动、发布批次计划以及预测储罐库存水平
测量与操作	捕获包含可观测数据的票据测量信息及产品质量信息，以计算毛量和净量；维护计量校验（Meter Proving）记录库
库存管理	将票据中记录的数量分配给托运人；管理对账流程，包括管道损耗与增益的核算以及账面与实物库存的对账

续表

项目	功能
计票与开票	管理和调整管道费率；根据特定场景和业务规则计算费用并生成账单
警报与报告	在工作流程中触发特定事件时通知用户；可查看、下载或发送报告包给客户和管理层

第三，相比科洛尼尔，巴克艾的管容分配制度更倾向于小型托运商。巴克艾规定，在上一个基期内连续进行12个月运输便可成为长期托运商，无运输量要求。对于非常规市场情况下的管容分配，如由于油气下游市场的不稳定波动，可能会导致某些月份的计划运输量临时超出管道输送能力，巴克艾公司规定，对于长期托运商，其分配的运输量将与其在上一基准期内的平均运输量相匹配。对于其他托运商，则依据市场常态下的分配准则来分配管道输送容量。在出现管道拥堵的情况下，会优先通知运输方调整其运输计划，以避免因供应短缺而造成的经济损失，确保托运商的基本权益不受损害，最终在调整后的运输计划基础上，根据历史运输数据按比例分配管道输送能力。对于新托运商，每个新托运商所分配得到的管容不超过总可用管容的2.5%，所有新托运商的总分配量不超过总管容的10%。并且以新托运商数量是否>4为界。

第四，凭借在运输和运营系统中的优势，巴克艾可以向客户提供产品转移服务（PRODUCT TRANSFER ORDERS）。即托运人在巴克艾收到产品后可以转让产品所有权。服务费率为巴克艾向卖方收取每桶7美分的转移费，每次PTO服务的最低费用限制为700美元。商品的买方和卖方均可在Transport4平台上向对方发起交易申请，申请批准后便会将指令发送给巴克艾予以最终批准和执行。

第五，巴克艾运输系统经过巧妙设计并已申请专利，该系统主要由采样站、泊位装载臂、仪表、源罐（Source Tank）、泵（Pump）、蒸汽燃烧单元（Steam Combustion Unit）六个部分组成。巴克艾运输系统首先接收关于指定流体产品、传输量和目标流速的输入。根据产品、设备可用性和其他参数，确定容积传输和流速的允许操作范围。控制信号被发送到阀门，以建立流体产品的流动路径。该方法包括监控设备状态变化以及应对这些变化的步骤。该系统可以支持多个源罐和泵，每个源罐存储不同类型的流体。此外，流动系统中的阀门可以有两个部分，分别负责关闭和打开，以建立或中断流动路径。

2.3 美国成品油管道公司金德摩根

金德摩根（Kinder Morgan）是北美最大的独立石油产品运输商，产品管道资产战略性地位于美国西部、东南部和中西部。主营业务是通过大约9500英里的管道输送汽油、喷气燃料、柴油、原油和凝析油，每天运输约240万桶石油，共拥有约65个液体终端，用于储存燃料并提供乙醇和生物燃料的混合服务。金德摩根将业务板块划分为太平洋地区业务和东南部业务。太平洋地区业务包含约3000英里的成品油管道。其历史可以追溯到1956年，是美国西部最大的产品管道业务，每天输送超过100万桶汽油、喷气燃料和柴油；东南业务服务于美国东南部的3100英里精炼石油产品系统，由佛罗里达中部管道系统（包含两条长110英里、85英里的管道）以及由遍布美国东南部的26个成品油码头组成的金德摩根东南码头（简称KMST）。

金德摩根管理着庞大而复杂的成品油管道网络，涵盖了多个地区和州，共运营五大系列产品管道系统，分别是：CALNEV管道、SFPP系统、波特兰机场管道、佛罗里达中部管道以及PPL管道，使其能够有效地运输和分发成品油产品。CALNEV管道主要输送汽油、柴油和喷气燃料；SFPP系列包括13个卡车装载码头，提供短期产品存储、卡车装载、蒸汽处理、添加剂注入、染料注入和含氧化合物混合；波特兰机场管道运输航煤；佛罗里达中部管道输汽油、工业酒精、柴油和航煤；PPL管道输送车用汽油、柴油（包括生物柴油）、煤油以及商用和军用航煤。

金德摩根的管容分配政策旨在处理拥塞情况下的容量分配公平性问题。为明确起见，金德摩根会逐线逐段分配容量。当金德摩根判定某月份的任何分段系统容量的提名超过了其容量上限时，首先将该分段总可用容量的1%分配给新托运商，所有新托运商分得的总容量不得超过该分段总可用容量的5%；其次，该月的剩余容量将提供给已提名容量的常规托运商。这些剩余容量将按照它们的基础运输百分比在常规托运商之间分配。如果按照基础运输百分比分配给托运商的容量大于其提名的容量，则超额容量将按照它们的基础运输百分比重新分配给所有其他常规托运商。

金德摩根规范提货和调度政策，确保及时交付产品，维护客户满意度。托运商必须通过Common Customer Interface（简称"CCI"）系统在每月的10日前提交每个管道段在随后月份提货的电子提名。未使用托运商CCI系统提交提名的必须以要求的书面格式提交，并在该月的5日前寄送给管道公司，如果发生信息变更，托运商必须至少提前7个工作日通过CCI系统提交变更的批量体积、目的地、供应商和收货人等信息。管道公司将从上述信息生成月度时间表，并在每月的20号将其分发给托运商、托运商指定的供应商和收货人。

2.4 法国成品油管道公司 TRAPIL

TRAPIL 公司成立于1950年，是欧洲最大的成品油管道公司之一，由道达尔能源、壳牌公司、埃克森美孚等公司共同拥有和运营。TRAPIL 的股权结构分散，主要股东包括 Noven SAS（49.72%）、Freesia Holding Participation BV（24.25%）、ESSO Société Anonyme Française（17.23%）、TotalEnergies Marketing France（5.50%）和员工持股计划（3.08%），其中 ESSO Société Anonyme Française 是埃克森美孚公司在法国的子公司。TRAPIL 公司管辖三条成品油管道，分别为勒阿弗尔—巴黎（LHP）管道、共和国国防管道（ODC）和地中海-罗讷河（PMR）管道网络，主要运输汽车燃料、喷气燃料和家用燃料。TRAPIL 公司负责输送法国约55%的成品油消费量，其管道里程达4700公里，拥有油罐850000立方米，还拥有160个泵站和交付设施。

在法国，成品油的管道并没有被强制要求第三方准入。第三方接入费率由运营商在开始运营时确定，并至少在其生效前提前两个月提交给能源部长进行审查。对这些接入费率的任何后续修改都可以通过至少在生效前提前一个月向能源部长提交声明来完成（《2012年5月2日第2012-615号法令》第6条）。

TRAPIL 的 LHP 管道网络所运输的成品油主要来自公司股东道达尔能源和埃克森美孚子公司的炼厂，运输采用点对点输送的方式，避免了两个上游炼厂运输要求的冲突问题。同时公司通过严格的批次管理和混油控制程序，管理不同石油产品的输送。

TRAPIL 公司负责统筹制定其管道网络的统一生产计划，各分支机构依据各自管理的管道系统进行相应的调节控制。在混油界面跟踪方面，TRAPIL 公司同时采用密度和荧光剂两种方法。当相邻产品的密度相差较大时，利用在线密度计监测密度的变化进行界面跟踪和产品区分。当相邻产品的密度相近时，在混油段注入荧光剂，通过监测荧光剂的位置来精确跟踪混油界面并区分产品。进行混油的切割和控制方时，当产品物性相近时采用直接顺序输送，每个批次切割的混油量控制在规定范围之内。而当产品物性相差较大时，采用橡胶球作为隔离球进行输送，以减少混油量。混油处理方式包括掺混、自建装置进行常压加热分离和送回就近炼厂处理三种方式。

TRAPIL 公司对输送产品质量的严格控制和管理，输送产品质量控制的好坏，对公司效益有很大的影响。为全面及时监控输送产品的质量，通过建立化验中心对产品在输送前、输送中和输送后三个阶段进行严格化验。同时，在每个站场和关键阀室都配备自动和手动的产品样本采集仪。该仪器在产品输送过程中按批次定期自动采集样品，并分成三份，分别为自己保存、移交客户保存和送化验中心进行质量化验。当客户对产品质量有异议时，将保存样品送第三方进行化验、验证。当有临时需要或客户要求时，则采用手动采集产品样本，进行分析化验，确保能及时准确的掌握产品质量。

3 我国成品油管网公平开放的现状与挑战

3.1 我国成品油管网设施发展现状研究

随着国家《"十四五"现代能源体系规划》的出台，政府着重指出了加强省际和区域间油气输送网络建设的重要性。这包括完善原油和成品油的长距离输送管道，优化东部沿海炼油厂的原油供应结构，以及改进成品油管道的布局，旨在提升成品油通过管道输送的比例。国家正致力于加速成品油管道网络体系的构建，主动推进原本属于三大国有石油公司的主要管道和社会上其他管道之间的互连互通。同时，逐步实施支线管道的规划和投资建设工作，长远目标是通过管网的互联互通，扩大管道油源和市场覆盖范围，实现管道之间、管道与储油站、以及储油站之间的全面连接，最终形成全国范围内统一的成品油管道运输网络。

3.1.1 成品油管道

国内成品油以管道输送为主，铁路、公路为辅，构建形成了西北、西南、珠三角、华北华东四大成品油骨干运输走廊。截至 2023 年底，中国共计成品油管道 58 条，成品油管道里程约 3.2 万公里。自国家管网集团公司成立以来，已高效构建起以"两纵两横"为主干的油气输送网络体系，并深化布局了华北、华东、华中、华南及西南五大核心区域的管网覆盖。2023 年底，该公司已成功运营成品油管道总里程达 2.66 万公里，约占全国成品油管道总里程的 87%。管网覆盖范围广泛，触达全国 27 个省市区，这一庞大的网络支持了全国 30% 的成品油消费量，为促进区域经济平衡发展、保障国家能源安全提供了坚实支撑。

3.1.2 炼化企业及成品油仓储

2023 年，国内炼油能力延续小幅增长态势，总炼化能力 9.5 亿吨/年，同比增长 1.3%，增速放缓 1.9 个百分点，约占全球炼油能力的 18.3%。其中（图 1），中石化炼化占比 32%，中石油占比 25%，其他主营占比 7%，地方炼厂占比 28%，民营炼厂占比 8%。国有炼厂仍占主导地位，但地方炼厂和民营炼厂的规模不断增加。全国千万吨级炼厂增至 36 家，合计炼油能力 5.22 亿吨/年，占国内炼油总能力的 55.8%。目前，成品油管网连接国内 37 家大型炼厂，覆盖全国 27 个省市区。在"十四五"规划期间，炼化产业正以整合资源、规模扩张和园区发展为支撑，提升其核心竞争能力。截至目前，以环渤海、长三角和珠三角三个主要炼化企业集群，以及东北、西北和沿江三个炼化产业带为标志的"三圈三带"产业布局已基本确立，炼化一体化将成为新建炼厂标配。自 2020 年以来，新一轮炼化扩能潮推动市场竞争白热化，中国成品油产量实现较快增长，国内成品油供应能力得到有力保障。

在油库建设领域，成品油企业正积极扩展其业务规模，其中中国石油和中国石化两大集团占据了市场的主导地位，同时，包括中国海油在内的新兴企业也在逐步增加其油库的持有量，共同助力中国油库向现代化转型。2015 年我国成品油商业油库容量约为 4710 万立方米，到 2022 年成品油商业油库容量达到 5067 万立方米。国家管网成立初期，成品油油库尚不具备公平开放的条件。国家管网成立后，陆续为所接收的油库新增地付设施，使其具备下载功能，从而初步具备了向第三方开放的物理条件，为进一步推动油库公平开放奠定了基础。

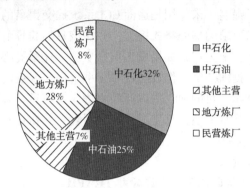

图 1　炼化能力市场份额

3.1.3 成品油接卸码头

根据《2023 年交通运输行业发展统计公报》，中国现有液体散货专业化泊位 1544 个，其中成品油泊位 161 个，较 2022 年增长 6 个。成品油的卸载码头主要分为两种类型：一种是供港口使用的公共码头，另一种则是专门为石化企业服务的专用码头，主要分布在环渤海地区、长三角地区、珠三角地区三个区域。随着国内能源需求的增长和港口基础设施的现代化，成品油接卸码头的容量和效率不断提升。这些码头不仅支持大宗成品油的进口，还兼顾了国内供应链的稳定与安全。尤其在长三角、珠三角和渤海湾等经济活跃区域，大型油气公司加强了对码头设施的投资，改善了接卸设备，提升了自动化和环保水平，确保了成品油的高效、安全运输。然而，随着国际油价波动和环保政策趋严，部分老旧码头面临升级改造压力，整体行业依然处于优化整合的过程中。

3.1.4 成品油用户侧工矿企业、加油站等

成品油零售市场竞争日趋激烈，民营企业竞争力不断增强。截止到 2023 年年末，我国加油站数量 12.3 万座左右。据卓创资讯统计数据显示，民营加油站占比为 52.13%，中石化加油站数量占比 26.20%，中石油加油站数量占比 19.64%，外资加油站数量占比 2.03%。数据对比，民营加油站数量庞大，占据中国加油站数量半壁江山。2024 年我国加油站数量已减至 11.06 万座，较上年下降 1.92%，预计 2025 年至 2030 年，国内加油站数量将持续下滑至 10 万座，至 2035 年可能降至 7 万座至 8 万座。

3.2 我国成品油管网公平开放的必要性

3.2.1 双碳背景下成品油需求逐渐饱和，公平开放是应对达峰挑战的现实选择

在碳达峰和碳中和愿景的推动下，交通能源电动化进程不断加快，引发了能源、交通和工业等领域的深刻变革（表4）。这些变革预计在2025至2030年间，将推动我国石油需求达到约8亿吨的峰值平台期，其中成品油需求预计在2025年触及峰值。在这样的背景下，新建成品油储运基础设施面临着严峻的经济效益挑战。为避免资源浪费和重复建设，迫切需要探索更为高效、可持续的发展路径。公平开放和共享存量基础设施成为关键所在。通过打破行业壁垒，促进不同企业之间的合作与共享，可以充分利用现有资源，提高资产利用效率。这不仅能够降低全社会的成品油储运成本，还有助于优化资源配置，推动产业链向双碳转型发展。

表4 石油石化需求分情景展望

分类	内容
全球石油需求预测	埃克森美孚：2050年石油需求将保持在每日1亿桶，与当前水平相当
	英国石油公司（BP）：2050年需求预计降至每日7500万桶
石化产品需求增长	2022—2028年，石化原料需求预计占石油总需求增长的40%以上
	主要增长区域：发展中国家（特别是亚洲），印度预计2027年超越中国成为石油消费增长主要来源
区域需求差异	中国：2024年上半年柴油、石脑油加工量下降，国际能源署（IEA）下调需求增长预期
	其他亚太国家：非经合组织国家需求增长，预计到2030年前每年增加约43万桶/天，但难以复制中国过去20年的增长表现
供需平衡与价格展望	OPEC增产与页岩油供应将使2025年全球石油供需趋于平衡或宽松
	页岩油边际平衡点：每桶62-70美元
	预期油价下降可能减缓柴油和石脑油在重卡和烯烃领域的替代趋势

3.2.2 成品油市场多元化发展，民营企业影响力不断增强，入网意愿强烈

成品油市场正呈现出多元化的发展态势，民营企业在这一进程中的影响力持续增强，展现出强烈的入网意愿。虽然国有炼厂在市场中仍占据主导地位，但地方炼厂和民营炼厂的规模正在不断扩大，逐渐形成了一股不可忽视的力量。在成品油零售市场方面，竞争日趋激烈，民营企业竞争力更是不断增强。

民营企业入网意愿强烈，但由于管网设施共享问题的限制，往往难以成功入网。具体来说，由于现有的沿线下载油库基本上隶属于中石油和中石化等大型国有企业，民营企业在资源入网后，往往难以找到公平开放的下载口进行利用。这种情况使得民营企业在成品油市场中处于相对被动的地位，难以充分发挥其竞争力和市场影响力。

3.2.3 引导供需两端规范运作，是推动成品油市场转型升级的关键举措

彻底整顿扰乱成品油市场乱象，需要成品油管网公平开放。成品油管网公平开放要求基础设施运营企业公开剩余容量等信息，同时对成品油质量实施检测和计量。此外，实现成品油管网公平开放，还将在供需两端形成积极的倒逼效应。为了适应市场变化，供需双方将不得不规范成品油台账管理制度，确保油品来源和流向的清晰透明。这将有效减少非法加工、调和成品油以及经营不符合产品质量标准或环保强制性标准的成品油等扰乱市场秩序的行为。

成品油行业数字化转型跨越式发展也要求成品油管网公平开放。2019年亚太国际管道会议上，中国工程院院士黄维和指出："我国油气管网智能化发展方向是形成感知、数据、知识、应用、决策五个层次的总体架构，建设智慧一体化数据库和专业性知识库，通过构建管道数字孪生体，支撑管网全方位感知、综合性预判、自适应优化、一体化管控。"随着数字化转型的持续推进，2022年，工业和信息化部联合国家发展改革委等多个部门共同发布了《关于"十四五"推动石化化工行业高质量发展的指导意见》，提到引导成品油行业发展向数字化、信息化、高端化发展，这必将依赖成品油管网的公平开放和信息公开披露。通过公平开放，能够吸引更多的市场主体参与市场竞争，推动成品油市场的多元化和高效发展；信息公开披露有助于增强市场的透明度和公信力，减少信息不对称，为数字化转型提供有力的数据支撑和决策依据。

3.3 我国成品油管网公平开放面临的关键问题

3.3.1 储油罐设施各为其主，多种油品的管理、接收、储存和混油处理存在困难

相较于美国科洛尼尔等成熟的成品油管道

公司，国内成品油在首站、末站储罐分类和"标签化"输送上有较为明显的困难（表5）。国际公司的成品油管道网络直接与多个炼油厂、铁路货运站、公路货运站和港口等关键节点相连，展现出强大的油品接收、转运、分配和卸载能力。这些系统在首站、末站、注入点和分输点等关键站场具备丰富的油品存储和混合油处理能力。

表5 中国与国外成品油情况对比

公平开放关键问题	中国	俄罗斯（国家主导模式）	欧美（市场主导模式）
油品的管理	资源企业坚持采用油品"标签化"运输方式，不允许资源互认，不同油源的、相同牌号油品，不能同罐储存、混合输送	油品的分类管理主要以油品的牌号为基础，依据不同炼油厂产品的特性，在具体的质量指标上有所差异。针对不同来源油品的硫含量、汽油的辛烷值、柴油的闪点等关键指标进行区分，并据此分类存储。这种管理策略允许相邻批次油品之间有较大的混合量，从而便于进行混合油的处理	分储分输管理方式。通过分别存储和输送不同来源和不同牌号的油品，能够确保每个油品批次的质量标准和物理特性保持一致性。同时设可分离批次和可替代批次，用户自主选择混输或不混输
油品的接收、储存	国家管网的成品油管道在首站的储罐种类较为有限，无法实现对多个来源油品的独立存储。在国内，大多数成品油管道的分输站和末站并未配备国家管网所属的储油设施，因此中途卸载和管道终点的油品一般直接被输送至国家管网以外的储油库	成品油管道网络直接与多个炼油厂、铁路货运站、公路货运站以及港口等关键节点相连接，展现出强大的油品接收、转运、分配和卸载功能。此外，该系统还配备了829个成品油储存设施	按照分储分输的原则，成品油管道系统的首站、末站、注入点和分输点等关键站场需要具备充足的油品存储能力，这意味着必须配备足够数量和容量的储罐。在成品油管道市场中，主导型公司通常拥有包括管道和铁路在内的多种油品输入方式，分输站的功能类似于油库，既可以供用户提取油品，也可以进行油品的分配输送
混油的处理	由于末站的混合油罐数量有限，无法实现对混合油的多阶段（例如四段或五段）分离，这增加了处理混合油的复杂性。由于末站缺乏足够的储油设施，无法采用油罐混合工艺，为了确保输出油品的质量，不得不增加混合油的卸载量。此外，在线混合和蒸馏分离技术的应用也受到限制	广泛应用的隔离液技术（PIG技术）通过引入隔离液形成物理屏障，有效隔离不同批次油品；多段输送与分批调度技术实现了油品的精准调度，确保不同种类的油品按顺序、分段运输；部分管道采用双层管道技术，通过内外管道层分隔运输油品，减少混油风险并提高管道的耐久性	管道系统的末站配备了完善的混合油接收与处理设备，而首站、末站、注入点和分输点等关键站场均具备充足的油品存储能力，配备了足够数量和合适容量的储罐

相较之下，国内的成品油管道系统存在一些局限性。在储油罐设施方面，我国成品油管网普遍存在储油罐设施分散、独立运营的现象。各大油品公司和油田公司往往根据自己的需求独立建设和管理储油设施，导致储油罐的布局缺乏整体规划和系统性，难以进行集中调度和高效管理。这种分散化的存储结构使得油品在储存和接收过程中缺乏统一的标准和操作规程，增加了油品混运的风险。此外，国内的末站大多仅配备两座混合油罐。尽管可以通过两段式切割处理汽油批次间的混合油，以及通过三段式切割处理汽油和柴油批次间的混合油，但缺乏对混合油进行多段（如四段、五段等）切割的能力。

3.3.2 成品油管网公平开放亟需打破资产所有权限制，发展基础设施共享经济

提高成品油管网公平开放程度，发展成品油管网共享经济，能全方位提升成品油管网信息化智能化管理水平、挖掘市场需求潜力。共享经济可以提高信息流动性、打破管网、储罐等设施的所有权壁垒，提高供需匹配效率和管网设施利用率，降低成品油运输成本。

共享经济强调所有权与使用权的相对分离，滴滴的商业模式就是共享经济的典型代表（图2）。滴滴的商业模式可以在一定程度上为管网公平开放遇到的难题提供思路。滴滴通过将车辆的所有权与使用权分离，解决了闲置车辆资源的

低效问题，这一经验可以应用于成品油管网设施的共享，通过平台化运营，将储罐、管道等基础设施的使用权开放给更多市场主体，提高资源利用率，打破资产所有权的限制。

滴滴的智能调度系统也是成品油管网管理的潜在参考方向。滴滴利用数据分析匹配司机与乘客，优化了资源分配效率。同样，成品油管网可以通过建立统一的信息共享与管理平台，实时更新管网容量、使用情况等数据，精准匹配供需双方的资源使用需求，减少信息不对称，提高管网设施的运行效率。此外，滴滴对用户需求的实时响应机制也能为成品油管网提供启发，通过动态调整管输计划，有效应对短期市场波动和突发需求。

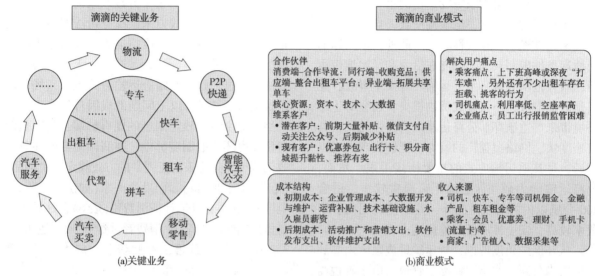

图 2　滴滴出行的关键业务和商业模式

滴滴的规范化管理制度有助于提升行业透明度，这一点在成品油市场尤为重要。当前，成品油市场存在变票销售、非法流通等问题，扰乱了市场秩序。借鉴滴滴的订单和支付管理系统，成品油管网可以建立完善的数字化交易和记录体系，将油品来源、运输路径和交易环节透明化，同时通过数据追踪加强监管，降低非法操作的空间。这种数字化管理不仅有助于构建更加规范的成品油市场，也能为行业的长远发展提供可靠保障。

3.3.3　相比交通领域共享经济，成品油管网基础设施共享还需要一些扶持政策

不同于分散的网约车市场，油气基础设施供应具有垄断竞争性。出自于市场战略的和品牌战略的考虑，一些石油公司的成品油储罐只接收自家炼厂来油。市场战略方面，由于市场竞争激烈、低碳政策导向，成品油市场基本饱和，部分石油公司缺少设施公平开放的动力；品牌战略方面，托运商坚持自家油品质量突出，不允许品牌间混输混储，要求"标签化"输送。

因此，发展管网基础设施共享经济必须在共享平台合法地位、公平开放主体清单、公平开放信息披露罚则规定等方面给予制度支持。在借鉴政府对网约车共享经济前期监管包容、多元化发展经验基础上，必须彻底疏通成品油公平开放的基础设施堵点，才能提高基础设施利用效率。

3.3.4　成品油管道运输面临铁路、水路等多种运输方式竞争

与欧美管道输送为主的运输方式不同，我国成品油运输方式多元。我国成品油运输以管道、铁路、水路、公路运输为主，其中管道与铁路是最主要的运输方式，水路主要承运部分沿海、沿江运输，公路以近距离集散为主。

不同的运输方式各有优缺点，管道运输以运量大、效率高为主要特点。其平稳、连续的输送模式受外界环境影响较小，具备较高的安全性，即使在极端天气条件下仍能保持稳定运行。此外，管道密闭输送的特点大幅减少运输过程中的油品损耗，同时高度机械化和自动化的运行方式提供了远程集中监控的优势，对环境的影响也较小。然而，管道运输灵活性不足，仅适用于大量、单向、定点的运输需求，不如车船灵活多样。不同油品的混输可能导致混油问题，需要额外处理或降级使用，增加输配成本。铁路运输在速度和计划性方面具有显著优势，其运输时间较为准确，且运送能力较大。铁路运输受自然条件

影响较小，能够全天候运行，安全性较高。然而，铁路运输受限于线路布局，不具备普遍适用性。同时，成品油铁路运输需占用双倍运力（油罐车空回），影响了运输连续性。与管道相比，其安全性略低，且由于运输过程中的散逸问题，油品损耗较高。水路运输成本低、运能大，并具有节能减排的优势。沿海和长江干线的运输成本显著低于铁路，是大宗货物运输的重要方式。然而，水路运输速度慢，运输周期长，且容易受自然气象条件影响，连续性较差。此外，水路运输受地理条件限制较大，仅能在沿海和江河地带运行，服务范围有限。水上运输一旦发生事故，尤其是溢油事故，不仅造成货物损失，还会对生态环境带来严重破坏。公路运输以灵活性见长，具有速度快、网络覆盖广的特点，能够实现点对点配送，是其他运输方式的重要衔接手段。然而，公路运输能力较小，运输距离通常较短，适合500公里以内的需求。此外，其运输费用高、事故率较高，安全性不及其他运输方式。

从环保的角度考虑，货运碳排放强度由高到低依次为公路、水运、铁路、管道。由于碳税尚未开征，成品油管道运输能耗低、碳排放少的特点无法在价格中体现。但是如果不将环保性纳入定价考量，管道运输则处于不利地位。在运输选项多样的成品油市场区域，运输服务提供商之间的竞争十分激烈，导致某些管道运输的每吨油品成本相较于其他运输途径显得较高。

3.3.5 变票或无票销售油品通过难以监管的公路、水路等途径销售，扰乱市场秩序

公路和水路运输具有高度灵活性，油品可以绕过正规管网设施进行销售，避开税收和监管义务。这使合法企业难以与非法操作竞争，形成不公平竞争环境，影响了行业整体规范化发展。变票或无票销售绕过管道运输和规范化储运设施，可能造成管网资源闲置，降低管网负荷率。这种行为导致油品流通效率降低，削弱了成品油管网的服务功能。同时，非法销售未能充分利用管道运输的低成本和高效率优势，加剧了运输成本的不平衡，进一步阻碍公平开放政策的落实。

自金税三期上线以来，税务司法部门查处石化企业偷逃消费税的力度空前大，但利益驱动下新型变票交易仍不断涌现。变名、无票、票货分离油品通过公路、水路等运输方式流入市场，严重扰乱市场秩序。以上问题的成因首先在于应税油品与免税油品界定困难的固有弊端，其次是消费税征税环节的制度弊端。在我国，生产环节征收消费税，只能采用正列举的方式规定应税油品，但正列举很难穷尽。而发达国家（如法国、荷兰、英国）是在零售环节计征消费税（表6），这样更容易做出判断，也符合"消费者付费"的底层逻辑。

3.3.6 混油处理涉及经验资质、成本分摊、重复征税等一系列问题

出现混油的情形主要有五种，分别是初始混油、过站混油、停输混油、意外混油和沿程混油。其中沿程混油主要由管道径向速度分布差异、紊流扩散和密度差引起。出现混油会设计经验资质、成本分摊、重复征税等一系列问题。

在混油经验资质方面，国家管网集团无成品油经营资质，混油难处理。当前管道内输送的成品油均为中石油、中石化资源，由于国家管网集团无成品油经营资质，社会独立炼厂资源入网后，企业间产生的混油如何处理是个难题。

表6 中美欧成品油消费税对比

地区	征税范围	税率设置	征税环节
中国	汽油、柴油、石脑油、溶剂油、航空煤油、润滑油、燃料油	全国统一	生产环节、进口环节
美国	汽油、柴油、石脑油、溶剂油、航空煤油、润滑油、燃料油、液化石油气、压缩天然气、醇类燃料、混合燃料等	全国统一的联邦消费税+各州自行制定的州消费	消费税主要集中在生产和进口环节，但实际缴税中可采取"消费税递延纳税安排"，借助"消费税移送和监控系统"实现实缴税环节的后移
欧盟	汽油、柴油、石脑油、溶剂油、航空煤油、润滑油、燃料油、煤油、蜡油、石油沥青等	仅规定成员国适用的最低税率	多环节征收，不同征收对象的征收环节不同，主要以零售环节征税为主

成本分摊方面，混油处理的成本（收益）分摊问题亟需明确。混油处理过程涉及多种利益纠纷情形，例如混油造成的托运商损失由谁负责，混油处理（返厂回炼、在线掺混、蒸馏分离等）过程产生的成本由谁承担，以及处理后混油的收益由谁享有。

重复征税方面，混油的税费问题仍待界定。目前成品油已经需要缴纳增值税、消费税、附加税等税费，如果混油处理后的成品油仍按一般成品油征收，会涉及重复征税的问题，需要进一步研究混油处理税率的界定。

4 关于进一步深化成品油管网公平开放的政策建议

4.1 发展成品油管道、储罐等基础设施共享经济

明确成品油储运设施公平开放的范围，要求中游油库基础设施向市场主体开放，各运营主体积极承担公平开放责任，推动建立成品油储运环节的信息报送、公开和披露制度。市场主体需按规定披露设施使用、运力分配等信息，为监管和市场交易提供透明的数据支持。同时，对现有成品油管道首站、注油站、分输站、末站储罐的产权归属进行全面摸底，加强与粮储局、民营企业等社会油库的物理互联和商务合作，逐步完善管道上下载点和中间油库设施，提升主干管网首末端覆盖率。通过与社会储罐所有者、社会终端销售企业合资建设等方式，打通成品油管网设施开放共享的堵点，缓解上下载基础设施不足问题，并提高现有设施的利用效率。此外，应进一步探索与社会管输资源一体化高效运作的新型商务合作模式，以市场化方式优化"商务一张网"的布局，构建更加灵活、高效的设施共享体系，推动构建全国统一大市场。

4.2 搭建公共信息服务平台，推动成品油多式联运

当前，中国已经建立了包括水运、铁路和公路在内的多式联运公共信息平台。例如，2016年启动的长江经济带多式联运公共信息与交易平台，整合了煤炭、钢铁等多种货物的物流信息，具备发布运输节点信息、在线交易处理、路线优化和提供多式联运方案等功能。借鉴长江经济带多式联运公共信息与交易平台的模式，通过整合管道、铁路、水路和公路四种运输方式的优势，可以进一步提升运输系统的综合效率，加速综合交通系统的整合。研究和开发信息共享机制、多式联运的定价和交易机制，有助于更有效地将管道运输整合到综合交通网络中，实现管道运输与其他运输方式的高效协作，构建一个更安全、便捷、高效、环保和经济的能源运输网络。

4.3 优化"标签化"输送，实行批次输送模式

建议由监管机构制定一套全国性的成品油管网接入规范，从国家层面对成品油的不同类型及油品的进出管道技术与管理体系进行统一，确保所有托运商在成品油进出管道时均遵循国家成品油质量标准。实行按油品质量的批次输送模式，分为"可替代批次"、"分离批次"。在管输价格和混油处理上，对"可替代批次"、"分离批次"输送予以区分，即"分离批次"输送等于VIP服务。

4.4 针对混油处理系列问题，完善混油处理的制度安排

针对成品油管道输送中混油问题，制定行业标准和规范，明确混油的范围及处理方法，合理界定国家管网集团在混油经营中的管理责任与权责。通过政策支持，推动混油税费减免或补贴安排，降低企业成本压力。完善混油管理制度，加强混油处理技术研究和推广，为行业提供统一的操作指引和技术支持，保障成品油管网的高效运行。

4.5 建立考虑环境效益的输送定价体系，以引导成品油流通行业低碳转型

管道是天然气长距离运输的最经济高效方式，也是最低碳环保的运输方式。管道运输面临着公路、铁路、水路等多种运输方式的竞争，但目前水路等高环保风险的运输方式定价未考虑环境成本。建议优化管输定价方式，考虑环境效益，参考铁路、水路等运输比价关系，建立科学合理的价格体系并核定运价率。

5 结论

本研究深入探讨了国外典型国家和管网公司的成品油管道运营机制，以及国内成品油管道的运营现状。对比发现目前我国成品油管网主要存在成品油基础设施分散、互联互通性差；在多种运输途径竞争激烈的情况下，价格优势不突出，难以有效吸引客户；市场监管不完善导致变票或无票销售油品行为频发；混油处理涉及经验资质、成本分摊、重复征税等一系列问题，极大地制约了我国成品油管网的公平开放。为完善我国成品油管网运营制度，本文提出打造新型管网设施商务合作模式，发展共享经济；搭建公共信息服务平台，实现成品油多式联运；优化"标签化"输送，实行批次输送模式；完善混油处理及

定价制度，指导成品油运输市场健康发展。希望可以建立一套符合我国国情的成品油管网公平开放机制，帮助我国成品油管网实现增输增效。

参 考 文 献

[1] 梁严,郭海涛,周淑慧,等.美国成品油管道公平开放现状及启示[J].油气储运,2019,38(06):609-616.

[2] 王果涛,廖绮,梁永图,等.美国成品油管道管输能力分配机制及其启示[J].国际石油经济,2022,30(01):85-93.

[3] 梁永图,廖绮,邱睿,等.市场化改革背景下成品油管网运营关键技术及展望[J].油气储运,2023,42(09):978-987+1008.

[4] 王大鹏.科洛尼尔管道油品质量管理经验与启示[J].油气储运,2018,37(03):291-294.

[5] 宫敬,于达.国家管网公司旗下成品油管道运营模式探讨[J].辽宁石油化工大学学报,2020,40(04):87-91.

[6] EMERSON. Brochure: Synthesis for Pipelines [EB/OL]. (2017-2-17) [2024-12-15]. https://www.emerson.com/documents/automation/brochure-synthesis-for-pipelines-en-68214.pdf.

[7] Michel Guénaire, et. al. Oil and gas regulation in France: overview [EB/OL]. (2020-11-1) [2024-12-15]. https://www.gide.com/sites/default/files/oil_and_gas_regulation_in_france_overview_4-629-73281.pdf.

[8] 李玲.油气基础设施建设再迎政策"东风"[N].中国能源报,2022-04-25(013).

[9] 李月清.炼油和零售终端利润稳步提升[J].中国石油企业,2023,(05):41-42.

[10] 李然,田磊,石洪宇,等.全球石油消费超疫情前水平我国成品油市场格局加速演变——2023年国内外石油市场发展形势及2024年展望[J].中国能源,2024,46(Z1):30-42.

[11] 本刊评论员.炼化未来已来——优胜劣汰VS深度重构[J].国企管理,2022,(16):7.

[12] 孙仁金,孙悦,于楠,等.2022年中国成品油行业运行特点分析与2023年展望[J].现代化工,2023,43(07):1-7.

[13] 程佳佳,王成金.我国港口原油码头布局及供需能力分析[J].综合运输,2017,39(04):22-27.

[14] 中共日照市委宣传部理论课题组,毛继春,郭长海,等.亿吨综合大港:海洋特色新兴城市的强力引擎[N].日照日报,2010-04-29(A04).

[15] 刘柏盛.中老铁路运输通道发展对策研究[J].铁道运输与经济,2017,39(08):20-24+47.

[16] 孙明华,王继勇,董雷,等.提高核心竞争力[J].国企管理,2023,(04):34-43.

一种音速喷嘴摩尔质量的修正方法

陈曦宇　王柯栩　裴勇涛

(国家管网集团西气东输分公司)

摘　要　在长输管道中,天然气的摩尔质量主要由站场气相色谱分析仪进行分析并进一步计算得出,而在实际应用场景中,气相色谱分析仪安装位置往往距离流量计量仪表的位置距离较远,途经工艺流程较为复杂,可能产生流速、温度等变化,对于使用音速喷嘴的场景影响更加明显,需要进一步对喷嘴处的气体摩尔质量进行修正以提高音速喷嘴的适用场景和计量准确性。本文建立了一种摩尔质量的修正模型,基于超声流量计的声速核查原理研发相关计算模型实现对在线气相色谱仪测量计算得出的天然气摩尔质量的实时修正,利用色谱仪组分信息计算的摩尔质量和声速值,对实际的摩尔质量值进行修正,得到喷嘴处的实际摩尔质量。通过实验涉及及结果分析,对同一支喷嘴在不同压力段进行分析,3支喷嘴随着计量系统入口压力段的降低,修正量的绝对值均呈现增大的趋势。这种趋势在喉径 $d=9.5838$mm 的音速喷嘴上表现得尤为明显,在入口压力为 5.60MPa 时甚至出现了修正台阶。通过结果分析得到一下结论:音速喷嘴的摩尔质量修正受到喉径尺寸和入口压力的共同影响。在设计和使用音速喷嘴时,应考虑到这些因素对摩尔质量修正的潜在影响,并采取相应的优化措施以提高喷嘴的性能和测量精度。同时,该方法已经在国家石油天然气大流量计量站武汉分站体积管流量标准装置得到了应用,实现了在 2.5~10MPa 压力条件下,高效校准音速喷嘴的技术能力,测量不确定度达到 0.11%($k=2$)。

关键词　音速喷嘴;摩尔质量;修正模型;准确度

Method for molar mass correction of a sonic nozzle

Chen Xiyu　Wang Kexu　Pei Yongtao

(Wuhan Metrology Research Center of West to East Gas Transmission Company of Pipe China)

Abstract　In long-distance pipelines, the molar mass of natural gas is mainly analyzed and further calculated by station gas chromatograph analyzers. In this paper, a correction model of molar mass is established, using the sound velocity measurement function of a stable ultrasonicflowmeter, and the molar mass and sound velocity values calculated by using the component information of the chromatograph are corrected to the actual molar mass values to obtain the actual molar mass at the nozzle.

The model utilizes the ultrasonicflowmeter, gas chromatograph, and temperature and pressure measurement equipment to correctly calculate the molar mass at the nozzle and achieve highly accurate metering.

Analyzing the same nozzle in different pressure segments, all three nozzles show an increasing trend in the absolute value of the correction as the inlet pressure segment of the metering system decreases. This trend is especially obvious for the sonic nozzle with throat diameter $d=9.5838$mm, and even a correction step occurs when the inlet pressure is 5.60MPa; while for the nozzle with $d=25.0785$mm, this change is less obvious.

The molar mass correction of a sonic nozzle is influenced by a combination of throat size and inlet pressure. When designing and using a sonic nozzle, the potential effects of these factors on the molar mass correction should be considered and optimized to improve nozzle performance and measurement accuracy. The method has been applied in the volume pipe flow standard device of Wuhan flow metering Station, and the technical ability to efficiently calibrate the sonic nozzle under the pressure condition of 2.5~10MPa has been realized, and the measurement uncertainty has reached 0.11% ($k=2$).

Key words　sonic nozzle; molar mass; correction model; accuracy

1 引言

喷嘴,作为流体动力学中的核心组件,在众多工业领域中已经得到了广泛地应用。特别是在天然气计量、燃油喷射等技术领域,喷嘴的性能直接影响着整个系统的效果和效率。为了适应这些应用领域对喷嘴性能的要求,全球范围内的研究人员一直积极致力于对喷嘴结构的改进优化以及对流出系数的补偿修正中。

早在1983年,孟明莘就认识到了在较高压力下,或者较高的温度的条件下使用喷嘴时,若使用喷嘴出厂证明书上的流量系数进行计算将会产生一定的误差,需要采取相应的修正来消减这一误差。2012年、2019年,胡志鹏、周晓亮分别对等人对临界流文丘里喷嘴质量流量计算影响参数之一湿度进行了实验,参照 ISO 9300:2005 的湿度修正方法,定义了湿度影响因子,同时通过实验证明湿度修正的影响规律。通过系列研究表明,喷嘴流出系数受到环境及介质的影响。

随着中国西气东输工程等大型输气管道项目的推进,天然气的需求量急剧上升,天然气应用进入了快速发展期,这一变化对气体流量计的检测范围和测量精度提出了更高的挑战。在这种背景下,音速喷嘴作为气体流量标准装置的一部分,对于确保气体流量计的准确校准变得尤为重要。摩尔质量作为音速喷嘴流出系数得出的一环,主要由站场气相色谱分析仪提供,而在实际应用场景中,气相色谱分析仪安装位置往往距离流量计量仪表的位置距离较远,途经工艺流程较为复杂,可能产生流速、温度等变化,导致计量准确性下降。本文建立了一种摩尔质量的修正模型,对实际的摩尔质量值进行修正,得到喷嘴处的准确摩尔质量,实现高准确度的音速喷嘴计量。

2 技术思路及研究方法

2.1 模型建立

本文建立了一种摩尔质量修正模型,使用性能稳定的超声流量计的声速测量功能,通过在音速喷嘴检定台位上游 9D 位置布置专用超声流量计,测量天然气介质中的声速,对标准装置上游安装的气相色谱仪分析计算的摩尔质量进行实时修正,消除了气相色谱仪分析时间滞后产生的天然气组成数据误差。通过分析气体组分与声速之间的关系,建立天然气介质中声速与摩尔质量的函数关系如下。

摩尔质量修正计算模型:

$$M_{NOZ} = \left(\frac{SOS_{GC}}{SOS_{USM}}\right)^2 M_{GC}$$

音速喷嘴流出系数计算模型:

$$C_d = \frac{q_v p_{ref}}{\frac{\pi d^2}{4} C_* p_0 T_{ref} Z_0} \sqrt{\frac{M_{NOZ}}{R} T_0}$$

式中,M_{NOZ} 为音速喷嘴处的摩尔质量,kg/mol;M_{GC} 为色谱仪组分计算的摩尔质量,kg/mol;SOS_{GC} 为色谱仪组分计算的声速,m/s;SOS_{USM} 为超声流量计测量的声速,m/s。p_{ref} 为高压活塞体积管天然气流量原级标准处的压力,Pa;T_{ref} 为 HPPP 标准装置处的温度,K;Z_0 为喷嘴滞止条件下的压缩因子,无量纲;d 为喷嘴的喉部直径,m;P_0 为临界流喷嘴上游入口处的滞止压力,Pa;T_0 为临界流喷嘴上游入口处的滞止温度,K;Z_0 为临界流喷嘴上游入口处滞止状态下的天然气的因子,无量纲;q_V 为工况条件下 HPPP 复现的体积流量,m³/s;C_* 为临界流函数。

2.2 实验设计

2019年,体积管法流量标准装置参与了由 NIM 组织的实验室间比对,采用 4 支音速喷嘴作为传递标准,考虑到不同实验室的压力范围和工作介质差异,拟合流出系数与雷诺数之间的关系曲线,以流出系数测量值与参考值的归一化偏差 E_n 值来评估装置间的一致性。在此次实验室比对中,武汉分站的最大归一化偏差 E_n 为 0.79,证明了实验室的音速喷嘴测量能力的一致性及准确性。

本文实验设计是依托于国家石油天然气大流量计量站武汉分站的体积管流量标准装置建立。该原级标准使用到了高压活塞体积管法,由体积管、活塞、四通阀组、过滤器、单向止回阀、传递涡轮、限流音速喷嘴组等部分组成,具体实验流程见图1,摩尔质量修正的软件实现方法见图2。

实验从两方面进行展开:1. 研究同一喉径的多支音速喷嘴在同一压力点下摩尔质量修正的表现情况。选取多支喉径一致的音速喷嘴在同一压力点下进行多次测量,分析摩尔质量的修正情况。2. 研究不同喉径的多支音速喷嘴的摩尔质量修正情况与计量系统入口压力的关联情况。选取不同喉径的多支音速喷嘴在不同压力点下进行多次测量,分析摩尔质量修正情况。

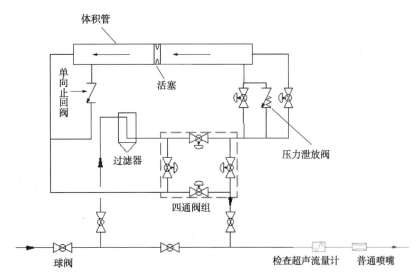

图 1　实验测试流程

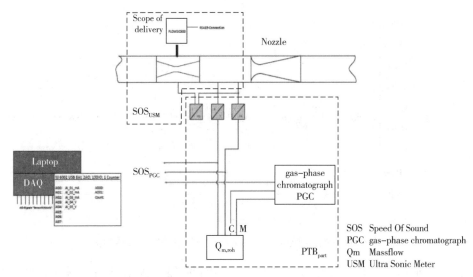

图 2　摩尔质量 M 修正的软件实现方法

实际喷嘴处摩尔质量和色谱测出的气体摩尔质量由于距离原因导致存在偏差，所以通过比较二者的差异进行摩尔质量修正，通过下式计算摩尔质量修正量。

$$修正量 = \frac{M_{NOZ} - M_{GC}}{M_{GC}} * 100\%$$

式中，M_{NOZ} 为音速喷嘴处的摩尔质量，即修正后摩尔质量，kg/mol；M_{GC} 为色谱仪组分计算的摩尔质量，即修正前摩尔质量，kg/mol。

3　实验结果分析

对于同一喉径的多支音速喷嘴在同一压力点下摩尔质量修正的表现情况，本次实验选取相同出场厂家按照同一标准生产的 3 支喷嘴，出厂喉径为 $d = 25.0894$ mm（参考其几何尺寸校准证书，极差为 0.0106mm，偏差在可接受范围内），编号为 1#、2#、3#，每支音速喷嘴均在计量系统入口压力为 5.60MPa 下重复测量 6 次，并对摩尔质量的修正情况进行统计分析，结果如图 3 所示。

通过图 3 看出，三支音速喷嘴的摩尔质量修正量集中在 -0.04% ~ 0.06% 之间，1# 和 3# 喷嘴的修正情况较为一致，3# 喷嘴的修正情况最稳定。这表明同一喉径音速喷嘴的修正情况较为集中，且在一定程度上表现一致。但单点表现和趋势波动并不完全一致，这可能与喷嘴自身加工情况、计量系统所处的环境条件有一定关联性。

对于不同喉径的多支音速喷嘴的摩尔质量修正情况与计量系统入口压力的关联情况的研究，

本次实验选取喉径 $d = 9.5838$mm、$d = 19.136$mm、$d = 25.0785$mm 的音速喷嘴三支，分别在计量系统入口压力为 6.70MPa、6.00MPa、5.60MPa 三个压力点下进行测量，每支音速喷嘴每个压力点分别测量 6 次，并对摩尔质量的修正情况进行统计分析，结果如图 4。

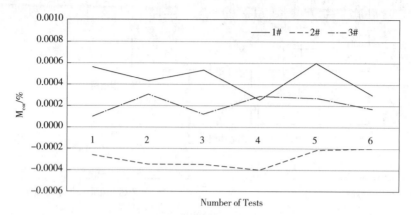

图 3 同喉径音速喷嘴在同一压力点摩尔质量修正情况

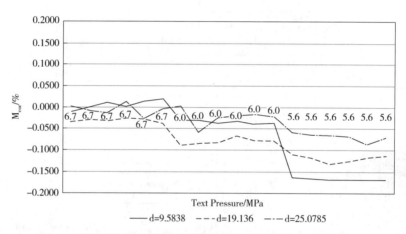

图 4 不同喉径音速喷嘴摩尔质量修正与计量系统入口压力关系

对计量系统不同入口压力段进行分析，入口压力为 6.70MPa 时，3 个口径的音速喷嘴摩尔质量修正量集中在 $-0.04\% \sim 0.02\%$ 之间，喉径 $d = 19.136$mm 的音速喷嘴测量点之间的变化最小；在入口压力为 6.00MPa 时，修正量集中在 $-0.04\% \sim 0.00\%$ 之间，喉径 $d = 9.5838$mm 的音速喷嘴测量点之间的变化最小；在入口压力为 5.60MPa 时，修正量集中在 $-0.17\% \sim -0.05\%$ 之间，喉径 $d = 9.5838$mm 的音速喷嘴测量点之间的变化最小。

对同一支喷嘴在不同压力段进行分析，3 支喷嘴随着计量系统入口压力段的降低，修正量的绝对值均呈现增大的趋势。这种趋势在喉径 $d = 9.5838$mm 的音速喷嘴上表现得尤为明显，在入口压力为 5.60MPa 时甚至出现了修正台阶；而对于 $d = 25.0785$mm 的喷嘴而言，这种变化就不太明显了。同时能够发现，在低压（5.6MPa）下，喷嘴喉径越大则流出系数越接近 0，即喷嘴处组分与色谱测出组分几乎没有差异，表明喉径越大对气体流速影响越小，受距限制也越小；高压（6.7）MPa 下三种喉径都差异几乎为 0，此时压力造成的高流速影响克服了喉径差异的影响。

4 结论

本文使用性能稳定的超声流量计的声速测量功能建立了一种摩尔质量的修正模型，同时涉及了系列实验验证了同喉径音速喷嘴修正量的一致性，修正量与喉径、压力的关系，得出结论如下：

（1）同喉径音速喷嘴摩尔质量修正的一致性：在相同喉径条件下，音速喷嘴的摩尔质量修正具有较好的集中性和一致性，但同时也存在一定的个体差异，这可能与喷嘴的加工精度和使用环境有关。

（2）不同喉径音速喷嘴摩尔质量修正与入口压力的关系：对于音速喷嘴而言，喉径大小与入口压力之间存在显著的相互作用，且在较低压力条件下，小喉径喷嘴的摩尔质量修正更为敏感。

（3）音速喷嘴的摩尔质量修正受到喉径尺寸和入口压力的共同影响。在设计和使用音速喷嘴时，应考虑到这些因素对摩尔质量修正的潜在影响，并采取相应的优化措施以提高喷嘴的性能和测量精度。此外，对于特定应用场景，选择合适的喉径和调整入口压力至适当水平，也是确保喷嘴稳定运行和精确测量的关键。

参 考 文 献

[1] 孟明莘. 高压下音速喷嘴流量系数的修正[J]. 自动化仪表，1983(05)：16.

[2] 胡志鹏，孙秀良，赵艳. 临界流文丘里喷嘴质量流量计算中临界流函数的取用及湿度修正的影响[J]. 工业计量，2012，22(S1)：59-60.

[3] 周晓亮，张立谦，刘锴，等. 临界流文丘里喷嘴湿度修正因子的研究[J]. 计量技术，2019(04)：15-18.

基于双向流固耦合的天然气科氏质量流量计模拟与优化

裴全斌

(国家管网集团西气东输分公司)

摘 要 【目的】随着天然气工业的迅速发展,对高精度流量测量技术的需求日益增加。大口径双U型科里奥利质量流量计在天然气流量测量中扮演着重要角色,但其在实际应用中存在测量精度不高、压力损失较大等问题。本研究旨在通过建立仿真模型并采用双向流固耦合方法,深入分析科氏流量计在不同工况下的流场特性和振动特性,以期解决这些问题,提高测量的准确性和可靠性。【方法】本文基于ANSYS软件构建了科氏流量计的仿真模型,并运用双向流固耦合技术进行模拟。通过模态分析和振动分析,计算了在不同工况下流量计的谐振频率、时间差等关键参数。研究中利用了有限元方法、计算流体力学以及结构力学等多物理场耦合计算方法,结合了NIST真实气体模型和Realizable可拓展壁面函数模型。【结果】研究发现,对于同一结构,入口质量流量和工作压力的变化对模态频率的影响不大。随着质量流量的增加,振动管的时间差和相位差呈线性增长,而拾振器处的振幅逐渐减小。在相同质量流量下,流体压力的增加导致测量管拾振器处的相位差、时间差减小,且振幅明显下降。此外,气体的可压缩性和弯管结构导致气体在测量管内的速度和密度分布不均匀,进而影响测量误差。这些发现与前人在科氏流量计的研究中有所不同,特别是在考虑气体可压缩性和结构影响方面。【结论】本研究的仿真结果为科氏流量计的设计和校准提供了新的见解,特别是在大口径天然气管道测量中的应用。研究结果表明,测量时间差和相位差随流量增加呈线性增长,而流体压力的增加则导致测量管振幅下降。这些发现有助于优化科氏流量计的设计,减少测量误差,提高其在天然气流量测量中的性能和准确性,对天然气工业的流量测量技术发展具有重要的参考和导向作用。

关键词 科氏质量流量计;流固耦合;天然气;流量测量

Simulation and Optimization of Coriolis Mass Flowmeters for Natural Gas Based on Bidirectional Fluid-Solid Coupling

Pei Quanbin

(State Oil and Gas Pipeline Network Group Co., Ltd. West-East Gas Transmission Branch Company)

Abstract [Purpose] With the rapid development of the natural gas industry, there is an increasing demand for high-precision flow measurement technology. Large-diameter double-U Coriolis mass flowmeters play a significant role in natural gas flow measurement, but they face issues such as low measurement accuracy and high pressure loss in practical applications. This study aims to build a simulation model and use bidirectional fluid-structure coupling methods to deeply analyze the flow field and vibration characteristics of Coriolis flowmeters under various working conditions, in order to address these problems and improve measurement accuracy and reliability. [Method] This paper constructs a simulation model of the Coriolis flowmeter using ANSYS software and employs bidirectional fluid-structure coupling techniques for simulation. Modal and vibration analyses were conducted to calculate key parameters such as resonant frequency and time difference under different working conditions. The study utilized multi-physics field coupling calculation methods such as the finite element method, computational fluid dynamics, and structural mechanics, combined with the NIST real gas model and the Realizable k-ε turbulence model. [Results] The study found that for the same structure, changes in inlet mass flow rate and working pressure have little effect on modal frequency. As the mass flow rate increases, the time difference and phase difference of the vibrating tube increase linearly, while the amplitude at the pick-up point gradually decreases. At the same mass flow rate, an increase in fluid pressure leads to a decrease in phase difference, time difference, and a significant drop in the amplitude of the measuring tube. Additionally, the compressibility of the gas and the presence of bend structures cause the velocity and density distribution of the gas in the measuring tube to become uneven, affecting measurement errors. These findings differ from previous studies on Coriolis flowmeters, especially in considering the effects of gas compressibility and structure. [Conclusion] The simulation results of this study provide new insights for the design and calibration of Coriolis flowmeters, especially in the application of large-diameter natural gas pipeline measurement. The results indicate that the measurement time difference and phase difference increase linearly with the flow rate, while an increase in fluid pressure leads to a decrease in the amplitude of the measuring tube. These findings help optimize the design of Coriolis flowmeters, reduce measurement errors, and improve their performance and accuracy in natural gas flow measurement, playing a significant role in guiding the development of flow measurement technology in the natural gas industry.

Key words Coriolis flowmeter; fluid-structure coupling; natural gas; flow measurement

科里奥利质量流量计适用于多种液体及气体的测量，被广泛接受并成为石油化工行业主流流量计。目前，天然气计量表多选用质量流量计、孔板流量计和气体涡轮流量计。孔板流量计存在磨损大准确度低，量程比小，压力损失大等缺点。气体涡轮流量计对被测介质的清洁度要求很高，涡轮流量计的叶轮容易损坏，导致涡轮流量计的后期维护量增加。科里奥利质量流量计其最大特点是可以直接测量流体的质量，能够测量高粘度流体和高压气体的流量，而在流体通道内没有阻流元件和可动部件，具有良好的可靠性、长使用寿命、准确性、重复性和稳定性等优点。科里奥利质量流量计在压缩天然气(CNG)、液化天然气(LNG)等能源及航天工业、科研部门中已经得到了广泛的应用。

在天然气管道工业计量领域，中小口径的气体质量流量计发展较快，但其在大口径管道流量测量领域存在测量精度不高、压力损失较大、安装固定要求高等缺点，限制了其在天然气长输管道中的计量应用。而且流量计在标定和校准装置方面受到了一定的限制，目前科氏流量计产品主要依赖进口，其测量性能和稳定性与国外产品存在一定的差距。

科氏流量计在进行测量时，压力、密度、温度、流速、压缩性等参数均会对测量结果产生影响。当测量管振动时，其能量在测量过程中将通过其边界结构传递到传感器外壳或管路，这将会导致传感器振动状态的不稳定。在许多工况下，介质的工作压力较低，通常小于1MPa，压力对流量的影响可以忽略。然而，在实际应用中，较高的流体压力可能会导致流量测量的偏差。Wang T等提出了利用线性阻尼模型进行分析的理论方法，对直管型科氏流量计进行了低压和高压流动实验测试，与理论分析结果具有较好的一致性。Cascetta等利用流量校准设备对弯管科氏流量计进行测试，实验结果表明当流体压力大于1.5MPa时，被测仪表的精度开始呈现出负误差。Büker O等以水和油为测试介质，在不同流量的测量过程中发现，质量流量一定时，温度随压力的增加而升高，但压力对质量流量测量精度没有发现明显的依赖性。

研究者采用数值模拟方法对科氏流量计的测量特性及影响规律进行了预测和研究，以指导流量计的设计和校准。Stack等引入有限元方法求解输送流体的Timoshenko梁的控制方程以来，采用数值方法对流量传感器进行建模已成为一种重要的开发工具。Belhadj等利用ANSYS模拟了一系列几何形状不同的科里奥利测量管(直管和弯管)内部的脉动流动。Bobovnik等首先使用CFD代码模拟了直束型科里奥利流管，表明了CFD仿真计算对科氏流量计模拟的可靠性。

随着计算流体力学、结构力学等多物理场耦合计算方法及软件的发展，数值模拟广泛应用在科氏流量计的流固耦合特性研究方面，并逐渐从单向流固耦合转向计算精度更高的双向流固耦合方法。针对大口径天然气管道测量用的双U型科氏流量计，本文基于ANSYS仿真环境搭建了流量计仿真模型，并采用双向流固耦合方法对其进行了模拟研究，计算分析了不同工况下该科氏流量计在测量天然气介质时的流场特性、振动模态频率及时间差等预测结果及影响规律。

1 测量原理

科氏流量计利用了流体与固体相互作用产生的科里奥利力引起的振动来测量流体的质量流量。在工作中，振动管在电磁驱动器和科氏惯性力的共同作用下发生弹性变形。该振荡取决于其测量管的弹性特性、几何结构（即流量传感器形状、管道的尺寸、横截面和厚度等）以及流体的流动特性等。

以双 U 型科氏流量计为例，其工作原理如图 1 所示。

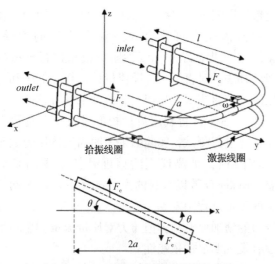

图 1 双 U 型科氏流量计测量原理图

如图 1 所示，流体由测量管入口端 inlet 流入，流通经过 U 型管后从另一出口端 outlet 流出，同时在该过程下，测量管将在驱振器的激振下处于共振状态。当管道中无流体流动时，可以看到测量管上各点的振动幅度相同，因此 A 与 B 两侧输出的是同相位信号。当管中有流体流动时，由于科里奥利力的存在，测量管在振动的同时还发生了扭转，则此时由 A 与 B 两侧输出的信号将存在相位差，与管内质量流量成正比。

当密度为 ρ 的流体在旋转管道中以恒定速度 V 流动时，任何一段长度 Δx 的管道都将受到一个 ΔF_c 的切向科里奥利力：

$$\Delta F_c = 2\omega V \rho A \Delta x \quad (1)$$

式中，A 为管道的流通截面积，ω 为旋转管道的旋转角速度。由于质量流量计流量即为 $q_m = \rho V A$，所以

$$\Delta F_c = 2\omega q_m \Delta x \quad (2)$$

则 U 形管中，直管上一微小长度 dx 受到的扭矩作用为：

$$dM = 2r dF_c = 4\omega R q_m dx \quad (3)$$

r 为弯管段半径，对上式两边求积分得：

$$M = \int dM = \int 4 q_m \omega r dx = 4 q_m \omega r L \quad (4)$$

其中 L 为 U 型管直管段的长度。

在该扭矩的作用下，U 形管产生的扭角 θ 很小则有：

$$M = K_s \theta \quad (5)$$

式中，K_s 为流量管的角弹性模量，由式 (4)、式 (5) 可得：

$$q_m = \frac{K_s \theta}{4\omega r L} \quad (6)$$

V_t 为流量管在振动方向的线速度，即 $V_t = \omega L$，则在形成扭角 θ 的 Δt 时间内

$$\sin\theta = \frac{V_t \Delta t}{2r} \quad (7)$$

由于 θ 很小，因此可近似为 $\theta = \sin\theta$，则

$$\theta = \frac{\omega L \Delta t}{2r} \quad (8)$$

所以由式 (6)、式 (8) 可得：

$$q_m = \frac{K_s}{8 r^2} \Delta t \quad (9)$$

从式 (9) 可以看出流体的质量流量与时间差成正比，时间差前的系数为仪表的标定系数。因此由拾振器位置测得的信号求取相位差，并由相位差与时间差之间的关系求出时间差 Δt，即可计算出质量流量 q_m。

2 数值模拟

2.1 仿真模型构建

本文以某一结构的双 U 型科氏流量计 (CMF) 为例建立有限元模型并进行数值模拟分析。通过双向流固耦合的模态及振动分析可得到反映其性能的谐振频率及时间差等参数，在其结构参数变化带来的测量影响以外，不同的工作条件也会对测量精度造成影响。

双 U 型科氏流量计传感器的结构主要包括测量管、减振板、拾振器、激振器、温度传感器、分流器、底座及外壳。考虑到测量管末端与分流器刚性连接，底座与外壳为刚性部件，拾振

器与激振器的质量仅为几十克,对测量管的振动特性分析影响较小,故在建立有限模型时可忽略分流器、底座、外壳、拾振器及激振器,使计算模型得到简化。因此,本文建立的双U型CMF传感器模型包括测量管与减振板,如图2所示。

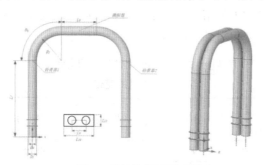

图2 流量计模型示意图

测量管与减振板的结构参数包括各管段长度、弯管半径、管径、壁厚、倾斜度、板厚、板长、板宽等,具体数值如表1所示。

表1 结构设计参数

符号	值	符号	值
L_7	420.00mm	θ_5	90°
L_8	100.50mm	D_3	15.00mm
L_9	58.00mm	D_4	12.00mm
L_{10}	121.00mm	R_3	165.88mm
L_{11}	63.00mm		

测量管与减振板的材质均为退火处理后的316的不锈钢,在仿真中需设置其弹性模量、泊松比和密度,其相关材料参数见于表2。

表2 双U型CMF固体材料参数316不锈钢

参数	数值
弹性模量	193GPa
泊松比	0.275
密度	7980kg/m³
拉伸屈服强度	252.1MPa
拉伸极限强度	565.1MPa

使用Mechanical软件对模型分别进行流体网格划分及固体网格划分,图3和图4分别展示了CMF的固体结构网格划分和内部的流体结构网格划分情况。经过网格无关性验证后,综合考虑仿真过程的计算精度及计算时间,将其中固体网格大小设置为1.5mm,生成网格数为434394,平均单元质量0.81617,最大偏度0.6127,横纵比1.5225。流体网格大小设置为3mm,生成网格数为680833,平均单元质量0.55792,横纵比3.6738。

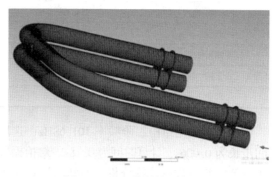

图3 CMF固体结构网格划分

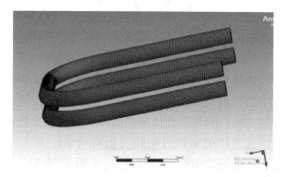

图4 CMF内部流体网格划分

2.2 仿真模型设置

CMF内部的流体部分使用Fluent进行仿真。介质模型使用NIST真实气体模型,粘性模型选用$k-\varepsilon$中Realizable可拓展壁面函数模型,入口边界条件为速度入口。介质设置为纯甲烷,工作温度为300K,工作压力和流量参数在模型计算过程中进行变化。

流固耦合振动分析需要给测量管施加正弦驱动力,其频率即测量管的谐振频率。通过模态分析求取合适的振型对应的固有频率是进行流固耦合振动分析的基础。模态是结构的固有振动特性,每一阶模态具有特定的固有频率、阻尼比和模态振型。通过Fluent软件对管路内流体流动情况进行计算求解后,将求解结果导入Mechanical静态结构模块进行预应力分析,而后进行预应力模态求解,计算完成得到测量管结构的一阶至六阶固有频率如表3所示。通过计算分析得到的各阶模态下的振型,在二阶模态101.68Hz下,其振型较符合CMF实际的振动状态,其测量管产生一个径向的角速度,符合流量计的运动规律,因此选取二阶模态所对应的固有频率对测量管的流固耦合激振频率进行设置。

表3 双U型CMF各阶模态频率

阶数	固有频率(Hz)
一阶	90.84
二阶	101.68
三阶	140.04
四阶	157.7
五阶	209.08
六阶	232.45

结构仿真使用结构二阶频率101.68Hz,选取时间步长为0.006s,计算时间为0.18s。对拾振器位置设置监测点进行计算分析,得到测量管振动图像如图5所示,其中激振力幅值F_0为0.001N,初始相位φ为0°,振动在0.05s处达到峰值后,下降幅度小于1%,即可判定振动基本达到稳定状态,并以某一固定频率及幅值保持稳定振动。

数据处理通过Matlab的曲线拟合工具箱功能中的插值(Interpolant)拟合方法来选择合适的插值方法。利用三次样条插值函数对监测点振动曲线进行插值可以得到较为精确的Δt,取位移值相同的振动曲线上的两点t_1、t_2做差得Δt,进一步可通过相位差与时间差的关系求出相位差$\Delta\theta$。因此,通过上述流固耦合振动分析可以研究CMF结构参数、流动参数等对测量时间差及相位差的影响。

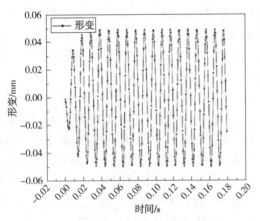

图5 拾振器监测点振动曲线图

3 数据分析

3.1 不同工况下的流固耦合分析

在考虑流体作用的情况下进行双向流固耦合仿真计算,选取300K甲烷作为测量介质,压强为2MPa,密度为13.3kg/m³。对如表1所示结构条件下的科氏流量计进行模拟计算,入口流量条件分别为2000、4000、8000、13000、16000m³/h的测量点进行模拟计算。表4给出了各阶固有频率随入口流量增大变化的趋势,在一定的工作压力条件下,该结构的各阶固有频率变化几乎不变,但在较高的流量点处仍然存在一阶固有频率略微增加的情况,通常在分析中可以认为质量流量的变化对结构固有频率的影响较小。

表4 压力为2MPa时测量管固有频率随入口流量的变化

质量流量/(kg/h)	一阶频率/Hz	二阶频率/Hz	三阶频率/Hz
2000	90.84	101.68	140.05
4000	90.84	101.68	140.05
6000	90.84	101.68	140.05
8000	90.84	101.68	140.05
12000	90.85	101.68	140.05
16000	90.85	101.68	140.05

图6给出了拾振器位置计算得到的相位差随流量变化的趋势。在一定的工作压力条件下,随着测量管入口质量流量的增大,其相位差由最小值0.000301293°增长至最大值0.003510614°,从曲线变化趋势上看,该结构科里奥利流量计的时间差、相位差随流量呈现较稳定的增长趋势。图7给出了不同流量点下,拾振器监测点获取的振幅的变化情况。如图7所示,随着流量计入口质量流量的增大,测量管内的流体质量增加,导致测量管拾振器位置处的振幅呈现出降低的趋势。在该流量范围内其振幅波动相对较小,最大值为0.049306mm,最小值为0.049229mm,其波动处于0.0001mm的范围内。

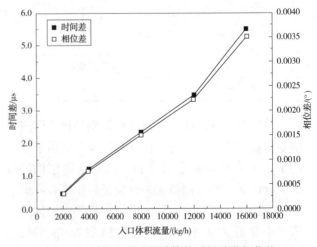

图6 不同流量点下所计算的时间差及相位差

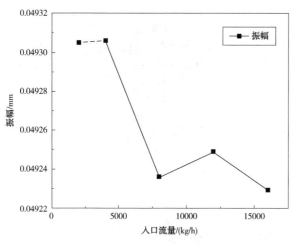

图7 不同流量点对应的测量管振幅

表5给出了相同流量点情况下采用不同工作压力时测量特性参数的变化。在相同流量点下，当流体工作压力越大，介质的流速也越小，这主要是气体被压缩，导致气体密度改变造成的。由表中数据可以看出，对2000kg/h和8000kg/h的流量点分别进行计算，当流体压力由8MPa提高至10MPa时，随压力增大，测量管拾振器处的相位差、时间差也随之减小，且当压力升高时，其测量管的振幅明显下降。在相同流量点下，压力增加所引起的时间差变化对比振幅的影响相对更大。

表5 不同压力条件下的仿真数据

流量点/ (KG/H)	压力/ MPa	入口 速度/ (m/s)	质量 流量/ (kg/h)	仿真 时间差/ s	仿真 相位差/ 度
2000	8	3.119	2068.56	0.000000921	0.000588254
	10	2.431	2067.84	0.000000699	0.000446828
8000	8	12.478	8074.08	8074.08	0.001783736
	10	9.725	8017.2	0.000002645	0.001689566

较高压力下产生的测量时间差的变化由流体压力和管道材料及结构变化等多种因素造成。工作压力对测量管的影响分为绷紧现象、波登管效应与轴向应变微弯效应，前两者导致刚度增加，后者导致刚度减小，在U形管中绷紧现象强于轴向应变微弯效应，故可能导致测量结果偏小。由于压力升高使得振荡管道的材质变硬，表现出与低压工况相比更大的刚度，惯性科里奥利力对管道的作用将产生更小的弹性变形，因此会受到固体结构的影响时间差信号将表现出变小的趋势，在高压下使用的科氏流量计需要进行一定的压力修正。

3.2 测量管内部流动特性分析

对流量计在不同流量点及不同压力条件下的测量管内部流场进行计算可以分析测量管内部的流动特性。图8给出了流量点8000kg/h、工作压力为2MPa时的测量管内布速度分布情况。

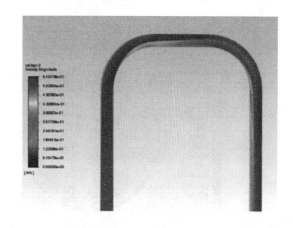

图8 8000kg/h、2MPa时的测量管内速度分布

在实际工作过程中，由于U型科氏流量计的测量管存在弯管结构等原因流体会产生局部的扰动。相对于密度较大的液体测量，特别在进行密度较低的气体流量测量时，结构扰动及测量管振动对管道内部气体流场分布的影响更大。如图8所示，当流体进入左侧的测量管入口后，速度分布符合均匀的圆管流速分布情况，经过两处弯管结构后，由于受到离心力的作用以及气体可压缩性影响，使得测量管内的流速分布逐渐呈现不均匀的趋势。

图9给出了当工作压力为2MPa时，在不同的质量流量点情况下测量管入口侧和出口侧的两个监测点位置处的横截面速度分布云图。图10给出了当测量点流量为8000kg/h时，在不同的工作压力下测量管入口侧和出口侧的两个监测点位置处的横截面速度分布云图。如图9和图10所示，测量管入口一侧管道上布置的监测点处的截面流速分布较为均匀，符合常规的圆管流速分布特性；同时，靠近测量管出口一侧管道上布置的监测点处的流速分布则明显呈现不均匀的状况，管外侧流体流速大于内侧，在内侧管道壁面处出现局部低速区域，这将使得流体振动场中科氏力分布不均匀并影响测量管上拾振器传感测点的受力情况，从而影响拾振器位置处的数据精度及流量计的整体测量效果。

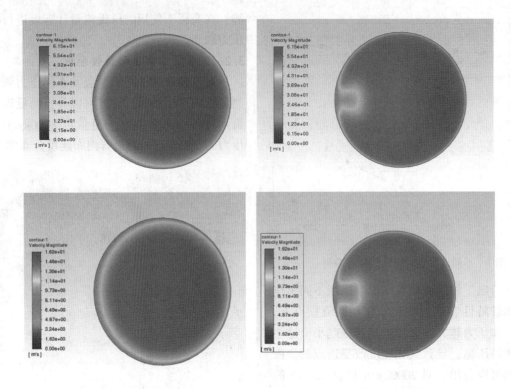

图 9　2MPa 时不同质量流量流场截面速度分布云图(上 8000kg/h 下 2000kg/h)

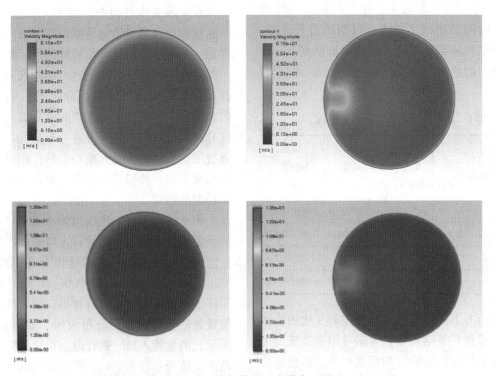

图 10　8000kg/h 时不同压力下流场截面速度分布云图(上 2MPa 下 8MPa)

表 6 给出了对应的不同流量点、不同压力下测量管的入口速度以及测量管内外侧拾振器位置处的速度场数据。表 7 给出了对应的不同工况下的各位置的密度数据。与液体介质相比，对于气体介质而言，经过弯管结构后产生的截面速度和密度差异更明显。由表中可以看出，拾振器 2 位置处的流体速度均比拾振器 1 位置处的流体速度偏大，同时拾振器 2 处的密度比拾振器 1 处的偏小。速度的增加将造成科氏力及相应的时间差增大，同时密度的降低将引起时间差减小，因此对于两者的同时变化所造成的影响需要综合考虑。

表6 不同工况下测量管内的速度场数据

质量流量/ (KG/H)	压力/ MPa	入口速度/ (m/s)	拾振器1 流速/(m/s)	拾振器2 流速/(m/s)
2000	2	13.735	9.7339	11.3562
	5	5.683	4.12089	5.88699
	8	3.119	2.81469	3.16663
	10	2.431	1.43727	2.29954
8000	2	54.941	43.0834	49.2382
	5	22.733	15.7782	22.5403
	8	12.478	9.46688	13.5241
	10	9.725	6.57	8.75

表7 不同工况下测量管内的密度场数据

质量流量/ (KG/H)	压力/ MPa	拾振器1 密度/(kg/m³)	拾振器2 密度/(kg/m³)	驱动频率/ Hz
2000	2	13.3072	13.3055	101.68
	5	34.9724	34.9715	101.68
	8	58.591	58.591	101.68
	10	75.177	75.177	101.68
8000	2	13.3059	13.2996	101.68
	5	34.9698	34.9723	101.68
	8	58.5910	58.5894	101.68
	10	75.1776	75.1751	101.68

4 结论

本文采用双向流固耦合方法对天然气管道用的双U型科里奥利质量流量计进行了理论分析和仿真研究，对流量计在不同工况下的测量特性和流场特性进行了分析，探究了流量计的振动特性和流动特性，为科里奥利质量流量计在不同介质下测量及量值溯源提供了参考。对于该流量计结构而言，通过湿模态分析得到了该流量计适用于采用二阶模态，且不同流量点和压力下的模态和固有频率几乎不变。测量时间差和相位差随流量的增加呈线性增长的趋势，同时拾振器处的振幅随流量的增加在小范围内逐渐减小。在相同的质量流量下，随着流体压力增大，测量管拾振器处的相位差、时间差也随之减小，且当压力升高时，其测量管的振幅明显下降。由于双U型流量计中的弯管结构和存在以及气体的可压缩性，使得气体在测量管内的沿程的速度及密度分布逐渐呈现不均匀的趋势，从而导致管道的受力不均，将进一步引起实际流量计中测量时间差的误差。

参 考 文 献

[1] 薛红梅，张东峰. 科里奥利质量流量计在天然气计量中的应用[J]. 石油仪器，2013，27(02)：36-38+9.

[2] 穆剑，任佳，闵伟，等. 天然气测量中科氏流量计计量性能影响因素探讨[J]. 中国计量，2008，(08)：86-87.

[3] 邢建文，关卫红，李伟. 质量流量计计量准确度的影响因素[J]. 中国仪器仪表，2013(9)：49-51

[4] 范尚春. 科氏质量流量计的若干干扰因素及其抑止[J]. 科学技术与工程，2004，4(4)：269-271.

[5] Wang T, Hussain Y. Pressure effects on Coriolis mass-flowmeters[J]. Flow Measurement and Instrumentation, 2010, 21(4)：504-510.

[6] Cascetta F. Effect of fluid pressure on Coriolis mass flowmeter's performance[J]. ISA Transactions, 1996, 35(4)：365-370.

[7] Büker O, Stolt K, Huu D M, et al. Investigations on pressure dependence of Coriolis Mass Flow Meters used at Hydrogen Refueling Stations[J]. Flow Measurement and Instrumentation, 2020, 76：101815.

[8] Stack C. P., Barnett R. B., Pawlas G. E. A Finite Element for the vibration analysis of a fluid-conveying Timoshenko beam. In：AIAA/ASME structures, structural dynamics and materials conference；1993. 2120-2129.

[9] Belhadj A, Cheesewright R, Clark C. The simulation of Coriolis meter response to pulsating flow using a general purpose FE code. J Fluid Struct 2000；14：613-34.

[10] Bobovnik G, Kutin J, Bajsić I. The effect of flow conditions on the sensitivity of the Coriolis flowmeter. Flow Meas Instrum 2004；15：69-76.

[11] 任建新，谭剑，熊亮，等. 基于刚度模型分析流体压力对直管科氏质量流量计的影响[J]. 机械科学与技术，2012，31(1)：67-70.

[12] 倪伟. 科里奥利质量流量计数字信号处理方法的研究[D]. 合肥工业大学，2004.

[13] Gao R, Wang Y, Xu G. Advanced signal processing techniques for Coriolis mass flowmeters：A review. Journal of Flow Measurement and Instrumentation, 2019, 52：1-12.

[14] Mokhtari M, Azzopardi G. Coriolis mass flowmeter for cryogenic fluids：A review of the state of the art. Measurement, 2020, 158：107749.

[15] Xu G, Gao R, Wang Y. Recent developments in Coriolis mass flowmeter technology：A review. Journal of

Flow Measurement and Instrumentation, 2021, 54: 1-15.

[16] Kottke V, Mokhtari M. Calibration of Coriolis mass flowmeters using a flexible hose: An experimental study. Flow Measurement and Instrumentation, 2022, 78: 101913.

[17] Liu Y, Xu G, Wang Y. A review on the advancements in Coriolis mass flowmeter calibration methods. Measurement, 2023, 192: 109982.

[18] Tijani M H, Mokhtari M. Coriolis flowmeters for multiphase flow measurement: A review. Flow Measurement and Instrumentation, 2020, 68: 101601.

[19] Lin Z, Huang Q, Wang W. Error analysis and compensation for Coriolis mass flowmeters in high-temperature applications. Flow Measurement and Instrumentation, 2021, 70: 101731.

[20] Gao R, Wang Y, Xu G. A novel approach to calibrate Coriolis mass flowmeters using a virtual mass method. Journal of Flow Measurement and Instrumentation, 2022, 55: 1-10.

[21] Kottke V, Mokhtari M. A new method for in-line calibration of Coriolis mass flowmeters. Flow Measurement and Instrumentation, 2023, 79: 102035.

[22] Mettler M D, Hamwi A A. Advances in Coriolis mass flowmeter technology for high-precision measurement. Measurement, 2024, 201: 109744.

Fatigue Failure Analysis of Tensile Armor Layers of Deep-sea Flexible Pipes at Joints

Liu Yu　Chen Yanfei　Zhang Ye　Zhong Ronfeng　Lu Shuntian　Xiang Tao

(National Engineering Laboratory for Pipeline Safety, China University of Petroleum)

Abstract　Dynamic flexible pipes are the essential equipment for deep-sea oil and gas transportation. In the complex dynamic marine environment influenced by waves and currents, the tensile armor layer at the joint is more susceptible to fatigue failure due to residual stresses introduced during the manufacturing process. To address this, a finite element model of the tensile armor layer at the joint was established, and fatigue failure analysis was performed. The results indicate that during the assembly of the tensile armor layer, plastic deformation occurs at all bending angles, but the material returns to its elastic state after relaxation. As the bending angle increases, the stress in the tensile armor layer gradually decreases. Additionally, as the amplitude and cyclic loading increase, the fatigue behavior of the tensile armor layer at the joint transitions from high-cycle to low-cycle fatigue. The traditional tensile armor layer at the joint is most prone to fatigue failure at the bend, suggesting that structural optimization should aim to minimize bending features to enhance fatigue resistance. These findings fill a gap in domestic research on the structural analysis and fatigue failure of tensile armor layers in marine flexible pipes joints.

Keywords　Dynamic Flexible Pipe; Tensile Armor Layer; Fatigue Life Prediction; Fatigue Failure

1 Introduction

The non-bonded flexible pipes consist primarily of the pipe body and the joint. The pipe body is made up of metal and polymer layers, without the need for chemical bonding agents. The number and type of layers, as well as their stacking order, depend on specific design requirements. Users can adjust the number of layers, such as adding an insulation layer, based on practical needs during the design process. The allowance for relative displacement and sliding between the layers enhances the pipe's bending performance, enabling non-bonded flexible pipes to better adapt to the complex marine environment.

Figure 1 shows a schematic diagram of a typical cross-section of a non-bonded flexible pipes.

Due to the differences in the structure of various flexible pipes bodies, the design of the joint needs to be specially tailored to the structure of each pipelines. Figure 2 illustrates a typical joint structure and its components. In the figure, label 4 refers to the tensile armor layer.

1) Stainless steel carcass
2) Polymer fluid barrier
3) Carbon steel pressure armour
4) Anti wear/birdcaging tapes
5) Carbon steel tensile armour
6) Polymer external sheath

Fig. 1　Non-bonded flexible pipe structure

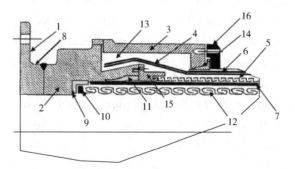

Fig. 2　Non-bonded flexible pipe end fitting structure

Fatigue refers to the progressive and localized structural damage that occurs in materials under cyclic loading. Indeepwater flexible pipes, common types of fatigue include metal layer fatigue, wear-related fatigue caused by relative movement between layers and interlayer contact pressure—such as the wear-related fatigue in the tensile and pressure armor layers—and corrosion fatigue, which arises from crack initiation and propagation due to metal corrosion. While non-metallic polymer layers rarely experience fatigue issues, fatigue failure of the inner lining can occur at the joints or due to pressure and temperature cycling in the pipelines body.

McCarthy et al. noted that the significant stiffness variation between the pipelines and the connector in deepwater environments can lead to progressive fatigue failure, with corrosion potentially accelerating this process. Elman et al. observed that the section of the pipelines near the elbow at the end is particularly prone to damage. Damage to the outer protective layer may result in annular leakage, corrosion of the armor layer, and degradation of the polymer layer, thereby exacerbating fatigue damage and reducing the service life of the pipelines. Nielsen et al. attributed pipelines fatigue to changes in tension and curvature caused by vessel motion and wave loading. Saunders et al. identified damage to the outer protective layer during installation as a key factor in pipelines fatigue, with seawater entering the annulus leading to varying degrees of corrosion in the armor layer. The acidic environment further reduces fatigue resistance. As a result, fatigue assessments should incorporate S-N curves that reflect different permeation components under varying partial pressures. Nielsen et al. also applied the Palmgren-Miner linear damage hypothesis to calculate cumulative fatigue damage. Clements et al. studied the fatigue process of materials under corrosive conditions, developing distinct S-N curves to evaluate the conservativeness of service life estimates. Zhao Lin et al., using a finite element model to analyze fatigue life, found that sections directly subjected to external pressure in interlock structures are most vulnerable to fatigue failure. Xi Yonghui et al. performed semi-physical fatigue simulation tests, analyzing the coordination between loading and deformation. Yang Chan et al. developed a fatigue load analysis model for a single steel strip spirally wound around a cylinder, examining the model's applicability and the influence of the steel strip's geometric parameters.

Campello et al. proposed a new type of joint design aimed at reducing stress concentration within the joint. The experimental results demonstrate that the maximum stress in the steel wire region inside the joint is approximately 1.4 times higher than in the pipe body region. While the stress concentration factor (SCF) is notably reduced, it still needs to be accounted for when calculating the fatigue life of the tensile armor layer steel strips. In another study, Campello et al. used analytical methods and a two-dimensional finite element model to investigate the assembly process of the tensile armor layer within the joint, assessing stress during installation. Unlike the study by Shen et al., the interaction between the tensile armor layer and the epoxy resin, caused by epoxy shrinkage, was also considered. Miyazaki et al. conducted a three-dimensional finite element analysis, incorporating material and geometric non-linearities to simulate the residual stress and geometry changes during the manufacturing process, and calculated the final residual stress under FAT testing and operational loads. Campello et al. employed a local analytical model to predict the stress state during joint assembly and operation, validating the method through comparison with prototype testing results. Based on experimental tests by Campello et al., the stress in the tensile armor layer inside the joint is 2.4 times greater than in the pipe body, making the joint's interior the most vulnerable point for fatigue failure.

Fig. 3 Fatigue fracture of tensile armor layer in endfitting

However, fatigue life assessments of the tensile armor steel strips typically focus on the cross-sectional area of the pipe body, without accounting for the concentrated stress occurring within the tensile armor layer inside the joint. This omission oftenresults in an overestimation of the pipeline's fatigue life. While there has been some accumulation of research abroad on the structural analysis and fatigue failure of tensile armor layers in dynamic flexible pipes, domestic research in China, which focus on the structural analysis and fatigue failure of tensile armor layers at the joints of marine flexible pipes, remains largely undeveloped.

This paper employs finite element software to create a two-dimensional model of an individual steel wire in the tensile armor layer at the joint, simulating its bending and folding during the manufacturing process as well as the testing process during installation and operation. The model's feasibility is verified through comparison with experimental data. Fatigue software is then used to simulate the fatigue failure of the tensile armor layer in the flexible pipes.

2 Finite Element Model and Validation

2.1 Finite Element Model

This study investigates anunbonded flexible pipe with an inner diameter of approximately 200 mm, which contains a pair of tensile armor layers. The inner layer consists of 49 tensile armor wires wound at a positive helix angle, while the outer layer is also made up of 49 tensile armor wires, but wound at a negative angle. The tensile armor wires have a rectangular cross-section and are not bonded to the epoxy resin.

The outer shell of the joint and the internal epoxy resin are both made of linear elastic isotropic materials. The detailed fundamental data for the tensile armor layer inside the joint are provided in Table 1.

Table 1 Basic Data

Name	Size and Unit
Inner diameter (d)	200mm
Number of inner and outer armor wires (n)	49
Armor wire laying angle (α)	26°

续表

Name	Size and Unit
Cross-sectional width of tensile armor layer (w)	12mm
Cross-sectional height of tensile armor layer (t)	6mm
Friction coefficient between epoxy resin and tensile armorwires (μ)	0.3
Bending guide radius during assembly (R_g)	50mm
Young's modulus of tensile armor wires (E)	200GPa
Poisson's ratio of tensile armor wires (ν)	0.3
Yield strength of tensile armor wires (Y_s)	1331.5MPa
Tensile strength of tensile armor wires (UTS)	1550MPa
Young's modulus of epoxy resin inside the joint (E_0)	3800MPa
Poisson's ratio of epoxy resin inside the joint (ν_0)	0.33

The model of the tensile armor layer at the joint is composed of two distinct finite element sub-models, known as the assembly model and the loading model.

2.1.1 Assembly Model

The first sub-model, referred to as the "assembly model" in this paper, is responsible for simulating all aspects related to the tensile armor steel wires during the joint assembly process, aiming to bring the wires to the configuration shown in Figure 4.

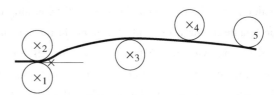

Fig. 4 Assembly model

The assembly model first simulates the post-bend behavior of the tensile armor wire at the entrance bend template of the joint, followed by the unloading and springback effects of the wire, and finally the fixing process at its final position.

The boundary conditions for the specific bending process are as follows: the left end of the tensile armor wire is fixed, template 2 is held in place, and template 1 is rotated counterclockwise around the center of template 2. Then, template1 is rotated clockwise back to its initial position to simulate the unloading and springback of the wire. Next, template 4 rotates clockwise by a certain angle around the origin to apply downward pressure on the wire, while the wire is fixed at template 3. Finally, template 5 moves downward to secure the wire's end in its designated position. Detailed operational steps are shown in Figure 5.

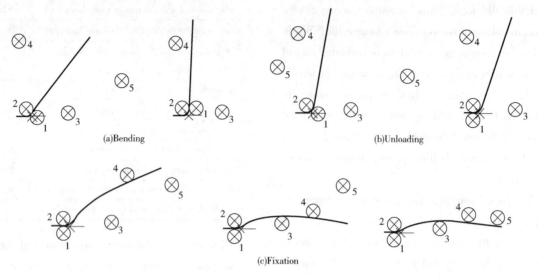

Fig. 5　Folding steps of tensile armor steel wire

2.1.2　Load Model

The second sub-model, referred to in this paper as the "loading model", simulates the application of tensile loads from Factory Acceptance Testing (FAT) and axial operational loads on the tensile armor wire.

The joint's outer shell and epoxy resin areboth treated as isotropic linear elastic materials. This model primarily evaluates the stress distribution in the tensile armor wire inside the joint during tensile load testing. General contact is established between the joint and the epoxy resin, while paired contact elements are used at the interface between the tensile armor wire (including its tip) and the epoxy resin to monitor and control the contact between any two components, as shown in Figure 6.

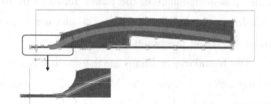

Fig. 6　Friction contact conditions

In the loading model, the tensile load is applied at the end of the tensile armor wire outside the joint, and the right end of the wire is mechanically anchored, as shown in Figure 7.

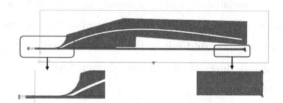

Fig. 7　Tensile and anchoring conditions

2.2　Validation of Validity

2.2.1　Assembly Model

During the joint assembly process, the wire is first bent and folded backward, then it undergoes springback, and finally, it is fixed in its final position inside the joint. Table 2 presents the results of the maximum stress values during the first springback and final shaping for different bending angles, simulated in the first assembly model to replicate manual processing.

Table 2　Stress values under different bending angles

NO.	Bend Angle/(°)	Radian	Maximum Springback Stress/MPa	Maximum Stress in Final Shaping/MPa
1	30	0.5236	983.8	961.7
2	35	0.6109	988.8	1101
3	40	0.6981	968.5	1423
4	45	0.7854	963.2	1503
5	50	0.8727	957.6	1501
6	60	1.0472	952.3	1535
7	70	1.2217	947.2	1547

续表

NO.	Bend Angle/(°)	Radian	Maximum Springback Stress/MPa	Maximum Stress in Final Shaping/MPa
8	80	1.3963	948.6	1547
9	90	1.5708	946.5	1548
10	100	1.7453	941.6	1549
11	110	1.9199	945.7	1550

After installation and final shaping, the stress distribution resulting from different bending and folding angles (along the upper and lower surfaces of the tensile armor wire) can be observed in Figures 8 and Figures 9.

As the bending angle increases, the maximum stress value after final shaping also increases accordingly. When the bending angle is less than 40°, this stress value remains below the yield limit, meaning the wire is still in the elastic stage. Once the bending angle exceeds 40°, a distinct region forms on the armor layer near the joint entrance. This region, known as the "plasticization region", extends from approximately 70mm to 270mm from the free end of the armor layer. The length of this plasticization region increases as the bending angle increases. Between 45° and 60°, the maximum stress value after final shaping continues to rise with increasing bending angles, and the stress peak enters the plastic stage. When the bending angle exceeds 70°, further increases in bending angle have a diminishing effect on the maximum stress value, which approaches the tensile limit and is accompanied by significant plastic deformation.

This trend is consistent with the findings of Shen et al., showing that the stress value is highest in the bend region. In the front section of the bend, the stress value continuously increases, while in the rear section, it gradually decreases. This validates the accuracy of the bending process model.

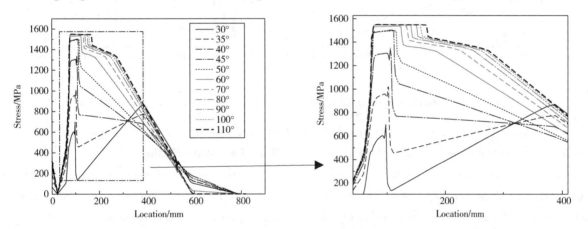

Fig. 8 Stress distribution along the upper surface of steel wire at different bending angles

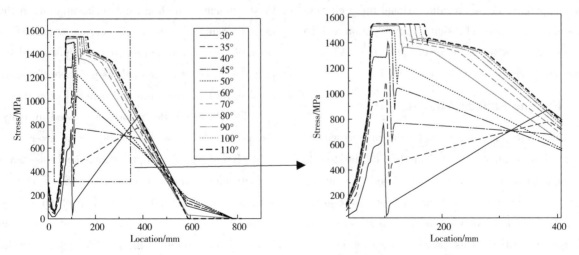

Fig. 9 Stress distribution along the lower surface of steel wire at different bending angles

The bending process induces residual stress in the plasticization region of the wire. During joint assembly, the upper surface of the wire experiences compression, while the lower surface is under tension. By comparing Figures 8 and Figures 9, it can be observed that after installation and shaping, the difference in stress values between the upper and lower surfaces is minimal. Considering the ease of operation for workers during the bending process and the extent of plastic deformation, a bending angle of 45° is selected for the simulation in the following sections.

2.2.2 Load Model

According to the API specifications, the design of unbonded flexible pipes must ensure that the utilization factor (UF) does not exceed 55% during normal operation. Under extreme conditions, the tensile limit is set at 85%. A comparison of the stress results along the upper and lower surfaces of the wire under a tensile load of UF = 55% and no load (UF = 0%) is presented in Figure 10.

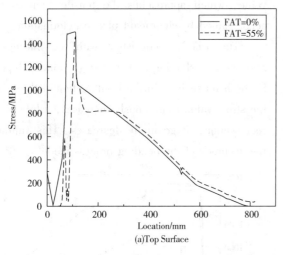 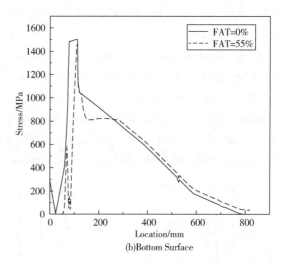

Fig. 10 Comparison of stress along the lower surface of steel wire between UF = 55% and UF = 0%

As shown in Figure 10, FAT acts as a mechanism for relieving residual stress, allowing the wire in the plasticization region to gradually return to an elastic state. However, significant stress concentration remains along the wire in the plasticization region.

The results are consistent with the trend observed in Campello et al.'s two-dimensional finite element model, which showed a significant reduction in high stress values in the bend region after the FAT test. This confirms the accuracy of the model.

3 Fatigue Failure Analysis

Currently, a stress-based fatigue analysis method is used to evaluate the fatigue life of the tensile armor layer at the joint. Therefore, when calculating the fatigue life of the tensile armor layer at the joint, it is essential to define the structural stress for fatigue evaluation. This process primarily involves configuring the fatigue analysis parameters, including the load and material fatigue properties.

3.1 Parameter Setting

In the analysis of high-cycle fatigue failure in marineflexible pipes, key research areas include the load spectrum, crack initiation and propagation, and fatigue life prediction. Given the complexity of these problems, finding a universal analysis method is challenging. To meet engineering requirements, research needs to focus on essential aspects and simplify the model. Fatigue analysis parameters primarily involve the load spectrum and material properties.

The Brown-Miller algorithm is used to analyze the steel material of the tensile armor layer at the joint. Based on the input data, such as tensile strength and Young's modulus, the S-N curve is automatically generated. The load spectrum follows a sinusoidal cyclic stress pattern, and the Stress Dataset is configured in the Loading Settings module for analysis. The results from the analysis software can be imported into the finite element software's visualization module for review.

3.2 Calculation Results

The Young's modulus of the epoxy resin is set to 3800MPa, with a friction coefficient of 0.30. The FAT load is based on a utilization factor (UF) of 55%. The fatigue life results from the analysis are presented on a logarithmic scale. After importing the results into the post-processing module of the finite element software, a contour plot of the fatigue life distribution can be analyzed.

Since the tensile armor layer at the joint in this study is subjected only to tensile loads during actual operation, the amplitude in the Stress Dataset is defined as the ratio of the external sinusoidal cyclic load to UF=55% of the FAT load.

When the amplitude is set to 0.10, 0.20, or 0.30, no damage is observed.

For amplitudes of 0.35, 0.45, 0.50, 0.55, 0.60, 0.65, 0.70, 0.80, 0.90, or 1.00, the maximum fatigue life reaches 1.00E+07, primarily concentrated on the inner side of the tensile armor layer outside the joint. The minimum fatigue life occurs on the upper surface of the bent tensile armor layer inside the joint, closely aligning with the location of fatigue fractures reported in recent joint failures both domestically and internationally. The LOG Life–Repeats contour plots for amplitudes of 0.5, 0.7, 0.9, and 1.0 are shown in Figure 11.

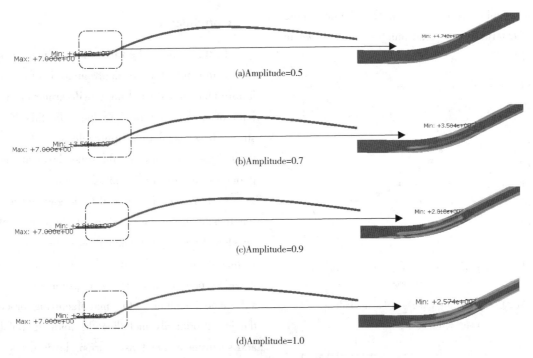

Fig. 11 LOG Life–Repeats diagram for different amplitudes

To enhance fatigue life and safety, it is recommended to reconsider the bend design at the joint entrance when shaping the tensile armor layer structure. The aim is to effectively reduce stress concentration caused by residual stresses.

3.3 Fatigue Life Analysis

The relationship between fatigue life and the analysis results is as follows:

$$S_x = \lg N \qquad (1)$$

Based on the contour plots, calculations indicate that after applying external sinusoidal cyclic loads with varying amplitudes to the tensile armor layer structure at the joint, the corresponding number of cycles to fatigue failure and the maximum damage can be obtained, as shown in Table 3.

Table 3 Results of worst life repeats under different amplitudes

NO.	Amplitude	Life Result	Number of Cycles to Fatigue Failure	Maximum Damage
1	0.10	—	—	0
2	0.20	—	—	0

续表

NO.	Amplitude	Life Result	Number of Cycles to Fatigue Failure	Maximum Damage
3	0.30	—	—	0
4	0.35	6.305	2426840	4.121E-7
5	0.40	5.746	556952.813	1.795E-6
6	0.45	5.202	159373.328	6.275E-6
7	0.50	4.742	55156.738	1.813E-5
8	0.55	4.352	22480.225	4.448E-5
9	0.60	4.023	10532.482	9.494E-5
10	0.65	3.743	5537.753	1.806E-4
11	0.70	3.504	3194.61	3.13E-4
12	0.80	3.120	1311.028	7.628E-4
13	0.90	2.810	655.185	1.526E-3
14	1.00	2.574	373.516	2.677E-3

The number of cycles to fatigue failure under different external cyclic load amplitudes is shown in Figure 12.

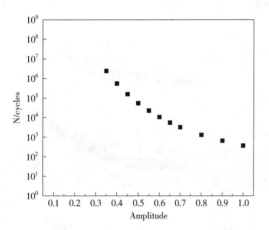

Fig. 12 Worst life repeats of fatigue failure under differentamplitudes

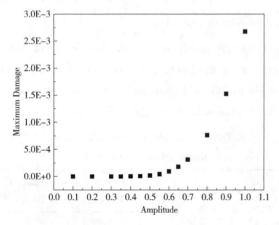

Fig. 13 The largest damage of fatigue failure under different amplitudes

When the cyclic load amplitude is below 0.30, the tensile armor layer at the joint does not experience fatigue failure and can theoretically withstand an infinite number of cycles. However, when the amplitude is between 0.35 and 0.45, the tensile armor layer enters the high-cycle fatigue stage. As the amplitude increases from 0.50 to 1.00, the layer gradually transitions from high-cycle to low-cycle fatigue. At this point, as the cyclic load increases, the variation in the number of cycles to failure in the worst-case scenario becomes progressively smaller. It is important to note that as the amplitude and cyclic load increase, the maximum damage continues to rise, and the rate of damage accumulation amplifies accordingly.

4 Conclusion

In this paper, the finite element method is employed to conduct an in-depth analysis of the fatigue failure characteristics of the tensile armor layer at the joint of a dynamic flexiblepipe. On this basis, a stress-based analytical approach is used to explore the fatigue life of the tensile armor layer at the joint under varying external load conditions.

(1) A two-dimensional finite element model of the tensile armor layer wires at thepipe joint was developed based on the actual assembly process used in manufacturing. The model consists of two sub-models, both treated as static problems. The first sub-model simulates the manufacturing process of the joint, primarily including the folding, unfolding, and securing of the tensile armor layer wires at the joint. The second sub-model is used to simulate the FAT. When compared with existing experimental results, the model demonstrates high accuracy.

(2) During the assembly of the tensile armor layer at the joint, complete plastic deformation occurs regardless of the bending angle, but it returns to an elastic state after springback. Under different simulated bending angles, the maximum stress generated by the initial springback is not significantly affected. However, as the bending angle increases, this maximum stress gradually decreases.

(3) When the operational load is relatively

small, specifically when the load amplitude is less than 0.3, the tensile armor layer at the joint does not experience fatigue failure. However, as the load increases, the failure mode gradually shifts from high-cycle fatigue to low-cycle fatigue, and the number of cycles to fatigue failure progressively decreases. However, this rate of decrease gradually slows down as the load continues to increase.

(4) By observing the fatigue life contour plot, it is evident that the locations with the shortest lifespan are situated at points of maximum curvature on the upper surface of the bend. When designing the structure and shape of the tensile armor layer at the joint, adjustments to the bending and fixation methods can be considered to effectively reduce stress concentrations caused by residual stress.

References

[1] CHEN Yanfei, LIU Hao, SUN Weidong, et al. A review on collapse failure mechanism and safety evaluation of flexible pipes carcass[J]. The Ocean Engineering, 2021, 39(4): 9.

[2] CARNEVAL R O, MARINHO M G, SANTOS J M. Flexible line inspection[C]//European Conference on Nondestructive Testing (ECNDT). 2006.

[3] ANDERSON K, MACLEOD I, O'KEEFFE B. In-service repair of flexible riser damage experience with the north sea galley field[C]//International Conference on Offshore Mechanics and Arctic Engineering. 2007, 4269: 355-362.

[4] SMITH R, O'BRIEN P, O'SULLIVAN T, et al. Fatigue analysis of unbonded flexible risers with irregular seas and hysteresis[C]//Offshore technology conference. OnePetro, 2007.

[5] WITZ J A. A case study in the cross-section analysis of flexiblerisers[J]. Marine Structures, 1996, 9(9): 885-904.

[6] SHEN Y, MA F, TAN Z, et al. Development of the end fitting tensile wires fatigue analysis model: sample tests and validation in an unbonded flexible pipe[C]// Offshore Technology Conference. OnePetro, 2008.

[7] CAMPELLO G, BERTORI F, DE SOUSA J R M, et al. A novel concept of flexible pipe endfitting: Tensile armor foldless assembly[C]//International Conference on Offshore Mechanics and Arctic Engineering. American Society of Mechanical Engineers, 2012, 44908: 413-421.

[8] DE SOUSA J R M, CAMPELLO G C, BERTONI F, et al. A FE model to predict thestress concentration factors in the tensile armor wires of flexible pipes inside end fittings[C]//International Conference on Offshore Mechanics and Arctic Engineering. American Society of Mechanical Engineers, 2013, 55379: V04BT04A016.

[9] CAMPELLO G C. Flexible pipe end fitting anchoring system design methodology and new technology proposal [J]. DSc thesis in Portuguese, Federal University of Rio de Janeiro, Civil Engineering Program (available in www.coc.ufrj.br), 2014.

[10] MIYAZAKI M N R. Tensile armor wire analysis inside the end-fitting of a flexible pipe during and after mounting process based on 3D finite element models [J]. Universidade Federal do Rio de Janeiro, Rio de Janeiro, 2015.

[11] MIYAZAKI M N R, DE SOUSA J R M, ELLWANGER G B, et al. A three-dimensional FE approach for the fatigue analysis of flexible pipes tensile armors inside end fittings[C]//International Conference on Offshore Mechanics and Arctic Engineering. American Society of Mechanical Engineers, 2020, 84355: V004T04A017.

[12] CHEN Yanfei, ZHANG Ye, FENG Wei, et al. Review on failure mechanism of un-bonded flexible pipes metal layers[J]. China Offshore Platform, 2022, 37 (01): 53-62+69.

[13] ELMANP, ALVIM R. Development of a failure detection system for flexible risers[C]//International Society of Offshore and Polar Engineers, 2008.

[14] NIELSEN N J, WEPPENAAR N, COUR D D L, et al. Managing fatigue in deepwater flexible risers[C]// Proceedings of Offshore Technology Conference. 2008.

[15] SAUNDERS, CHRIS, O'SULLIVAN, et al. Integrity management and life extension of flexible pipe[J]. Offshore, 2008, 68(1).

[16] CLEMENTS R A, JAMAL N, SHELDRAKE T. Riser strategies: fatigue testing and analysis methodologies for flexible risers[C]. //Offshore Technology Conference 2006: New Depths. New Horizons vol. 1. 2006: 19-24.

[17] ZHAO Lin, YIN Xiao, DUAN Wenjing, et al. Marine flexibility based on ABAQUS/fe-safe fatigue analysis of piping skeleton layers[J]. China Water Transport, 2015, 15(10): 283-285.

[18] XI Yonghui. Semi-physical Simulation of Marine Flexible Pipe and Cable's Fatigue[D]. Dalian University of Technology, 2012.

[19] YANG Chan. Some problems of fatigue resistance design of flexible risers used in deep water[D]. Dalian University of Technology, 2015.

[20] MCCARTHY J C, BUTTLE D J. MAPS-FR structural integrity monitoring for flexible risers[C]//Proceedings of the 22th International Offshore and Polar Engineering Conference. 2012.

[21] CAMPELLO G C, DE SOUSA J R M, VARDARO E. An analytical approach to predict the fatigue life of flexible pipes inside end fittings[C]//Offshore Technology Conference. OnePetro, 2016.

[22] SHEN Y, MA F, TAN Z, et al. Development of the end fitting tensile wires fatigue analysis model: sample tests and validation in an unbonded flexible pipe[C]//Offshore Technology Conference. OnePetro, 2008.

[23] API RP 17B, 2008, Recommended Practice for Flexible Pipe, 4^a ed., Washington, American Petroleum Institute.

[24] API RP 17J, 2008, Specification for Unbonded Flexible Pipe, 3^a ed., Washington, American Petroleum Institute.

油气管道智能工地建设质量问题剖析及技术提升策略

刘海春[1]　郭　旭[2]　苏维刚[2]　李庆生[2]　姜　鹏[1]　陈　群[1]　李增材[1]

(1. 国家管网集团工程部；2. 国家管网集团工程质量监督检验有限公司)

摘　要　数字化转型是油气管道工程建设在"全国一张网"背景下践行高质量发展的重大举措，智能工地以精细化管理为规则标准，搭建各方广泛参与、汇聚管理共识、凝聚高水平建设力量的数字化工程管理实践平台，是工程业务域数字化转型的创新深化路径和重要实践典范。质量监督机构历来重视智能工地建设问题的专项检查，在年度检查清单中智能工地问题占比为2.16%，抽检合格率94.82%，频发问题和共性问题包括数据丢失、数据乱码、数据失真等。结合"工程信息管理系统(PIM一期)"的运行经验，围绕自动焊数据采集等关键技术应用现状、现存技术瓶颈及各项使用难题，提出不同规模、不同场景智能工地关键技术提升策略，稳步推进在建管道工程智能工地建设水平，充分彰显数字化转型在工程建设业务领域的价值意蕴。

关键词　工程建设；智能工地；数据采集；质量监督；自动焊

Analysis of QualityProblems and Technical Improvement Strategies for Intelligent Construction Sites of Oil and Gas Pipelines

Liu Haichun[1]　Guo Xu[2]　Su Weigang[2]
Li Qingsheng[2]　Jiang Peng[1]　Chen Qun[1]　Li Zengcai[1]

(1. PipeChina Engineering Department; 2. PipeChina Engineering Quality Supervision and Inspection Company)

Abstract　"Digital Transformation" is major measure for the high-quality development of oil and gas pipeline construction in the context of "one national network". "Intelligent Construction Cites" are based on the principle of meticulous management, establish a digital engineering management practice platform that involves all parties, gathers management consensus, and unites high-level construction forces, It is an innovation deepening path and an important practice model for the digital transformation of engineering construction. The quality supervision organization has always attached importance to the special inspection of intelligent construction site problems. in the annual checklist, the proportion of intelligent construction site problems is 2.16%, and the sampling inspection pass rate is 94.82%. Frequent problems and common problems include data loss, data scrambling, data distortion, etc. Combined with the operation experience of "Project Information Management System (PIM I)", focusing on the application status of key technologies such as automatic welding data acquisition, existing technical bottlenecks and various use problems, this paper puts forward key technology improvement strategies for intelligent construction sites with different scales and scenarios, steadily promotes the construction level of intelligent construction sites of pipeline projects under construction, and fully demonstrates the value of digital transformation in the field of engineering construction business.

Key words　Engineering Construction; Intelligent Construction Site; Data Acquisition; Quality Supervision; Automatic Welding

1 前言

当前我国能源行业基于数字化转型升级工程的形势、走势不断深化和加强，管理制度化、制度表单流程化、表达流程信息化等多维度的生产经营和工程建设业务管理工作全面走深走实，带来深刻的制度化、规范化和标准化价值意蕴。数字化转型是油气储运企业应对数字化时代挑战、贯彻落实能源安全新战略、推动能源高质量发展的深化实践路径。"十四五"期间，在国家能源供给规模持续增强和"全国一张网"油气储运设施加快构建的背景下，油气管道工程建设业务领域从战略层面出发，抓住数字化和智能化重要发展机遇，明确工程建设数字化转型的目标和路径，制定详细精确的实施计划，对工程项目的业务流程、组织结构、文化理念等进行全面深入的改造和升级，以实现更高效、更智能、更灵活的建设施工模式，确保数字化转型与工程建设整体战略保持一致。

数字化转型是一个全面、深入、系统的过程，涉及技术、组织等多个层面。从技术层面来看，数字化转型是利用数字技术对传统业务进行改造和升级的过程，包括数据采集、存储、处理、分析和应用等各个环节，通过技术手段提升业务效率和质量。从组织层面来看，数字化转型要求企业打破传统的组织结构和管理模式，推动跨部门、跨层级的协作和创新，实现组织文化的数字化转型，建立更加扁平化、灵活和高效的运营体系。具体到油气管道行业来说，中国油气管道行业的市场规模不断扩大，投资规模和管道网络覆盖率在"十四五"期间显著提升，政府政策扶持亦持续加强，为油气管道行业的发展带来大量的机遇和挑战。

油气管道工程建设业务领域中，智能工地是落实践行高质量发展的重大举措，是工程业务域数字化转型的典型深化路径和重要实践典范。智能工地以精细化管理为规则标准，通过软硬件结合的方式在现场部署各类智能终端，实现施工安全监管和质量监控，搭建各方广泛参与、汇聚管理共识、凝聚高水平建设力量的数字化工程管理实践平台。智能工地能够解决管理人员对施工现场不掌握、现场检查没有抓手的问题，使管理人员实时掌握现场情况和工程进展，在AI、大数据、边缘计算等智能技术加持下，助力工程建设智能化管理持续提升，深化国家能源建设新机制发展，纵深推进油气管道工程建设技术革命，加快构建支撑高水平自立自强的新型智能工地创新科技。

2 智能工地建设的质量问题分析

油气储运工程智能工地建设起始于中俄东线天然气管道工程(黑河-长岭)段，本质上是施工要素的数字化，围绕施工现场项目管理的"人、机、料、法、环、测"六大核心要素，将物联网、人工智能、大数据等新型信息化技术，与传统施工项目管理深度融合，把智能感知、施工监测与过程管理等汇聚到一个平台上，为油气储运工程传统施工模式提供助力，辅助管理人员进行项目决策、监督与管理。

工程质量监督在推进智能工地建设质量管控过程中发挥了重要作用。质量监督是油气管道工程建设质量管控的重要形式之一，质量监督机构依据《中华人民共和国建筑法》、《建设工程质量管理条例》等国家法律法规和规章制度，以及国行企标、施工图设计等文件要求，履行政府赋予的行政职责，严格监管项目实施全过程，把控工程施工质量目标，对工程建设各方责任主体质量行为及实体质量实施的监督。

2.1 智能工地问题分布

质量监督机构历来重视智能工地建设问题的专项检查，以2023年全年检查问题清单为例，2023年国家管网集团公司针对所有投资建设的油气储运项目，包括原油、成品油及天然气管道，原油、成品油库，储气库，液化天然气接收站及相应配套设施的工程建设项目，开展了766次现场检查工作，发现17530项质量安全问题，其中智能工地抽测3415点，不合格177点，抽检合格率94.82%。测点位置主要包括摄像头、一体机、现场组网、焊接参数实时上传系统、焊口二维码、设备二维码、人员入场二维码等。在全部17530项检查问题清单中，智能工地问题为378项，占比为2.16%。检查问题的类型分布如图1所示。

378项智能工地问题中，按责任主体划分，施工单位问题数量为237项，占比62.70%；无损检测单位79项，第四方复评单位10项，共占比23.54%；监理单位27项，占比7.14%；建设单位21项，占比5.56%；设计单位4项，占比

1.06%。按问题类型划分，质量行为问题 310 项，占比 82.01%；实体质量问题 68 项，占比 17.99%。智能工地检查问题的类型分布如图 2 和图 3 所示。

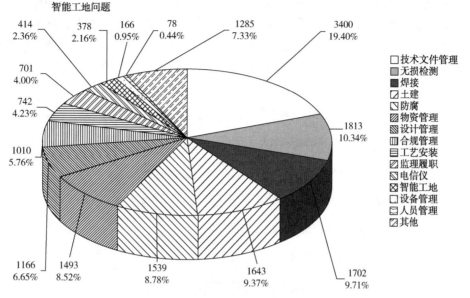

图 1　2023 年度质量监督问题统计与分析

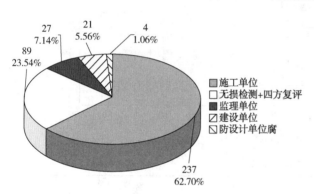

图 2　智能工地问题分类-按责任主体划分

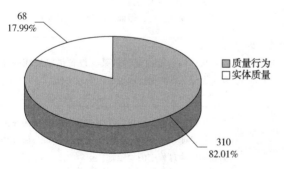

图 3　智能工地问题分类-按问题类型划分

按照问题性质划分，378 项智能工地问题中，施工数据未实时采集 78 项，占比 20.63%；上传数据与实际数据偏差较大 56 项，占比 14.81%；无损检测底片未上传 51 项，占比 13.49%；设计文件或施工方案未编制智能工地专篇 45 项，占比 11.90%；摄像头位置未正对施工作业面 37 项，占比 9.79%；人员信息未及时录入 29 项，占比 7.67%；数据审核不及时 21 项，占比 5.56%；硬件架设数量不足 14 项，占比 3.70%；硬件损毁 13 项，占比 3.44%；二维码电子标签未投入使用 13 项，占比 3.44%；未及时完成智能工地施工进度、技术规格书未录入、监理履职不到位、数据格式不正确、数据单位不正确等其他问题 21 项，占比 5.56%。智能工地检查问题按性质分布如图 4 所示。

根据 2023 年智能工地质量监督问题清单的数据分析可知，因施工单位、无损检测单位对智能工地建设相关标准规范不熟悉或理解不透彻，智能工地搭建、施工数据采集上传、无损检测底片数字化等问题较为突出，数据丢失、数据乱码、数据失真等问题频发，亟需项目各方责任单位进行重点关注和专项整改。后续的质量检查和监督过程中，应重点针对施工单位、无损检测单位的智能工地和工程信息管理系统要求落实情况开展督查，就其相关质量行为和实体质量问题进一步分析研判，制定专项整改措施。

2.2　典型问题分析

通过全面总结 2023 年各类工程质量监督检查、专项检查、设计督查过程中发现的智能工地典型问题，结合近年来油气管道质量监督机构各类型检查工作情况通报文件，深入分析典型问题发生的原因，对于管理行为方面的典型问题，如智能工地技术文件管理不规范、智慧管网设计专篇深度不够、人员履职不到位、管理规定执行不到位等，应认真分析问题背后的体制机制缺陷和管理漏洞，举一反三、强化管理，建立健全长效管控机制，有效减少智能工地典型问题屡查屡犯的情况。对于智能工地现场工程作业实体质量问题，如数据采集设备精度不符合要求、数据录入不及时、硬件损毁、无损检测底片数字化要求不落实等问题，从严管理、细化措施、切实整改，通过加强对员工的宣贯培训和正面引导，增强现场施工作业员工职业素养，以基层为着力点做好智能工地管理工作。智能工地建设问题分类分析及建议措施见表1所示。

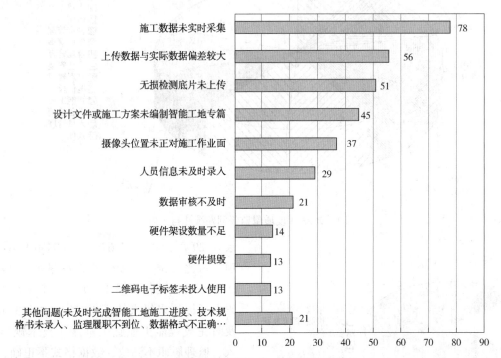

图 4　智能工地问题分类−按问题性质划分

表 1　管道施工现场智能工地建设问题汇总

序号	类型	问题	描述	建议
1	数据丢失	焊层数据错乱	填充焊和盖面焊工艺中，同一台焊机两把焊枪应为不同焊层，但部分厂家的焊机反馈的数据出现两把焊枪传回同一焊层数据，严重影响使用效果	焊机厂商需要加强焊机采集设备的软件稳定性，及时升级焊机设备，保证现场数据的真实可靠
2		线缆受干扰	部分厂家的焊机设备数据传输线缆采用非屏蔽的普通线缆，使用一段时间后，存在线缆被干扰的问题，导致数据在线缆传输过程中丢失，严重影响使用效果	焊机厂商需要加强焊机采集设备传输线缆的品质，保证现场数据传输过程中的稳定性
3		USB口松动或接触不良	部分厂家的外焊机仅有1个无固定措施的USB口做为数据传输口，在现场施工作业过程中，常因USB线松动或脱落，造成数据无法正常传输，严重影响使用效果	焊机厂商需要加强焊机采集设备传输接口的稳固性，尽量使用带有自锁功能的工业级接口，保证现场数据传输过程中的稳定性
4		不能读取历史数据	目前的焊机厂家只提供了实时数据采集的接口，未扩展出历史数据采集接口	焊机或焊机改造厂家需要支持实时数据传输和历史数据读取
5		未做线缆检查	现场进行焊接作业前，相关人员没有对焊机和焊机盒子的线缆连通情况做确认	现场进行焊接作业前，相关人员需要按照检查表进行各设备情况的确认
6		未做供电检查	现场进行焊接作业前，相关人员没有对焊机采集盒和焊机采集智能网关的供电情况进行检查	现场进行焊接作业前，相关人员需要按照检查表进行各设备情况的确认
7		未做扫码检查	现场进行焊接作业前，相关人员进行扫码工作的时候忽略了二次确认，导致二维码未成功扫入到设备中	现场进行焊接作业前，相关人员需要按照检查表进行各设备情况的确认

续表

序号	类型	问题	描述	建议
8	数据乱码	焊机程序出错	焊机作业一段时间内出现部分数据显示为乱码的情况,经过排查发现是焊机设备编译程序出错,传输了乱码数据	焊机厂商需要加强焊机采集设备的软件稳定性,及时升级焊机设备,保证现场数据的真实可靠
9	数据乱码	线缆受干扰	部分厂家的焊机在一段时间内的数据会出现随机乱码现象,经过排查,存在线缆被干扰的问题,导致数据在线缆传输过程中出现乱码,更换线缆后可以正常稳定传输	焊机厂商需要加强焊机采集设备传输线缆的品质,保证现场数据传输过程中的稳定性
10	数据乱码	扫码枪损坏	现场人员使用的扫码枪出现问题后,没有及时的去维修扫码枪,导致二维码扫码出现乱码现象	现场进行焊接作业前,相关人员需要按照检查表进行各设备情况的确认,并且及时的维护维修设备
11	数据失真	人工补录数据	现场人员通过一体机确认今天的焊机数据量是否完整,如果不完整,现场人员会补录数据,补录数据采用手工电脑补录方式,这种方式存在假数据补录	现场采集数据尽量通过一体机和智能网关之间的无人值守数据采集和数据补传,避免人工填写
12	数据失真	现场修改数据	现场有权限的人员可以进行数据库的登入操作,登入后可以对一些数据进行修改	现场数字化管理人员加强密码安全意识
13	数据失真	焊机程序出现异常卡顿	部分厂家的焊机发现采集过程中的数据一直未出现正常变动,一直是同样的一条数据,之后又恢复正常,经排查发现,是焊机采集程序出现bug,修复后正常	焊机厂商需要加强焊机采集设备的软件稳定性,及时升级焊机设备,保证现场数据的真实可靠
14	数据失真	部分数据依赖人工填写	现场人员还有很多设备需要进行手工填写或者APP录入,如焊层温度、预热温度等,这个填写过程容易出现假数据	现场关键性的数据尽量升级成自动采集,避免人为因素干扰,以便有效地、真实地记录

2.3 智能工地质量管理策略

结合智能工地发现的质量问题,制定专项质量监督和管控策略,以智能工地专项质监点为抓手,强化智能工地应用管理,确保数字化转型有效运行。质量监督机构应严格按照设计要求、现场实际情况开展智能工地检查,彻底督查已建成的智能工地设备设施的使用效能,将智能工地应用落到实处。各参建单位自上而下层层把关、层层负责,严格落实智能工地全过程应用管理,全面提高工程建设数字化工作质量和水平,推进各级各部门认真履职、担当负责。智能工地的质监点设置建议如表2所示。

以施工现场关键工序之一的管道线路焊接为例,随着全自动焊接在管道建设中的规模化应用,以及管道全生命周期管理的开展,对管道建设期各阶段采集的多种类数据加以梳理,构建基于大数据的焊接质量分析模型,改变传统的管道焊接质量管理模式,对于预警管道焊接中存在的问题,准确评估管道焊接质量提供决策支持。采集与管道焊接质量相关的各类数据,围绕"人、机、料、法、环、测"进行数据的采集与整合,建议智能工地焊接主要数据采集内容可采用表3的形式进行全面录入。

表2 智能工地的质监点设置建议

序号	质监点名称	涉及单位	质监点性质	检查内容
1	自动焊焊接数据核查	施工单位 监理单位 建设单位	必监点	根焊与热焊间隔时间、焊接电流、焊接电压、层间温度、焊接速度、层道数等关键参数抽查
2	视频监控影像质量核查	施工单位	巡监点	开机率抽查、图像角度、图像清晰度、视频传输可靠性、视频影像链路稳定性等
3	数据一致性核查	施工单位	巡监点	现场数据与智能工地采集系统数据一致性核对

续表

序号	质监点名称	涉及单位	质监点性质	检查内容
4	数据真实性核查	施工单位 技术服务商	巡监点	施工现场抽查人工补录数据、现场修改数据、部分数据依赖人工填写、现场违规操作导致的施工数据失真等
5	智能工地数据审核核查	施工单位 监理单位	巡监点	检查施工单位、监理单位对智能工地数据审核履职情况
6	人员设备入场报验核查	监理单位	巡监点	检查各参建单位智能工地相关的人员、设备入场报验情况
7	现场人员不安全行为和物的不安全状态核查	施工单位 监理单位 建设单位	巡监点	通过智能工地视频影像监控系统检查现场人员不安全行为和物的不安全状态

表3 主要数据采集内容

序号	类别	数据项	数据内容
1	人员	焊接技术人员、检查检验人员、辅助操作人员、辅助操作人员等管道焊接施工相关的各类人员	项目、标段、员工编号、姓名、年龄、性别、资质证书、所属项目、机组、单位、工艺培训、焊接考核、技术评价等
2	设备	焊机、对口器、坡口机、加热设备、计量器具等各类设备设施	项目、标段、设备编号、设备名称、类型、型号、出厂合格证、质量合格证、所属项目、使用时间、归还时间、使用人员、维护时间、维护人员等
3	工艺	焊接工艺规程	项目、标段、施工验收规范、管材规格范围、焊接方法、焊接材料、焊接技术要求等
4		坡口	项目、标段、前钢管编号、后钢管编号、破坏形状、坡口尺寸、是否内卷边、使用设备编号、施工人员编号等
5		根焊、热焊、填充焊、盖面焊	项目、标段、电源极性、焊接方向、焊接电流、电弧电压、焊接速度、送丝速度、气体流量、设备编号、焊工编号、焊接时间等
6	材料	管材	项目、标段、编号、材质、壁厚、长度、管径、重量、制管形式、制管标准、炉批号、压力等级、管材等级、防腐类型、防腐等级、出厂日期、制造厂商等
7		焊材	项目、标段、规格、型号、数量、类别、牌号、依据标准、生产日期、生产厂家、出厂合格证、质量合格证、入库时间、领用时间、烘烤时间、烘烤温度、经手人等
8		保护气体	项目、标段、气体名称、密度、纯度、温度、压力、生产厂家、生产日期等
9	环境	天气环境	项目、标段、机组、温度、湿度、风速、天气状况、时间
10		地理环境	项目、标段、机组、地形类型、坡度、海拔、时间
11	检测	包括无损检测（射线检验（RT）、超声检测（UT）、磁粉检测（MT）和液体渗透检测（PT））、外观检验、硬度测定	项目、标段、焊缝编号、检测单位、检测时间、检测方法、是否存在缺陷、缺陷类型、缺陷参数、检测设备编号、检测设备名称、执行标准、检测人编号、检测人姓名
12		其他相关焊接检测数据	

3 智能工地关键技术提升思路

围绕质量监督机构在智能工地发现的问题清单分析总结情况，结合油气管道行业智能工地近年来的引用现状和发展情况综合研判，当前由于受到各种技术成本和行业条件的限制，且油气管道从业单位的数字化管理工作起步较晚，各项工作尚在摸索阶段，智能工地建设从顶层设计和宏观层面仍然存在一定的问题和不足。

（1）智能工地设计专篇编制深度不足，未针对具体项目进行专项细化；

（2）智能工地划分原则不清晰，施工场景划分标准不明确；

（3）智能工地建设的标准化工作未全面开展，没有形成成熟的标准体系；

（4）关键参数采集真实性和实效性有待提高，数据缺失、逻辑错误等现象时有发生；

（5）半自动焊、手工焊参数未实现自动采集，焊接参数采集不全面；

（6）无人机管控手段缺失，飞控成果未充分利用；

（7）现场数据填报工作量大，线上线下数据填报存在一定程度的重叠重复，人工填报数据多，审核工作量大，数据准确性及真实性难以全面保证；

（8）现场组网效果差，数据采集完整性和传输及时性难以保证；

（9）管道焊接底片评定成本高、效率底。目前管道焊缝射线检测电子化底片近百万张，传统底片评片、扫描等产生较大工作量，且效率低下，常出现误评等问题。如何有效利用海量电子化底片数据，发挥其有效价值，是当前亟需解决的问题；

（10）其他人员能力、设备管理、重视程度、整体性和系统性等问题较为突出，例如建设单位、监理单位、施工单位等各责任主体主要以工程管理人才和工程技术人才为主，数字化建设能力普遍不足，未能提供智能工地建设的有力支撑。视频监控开机率不足、遮挡摄像头、安装位置不当、焊接改造维护不系统、信息化小屋管理不善、一体机使用不力，导致智能工地系统整体运行不稳定的情况时有发生。个别建设单位、施工单位存在错误思想，认为设备搭建起来即可完成任务，无法保障流程运行，未深入理解数字化转型的重大意义。智能工地建设、运行缺乏顶层设计和系统管理的指导，智能工地建设和运行存在秩序混乱、效率低下的问题，缺乏整体性和系统性规划。

在总结梳理智能工地建设需求基础上，建议后续油气管道智能工地的研究方向应聚焦6项以下关键技术问题：

（1）不同规模、不同场景智能工地建设需求和划分方法，形成施工场景划分标准，调研焊接等关键施工设备的参数采集的技术现状；

（2）研究现有和相关行业智能工地建设成功案例，精细化构建适用于不同规模和不同场景的智能工地建设标准体系；

（3）研究关键施工设备参数自动采集与传输技术，形成对应半自动焊机改造方案以及全自动焊机的扩展方案；

（4）研究无人机航飞需求、高危场景等，搭建无人机航飞场景和高危场景的自动识别模型，收集航飞影像训练样本，结合国家规定以及行业情况，形成管网的无人机飞控管理规定。研究航飞成果与进度测量模型结合技术，实现进度测量的自动反馈。研究根据无人机航飞成果站场自动三维建模技术；

（5）研究喷版图像、纸质表单资料照片等资料的算法，依托智能一体机进行现场部署，形成"无人力参与（干扰）"的工地数据自动采集模式；

（6）研究现场组网优化提升方案及关键技术，优化现场组网效果，形成统一数据接口，支持多种类型物联网数据传输方式。

4 结论

在油气管网规模化工程建设过程中，管道工程标准化、模块化、信息化水平不断提高，完成从传统管道向数字管道的转变，实现设计数字化、施工机械化、物采电子化、管理信息化。随着大数据、云计算、物联网等信息技术的日趋成熟，国家管网等央企智慧管道建设具备了坚实基础，智能工地建设正在快速稳步发展。但由于受到硬件、软件等方面的限制，要实现从交接桩、测量放线、焊接施工、防腐补口、清管测径、下沟回填等长输管道施工全过程数据的存储和实时传输，最终实现数字化移交还有很长的路要走，尚需要建设单位、设计单位、设备生产厂家、各责任主体共同研究、开发和探索。

下一步，油气管道工程建设业务领域应立足长远，坚持价值导向，强化科技创新，通过加强科研产业融合发展，重大风险防控技术攻关，加快数字智能技术应用等一系列措施，推动油气管道工程建设发生深刻变革。特别要通过大数据、云计算、物联网、人工智能等科技创新手段赋能增效，推动石油石化行业高质量发展。未来要加快石油化工安全生产智能化建设，特别是在加快发展新质生产力的过程中，需深度融合新一代信息技术、数字孪生技术、智能化装备，迫切需要大数据、云计算、物联网、人工智能等科技创新手段赋能增效。

参 考 文 献

[1] 王振声；陈朋超；王巨洪. 中俄东线天然气管道智能化关键技术创新与思考[J]. 油气储运，2020，39(07)：730-739.

[2] 刘袁辰；董锦坤；贾君；张玉桂. 智能技术在智慧工地中的应用研究——以顶管工程为例[J]. 辽宁工业大学学报(自然科学版)，2024，44(03)：181-187.

[3] 李久林；王忠铖；田军；陈利敏；徐浩；刘廷勇. 智能建造背景下的智慧工地发展与实践研究[J]. 建筑技术，2023，54(06)：645-648.

[4] 邹骅. 智能建造背景下的智慧工地集成与协同优化策略研究[J]. 建筑工人，2024，45(04)：17-20.

[5] 铁明亮；詹胜文；李杰. 长输油气管道智慧工地数字化管理流程浅析[J]. 中国勘察设计，2024(01)：93-96.

[6] 王永胜；张朝辉；赵亮. 石油石化场站工程标准化智能工地的创新和应用[J]. 石油化工安全环保技术，2023，39(06)：6-8+5.

[7] 赵光辉；石阳；郭鹏；孙皓；石志伟；乔晓亮. 智能工地技术在北京燃气天津南港LNG应急储备项目中的应用与实践[J]. 中国石油和化工标准与质量，2023，43(08)：65-67.

[8] 谢武军；谌杨；褚荣光；王兆坤；马睿；刘梓舟. 智能管网时代油气管道施工现场的标准化建设[J]. 石油工程建设，2021，47(S1)：1121-1126.

[9] 杨赫；牟晓亮. 深圳LNG外输管道工程智能监控的建立与探索[J]. 科技和产业，2021，21(04)：215-222.

[10] 吴仁祖. 建设工程质量监督管理中问题解决策略探讨[J]. 产品可靠性报告，2024(06)：148-149.

[11] 刘斌. 基于质量监督管理的油气田地面工程建设标准化方法分析研究[J]. 中国石油和化工标准与质量，2024，44(12)：13-15.

[12] 宋博远昌. 油气田工程质量监督闭环管理的实践与认识[J]. 中国石油和化工标准与质量，2024，44(11)：43-47.

[13] 郑健；刘欢；周子凯. 管道全自动焊接质量智能化管理模式探析[J]. 中国石油和化工标准与质量，2022，42(19)：56-58.

[14] 赵云峰；姜有文；王巨洪. 基于卷积神经网络的油气管道智能工地监控实现[J]. 管道技术与设备，2021(06)：23-26.

综合法天然气管道泄漏监测系统研发与应用

张 春　曹旦夫　王军防　王浩霖

（国家管网集团东部原油储运有限公司）

摘　要　管道泄漏监测技术是及时发现泄漏事故并对泄漏点进行定位的重要手段，当前天然气管道泄漏监测技术多采用单一监测方法，存在检测灵敏度低、容错率低、对微小泄漏检测能力不足等问题。针对当前天然气管道泄漏监测存在的问题展开研究，从"感"和"知"两方面开展研究。研制了基于压电原理的动态压力变送器，解决了微弱泄漏信号的检测难题；研究了泄漏信号特征提取及增强方法，引入非周期随机共振方法与快速累计差分方法相结合的双增强策略，实现了强干扰背景下泄漏信号的增强处理，提高了泄漏信号的信噪比，采用动态迭代时间规整算法实现了上下游信号的有效匹配，解决了微弱泄漏信号难以被传统检测手段捕捉的难题；针对单一泄漏监测方法存在的不足，综合各类传感信号及SCADA数据等多源数据，基于不同的信号特征构建泄漏多源感知模型，设计开发了综合法天然气管道泄漏监测系统，并在中俄东线唐山至宝坻段完成现场应用测试。现场测试表明，研制的动态压力变送器具有较好的灵敏度、低频响应特性等性能；针对63组现场放气测试，研发的管道泄漏监测系统均能识别并自动报警，报警响应时间在60秒以内，最小可监测泄漏孔径达1mm，针对2mm、3mm、4mm和6mm泄漏孔径的放气实验平均定位误差78~87m，针对1mm泄漏孔径的放气实验，平均定位误差为152m，总体与国外先进技术水平相当。

关键词　天然气管道；泄漏监测；综合法；动态压力变送器；定位

Research and application of comprehensive natural gas pipeline leakage monitoring system

Zhang Chun　Cao Danfu　Wang Junfang　Wang Haolin

(Pipe China Eastern Oil Storage and Transportation Co. Ltd)

Abstract　Pipeline leakage monitoring technology is an important means to detect leakage accidents in a timely manner and locate the leakage points. Currently, most natural gas pipeline leakage monitoring technologies adopt single monitoring methods, which have problems such as low detection sensitivity, low fault tolerance, and insufficient ability to detect tiny leaks. To tackle the existing problems in natural gas pipeline leakage monitoring, research has been initiated from the perspectives of "sensing" and "perceiving". A dynamic pressure transmitter founded on the piezoelectric technology principle has been devised, resolving the challenge of detecting feeble leakage signals. The techniques for extracting and augmenting the characteristics of leakage signals have been explored. A dual enhancement strategy, integrating the aperiodic stochastic resonance method and the fast cumulative difference method, has been introduced to bolster leakage signals against a backdrop of strong interference, thereby enhancing the signal-to-noise ratio of these signals. The dynamic iterative time warping algorithm has been utilized to actualize the effective matching of upstream and downstream signals, surmounting the obstacle of weak leakage signals being elusive to traditional detection methods. In response to the shortcomings of a single leakage monitoring method, diverse source data, including various sensor signals and SCADA data, have been consolidated. Based on distinct signal features, a multi-source perception model for leakage has been established, and a comprehensive natural gas pipeline leakage monitoring system has been designed and developed. Field application tests have been concluded in the Tangshan to Baodi section of the China-Russia Eastern Gas Pipeline. Field tests demonstrate that the developed dynamic pressure transmitter exhibits favorable performance in terms of sensitivity, low-frequency response characteristics, and the like. For 63 sets of on-site venting tests, the developed pipeline leakage monitoring system can recognize and automatically alarm, with an alarm response time of less than 60 seconds. The minimum detectable leakage aperture is 1mm. The average positioning error for venting experiments with 2mm, 3mm, 4mm, and 6mm leakage apertures is 78–87m, and for venting experiments with 1mm leakage aperture, the average positioning error is 152m. Overall, it is comparable to the advanced technology level abroad.

Key words　natural gas pipeline; leakage monitoring; comprehensive method; dynamic pressure transmitter; location

1　引言

油气管道作为能源运输的重要基础设施，其安全稳定运行直接关系到国家能源安全、经济发展及社会稳定。然而由于油气管道覆盖地域广、高后果区密集，一旦发生泄漏，极易引发火灾、爆炸等重大伤亡事故，造成难以估量的损失。管道泄漏监测系统能够实时检测管道运行状态，及早发现泄漏风险并报警，是保障油气管道安全运行、降低事故危害的重要环节。国内外对长输管道泄漏监测技术的研究已经有几十年的历史，根据工作原理的不同，管道泄漏监测方法可分为负压波法、流量平衡法、光纤法、瞬态模拟法、声波法、内检测法等。目前输油管道泄漏监测技术相对较为成熟，其原理多以负压波法或负压波+流量平衡法为主，代表性厂家有中加诚信、天津精仪精测、管道科技中心等，其应用管道里程均在4000公里以上。由于天然气具有可压缩性，管道泄漏后压力、流量等参数变化幅度不明显，难以将输油管道上相对成熟的负压波法等技术直接用于天然气管道的泄漏监测。

天然气管道泄漏后会产生声波、应力波、泄漏点附近温度下降等物理现象，上述变化将从泄漏点处沿管道介质、管壁向上下游传播，同时泄漏点附近温度也会因气体的迅速膨胀而发生下降。因此，国内外对天然气管道泄漏监测的研究普遍以声波法和光纤测温法为主。声波法管道泄漏监测方面，国外以美国ASI音波测漏系统最为成熟，已在包括气体、液体、多相流等输送管道得到广泛应用。国内北方管道公司、西部管道公司、北京寰宇等开展了基于声波法的天然气管道泄漏监测技术研究，在大沈线、西气东输部分管段进行了小范围试用，验证了声波法在天然气管道泄漏监测方面的可行性，但灵敏度、定位精度等性能指标尚不能完全满足生产需求。光纤测温泄漏监测是天然气管道泄漏监测的新兴技术，德国GESO公司、法国METRAVIB RDS公司、瑞士Omnisens公司等均开展了相关研究应用，如瑞士Omnisens公司的DiTEST-AIM产品在秘鲁某液化天然气公司的一条408km高压输气管道进行了应用，取得了较好的效果。国内北方管道公司、天津大学、中国计量院等企业高校也开展了分布式光纤测温泄漏监测技术研究，取得了一定进展，但尚未正式应用。总体而言，与输油管道泄漏监测相比，天然气管道泄漏监测技术还不够成熟，目前在用的管道泄漏监测系统多采用单一的监测技术，系统的灵敏性受限于单一传感器的检测灵敏度和可靠性，存在微小泄漏监测能力不足、单一原理的泄漏监测技术稳定性差等问题。本文采用硬件研制与软件开发相结合的方式，研制高灵敏度的动态压力变送器，解决微小泄漏的监测难题，同时融合多源传感信号及SCADA数据综合进行泄漏监测，开发综合法管道泄漏监测系统，并在现场管道应用，测试验证技术的可行性。

2　技术路线和研究方法

围绕天然气管道泄漏监测存在的问题，从"感"和"知"两方面开展研究。在"感"的方面，设计并研制灵敏度高、抗电磁干扰能力强的动态压力传感器，同时采集站场声振信号及SCADA系统生产数据（如压缩机启停、阀门开关等），采用不同的采样策略，形成多源异构异步数据源；在"知"的方面，针对采集的多源数据，采用信号增强、动态特征提取、异构数据融合等多种技术，建立基于多源数据的管道泄漏监测模型，开发天然气管道泄漏监测系统，并选取测试管道进行示范应用（图1）。

3　结果和效果

3.1　动态压力变送器设计与研发

管道微小泄漏在泄漏事件中的占比越来越大，解决小泄漏的检测问题显得尤为重要，高灵敏度的传感器是提高泄漏检测准确率和信噪比的关键环节。天然气管道泄漏后，除了管内介质流动规律的变化外，还伴随有声波、应力波等物理现象，上述变化会从泄漏点处向管道上下游传播，通过在管道两端安装高灵敏度的变送器可实现异常信号的监听与捕捉。声波法是当前管道泄漏监测研究的热点，与传统的负压波法、流量平衡法等相比，具有灵敏度高、定位精度高、适应性强等特点。声波中的动态压力波，具有幅值大、频响范围宽、监测管段长等特点，基于动态压力波监测技术的变送器能够直接检测管道内微弱声压信号变化，与普通压力变送器相比具有更高的灵敏度和泄漏判别效果，是长距离监测管道泄漏的最佳选择。按照工作原理的不同，动态压力变送器分为压电式、压阻式、压变式及电容式等类型，其中压电式应用较多，其工作原理是基于压电效应，当压电材料受到外部压力或拉力时，其表面会产生电荷信号，信号的幅度与压力或拉力的变化成正比。基于压电原理的动态压力变送器，具有较宽的频响范围和较强的适应性，不仅能够准确捕捉短时间内的幅值波动，满足对复杂次声波信号监测的需求，还能适应各种强度的变化，可为检漏系统提供准确的信号。

基于此，研制了基于压电原理的动态压力变送器，变送器整体由防爆壳体、变送器电路、压电膜片及接头等部分构成，其监测灵敏度与准确性由传感器（测量部分）及变送器电路决定。为提高低频段信号的响应能力及抗干扰能力，经调研膜片选用 PCB 公司压电膜片。结合变送器总体电性能要求，创新性地设计了动态压力变送器的电路。其变送器电路（见图 2）由电源电路、电压基准电路 VREF、恒流源电路 CIS、交流放大电路 Aac 和电压电流转换电路 V2I 等部分组成，CIS 和 Aac 共用 1 个双运算放大器（分别记为 U2A 和 U2B）。变送器电路采用二点式或三点式安装，能够较好地消除来自管道引入的电气干扰，同时去除了动态压力传感器自身的输出偏置成分，保证了输出零位的准确性，通过动态压力转换电流和输出零位电流线性叠加，可以方便的设置整个传感器的灵敏度和输出零位。动态压力变送器整体通过了防爆防护认证（防爆标志 Exia IIC T4），满足现场使用环境对设备防爆防护等级的要求。

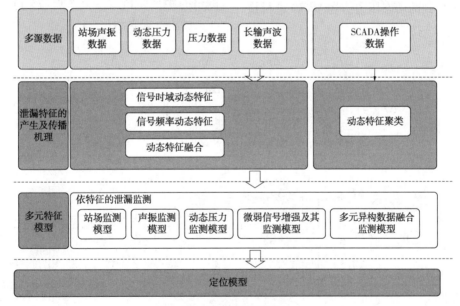

图 1　课题技术路线

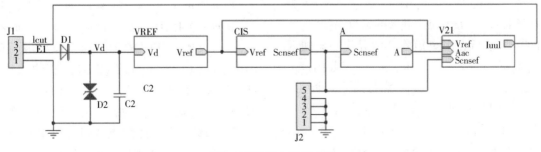

图 2　变送器电路示意图

3.2 泄漏信号提取方法应用研究

受现场设备电磁干扰、振动等因素的影响，传感器采集的原始信号信噪比较低，采取有效的滤波方法对信号进行降噪处理，提取微弱特征信号，找到信号突变拐点，计算准确时间差是管道泄漏监测的难点。

对于微弱泄漏信号，可采用随机共振方法实现信号的增强，但随机共振后的信号波动较大，很难寻找到统一的基准来检测泄漏信号，因此不仅需要考虑信号的前后关联性，还需要考虑信号的归一性。为此，提出了基于非周期随机共振的快速累计差分算法，利用自适应随机共振算法（下称 ASR 算法）实现强干扰背景下泄漏信号的增强，引入快速累计差分算法（下称 FAD）对信号进行连续化处理，达到了微弱信号的双增强效果，提高了泄漏信号的信噪比。基于 ASR-FAD 泄漏检测技术，其实现流程可以描述为：

（1）使用动态压力变送器采集待监测管段两端的实时动态压力信号 $X_1(k)$；

（2）对采集到的动态压力信号 $X_1(k)$ 进行预处理，得到 $X_2(k)$；

（3）根据当前时刻的信号 $X_2(k)$ 和下一时刻的信号 $X_2(k+1)$ 的值，使用 ASR 算法求得下一时刻的输出信号 $W_1(k+1)$；

（4）使用 FAD 算法对 $W_1(k+1)$ 进行处理，得到 $\Delta P(k+1)$；

（5）对信号 $\Delta P(k+1)$ 进行后续处理。

信号经 ASR-FAD 算法处理后，还需进行上下游信号的有效匹配。动态时间规整算法（下称 DTW）能够实现计算管道两端快速差分序列相似波峰时间间隔的问题，但有一个问题需要解决，传统的 DTW 算法只适用于有限长度序列，而对于泄漏定位来说其信号为未知长度的时间序列信号。因此在传统 DTW 算法的基础上，利用结果复用的思想提出了可用于计算实时数据的动态迭代时间规整算法。

管道两个监测点的实时快速累积差分序列分别为：$\overline{\Delta P}_{up} = \overline{\Delta p}_{up,1}$, $\overline{\Delta p}_{up,2}$, $\cdots$, $\overline{\Delta p}_{up,N}$ 和 $\overline{\Delta P}_{down} = \overline{\Delta p}_{down,1}$, $\overline{\Delta p}_{down,2}$, $\cdots$, $\overline{\Delta p}_{down,N}$，其中 N 为差分序列长度。以 $\overline{\Delta P}_{up}$ 为纵轴、$\overline{\Delta P}_{down}$ 为横轴构造 $N \times N$ 网格的距离矩阵 d，交叉点 (i,j) 元素为 $d(i,j)$，可用下式表示：

$$\overline{\Delta P}_{up} = \begin{bmatrix} \overline{\Delta P}_{up,1} \\ \overline{\Delta P}_{up,2} \\ \vdots \\ \overline{\Delta P}_{up,k} \\ \vdots \end{bmatrix} = \begin{bmatrix} \overline{\Delta p}_{up,1} & \overline{\Delta p}_{up,2} & \cdots & \overline{\Delta p}_{up,L} \\ \overline{\Delta p}_{up,\Delta L+1} & \overline{\Delta p}_{up,\Delta L+2} & \cdots & \overline{\Delta p}_{up,\Delta L+L} \\ \vdots & \vdots & & \vdots \\ \overline{\Delta p}_{up,k \times \Delta L+1} & \overline{\Delta p}_{up,k \times \Delta L+2} & \cdots & \overline{\Delta p}_{up,k \times \Delta L+L} \\ \vdots & \vdots & & \vdots \end{bmatrix} \quad (1)$$

$$\overline{\Delta P}_{down} = \begin{bmatrix} \overline{\Delta P}_{down,1} \\ \overline{\Delta P}_{down,2} \\ \vdots \\ \overline{\Delta P}_{down,k} \\ \vdots \end{bmatrix} = \begin{bmatrix} \overline{\Delta p}_{down,1} & \overline{\Delta p}_{down,2} & \cdots & \overline{\Delta p}_{down,L} \\ \overline{\Delta p}_{down,\Delta L+1} & \overline{\Delta p}_{down,\Delta L+2} & \cdots & \overline{\Delta p}_{down,\Delta L+L} \\ \vdots & \vdots & & \vdots \\ \overline{\Delta p}_{down,k \times \Delta L+1} & \overline{\Delta p}_{down,k \times \Delta L+2} & \cdots & \overline{\Delta p}_{down,k \times \Delta L+L} \\ \vdots & \vdots & & \vdots \end{bmatrix} \quad (2)$$

$$d(i,j) = |\overline{\Delta p}_{up,i} - \overline{\Delta p}_{down,j}| \quad (3)$$

其中，$i=1, 2, \cdots N$，$j=1, 2,, \cdots N$，L 为序列初始长度，ΔL 为每次载入的新数据量。

DTW 算法旨在距离矩阵 d 中寻找一条映射路径 $w_p = (w_{p,1}, \cdots, w_{p,l}, \cdots, w_{p,L})$，其中 $w_{p,l} = (n_l, m_l) \in [1:N] \times [1:N]$。满足如下 4 个约束条件：(1) 边界性，$w_{p,1} = (1, 1)$，$w_{p,L} = (N, N)$；(2) 单调性，若 $w_{p,l_1} = (n_{l_1}, m_{l_1})$，$w_{p,l_2} = (n_{l_2}, m_{l_2})$，$l_2 > l_1$ 则 $n_{l_1} > n_{l_2}$，$m_{l_1} > m_{l_2}$；(3) 连续性，若 $w_{p,l} = (n_l, m_l)$，$w_{p,l+1} = (n_{l+1}, m_{l+1})$ 则 $n_{l+1} \geqslant n_l$，$m_{l+1} \geqslant m_l$；(4) 有界性，若 $w_{p,l} = (n_l, m_l)$，则 $|n_l, m_l| < M_0$，其中 M_0 为路径扭曲界限。其中，约束条件(1)至(3)为传统 DTW 算法的约束条件，约束条件 4 为快速累积差分序列需增加映射路径界限约束。基于动态压力的泄漏监测技术，动态压力波在两个监测点之间的传播时间有限，因此泄漏信号传输到两个站点的时间差也有限，

其最大值为泄漏信号从管道一端传递到另一端所需要的时间,同理,基于快速累积差分信号也有相同的时间差上限。

为保证对泄漏监测的实时性,ΔL值设置为较小值,即差分序列中的历史数据远大于新数据量,若每次使用传统DTW算法对序列进行计算,距离矩阵d和累积距离D包含了大量重复信息,对历史数据重复计算增加了计算时间;ΔL设置太小则会出现待计算数据累积的情况,难以保证实时性。为此,使用结果复用的思想提出了动态迭代时间规整算法。对于映射路径矩阵的信息复用有两种情况。根据式计算D_k中的映射路径:当映射路径$w_{p,k}$与映射路径$w_{p,k-1}$相遇时,$w_{p,k}=[w'_{p,k-1};w'_{p,k}]$,其中$w'_{p,k-1}$为映射路径$w_{p,k-1}$相遇点之前的映射关系,$w'_{p,k}$为根据累积距离矩阵计算的从相遇点到序列末端的路径;若不存在相遇点,则根据传统的动态时间规整算法求解相应的新的映射路径。

3.3 泄漏多源感知模型的建立

目前大多数管道泄漏监测系统采用单一的监测技术,系统的灵敏性受限于单一变送器的灵敏度,存在容错能力较低、稳定性差等问题。针对单一监测技术的不足,本文研究综合多种传感信号(包括动态压力信号和振动信号等)、SCADA数据等多源数据,基于不同的信号特征构建泄漏多源感知模型,提高泄漏监测的准确性和效率。模型的诊断流程见图3。

动态压力变送器采集的次声信号是实现管道泄漏监测的关键,当天然气管道发生泄漏时,泄漏点处会激发出一系列异常信号,这些信号会沿着管道传播。通过在监测管段两段安装动态压力变送器实时采集动态压力信号,并借助傅里叶变换、ASR-FAD算法、DTW算法等信号处理算法,对原始信号进行一系列处理,提取与泄漏相关的频率、幅度等特征,实现管道泄漏的监测与定位。其中,傅里叶变换实现泄漏信号由时域至频域的转换,ASR-FAD算法实现信号的增强及奇异点的检测,动态迭代时间规整算法用于异常信号的对齐,通过动态匹配找到与已知泄漏特征相符的信号段。

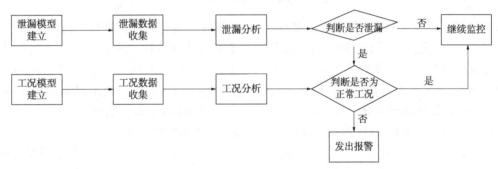

图3 多源泄漏感知模型诊断流程

振动传感器主要用于采集站内工况操作产生的振动信号,辅助进行工况判断。当振动传感器检测到信号异常波动时,会立即触发工况模型进行进一步分析,工况模型则会根据振动信号的特征,结合管道的实时运行数据及设备状态数据,进一步确认是否由工况引起的异常,通过这种方式,可以有效区分正常工况变化与泄漏事件,提高泄漏检测的准确性。

SCADA数据主要用于工况判断,通过采集的实时运行数据及关键设备运行状态,分析管道是否存在工况操作,确认异常信号是否为工况操作所致。模型综合工况识别结果与泄漏诊断结果,给出最终报警信息。

3.4 泄漏监测系统开发

结合天然气管道泄漏监测需求,设计了天然气管道泄漏监测系统架构。系统采用五层架构模式(见图4),包括仪表层、数据实时采集层、数据中心、诊断算法层和诊断结果发布层等。仪表层:为泄漏监测系统的数据基础,借助各类传感器实时采集和检测管道运行状态及生产过程参数;数据采集层负责接收、处理来自仪表层的数据,并将数据传输给数据中心;数据中心负责接收来自数据采集层的大量数据,并进行数据的集中处理、储存;诊断层将采集到传感信号通过数据通道流向各方法的检测程序,并汇总多种方法的检测结果,采用一种组合算法输出最终报警结果;诊断结果发布层将各算法的报警结果以及综合决策的报警结果均存入数据库,并提供数据接口用以报警结果的发布。系统可读取实时与历史数据,将实时报警事件展示在用户界面上,并可提供历史报警记录查询模块。

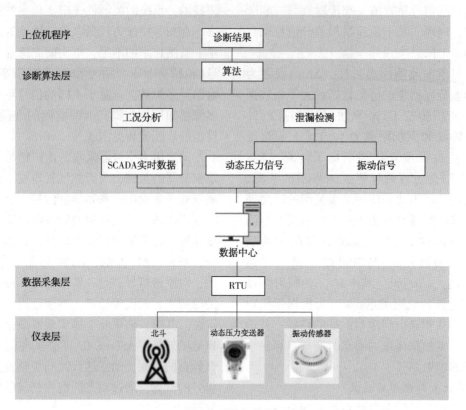

图 4 系统架构图

管道泄漏监测系统采用"成熟的硬件架构+自主模型"的开发模式,保障了系统运行的稳定性、可靠性及高效性,开发框架见图 5。系统基于 B/S 架构开发,用户无需安装专门的客户端软件,通过浏览器即可访问系统,简化了部署和维护,并提供了更好的可访问性、系统拓展性及兼容性。系统服务端采用 Java 语言结合 Spring Boot+MyBatis 框架开发,实现了 restful 风格化服务,凭借 Java 语言的 JVM 运行机制带来的跨平台优势,支持跨平台部署。数据库采用 PostgreSQL 数据库,其强大的扩展性、高可靠性、安全性、丰富的功能特性,成为了适用于处理大型数据集、高并发请求和复杂业务逻辑的优选开源关系型数据库管理系统。前端界面采用 VUE 框架开发,界面美观、操作简便。

3.5 系统测试分析

开发的综合法管道泄漏监测系统在中俄东线唐山至宝坻段进行了应用测试。该管道规格 $\varphi1219×22mm$,管道长度 79.5km,阀室或站场间的最大间距为 25.4km,最小间距 14.6km;管道设计压力 10MPa,唐山至宝坻段运行压力为 7MPa 左右。各站点设备配置情况见表 1 和图 6。

表 1 试点管线站场设备配置表

位置	设备名称	泄漏监测服务器	数据采集RTU	动态压力传感器	振动传感器
唐山站	出站端	—	1	1	1
40#阀室	出站端	—	1	1	1
41#阀室	出站端	—	1	1	1
42#阀室	出站端	—	1	1	1
宝坻站	进站端	—	1	1	1
监测中心		1	—	—	—
总计		1	5	5	5

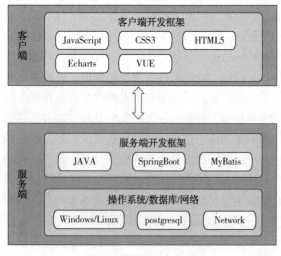

图 5 系统开发框架图

为充分测试系统的性能指标，在42#阀室开展了现场放气测试。放气模拟管路管径DN50，耐压等级10MPa，可通过改变法兰板的过流孔径实现不同口径泄漏量的模拟，共制作了1mm、2mm、3mm、4mm、6mm、9mm等不同口径的过流孔，可实现不同泄漏孔径的泄放模拟。

通过缓开缓关、快开快关、调整泄漏孔径等方式调节管道泄漏量，共开展了63组放气测试，波形图及系统报警情况见图8～图13。

图6 阀室现场实施照片

图7 放气模拟管路现场照片

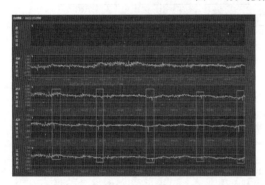

图8 4mm泄漏孔径测试波形图

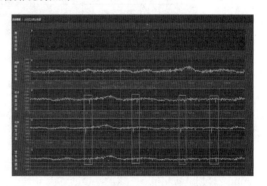

图9 3mm泄漏孔径测试波形图

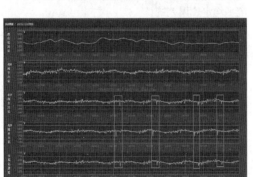

图10 2mm泄漏孔径测试波形图

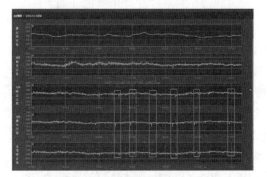

图11 1mm泄漏孔径测试波形图

·353·

识别报警 / 泄漏报警

泄漏报警列表

	泄漏管道	泄漏时间	泄漏位置	定位方式	级别	操作
1	中俄东线 42#阀室出站 至 宝坻站进站 波动异常	2024-05-20 11:14:21	距离42#阀室 0.173公里	自动定位	1	查看详情
2	中俄东线 42#阀室出站 至 宝坻站进站 波动异常	2024-05-20 11:28:47	距离42#阀室 -0.011公里	自动定位	1	查看详情
3	中俄东线 42#阀室出站 至 宝坻站进站 波动异常	2024-05-20 11:42:00	距离42#阀室 0.021公里	自动定位	1	查看详情
4	中俄东线 42#阀室出站 至 宝坻站进站 波动异常	2024-05-20 11:50:49	距离42#阀室 0.029公里	自动定位	1	查看详情
5	中俄东线 42#阀室出站 至 宝坻站进站 波动异常	2024-05-20 12:03:06	距离42#阀室 -0.091公里	自动定位	1	查看详情
6	中俄东线 42#阀室出站 至 宝坻站进站 波动异常	2024-05-20 12:14:12	距离42#阀室 0.132公里	自动定位	1	查看详情
7	中俄东线 42#阀室出站 至 宝坻站进站 波动异常	2024-05-20 12:25:49	距离42#阀室 -0.133公里	自动定位	1	查看详情
8	中俄东线 42#阀室出站 至 宝坻站进站 波动异常	2024-05-20 13:09:05	距离42#阀室 -0.110公里	自动定位	1	查看详情
9	中俄东线 42#阀室出站 至 宝坻站进站 波动异常	2024-05-20 13:25:29	距离42#阀室 0.007公里	自动定位	1	查看详情
10	中俄东线 42#阀室出站 至 宝坻站进站 波动异常	2024-05-20 13:42:21	距离42#阀室 0.125公里	自动定位	1	查看详情

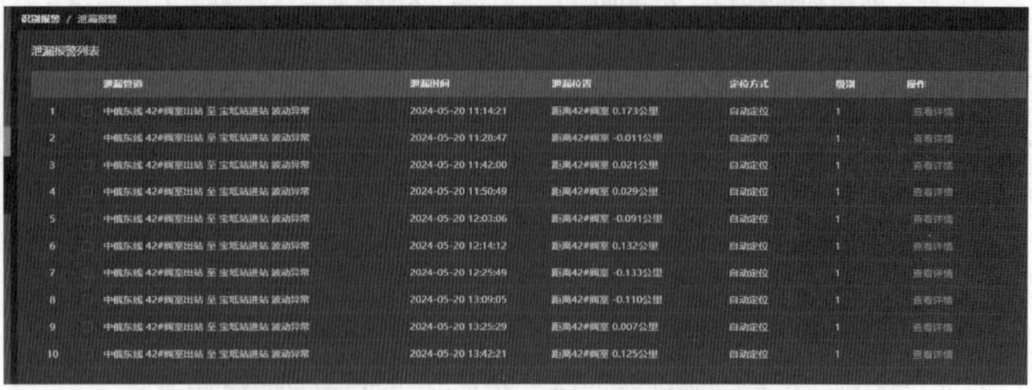

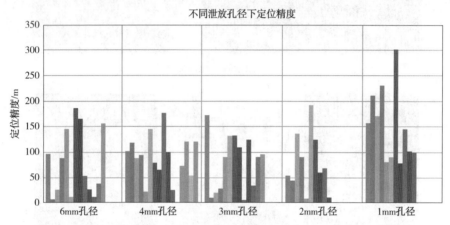

图 12　2024 年 5 月 20 日及 22 日部分系统报警截图 1

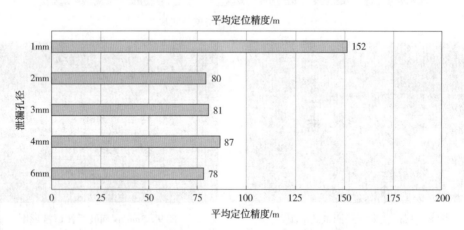

图 13　不同泄漏孔径下定位精度统计

图 14　不同泄漏孔径下系统平均定位误差统计

现场实际测试结果表明：

（1）在微小泄漏监测及报警识别方面，针对 63 组放气测试，系统均能识别并自动报警，系统可检测的最小孔径为 1mm，报警响应时间在 60 秒之内。

（2）在定位误差方面，针对 2mm、3mm、4mm 和 6mm 泄漏孔径的放气实验，系统自动定位误差均在 200m 以内，平均定位误差为 78～87m，部分测试定位误差可达 30 米以内；针对 1mm 泄漏孔径的放气实验，系统定位误差在 79～301m，平均误差为 152m。

文献资料显示，国内类似的天然气管道泄漏监测系统最小可监测孔径多在 6～7mm，最小定位误差在 50 米以内；国外以美国 ASI 公司为代表的音波泄漏监测系统，可监测最小孔径为 2mm，最小定位误差小于 30 米，响应时间在 60 秒以内。本文融合动态压力变送器、振动传感器及 SCADA 系统等多源数据，开发形成的综合法天然气管道泄漏监测系统灵敏度高、响应迅速、定位准确，在微小泄漏报警、定位误差及响应时间等方面与国外先进技术水平相当。

4 结论及建议

针对当前天然气管道泄漏监测存在的问题，从"感"和"知"两方面开展研究。在"感"的方面，研制了基于压电原理的动态压力变送器，解决了微弱泄漏信号的检测难题；在"知"的方面，通过融合多种传感信号、SCADA 数据等多源数据，基于不同的信号特征构建泄漏多源感知模型，开发了综合法管道泄漏监测系统，并在中俄东线唐山至宝坻段开展了现场放气测试，得到的主要结论如下：

（1）研制的动态压力变送器具有灵敏度高、频带响应宽、响应时间短等性能，具备微弱天然气泄漏信号的检测能力。针对 1mm 泄漏孔径的放气试验，波形信号可看出明显的响应。

（2）融合各类传感信号及 SCADA 数据等多源数据，研发的综合法天然气管道泄漏监测系统在泄漏监测报警、定位及响应方面满足预期要求。针对 63 组放气测试，泄漏监测系统均能识别并自动报警，报警响应时间均在 60 秒之内，系统最小可监测孔径达 1mm，针对 2mm、3mm、4mm 和 6mm 泄漏孔径的放气实验平均定位误差 78~87m，针对 1mm 泄漏孔径的放气实验，平均定位误差为 152m，总体与国外先进技术水平相当。

随着大数据、人工智能等技术的不断发展，下一步将深化大数据、深度学习等新型技术在管道泄漏监测方面的应用研究，提高模型对复杂泄漏场景的识别能力，进一步增强泄漏监测系统的准确性和鲁棒性。系统采集的传感数据、SCADA 数据、GIS 系统等多途径海量数据具有时序性、多样性、关联性和复杂性等特点，通过大数据平台可将上述数据进行集成和整合，形成全面的管道运行状态数据库，并利用大数据分析、深度学习模型等，对管道泄漏数据进行深度挖掘分析、特征提取及模式识别，实现管道泄漏故障的智能诊断和趋势预测，为管道运行及维护提供决策依据。

参 考 文 献

[1] 李昕. 油气管道泄漏安全对策措施研究[J]. 中国化工贸易，2023，15：127-129.

[2] 王洪超，吴琼，王宁等. PLC/RTU 在负压波法管道泄漏监测中的应用[J]. 物联网技术，2023，13(6)：22-24.

[3] 秦程. 基于负压波与流量平衡法的管道泄漏监测系统研究[D]. 大连理工大学，2021：8-30.

[4] 马文静，任远等. 天然气管道泄漏监测技术发展现状研究[J]. 中国科技期刊数据库，2022：112-115.

[5] 袁文强，郎宪明，曹江涛等. 基于声波法的管道泄漏监测技术研究进展[J]. 油气储运，2023，42(2)：141-151.

[6] 刘良果，梅茜迪等. 基于次声波传感的输气管道泄漏监测技术应用研究[J]. 2018 年全国天然气学术年会.

[7] 王洪超，马云宾等. 基于光纤声波传感器的管道泄漏监测技术研究[J]. 山东工业技术，2024，5：40-45.

[8] 王江伟，王红义等. 光纤法泄漏监测技术的有效性与局限性[J]. 中国土木工程学会燃气分会 2021 年学术年会.

[9] 李玉星，刘翠伟. 基于声波的输气管道泄漏监测技术研究进展[J]. 中国科学，2017，62(7)：650-658.

[10] 刘翠伟. 输气管道泄漏声波产生及传播特性研究[D]. 中国石油大学(华东). 2016，1-34.

[11] 张宇，靳世久等. 基于动态压力信号的管道泄漏特征提取方法研究[J]. 石油学报，2010，31(2)：338-342.

[12] 周巍，罗润等. 智能音波测漏系统在长输天然气管道上的实践[J]. 传感器技术与应用，2020，8(3)：96-106.

[13] ASI 音波测漏系统[J]. 石油化工自动化，2009，88.

燃气轮机伺服控制系统的关键技术分析与架构设计

郑 明 关 睿 刘 超 姚 珺

(国家管网集团储运技术发展有限公司)

摘 要 随着能源领域国产化进程的加速,燃机进口导叶的伺服控制系统国产化替换迫在眉睫。本文深入剖析了燃机进口导叶伺服控制的技术特质、棘手难点以及实际应用场景,据此精心策划并提出一套完整的燃机伺服控制系统设计蓝图,涵盖系统整体架构的搭建、核心伺服控制模块的研发以及配套软件的定制化设计。同时,紧密结合该系统常见的故障类别与行之有效的诊断处理手段,将此设计方案成功落地于国产 T9100 控制系统平台之上。经 GE LM2500+ 型号燃机的伺服控制功能严格测试,各项试验数据精准契合燃机伺服控制系统预设指标,充分验证了本方案能够圆满达成燃机伺服控制系统国产化替代的既定目标,为推动我国燃机技术的自主可控发展提供了有力支撑。

关键词 燃气轮机;进口导叶;伺服控制;国产化

Key technology analysis and architecture design of gas turbine servo control system

Zheng Ming   Guan Rui   Liu Chao   Yao Jun

(Pipechina Pipeline Technology Development Co., Ltd)

Abstract With the acceleratI/On of the localizatI/On process in the energy field, the localizatI/On and replacement of the servo control system of the imported guide vane of the gas turbine is imminent. This paper deeply analyzes the technical characteristics, difficulties and practical applicatI/On scenarI/Os of the servo control of the imported guide vane of the gas turbine, and proposes a complete set of design blueprint for the servo control system of the gas turbine, covering the constructI/On of the overall system architecture, the research and development of the core servo control module, and the customized design of the supporting software. At the same time, the common fault categories of the system and the effective diagnosis and treatment methods are closely combined, and the design scheme is successfully implemented on the domestic T9100 control system platform. After rigorous testing of the servo control functI/On of the GE LM2500+ gas turbine, the test data accurately fit the preset indicators of the gas turbine servo control system, which fully verified that the solutI/On can successfully achieve the established goal of localizing the gas turbine servo control system, and provided strong support for promoting the independent and controllable development of gas turbine technology in China.

Key words Gas Turbine; Inlet Guide Vane; Servo Control; Domestic

在燃气轮机的运行体系中,进口导叶控制系统无疑占据着关键地位,其性能优劣直接关乎燃机的整体运行效能。储运技术公司压缩机组维检修分公司试车台所采用的美国 AB Rockwell 公司的 ControlLogix 燃机试车控制系统,自投入使用至今已近十载,随着时间的推移,其老化磨损问题逐渐凸显,目前已临近必须替换更新的关键节点。

该套控制系统的硬件设备与软件程序均依赖国外进口,这不仅使得系统的日常维护工作面临重重困难,而且在进行版本升级时,往往需要承担高昂的费用,同时还不得不忍受漫长的响应周期，这无疑极大地增加了企业的运营成本和时间成本，严重制约了公司的发展效率。

值得注意的是，当下燃气轮机在国内众多行业领域的应用呈现出愈发广泛的趋势，然而令人遗憾的是，无论是燃机的实际运行环节，还是试车测试阶段所运用的控制系统，几乎全部被国外产品所垄断。特别是燃机测试过程中所使用的进口导叶系统伺服控制模块，由于其并非标准通用件，市场供应渠道狭窄，导致其价格居高不下，而且频繁出现故障，严重阻碍了燃机的正常起机测试流程以及稳定运行状态，给企业的生产运营带来了极大的困扰和损失。

面对如此严峻的形势，为了坚决打破国外的技术封锁壁垒，储运技术公司压检分公司积极开展试车台国产控制系统的硬件开发工作，并进行全面深入的对比测试，已然成为当前燃机技术发展进程中亟待解决的首要任务，这对于提升我国燃气轮机产业的自主可控能力和核心竞争力具有极其重要的战略意义和现实价值。

1 燃气轮机导叶伺服控制原理

1.1 导叶控制系统结构

伺服控制模块依据反馈值与设定值的偏差进行控制计算，输出相应的控制信号。控制信号经过运算放大，进入到伺服阀的各个线圈中，线圈中流过的电流大小和方向，决定了液压油的流速和方向，流动的液压油推动作动筒伸缩，从而控制导叶的角度。因为仪表在实际测量中可能存在误差，所以进口导叶伺服控制系统采用多个 LVDT 传感器测量进口导叶的实际角度，以增加测量结果的可信度。使用多个传感器的另一个好处是增加了系统的可用性。对于伺服阀控制同样采用冗余设计，使用多个线圈驱动一个伺服阀，即使某个线圈断线，剩下的正常线圈依然可以驱动，增加了伺服阀的故障容忍能力。

1.2 角度与转速的关系

燃机运行过程中需要根据转子的修正转速进行燃机进口导叶系统调节，30MW 级燃机的导叶角度与燃机转速的对应关系曲线图 1 所示。

进口导叶伺服控制模块作为燃机控制系统中的重要元件，其本身的控制精度、响应实时性、可靠性直接影响燃机进口导叶控制的准确度，进而影响燃机的效率。此外，伺服控制模块对各种信号故障类型的精确检测，以及对故障处理的及时性将直接关系到燃机设备的安全。因此，燃机对进口导叶系统的控制要求极高，其对应的控制模块要求也非常严格。

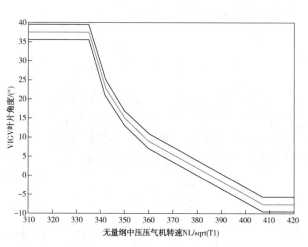
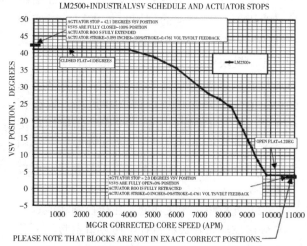

图 1 RB211 和 LM2500 进口导叶角度与转速关系曲线

2 燃机伺服控制功能设计

针对燃机伺服控制的高控制性能、高可用性要求，综合 T9100 控制系统的结构特点，重点从系统架构、伺服控制模块、软件这三大方面对燃机伺服控制系统功能进行设计。

2.1 系统架构设计

中控技术的 T9100 系统为完全自主研发的大型动设备控制系统，基于中控技术的大型安全仪表系统 TCS-900，具有 TÜV SIL3 认证。功能齐全，控制算法丰富，精度高，组态方便，具有动设备控制的丰富应用经验，能胜任燃气轮机的控制工作。

T9100系统的常用组件包括组态软件、监控软件、控制器、通信模块、各类I/O模块。控制器作为整个控制系统的主要控制核心，它通过安全通信总线接收各类输入型I/O模块的采样数据，经过程序运算后，输出控制数据到输出型I/O模块。伺服控制模块作为兼具输入和输出功能的I/O模块，在模块级实现闭环控制，可提高控制速度。可使用T9100系统搭建图2所示的燃机伺服控制系统。

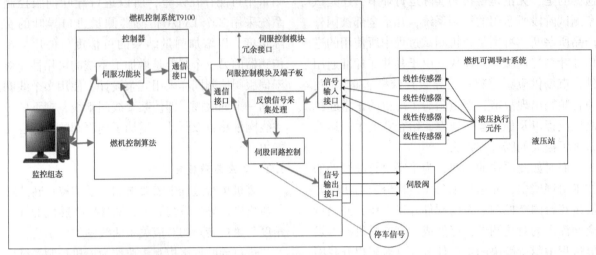

图2 燃机伺服控制系统整体架构图

2.2 伺服控制模块功能设计

进口导叶伺服控制的主要功能在伺服控制模块中实现。在伺服控制模块功能设计中，应确保：

（1）1个伺服控制模块可同时控制2个伺服回路，并采用三重化冗余设计。

具备16位的输入采样精度，以及多种输出信号范围。具备5ms的伺服扫描周期，以实现块速的伺服控制。对于每个回路，伺服控制模块伺服控制模块可以采集至多4路LVDT反馈信号再通过表决得到反馈值Pv，并根据控制器下发的设定值Sv，经过模块内部PID运算，输出至多3路AO信号控制伺服阀。伺服控制模块可按冗余或非冗余模式配置。冗余配置情况下，备用卡处于热备状态，在主卡发生故障时能快速进行主备切换。

（2）支持组态接收功能、在线控制功能。

伺服模块可在运行过程中接收组态数据，并根据组态信息执行相应的伺服控制功能。

伺服模块支持在线调整参数和在线控制功能，包括PID参数在线调整、输入输出信号投切控制、LVDT信号零幅值校准控制等。

（3）具备内外部故障检测与处理能力。

具备系统内部故障的检测与处理能力，包括伺服模块自身的故障、控制器的故障、通信故障等。

具备外部故障的检测与处理能力，包括输入信号、输出信号故障等。

（4）采用微分先行带抗积分饱和的PID算法。

燃机导叶控制速度以其快速性准确性为特征，采用微分先行的PID算法是非常合适的，该算法在设定值变化时只执行比例和积分控制，没有微分作用。该算法适合应用在希望对设定值变化有较好的跟踪特性的场合。

其算法为：

$$\Delta MV_n = \frac{100}{PB}\left(\Delta E_n + \frac{T_S}{T_I}E_n + \Delta U_n\right) \quad (1)$$

式中，E_n为设定值与测量值的偏差；PV_n、PV_{n-1}为前后2次测量值；PB为比例度；T_I为积分时间；T_D为微分时间；T_S为控制周期；微分项采用不完全微分，其系数为$K_D=8$；

$$\Delta E_n = E_n - E_{n-1} \quad (2)$$

$$\Delta U_n = U_n - U_{n-1} \quad (3)$$

$$U_n = \frac{T_D}{K_D T_S + T_D}[U_{n-1} + K_D(PV_n - PV_{n-1})] \quad (4)$$

本周期控制量=上周期控制量+ΔMV_n。

PID算法中采用了积分切除方式抗积分饱和，所谓积分饱和是指当系统存在一个方向的偏差，则由于积分作用的不断累积会使PID输出值不断扩大并超出正常范围进入饱和区，而当系统出现反向的偏差时，需要首先从饱和退出，无法立即对反向的偏差做出响应。为解决这个问题，使用积分切除方法，当上一周期的控制量输出超过输出上限或低于输出下限时，积分项不起作用。

（5）自动LVDT零幅值校准功能。

该功能用于自动获取实际应用时LVDT的两个极限位置，作为反馈信号的零值和幅值。

2.3 软件设计

（1）用户组态的编辑功能。

支持用户进行组态建立、硬件组态编辑、变量组态编辑、通信组态编辑、用户程序编写，并支持组态保存与导入导出功能。

（2）组态的全体和增量编译功能。

支持增量编译功能，仅对用户操作过的、发生了更改的组态内容进行单独编译，单独改变组态内容，从编译环节确保了组态更改的局部影响性。

（3）组态的全体和增量下载功能。

增量下载与增量编译相似，仅对发生了更改的组态内容进行下载，从组态下载操作环节确保了组态更改的局部影响性。

（4）实时监测与手动控制功能。

支持以趋势图、流程图等形式实时显示系统的导叶角度、转速、压力等数据，支持对相关的变量位号进行手动控制的命令下发。

3 常见故障类型与诊断处理方式

3.1 LVDT线路故障诊断与恢复操作

在实际运行过程中，LVDT线路故障属于较为常见的故障类型。一旦发生LVDT线路故障，会致使反馈信号失准，进而对控制回路的正常运行产生不良影响。伺服控制模块具备对每路LVDT传感器的线路故障检测、切除以及投入等功能。当某一个LVDT传感器出现断线故障时，该传感器会自动被切除，不再参与反馈值的表决环节。

对于出现断线故障的LVDT传感器，可以进行在线修复操作。只有在伺服控制模块判定其现场已恢复正常之后，才能够借助软件将对应的LVDT传感器重新投入使用。倘若所有输入的LVDT均发生故障，那么系统将会依据安全预设输出值来执行输出动作，以此保障整体运行在特殊情况下仍能处于相对安全稳定的状态，避免因传感器故障而引发不可控的风险或事故，最大程度降低故障对整个运行体系的冲击和破坏。

3.2 LVDT偏差大故障诊断与处理方式

LVDT在长期使用过程中可能出现例如精度变差的情况，伺服控制模块具备LVDT间偏差诊断功能，当多个LVDT信号的差值超过所设置的阈值并经故障确认后，则剔除该路LVDT，不再参与PV表决。

3.3 AO故障检测与恢复操作

伺服输出故障也是常见的一种故障类型，故障因素有触点接触不良、输出线路断线、伺服阀内部线路损坏、环境信号干扰等。伺服控制模块可对AO线路进行开路、短路、输出异常等检测，检测为故障的线路可被切除。

可在线排除AO外部故障，恢复AO线路输出功能，通过AO输出电路故障状态指示信息，确认故障已恢复。

3.4 位置控制偏差故障诊断处理方式

伺服控制模块使用两种方式进行位置控制偏差诊断。方式一是在检测到PV与SV偏差大于5%时启动故障确认机制，周期性判断偏差是否收敛。方式二是在PV与SV之间偏差超过组态配置值的情况下，若持续超过组态配置的确认时间，则报位置控制偏差故障。常见的位置控制偏差故障因素有伺服控制PID参数设置不合理、油压不足、伺服阀卡涩、导叶机构卡住等。

3.5 外部停车功能设计

为保障设备安全性，在某些紧急情况下需绕过控制逻辑，直接要求伺服控制输出到安全位置。为了获得更快响应速度，伺服控制模块内置了外部停车信号(DI)的联锁功能及可组态的联锁AO输出安全值功能。当检测到停车信号(DI)触发并经确认后，由伺服模块直接触发停车，AO输出预设的安全值。

4 燃机伺服控制验证

为了验证本次开发的燃机导叶伺服控制系统是否满足要求，在GE LM2500+型燃气轮机上进行了伺服控制功能测试。

4.1 原有燃机控制系统改造

储运技术公司压缩机组维检修分公司的试车台燃机试车控制系统，采用的是由AB Rockwell公司所生产的ControlLogix。此系统并未配备专用的伺服控制模块，其输入与输出信号均需借助变送器来完成信号转换过程，而后依靠1756-L61控制器达成导叶闭环控制。而T9100的伺服控制模块则将 LVDT 信号采集功能以及伺服控制电流输出功能进行了有机整合，有效省略了中间的信号变送单元。在本次测试中，最大程度地保留了原控制系统的燃机控制主体功能，达成了 T9100 系统伺服控制系统对燃机进口导叶的控制目标，从而对国产化替换的可行性展开验证工作，为后续相关技术应用与系统升级改造提供了极具价值的实践依据与数据支撑，有力推动了该领域技术的国产化进程与自主创新发展。

4.2 燃机伺服控制功能测试

试验平台搭建完成后，燃机进行干转启动，达到盘车转速后，先进行两个反馈信号的零幅值校准，然后通过改变导叶设定目标，使用 T9100 伺服控制系统专用快速趋势记录软件，观察导叶控制的阶跃响应曲线，经调节 PID 参数后，可使阶跃响应特性达到原有效果，运行过程中能很好的表现出对导叶角度设定值的跟随性，改造前后的控制效果见图 3 与图 4。

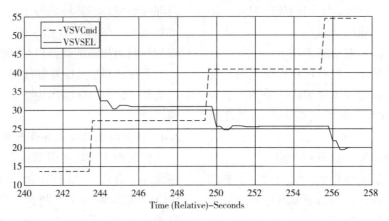

图 3　进口导叶伺服控制回路改造前阶跃响应曲线

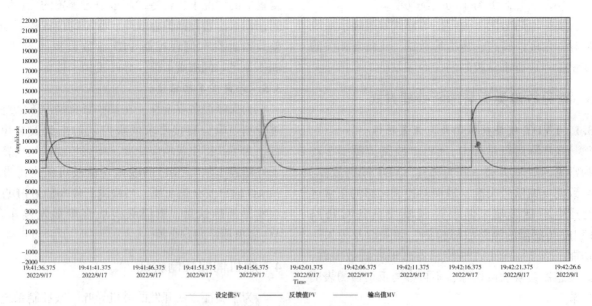

图 4　进口导叶伺服控制回路改造后阶跃响应曲线

T9100 伺服控制是 I/O 模块级的闭环控制，其扫描速度高于原有控制系统，且无需信号中转变送，经调整合适的 PID 参数后可超越原有系统的调节效果。

5　结束语

本文深入且系统地剖析了燃机进口导叶伺服控制系统的技术特性与国产化替换需求，详细解析了燃机伺服控制系统的常见故障类别及其诊断处理手段，进而提出一套燃机伺服控制功能设计方案，并成功应用于 T9100 控制系统之中。通过对燃机控制系统平台实施适度改造，对燃机导叶伺服控制功能展开验证，结果显示该设计方案契合燃机伺服控制系统的各项标准，有力证实了试车台控制系统在伺服控制环节全面推行国产化改造的可行性，为后续试车台控制系统的国产化变革筑牢了技术根基。鉴于燃驱压缩机组控制系统与试车台控制系统皆以燃机控制为核心,且控制原理大体相同,故而在试车台控制系统国产化替代研究取得成功后,能够在燃驱压气站、燃机发电等燃机控制相关领域广泛推广运用。

参 考 文 献

[1] IEC 61508-3, FunctI/Onal safety of electrical/electronic/programmable electronic safety-related systems-Part 1: General requirements[S]. CENELEC, 2010.

[2] 刘志勇,黄文君,金建祥. 工业控制器跨平台特性设计[J]. 计算机工程, 2012. 第38卷 第14期: 227-228.

[3] 刘志勇. 工业控制器的关键技术研究与开发[D]. 杭州: 浙江大学, 2012.

[4] 方珂琦,朱杰,张则立,等. TCS-900安全控制系统的功能安全通信设计[J]. 工业控制计算机, 2021, 第34卷 第8期: 9-11.

[5] 丁建南,童蕴真. 6FA燃机伺服阀控制原理及故障分析[J]. 中国科技信息, 2021(15): 36-38.

[6] 闫桂山,金振林,赵鹏辉. 燃气轮机电液伺服泵控IGV位置补偿控制研究[J]. 船舶工程, 2020, 42(12): 85-92.

Optimization analysis of shutdown and restart of multiphase transportation operations in the South China Sea

Han Dong　Zheng Chengming　Jiang Ling　Zhang Meng　Qi Baobao
Chen Zhu　Zhang Lin　Guo Yingzhen　Sun Xinghua　Wang Bing

(China National Offshore Oil Corporation)

Abstract　Currently, a South China Sea subsea pipeline performs oil & gas multiphase transportation. To understand the heat, mass transfer in shutdown and restart scenario under different upstream and downstream working conditions, this paper presents engineering practice of multiphase transportation for shutdown and restart scenarios in a gas condensate field. The paper also presents an innovative process in which an intermittent restart process is used to maximize production and analyzes the shutdown and intermittent restart of a subsea pipeline using a dynamic multiphase flow model. Several intermittent restart processes were compared and assessed considering the operation parameters of the inland process terminal system. The results showed that flow instability under different intermittent restart processes will not threaten the inland process terminal system. When the gas flow rate is less than $8\times10^4 m^3/h$, gas has no liquid-carrying ability. The shorter the intermittent restart time, the smaller the liquid flow rate to the inland process terminal system. If the flow is restarted intermittently with higher gas flow rate, liquid flow rate when landing the terminal is higher, up to $400m^3/h$. This scenario can be prevented by an intermittent restart process with a maximum value of $282m^3/h$. A larger intermittent restart reduces the gas flow rate by 30%. This paper's conclusions can help engineers develop process plans for transient situations, as well as being a reference for the production system's safe operation and production capacity release.

Keywords　Subsea pipeline; Multiphase transportation; Flow assurance; Shutdown and restart

1 Introduction

Lingshui, Dongfang, Ledong and Liwan gas field in the South China Sea accounts for 90% of the Hainan province's home and industrial gas usage. With the continued development of deep and ultradeep sea gas fields such as the Dongfang Gas Field, Lingshui Gas Field, and Shenhai No. 1 (deepwater semisubmersible production and storage platform), subsea pipelines connect the platform to an inland process terminal where multiphase transportation is commonly used, and pigging is an important method for daily operation and maintenance of subsea pipelines (Det Norske Veritas 2000). Pigging can effectively avoid multiphase transportation of subsea pipelines due to the presence of U-shaped pipe sections, which can cause liquid accumulation, corrosion, and other conditions that compromise pipeline integrity (Petroleum Industry Press 2007; Chunzhi Cai 1994). CNOOC requires that pigging operations be performed at least twice a year on subsea pipelines that fulfill the pigging criteria. Furthermore, shutdown and restart are necessary working conditions for daily operations. The safety of both upstream and downstream is greatly threatened by flow instability if improperly managed (Yu 2017). Thus, to boost gas field production capacity and enhance energy supply assurance, numerical simulation is necessary as a reference for pigging, shut down and restart.

In terms of pigging, LiuHuaizeng et al. (2015) used a fault tree quantitative analysis method combined with a rough comprehensive evaluation mathematical model of pigging risk to quantify the risk assessment of subsea pipeline pigging. Wang Xuecheng et al. (2023); Gangsheng et al. (2021); Hongjun at al. (2019) and others conducted theoretical calculations and simulation analysis on the shape of pigging and the flow field under different bypass rates during the pigging process, providing guidance for a more economical and efficient pigging design. Li Panke (2020) analyzed the risks, advantages, and disadvantages of the landing process in the Dongfang gas field inland terminal and proposed optimization measures. Changliang (2021), Houbin et al. (2020); Zhaoting et al. (2008); Yang (2022); and Pengqun (2021) and others focused on the practice of pigging operations in offshore oil and gas fields, including subsea mixed-transport flexible hoses, ball-threading jamming, forward and reverse pig-threading, and pig-threading of the inner loop of oil and gas subsea pipelines on offshore unmanned platforms. Research and improvement analysis was conducted by these authors. Yuhang and Wei (2015); Hichong and Jiujun (2005) and others used OLGA to conduct numerical simulation analysis of pressure and flow fluctuations during the pigging process of an offshore natural gas condensate pipeline.

In terms of shutdown and restart, YuXichong et al. (2008) used a combination of a large-scale experimental loop test and OLGA to study the transient flow rules of subsea pipelines during shutdown and restart. Chen Bing et al. (2022) studied the changes in process parameters of supercritical impurity CO_2 pipelines during shutdown and restart. Li Yuxing et al. (2022); Zhenpeng and Peng (2017); Yuxing et al. (2017) carried out transient flow law and process research on gas-liquid, two-phase mixed transportation pipelines.

The preceding study does not provide significant insight into the practical engineering use of multiphase transient situations in conjunction with actual-production. While some academics have conducted numerical simulation analyses using commercial software or theories such as fluid mechanics, most simulation studies are specific to subsea pipelines and have limited generalization. Additionally, there has been a lack of research on numerical simulation of transient working conditions of mixed gas-liquid pipelines in recent years.

This paper presents actual engineering practice used in multiphase transportation for subsea pipelines, combined with the actual inland process terminals, using OLGA. The goal is to establish the transient working conditions, modelling the subsea pipeline, and revise the numerical model based on practical production data. The engineering practice proposes an optimization plan, provides technical support for the formulation of transient operating conditions plans for the gas field, and ensures safe production.

2 Multiphase Flow Mechanism Model

The dynamic multiphase flow simulator is a commercial modelling software tool for transient simulation of multiphase flow. In the oil and gas industry, this simulator is extensively used, particularly for dynamic simulation and optimization in both onshore and offshore oil and gas fields. This dynamic multiphase flow simulator offers the following features and capabilities:

● Multiphase dynamic flow simulation: in wellbores, pipelines, and mid-and downstream process equipment. This simulator can capture the multiphase transient flow behavior, e. g., it is possible to replicate real-world operatingconditions, including packing, shutdown, restart, and valve operations in a multiphase transportation system.

● Flow assurance challenges, including backpressure in subsea pipelines will be impacted by wax formation, hydrates, etc., and in extreme situations, it might block the entire pipeline. This simulator can predict the thickness of wax deposition.

The laminar flow model in the simulator is based on the average mass conservation equation of the pipeline cross section. Its mass conservation equation is given by Eq. (1) (Biberg et al. 2015):

$$\frac{\partial A_f}{\partial t}+\frac{\partial A_f U_f}{\partial z}=0 \qquad (1)$$

where f is g, h, a indicates gas, condensate, and aqueous; t is time, s; z is axial coordinate, m; A_f is the flow areas, m^2; U_f is bulk velocity, m/s.

The momentum equation is described by Eq.

(2)-(4) (OLGA User manual 2024)

$$\rho_g\left(\frac{\partial A_g U_g}{\partial t}+\frac{\partial A_g U_g}{\partial z}\right)=-A_g\left(\frac{\partial P_g}{\partial z}+\rho_g g\sin\theta\right)-S_g\tau_g-S_i\tau_i \quad (2)$$

$$\rho_h\left(\frac{\partial A_h U_h}{\partial t}+\frac{\partial A_h U_h}{\partial z}\right)=-A_h\left(\frac{\partial P_h}{\partial z}+\rho_h g\sin\theta\right)-S_h\tau_h+S_i\tau_i-S_j\tau_j \quad (3)$$

$$\rho_a\left(\frac{\partial A_a U_a}{\partial t}+\frac{\partial A_a U_a}{\partial z}\right)=-A_a\left(\frac{\partial P_a}{\partial z}+\rho_a g\sin\theta\right)-S_a\tau_a+S_j\tau_j \quad (4)$$

where P = phase pressure, Pa; i and j denote gas/condensate and water/condensate interfaces.

The energy equation Eq. (5) is described by the followingequation:

$$\frac{\partial m_i E_i}{\partial t}+\frac{\partial m_i U_i H_i}{\partial z}=S+Q+\frac{\partial P}{\partial t} \quad (5)$$

where m_i is mass for different phases; E_i is denotes the field energy; S is denotes enthalpy source/sink; Q is heat flux through pipe wall.

To capture the slug behavior, a hydrodynamic slugging model is added in the model, and the hydrate slug model is described by Eq. (6) (SLB user manual 2024):

$$\frac{\partial N}{\partial t}+\frac{\partial}{\partial z}(NU_A)=B-D \quad (6)$$

where N is the density of slug unit cells in the pipeline (1/m); U_A is the advection velocity; B is the birth rate of slug unit cells; N is the death rate of slug unit cells.

The solution equation used to describe the 3D velocity field distribution is described by Eq. (7) by (Biberg et al. 2015).

$$\int U_f^2 dA \approx U_f A_f^2 \quad (7)$$

When modeling and simulating the subsea pipeline, the simulator software model was selected with oil, gas, and water three-phase velocity distribution as shown in Fig. 1.

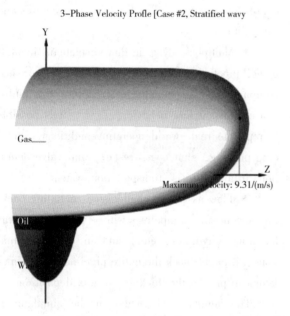
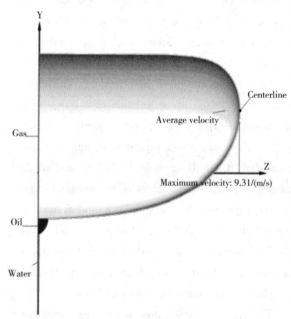

Fig. 1 Oil, gas, and water three-phase velocity distribution based on the dynamic multiphase flow simulator

3 Basic Parameters of Subsea Pipelines

3.1 Fluid Components

With the development of China's offshore oil and gas fields, fluid from Gas Field A to the inland process terminal changed from natural gas to condensate oil and natural gas. Condensate from the Gas Field B, together with natural gas from Gas Field A was transported to the inland process terminal. Table 1 summarizes the components of condensate and natural gas.

Table 1 Condensate and natural gas components

Component	Condensate (mol%)	Natural Gas (mol%)
CO_2	—	40
N_2	—	10
C_1	—	50
C_2	0.001	—

续表

Component	Condensate (mol%)	Natural Gas (mol%)
C_3	0.017	—
iC_4	0.035	—
nC_4	0.037	—
iC_5	0.38	—
nC_5	0.3	—
C_6	1.167	—
C_7	3.551	—
C_8	11.35	—
C_9	26.858	—
C_{10}	26.641	—
C_{11+}	29.663	—

For fluid physical property description, Multiflash® modeling software is used for fluid physical property modeling, and the selection of equation of state (EOS) is PRA[PR-Advanced]. Phase envelopes of natural gas and condensate are shown in Fig. 2.

3.2 Parameters of Subsea Pipeline

The subsea pipeline from Gas Field A to the inland process terminal has a total length of 106km with an outer diameter of 559mm, pipeline wall thickness of 12.7mm, flow rate of $350m^3/d$ for condensate oil and $535 \times 10^4 m^3/d$ for dry gas. The pipeline has an inlet temperature of 37℃ and inlet pressure of 5.53MPa. The outlet temperature and pressure are 27℃ and 4MPa, respectively, and total heat transfer coefficient is $10W/m^2 \cdot ℃$. Fig. 3 shows the geometry of the subsea pipeline from Gas Field A to the inland process terminal.

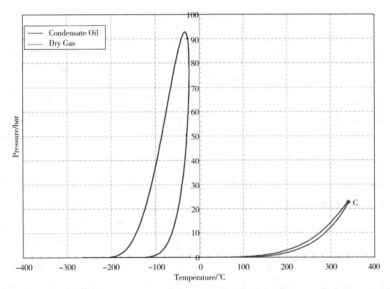

Fig. 2　Phase envelopes for condensate oil and dry gas

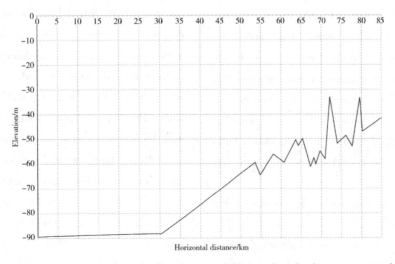

Fig. 3　Geometry for subsea pipeline from gas field A to the inland process terminal

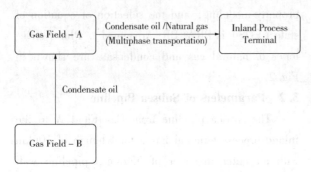

Fig. 4 Gas field development strategy

Emergency shutdown (ESD) signals are available at the inlet of the inland process terminal to guarantee pressure stability in processing systems. Furthermore, slug catcher and condensate oil systems processing capacities determine the processing limit of the incoming liquid volume. The condensate oil stabilization system can operate at maximum load capacity of 700m³/d, while the slug flow catcher has a storage limit of 1268m³. Consequently, accumulative flow rate after pigging should be less than 1968m³/d, and the pressure fluctuation range should be between 3.4 and 5.5MPa.

3.3 Case Study and Analysis

Currently, there is no pertinent reference for guidance on gas production resumption, which might result in uncontrollable on-site risks and severely restrict the release of production capacity. This multiphase transportation pipeline only has practical engineering use in complex transient conditions, has had only two shutdown and restart cases. The two scenarios are compared between infield production data with numerical simulation.

Scenario 1—Gas Field A: Shutdown and Gas Field B: Production. A description of operating conditions for shutdown and restart are shown, in Fig. 5.

- Gas Field A's production was progressively shut down. The inland terminal's condensate oil stabilization system was turned off, Gas Field B's external gas flow rate decreased from 200000m³/h to 130000m³/h, and the condensate oil flow rate was zero.
- Gas supply in Gas FieldA resumed as usual after a 4h shut down.
- Gas Field B produced condensate during Gas Field A shut down, yielding approximately 1800m³ of condensate oil overall.
- The condensate storage tank and four slug catcher's liquid contents decreased in the inland process terminal. When the condensate stabilization system reached maximum efficiency, a significant volume ofliquid began to accumulate. A total of 1600m³ of condensate oil were obtained following completion of the liquid collection. The receiving capacity of the terminal was surpassed by the liquid collecting volume.

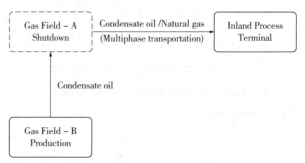

Fig. 5 Logistic diagram for Gas Field A Shutdown and Gas Field B Production

Scenario 2—GasField A: Shutdown and Restart with Pig Injection and Gas Field B: Production.

The working conditions of shutdown and restart are shown in Fig. 6.

- Gas Field A shutdown and restart on the 6th day. A pig is launched concurrently because the restart process starts, and the gas supply is subsequently increased to $20 \times 10^4 m^3/h$.
- After the gas supply was restored, there was basically no condensate coming inland when the onshore terminal gas consumption was 10000m³/h. After approximately 3 hours when the gas consumption was increased to 20000m³/h, a small amount of condensate came inland. After about 9 hours, the condensate oil came inland. The liquid volume increased to 50m³/h and stabilized about 4 hours later. The liquid carrying volume began to decrease, and the liquid carrying volume dropped to 0.
- The land terminal condensateprocessing unit was operating at full capacity. During the shutdown of Gas Field A, Gas Field B exported 1377m³ of measured condensate. The total oil recovery volume was approximately 1000m³. There was currently about 400m³ of liquid accumulation in the subsea

pipeline. Compared with Gas Field B, the exported condensate data are essentially consistent.

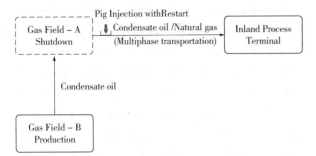

Fig. 6 Logistic diagram for Gas Field A Shutdown and restart with pig injection and Gas Field B Production

A numerical simulation model in the dynamic multiphase flow simulator for the subsea pipeline was created based on fundamental data and engineering practice for transient operating situations. The in-field data and the numerical simulation results were compared as shown in Fig. 7 and Fig. 8.

As shown in Fig. 7, the simulation results for Scenario 2 show a peak of incoming liquid flow when the pig arrives at the trap position. Also, while the flow rate decreases rapidly, the maximum flow rate is slightly larger than the actual process, but after the pig arrived in the actual process, the liquid flow rate shows a slow downward trend. In contrast, the simulation results of Scenario 1 show two peaks of incoming liquid flow, while the actual process only shows one peak.

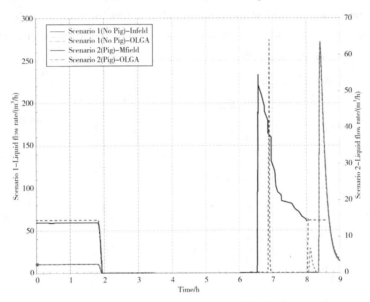

Fig. 7 Comparison of liquid flowrate on shore between infield and numerical simulation

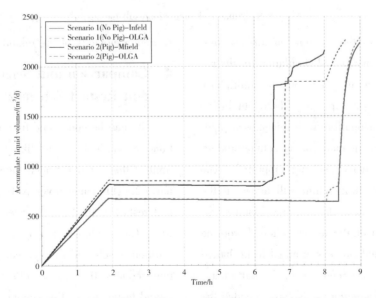

Fig. 8 Comparison of accumulated liquid volume on shore between infield and numerical simulation

Combining the two working conditions, the simulation results arein agreement with the actual process in the steady-state stage prior to the pig arriving or restarting. However, in the transient working condition following the pig's arrival or restart, the simulation results are different from the actual process in that the peak arrival time of the incoming liquid flow rate in the actual process is accurately predicted. Furthermore, the simulation results are more complex than the actual process, which can assist the site in leaving a certain amount of redundancy when creating the process plan to ensure pig pass. The inland terminal's process conditions are stable.

Fig. 8 shows in the actual process of Scenario 2, the incoming liquid flow rate slowly decreases after the pig arrives at the shore; thus, the accumulated liquid volume ashore shows a slowly rising trend, while in the simulation results, the liquid volume after pig trapped is once on landing, the liquid accumulation in the pipe had not yet been established, resulting in no condensate coming inland for about 0.5 days. Judging from the final accumulated liquid volume inland, the total liquid volume in the simulation and the actual process are basically consistent, slightly larger than the actual process, but the errors of both are within the scope of the project.

Fig. 9 shows the pig velocity and total travel distance in numerical model. In the OLGA model, default settings were adopted in pig settings (mass, static force, wall friction coefficient, fluid friction, etc.), and the pig's properties are different from the actual pig's property onsite, which is greatly affect the velocity of pigging.

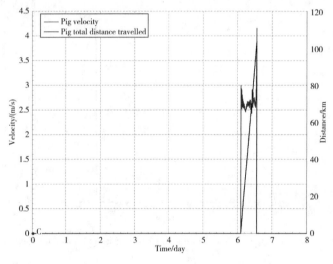

Fig. 9 Numerical simulationresult for pig

It is evident that the dynamic multiphase flow simulator numerical model and simulation results are appropriate for operating conditions of multiphase transportation in subsea pipeline. These conditions have become more compatible with this subsea pipeline with the adjustment made using actual production data (fluid property and pig property). The entire volume of incoming liquid will arrive at the coast in 1 to 2 days under the aforementioned working conditions. Both of the aforementioned working conditions for the land process terminal to be halted and vented because in Scenario 1, in particular, the total volume of incoming liquid has exceeded the receiving capacity of the inland process terminal.

4 Comparison and Selection of Intermittent Restart Schemes

As can be concluded from the previous information, Scenario 1 is more likely to threaten the upper limit of the terminal's acceptance capacity than Scenario 2. This paper inventively proposes an intermittent restart method for the special working condition when Gas Field A is shutdown, and the optimal solution is selected through comparison. This condition reduces the liquid collection pressure on the inland terminal. In this section, two types of restart methods are proposed.

When the total gas volume remains unchanged, the first type of intermittent restart process increases the gas volume, and the second type is intermittent large volume recovery with different intermittent time intervals. There are six types of inter in total as shown in Table 2. Assume that Gas Field A and Gas Field B operate in a steady state for 2 days before shutdown occurs in each plan. At that time, Gas Field A is shut down for 5 days (on average, it takes 5 days for shutdown and overhaul), during which Gas Field B produces condensate normally.

Table 2 Different intermittent restart processes

Gas Flowrate/ ($10^4 m^3/h$)	Duration of each stage/h			Gas Flowrate/ ($10^4 m^3/h$)	Duration of each stage/h		
	Case 1	Case 2	Case 3		Case 4	Case 5	Case 6
22	10	10	10	8	27.5	27.5	27.5
0	4	7	10	0	4	7	10
22	10	10	10	16	13.75	13.75	13.75
0	4	7	10	0	4	7	10
22	10	10	10	22	10	10	10

Fig. 10 shows the total liquid volume of subsea pipelines that were landed according to different restart plans. Three liquid – collecting procedures (measuring liquid flow volume) are included in Case 1 to 3, with corresponding liquid volumes of $1086m^3$, $1023m^3$, and $260m^3$. There are two liquid-collecting procedures in each of Case 4 to 6, and there are $1951m^3$ and $596m^3$ of liquid in each case. It is evident that no liquid is carried inland when the gas supply volume is smaller than $8 \times 10^4 m^3/h$. This condition indicates that the gas lacks the potential to carry liquids. While increasing the interval time can provide the inland terminal enough time to handle the liquid volume, it has minimal influence on the total amount of liquid deposited inland.

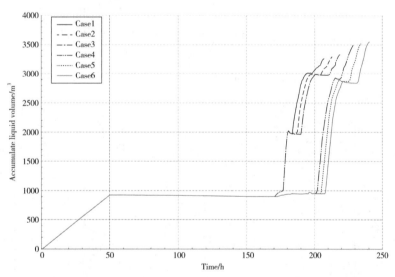

Fig. 10 Difference in accumulated liquid volume inland for different restart cases

Irrespective of whether it is the two intermittent restart methods of increasing gas volume in steps, restoring large volume, or the interval time, and other factors, there will be no slug flow inland.

The difference in offshore pressure of the subsea pipeline under different restart processes are shown in Fig. 11. During each intermittent restart of Case 1 to 6, pressure inland will have two peaks in which the 2nd peak is greater than the first peak. Among the different processes, the maximum value of the pressure fluctuation appears. During the first intermittent restart of Case 1 to 3, the maximum pressure is 0.4 bar. The reason for the two peaks during intermittent restart is that the largest amount of liquid accumulation during the shutdown period is usually at the bottom of the platform under the subsea pipeline.

When each intermittent restart occurs, a large amount of accumulated liquid will be pushed out, causing the first pressure on the subsea pipeline to go inland. The fluctuation then gradually calms down. When the gas volume is about to reach the inland process terminal, the instantaneous two-phase flow rate increases, resulting in a 2nd pressure fluctuation, and the second pressure fluctuation is larger than the first pressure fluctuation. But overall, irrespective of which method is used, the pressure fluctuations caused by the landing are small and will not pose a threat to the stability of the inland terminal process system.

The changes in the inland liquid phase flow rate of the subsea pipe under different schemes are shown in Fig. 12. During each intermittent restart of Cases 1 to 6, the inland will also have two peaks. Except for the first restart, the second flow peak is greater than the first and the maximum flow rate reached is 400m³/h, and the second flow peak value during another restart was lower than the first time. In addition, as the interval time increases, the inland instantaneous liquid flow rate gradually increases.

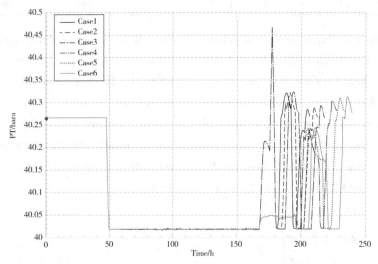

Fig. 11 Differences in inland pressure under different restart cases

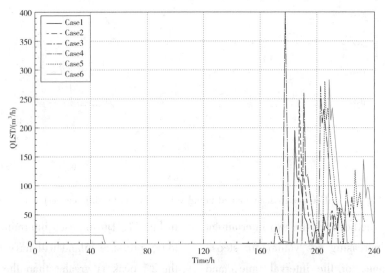

Fig. 12 Differences in inland liquid flow rate under different restart cases

From the analysis of the inland accumulated liquid volume, if the first gas volume setting in the stepped reproduction gas volume is unreasonable, resulting in no liquid carrying capacity, the inland liquid volume will increase during the second restart. From the inland pressure and slug flow analysis, irrespective of the method, will not affect the stability of the inland terminal process system; from the inland liquid phase flow analysis, the intermittent large-volume recovery method has a larger instantaneous flow rate at the first restart. The risk is that the shorter the intermittent time, the smaller the inland instantaneous liquid flow.

Therefore, according to the analysis results, the intermittent restart process is feasible and can effectively reduce the risk to the inland terminal process system. Intermittent steps of nearly large transportation volumes can be adopted to increase the gas volume. For example, the gas volume ladder can be shortened from 8→16→22 in this paper to 16→19→22; thereby, reducing the inland instantaneous liquid flow during the initial restart, and a shorter interval can be selected to shorten the production resumption time limit and fully release the upstream and downstream production capacity of the subsea pipeline.

5 Conclusions

Operating procedures vary with the system operating conditions during shutdown and restart. OLGA could provide necessary engineeringreference for complexity onsite operations.

In the special transient conditions of multiphase transportation, shutdown and restart, and pig, the inland process terminal needs to stop production and vent to cope with these conditions. The total amount of incoming liquid is the largest in the inland process terminal.

This paper innovatively proposes intermittent restart and terminal intermittent liquid collection methods, which can effectively improve the safety margin of the terminal process system under complex transient conditions of gas-liquid two-phase mixed transport in subsea pipelines.

The intermittent step-by-step gas volume increase method is safer than the intermittent large-volume method. This condition mainly implies that the intermittent step-by-step gas volume increase method has smaller liquid phase flow rate for the first time and has less impact on the terminal. However, it should be noted that the minimum step gas volume method is selected to avoid the gas having no liquid-carrying ability due to too small a gas volume; thus, increasing the total liquid volume landed during the second resumption of production.

The interval can be selected according to the requirements of the onshore terminal process system. If the interval is long, sufficient liquid processing time can be reserved for the terminal (mainly depends on the daily processing volume and slug of the condensate stabilization system the total volume of the flow trap). If the interval is short, the inland liquid phase flow during each restart can be reduced to avoid water hammering and erosion of pipelines and equipment.

References

[1] Bai, Gangsheng, Li, Weiquan, Zhang, Yuan, et al. 2021. Performance Design of Oil and Gas Pipeline Pigs [J]. Oil-Gas Field Surface Engineering, volume: 40 (04): 71-76.

[2] Cai, Chunzhi 1994. Oil and gas storage and transportation technology [M]. Petroleum Industry Press. volume: 384-385.

[3] Chen, Yang 2022. Practice and Effect of the First Ball Pigging Operation of Marine Pipe[J]. Chemical Management, volume: (28): 134-137.

[4] Chen, Bing, Xu Menglin, Fang Qichao, et al. 2022. Effect of supercritical impurity-containing CO_2 pipe network shutdown and restarting conditions on hydrate generation[J]. Chemical Engineering of Oil & Gas, volume: 51 (04): 43-50.

[5] D. Biberg, G. Staff, N. Hoyer, and H. Holm 2015. Accounting for Flow Model Uncertainties in Gas-Condensate Field Design Using the OLGA High Definition Stratified Flow Model[J]. BHR-2015-G5.

[6] Det Norske Veritas 2000. DNV-OS-F101-2000 Subsea pipeline systems[S]. Oslo: DNV Publishing Services.

[7] Fan, Zhaoting, Qu Huihui, Yuan Zongming, et al. 2008. Calculation and Analysis of Ball Passing Scheme for Ocean Gas Pipeline[J]. Journal of Chongqing University of Science and Technology (Natural Science Edition), volume: (05): 43-45.

[8] Gong, Pengqun 2021. The problem and solution of ball

access of oil and gas inner loop of offshore oil and gas unmanned platform[J]. Petrochemical Technology, volume: 28 (09): 42-43.

[9] Liu, Huaizeng, Wang, Yuchuan, and Huang, Gang 2015. Quantitative analysis on failure tree of subsea pipeline pigging risk[J]. Oil & Gas Storage and Transportation. volume: 34 (04): 442-446.

[10] Li, Panke 2020. Design and Risk Response of Ball Receiving Process at Dongfang Terminal[J]. Chemical Management, volume: (06): 68-69.

[11] Lu, Changliang 2021. Analysis and Research on the First Online Ball Connection of Flexible Flexible Hoses for Mixed Transportation in an Offshore Oilfield[J]. Petroleum and Chemical Equipment, volume: 24 (09): 20-22.

[12] Li, Jianing, Du Shengnan, Fan Kaifeng, et al. 2022. Study on the Transient Flow Law of Shutdown and Restart of Oil and Gas Mixed Pipelines with Multiple Fluctuations and Large Height Differences[J]. Journal of Liaoning Petrochemical University, volume: 42 (02): 55-60.

[13] Li, Yuxing, Liu, Youchao, Yang, Fan, et al. 2017. Research on Shutdown and Restart Process of the Long-distance Wet Gas Pipeline[J]. Oil-Gas Field Surface Engineering, volume: 36 (04): 40-46.

[14] SLB User Manual 2024.

[15] Subsea pipeline design of offshore petroleum engineering[M] 2007. Petroleum Industry Press, 2007: 85.

[16] Tian, Hongjun, Wei, Yungang, Cao, Yuguang, et al. 2019. Three-dimensional numerical simulation analysis of bypass flow field of pipeline cleaner[J]. Petroleum Machinery, volume: 47 (01): 120-129.

[17] Wang, Xuecheng, Zhang, Leilei, Gong, Zhihai, et al. 2023. Summary of Research on the Motion Mechanism of Leather Bowl Ball during Ball Passing[J]. Liaoning Chemical Industry, volume: 52 (01): 117-120.

[18] Xie, Houbin, Qi, Baobao, Li, richeng, et al. 2020. Reason Analysis and Improvement Measures of Pig Blockage during Pigging in Subsea Pipeline of Ledong Gas Field[J]. Guangdong Chemical Industry, volume: 47 (16): 153-155+146.

[19] Yang, Yuhang and Li Wei 2015. Simulation analysis of natural gas condensate offshore pipeline cleaning [J]. Inner Mongolia Petrochemical Industry, volume: 41 (08): 37-39.

[20] Yu, Xichong and Wu, Jiujun 2005. Numerical Simulation Study on Cleaning Process of Subsea Mixed Pipeline[J]. China Offshore Oil and Gas, volume: 03): 203-208.

[21] Yu, Xichong, Li Qingping, An Weijie, et al. 2008. Shutdown and restart transient flow characteristic study in offshore multiphase pipeline[J]. Journal of Engineering Thermophysics, volume: 02): 251-255.

[22] Zou, Yu 2017. Research on Slug Flow Control Method for M Gas Field Pigging and Shutdown Restart Process [D]. Southwest Petroleum University.

[23] Zhang, Zhenpeng and Luo Peng 2017. Simulation Analysis of Shutdown and Restart of Natural Gas Condensate Pipeline[J]. Petroleum and Chemical Equipment, volume: 20 (02): 57-60.

Study on the effect law of microcrystalline wax on EVA to improve the low temperature rheology of model waxy oil

Xia Xue　Yan Feng　Li Qifu　Yu Hongmei　Li Zhengbin

(PipeChina Research Institute of Science and Technology)

Abstract　Pipeline is a crucial link between crude oil production, transportation and sales, and is the lifeline to assure the safe and effective development of petroleum energy. However, at this moment, most of China's domestically produced crude oils are easily condensable and extremely viscous crude oils, which have high pour point and high viscosity at room temperature, gravely compromising the safety of crude oil pipeline transportation. Pour point depressant (PPD) is frequently used in engineering to increase the fluidity of waxy crude oil, but there is an obvious mismatch between the effect of PPD and the components of crude oil. Among these, waxes are a major factor that alters the viscosity of crude oil and can have a substantial impact on the modifying effect of PPDs. But at this point, the majority of the study is focused on paraffin waxes, and there are still many unanswered questions about microcrystalline waxes. In order to examine the mechanism by which microcrystalline waxes affect the effect of EVA PPD, this paper combines a number of techniques, such as macro rheological testing, differential scanning calorimetry (DSC) and polarized light microscopy observation. It also offers recommendations for the investigation and development of new PPDs for typical crude oils.

Keywords　microcrystalline wax; rheology; model oil; pour point depressant

1　Introduction

Wax precipitation is a non-negligible phenomenon in the pipeline transportation of waxy crude oil. This phenomena can lower the pipeline's effective flow area, dramatically raise the energy consumption of pipeline transportation, and in extreme circumstances, even cause a "condensation" accident. Globally, "additive transportation" has been used as a cost-effective and efficient method to reduce the risks associated with pipeline transmission. Nonetheless, PPD and waxy crude oil are highly compatible, and the component s of waxy crude oil have a direct impact on its modifying effect. Among them, the composition and carbon number distribution of waxes in crude oil are the key factors affecting the PPD effect.

The wax fractions in crude oil can be classified as paraffin or ground wax based on the distribution of carbon numbers. The primary constituents of paraffin wax, often referred to as large crystal wax, are straight chain alkanes (approximately 80% to 95%), with trace amounts of alkanes with branched chains and monocyclic cycloalkanes with long side chains; paraffin wax is a mixture of hydrocarbons with an atomic number of roughly 18 to 30. Ground wax, also known as microcrystalline wax, is made up of hydrocarbons with carbon atoms numbering 35~60, and contains a large number of high molecular weight isomerized alkanes and cycloalkanes. According to fundamental chemical principles, performance is determined by characteristics, which are determined by molecules. Paraffin wax and microcrystalline wax have quite different crystallization behaviors because of their various compositions. Generally speaking, paraffin wax crystallizes as bigger flakes, whereas microcrystalline waxes typically form finer granular crystals.

As the primary cause of viscosity in waxy crude oil, wax has a significant impact on PPD performance. Current research on this topic primarily focuses on the mechanism and interaction law between paraffin wax and PPD, based on which a series of traditional PPD have been developed. However, the presence of numerous microcrystalline waxes significantly limits the impact of PPD in the transportation of some common waxy crude oils (such Nanyang oil delivered on the Weigang-Jingmen line). Nevertheless, at this point, experts both domestically and internationally lack significant expertise on this topic. Therefore, it is necessary to deeply reveal the influence of microcrystalline wax on PPD performance and its mechanism, so as to provide theoretical guidance for the research and development of new PPD for typical crude oils.

2 Experiment

2.1 Material

Sliced paraffin Ⅰ (CP, 62~64℃) and sliced paraffin Ⅱ (CP, 50~52℃) were purchased from Sinopharm Chemical Reagent Co. The sliced paraffin Ⅰ and Ⅱ were prepared into mixed paraffin waxes at a mass ratio of 3 : 1, and the carbon number distribution of the paraffin waxes before and after mixing was measured by GPC, as shown in Figure 1 (a), (b) and (c). It can be seen that the mixed paraffin has a wide carbon number distribution, which is similar to the real waxy crude oil.

The microcrystalline wax (Sasolwax 3971) was purchased from Sasolwax GmbH, and the carbon number distribution was obtained by high temperature gas chromatography (HTGC) performed by Sasol Wax GmbH using a proprietary HTGC protocol. Sasolwax 3971 (78~84℃) consists of 16wt% of linear alkanes (carbon number range $C_{22} \sim C_{58}$) and 84 wt% of nonlinear alkanes (carbon number range $C_{25} \sim C_{80}$). The carbon number distribution is shown in Figure 1(d). From Figure 1(d), it can be seen that the carbon number distribution of this microcrystalline wax is wide, and the dominant carbon number distribution is around $C_{50} \sim C_{60}$. In addition, the PPD used in the test was EVA-28, a commercialized PPD purchased from ARKEMA, France. In this paper, the dosage of EVA is fixed at 100ppm.

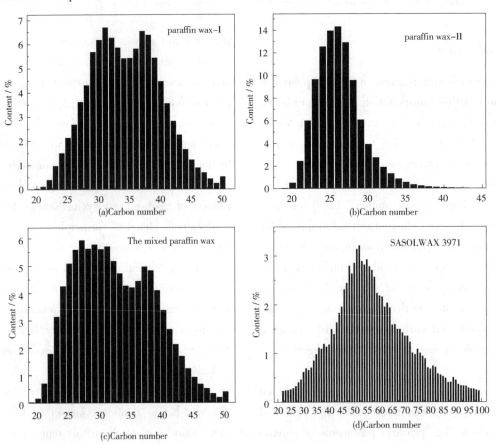

Figure 1 Carbon number distribution of experimental waxes

2.2 Method

2.2.1 Preparation of model waxy oil

The model wax oil consists of solute (mixed sliced paraffin) and solvent (liquid paraffin with xylene). Among them, the solute in the model wax oil consists of sliced paraffin wax (0~15 wt%) and microcrystalline wax (0~15 wt%), the total amount of which is fixed at 15 wt%. For the convenience of description, the oil samples with different wax ratios were named OS-1~OS-5 (details are shown in Table 1).

Table 1 Nomenclature of different model wax oils

Oil sample	Wax content/wt%	
	Paraffin wax	Microcrystalline wax
OS-1	15	0
OS-2	14	1
OS-3	10	5
OS-4	5	10
OS-5	0	15

2.2.2 Pour point test

The pour point of undoped/doped oil samples were measured according to the Chinese oil and gas industry standard SY/T 0541—2009.

2.2.3 Viscosity test

The variation rule of apparent viscosity with shear rate can be obtained by scanning the shear rate of oil samples at a certain temperature, and the range of shear rate was $1\sim200s^{-1}$. In order to ensure that the oil samples to be tested are completely gelled, the test temperature was set at 10℃.

2.2.4 DSC experiments

The DSC test was carried out according to the Chinese Petroleum Industry Standard SY/T 0545—2012 by using the differential scanning calorimeter (821e).

2.2.5 Microscopic observation

The BX51 microscope was selected to obtain micrographs of wax crystals. Moreover, it is worth noting that the photographs were captured after the temperature dropped from 70℃ to 20℃.

3 Results and Discussion

3.1 Pour Point tests

Figure 2 shows the pour point of model wax oil containing different concentrations of microcrystalline wax. It can be observed that the pour point of model oil increased gradually with the increase of microcrystalline wax content. The pour point of pure paraffin oil was 25℃, and with the gradual increase of microcrystalline wax content from 1 wt% to 15 wt%, the pour point monotonically increases from 27℃ to 41℃. After the addition of EVA PPD, the pour point still showed a similar trend, but the pour point reduction showed a phenomenon of decreasing first and then increasing: with the microcrystalline wax content gradually increased from 0wt% to 15wt%, the pour point reduction decreased from 5℃ to 2℃, and then rebounded to 12℃. Obviously, the coexistence of paraffin wax and microcrystalline wax is not conducive to the condensation reduction effect of EVA.

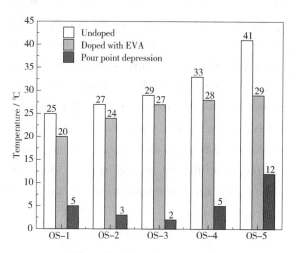

Figure 2 Pour point test results

3.2 Viscosity Test

The results of viscosity tests of OS-1~OS-5 are shown in Table 3. From the table, it can be seen that under the temperature conditions of model wax oils completely gelled, when paraffin wax or microcrystalline wax exists alone (OS-1, OS-5), the gelled structure formed by the oil samples has a weaker ability to shear deformation, which is reflected in the lower viscosity value. In contrast, when microcrystalline and paraffin waxes coexisted, the strength of the gelling structure of waxes in the system was high and difficult to be destroyed by shear. At the same time, when EVA was added, the phenomenon did not change, and the viscosity reduction ability of EVA and its capacity to reduce pour point showed a similar pattern of change.

Table 3 Statistical table of viscosity test results

Oil sample	Viscosity ($10s^{-1}$)/mPa·s		Viscosity ($100s^{-1}$)/mPa·s	
	Undoped	Doped with EVA	Undoped	Doped with EVA
OS-1	972.36	180.323	347.19	88.82
OS-2	4259.09	2307.08	1118.03	885.28
OS-3	12450.8	7628.14	2689.17	1504.83
OS-4	953.33	780.95	494.56	322.57
OS-5	356.38	230.57	155.87	77.93

3.3 DSC tests

The DSC curves of OS-1 ~ OS-5 before and after the addition of EVA are shown in Figure 2. As can be seen from the figure, the wax appearance temperature (WAT) of OS-1 was 32.76℃, which decreased to 30.26℃ after the addition of 100 ppm EVA. The reason for the decrease is that the nonpolar groups of EVA molecules can eutectically precipitate with the wax molecules during the cooling process, so that the effective solubility of the wax molecules can be increased. In addition, during the process of eutectic precipitation of EVA and wax molecules, the polar VA group increased the interfacial tension between wax crystals and oil phase, and increased the nucleation barrier, which increased the critical nucleation radius of wax crystal precipitation, making it more difficult for wax crystals to precipitate, and decreased the WAT of the oil samples. With the increase of the content of microcrystalline waxes, the DSC curve goes through the changes of single peak - bimodal peak - single peak, and the WAT is also increased from 33.79℃ to 54.77℃. After the addition of EVA, the peak shape of the curve had a similar course of change, but the area of the small peak in the high-temperature section increased significantly, which implied that EVA would also participate in the crystallization process of the microcrystalline waxes to make them concentrate on the exothermic crystallization. Meanwhile, the decrease of WAT also has the same trend with the decrease of pour point, which reflects the difference of EVA's eutectic ability in different environments. Meanwhile, the decrease of WAT also has the same trend with the decrease of pour point, which reflects the difference of EVA in eutectic ability.

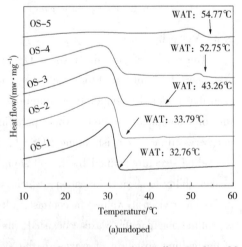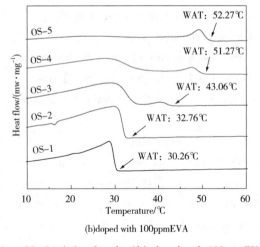

Figure 3　Exothermic curves of crystallization from OS-1 to OS-5: (a) undoped; (b) doped with 100ppmEVA

3.4 Micrograph of wax crystal

The polarized light micrographs of OS-1, OS-3 and OS-5 at 20℃ are shown in Figure 3. As observed from Figure 3, the presence of paraffin wax caused a large number of needle-like or point-like wax crystals to appear in the field of view, and it is easy for these wax crystals to overlap with each other to form a network structure, which significantly deteriorated the fluidity of the oil samples. After the addition of EVA, some large bar crystals appeared in the view, which meant that EVA was involved in the crystallization process of paraffin waxes, leading to the change of wax crystal morphology. However, when paraffin wax and microcrystalline wax coexisted, there was no significant change in the morphology of wax crystals before and after the addition of EVA, and the wax crystals appeared to be disorganized clusters. But when only microcrystalline waxes were present in the oil phase, the addition of PPD significantly increased the number of visible wax crystals, which implies that EVA guides the microcrystalline waxes to form denser wax crystals, thus improving the low-temperature rheology of the model wax oils.

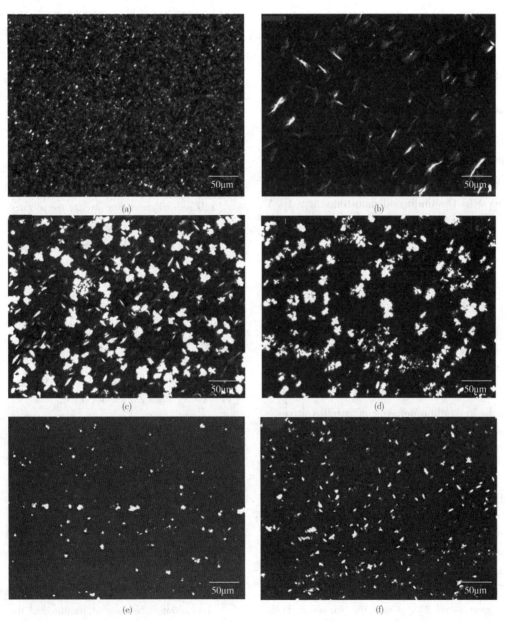

Figure 4 The polarized light micrographs of OS-1(a)/(b), OS-3(c)/(d) and OS-5(e)/(f)

3.5 Mechanism Discussion

From the above experimental phenomena, it can be concluded that when paraffin or microcrystalline waxes are present alone, the EVA PPD has an excellent modifying effect, but when paraffin and microcrystalline waxes coexist, the effect of EVA is significantly inhibited. Research has reported that the crystallization behavior of low carbon number waxes in oil samples is interfered by high carbon number waxes. When paraffin wax or microcrystalline wax exists alone, the carbon number distribution of wax is more concentrated, and the interference ability to the eutectic effect of EVA PPD is relatively weak, so that EVA can give full play to its modification effect. However, when microcrystalline wax and paraffin wax coexist, the presence of microcrystalline wax changes the crystallization process of paraffin wax, and the influence of EVA PPD cannot reverse the phenomenon, making the effect of EVA weaker.

4 Conclusion and recommendation

In this work, we examined the mechanisms and laws governing the impact of microcrystalline waxes on the low-temperature rheology of undoped/doped model waxy oil. The findings demonstrated that EVA PPD had a great modifying effect when paraffin or microcrystalline waxes were present alone, but that the effect of EVA was greatly reduced when paraffin and microcrystalline waxes were present together. It was thus hypothesized that, for crudes containing microcrystalline waxes, the research of PPDs should focus on microcrystalline waxes, and the low-temperature rheology of the oil samples should be improved by weakening the influence of microcrystalline waxes on the crystalline behavior of paraffin waxes.

References

[1] Alnaimat F, Ziauddin M. Wax deposition and prediction in petroleum pipelines. Journal of Petroleum Science and Engineering 2020; 184.

[2] Al-Sabagh AM, Betiha MA, Osman DI, Hashim AI, El-Sukkary MM, Mahmoud T. Preparation and Evaluation of Poly(methyl methacrylate)-Graphene Oxide Nanohybrid Polymers as Pour Point Depressants and Flow Improvers for Waxy Crude Oil. Energy & Fuels 2016; 30(9): 7610-21.

[3] Xie Y, Zhang J, Ma C, Chen C, Huang Q, Li Z, et al. Combined treatment of electrical and ethylene-vinyl acetate copolymer (EVA) to improve the cold flowability of waxy crude oils. Fuel 2020; 267.

[4] Joonaki E, Hassanpouryouzband A, Burgass R, Hase A, Tohidi B. Effects of Waxes and the Related Chemicals on Asphaltene Aggregation and Deposition Phenomena: Experimental and Modeling Studies. ACS Omega 2020; 5(13): 7124-34.

[5] Binks BP, Fletcher PD, Roberts NA, Dunkerley J, Greenfield H, Mastrangelo A, et al. How polymer additives reduce the pour point of hydrocarbon solvents containing wax crystals. Phys Chem Chem Phys 2015; 17(6): 4107-17.

[6] Zhao Z, Xia X, Li Y, Liu D, Cai W, Li C, et al. Effect of Dodecylbenzenesulfonic Acid as an Asphaltene Dispersant on the W/O Emulsion Stabilized by Asphaltenes and Paraffin Wax. Energy & Fuels 2023; 37(6): 4244-55.

[7] Chen X, Hou L, Li W, Li S. Influence of electric field on the viscosity of waxy crude oil and micro property of paraffin: A molecular dynamics simulation study. Journal of Molecular Liquids 2018; 272: 973-81.

[8] Ashbaugh HS, Guo XH, Schwahn D, Prud'homme RK, Richter D, Fetters LJ. Interaction of paraffin wax gels with ethylene/vinyl acetate co-polymers. Energy & Fuels 2005; 19(1): 138-44.

[9] Kurniawan M, Subramanian S, Norrman J, Paso K. Influence of Microcrystalline Wax on the Properties of Model Wax-Oil Gels. Energy & Fuels 2018; 32(5): 5857-67.

[10] Guo XH, Pethica BA, Huang JS, Adamson DH, Prud'homme RK. Effect of cooling rate on crystallization of model waxy oils with microcrystalline poly(ethylene butene). Energy & Fuels 2006; 20(1): 250-6.

[11] Xia X, Zhao Z, Cai W, Li C, Yang F, Yao B, et al. Effects of paraffin wax content and test temperature on the stability of water-in-model waxy crude oil emulsions. Colloids and Surfaces A: Physicochemical and Engineering Aspects 2022; 652.

[12] Xia X, Lian W, Li C, Sun G, Yao B, Ma W, et al. Exploration of the fundamental factor for the synergistic modification of model waxy oil by EVA and asphaltene: An experiment and simulation study. Geoenergy Science and Engineering 2023; 223.

[13] Wang CH, Gao LY, Liu MH, Xia SQ, Han Y. Self-crystallization behavior of paraffin and the mechanism study of SiO_2 nanoparticles affecting paraffin crystallization. Chemical Engineering Journal 2023; 452.

[14] Xia X, Li C, Qi Y, Shi H, Sun G, Yao B, et al. Asphaltene dispersants weaken the synergistic modification effect of ethylene-vinyl acetate and asphaltene for

model waxy oil. Fuel 2023; 341.

[15] Xue J, Li C, He Q. Modeling of wax and asphaltene precipitation in crude oils using four-phase equilibrium. Fluid Phase Equilibria 2019; 497: 122-32.

[16] Ruwoldt J, Kurniawan M, Humborstad Sørland G, Simon S, Sjöblom J. Influence of wax inhibitor molecular weight: Fractionation and effect on crystallization of polydisperse waxes. Journal of Dispersion Science and Technology 2019; 41(8): 1201-16.

[17] Wang C, Gao L, Liu M, Xia S, Han Y. Self-crystallization behavior of paraffin and the mechanism study of SiO_2 nanoparticles affecting paraffin crystallization. Chemical Engineering Journal 2023; 452: 138287-301.

掺氢长输管道气体传质规律研究

柴 冲　张瀚文　彭世垚　裴业斌

（国家石油天然气管网集团有限公司科学技术研究总院分公司）

摘 要 【目的】利用在役天然气管道进行掺氢输送是实现氢能大规模输运的有效方式，当前我国长输天然气管道最大高程差可达千米，掺氢天然气在大落差管道内是否发生分层备受业内关注。【方法】本文基于经典的玻尔兹曼分布律将分子动力学模拟与热力学和扩散理论相结合，建立了长输天然气管道静置工况下的数值模型，系统研究氢-甲烷混合物在重力场中的平衡分布。并建立了国际首座全尺寸掺氢天然气静置分层试验平台进行试验验证。【结果】研究结果表明：两种气体的分压都随着海拔高度的增加而降低，而氢气由于摩尔质量较小，因此下降速度较慢，对于竖直高度为百米级的掺氢天然气管道，静置100天内，氢气组分的最大变化比例小于0.04%，氢气组分变化对管道的影响可完全忽略；对于竖直高度小于千米的掺氢天然气管道，静置100天内，氢气组分的最大变化比例小于0.22%，氢气组分变化对管道的影响可基本忽略；【结论】本文中提出的掺氢天然气分层计算模型与符合试验结果符合，可用于指导掺氢天然气管道停输时间的确定，对管道安全运维具有重要指导意义。

关键词 掺氢天然气；玻尔兹曼分布；分层

Study on the law of gas mass transfer in long-distance hydrogen pipeline

Chai Chong　Zhang Hanwen　Peng Shiyao　Pei Yebin

(Pipe China Institute of Science and Technology)

Abstract It is an effective way to realize the large-scale transportation of hydrogen energy by using in-service natural gas pipelines to carry out hydrogen-doped transportation. At present, the maximum elevation difference of long-distance natural gas pipelines in China can reach kilometers. Whether hydrogen-doped natural gas is stratified in large drop pipelines has attracted much attention in the industry. In this paper, based on the classical Boltzmann distribution law, a numerical model of long-distance natural gas pipeline under static conditions is established, and an international first full-scale hydrogen-doped natural gas static stratification test platform is established for experimental research. The results show that for the hydrogen-doped natural gas pipeline with a vertical height of 100 meters, the maximum change ratio of hydrogen component is less than 0.04% within 100 days, and the influence of hydrogen component change on the pipeline can be completely ignored. For the hydrogen-doped natural gas pipeline with a vertical height of less than km, the maximum change ratio of hydrogen composition is less than 0.22% within 100 days of standing, and the influence of hydrogen composition change on the pipeline can be basically ignored. The layered calculation model of hydrogen-doped natural gas proposed in this paper is consistent with the test results, which can be used to guide the determination of shutdown time of hydrogen-doped natural gas pipeline, and has important guiding significance for pipeline safety operation and maintenance.

Key words hydrogen-doped natural gas; Boltzmann distribution; delamination

氢能是一种清洁的二次能源，具有来源广、可再生、可电可燃、零碳排等优点，我国氢能资源分布不均，利用在役天然气管道进行掺氢输送是实现氢能资源转运的有效手段。目前，掺氢天然气对钢制天然气长输管道作用机理尚不明确，为保证掺氢天然气管道安全运行，需保证管道沿线各处掺氢比例低于临界值。长输天然气管道沿程落差较大，管道处于事故停输工况时，掺氢天然气被封闭在存在高程落差的封闭空间内，存在分层风险，可能导致管道局部位置掺氢比例升高，增加管道氢损伤风险，影响管道运行寿命。

为保证掺氢天然气管道安全、平稳运行，需严格控制氢安全比例。目前国内外针对掺氢天然气静置分层的研究相对较少，2022 年，朱红钧等人利用数值模拟的方式，分析了起伏高程对掺氢天然气静置分层的影响，发现气流剪切作用越强，静置分层达到稳定所需的时间越长；刘翠伟等人采用计算流体力学的方法研究了储氢瓶静置、管道停输与流动工况下的氢气体积分数变化，研究表明在管道重力方式氢气与天然气出现明显分层，且氢气分布梯度随压力和掺氢比的增大而增大，随温度和管径的增大而减小。2024 年李敬法等人基于热力学能量最小原理，推到了掺氢天然气浓度分布模型，模拟结果表明，掺氢天然气分层现象可基本忽略。2023 年李芮芮等人通过实验手段探究了掺氢天然气泄漏后的氢气浓度分布，发现掺氢天然气静置后厨房内甲烷和氢气分层现象不明显，在居民住宅内的掺氢天然气泄漏燃爆特性研究中，可忽略掺氢天然气分层的影响；2017 年，利兹大学等开展了实验条件为直径 150mm，高度 7m 的掺氢天然气分层实验，实验周期为 1 年，每周测量顶/底部氢含量，实验发现未发生重力分层现象。目前，国内外学者对掺氢天然气是否发生重力分层问题仍存在争议，且当前研究结论均依托于数值计算或小尺度实验等方式，对长输管道工程指导作用受限。因此，本研究建立了国际首座全尺寸掺氢分层试验平台，研究长输管道停输工况下的掺氢天然气体积分数变化规律。

1 实验部分

1.1 实验原理

在重力场中，气体在分子间的作用力与重力的作用下进行运动、扩散。在重力场作用下，气体热运动达到平衡时的分布状态符合玻尔兹曼分布。

在一个充满理想气体的刚性垂直管道中，气体分子的归一化概率密度 i ($i=H_2$ 或 CH_4) 随高度 h 的变化由玻尔兹曼分布给出：

$$\rho_i(h) = \frac{\exp\left(-\frac{m_i g h}{kT}\right)}{\int_0^H \exp\left(-\frac{m_i g h}{kT}\right) dh}$$

其中 m_i 是气体分子的质量；g 是重力加速度；k 是玻尔兹曼常数；T 是绝对温度；H 是管道的最大高度。

在得到概率密度后，基于理想气体近似，我们可以通过以下方式计算气体分压分布：

$$P_i(h) = P^{tot} F_i^{tot} \rho_i(h) H$$

其中 P^{tot} 是整个系统的平均压力；F_i^{tot} 是气体 i 在整个系统中的体积分数。

在下面我们使用更方便讨论的归一化分压：

$$p_i(h) = P_i(h)/P^{tot} = F_i^{tot} \rho_i(h) H$$

在上述基础上，可得出气体 i 的体积分数随高度的变化：

$$F_i(h) = \frac{F_i^{tot} \rho_i(h) H}{\sum_j F_j^{tot} \rho_j(h) H} = \frac{p_i(h)}{\sum_j p_j(h)}$$

基于玻尔兹曼分布律分别对高度为 100m，初始掺混比例为 1∶1 条件下的 H_2/CH_4 与 H_2/FS_6 的气体分布状态进行数值计算，结果如下：图 1 所示为 CH_4 与 H_2 在竖直高度 100m 的管道内静置 100 天后不同高度时的浓度分布，静置 100 天后，在高度为 100m 的位置，氢气的体积分数为 50.0299%，甲烷的体积分数为 49.9701%，氢气组分变化比例为 0.03%。图 2 所示为 SF_6 与 H_2 在竖直高度 100m 的管道内静置 100 天后不同高度时的浓度分布，静置 100 天后，在高度为 100m 的位置，氢气的体积分数为 50.3108%，甲烷的体积分数为 49.6892%，氢气组分变化比例为 0.31%。图 3 所示为 CH_4 与 H_2 在倾斜角为 5°，管道长度为 100m 的管道内静置 100 天后不同位置处的浓度分布，静置 100 天后，在管道顶部位置，氢气的体积分数为 50.0026%，氢气组分变化比例为 0.0026%。

1.2 实验平台

基于玻尔兹曼分布律得到的气体分布规律，建立两座竖直管道井平台和一座倾斜试验管沟平台，竖直管道井开口直径为 300mm，管井深度为 100m，井内放置直径为 160mm 的实验管道一根，在管道顶部设置取样口，如图 4 所示；倾斜试验管沟倾角为 5°，管沟长度为 100m，管沟内放置直径为 160mm 的实验管道一根，管道顶部与底部分别设置取样口，如图 5 所示。

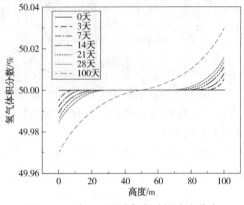
图 1　CH_4 与 H_2 不同高度时的浓度分布

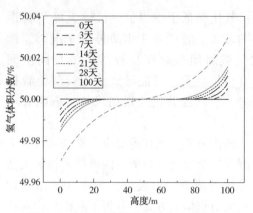
图 2　SF_6 与 H_2 不同高度时的浓度分布

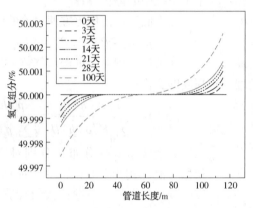
图 3　CH_4 与 H_2 在管道不同位置处的浓度分布

图 4　竖直管道井管道按照俯视图

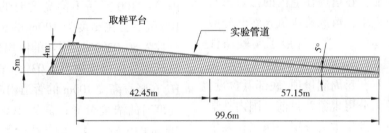
图 5　倾斜试验管沟管道安装正视图

考虑实验平台施工难度，实验采用高分子聚合物材质的 PE 管，PE 管按照静液压强等级分为 PE40、PE60、PE80、PE100，PE 等级越高，强度越高，防气体渗透性越好，考虑实验介质中氢气分子的易渗漏特性，试验管道 PE100、SDR17 的管道，所有试验管道依次经过管道焊接、管道严密性测试、管道强度试验、气体置换等测试合格后投入使用。

1.3　实验方案

以 50∶50 比例的 CH_4—H_2 和 50∶50 比例的 SF_6—H_2 等混合气体作为分层实验介质，依据玻尔兹曼气体分布定律，重力场作用下的混合气体浓度分布，气体的浓度分布梯度与高度和分子质量的乘积存在正相关关系。六氟化硫常温常压下是一种无色无味无毒且不可燃的惰性气体，常温常压下其分子质量为 146g/mol，是 CH_4 分子量的约 9 倍。利用 SF_6—H_2 混合气体进行竖直高度 100m 场景下的分层实验，可定量表征在千米落差长输管道中，CH_4—H_2 混合气体的分层规律。

本实验共设置三根实验管道，试验管编号及充装介质见表 1。

表 1　实验管道及实验介质

实验平台	管道直径	实验介质	取样口位置
1#竖直试验井		SF_6—H_2	顶部
2#竖直试验井	160mm	SF_6—H_2	顶部
3#倾斜试验管沟		CH_4—H_2	顶部+底部

1.4 取样流程与检测方案

1）气体取样流程：打开取样口阀门对取样口管道内空气进行吹扫；连接取样袋，打开阀门取少量气后，将取样袋抽真空对取样袋进行冲洗，重复两次洗袋操作，清除取样袋内杂质气体；取气，取样结束后关闭取样口管道阀门，在管道取样口处缠绕氢敏胶带，观察是否存在氢气泄漏；氢气检测。通过以上步骤，可消除取样过程中杂质气体的影响。

2）氢气检测方案：用标气吹扫色谱仪器的进气管路；通入标气进行检测，以标气检测成分峰面积结果作为基准进行样品成分分析，重复标气检测流程至出现两次峰面积结果误差≤0.5%；计算检测结果。利用该方法对标准气进行检测，本实验中采用仪器的测试误差小于0.2%。

被测气体浓度计算公式如下：

$$C_x = C_s \cdot \frac{Ax}{As}$$

式中，C_x 为被测气体的浓度；C_s 为标准气体的浓度；Ax 为被测气体的峰面积，As 为标准气体的峰面积。

依据长输管道极端停输工况，本实验开展周期为4个月，实验开始首个月，气体成分检测频次为1次/周，共计检测5次，实验开始后3个月，气体成分检测频次为1月/次，共计检测4次。实验周期内共进行9次气体检测实验。

2 实验结果与讨论

2.1 H_2 与 SF_6 混合体系分层实验结果

竖直试验管道开始试验首天对试验气体进行初次检测，为排除杂质组分的干扰，分别检测六氟化硫和氢气的组分比例后进行归一化处理，得到混合气体的组分比例。如图6与图7所示，竖井1试验管内初始 H_2 组分占比为50.00%，静置128天后，试验管顶部 H_2 组分比例为50.22%，氢气组分变化为0.22%。竖井2试验管内初始 H_2 与 SF_6 掺混比例为50.06%，静置128天后，试验管顶部 H_2 组分比例为50.10%，氢气组分变化为0.04%。每次取样检测过程中，由于操作和测试误差导致测试结果不完成符合仿真结果，综合整个试验周期所有测试结果，该试验结果与基于玻尔兹曼分布律的数值计算结果接近。

2.2 H_2 与 CH_4 混合体系分层实验结果

倾斜井的实验介质为 H_2 与 CH_4 混合气体。如图8与图9所示，倾斜井顶部初始 H_2 组分占比为50.56%，静置128天后，试验管顶部 H_2 组分比例为50.66%，氢气组分变化为0.1%。倾斜井底部初始 H_2 与 CH_4 掺混比例为49.75%，静置128天后，试验管底部 H_2 组分比例为49.47%，氢气组分变化为-0.28%。由于玻尔兹曼分布律得出的氢气组分变化比例为0.0026%，远小于仪器的测试误差0.1%，但试验结果仍能定量表明在本实验工况条件下，氢气与天然气未发生明显分层特征。

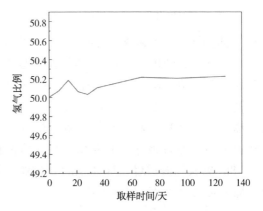

图6 #竖直井1-DN160管氢气体积分数变化

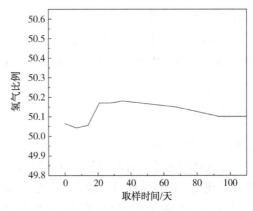

图7 #竖直井1-DN160管氢气体积分数变化

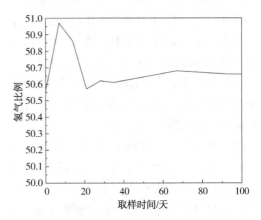

图8 #倾斜井顶部氢气体积分数变化

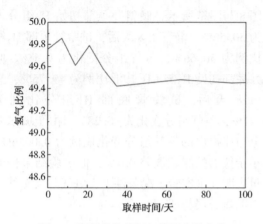

图9 #倾斜井底部氢气体积分数变化

2.3 氢气在PE管中渗透的影响

氢气与甲烷在PE管中的渗透量计算参考如下公式，在实验工况下，氢气在PE100材料中的渗透系数为6×10^{-7} cm³·cm/cm²·s·MPa，可计算出100内DN160实验管的氢气渗透量为9.66×10^{-4} m³，DN160实验管初始充装气体体积为2.18 m³，氢气渗透量占原体积的0.04%，氢气渗漏量对试验结果影响可忽略。

$$Q=\frac{P_1\cdot S\cdot t\cdot P}{L}$$

式中，Q为渗透量，cm³；P_1为渗透系数，cm³·cm/cm²·s·MPa；S为渗透表面积，cm²；t为渗透时间，s；P为内外压差，MPa；L为厚度，cm。

3 结论

本文基于玻尔兹曼气体分布律建立了氢气与甲烷二元气体分布数值计算模型，基于数值计算结果设计建设了两座全尺寸气体静置分层试验平台，得出分层试验结果并验证了模型的准确性。

（1）针对高度为100m的竖直试验管，H_2与SF_6混合体系，SF_6分子量为CH_4的近9倍，静置试验进行100天后，DN160试验管顶部的氢气组分变化为0.22%。H_2与CH_4混合体系，静置试验进行100天后，DN160试验管顶部的氢气组分变化为0.04%。竖直试验管静置分层试验结果基本符合玻尔兹曼分布模型。

（2）针对长度为100m，倾斜角为5°的试验管，H_2与CH_4混合体系，静置试验进行100天后，倾斜试验管顶部的氢气组分变化为0.08%，未发生明显分层现象。倾斜试验管掺氢天然气静置分层试验结果基本符合玻尔兹曼分布模型。

（3）通过实验结果可知，掺氢比例小于50%、高程小于100m的天然气管道在停输静置工况下，100天内管道中氢气组分不会发生明显变化，可基本忽略由于氢气浓度变化对管道造成的材料损伤。

参 考 文 献

[1] 程德宝，蔺建刚，魏乃腾，于雯霞，党春蕾，田华峰. 氢气输送管道技术发展现状[J/OL]. 油气储运，1-11[2024-04-15].

[2] 刘翠伟，裴业斌，韩辉，周慧，张睿，李玉星等. 氢能产业链及储运技术研究现状与发展趋势[J]. 油气储运，2022，41（05）：498-514.

[3] 王玮，王秋岩，邓海全，程光旭，李云. 天然气管道输送混氢天然气的可行性[J]. 天然气工业，2020，40（03）：130-136.

[4] 李玉星，张睿，刘翠伟，王财林，杨宏超，胡其会等. 掺氢天然气管道典型管线钢氢脆行为[J]. 油气储运，2022，41（06）：732-742.

[5] 程玉峰. 高压氢气管道氢脆问题明晰[J]. 油气储运，2023，42（01）：1-8.

[6] 刘刚，崔振莹，魏甲强，陈雷，姜堡垒，李东泽等. 掺氢天然气环境CH_4对管线钢氢脆的抑制行为[J]. 油气储运，2023，42（01）：16-23.

[7] 朱红钧，陈俊文，粟华忠，唐堂，何山. 起伏天然气掺氢管道气体静置分层过程数值研究[J]. 西南石油大学学报（自然科学版），2022，44（06）：132-140.

[8] 王帅，杨成功，周海，牛蓓媛，李建宁，罗婷婷. T型天然气管道掺氢输送流动特性数值模拟研究[J]. 延安大学学报（自然科学版），2023，42（04）：1-8.

[9] 安永伟，孙晨，冀守虎，贾冠伟，许未晴，刘伟等. 天然气掺氢在管道流动中的氢浓度分布[J]. 力学与实践，2022，44（04）：767-775.

[10] 刘翠伟，崔兆雪，张家轩，裴业斌，段鹏飞，李璐伶等. 掺氢天然气管道的分层现象[J]. 中国石油大学学报（自然科学版），2022，46（05）：153-161.

[11] 李敬法，宇波，苏越，刘翠伟，李玉星等. 静置工况条件下掺氢天然气浓度分布规律[J]. 天然气工业，2024，44（02）：145-155.

[12] 李莘莘，刘振翼，李鹏亮. 室内掺氢天然气泄漏分层现象模拟研究[J]. 中国安全生产科学技术，2023，19（09）：129-135.

[13] Badino, Giovanni. "The legend of carbon dioxide heaviness." Journal of Cave and Karst Studies 71.1 (2009): 100-107.

[14] Azatyan, V. V., et al. "Possible Gravitational Strati-

fication of Components in Mixtures of Reaction Gases." Russian Journal of Physical Chemistry A 93. 5 (2019): 986-987.

[15] 丁欣. 基于超声波相位差法的氢气检测仪设计[D]. 哈尔滨理工大学, 2017.

[16] 吴红志. 氢气含量的几种检测方法[J]. 氯碱工业, 2011, 47(04): 34-36.

[17] 张兴磊, 花榕, 陈双喜, 周跃明, 陈焕文. 低浓度氢气检测方法研究进展[J]. 分析仪器, 2009, (05): 6-12.

[18] 郑度奎, 李敬法, 宇波, 张引弟, 赵杰, 韩东旭等. 非金属PE管材氢气-甲烷渗透研究进展[J/OL]. 材料导报, 1-22[2024-04-15].

[19] 郭淑芬, 吕家龙, 高智惠, 薛海龙, 梁军辉, 孙福龙. 有关非金属内胆全缠绕储氢气瓶渗透参数的探讨[J]. 低温与特气, 2020, 38(05): 16-18.

[20] 熊耀强, 吴平, 赵建平. Ⅳ型储氢气瓶内衬材料的氢渗透行为分子动力学模拟[J]. 压力容器, 2023, 40(07): 19-28.

站场典型燃气锅炉对掺氢天然气的适应性研究

程 磊[1]　张瀚文[1]　张 扬[2]

(1. 国家石油天然气管网集团有限公司科学技术研究总院分公司；2. 清华大学山西清洁能源研究院)

摘　要　通过长输天然气管道掺混输氢是打破氢能产业发展运输瓶颈的关键措施，并通过提高零碳燃料的比例显著优化了能源燃料结构，对中国顺利实现碳达峰、碳中和"3060"目标具有意义。掺氢天然气的基础燃烧特性，如燃烧速度、燃烧温度、火焰形态、回火特性、压力波动特性和氮氧化物生成特性等较常规天然气有显著变化，对站场燃气轮机、燃气锅炉等重要管输站场设备的运行效率、安全性、稳定性和污染物排放特性有重要影响。本研究通过在常规天然气中掺混0~30%的氢气，开展了混气后的物性参数变化规律、流动与燃烧特性及规律的研究，明确了掺氢比和当量参数等对火焰特征与火焰稳定特性等的影响规律。基于掺氢天然气燃烧特性变化规律，通过现场原机试验开展了燃气锅炉在不同掺氢比例燃气的相关燃烧试验研究，进行了燃烧稳定性、燃烧效率、燃烧污染物与水汽排放特性，测试和评估额定输入压力或功率下设备的燃烧与传热效率、热负荷、烟气排放指标、温度场分布等。通过研究提出了适应典型站场燃气锅炉工作性能要求的掺氢比例，可指导管输天然气掺氢控制方案以及终端设备改造措施的制定和实施。

关键词　掺混输氢、燃气锅炉、燃烧、污染物排放

Study on the adaptability of typical gas boilers in stations to hydrogen blended natural gas

Cheng Lei[1]　Zhang Hanwen[1]　Zhang yang[2]

(1. PipeChina Institute of Science and Technology; 2. Tsinghua University Shanxi Clean Energy Research Institute)

Abstract　Mixing and transporting hydrogen through long-distance natural gas pipelines is a key measure to break the transportation bottleneck in the development of the hydrogen energy industry. By increasing the proportion of zero carbon fuels, the energy fuel structure has been significantly optimized, which is significant for China to successfully achieve the "3060" goal of carbon peak and carbon neutrality. The basic combustion characteristics of hydrogen blended natural gas, such as combustion speed, combustion temperature, flame shape, tempering characteristics, pressure fluctuation characteristics, and nitrogen oxide generation characteristics, have significant changes compared to conventional natural gas, which have a significant impact on the operational efficiency, safety, stability, and pollutant emission characteristics of important pipeline equipment such as gas turbines and gas boilers in power stations. This study investigated the changes in physical properties, flow and combustion characteristics, and laws of conventional natural gas by blending 0~30% hydrogen gas. The study clarified the influence of hydrogen blending ratio and equivalent parameters on flame characteristics and flame stability. Based on the variation law of combustion characteristics of hydrogen blended natural gas, relevant combustion tests of gas boilers with different hydrogen blending ratios were carried out through on-site original machine experiments. The combustion stability, combustion efficiency, combustion pollutants and water vapor emission characteristics were tested and evaluated, and the combustion and heat transfer efficiency, heat load, flue gas emission indicators, temperature field distribution, etc. of the equipment were tested and evaluated under rated input pressure or power. A hydrogen blending ratio suitable for the performance requirements of typical station gas boilers has been proposed through research, which can guide the development and implementation of hydrogen blending control schemes for pipeline natural gas and terminal equipment renovation measures.

Key words　Mixing and transporting hydrogen; gas boilers; combustion; pollutant emissions

利用已有的天然气管网输送基础设施，可将上游氢气通过天然气管网输送到下游终端用户，因此天然气掺氢技术既能解决大规模可再生能源制氢、工业副产氢等气源消纳的问题，又能高效低成本地输送氢气，实现"氢进万家"，是降低天然气利用过程碳排放强度以及保障燃气供应安全的有效途径。为了更好地掌握掺氢天然气的基础燃烧特性以及掺混燃气应用于站场典型燃气设备的适应性，开展了掺氢天然气的基础燃烧特性、不同工况下设备掺氢天然气燃烧数值模拟研究和典型设备掺氢天然气现场验证燃烧试验等研究。

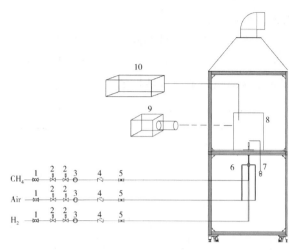

图1 实验系统原理图

1—手动球阀；2—电磁阀；3—质量流量计；
4—单向阀；5—阻火器；6—混气管；7—点火电极；
8—耐火隔热罩；9—高速相机；10—水露点检测仪/烟气分析仪

1 掺氢天然气的基础燃烧特性研究

通过搭建射流火焰实验台对掺氢天然气的基础燃烧特性开展了实验研究，获得了不同掺氢比和当量比下的燃烧火焰形状、燃烧温度、传播速度、回火指数、烟气水露点以及NO_x、CO污染物排放特性，明确了掺氢天然气与普通燃气互换性、火焰特征、稳定性、燃烧产物特征等。

1.1 射流火焰实验系统

图1为搭建的射流火焰实验系统图，实验系统主要由气体供给系统、流量调节系统、混合装置和数据采集系统组成。压缩空气罐供给燃料气体，气体纯度99.9%。流量调节系统由压力调节阀、流量调节阀及高压通气管组成，连续地向实验测试模块供给足够可调流量的燃气和空气。采用H_2量程5L、CH_4量程5L、空气量程50L的质量流量计。实验系统可采用混合燃料气体，气瓶出口安装安全防爆阀门，管路上安装单向阀和阻火器，防止气体倒流引发爆炸事故。燃料与空气经质量流量计调节进入混气管中进行气体混合。

试验台结构如图2所示，尺寸为1m×1m×2.3m；混气结构材料采用304制作；气路入口1为甲烷，外径尺寸6mm，空气入口管路2外径尺寸6mm，氢气入口管路3外径尺寸12mm；保温层材料采用硅酸铝，前侧留有2cm长槽，拍摄火焰形态，左右留有1cm长槽，为激光预留口。实验台中的电子控制柜所需材料包括PLC、触摸屏、继电器、空气开关和电缆等。此外，实验配置仪器还包括高速相机、水露点检测仪和MRU红外烟气分析仪。

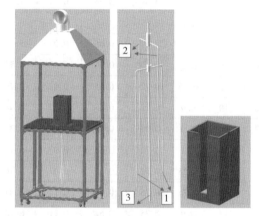

图2 实验台结构示意图

1.2 实验方法

实验装置可进行不同情况的火焰现象实验，实验时先开启控制面板开关以打开空气供应，并调节所需空气流量，然后打开燃气阀门并在燃烧喷口处进行点火，点火成功后调节实验工况下所需燃料气体流量。燃烧稳定后，燃气与空气充分混合后在喷口处形成稳定的正锥形层流火焰。利用高速相机获取燃烧火焰形状，同时根据燃烧结构使用面积法计算传播速度，计算方法如公式1所示：

$$u_0 = \frac{Q}{F} \quad (1)$$

式中，u_0为传播速度，m/s；F为焰峰面积，m^2；Q为气体流量，m^3/s。

利用热电偶及无纸记录仪获取燃烧内焰温度，即为此时燃烧温度，同时在烟气出口处利用水露点测试仪和MRU红外测试仪器测量污染物排放。

1.3 研究结果

实验共开展了7种不同掺氢比例、5种不同当量比下共计35组掺氢天然气燃烧实验，获得了掺氢天然气基础燃烧特性的变化规律。

表1 掺氢天然气基础燃烧实验工况表

	掺氢比/%	当量比	燃料流量/(L/min)	甲烷流量/(L/min)	氢气流量/(L/min)	空气流量/(L/min)
(1)	0	0.75	0.67	0.67	0	7.98
(2)	0	0.9	0.67	0.67	0	7.02
(3)	0	1.0	0.67	0.67	0	6.38
(4)	0	1.1	0.67	0.67	0	5.74
(5)	0	1.25	0.67	0.67	0	4.79
(6)	5	0.75	0.69	0.66	0.03	7.96
(7)	5	0.9	0.69	0.66	0.03	7.00
(8)	5	1.0	0.69	0.66	0.03	6.37
(9)	5	1.1	0.69	0.66	0.03	5.73
(10)	5	1.25	0.69	0.66	0.03	4.77
(11)	10	0.75	0.72	0.65	0.07	3.79
(12)	10	0.9	0.72	0.65	0.07	6.98
(13)	10	1.0	0.72	0.65	0.07	6.35
(14)	10	1.1	0.72	0.65	0.07	5.71
(15)	10	1.25	0.72	0.65	0.07	4.76
(16)	15	0.75	0.75	0.64	0.11	7.91
(17)	15	0.9	0.75	0.64	0.11	6.96
(18)	15	1.0	0.75	0.64	0.11	6.33
(19)	15	1.1	0.75	0.64	0.11	5.69
(20)	15	1.25	0.75	0.64	0.11	4.75
(21)	20	0.75	0.78	0.62	0.16	7.88
(22)	20	0.9	0.78	0.62	0.16	6.94
(23)	20	1.0	0.78	0.62	0.16	6.31
(24)	20	1.1	0.78	0.62	0.16	5.67
(25)	20	1.25	0.78	0.62	0.16	4.73
(26)	25	0.75	0.81	0.61	0.20	7.85
(27)	25	0.9	0.81	0.61	0.20	6.91
(28)	25	1.0	0.81	0.61	0.20	6.28
(29)	25	1.1	0.81	0.61	0.20	5.65
(30)	25	1.25	0.81	0.61	0.20	4.71
(31)	30	0.75	0.85	0.59	0.25	7.82
(32)	30	0.9	0.85	0.59	0.25	6.88
(33)	30	1.0	0.85	0.59	0.25	6.26
(34)	30	1.1	0.85	0.59	0.25	5.63
(35)	30	1.25	0.85	0.59	0.25	4.69

1) 火焰外观

不同工况下火焰外观变化规律如图3所示，可知随着掺氢比的增加，火焰由狭长变得宽短，这也是由于掺入氢气后，加快了混合燃气的火焰传播速度，使燃烧变得更快。相同掺氢比下，当量比减小，火焰逐渐狭长，火焰亮度变暗，且$\alpha=0.75$和$\alpha=1.25$时的燃烧火焰变化明显。$\alpha=1.25$时，火焰分层现象明显，随着当量比的减小，分层逐渐减弱。

2) 燃烧温度

燃烧温度是表征燃烧反应的重要参数，对判定燃烧效率，控制燃烧过程以及优化燃烧条件具有重要意义。实验研究了不同掺氢比例工况下，燃烧温度的变化规律，考虑到温度实测值均<1000℃，绝对值参考意义不大，为直观对比温度变化规律，以$\alpha=1$时的实测值作为基准，进行数据处理，测得的实验温度相对比值如图4所示。

从图4可知，同一掺氢比下，$\alpha=1$时燃烧强度最大，此时温度达到最高，当量比稍高或稍低($\alpha=1.1$和0.9)时，温度变化呈现增大趋势，但波动相差不大，并且接近$\alpha=1$时的燃烧温度。当量比为0.75和1.25时，温度变化呈现增大趋势，但相较$\alpha=1$时，燃烧温度显著降低。燃烧实测时会有散热损失，实测温度相较于绝热火焰温度降低，因此同步采用Chemkin中一维PSR反应器计算了不同掺氢比下掺氢天然气绝热火焰温度(在绝热条件下完全燃烧时燃烧产物的温度)进行对比，可以看出同一当量比下，随着掺氢比的增加，绝热火焰温度逐渐增大。在当量比1，掺氢比30%时，绝热火焰温度最高可达2237K，当量比在1附近时，绝热燃烧温度也相对较高。随着当量比的减小，绝热火焰温度呈现出先增大后减小的变化趋势。

3) 火焰传播速度

实验得到的火焰传播速度如图5所示，随着掺氢比例的增加，预混燃料的火焰传播速度显著增加。火焰传播速度随着当量比的减小，呈现先增后减的趋势，并且传播速度的峰值出现在当量比为1.1处，当量比数在0.9~1.1范围内，掺氢天然气的火焰传播速度相对接近。

4) 烟气水露点

烟气水露点为燃气燃烧所产生的烟气中水蒸汽开始凝结时的温度。图6为当量比在0.75~1.25范围内，不同掺氢比对烟气水露点的影响规律。同时采集烟气中水分含量,利用水蒸气压力和饱和温度曲线进行验证校核。随着掺氢比增大,烟气中水分含量增大,同时水露点温度逐渐升高。不同当量比下,水露点温度变化范围在 2℃ 以内,温度峰值出现在 $\alpha=1$ 附近;$\alpha=0.75$ 时,水分含量最低,水露点温度也最低。

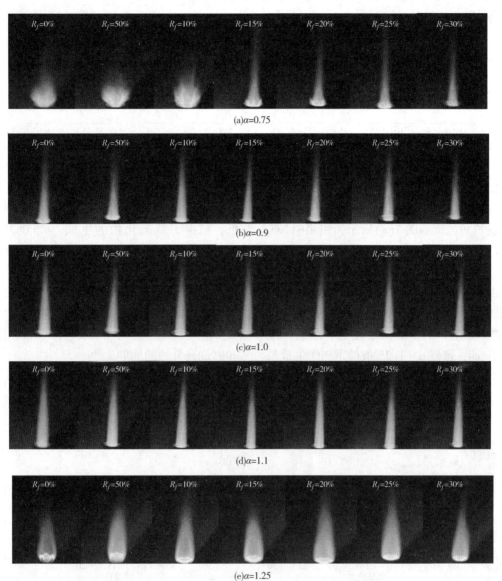

图 3　不同工况下火焰外观

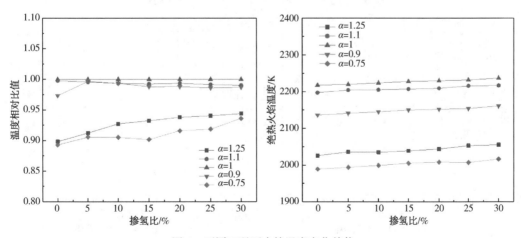

图 4　不同工况下火焰温度变化趋势

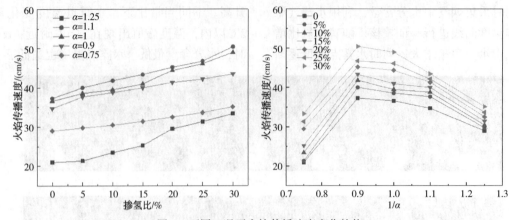

图 5 不同工况下火焰传播速度变化趋势

5）燃烧污染物特性

本研究测试的不同当量比和不同掺氢比下的 NO_x 污染物排放浓度如图7(a)所示，实测中 NO_2 浓度基本为 0，因此主要以 NO 代表 NO_x。同一当量比下，随着掺氢比例的增加，NO 的排放量呈现上升趋势，这是因为氢气加入后提高了火焰温度，温度的升高促进了 NO 的生成。相同掺氢比下，排放峰值在贫燃 $\alpha=0.9$ 附近，最高浓度可达 42.7ppm；富燃 $\alpha=1.25$ 条件下 NO 生成最少，随着掺氢比的增大，NO 浓度由 6.4ppm 增加到 15.7ppm，这是因为富燃条件下，氧气浓度低，减少了氮氧化物的生成。$\alpha=0.75$ 条件下 NO 生成也相对较少，这是因为氧气的过量加入，降低了燃烧温度，使得氮氧化物生成减少。因此，在偏离完全燃烧状态时的当量比条件下，都能在一定程度上降低 NO_x 污染物的生成，但是实际燃烧技术应用时还需要考虑燃料燃烧稳定性和燃烧充分性的问题。

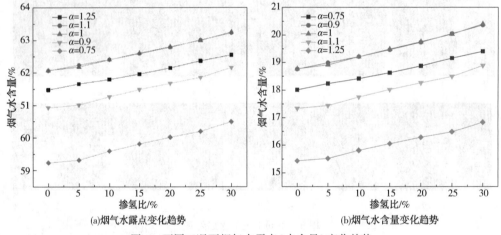

(a)烟气水露点变化趋势　　　(b)烟气水含量变化趋势

图 6 不同工况下烟气水露点(水含量)变化趋势

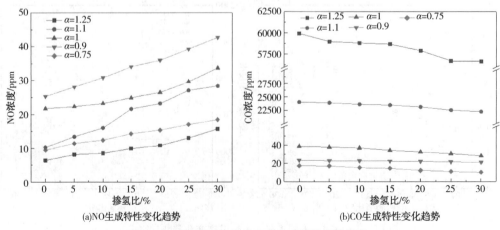

(a)NO生成特性变化趋势　　　(b)CO生成特性变化趋势

图 7 不同工况下 NO 和 CO 生成特性变化趋势

不同当量比和不同掺氢比下的 CO 排放浓度见图 7(b)所示，同一当量比下，CO 排放浓度随着掺氢比例的增加而降低，这是因为掺氢量增加，甲烷含量相对减少，降低了燃料含碳量。氢的体积分数增加，使反应过程中的 OH 浓度变大，加速反应进行。相同掺氢比下，富燃条件下 CO 浓度均超标，其中 $\alpha=0.75$ 时 CO 排放量最高，此时 CO 浓度变化区间为 56651.5~59945.9 ppm，$\alpha=1.1$ 时 CO 浓度变化区间为 22225.2~24028.5 ppm。α 为 0.75、0.9 和 1 时，CO 浓度急剧下降，α 为 0.9~1.0 时排放范围在 40 ppm 以内，α 为 0.75 时排放浓度在 20 ppm 以内，这是由于富燃 $\alpha>1$ 条件下，氧气含量严重不足，燃烧极不充分，造成 CO 大量生成，而贫燃条件下，氧气供给过量，燃料能够实现完全燃烧，因此 CO 含量极低。

2 掺氢天然气锅炉适应性试验研究

基于上述掺氢天然气的基础燃烧特性变化规律研究，在典型燃气锅炉上开展了掺氢天然气燃烧试验，分别测量了不同掺氢比例下锅炉的实际热负荷、炉膛温度和烟气污染物排放浓度等相关参数，明确了燃烧震荡、回火等锅炉掺氢燃烧运行风险现象发生的可能性，并开展了掺氢对锅炉整机性能、燃烧稳定性和污染物排放的影响规律研究。

2.1 试验原理

通过设置天然气与氢气的掺混装置，将按照一定比例掺混的掺氢混合燃气，在典型燃气锅炉上直接进行燃烧试验，燃烧期间分别使用各仪器测量不同掺氢比例下锅炉的实际热负荷、炉膛温度、烟气再循环量、排烟温度、烟气污染物排放浓度和水含量、工质侧温升变化、热效率、燃烧效率、炉内燃烧情况和噪音等相关参数。

2.2 试验平台

典型锅炉掺氢试验基于某专用燃气试验台进行，场地具有高效的通风系统，能够确保氢气和天然气的快速稀释和排放，防止气体积聚。试验场地安装有燃料供应系统、燃料混合装置、百得 TBG 80 燃烧器、型号为 CWNS1.4-85/60-Q 的常压燃气热水锅炉、温度和压力传感器、数据采集系统等。

1) 燃料供应系统

本次使用移动式高压氢气瓶作为氢气源，用于提供试验所需的氢气，并将高压氢气减压至适合试验的压力范围。天然气采用常规市供城燃天然气。

2) 燃料混合装置

通过静态流动混合器确保氢气和燃料充分混合，形成均匀的混合气体。

3) 锅炉系统

选用型号为 CWNS1.4-85/60-Q 的常压燃气热水锅炉，并设置有观察窗口，用于观察火焰形态和燃烧过程。

4) 测量与控制系统

温度测量：使用热电偶测量炉膛、介质的温度；压力测量：监测炉膛内的压力变化；流量控制：通过质量流量控制器（MFC）精确控制氢气和天然气的流量。

5) 数据采集系统

记录和分析试验数据，如温度、压力、流量等。

6) 点火系统

点火装置由点火变压器和点火探针组成，用于点燃混合气体，并自动控制点火时机。

7) 排放处理单元

使用便携式烟气分析仪分析燃烧产生的烟气成分，如 CO、CO_2、NO_x 等。

2.3 试验步骤

每项试验的主要步骤如下：

1) 天然气管线以及氢气管线用氮气按要求置换完成，并按照试验燃烧器设计的供气压力完成燃料供应；

2) 将待测试的燃气燃烧器按要求安装完成，送电检查所有在线测点和便携式测量仪器是否安装到位；

3) 燃烧器天然气点火启动，调整好天然气的流量后根据氢气管线上的流量计按照试验要求掺入不同比例的氢气；

4) 每个工况下在排烟处开展 CO、CO_2、H_2O 和 NO_x 等烟气组分浓度测量；

5) 在排烟出口测量排烟温度；

6) 记录收据相关表盘数据（燃烧器运行模式、炉膛温度、进出口热水参数、甲烷和氢气流量参数、供气压力、供风温度等）。

2.4 试验内容

在典型燃气锅炉 100% 负荷下进行空燃当量比为 0.75 和 0.9 的不同掺氢比例的天然气燃烧试验，试验工况如下：

表2 掺氢天然气锅炉试验工况表

掺氢比例	空燃当量比	测试参数	测试时间	稳定条件
0 5% 10% 15% 20% 25% 30%	0.75 0.90	实际热负荷	测给气量3min，	每个工况需稳定运行半小时
		炉膛温度	稳定后表盘读取，取三组	
		排烟温度	试验测量	
		烟气露点	试验测量	
		烟气污染物排放浓度	稳定后测试3min	
		烟气水含量	稳定后测试3min	
		工质侧温升变化	表盘得出	
		热效率	计算得出	
		燃烧效率	计算得出	
		炉内燃烧情况视频	拍摄30s	
		燃烧侧噪音分贝	测试1min	

2.5 试验结果

通过在100%燃烧负荷运行条件下，设置不同的当量比（当量比分别为0.75、0.9），在天然气中掺入不同比例（0、5%、10%、15%、20%、25%、30%）的氢气进行燃烧试验，具体结论分析如下：

1）锅炉炉膛温度及热效率变化

通过对锅炉炉膛温度及热效率变化可以看出，不论在当量比为0.9和0.75，随着掺氢比增大，炉膛温度整体呈现上升趋势，锅炉热效率呈现下降趋势。主要由于氢气掺入到天然气中，燃烧器根部燃烧加剧，火焰变短，局部高温区域增大，炉内火焰温度升高，而掺入氢气燃烧产生的高温烟气中水蒸汽增大，相比甲烷燃烧产生的CO_2，分子量小，炉内辐射换热降低，导致炉膛出口烟气温度升高，从而排烟温度升高，锅炉热效率下降（图8）。

2）锅炉炉膛压力及噪声变化

在天然气掺入氢气后，整个燃烧过程火焰相对稳定，炉膛压力波动在合理范围内（±30Pa），燃烧噪音参数与纯烧天然气基本相同（图9）。

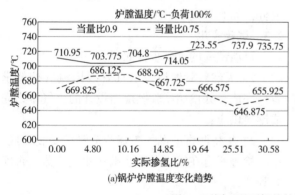
(a)锅炉炉膛温度变化趋势

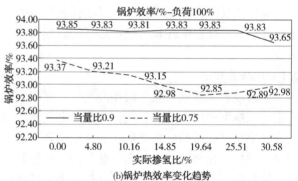
(b)锅炉热效率变化趋势

图8 不同工况下锅炉炉膛温度及热效率变化趋势

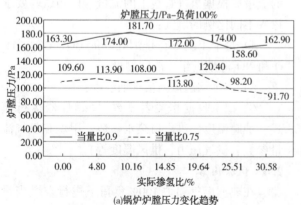
(a)锅炉炉膛压力变化趋势

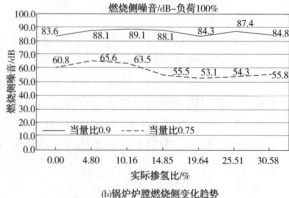
(b)锅炉炉膛燃烧侧变化趋势

图9 不同工况下锅炉炉膛压力及噪声变化趋势

3）烟气 NO_x、CO_2 含量和水露点变化

通过对尾部烟气中的 NO_x 和 CO_2 含量变化可以看出，不论在当量比为 0.9 和 0.75，随着掺氢比增大，尾部烟气中的 NO_x 整体呈现上升趋势，CO_2 整体呈现下降趋势。主要原因天然气掺入氢气后，燃烧加剧，燃烧器根部火焰燃烧剧烈，局部高温区域增大，热力型 NO_x 生成量增大，导致尾部烟气中 NO_x 含量随着掺氢比升高而升高。由于氢气掺入后，替代了部分天然气燃烧，所有尾部烟气中的 CO_2 含量降低。由于烟气中的水蒸汽占比增大，水蒸气分压比提升，烟气中的水露点温度升高，烟气相对湿度升高（图10、图11）。

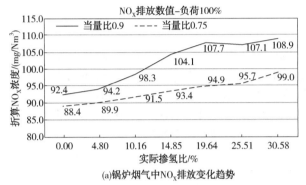 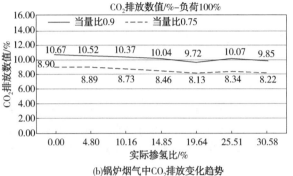

图10 不同工况下锅炉炉膛烟气 NO_x、CO_2 变化趋势

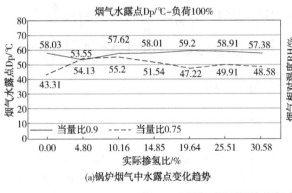 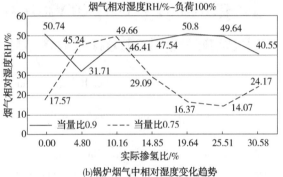

图11 不同工况下锅炉烟气水露点和相对湿度变化趋势

3 结论

本研究通过设置不同的当量比（当量比分别为 0.75、0.9），在天然气中掺入不同比例（0、5%、10%、15%、20%、25%、30%）的氢气进行燃烧试验。试验表明：随着掺氢比增大，炉膛温度整体呈现上升趋势，锅炉热效率呈现下降趋势。尾部烟气中的 NOx 整体呈现上升趋势，CO_2 整体呈现下降趋势。在天然气掺入氢气后，整个燃烧过程火焰相对稳定，无火焰震荡及回火现象出现，炉膛压力波动在合理范围内（±30Pa），燃烧噪音参数与纯烧天然气基本相同。通过本次天然气掺氢燃烧试验研究，在燃烧方面，火焰整体稳定，无火焰震荡及回火现象出现，炉膛温度虽然有一定升高，但炉内金属材料能够满足要求。在燃烧效率方面，整个过程燃尽率为100%。在尾部烟气成分方面，烟气中的 NOx 升高、烟气中的含水量升高，CO_2 下降，当掺氢比达到30%时，烟气中的 NOx 升高约20%。在锅炉热效率方面，随着掺氢比增大，排烟温度升高，锅炉热效率降低，当掺氢比达到30%时，锅炉热效率降低约0.3%。

通过本项目试验表明，天然气掺氢燃烧稳定，在30%掺氢比以下，易于实现锅炉等站场典型燃气设备高效利用，无安全风险，仅需考虑环保、节能方面所带来的影响。

国外成品油管输定价监管现状及对我国的启示

牛国富[1]　石博涵[2]　张仲蓥[1]　温文[1]　阴佳乐[1]

（1. 国家石油天然气管网集团有限公司科学技术研究总院分公司；2. 国家石油天然气管网集团有限公司）

摘要　国内成品油管道受限于市场条件等因素未能有效实现公平开放，安全稳定、高效低价的优势未能充分发挥，亟需系统研究国外成熟的成品油管输定价机制，结合我国实际制定科学合理的价格机制，促进成品油管道行业健康持续发展。本文详细梳理了美国、加拿大和俄罗斯的成品油管输定价与监管现状，包括监管机构、定价方法、调价规则、定价示例、监管规则等方面内容及特点。结合国内成品油管输价格机制最新政策现状，借鉴国外定价与监管做法，提出了以下构想：①借鉴参数制定规则；②探索新的管输价格形式；③研究将管道弃置费用纳入定价成本；④做好成本监审和信息公开。

关键词　成品油管道；监管；价格

The Current Situation of Pricing and Regulatory Supervision of Foreign Oil Pipeline Transportation and Its Implications for China

Niu Guofu[1]　Shi Bohan[2]　Zhang Zhongli[1]　Wen Wen[1]　Yin Jiale[1]

(1. PipeChina Institute of Science and Technology; 2. China Oil & Gas Pipeline Network Corporation)

Abstract　Due to market conditions and other factors, domestic refined oil pipelines have not effectively achieved fair and open access, and their advantages of safety, stability, high efficiency and low cost have not been fully utilized. It is urgent to systematically study the mature refined oil pipeline transportation pricing mechanisms in foreign countries, and formulate a scientific and reasonable pricing mechanism in line with China's actual situation to promote the healthy and sustainable development of the refined oil pipeline industry. This paper elaborately sorts out the current situation of refined oil pipeline transportation pricing and supervision in the United States, Canada and Russia, including regulatory institutions, pricing methods, price adjustment rules, pricing examples, and regulatory rules, as well as their characteristics. Combining the latest policy status of domestic refined oil pipeline transportation pricing mechanisms and drawing on foreign pricing and supervision practices, the following ideas are proposed: ①draw on the parameter formulation rules; ②explore new pipeline transportation pricing forms; ③study the inclusion of pipeline abandonment costs in pricing costs; ④do a good job in cost supervision and information disclosure.

Key words　Finished oil pipeline; Regulation; Price

国家管网成立之后，原所属三大石油公司的相应成品油管道纳入国家管网集团正式运营。受限于物理设施、市场条件等客观因素，成品油管道还未能有效实现公平开放，管输量占消费量比重远低于欧美发达国家水平。与铁路等其他运输方式相比，成品油管道运输价格较低，短途管输同等占用管容但价格显著偏低，未能发挥其安全稳定、高效低价的优势，导致国内部分成品油管道负荷率较低，亟需研究制定科学合理的价格机制，促进成品油管道行业健康持续发展。北美、俄罗斯等地区成品油管道发展成熟，已形成了相对完善的政府监管体系和定价机制。本文旨在通过研究总结国外成品油管道定价与监管经验，为国内提供参考借鉴。

1 美国成品油管输定价及监管

美国成品油管道约10万公里，集中于中部和南部的得克萨斯州、俄克拉荷马州以及墨西哥湾沿岸。主要的成品油管道公司有麦哲伦（Magellan）、科洛尼尔（Colonial）、金德摩根（Kinder-Morgan）等。

1.1 联邦、各州监管机构和职责

美国成品油管道监管法规与体系比较完善，全美跨国和跨州成品油天然气运输管道的管理工作由联邦能源管理委员会（简称FERC）承担。FERC隶属于能源部，但独立行使职权，负责监督审查管网运输公司制定运输服务质量与输量标准和服务费率，确保第三方公平获得管输服务，实现企业和消费者利益的有效平衡。委员会每月召开2次会议，研究许可证申请、管输费率、建设项目等问题。

完全位于一州境内的油气运输管道由各州专门的输油管道监管机构监管，大多数州与公共事业的监管职能放在一起，例如德克萨斯州铁路委员会、阿拉斯加监管委员会。

1.2 成品油管输价格管理

1906年–20世纪90年代，管道运输费率采用服务成本法制定。1993年后，经过广泛的行业咨询和征求意见，监管部门采纳了指数法作为监管成品油管道费率的主要方法，同时也允许某些情况下使用服务成本法、市场法和协商法。管道价格初始核定及后续调整的信息均在FERC官网发布（图1）。

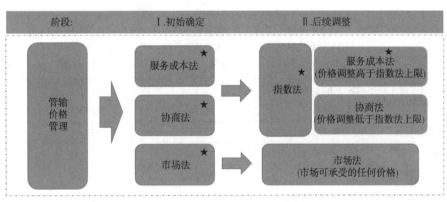

图1 美国成品油管道价格管理模式

1.2.1 管输价格首次核定以服务成本法为主

美国成品油管道首次运价确定分为服务成本法核定、双方协商、基于市场确定三种类型，均需FERC监管。

（1）由政府按照服务成本法核定管输费。1985年FERC发布了《154-B号条例》，确定了以服务成本法来制定管输运价，管道运营商需提交服务成本详细情况（在官方网站进行电子申报），以证明其价格合理性，报美国联邦能源委员会审批。目前大多数管道公司采用该方法确定初始价格。准许收益率为加权平均资本成本，即将公司债权成本、股权收益率按相应的资本结构加权求和得出，其中债权成本据实计算，股权收益率最高不超过14%。

（2）双方协商确定。在管道托运商唯一或数量少的情况下，FERC允许管道运营商通过协商一致并报FERC的方式确定首次管输价格。该方式需托运双方协商达成一致价格意见，管道运营商向FERC提交至少一个拟使用其管输服务的非关联托运商同意的价格承诺书，以该价格申请确定为首次管输价格。

（3）基于市场确定。管道运营商在成品油运输的上载点市场及目的地市场均不存在显著的市场势力的情况下，可以向FERC提交相关申请，申请审核周期可能较长，但如果申请获得批准，运营商可以将费率设定在市场能够承受的任何水平。虽然美国成品油管道市场化程度较高，但采用该方式定价的管道数量极少。

1.2.2 后续价格主要按照指数法调整

指数法是美国联邦能源委员会为了简化成品油管道费率监管，制定的简单且公平合理的费率调整方法。FERC将每年发布的费率调整指数作为价格调整上限。对于政府采用服务成本法和协商一致确定首次管输价格的管道，若价格调整不超过指数法确定的价格上限，仅需取得托运商的书面同意，不必向FERC提交成本等数据和信息；若上限价格无法涵盖管道运营商实际成本费用，其可以向 FERC 申请服务成本法重新核定。对于采用市场法确定首次管输价格的管道，后续价格可根据市场情况自由调整。

1993 年 10 月 22 日发布了《561 号条例》，采用指数法制定管输上限运价，该指数是在工业生产者出厂价格指数（PPI）基础上加或减一个百分比来确定，FERC 每五年对该百分比进行一次修正。例如 2021 年 7 月—2026 年 6 月价格指数调整规则为"PPI-0.21%"，近五年价格调整指数详见表 1。

表 1 近五年价格调整指数

时间	价格指数
2024 年 7 月 1 日—2025 年 6 月 30 日	1.012647
2023 年 7 月 1 日—2024 年 6 月 30 日	1.133194
2022 年 7 月 1 日—2023 年 6 月 30 日	1.087107
2021 年 7 月 1 日—2022 年 6 月 30 日	0.984288
2020 年 7 月 1 日—2021 年 6 月 30 日	1.020139

1.3 成品油管道运价示例

根据科洛尼尔成品油管道公布的运价（详见表 2），其 2024 年 7 月 1 日之后调整的运价是按照 FREC 公布的价格指数 1.012647 乘上期价格得出。

表 2 科洛尼尔成品油管道运价节选（美分/桶）

目的地	起点			
	德克萨斯州 哈里斯县休斯顿 （帕萨迪纳，锡达湾）	德克萨斯州 杰斐逊县赫伯特 （博蒙特-波特阿瑟）	路易斯安那州 卡尔卡西厄县 莱克查尔斯	路易斯安那州 圣兰德里县 克罗茨斯普林斯
阿拉巴马州（2023 年 7 月 1 日—2024 年 6 月 30 日价格）				
伯明翰（杰斐逊县）	138.41	127.46	123.43	109.99
伯明翰-赫勒纳（谢尔比县）	138.41	127.46	123.43	109.99
博利吉（格林县）	138.41	127.46	123.43	109.99
牛津（卡尔霍恩县）	150.34	139.42	135.36	121.92
阿拉巴马州（2024 年 7 月 1 日—2025 年 6 月 30 日价格）				
伯明翰（杰斐逊县）	140.16	129.07	124.99	111.38
伯明翰-赫勒纳（谢尔比县）	140.16	129.07	124.99	111.38
博利吉（格林县）	140.16	129.07	124.99	111.38
牛津（卡尔霍恩县）	152.24	141.18	137.07	123.46

美国所有跨州成品油管道运价均由 FERC 统一监管，除少数通过双方协商或基于市场确定之外，大多数采用服务成本法由 FERC 定价，后续按指数法调整，以确保第三方公平获得管输服务，实现企业和消费者利益的有效平衡。

2 加拿大成品油管输定价及监管

加拿大幅员辽阔且石油储量丰富，但因为环境问题、当地土著反对以及司法流程等方面的原因导致管道设施的建设滞后。加拿大成品油管道总长约 4000 公里，主要的成品油管道运营商包括 TC Energy、安桥（Enbridge）等。

2.1 监管机构及职责

加拿大能源监管机构（简称 CER）成立于 2019 年 8 月 28 日，与之伴随的是将《加拿大能源监管法案》（CER 法案）定为法律。该机构的前身是加拿大能源局（NEB，成立于 1959 年），因此 CER 在监管加拿大的能源公司和项目方面有近 60 年的经验。CER 的作用是确保在存在市场垄断的地方，市场垄断不被滥用，规范其管辖范围内管道公司的服务价格、服务条款和条件，以确保管输费公正合理，并且在管输费或服务方面不存在不公正的歧视。其主要职责是在石油、天然气和电力领域实施监管，以保护公众利益。如果管道系统跨越省或国际边界，则由 CER 管理。完全位于一个省内的管道由该省的监管机构监管。

CER 监管的管道公司分为两组：第一组包括 13 家大型管道公司，其拥有广泛管道系统，并向第三方托运商提供管输服务；第二组为运营规模较小、管道系统简单、不提供第三方托运服务的小型管道公司，共85家，见表3。管输价格核定及后续调整的信息（企业申请和政府批复）均在CER官网发布。

表3 加拿大被监管管道公司

类别	受监管管道公司
第一组	Alliance Pipeline Ltd.
	PKM Cochin ULC
	Enbridge Pipelines（NW）Inc.
	Enbridge Pipelines Inc.
	Foothills Pipe Lines Ltd.
	Maritimes & Northeast Pipeline Management Ltd.
	NGTL GP Ltd., on behalf of NGTL Limited Partnership
	Trans Mountain Pipeline ULC
	Trans Québec and Maritimes Pipeline Inc.
	TransCanada Keystone Pipeline GP Ltd.
	TransCanada PipeLines Limited
	Trans-Northern Pipelines Inc.
	Westcoast Energy Inc.
第二组	85家小型管道公司

2.2 成品油管输价格管理

首次价格确定时，大型管道公司主要由政府按照服务成本法核定，小型管道（集团内部使用）公司由双方协商确定。

在1970-1995年间，NEB几乎全面使用服务成本法确定运输费，运费的调整需按照固定格式内容提出申请，NEB征求各方意见后发布最终运费文件。服务成本由运营费用、折旧、税金等组成。准许股权收益率计算公式为：下年准许股权收益率=当年准许股权收益率+（当年预测30年期政府债券收益率-计算当年准许股权收益率采用的30年期政府债券收益率）*0.75。近几年，准许股权收益率从2021年的6.63%逐步提升到2024年的7.88%。

为了提高监管程序的有效性和便捷性，自1980年代中期以来支持使用政府磋商作为收费听证会的替代方案。1988年9月颁布了第一份《托运、服务价格和条款协商解决准则》，1994年8月更新，并于2002年6月再次修订，以便在处理有争议时提供灵活性的解决办法。以上定价方式均需CER审批或备案。

（1）第一组公司监管

对于第一组公司（不含Kinder Morgan Cochin和Enbridge Norman Wells管道）的管输费审批，CER通常采用以下两种方式之一：一是对其提交的详细服务成本价格申请进行评估，在大多数情况下，CER会举行正式的公开听证会，听取各方的证据和论点后发布最终决定。二是根据2002年6月12日修订的《协商解决托运、服务价格和条款的指南》（简称《指南》），为提高监管有效性和便捷性，允许少数管道由管道公司、托运商和其他利益相关者进行协商确定价格，管道公司将提交一份文件，详细说明协商条款报CER审批。CER根据《指南》评估管输费协议，以确管输费公正合理，且所有条款均不违反《指南》规定，还将征求托运商和其他利益相关者的意见。

根据《指南》，政府对管输费方案确定的步骤为：①管道公司向CER提交定价方案申请，包括收入确定、管输费计算、调整、与托运商沟通情况等内容。当方案未获得所有托运商认同时，需说明CER接受该方案的理由。同时将申请提交给所有相关方。②CER收到申请后，邀请相关方评价定价方案。③若无任何相关方反对该方案，CER视同其公正合理，无需公开听证。④若相关方反对定价方案，应在规定时限向CER提交反对声明。收到反对声明后，管道公司和支持定价方案的相关方可在规定时限内递交针对反对声明的回应。CER基于收到的所有信息做出决定，包括：驳回反对意见，批准定价方案；否定价方案，移交听证；暂时批准定价方案，举行听证会解决反对方提出的问题。

（2）第二组公司监管

对于第二组公司和第一组中的两条输油管道（Kinder Morgan Cochin和Enbridge Norman Wells），目前采用管道运营商和托运商、其他利益相关方协商确定，有争议向CER申诉的管输费监管方式。管道公司有责任向托运商和其他利害关系人提供足够的信息，以确定管输费的合理性，并提交给CER备案。如果托运商没有提出申诉，CER则认为该管输费公正合理，自动生效。如果托运商与管道公司存在无法解决的争议事项，可以向CER提起申诉，CER通过特定的程序来审查管输费和条款，直到消除异议。

后续价格调整时，以服务成本法定价的大型管道公司定期向CER申请新的管输价格，双方协商的小型管道公司到期后重新向CER报备。

对于服务成本法核定管输费，能源委员会要求第一组中的管道公司，每季度按照规定内容与格式提交监督报告，包括公司财务业绩、重大变更情况分析等，这些报告作为是否调整运输费的依据。成本数据按照《天然气管道统一会计准则》、《输油管道统一会计准则》要求执行，并与之保持一致；第二组的管道公司每年提交一份审计的财务状况报告。

在服务成本监管的背景下，管输费变更要求管道公司定期提交申请，并提供CER申报手册指南中规定的文件。然后举行公开听证会，以听取有关各方的意见。在考虑了各方的证据和论点后，批准最终管输费。

采用协商方式确定的管输费协议有些包括明确的终止日期，例如Trans Mountain管道公司的2019—2021年管输费结算，有些则规定无限期地执行。对于到期的管输费，由管道公司重新申请报备，例如Enbridge加拿大干线管输费协议2021年7月1日到期后，由其向CER提交新的管输费申请。

2.3 成品油管道运价示例

Trans-Northern成品油管道由Trans Northern公司于1952年建成投用，约850公里，将成品油魁北克省蒙特利尔向西输送到安大略省多伦多，并从帝国石油有限公司位于安大略省东部南蒂科克的炼油厂运往多伦多。以蒙特利尔-多伦多段管输费为例，各年CER批准调整见表4。加拿大NEB（国家能源局）要求从2015年开始提取弃置费用，并允许其通过管输价格回收，相应的管输费包含基础运费和弃置费用运费。从2014年至2021年运费逐年增加。

加拿大所有跨省成品油管道运价均由CER统一监管。对于拥有广泛管道系统，并向第三方托运商提供管输服务的13家大型管道公司，主要由政府按照服务成本法核定，每季度向CER提交成本等监测信息，定期向CER申请核定新的管输价格；同时为提高监管有效性和便捷性，允许少数管道通过"双方协商、政府审批"的方式确定管输价格；对于运营规模较小、管道系统简单、不提供第三方托运服务（主要集团内部使用）的85家小型管道公司，此类公司管输费主要由双方协商确定并向CER报备，若双方存在无法解决的争议，向CER申诉解决。CER通过这些监管确保市场垄断不被滥用，管输费公正合理，管输服务不存在不公正歧视，以保护公众利益。

表4 Trans-Northern蒙特利尔-多伦多段管输费
（美元/立方米）

年费	最终运费	基础运费	弃置费用运费
2014	11.942	—	—
2015	14.493	14.011	0.482
2016	15.753	15.264	0.489
2017	21.739	21.086	0.653
2018	43.747	42.922	0.825
2019	43.656	42.722	0.934
2020	44.674	43.865	0.809
2021	50.093	48.784	1.309

注：最终运费=基础运费+弃置费用运费。

3 俄罗斯成品油管输定价及监管

俄罗斯成品油输送管网基本由隶属于Transneft公司的几家成品油运输公司负责，其中俄罗斯成品油运输股份公司（Transnefteproduct）占比最大，目前运营超过1.9万公里的成品油管道，其中俄罗斯境内管网1.64万公里，其余部分管网主要分布在乌克兰、白俄罗斯、哈萨克斯坦等地区。

3.1 监管机构及职责

俄罗斯成品油管道由能源署（原俄罗斯联邦能源委员会）进行监管，其主要职能之一就是重新审查和批准俄罗斯国家石油运输公司和成品油运输公司所属干线管道输油的运价。代表国家调整管道运价制定的方法并给出指标参数，具体运价的数据则可由地方能源委员会根据联邦能源署制定的方法参数来确定。地方能源委员会与联邦能源署使用统一的标准规范，并可以在一定的权限范围内自主做出决定。俄联邦价格办公室是俄联邦政府价格制定标准的具体执行机构之一，由它来全权代表国家对自然垄断主体所产商品或所提供服务的价格进行制定、调整及监管。

3.2 成品油管输价格管理

俄罗斯成品油管输价格参照铁路运费作为标杆定价，上限不得超过对应的铁路运价，对于唯一托运商可双方协商。为确定成品油运输费率，俄联邦价格办公室针对每个成品油管道运输公司批准其最高费率。成品油管输价格制定依据的原则如下：1）计算成品油管道运输公司的计划性日常和基本开支需求；2)保证成品油管道运输在运输服务市场上的竞争力(和铁路相比的竞争力)；3)俄罗斯联邦价格办公室对各成品油管道运输平等要求，即所有发货人和运输方向(出口和内需)执行相同费率。

俄罗斯成品油运输股份公司每个运价区段的管输费率，不超过同一类型成品油铁路运输费率确定。如果唯一托运商的成品油沿此费率段(装油、转运、卸油、交油站)运输，也可以应用协议费率。在此费率段发货人与成品油运输公司之间产生严重分歧的情况下，俄联邦价格办公室单独对此费率段成品油运输费率进行研究，并在相关发货公司的参与下，根据成品油运输公司的报告，予以研究批准。

后续价格根据铁路运费和管输企业成本变化情况调整，申请价格修订的因素主要有：1)成品油管道运输企业费用要素价格变化；2)拟定的结算年度通货膨胀变化；3)铁路运输服务费率发生变化。

3.3 成品油管道运价示例

根据俄联邦能源委员会 2002 年 12 月 18 日 93-э/8 号决议，批准了俄罗斯成品油运输股份公司长输管道系统输油服务费率，自 2003 年 1 月 1 日起生效实施，见表5。

俄罗斯成品油管输价格采用标杆法定价，价格参照铁路，上限不超过铁路运价。对于唯一托运商的管输价格，可由双方协商确定，存在分歧时由俄联邦价格办公室研究批准。后续管输价格根据铁路运费和管输企业成本变化情况相应申请调整。

表5 俄罗斯成品油运输股份公司长输管道系统输油服务费率表

公司名称	管输费率(卢布/100吨公里)
"海洋成品油运输"股份公司	27.51
"圣彼得堡成品油运输"股份公司	30.86

4 国外成品油管输价格管理模式总结

从上述国家成品油管道运输价格监管情况看，主要有服务成本法、协商法、标杆法和市场法四种定价方法，其适用条件和做法见表6。国外成品油管道运输价格主要由政府监管定价，政府机构主要采用服务成本法、市场标杆法进行公开透明的定价，制定了包括定价程序、首次定价方法、后续价格调整等一整套规则；也规定了在一定情形下，管道运营商可以与托运商协商价格，但协商不一致的仍由政府定价。美国极少管道也可以通过市场竞争确定价格，但这种情况仅适用于成品油管道存在多运营主体、多种替代运输方式的非垄断市场中，允许托运双方在充分竞争的市场中根据自由竞争原则参与托运、承运选择并确定管输价格。

表6 典型国家成品油管道价格管理模式简表

序号	定价方法	适用条件	使用此方法的国家及做法
1	服务成本法	适用于有垄断地位、管道系统复杂、向第三方提供托运服务的管道	• 美国：大部分管输价格首次核定以服务成本法为主。 • 加拿大：大型管道公司主要按服务成本法定价
2	协商法	主要适用于内部使用、不提供第三方托运或第三方托运商较少的管道，托运双方能就价格达成一致意见	• 美国：少数管道在托运商数量较少情况下，允许双方协商一致，报政府审批的方式定价。 • 加拿大：运营规模较小、管道系统简单、不提供第三方托运服务的小型管道公司的管输费由双方协商；少数大型管道公司允许协商并报政府审批。 • 俄罗斯：少数只有唯一托运商的管道可以双方协商定价
3	标杆法	适用于一家或少数几家公司垄断，且与铁路相竞争的管道	• 俄罗斯：大部分管输价格参照铁路运费确定
4	市场法	适用于多管道主体、市场充分竞争，不存在垄断的管道	• 美国：极少管道适用，需在政府严格审批不存在显著市场势力情况下，允许依据市场情况自由定价

5 对我国的启示

国家发改委 2024 年 11 月发布了《关于完善成品油管道运输价格形成机制的通知》(发改价格〔2024〕1703 号)，标志着国内成品油管输价格领域的一次重要改革。文件明确对跨省成品油管道运输价格实行弹性监管机制，在不超过国家核定的最高准许收入前提下，管输企业和成品油生产经营企业公平协商运输价格，并提出协商价格的具体原则，即对于其他运输方式可替代的，管道运输价格应不高于替代运输方式价格；对于其他运输方式无法替代的，管道运输价格可参照但不高于所在地区或邻近地区铁路运输价格。新机制的出台有利于充分发挥成品油管道运输竞争优势，深化石油天然气市场体系改革，对促进成品油行业发展具有积极重要意义。在未来的价格实践中，可以结合国外定价与监管经验做法，在参数确定、价格形式、定价成本范围等方面进行借鉴或探讨。

（1）借鉴参数制定规则。现有的定价机制规定了"设置最低负荷率要求"和准许收益率，明确了参数设定原则，如"准许收益率综合考虑企业加权平均资本成本、行业发展需要、用户承受能力等因素确定"、"最低负荷率根据实际负荷率、公平开放情况、其他运输方式替代程度等因素确定"，下一步在价格核定时，可以借鉴国外参数设定规则确定合理的准许收益率等参数。

（2）探索新的管输价格形式。目前价格机制明确与铁路对比的价格机制，为进一步促进成品油管道公平开放、提高管道使用效率，增强与铁路价格形式"基准价+运价率×运输距离"的可比性，解决管道短途价格显著低于铁路、难以回收管道成本的问题，建议在现行统一运价率模式之外，探索"管输基准价＝起步价+运价率×运距"价格模式，设置成品油管道运输起步价，充分发挥管道安全稳定、高效经济的优势。

（3）研究将管道弃置费用纳入定价成本。管道报废后需采用拆除、蒸汽吹扫、清洗、注浆等措施进行无害化处置，消除对环境和安全的不利影响，根据《企业产品成本核算制度——油气管网行业》等要求，油气管输企业近年来已开始计提弃置费用，确保未来管道处置时有充足的资金保障。建议借鉴加拿大做法，在以后定价及调整时，将弃置费用纳入定价成本，通过价格进行资金回收。

（4）做好成本监审和信息公开。为推动成品油管输价格形成机制尽早落地见效，建议参照国外管输企业主动申报、监管部门审核等成本监审方法和定价协调机制，尽快启动首个周期成本监审，成品油管输企业要规范成本核算，及时准确向监管部门提供成本监审数据和报表，配合监管部门高效完成成本监审。成本监审完成后，管输企业应在核定的最高准许收入下，细致梳理可比价格信息，加快制定具体价格协商方案，及时将与用户协商确定和调整管道运输价格情况向监管部门报告；若确难以协商一致的，各方应及时监管部门报告，并由监管部门协调确定，共同推动成品油行业稳定健康发展。同时，管输企业要做好信息报送和公开，在公司门户网站或指定平台向社会公开成品油管道的价格水平，促进社会各方及时获取准确运价信息。

参 考 文 献

[1] 国家管网北方管道有限责任公司. 世界管道概览（2020）[M]. 北京：石油工业出版社，2020：100-102.

[2] 梁严，郭海涛，周淑慧，等. 美国成品油管道公平开放现状及启示[J]. 油气储运，2019，38(06)：609-616.

[3] 张琦，高爱茹，徐强. 国外石油管道运输价格体系研究[J]. 国际石油经济，2006，14(7)：23-27.

[4] Federal Energy Regulatory Commission. Opinion No. 154-B[EB/OL]. (1985-06-28)[2024-12-03]. https://www.ferc.gov/sites/default/files/2020-04/opinion-154b_0.pdf.

[5] Federal Energy Regulatory Commission. Cost-of-Service Rates Manual[EB/OL]. (1999-06)[2024-12-03]. https://www.ferc.gov/sites/default/files/2020-08/cost-of-service-manual.pdf.

[6] Federal Energy Regulatory Commission. Market-based Ratemaking for Oil Pipelines(Final Rule)[EB/OL]. (1994-10-28)[2024-12-03]. https://www.govinfo.gov/content/pkg/FR-1994-11-16/html/94-27620.htm.

[7] Federal Energy Regulatory Commission. Order No. 561[EB/OL]. (1993-10-22)[2024-12-04]. https://www.ferc.gov/sites/default/files/2020-08/cost-of-service-manual.pdf.

[8] Canada Energy Regulator. Regulation of Pipelines and Power Lines. [EB/OL]. (2023-05-12)[2024-12-05]. https://www.cer-rec.gc.ca/en/about/who-we-are-what-we-do/responsibility/regulation-pipelines-power-lines.html.

[9] Canada Energy Regulator. Regulation of pipeline traffic, tolls and tariffs[EB/OL]. (2021-02-12)[2024-12-05]. https://www.cer-rec.gc.ca/en/about/who-we-are-what-we-do/responsibility/regulation-pipeline-traffic-tolls-tariffs.html.

[10] National Energy Board. RH-R-2-94[EB/OL]. (2009-10-08)[2024-12-05]. https://docs2.cer-rec.gc.ca/ll-eng/llisapi.dll/fetch/2000/90465/92833/553405/93101/554572/590939/590600/A24154%2D1_%E2%80%93_NEB_%E2%80%93_Reasons_for_Decision_%E2%80%93_Multi%2DClient_%E2%80%93_Review_of_RH%2D2%2D94_Cost_of_Capital_%E2%80%93_RH%2DR%2D2%2D.

[11] National Energy Board. 2010 Rate of Return on Common Equity (ROE) per Discontinued RH-2-94 Formula[EB/OL]. (2009-12-11)[2024-12-05]. https://docs2.cer-rec.gc.ca/ll-eng/llisapi.dll/fetch/2000/90465/92833/553405/93101/554572/586606/A1R0H8_%2D_2010_Rate_of_Return_on_Common_Equity_%28ROE%29_per_Discontinued_RH%2D2%2D94_Formula.pdf?nodeid=586607&vernum=-2.

[12] National Energy Board. Guidelines for Negotiated Settlements of Traffic, Tolls and Tariffs (Guidelines)[EB/OL]. (2002-06-12)[2024-12-05]. https://docs2.cer-rec.gc.ca/ll-eng/llisapi.dll/fetch/2000/90463/157025/208496/A02885%2D1_NEB_Decision_%E2%80%93_Guidelines_for_Negotiated_Settlements_of_Traffic%2C_Tolls_and_Tariffs_%28A0E4C1%29.pdf?nodeid=208497&vernum=-2.

[13] National Energy Board. Oil Pipeline Uniform Accounting Regulations[EB/OL]. (2020-03-16)[2024-12-05]. https://laws-lois.justice.gc.ca/eng/regulations/C.R.C.,_c._1058/index.html.

弯管相连腐蚀缺陷力学—电化学相互作用规律研究

张鹏[1]　赵明[2]　黄云飞[3]　许田[4]

(1. 西南石油大学土木工程与测绘学院；2. 西南石油大学石油与天然气工程学院；
3. 中石油西南油气田分公司安全环保与技术监督研究院；4. 西南石油大学机电工程学院)

摘　要　【目的】弯管作为管道系统的重要一环，外表面腐蚀缺陷存在会严重威胁管体安全。现有弯管腐蚀缺陷评估模型几乎仅适用含单一缺陷的弯管评估，因此对含复杂腐蚀缺陷弯管评估至关重要。【方法】为解决上述问题，建立三维(3D)多物理场耦合模型，研究含相连腐蚀缺陷X80弯管在近中性pH厌氧土壤环境中力学—电化学(M-E)相互作用规律，分析内压作用下含相连腐蚀缺陷弯管M-E相互作用规律和失效过程，研究缺陷几何形状变化对应力分布和电化学腐蚀进程影响规律。【结果】弯管内侧出现多个腐蚀缺陷且腐蚀缺陷在纵向相连时，会出现局部应力集中和M-E相互作用效应增强，进而促进腐蚀缺陷生长；与原生缺陷相比，附着缺陷表现出更高的应力集中和更快的腐蚀速度；M-E相互作用导致附着缺陷长度及边角处的腐蚀加剧，应力集中出现在缺陷长度处，最大局部应力出现在附着缺陷转角处，含相连腐蚀缺陷弯管在内压18MPa左右时发生爆裂失效；相连腐蚀缺陷的几何形状变化导致的机械和电化学参数变化基本一致。原生缺陷和附着缺陷的长度变化引起的M-E相互作用规律与缺陷深度变化时影响规律几乎相同。【结论】研究结果进一步完善含腐蚀缺陷弯管完整性评价模型，为管道系统完整性管理提供理论指导。

关键词　弯管；相连缺陷；多物理场耦合模型；腐蚀；有限元分析

Study onmechano-electrochemical interaction laws of connected corrosion defects on elbows

Zhang Peng[1]　Zhao Ming[2]　Huang Yunfei[3]　Xu Tian[3]

(1. School of Civil Engineering andGeomatics, Southwest Petroleum University;
2. School of Petroleum Engineering, Southwest Petroleum University;
3. School of Mechanical and Electrical Engineering, Southwest Petroleum University)

Abstract　[Objective] Elbows are important parts of pipeline system, corrosion defects on the external surface will seriously threaten the safety of pipe body. The existing corrosion defect evaluation models are almost only applicable to the evaluation of the elbow with a single defect, so it is very important for the evaluation of the elbow with complex corrosion defects. [Methods] In order to solve the above problems, a three-dimensional (3D) multi-physics field coupling model was developed to investigate the mechano-electrochemical (M-E) interaction of X80 elbows with connected corrosion defects in a near-neutral pH anaerobic soil environment. The M-E interaction law and failure process of elbows with connected corrosion defects under internal pressure was analyzed. The effect of defect geometry on stress distribution and electrochemical corrosion progression was studied. [Results] When multiple corrosion defects are present on the internal side of the elbow and are connected longitudinally, local stress concentration and M-E interaction effect are intensified, accelerating the growth of these defects. The attached defects exhibit higher stress concentration and a faster corrosion rate compared to the primary defects. The M-E interaction leads to increased corrosion at the defect's length and corners. Stress concentration is observed along the defect length, with the maximum local stress occurring at the attached defect corners. Failure of the elbow with connected corrosion defects occurs at an internal pressure of approximately 18MPa. The changes in mechanical and electrochemical parameters due to variations in the geometric shape of the connected defects are largely consistent. The M-E interaction law resulting from changes in the length of the primary and attached defects is similar to that caused by variations in defect depth. [Conclusion] The research results further improve the integrity evaluation model of elbows with corrosion defects, and provide theoretical guidance for the integrity management of pipeline system.

Key words elbow; connected defects; multi-physics field coupling model; corrosion; finite element analysis

油气管道腐蚀缺陷严重威胁管道安全平稳运行，腐蚀会导致管壁受损，甚至发生穿孔泄漏等危险，往往会造成经济损失，人员伤亡，环境污染等。与直管段不同，由于弯管具有一定的曲率，内压作用下弯管上应力分布不均，弯管的防腐涂层更容易脱落，导致弯管局部发生力学—电化学腐蚀。Gutman 理论认为机械应力加速了钢表面的腐蚀，而腐蚀缺陷的生长又反作用于结构的机械应力，被称为力学—电化学(M-E)协同作用。Xu 和 Cheng 基于该理论，将 M-E 效应概念引入管道腐蚀领域研究中。目前，M-E 相互作用概念作为应力腐蚀的基础研究方法已被广泛认可。

目前，用于腐蚀管道缺陷评估的标准和模型有很多，但这些标准和模型仅适用于直管段。近年来，对腐蚀弯管缺陷评估有一定研究。Kim 等人通过各种壁厚减薄几何形状以及不同的壁厚减薄位置对弯管破坏压力影响进行了实验研究。Shuai 等人研究了含单个腐蚀缺陷的弯管在近中性 pH 溶液中的力学—电化学(M-E)相互作用。但以上研究均是针对完整弯管或单缺陷弯管开展，而腐蚀缺陷在弯管上形状和排列方式各异，进行含复杂腐蚀缺陷弯管研究对管道系统完整性管理至关重要。根据 Shuai 等人和 Lee 等人的研究，发现弯管内侧的失效压力更低，含腐蚀缺陷时弯管内侧更容易失效，所以本文的腐蚀缺陷也建立在弯管内侧。

本文针对典型的 90°弯管，研究弯管相连腐蚀缺陷在不同内压、不同几何形状(长度和深度)下的 M-E 相互作用及对弯管影响，包括弯管失效压力、von Mises 应力和电化学腐蚀速率。通过参数变化的规律性研究，旨在为含复杂腐蚀缺陷的弯管完整性评价提供指导。

1 力学及电化学特性

1.1 管材力学性能及失效判据

本文采用各向同性硬化塑性的 X80 管材钢的本构关系模型。通过单轴拉伸试验得到的 X80 钢真实应力应变曲线(图 1)，杨氏模量(E)为 210GPa，泊松比(μ)为 0.3，屈服强度(σ_{ys})为 589MPa，极限抗拉强度(σ_u)为 786MPa。硬化函数(σ_{yhard})定义如式(1)所示：

$$\sigma_{yhard} = \sigma_{exp}\left(\varepsilon_p + \frac{\sigma_e}{E}\right) - \sigma_{ys} \quad (1)$$

式中，σ_{exp} 为拉伸试验获得的硬化函数；ε_p 为塑性应变；σ_e 为 von Mises 应力，MPa。

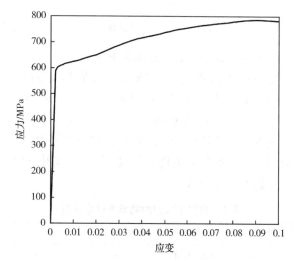

图 1 单轴拉伸试验得到的 X80 钢的真实应力—应变曲线

采用行业内通用的管道失效压力评估准则，认为当管道上某点最大 von Mises 应力达到管材极限抗拉强度时管道发生失效。该准则由于精度较高，适用于评估含有单一缺陷或复杂缺陷的管道，被广泛认可并应用于管道的失效压力预测。

1.2 电化学特性

假设弯管外表面防腐涂层脱落导致缺陷区暴露于近中性 pH 的土壤环境(NS4 溶液)中，在溶液中腐蚀表面发生的反应包括阳极铁的氧化反应和阴极析氢反应，如式(2)和(3)：

$$Fe \longrightarrow Fe^{2+} + 2e \quad (2)$$

$$2H^+ + 2e \longrightarrow H_2 \quad (3)$$

钢在溶液中腐蚀表面反应受活化控制，电化学动力学方程如式(4)-(6)：

$$i_a = i_{0,a} \exp\left(\frac{\eta_a}{b_a}\right) \quad (4)$$

$$i_c = i_{0,c} \exp\left(\frac{\eta_c}{b_c}\right) \quad (5)$$

$$\eta = \varphi - \varphi_{eq} \quad (6)$$

式中，i_a 和 i_c 分别为阳极和阴极电荷转移电流密度，A/cm²；$i_{0,a}$ 和 $i_{0,c}$ 分别为阳极和阴极交换电流密度，A/cm²；η_a 和 η_c 分别为阳极和阴极过电位，V；b_a 和 b_c 分别为阳极和阴极 Tafel 斜率，V/dec；φ 为电极电位；φ_{eq} 为平衡电极电位。

根据能斯特方程，阳极和阴极平衡电极电位如式（7）和（8）：

$$\varphi_{a,eq}=\varphi_{a,eq}^0+\frac{RT}{2F}\ln(\alpha_{Fe^{2+}}) \tag{7}$$

$$\varphi_{c,eq}=\varphi_{c,eq}^0+\frac{RT}{F}\ln(\alpha_{H^+}) \tag{8}$$

式中，$\varphi_{a,eq}^0$ 和 $\varphi_{c,eq}^0$ 分别为阳极和阴极标准平衡电位；R 为理想气体常数，其值为 8.314J/Kmol；T 为环境温度，其值为 298.15K；F 为法拉第常数，其值为 96485.34C/mol；$\alpha_{Fe^{2+}}$ 和 α_{H^+} 分别为 NS4 溶液中的亚铁离子和氢离子浓度。有限元模型的初始电化学参数（表1），其中 SCE 表示饱和甘汞电极。

表1 有限元模型的初始电化学参数

电化学反应	i_0(A/cm²)	b(V/dec)	φ_{eq}(V, SCE)
阳极	2.353×10⁻⁷	−0.207	−0.644
阴极	1.457×10⁻⁶	0.118	−0.859

1.3 力学—电化学耦合

管道钢结构的弹塑性变形和电化学的协同作用导致阳极反应的平衡电位变化如式（9）：

$$\varphi_{a,eq}=\varphi_{a,eq}^0-\frac{\Delta P_m V_m}{zF}-\frac{RT}{zF}\ln\left(\frac{v\alpha}{N_0}\varepsilon_p+1\right) \tag{9}$$

式中，ΔP_m 为引起弹性变形的超压，MPa，其值为钢结构屈服强度的 1/3；V_m 为钢的摩尔体积，其值为 7.13×10⁻⁶ m³/mol；z 为电荷数，其值为 2；v 为方向相关因子，其值为 0.45；α 为系数，其值为 1.67×10¹⁵ m⁻²；N_0 为初始位错密度，其值为 1×10⁸ cm⁻²。

2 有限元模型的建立

2.1 模型参数

利用有限元软件 COMSOL® 5.6 建立了含相连腐蚀缺陷的 90°弯管三维 M-E 相互作用模型（图2）。弯管外径 D 为 762mm，壁厚 t 为 11.72mm，弯曲半径 $R=5D$，为了解决非线性问题，选择 MUMPS 求解器进行求解，并采用 0.0001 的相对容差来保证数据收敛性和准确性。考虑到模型几何形状和载荷的对称性，同时为节省计算成本，对含相连腐蚀缺陷弯管的二分之一进行模拟。弯管上深度较大的缺陷为原生缺陷，参数包括长度 l_1 和深度 d_1。附着缺陷与原生缺陷纵向相连，深度小于原生缺陷深度，参数包括长度 l_2 和深度 d_2。为避免应力集中，在缺陷的边角处进行倒圆角设计（图3）。

2.2 边界条件

在固体力学和电化学边界条件设置中，弯管两端设为固定约束条件，环向截面设为对称约束条件，外表面设为自由边界条件，内表面设为电隔离边界条件，弯管内表面施加值为 p 的均匀压力载荷（图4）。

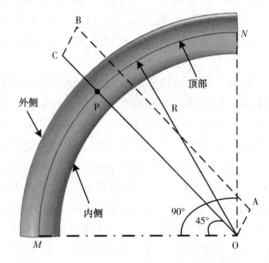

图2 弯管三维模型

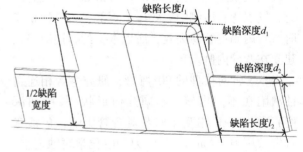

图3 相连腐蚀缺陷几何形状

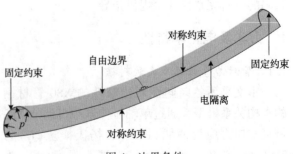

图4 边界条件

2.3 网格划分与验证

在网格划分中，为提高计算精度同时节省计算成本，在腐蚀缺陷及缺陷邻近区域进行局部网格细化处理，其余区域进行网格稀疏处理。由于二阶四面体单元适合用于处理复杂的三维几何图形，因此网格基于二阶四面体单元构建。根据网格无关性验证结果（图5），当元素个数超过70000后，单元数对最大von Mises应力和最大阳极电流密度的影响不再显著，得到腐蚀缺陷及缺陷邻近区域网格细化结果，最小网格尺寸为1mm，最大网格尺寸为10mm。

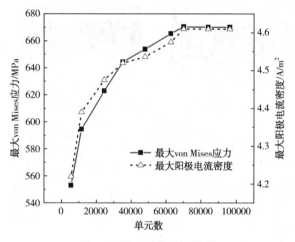

图5 网格无关性验证结果

2.4 模型验证

采用理论结果和模拟结果进行对比验证方法。由于弯管的特殊性，具有一定的曲率，导致在内压下应力分布不均匀，但在对称截面O-A-B-C与中心线M-N相交处存在一个特殊节点P，如图2所示。在该点处，不存在因应力分布不均而引起的额外变形，因此通常采用直管段的理论模型计算P点的环向应力来验证弯管的应力条件。直管段环向应力计算如式（10）：

$$\sigma_\theta = \frac{pD}{2t} \quad (10)$$

式中，σ_θ 为环向应力，MPa。

根据Gutman的研究，弹性应力对弯管腐蚀缺陷区阳极电流密度影响的理论结果如式（11）：

$$i'_a = i_a \exp\left(\frac{\Delta P V_m}{RT}\right) \quad (11)$$

式中，i_a 和 i'_a 分别为无应力钢和有应力钢的阳极电流密度，ΔP 为弹性变形阶段的超压，约为管道在内压作用下单轴拉伸应力的1/3，计算如式（10, 12, 13）：

$$\Delta P = (\sigma_\theta + \sigma_z)/3 \quad (12)$$

$$\sigma_z = \frac{pD}{4t} \quad (13)$$

式中，σ_z 为轴向应力，MPa。

表2列出了不同内压下位置P处理论计算和数值模拟得到的环向应力和阳极电流密度对比结果。可以看出，每种工况下模拟结果与理论值的差异均在5%以内，表明有限元模型是可靠的。

表2 模拟结果与理论值对比结果

内压/MPa		4	6	8	10	12
环向应力/MPa	理论值	130.04	195.05	260.07	325.1	390.1
	模拟值	128.15	191.45	257.66	319.79	383.78
	差异/%	−1.45	−1.85	−0.93	−1.63	−1.62
阳极电流密度/($\times 10^{-2}$ A/m^2)	理论值	2.663	2.923	3.21	3.525	3.87
	模拟值	2.542	2.78	3.067	3.446	3.747
	差异/%	−4.54	−4.89	−4.46	−2.24	−3.19

3 结果分析

3.1 内压对弯管的影响

研究了内压作用下相连腐蚀缺陷处von Mises应力及阳极电流密度分布（图6）和最大von Mises应力及最大阳极电流密度变化（图7）。可见，随着内压的增加，原生缺陷和附着缺陷内部应力水平逐渐增大，应力从原生缺陷和附着缺陷纵向边缘处逐渐向缺陷内扩展，应力集中出现在原生缺陷及附着缺陷纵向边缘处，最大局部应力出现在附着缺陷边角处。在内压18MPa时，附着缺陷转角处von Mises接近于极限抗拉强度786MPa，说明当附着缺陷与原生缺陷纵向相连时，当内压达到18MPa左右时弯管将失效。阳极电流密度（即腐蚀溶解速率）分布云图也呈现与von Mises应力相同的变化规律。最大von Mises应力和最大腐蚀速率均随着内压的增加而增大。值得注意的是18MPa时的最大阳极电流密度接近7.264×10^{-2}A/m^2，大约为未腐蚀区域的三倍，说明附着缺陷纵向边缘处腐蚀速率最快，管道管理人员应特别关注这一现象。

通过内压对含相连腐蚀缺陷弯管的研究，可以确定弯管上纵向相连腐蚀缺陷的失效压力和腐蚀过程。此外，随着内压的增加，附着缺陷纵向边缘处应力水平较大且腐蚀速率最快，弯管在此位置更容易发生局部壁厚减薄，甚至发生穿孔泄漏等问题，严重威胁弯管的安全性和可靠性。

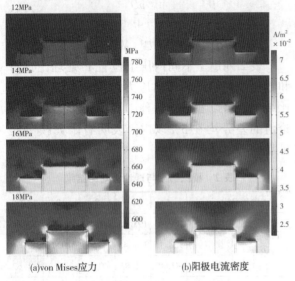

图6　内压作用下相连腐蚀缺陷处

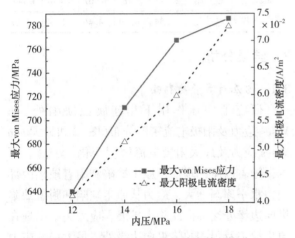

图7　内压对含相连腐蚀缺陷弯管的影响

3.2　缺陷几何形状对弯管的影响
3.2.1　缺陷长度

研究了原生缺陷长度变化时相连腐蚀缺陷处von Mises应力及阳极电流密度分布（图8）和最大von Mises应力及最大阳极电流密度变化（图9）。可见，原生缺陷长度变化对两个缺陷内部及周围的应力场和电化学腐蚀场同时产生影响，这种影响在附着缺陷上更加显著。随着原生缺陷长度的增加，应力和腐蚀从原生缺陷和附着缺陷纵向边缘处逐渐向各自内部延伸，原生缺陷和附着缺陷边角处应力和腐蚀向外扩展。应力集中出现在缺陷边角处和附着缺陷纵向边缘处，最大局部应力出现在附着缺陷边角处，在这里腐蚀速率也最快。最大von Mises应力和最大腐蚀速率与原生缺陷长度正相关的一致性较好，二者增长趋势相同。

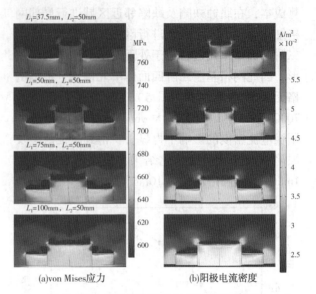

图8　原生缺陷长度变化时相连腐蚀缺陷处

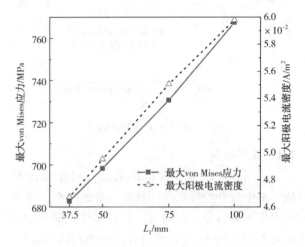

图9　原生缺陷长度对含相连腐蚀缺陷弯管的影响

研究了附着缺陷长度变化时相连腐蚀缺陷处von Mises应力及阳极电流密度分布（图10）和最大von Mises应力及最大阳极电流密度变化（图11）。可见，附着缺陷长度变化同样对原生缺陷产生影响。随着附着缺陷长度的增加，应力和腐蚀速率也在附着缺陷纵向边缘处增长。与原生缺陷长度增加变化规律相同，应力和腐蚀同样从原生缺陷和附着缺陷纵向边缘处逐渐向各自内部延伸，附着缺陷边角处应力和腐蚀向外扩展。应力集中和腐蚀出现在缺陷边角处和附着缺陷纵向边缘处，最大局部应力且腐蚀速率最快出现在附着缺陷边角处，但原生缺陷边角处应力几乎不发生变化。最大von Mises应力和最大腐蚀速率增长速率一致。

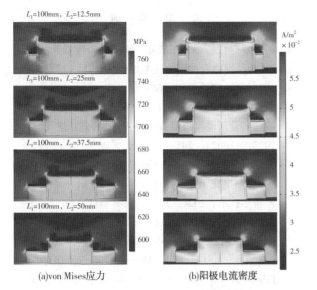

(a) von Mises应力　　　(b) 阳极电流密度

图 10　附着缺陷长度变化时相连腐蚀缺陷处

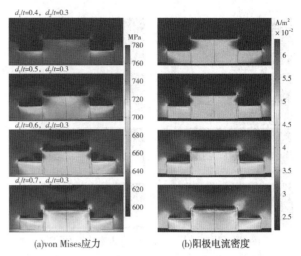

(a) von Mises应力　　　(b) 阳极电流密度

图 12　原生缺陷深度变化时相连腐蚀缺陷处

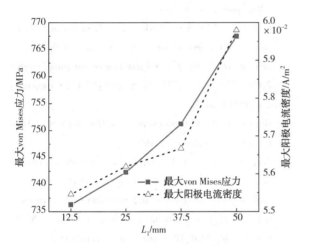

图 11　附着缺陷长度对含相连腐蚀缺陷弯管的影响

3.2.2　缺陷深度

研究了原生缺陷深度变化时相连腐蚀缺陷处 von Mises 应力及阳极电流密度分布（图 12）和最大 von Mises 应力及最大阳极电流密度变化（图 13）。可见，原生缺陷深度变化规律与原生缺陷长度变化规律相同，原生缺陷深度变化对附着缺陷影响显著。随着原生缺陷深度的增加，应力和腐蚀从原生缺陷和附着缺陷纵向边缘处逐渐向各自内部延伸，原生缺陷和附着缺陷边角处应力和腐蚀向外扩展。应力集中出现在原生缺陷边角处和附着缺陷纵向边缘处，最大局部应力且腐蚀最快出现在原生缺陷与附着缺陷相邻的边角处。在 $d_1/t = 0.6$ 以后最大 von Mises 应力和最大阳极电流密度增加变缓（图 13），说明 $d_1/t = 0.6$ 是一个临界值，管道管理者应特别注意原生缺陷深度达到壁厚的 60% 时弯管的情况。

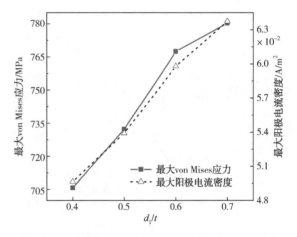

图 13　原生缺陷深度对含相连腐蚀缺陷弯管的影响

研究了附着缺陷深度变化时相连腐蚀缺陷处 von Mises 应力及阳极电流密度分布（图 14）和最大 von Mises 应力及最大阳极电流密度变化（图 15）。可见，附着缺陷深度变化规律与附着缺陷长度变化规律相同。随着附着缺陷深度增加，最大 von Mises 应力和最大腐蚀速率变化趋势相同（图 15），二者增长速率由慢变快在变慢，当 d_2/t 从 0.2 变化到 0.3 时，曲线斜率最大，这时应力水平和腐蚀速率骤增，管道管理者应特别注意附着缺陷深度达到弯管壁厚 20% 时的情况。

4　结论

建立了多物理场耦合模型，研究了含相连腐蚀缺陷 X80 弯管在碳酸氢盐土壤溶液中力学—电化学相互作用规律及对弯管的影响，得到如下结论：

（1）当弯管内侧出现多个腐蚀缺陷且腐蚀缺陷在纵向相连时，会导致局部应力集中和 M-E

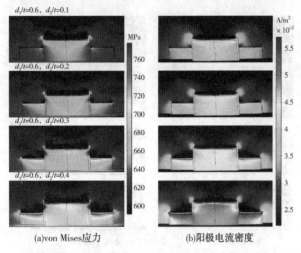

图14 附着缺陷深度变化时相连腐蚀缺陷处

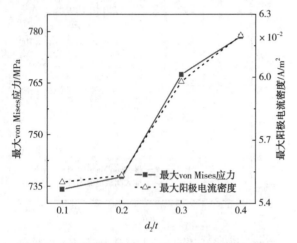

图15 附着缺陷深度对含相连腐蚀缺陷弯管的影响

相互作用效应增强。最大应力和腐蚀速率总是出现在附着缺陷纵向边缘处和附着缺陷边角处，由此产生的M-E相互作用导致腐蚀加剧，弯管附着缺陷上更容易发生失效，影响管道系统的完整性和可靠性。

（2）相连腐蚀缺陷处的局部应力集中和腐蚀溶解速率随着内压的增大而增大，附着缺陷比原生缺陷具有更高的应力和腐蚀速率。随着附着缺陷和原生缺陷长度和深度的增加，M-E相互作用同样在附着缺陷上更加显著。原生缺陷和附着缺陷的长度变化引起的M-E相互作用规律与深度变化时相同。

（3）弯管上腐蚀缺陷具有排列随机性和数量不确定性，本文仅针对X80弯管上相连腐蚀缺陷进行数值模拟研究，未来的研究建议围绕弯管上不同排列方式的腐蚀缺陷和不同数量的腐蚀缺陷开展，以丰富和完善管道系统完整性管理体系。

参考文献

[1] 肖述辉, 杜传甲, 王成军. 改进麻雀搜索算法优化BP神经网络管道腐蚀速率预测模型[J]. 油气储运, 2024, 43(07): 760-768+795.

[2] 袁文强, 郎宪明, 曹江涛, 蔡再洪, 郑浩. 基于声波法的管道泄漏检测技术研究进展[J]. 油气储运, 2023, 42(02): 141-151.

[3] 张鹏, 赵明, 罗梓洋, 许田. 弯管内外表面腐蚀缺陷力学-电化学相互作用规律研究[J]. 中国安全生产科学技术, 2024, 20(8): 104-111.

[4] ZHANG P, ZHAO M, LIU S M, XU T. Finite-Element Modeling of the Mechanoelectrochemical Interaction of Circumferentially Aligned Corrosion Defects on Elbows of Pipelines[J]. Journal of Pipeline Systems Engineering and Practice, 2025, 16(1). DOI: 10.1061/jpsea2.pseng-1630.

[5] KIM J W, LEE S H, PARK C Y. Experimental evaluation of the effect of local wall thinning on the failure pressure of elbows[J]. Nuclear Engineering and Design, 2009, 239(12): 2737-2746. DOI: 10.1016/j.nucengdes.2009.10.003.

[6] Gutman E M. Mechanochemistry of Materials[M]. UK: Cambridge Interscience. 1998.

[7] Xu L Y, Cheng Y F. Reliability and failure pressure prediction of various grades of pipeline steel in the presence of corrosion defects and pre-strain[J]. International Journal of Pressure Vessels and Piping, 2012, 89: 75-84. DOI: 10.1016/j.ijpvp.2011.09.008.

[8] SHUAI Y, WANG X H, LI J, WANG J Q, WANG T, HAN J Y, et al. Assessment by finite element modelling of the mechano-electrochemical interaction at corrosion defect on elbows of oil/gas pipelines[J]. Ocean Engineering, 2021, 234: 109228. DOI: 10.1016/j.oceaneng.2021.109228.

[9] SHUAI Y, ZHANG X, HUANG H, FENG C, CHENG Y F. Development of an empirical model to predict the burst pressure of corroded elbows of pipelines by finite element modelling[J]. International Journal of Pressure Vessels and Piping, 2022, 195: 104602. DOI: 10.1016/j.ijpvp.2021.104602.

[10] LEE G H, POURARIA H, SEO J K. Burst strength behaviour of an aging subsea gas pipeline elbow in different external and internal corrosion-damaged positions[J]. International Journal of Naval Architecture and Ocean Engineering, 2015, 7(3): 435-451. DOI: 10.1515/ijnaoe-2015-0031.

[11] WU Y, DU Z H, LI L Y, TIAN Z X. A new evalua-

tion method of dented natural gas pipeline based on ductile damage[J]. Applied Ocean Research, 2023, 135: 103533. DOI:10.1016/j.apor.2023.103533.

[12] XU L Y, CHENG Y F. Corrosion of X100 pipeline steel under plastic strain in a neutral pH bicarbonate solution[J]. Corrosion Science, 2012, 64: 145-152. DOI:10.1016/j.corsci.2012.07.012.

[13] 李明, 李秉军, 何永志, 王雷, 梁昌晶. 某埋地碳钢管道腐蚀失效分析[J]. 焊管, 2021, 44(08): 30-35.

[14] YANG Y, CHENG Y F. Stress Enhanced Corrosion at the Tip of Near-Neutral pH Stress Corrosion Cracks on Pipelines[J]. Corrosion, 2016, 72(8): 1035-1043. DOI:10.5006/2045.

[15] Bagotsky V S. Fundamentals of electrochemistry[M]. United States: Wiley-Interscience. 2006.

[16] XU L Y, CHENG Y F. Development of a finite element model for simulation and prediction of mechano-electrochemical effect of pipeline corrosion[J]. Corrosion Science, 2013, 73: 150-160. DOI:10.1016/j.corsci.2013.04.004.

[17] QIN G J, CHENG Y F. A review on defect assessment of pipelines: Principles, numerical solutions, and applications[J]. International Journal of Pressure Vessels and Piping, 2021, 191: 104329. DOI:10.1016/j.ijpvp.2021.104329.

[18] WANG Y C, XU L Y, SUN J L, CHENG Y F. Mechano-electrochemical interaction for pipeline corrosion: A review[J]. Journal of Pipeline Science and Engineering, 2021, 1(1): 1-16. DOI:10.1016/j.jpse.2021.01.002.

基于改进门控循环单元的原油储罐关键参数预测方法

刘鹏涛

(国家管网集团甘肃分公司)

摘要 传统的原油储罐关键参数预测方法在处理多因素交互和复杂非线性关系时存在一定局限性，尤其在面对大量连续、重复数据时，容易造成计算资源浪费并影响模型性能。本文提出一种基于改进门控循环单元神经网络的高效预测方法(I-GRU)。通过引入数据去重和降噪策略，显著减少冗余信息，从而提高计算效率并避免了过拟合。结合动态学习率调整机制，提出改进控循环单元网络结构，进一步增强模型的泛化能力和预测精度。实验结果表明，与传统回归模型相比，所提方法在预测精度、计算效率和实时性方面均表现出显著优势，能够有效满足原油储罐实时监控对快速响应和高精度预测的需求，为原油储罐压力预测提供了创新的解决方案，具有重要的应用价值。

关键词 原油储罐，I-GRU 神经网络，时序预测，数据去重，实时监控，深度学习

Key Parameter Prediction Method for Crude Oil Storage Tanks Based on Improved Gated Recurrent Unit

Liu Pengtao

(PipeChina Gansu Branch)

Abstract Traditional methods for predicting key parameters of crude oil storage tanks exhibit certain limitations in handling multi-factor interactions and complex nonlinear relationships. Particularly when dealing with large volumes of continuous and repetitive data, these methods often lead to wasted computational resources and degraded model performance. This paper proposes an efficient prediction method based on an improved gated recurrent unit neural network (I-GRU). By introducing data deduplication and denoising strategies, the method significantly reduces redundant information, thereby improving computational efficiency and avoiding overfitting. Additionally, a dynamic learning rate adjustment mechanism is integrated to enhance the network structure of the gated recurrent unit, further boosting the model's generalization ability and prediction accuracy. Experimental results demonstrate that, compared to traditional regression models, the proposed method achieves significant improvements in prediction accuracy, computational efficiency, and real-time performance. It effectively meets the demands of real-time monitoring for crude oil storage tanks, providing a rapid response and high-precision prediction solution. This innovative approach offers important applications in crude oil storage tank pressure prediction.

Key words crude oil tank; I-GRU neural network; time-series prediction; data deduplication; real-time monitoring; deep learning

1 引言

随着全球能源需求的不断增长，原油储罐作为石油工业的重要基础设施，其运行状态的监测和分析对确保生产安全、提升运营效率以及延长设备使用寿命至关重要。原油储罐的关键参数（如压力、温度、液位等）直接影响其正常运行和安全性，因此，准确预测和评估这些参数成为

石油行业亟待解决的重要问题。

传统的原油储罐压力预测方法多依赖于物理建模和统计回归模型，这些方法通常通过模拟与公式推导来建立储罐压力与各个变量之间的关系。然而，物理建模方法过于依赖理想化假设，难以应对多因素复杂交互的实际情况；而统计回归方法则忽略了数据中潜在的高维交互和非线性特征，无法有效捕捉复杂的动态变化。随着储罐运行环境变得日益复杂，传统方法的局限性逐渐显现，尤其是在处理大量实时监控数据时，难以满足高精度和快速响应的要求。

近年来，随着人工智能和大数据技术的飞速发展，基于机器学习和深度学习的预测方法逐渐成为解决这一问题的有效手段。尤其是循环神经网络（RNN）及其变种，如长短时记忆网络（LSTM）和门控循环单元（GRU），在处理复杂时序数据和多因素交互问题时表现出了显著优势。尽管 LSTM 能有效捕捉长时间序列中的依赖关系，但其计算复杂度较高，尤其在实时预测场景中，可能不适合高频、低延迟的要求。与之相比，GRU 通过简化网络结构并提高计算效率，在许多时序预测任务中表现出更优的性能。

目前，基于深度学习的原油储罐压力预测方法已取得了一定的进展。例如，Wu 等使用 GRU 模型对油气储罐的压力进行了有效预测，验证了 GRU 在处理复杂时序数据时的优势。Tang 等提出了基于神经图网络和门控循环单元（GRU）的可解释多变量时间序列异常检测方法，实验结果表明，GRU 在面对非线性和多因素影响的复杂数据时，比传统回归方法表现出更高的预测精度和效率。Selbekk 等提出的 GRU 混合模型通过集成学习进一步提高了预测精度。以上研究表明，深度学习方法能在处理多维度、非线性时序数据时，充分捕捉储罐压力与温度、液位、流量等多种因素之间复杂的动态关系，提供了更为精准、高效的解决方案。

然而，现有研究仍面临若干挑战。首先，原油储罐压力受多种因素的影响，如储罐内外部温度、液位、环境压力和气体流量等，且这些因素之间的相互作用复杂，具有明显的非线性特征。其次，当处理大量连续、重复的数据时，深度学习模型可能会浪费计算资源，并影响实验效果。冗余信息容易导致模型收敛速度变慢，甚至出现过拟合现象。此外，如何进一步提升深度学习模型在复杂工业环境中的泛化能力，尤其是在实时监控和故障预警中的应用，依然是一个亟待解决的问题。

为此，本文提出了一种改进门控循环单元的原油储罐关键参数预测方法（An Improved Gated Recurrent Unit Method for Predicting Key Parameters of Oil Storage Tanks，I-GRU），旨在解决现有方法在处理大量冗余数据时的效率问题，并提高预测精度。具体来说，本文在数据预处理阶段结合数据去重和降噪技术，有效消除冗余信息，进一步提高计算效率，并降低了过拟合风险。同时优化 GRU 网络结构，通过引入动态学习率调整机制，以增强模型的泛化能力和预测精度。实验结果表明，所提出的方法在预测精度、计算效率和实时性方面，均显著优于传统回归模型，能够满足原油储罐实时监控系统对快速响应和高精度预测的需求。本文的研究为原油储罐压力预测提供了创新的解决方案，具备重要的实际应用价值

2 I-GRU 模型设计

本文提出了一种改进的门控循环单元（I-GRU）模型，用于原油储罐关键参数的预测，旨在通过深度学习技术克服传统方法在处理复杂非线性时序数据时的局限性。I-GRU 模型通过自适应学习数据中的时序依赖性，能够准确捕捉原油储罐压力的动态变化，从而提高预测精度。如图 1 所示，该方法的主要流程包括以下几个步骤：（1）数据预处理：对原始数据进行缺失值填补、异常值剔除、时间窗口去重和标准化处理，以确保输入数据的质量和一致性。（2）模型设计：设计 I-GRU 模型，通过引入更新门和重置门机制，控制信息流动与遗忘，有效捕捉时序数据中的长期依赖关系。（3）训练与评估：使用处理后的数据进行模型训练，采用均方误差（MSE）和平均绝对误差（MAE）等指标进行评估，确保模型在原油储罐压力预测中的准确性和稳定性。

2.1 数据预处理

在进行原油储罐压力预测之前，首先对原始数据进行预处理。数据集来源于某石油公司提供的实际监控数据，包含储罐的液位、温度、流量、外部气压等多个与压力相关的因素。预处理步骤包括以下几个方面：

缺失值处理：对于数据集中的缺失值，采用

线性插值法填补缺失数据点,以确保数据的完整性和一致性。线性插值法通过连接缺失数据点前后的已知数据点来估算缺失值,从而减少因缺失值带来的数据偏差。

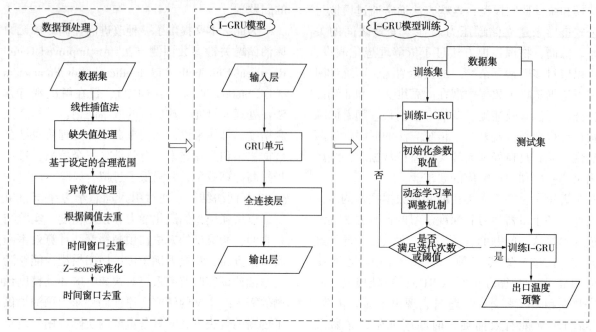

图 1 算法流程图

异常值剔除:通过对传感器采集的数据进行异常值检测,剔除那些偏离正常范围的异常值。异常值剔除方法基于设定的合理范围,如根据每个特征的历史数据分布来识别异常数据,从而减少噪声对模型训练的干扰。

时间窗口去重:为消除冗余数据对模型训练的影响,采用时间窗口方法进行数据去重。具体地,设定一个时间窗口 w,在每个时间窗口内选择一个代表性数据点。如果相邻数据点的差异小于预设的阈值 ϵ,则认为该数据点为冗余数据,予以去除。公式表示为:

$$if\ |x_i-x_{i+1}|<\epsilon\ then\ remove\ x_{i+1} \tag{1}$$

其中,$|x_i-x_{i+1}|$ 为相邻数据点之间的差异,若该差异小于阈值 ϵ,则认为该数据点冗余。通过此方法,显著减少冗余数据,提升模型训练效率,避免冗余信息对模型的负面影响。

标准化处理:采用 Z-score 标准化方法,使每个特征的均值为 0,方差为 1,从而保证数据的统一性,有助于提升模型的收敛速度,避免不同尺度的特征对模型训练产生不均衡的影响。

$$z_i=\frac{x_i-\mu}{\sigma} \tag{2}$$

其中:z_i 为标准化后的数据值;x_i 为原始数据值;μ 为该特征的均值;σ 为该特征的标准差。

2.2 I-GRU 模型设计

I-GRU(Gated Recurrent Unit)是门控神经网络(GRU)结构的变种,依据更新门和重置门来控制信息流动和遗忘,相较于传统 GRU,I-GRU 在长序列训练中具有更高的效率和较低的计算复杂度。本文中,I-GRU 模型包含以下几个主要部分:

输入层:输入层接收来自预处理后的数据集,包括温度、液位、流量和外部气压等特征变量。每个时间步的输入是一个多维特征向量。

GRU 单元:GRU 单元通过重置门和更新门控制信息的流动与遗忘,从而有效捕捉时间序列中的长期依赖信息。具体地:

重置门(Reset Gate):决定了当前输入信息对上一时间步隐藏状态的影响,公式如下:

$$\boldsymbol{r}_t=\sigma(\boldsymbol{W}_r\cdot[\boldsymbol{h}_{t-1},\boldsymbol{x}_t]+\boldsymbol{b}_r) \tag{3}$$

其中,$\boldsymbol{r}_t$ 为重置门;$\boldsymbol{W}_r$ 是权重矩阵;$\boldsymbol{b}_r$ 是偏置项;$\boldsymbol{h}_{t-1}$ 是上一个时间步的隐藏状态;$\boldsymbol{x}_t$ 是当前时间步的输入;σ 是 Sigmoid 激活函数。

更新门(Update Gate):决定当前隐藏状态与过去状态的结合程度,公式如下:

$$\boldsymbol{z}_t=\sigma(\boldsymbol{W}_z\cdot[\boldsymbol{h}_{t-1},\boldsymbol{x}_t]+\boldsymbol{b}_z) \tag{4}$$

其中,$\boldsymbol{z}_t$ 为更新门;$\boldsymbol{W}_z$ 是权重矩阵;$\boldsymbol{b}_z$ 是偏置项。

候选隐藏状态(Candidate Hidden State):候

选隐藏状态通过重置门和当前输入计算得到，公式如下：

$$\tilde{h}_t = \sigma(W_h \cdot [r_t \odot h_{t-1}, x_t] + b_h) \quad (5)$$

其中，$\tilde{h}_t$为候选隐藏状态；$\odot$表示按元素相乘。

当前隐藏状态（Current Hidden State）：当前隐藏状态由更新门和候选隐藏状态加权得到，公式如下：

$$h_t = (1-z_t) \odot \tilde{h}_t + z_t \odot h_{t-1}) \quad (6)$$

其中，h_t为当前隐藏状态；z_t是更新门；$\tilde{h}_t$是候选隐藏状态；h_{t-1}是上一个时间步的隐藏状态。

全连接层：GRU单元的输出通过全连接层进行线性变换，将高维特征映射到预测值空间，以获得储罐压力的最终预测值。

输出层：输出层由一个神经元组成，用于预测原油储罐的压力值，输出为连续型数值。

2.3 动态学习率调整机制

为了提高模型的训练效率并避免在训练过程中出现过拟合或梯度消失等问题，本文采用了指数衰减和基于验证集误差的动态调整两种学习率调整机制。

指数衰减：在训练过程中，学习率随迭代次数的增加而逐步降低，从而使模型在后期训练时能够更加精细地调整参数。具体的学习率调整公式如下：

$$\eta_t = \eta_0 \cdot e^{-\lambda t} \quad (7)$$

其中，η_t表示第t次迭代时的学习率；η_0为初始学习率；λ为衰减率（控制衰减速度）。

基于验证集误差的调整：当验证集的损失在一定轮次内未出现显著改善时，模型将自动降低学习率，从而避免在训练过程中发生过拟合。具体的调整公式如下：

$$\eta_t = \eta_0 \cdot \left(1 + \lambda \cdot \frac{val\_loss_t - val\_loss_{t-1}}{val\_loss_{t-1}}\right) e^{-\lambda t} \quad (8)$$

其中，η_t为当前学习率；η_0是初始学习率；λ是调整因子（通常设为0.1）；val_loss_t和val_loss_{t-1}分别为当前轮次和上一轮次的验证集损失。

3 案例分析

为了验证基于I-GRU模型的原油储罐关键参数预测方法的有效性，本文进行了大量的实验，涵盖了不同数据集的训练与测试，以及与传统回归模型和其他深度学习模型的对比实验。实验的主要目标是评估I-GRU模型在预测精度、计算效率和实时性方面的优势，验证其在复杂时序数据处理中的适用性。

3.1 数据集

实验所使用的数据集来自于某石油公司提供的原油储罐监控系统。该系统包含多个原油储罐的历史数据，涵盖温度、液位、流量、外部气压等多个特征。数据集记录了过去三年内的小时级时序数据，数据量达到10万条，包含不同工作状态下的储罐压力变化。这些数据集不仅包含了常规工作状态下的数据，还包括了一些异常工况下的记录，具有较高的代表性和广泛的适用性。如表1所示，数据集中的每条记录包括以下特征。

表1 数据记录特征

数据特征	数据描述
温度/℃	储罐内部的温度数据
液位/m	储罐内的液位信息
流量/(m³/h)	储罐的进出口油气流量
外部气压/Pa	储罐外部的环境气压
压力/Pa	储罐内部的压力数据（目标变量）

3.2 模型训练与评估方法

为了验证I-GRU模型的有效性，本文将数据集随机划分为训练集、验证集和测试集，其中训练集占比70%，验证集占比15%，测试集占比15%。训练集用于模型的训练，验证集用于调整超参数，测试集用于评估模型的性能。为了全面评估模型的预测性能，本文采用以下两种常用的回归评估指标：

均方误差（MSE，Mean Squared Error）：用于衡量模型预测值$\hat{s}_i^z$与真实值s_i^z之间的差异，MSE值越小，表示模型的预测精度越高。

$$MSE = \sqrt{\frac{\sum_z (\hat{s}_i^z - s_i^z)^2}{z}} \quad (9)$$

平均绝对误差（MAE，Mean Absolute Error）：通过计算预测值$\hat{s}_i^z$与真实值s_i^z的绝对误差的平均值，MAE同样用于衡量预测精度。

$$MAE = \frac{\sum_z |\hat{s}_i^z - s_i^z|}{z} \quad (10)$$

使用 Adam 优化器对 I-GRU 模型进行训练，采用批量梯度下降(Batch Gradient Descent)策略，以提高训练的稳定性和效率。训练过程中使用的学习率为 0.001，训练周期设置为 50 轮，批量大小为 64。在训练过程中，采用验证集来调整模型的超参数。具体而言，选择了不同的 GRU 单元数量和层数进行实验，最终选择了两层 GRU，每层 256 个单元的结构。为了防止过拟合，模型训练过程中还引入了 Dropout 正则化，Dropout 率设为 0.2。

3.3 对比模型

为了全面评估 I-GRU 模型的性能，本研究选择了以下几种常见的回归模型进行对比实验。传统回归模型(线性回归、支持向量回归(SVR))、LSTM(长短时记忆网络)、GRU(门控循环单元)、GRU-M(GRU 混合模型)。所有模型在相同的数据集上进行训练和测试，采用相同的评估指标，以确保对比的公平性。

实验组 1：GRU 与传统回归模型对比

在第一个实验组中，比较了 I-GRU 模型与传统线性回归模型和支持向量回归(SVR)模型在预测精度上的差异。预测精度使用均方误差(MSE)和平均绝对误差(MAE)作为评估指标。实验结果如表 2 所示：

表 2　I-GRU 与传统回归模型的对比

模型	MSE	MAE
线性回归	0.245	0.358
支持向量回归(SVR)	0.211	0.319
GRU(本研究)	0.128	0.213

从表 2 可以看出，I-GRU 模型在预测精度上明显优于线性回归和 SVR 模型，MSE 和 MAE 分别降低了 47.8% 和 40.5%。这表明，所提模型能够有效捕捉时序数据中的非线性关系，从而提高了预测精度。

实验组 2：I-GRU 与 LSTM 对比

在第二个实验组中，本文对比了 I-GRU 与 LSTM 模型在原油储罐压力预测中的表现，并进一步对比了 GRU(门控循环单元)和 GRU-M(GRU 混合模型)的性能。LSTM 模型采用与 GRU 相同的网络结构和训练设置。实验结果如表 3 所示：

表 3　I-GRU 与 LSTM 模型的对比

模型	MSE	MAE	训练时间/分钟
I-GRU	0.128	0.213	45
LSTM	0.147	0.238	58
GRU	0.135	0.220	47
GRU-M	0.125	0.210	50

从表 3 可以看出，所提模型在预测精度上优于 LSTM 模型，MSE 和 MAE 分别降低了 12.9% 和 10.5%。同时，也在计算效率上优于 LSTM，训练时间减少了 22.4%。相比之下，GRU(门控循环单元)的性能略逊色于所以模型，MSE 和 MAE 分别增加了 5.4% 和 3.3%。而 GRU-M(GRU 混合模型)在精度上略优于所提模型，MSE 和 MAE 分别降低了 2.4% 和 1.4%，但训练时间相对增加了 10%。这些结果表明，所提模型在预测精度和计算效率之间取得了较好的平衡，GRU-M 则在精度上稍有提升，但计算效率有所下降。

实验组 3：不同特征组合对 I-GRU 性能的影响

为了评估不同输入特征组合对 I-GRU 模型性能的影响，本文进行了多组实验，分别使用不同的特征组合进行训练与预测。实验组包括：使用单一特征(温度)训练 I-GRU。使用两种特征(温度、液位)训练 I-GRU。使用三种特征(温度、液位、流量)训练 I-GRU。使用所有特征(温度、液位、流量、外部环境压力)训练 I-GRU。实验结果如表 4 所示：

表 4　不同特征组合对 I-GRU 性能的影响

特征组合	MSE	MAE
仅温度	0.182	0.276
温度+液位	0.155	0.245
温度+液位+流量	0.136	0.221
温度+液位+流量+环境压力	0.128	0.213

从表 4 可以看出，随着输入特征的增加，所提模型的预测精度逐渐提高。使用所有特征组合时，所提模型的 MSE 和 MAE 分别降低了 29.8% 和 22.9%。这表明，更多的输入特征能够提高 I-GRU 模型更全面地捕捉储罐压力变化的影响因素，从而提升预测性能。

实验组 4：I-GRU 模型的稳定性与泛化能力分析

为了评估所提模型的稳定性和泛化能力，本

文采用不同的随机种子进行多次训练。每次实验使用不同的随机初始化，对模型进行训练，并记录每次训练的 MSE 和 MAE。实验结果图 2 所示：

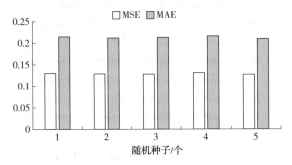

图 2　I-GRU 模型的稳定性与泛化能力

实验结果表明，所提模型在不同的随机初始化下，预测结果非常一致，MSE 和 MAE 的标准差分别为 0.002 和 0.003，显示了模型的高稳定性和较强的泛化能力。

实验组 5：不同训练周期对 I-GRU 模型性能的影响

为了研究不同训练周期对所提模型性能的影响，本文设置了 5 个不同的训练周期（10、20、30、40、50 轮），并评估每个周期的 MSE 和 MAE。实验结果如图 3 所示：

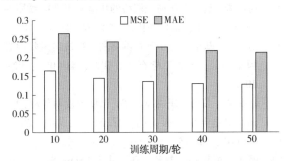

图 3　不同训练周期对 I-GRU 模型性能的影响

从表 5 可以看出，随着训练周期的增加，所提模型的预测精度逐渐提高。50 个训练周期时，模型的 MSE 和 MAE 分别达到了最优值。表明 I-GRU 模型能够通过充分训练，逐步优化参数，提高预测精度。

3.4　结果讨论

通过多组实验结果分析可以得出以下几点结论：

（1）I-GRU 优于传统回归模型和 SVR：所提模型 I-GRU 在处理原油储罐压力预测问题时，能够有效捕捉时序数据中的复杂非线性关系，表现出显著的预测精度优势。

（2）I-GRU 与 LSTM（长短时记忆网络），GRU（门控循环单元），GRU-M（GRU 混合模型）对比：尽管 LSTM、GRU、GRU-M 在某些长时序数据的建模上有所优势，但所提模型在精度和计算效率上的综合表现更佳，尤其适合实时监控应用。

（3）特征选择的影响：增加输入特征能够显著提升 I-GRU 模型的预测精度，表明更全面的特征能够帮助模型更好地捕捉储罐压力变化的规律。

（4）模型稳定性与泛化能力：所提模型 I-GRU 在不同的随机种子和训练周期下表现出高度稳定性和良好的泛化能力，适合在实际应用中进行大规模部署。

综上所述，I-GRU 模型在原油储罐压力预测任务中展现了较高的精度、计算效率以及稳定性，具有良好的应用前景。

4　结论与展望

本文提出了一种基于改进门控循环单元神经网络的高效预测方法（I-GRU），并通过实验验证了该方法的有效性。实验结果表明，所提模型在处理多因素、非线性时序数据方面具有显著优势，相较于传统回归模型和支持向量机（SVR），I-GRU 在预测精度和计算效率上表现更为出色，能够准确捕捉储罐压力的动态变化。与 LSTM（长短时记忆网络），GRU（门控循环单元），GRU-M（GRU 混合模型）相比，I-GRU 通过简化结构有效降低了计算复杂度，同时保持了较高的预测性能，适应了实时监控系统的需求。通过多维度特征（如温度、液位、流量等）的综合建模，I-GRU 能够较好地捕捉各因素间的时序依赖关系，进一步提升了预测精度。尽管取得了较好的实验效果，未来的研究可通过多模型融合进一步提升预测精度，并考虑更多外部环境因素的影响，以进一步优化模型表现。

参 考 文 献

[1] 孙东亮，蒋军成，张明广，等. 基于最大熵原理的卧罐爆炸碎片数量概率分布[J]. 化工学报，2011，62(S1)：219-224.

[2] 贾文龙，肖欢，冷翔宇，等. 原油储罐重质沉积物超声波空化微射流清洗实验及数值模拟[J/OL]. 化工学报，1-14[2024-12]

[3] 张启波，张护国，谭清磊，等. 油罐池火灾热辐射临界距离研究[J]. 安全与环境学报，2021，21

(06): 2533-2540.

[4] 姜良芹, 尉晨煜, 林钰博, 等. 锚固式方型储液罐动力特性与基频解析解[J]. 低温建筑技术, 2021, 43(11): 93-97.

[5] Cho K, Van Merrienboer B, Gulcehre C, et al. "Learning phrase representations using RNN encoder-decoder for statistical machine translation." Proceedings of the Conference on Empirical Methods in Natural Language Processing(EMNLP), 2014: 1724-1734.

[6] Khan S, Kumar V. "A novel hybrid GRU-CNN and residual bias(RB) based RB-GRU-CNN models for prediction of PTB Diagnostic ECG time series data." Biomedical Signal Processing and Control, 94 (2024): 106262.

[7] Sheikh R, Gupta V K, Yadav T, et al. "Temporal Dependency Analysis in Predicting RUL of Aircraft Structures Using Recurrent Neural Networks." In: Fracture Behavior of Nanocomposites and Reinforced Laminate Structures, Cham: Springer Nature Switzerland, 2024: 329-361.

[8] Qin C, Shi G, Tao J, et al. "RCLSTMNet: A Residual-convolutional-LSTM Neural Network for Forecasting Cutterhead Torque in Shield Machine." International Journal of Control, Automation and Systems, 22 (2) (2024): 705-721.

[9] Prabakar S. "Strategic Integration for Future Selection-LSTM Stock Prediction Algorithm based on the Internet of Things(IoT)." In: 2024 1st International Conference on Advanced Computing and Emerging Technologies (ACET), IEEE, 2024: 1-6.

[10] Qin C, Chen L, Cai Z, et al. "Long short-term memory with activation on gradient." Neural Networks, 164(2023): 135-145.

[11] Cahuantzi R, Chen X, Güttel S. "A comparison of LSTM and GRU networks for learning symbolic sequences." In: Science and Information Conference, Cham: Springer Nature Switzerland, 2023: 771-785.

[12] Hou X, Ge F, Chen D, et al. "Temporal distribution-based prediction strategy for dynamic multi-objective optimization assisted by GRU neural network." Information Sciences, 649(2023): 119627.

[13] Wu J, Tao Y, Wang X "A Comparative Study of Machine Learning Based Tank Pressure Prediction for Ships." In: 2024 International Conference on Artificial Intelligence and Power Systems (AIPS), IEEE, 2024.

[14] Tang C, Xu L, Yang B, et al. "GRU-based interpretable multivariate time series anomaly detection in industrial control system." Computers & Security 127 (2023): 103094.

[15] Selbekk A, Indrevoll P S. "Prediction of Tank Parameters in LNG Carriers Using a Hybrid Modeling Approach." MS thesis, NTNU, 2024.

[16] Owa K O, Sharma S K, Sutton R. "Optimised multivariable nonlinear predictive control for coupled tank applications." (2013): 14-14.

[17] 苏怀, 张劲军. 天然气管网大数据分析方法及发展建议[J]. 油气储运, 2020, 39(10): 1081-1095. 8136A 4 3 7

[18] 刘鹏涛. 基于循环神经网络的压缩机组性能预测模型[J]. 石油化工自动化, 2024, 60(01): 6-12.

[19] 刘鹏涛. 基于图卷积神经网络的压缩机组风险预警模型[J]. 天然气与石油, 2023, 41(05): 92-100.

[20] 赵继飞, 苏扬, 王海堂, 等. 温度湿度对储罐油面最高静电电位的影响[J]. 石油化工设备, 2011, 40(06): 12-15.

[21] 程迪, 黄松岭, 赵伟, 等. 基于PSO-LS-SVM的储罐底板缺陷量化方法研究[J]. 电测与仪表, 2018, 55(04): 87-92.

内压-缺陷尺寸耦合下管道极限弯曲应变特性分析

裴迎举　王　聪　凌瑜基　蒋程晨　薛喆中

（成都理工大学）

摘　要　（目的）管道运输是国家能源供给革命中重要的一环，管道的安全性能备受瞩目。本研究就内压-缺陷尺寸耦合作用下含缺陷管道极限弯曲应变特性进行探究。（方法）针对目前多聚焦于缺陷管道爆破压力而缺乏综合各类失效因素分析的问题，本研究充分考虑内压-缺陷尺寸耦合作用，基于所建立的三维仿真管道实体域模型进行多重载荷下缺陷管道极限弯曲应变特性的多维度综合分析。主要探究管道径厚比，缺陷长度，宽度，高度及组合缺陷对管道极限弯曲应变的影响。（结果）研究结果显示：管道径厚比与管道极限弯曲应变呈负相关，但在径厚比约为75处出现驼峰；缺陷长度，宽度，深度的增长引起极限弯曲应变的降低，且降幅呈现出先大后小的现象，同时三种缺陷尺寸中，深度系数变化引起应变变化的方差为长度系数的3倍，宽度系数的30倍，影响权重更大；内压-缺陷尺寸耦合作用时，组合缺陷长度，宽度尺寸大小的比值较小时管道极限弯曲应变与内压载荷为正相关，随着比值增大，相关性由正相关变为负相关。同时提出BP神经网络应用于缺陷管道极限弯曲应变的预测，预测值与计算值的最大误差仅为7.18%。（结论）研究结果可在管道日常检修与维护中为运行中含缺陷管道的适用性评价提供理论指导，同时利用BP神经网络预测模型可助力降低管道发生爆破的风险。

关键词　多重载荷；缺陷管道；极限弯曲应变；数值模拟；内压-缺陷尺寸耦合

Analysis of ultimate bending strain characteristics of pipelines under coupling of internal pressure and defect size

Pei Yingju[1]　Wang Cong[1]　Ling Yuji[1]　Jiang Chengchen[1]　Xue Zhezhong[2]

(1. School of Energy, Chengdu University of Technology;
2. School of Mechanical and Electrical Engineering, Chengdu University of Technology)

Abstract　[Objective] Pipeline transportation is an important part of the national energy supply revolution, and the safety performance of pipelines has attracted much attention. This study explores the ultimate bending strain characteristics of defective pipelines under the coupling effect of internal pressure and defect size. [method] In response to the current focus on the burst pressure of defective pipelines and the lack of comprehensive analysis of various failure factors, this study fully considers the coupling effect of internal pressure and defect size. Based on the established three-dimensional simulation pipeline solid domain model, a multidimensional comprehensive analysis of the ultimate bending strain characteristics of defective pipelines under multiple heavy loads is carried out. The main focus is on exploring the effects of pipeline diameter thickness ratio, defect length, width, height, and combined defects on the ultimate bending strain of pipelines. [Results] The research results show that the diameter to thickness ratio of the pipeline is negatively correlated with the ultimate bending strain of the pipeline, but a hump appears at a diameter to thickness ratio of about 75; The increase in defect length, width, and depth causes a decrease in the ultimate bending strain, and the decrease shows a phenomenon of first increasing and then decreasing. At the same time, among the three defect sizes, the variance of strain change caused by the change in depth coefficient is 3 times that of length coefficient and 30 times that of width coefficient, with a greater impact weight; When the coupling effect of internal pressure and defect size is applied, when the ratio of the length and width dimensions of the combined defect is small, the ultimate bending strain of the pipeline is positively correlated with the internal pressure load. As the ratio increases, the correlation changes from positive to negative. At the same time, the application of BP neural network in predicting the ultimate bending strain of defective pipelines was proposed, and the maximum error between the predicted value and the calculated value was only 7.18%. [Conclusion] The research results can provide theoretical guidance for the applicability evaluation of defective pipelines in daily pipeline maintenance and repair, and the use of BP neural network prediction models can help reduce the risk of pipeline explosion.

Key words Multiple payloads; Defective pipeline; Ultimate bending strain; Numerical simulation; Internal pressure defect size coupling

1 引言

管道运输是油气能源主要的运输方式。随着我国经济的快速发展，对油气运输的要求逐年提高。截至到 2020 年底，中国油气管道总长 14.40×10^4 km，对大口径，高碳钢制作的管道具有巨大的需求。X80 钢管具有高强度，高韧性，耐腐蚀，易焊接等优点，在国内管道建设中被大量使用。中国 X80 管材的生产及管道建设技术已进入国际先进行列。由于管道常年处于承压状态下，管道外部会因环境中的风沙、雨水、机械腐蚀或人为因素产生缺陷。这些缺陷会削弱管道的结构完整性，增加泄漏及破裂的风险，对管道的安全运行构成威胁。因此开展多重载荷作用下含缺陷运行管道的失效影响因素以减小管道失效概率的研究至关重要。

含缺陷管道运行状态下失效形式及影响因素引起国内外学者的广泛关注。同时大量学者针对缺陷管道的爆破压力提出了预测模型。王国庆等分析了弯曲对薄壁无缺陷和局部壁薄化弯头在内压和弯矩组合作用下爆破能力的影响。比普尔·钱德拉·蒙达尔等考虑轴向力与弯矩，针对不同轴力制定了组合弯矩和内压的失效位点。朱丽等对内压和弯矩条件下含环焊缝缺陷X80 管道的断裂响应进行分析。冯欣润等分析了内压与弯矩载荷联合作用下，凹槽、凹陷和组合缺陷对其极限弯矩承载能力及变形性能的影响效应。同时易帅等基于有限元实例，拟合开发了受弯曲和轴向载荷作用下腐蚀管道的爆破预测模型。帅健等采用非线性有限元法建立了含缺陷管道爆破失效的数值模型，为预测管道爆破压力提供理论支持。

综上所述，多重载荷作用下含缺陷管道的失效分析已成为研究热点。同时也存在多维度失效分析不够全面，预测模型的发展不够成熟等问题。因此本研究综合考虑管道在实际运行过程中所受载荷作用，并基于已公开研究成果中的实验数据，开展多重载荷作用下含腐蚀缺陷 X80 管道运行过程中极限弯曲应变（下文简称为"应变"）的多维度影响因素研究。近些年大量学者将 BP 神经网络应用于各个领域并取得较好的效果。本研究基于数值模拟所得数据，提出将 BP 神经网络应用于预测缺陷管道应变。通过大量重复性的计算所得的准确结果与预测结果相对比，验证此方案的可行性与准确性，为含缺陷管道适用性评价提供帮助。

2 管道实体域模型

2.1 模型及边界条件

本研究综合考虑到实体管道模型及管道载荷具有对称性，为降低数值模拟时间成本及提高计算准确性，对管道模型进行简化，建立 1/4 实体管道模型。管道腐蚀具有复杂性，其腐蚀缺陷呈不规则状，同时为避免应力集中现象影响计算精确性，对腐蚀区域进行倒角处理。

对实体管道模型采用对称约束，即对称约束面不可发生沿该面法线方向的位移，对称约束面仅可绕其法线进行旋转。同时采用 MPC 耦合约束，在无缺陷轴向面的中点建立节点，由节点引出大量刚性梁与无缺陷轴向面连接，在节点处施加力矩与轴力，如图 1 所示。管道相关数据如表 1 所示。

笔者使用挪威船级社发布的 DNV-RP-F101 Corroded Pipeline 中设置腐蚀缺陷的几何参数，以实现参数无量纲化。研究工况选取范围：长度系数 $L/\sqrt{Dt}$ 为 $0.6 \sim 5.1$；宽度系数 $\frac{j}{t}$ 为 $9.0 \sim 60$；深度系数 $\frac{d}{t}$ 为 $0.1 \sim 1.0$。同时定义内压系数 $\frac{p}{P}$，研究范围为 $0.04 \sim 1.4$。各个参数定义如图 1，图 2 及表 1 所示。

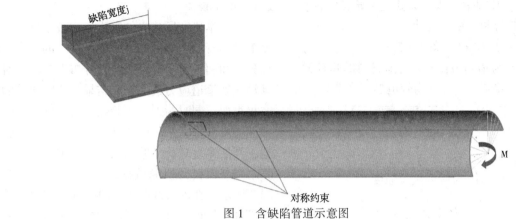

图 1 含缺陷管道示意图

表格 1 管道基本数据

管材	管道外径/D	管道长度	管道壁厚	运行内压/P	屈服强度	抗拉强度	泊松比	弹性模量/E
X80	711mm	6m	8.8mm	6.3MPa	450MPa	535MPa	0.3	205GPa

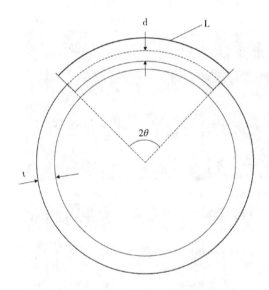

图 2 完整管道缺陷截面示意图

2.2 模型网格划分

本研究可选取 4 节点四面体网格单元和 10 节点四面体网格单元两种网格剖分形式。10 节点四面体网格单元模式下的单元网格有 10 个节点,其计算量较大,但在复杂几何模型及几何模型中存在尺寸相差较大等情况下具有更强的收敛性。因此基于研究建立的实体管道模型所具有的特点,选用 10 节点四面体网格单元网格划分模式。本研究针对建立的管道实体域模型,开展了多种网格尺寸划分综合分析。设置网格划分最大尺寸为 100mm、300mm 及 500mm,读取不同网格尺寸划分下,不同弯矩载荷作用下缺陷最深处沿壁厚方向中间节点的 Von Mises 等效应力,结果如图 3 所示。研究结果表明,同一弯矩下不同网格尺寸最大等效应力值的最大偏移量为 4.72%。故在此模型中采用最大尺寸为 500mm 的网格划分方案,划分结果如图 4 所示。

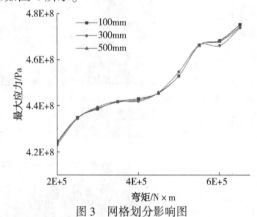

图 3 网格划分影响图

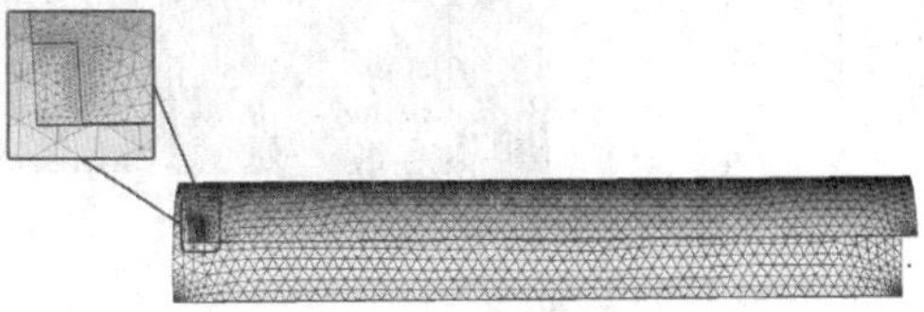

图 4 模型网格图

2.3 多重载荷及失效准则

本研究充分考虑运行中管道承受载荷主要为运行内压,外部载荷引起的弯矩,运行温度改变引起的热应力。式(1)为热应力公式。弯矩及热应力载荷于节点处施加,内压载荷于管道内壁处施加。含缺陷管道的失效准则主要为弹性失效准则和塑性失效准则。弹性失效准则假设材料在受力过程中始终处于弹性状态,而塑性失效准则考虑受力过程中会发生塑性形变。考虑到 X80 管材输油管道具有良好的弹性,综合分析下选用塑性失效准则,其判定表达式如式(2)所示。基于式(3)对管道弯曲应变进行计算。

$$P = E \times A \times \alpha \times (T_2 - T_1) \quad (1)$$

其中,P 为锚固力;E 为管道材料的弹性模量;A 为管道横截面积,α 为管道热膨胀系数。

$$\sigma_{eq} = \sqrt{[(\sigma_1-\sigma_2)^2+(\sigma_2-\sigma_3)^2+(\sigma_3-\sigma_1)^2]/2} \leq \sigma_u \quad (2)$$

σ_1、σ_2、σ_3 为三个方向上的主应力,σ_{eq} 为 Mises 等效应力,σ_u 为材料抗拉强度。

$$\varepsilon = K \times D \times \left(\frac{1}{R} - \frac{1}{R_0}\right) \quad (3)$$

其中,ε 为弯曲应变;K 为形状与材料的相关系数;D 为管道外径;R 为弯曲后的曲率半径;R_0 为初始曲率半径。

2.4 模型验证

基于本研究所建立的管道实体域模型,数值模拟得到管道在仅受弯矩载荷作用下的应力与位移分布如图5所示。管道应力分布沿其轴线对称且轴线处管道应力最小,此结果与李牧之等的研究相类似。同时为近一步验证此实体域模型的准确性与可靠性,基于已公开研究成果中的实验数据,开展多重载荷下的缺陷管道多维度分析。计算结果与已公开研究数据之间的对照图如图6,图7所示。在轴力载荷改变时,计算结果与研究数据的最小误差仅为 0.02%,平均误差为 4.14%。在弯矩载荷改变时,计算结果与研究数据的最小误差仅为 5.88%,平均误差为 8.23%。数据误差较小,验证了此实体域模型在本研究设定的多重载荷下具有准确性与可靠性。

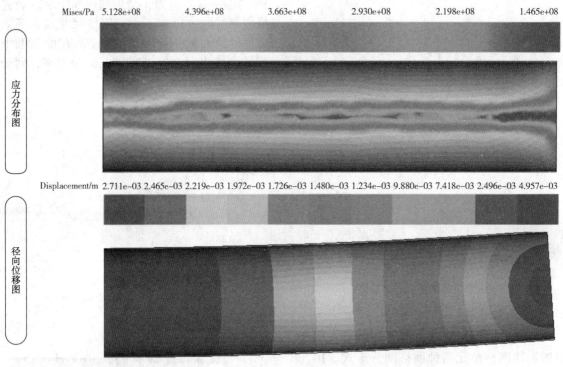

图5 应力位移图

3 数值模拟及结果分析

3.1 径厚比对极限弯曲应变的影响

研究多重载荷作用下含缺陷管道应变随径厚比改变的变化规律。选定缺陷长度系数为3.76,宽度系数为13.64,深度系数为0.5。考虑内压载荷作用并分别选定内压系数为1、0.68、0.36的三种工况。计算得不同径厚比下管道的应变结果如图8所示。在多重载荷及缺陷的作用下,管道应变与径厚比的相关性呈负相关,且随着径厚比的增加,内压载荷对管道应变的影响逐渐增加。管道应变随径厚比改变的变化曲线会于径厚比约为75处出现驼峰,且驼峰位置不受内压载荷影响。

3.2 缺陷长度对极限弯曲应变的影响

研究多重载荷作用下含缺陷管道应变随缺陷长度改变的变化规律。选定缺陷宽度系数为11.36,深度系数为0.57,考虑内压载荷作用并分别选定内压系数为1、0.68、0.36的三种工况。选定缺陷长度系数为0.6~4.0的工况以观察在内压载荷作用下缺陷长度对应变的影响规律,数值模拟结果如图9所示。在多重载荷及缺陷的作用下,管道应变与长度系数的相关性呈负相关。在长度系数小于2.04时,应变随单位长度系数改变的降幅大于长度系数大于2.04时的降幅。这是因为缺陷的延长方向为横向圆周方向,而弯矩载荷的作用方向为轴向,两方向所在平面相互垂直。而管道在多重载荷作用下,应力分布如图10所示。轴向垂直面应力最大,随着与轴向垂直面的角度增大,其上应力呈现为先减小后增大,且最小应力处与轴向垂直面夹角超过90度。缺陷在长度方向延长同一长度时,延长后长度系数小于2.04的情况下缺陷处所增加的应力大于延长后长度系数大于2.04的情况。对照不同运行压力下管道应变随缺陷长度的变化曲线,当长度系数小于1.57时,相同缺陷尺寸下,内压载荷大小与应变呈负相关。当长度系数大于2.04时,内压载荷大小与应变呈正相关。

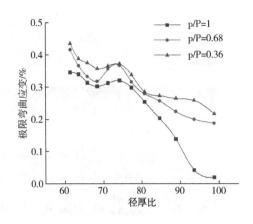

图8 径厚比影响

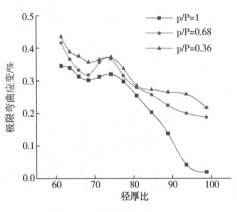

图9 缺陷长度影响

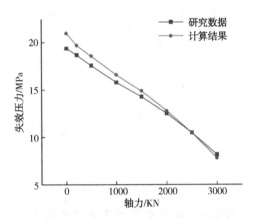

图6 失效压力随轴力变化对比图

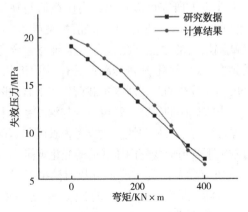

图7 失效压力随弯矩变化对比图

3.3 缺陷宽度对极限弯曲应变的影响

研究多重载荷作用下含缺陷管道应变随缺陷宽度改变的变化规律。选定缺陷长度系数为3.92,深度系数为0.23。选定缺陷宽度系数变化范围为9.0~60。考虑内压载荷作用并分别选定内压系数为1、0.68、0.36的三种工况,计算结果如图11所示。在多重载荷及缺陷的作用下,管道应变与宽度系数的相关性呈负相关,这与侯富恒等的研究结果类似。宽度系数小于13.64时,应变随单位宽度系数改变的降幅大于宽度系数大于13.64时的降幅。且宽度系数大于36.36后,管道的应变变化曲线呈现为上下波动,最后趋近于一条直线。这表明在缺陷宽度过大时,缺陷的宽度不再对管道应变造成明显影响。对照不同内压载荷下管道应变随宽度系数的变化规律,当宽度系数小于22.72时,相同缺陷尺寸下,内压载荷大小与应变呈正相关。而当宽度系数大于31.82时,内压载荷大小与应变呈负相关。

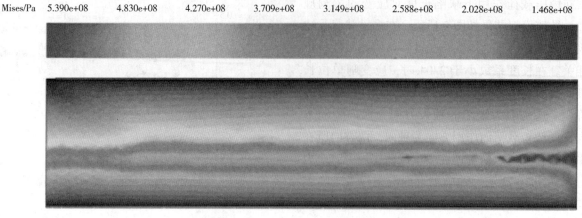

图10 联合载荷下应力分布图

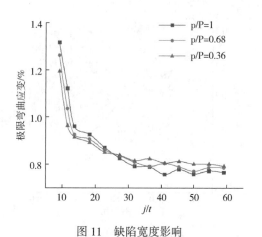

图11 缺陷宽度影响

3.4 缺陷深度对极限弯曲应变的影响

研究多重载荷作用下含缺陷管道应变随缺陷深度改变的变化规律。选定缺陷长度系数为2.51，缺陷宽度系数为9.01，考虑内压载荷作用并分别选定内压系数为1、0.68、0.36的三种工况。选定深度系数0~1的工况以观察在内压载荷作用下缺陷长度对应变的影响规律，计算的结果如图12所示。在多重载荷及缺陷的作用下，管道应变与深度系数的相关性呈负相关。$p/P=1$的内压载荷作用下，深度系数小于0.3864时应变随单位深度系数改变的降幅较大。$p/P=0.68$的内压载荷作用下，深度系数小于0.3409时应变随单位深度系数改变的降幅较大。而$p/P=0.36$的内压载荷作用下，深度系数小于0.2955时应变随单位深度系数改变的降幅较大。对照得知，内压载荷与应变降幅大小转折点的相关性呈正相关。如表2所示，三者影响因素中深度系数变化引起应变变化的方差为长度系数的3倍，宽度系数的30倍，影响权重更大，这与马明利等人的研究结果类似。对照同深度系数下应变随内压载荷大小的变化。研究结果显示在缺陷尺寸相同的情况下，应变与内压载荷呈正相关。

表2 缺陷尺寸影响方差

改变量	p/P			平均方差
	0.36	0.64	1	
长度	0.123	0.1	0.047	0.09
宽度	0.0112	0.0177	0.0262	0.018367
深度	0.2858	0.3155	0.5557	0.385667

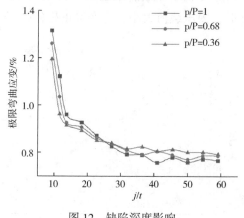

图12 缺陷深度影响

3.5 内压-缺陷尺寸耦合作用对极限弯曲应变的影响

基于以上研究可以得出，内压载荷与缺陷管道应变的相关性并不是单一的正相关或负相关，会随缺陷尺寸的改变产生变化。本研究综合考虑到管道实际运行过程中内压载荷具有不恒定性，因此开展内压载荷的改变对多重载荷作用下含缺陷管道应变影响规律的研究。管道应变与深度系数及管道径厚比具有单一相关性。故只考虑长度与宽度的组合缺陷下管道应变随内压载荷的变化规律。

本研究设定四类工况进行数值模拟以降低计算结果的偶然性。工况一的缺陷深度系数为0.34，长度系数为0.63，宽度系数为68.18。工况二的缺陷深度系数为0.34，长度系数为1.88，宽度系数为34.09。而工况三的缺陷深度系数为

0.34，长度系数为3.76，宽度系数为11.36。工况四的缺陷深度系数为0.34，长度系数为5.02，宽度系数为11.36。定义A为缺陷长度系数与缺陷宽度系数之比，计算得工况一中A=0.0092，工况二中A=0.0552，工况三中A=0.3312，工况四中A=0.4416。对四种工况进行数值模拟，结果如图13所示。

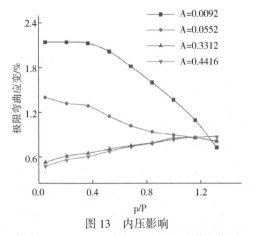

图13 内压影响

选取工况一与工况四两种具有代表性的缺陷管道实体域模型计算结果，如图14(a)，图14(b)所示。工况一中缺陷宽度大于缺陷长度，内压载荷与管道应变呈负相关。而工况二的缺陷长度与缺陷宽度的大小相近，缺陷径向方向上的投影呈类正方形，内压载荷与管道应变呈负相关。但对照工况一，减小幅度与降低速率更小。工况三中缺陷长度大于缺陷宽度，内压载荷与管道应变呈正相关。而在工况四中缺陷长度大于缺陷宽度，内压载荷与管道应变呈正相关。且对照工况三，增加幅度与增加速率更大。综上所述，随着比值A的增大，管道应变与内压载荷的相关性由负相关转变为正相关。且负相关时，A值越小，应变随内压载荷的变化幅度与变化速率越大。反之正相关时，A值越大，应变随内压载荷的变化幅度与变化速率越大。

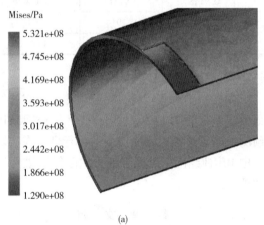
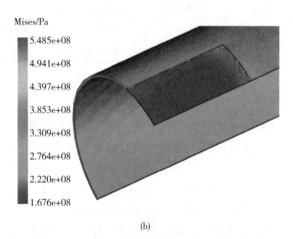

(a)　　　　　　　　　(b)

图14 工况图

4 BP神经网络预测极限弯曲应变

本研究提出将BP神经网络与多重载荷作用下含缺陷管道的多维度分析相结合，基于此前数值模拟所得数据调用trainlm训练函数进行BP网络训练，通过调整权重和偏置最小化预测误差。在验证和测试阶段，评估了网络在独立数据集上的泛化能力。最后对新数据进行预测，并通过反标准化将预测值转换回原始尺度，以分析和解释结果。运算逻辑如图15所示，图中各参数分别为：输入层i(缺陷长度、宽度、深度以及管道径厚比)；隐藏层h；输出层o(预测极限弯曲应变)，权重W；b1(隐藏层的偏置项)b2(输出层的偏置项)；x1, x2, x3, x4(输入信号：缺陷长度、宽度、深度以及径厚比)；$\sum_i w_i x_i + b$(加权求和公式)；W0，W1，W2，W4(影响权重)；f(x)(激活函数)。

本研究设置缺陷长度、宽度、深度及管道径厚比四个自由变量，基于此前研究所得数据对不同缺陷尺寸与径厚比管道的应变进行预测，预测结果如表3所示。同时对相同工况下管道进行计算分析，并将计算结果与预测结果相对照，误差值如表3所示。预测结果与计算结果误差最大仅为7.17%，证明通过BP神经网络可以有效预测多重载荷作用下含缺陷管道的应变及在实际生产过程中运用此方法快速准确的对缺陷管道进行适用性评价的可行性。同时此方法需要大量原始数据作为支撑以提供准确的预测结果。

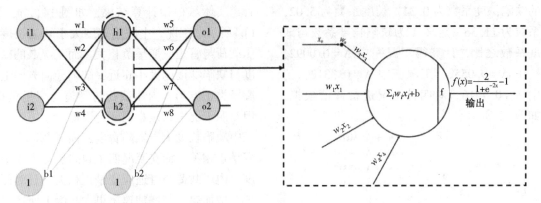

图 15 神经网络正反向传播及信息处理机制

表 3 预测与仿真数据

$L/\sqrt{Dt}$	$\dfrac{j}{t}$	$\dfrac{d}{t}$	径厚比	预测值/%	计算结果/%	误差/%
4.076549	34.090909	0.3333	80.795	0.4791	0.47636	0.575195
2.665431	27.272727	0.46666	78.563	0.496	0.5001	0.81984
3.620363	47.727273	0.4777	71.7	0.4473	0.4819	7.17991
2.508628	54.545455	0.431818	69.566	0.9148	0.91317	0.178499
3.135813	20.454545	0.5681	84.524	0.3081	0.2914	5.730954
3.76297	6.8181818	0.02222	88.659	1.8594	1.94869	4.58205

5 总结

本研究通过建立多重载荷作用下缺陷管道模型进行多维度分析，开展了对缺陷管道极限弯曲应变特性的研究，增加了多维度失效分析的全面性。同时提出并验证了可将 BP 神经网络与研究缺陷管道适用性相结合。主要结论如下：

（1）载荷及管道缺陷不变时，管道极限弯曲应变与径厚比呈负相关。在径厚比约为 75 处出现驼峰且驼峰峰顶值较大，该径厚比可在实际工程应用，以降低管道缺陷对管道强度的影响及减小经济投入。

（2）缺陷长度，缺陷宽度及缺陷深度与管道极限弯曲应变呈负相关，且增加单位尺寸的降幅呈现为先大后小。同时缺陷深度相较于其余两尺寸参数，其影响权重更大，在进行强度计算时应重点考虑这一因素。

（3）多重载荷作用下，含缺陷管道受不同内压载荷作用时，组合缺陷的改变会引起结果的改变。随着 A 值得增大，管道极限弯曲应变与内压载荷由正相关变为负相关，且管道极限弯曲应变的变化范围及变化速率呈现出先减小后增大的趋势。

（4）基于对照实验，得到预测值与实验结果最大误差仅为 7.17%。验证了 BP 神经网络用于预测缺陷管道极限弯曲应变的可行性。在实际生产中 BP 神经网络可用于助力缺陷管道进行适用性评价。

参 考 文 献

[1] 李秋扬,赵明华,张斌,等.2020 年全球油气管道建设现状及发展趋势[J].油气储运,2021,40(12):1330-1337+1348.

[2] 冯耀荣,吉玲康,李为卫,等.中国 X80 管线钢和钢管研发应用进展及展望[J].油气储运,2020,39(06):612-622.

[3] Wang Q, Zhou W. Burst capacity analysis of thin-walled pipe elbows under combined internal pressure and bending moment[J]. International Journal of Pressure Vessels and Piping, 2021, 194: 104562.

[4] Mondal B C, Dhar A S. Burst pressure of corroded pipelines considering combined axial forces and bending moments[J]. Engineering Structures, 2019, 186: 43-51.

[5] Zhu L, Li N, Jia B, et al. Fracture response of X80 pipe girth welds under combined internal pressure and

bending moment[J]. Materials, 2023, 16(9): 3588.

[6] 冯欣润. 复合载荷下缺陷海管剩余极限强度研究[D]. 武汉理工大学, 2019.

[7] Shuai Y, Zhang X, Feng C, et al. A novel model for prediction of burst capacity of corroded pipelines subjected to combined loads of bending moment and axial compression[J]. International Journal of Pressure Vessels and Piping, 2022, 196: 104621.

[8] 帅健, 张春娥, 陈福来. 非线性有限元法用于腐蚀管道失效压力预测[J]. 石油学报, 2008, (06): 933-937.

[9] 刘海峰, 刘浩天, 李罗胤, 等. 基于改进BP模型的沙漠砂混凝土高温后抗压强度预测[J/OL]. 河南理工大学学报(自然科学版), 1-12[2024-09-28]. http://kns.cnki.net/kcms/detail/41.1384.N.20240909.1332.004.html.

[10] Zhang J. [Retracted] Application of BP Neural Network in Matching Algorithm of Network E-Commerce Platform[J]. Journal of Sensors, 2022, 2022(1): 2045811.

[11] Ye H, Qian J, Yan S, et al. Limit bending moment for pipes with two circumferential flaws under combined internal pressure and bending[J]. International Journal of Mechanical Sciences, 2016, 106: 319-330.

[12] Zhou R, Gu X, Bi S, et al. Finite element analysis of the failure of high-strength steel pipelines containing group corrosion defects[J]. Engineering Failure Analysis, 2022, 136: 106203.

[13] Tian X, Zhang H, Lu M. Effect of axial force and bending moment on the limit internal pressure of dented pipelines[J]. Engineering failure analysis, 2019, 106: 104168.

[14] 陈严飞, 阎宇峰, 夏通璟, 等. 考虑弯矩影响的含腐蚀缺陷X80管道失效内压的计算方法[J]. 腐蚀与防护, 2024, 45(07): 78-83.

[15] Wang Y S. A plastic limit criterion for the remaining strength of corroded pipe: Proceedings of the International Conference on Offshore Mechanics and Arctic Engineering(10th) Stavanger Norway June23-281991[C]. Stavanger: ASME 1991.

[16] 张昊宇, 张引弟, 贺翔, 等. 外载荷作用下含腐蚀缺陷高钢级管道失效研究[J]. 当代化工, 2023, 52(10): 2511-2515+2520.

[17] 李牧之. 海底管道在弯矩和水压作用下的屈曲失效机理研究[D]. 天津大学, 2019.

[18] 夏通璟. 含腐蚀缺陷油气管道失效内压预测方法和可靠性评估[D]. 中国石油大学(北京), 2021.

[19] 侯富恒, 陈严飞, 贺国晏, 等. 含腐蚀-凹陷组合缺陷的海底管道极限弯矩承载力研究[J]. 中国造船, 2023, 64(06): 24-34.

[20] 马明利, 刘扬, 于心泷, 等. 面缺陷尺寸对管道承压能力的影响分析[J]. 科学技术与工程, 2011, 11(15): 3576-3579.

[21] 陆武慧. 基于改进BP神经网络的学生职业素质能力评价模型[J]. 自动化技术与应用, 2024, 43(09): 21-24.

油气田在役玻璃钢管道老化规律研究

熊新强[1] 宫敬[1] 刘杰[2] 唐德志[2] 廖丹丹[3]

[1. 中国石油大学(北京)机械与储运工程学院·城市油气输配技术北京市重点实验室·油气管道输送安全国家工程研究中心;2. 中国石油天然气股份有限公司规划总院;3. 四川大学机械工程学院]

摘 要 油气田玻璃钢管道存量大,随着服役时间的延长,老化失效问题逐渐凸显,但当前缺乏对老化过程的规律分析研究。为了探究油气田工况下玻璃钢管老化过程中各项性能的衰减规律,本文基于时温等效原理在实验室条件下开展了热空气、水和油三种不同介质环境的加速老化实验,采用宏观实验数据与微观结构相结合的分析方法,从老化前后玻璃钢样品的微观形貌与组织结构、化学性质及物理特性等几个方面对比了三种环境下不同老化时间的玻璃钢管老化行为,探究了老化机理的差异性。结果表明,玻璃钢管道老化机理分为三个阶段,第一阶段是介质从树脂层内表面破损处扩散侵入,产生吸湿增重和树脂降解,此过程主要取决于内表面固有缺陷、温度和介质类型。第二阶段是介质侵入增强层,在内部界面上造成了局部纤维损伤和基体收缩,界面脱粘,强度快速降低。第三阶段是介质的进一步扩散使树脂基体发生降解形成小分子气体,小孔洞融合成大空洞,整体强度降低,最终无法继续使用。另外,温度是玻璃钢管道性能影响最大的因素,对介质的扩散、树脂溶解和降解等热学过程有显著影响。油介质会极大提升介质侵入增强层的速度,因此含油介质比水介质具有更强烈的老化效应。基于以上结果提出了玻璃钢管道设计、施工、运行各阶段的管理建议,研究结论有利于油田管理单位明确玻璃钢管道老化影响因素和规律,及时开展现场管道的风险评估和预判,保障管道安全平稳运行。

关键词 玻璃钢管;老化因素;加速老化;影响规律

Research on the main aging factors and influence law of in-service GFRP pipes in oil and gas fields

Xiong Xinqiang[1] Gong Jin[1] Liu Jie[2] Tang Dezhi[2] Gu Tan[2] Liao Dandan[3]

(1. College of Mechanical and Transportation Engineering, China University of Petroleum;
2. Petro China Planning & Engineering Institute;
3. School of Mechanical Engineering, Sichuan University, Chengdu)

Abstract The stock of fiberglass pipelines in oil and gas fields is substantial, with aging and failure issues becoming increasingly prominent as service life extends. However, there is a lack of research on the patterns of the aging process. To investigate the degradation laws of various properties of fiberglass pipes under oilfield conditions during the aging process, this study conducted accelerated aging experiments using hot air, water, and oil environments based on the principle of time-temperature equivalence. An analytical method combining macroscopic experimental data with microscopic structure was employed to compare the aging behavior of fiberglass samples across different aging times in these three environments from aspects such as microstructure and organization, chemical properties, and physical characteristics before and after aging. The results indicate that the aging mechanism of fiberglass pipelines can be divided into three stages. Stage one involves the diffusion of media through surface defects of the resin layer, leading to moisture absorption, weight gain, and resin degradation, which primarily depends on inherent surface defects, temperature, and media type. Stage two sees enhanced media penetration into the reinforcement layer, causing localized fiber damage, matrix shrinkage, and interfacial debonding at internal interfaces, resulting in a rapid decrease in strength. Stage three is characterized by further diffusion of media causing the degradation of the resin matrix into small molecule gases, the fusion of small cavities into larger voids, overall strength reduction, and eventual unusability. Temperature is identified as the most influential factor affecting pipeline performance, significantly impacting the diffusion of media, resin solubility, and degradation. Oil media notably accelerate the speed of media intrusion into the reinforcement layer compared to water media, thus exhibiting more pronounced aging effects. Based on these findings, management recommendations for the design, construction, and operation phases of fiberglass pipelines are proposed. This study's conclusions aid oilfield management units in understanding the factors and patterns influencing fiberglass pipeline aging, enabling timely risk assessment and prediction of on-site pipelines, ensuring their safe and stable operation.

Key words Fiberglass pipe; Aging factors; Accelerated aging; Influence patterns

非金属管具有耐腐蚀、抗结垢结蜡、流体摩阻低、电绝缘、质量轻、使用寿命长等优势，在油气田管道选材中逐渐被重视，得到较为广泛的应用。

目前中国石油已应用非金属管道约占地面集输管道总量的15%，主要分为玻璃钢、柔性复合管、钢骨架复合管、塑料合金管以及聚乙烯管等五大类，其中玻璃钢管道占比52%，柔性复合管占比22%，是当前应用最广泛的两大类非金属管道。但由于复合材料复杂的结构特点以及现场严苛的工况条件，随着服役年限的增加，玻璃钢管道的失效率已经高于金属管道，管道长时间服役的老化因素与影响规律亟待研究。

玻璃钢管道主要应用与集油输油管道和输水注水管道，多年来，玻璃钢管道常常被认为是不会发生腐蚀的，行业内主要侧重于研究玻璃钢管道的位置探测、仿真计算、无损检测等方面，对与材料的的老化因素常常被忽略，对整管的性能衰减分析的研究较少，缺少油田真实模拟环境下的老化规律研究。玻璃钢材料的老化试验常常基于高分子聚合物的时温等效机理，在实验室利用短时高温模拟长时间的低温服役条件，研究发现高温下水介质会对纤维-基体界面造成较为严重的破坏，表现为层间剪切强度、拉伸强度、爆破压力和耐冲击性能发生衰减。Stocchi研究了玻璃钢管道在80℃温度下蒸馏水环境老化2周后的性能变化，发现纤维和基体都形成了明显损伤，弯曲性能和冲击性能显著降低。钱熙文等人发现95℃高温下，玻璃钢管道的颜色发生明显变化、基体产生孔洞缺陷，失重率显著上升，氧化基团增加。杜增智等人发现玻璃钢管道在油田模拟水环境下发生的表面树脂溶解和纤维水解损伤。吴瑞等人对比了玻纤/碳纤/亚麻纤维-树脂基复合材料在不同温度下的吸水性，结果表明玻纤/树脂复合材料的拉伸强度和层间剪切强度下降幅度较大，长期力学性能保持率较差。此外，对不同湿度环境和不同介质条件的研究表明，高温加速了液相介质的扩散和渗透，并显著降低了纤维增强复合材料的残余强度。同样，热老化和光化学老化对纤维增强复合材料性能的影响显示出一致的模式，导致聚合物骨架的降解和氧化。一些研究人员发现，在热氧老化后，复合材料中的基体和纤维/基体界面逐渐退化，表现为断链、重量损失和纤维/基质脱粘。然而，环氧树脂的降解和衰变会因纤维或其他增强项的作用而延迟。Lafarie Frenot系统地研究了在中性和氧化气氛中热循环对环氧树脂层压板的损伤，发现氧的加速作用和热循环的耦合对微裂纹扩展有加速作用。虽然关于玻璃钢复合材料老化的研究众多，但对于输油介质影响下的老化规律缺乏对比研究，对老化机理的研究不够深入。

为了解决油气田环境中以玻璃钢为代表的非金属管普遍存在的老化失效问题，本文基于时温等效原理设计了实验方案。在经过循环热烘箱加热的热空气、恒温水浴箱加热的水介质和油介质中，对玻璃钢管开展了加速老化实验。采用宏观实验数据与微观结构相结合的分析方法，从老化前后玻璃钢样品的微观形貌与组织结构，化学性质及物理特性等几个方面对比了三种环境下不同老化时间的玻璃钢管老化行为，探究了老化机理的差异性。有利于油田管理单位明确玻璃钢管道老化影响因素和规律，及时开展现场管道的风险评估和预判，保障管道安全平稳运行。

1 老化实验方法

在本研究中采用空气、水和油三种介质进行模拟实验，实验前对管材进行清洗烘干后将管道分别切割成管节试样，每组三个对照试样以确保实验数据的准确性。水介质采用某油田的典型采出水成分进行模拟，其主要成分及离子浓度如表1。

在实验室条件下为了加速老化实验的需要，依据API 15S标准中时温等效原理对老化温度进行适当提升，恒温水浴槽分别在65℃，80℃和95℃三个温度下进行水热老化实验。热氧老化和油浴老化实验在95℃的温度条件开展，最长热氧老化时间达到5000h。试验装置示意图如图1所示。

老化完成之后对试样的形貌特征、红外光谱、内部缺陷、环向强度等性能进行了测试。

表 1 某油田采出水模拟液成分(离子含量单位/mg·L^{-1})

pH 值	Ca^{2+}	Mg^{2+}	SO$_4^{2-}$	HCO$_3^-$	Cl$^-$	Fe^{2+}	Na$^+$/K$^+$
5.7	6300	290	635	325	85000	20	5000

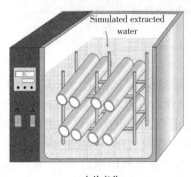
(a)水热老化

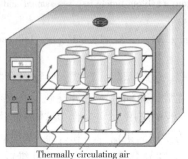
(b)热空气老化

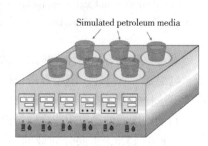
(c)油浴老化

图 1 试验装置示意图

2 实验结果

2.1 老化玻璃钢管的形貌分析

图2显示了玻璃钢管在95℃热氧条件下老化不同时间后颜色与外形变化。结果表明，初始状态玻璃纤维管表现出黄绿色，随老化时间的延长颜色逐渐加深，氧化1000小时观察到外层颜色由黄绿色转变成棕色，5000小时后再进一步向黑色转变。颜色变化表明环氧树脂由于热氧化而发生了化学反应，老化后环氧树脂的颜色变化的原因与聚烯结构以及醌类或环化共轭氮化合物形成有关。图4也进一步显示了热老化后GFRP管截面上氧化层的变化情况，氧化降解先发生在表面的暴露区域，氧化层随老化时间增加而颜色变暗且氧化深度进一步增加。相比较而言，同一老化时间下水热老化样品表面颜色比热氧老化样品颜色更浅。

图3是热空气介质下，不同老化时间后的微观形貌，随着老化时间的延长，在500h后截面上出现纤维脱出孔，内表面树脂开始收缩变形，纤维暴露。老化时间达到3000h后，截面上纤维脱出孔数量显著增加，纤维间树脂基体出现明显的微孔，内表面树脂呈现严重的脱落现象，产生线状缺陷。老化时间达到5000h后，在内表面和截面的树脂层都发现了大量的微孔，最大尺寸接近500nm，玻璃纤维界面严重损伤破坏。

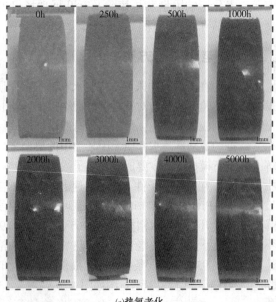
(a)热氧老化

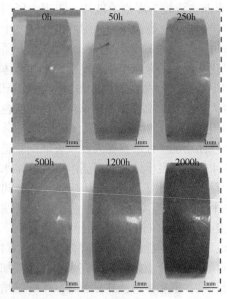
(b)水热老化

图 2 加速热老化玻璃钢管样品的宏观形貌图

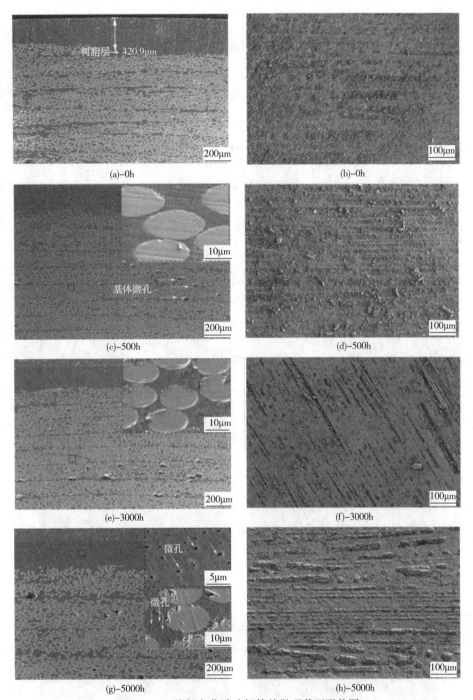

图 3 95℃热氧老化玻璃钢管的微观截面形貌图

图 4 是水介质下,不同老化时间后的微观形貌,随老化程度的加剧纤维脱出孔和树脂上的微孔越来越明显。同热氧老化相比,水热老化的不同之处在于它包含了复杂的化学反应与物理溶解两个过程。通过形貌分析发现水热环境对纤维界面造成的损伤比热氧更严重,观察到大范围的纤维损伤也佐证了这一结论。此外,水热老化3000h 后在管道内外表面还观察到大量均匀分布的圆形的腐蚀坑,最大直径达到 20μm。这些现象都表明玻璃纤维的溶解损伤和树脂表面的腐蚀坑都与水解过程有关。

图 5 是油介质下,不同老化时间后的微观形貌,通过观察发现内表面在老化过程中有更多的线性沟槽和树脂脱落,随着老化时间的增加而显著增加。内表面优先接触热介质,在热输入的作用下,树脂和表面玻纤油浴导热系数不同产生热应力,导致树脂收缩形成沟槽。相比较而言,同时伴随老化过程的延续,管道截面玻纤的损伤加强,玻纤的损伤面积增加,这与热输入的持续破坏有关。

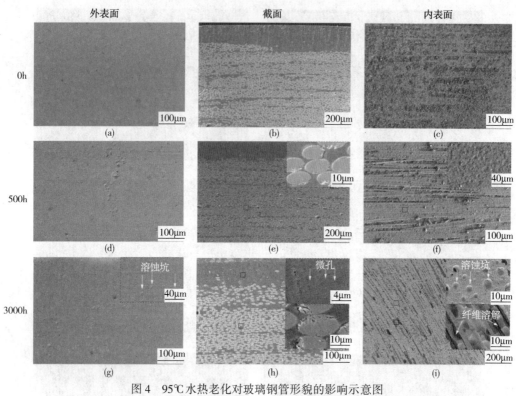

图 4　95℃水热老化对玻璃钢管形貌的影响示意图

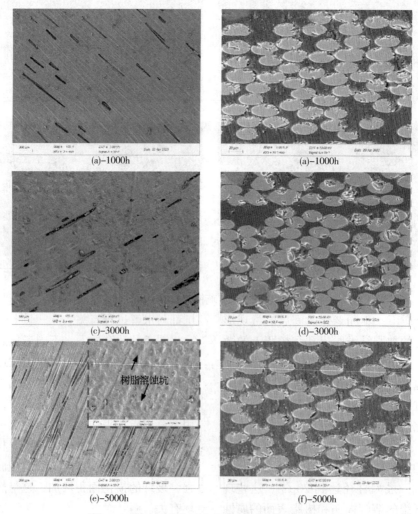

图 5　95℃油浴老化玻璃钢管的微观截面形貌图

2.2 热老化对成分的影响

对老化后的样品进行红外检测，结果如图6-图8所示。有如下结果：（1）随C-O-C键的减少可以明显的观察到1224cm^{-1}处C-O和1738cm^{-1}处C=O特征带的增加，红外光谱中的羰基指数和醚基指数增加，环氧基指数减小，表明主链上的环氧基团发生了氧化。（2）480cm^{-1}处的吸收峰与玻璃纤维中的Si-O键有关，Si-O峰的强度发生了明显衰减，玻璃纤维在水热老化过程中发生了水解。

油介质中老化1000h到4000h后，3200cm^{-1}～3600cm^{-1}区间的N-H、O-H峰强度增加1730cm^{-1}的酯基C=O键和1660cm^{-1}的酰胺C=O键强度增大，1482cm^{-1}～1535cm^{-1}区间的C-O峰和912cm^{-1}的C-O-C强度发生衰减。同时定量分析计算了老化过程成中的主要官能团含量，C=O强度从2.14%增加2.66%，C-O强度从1.53%减小到1.73%，C-O-C强度从3.33%减小到3.11%。

2.3 缺陷CT扫描

CT扫描的结果如图9、图10所示，从孔隙率的分析结果来看，三种老化环境下老化后GFRP样品的孔隙率明显高于未老化样品的0.1061%。且随着老化时间的增加，孔隙率有明显的增加趋势。在三种环境下长时间老化后，热氧，水热和油浴环境下的孔隙率依次增大，分别为0.6374%，0.959%和1.4954%。根据缺陷数量的统计结果分析发现老化后缺陷数量变少，由小体积缺陷融合成大体积缺陷，随老化时间延长缺陷尺寸增大，缺陷体积增加。

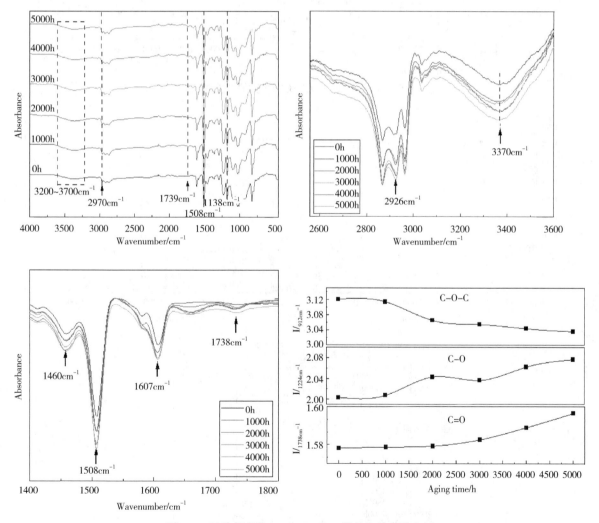

图6 95℃热氧老化(5000h)GFRP管的红外光谱分析

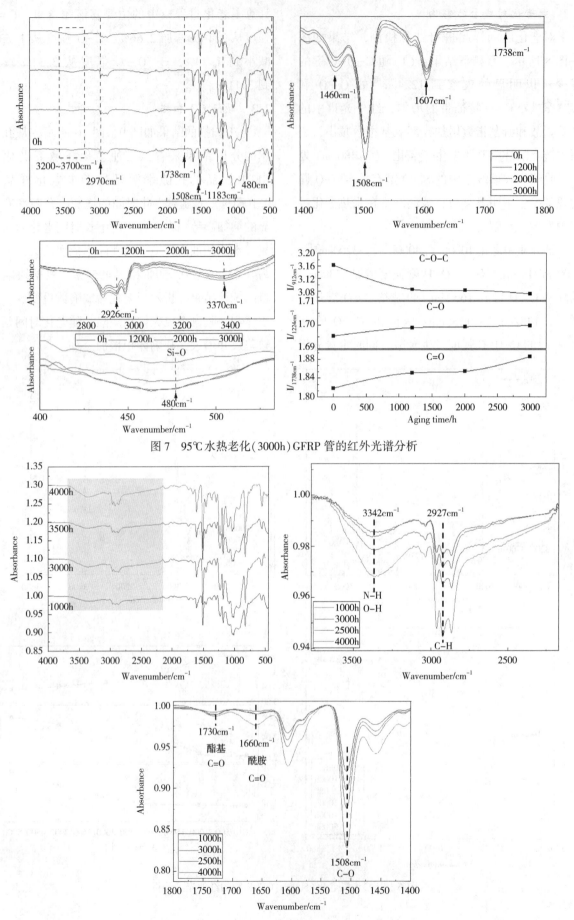

图 7 95℃水热老化(3000h)GFRP 管的红外光谱分析

图 8 95℃油热老化(4000h)GFRP 管的红外光谱分析

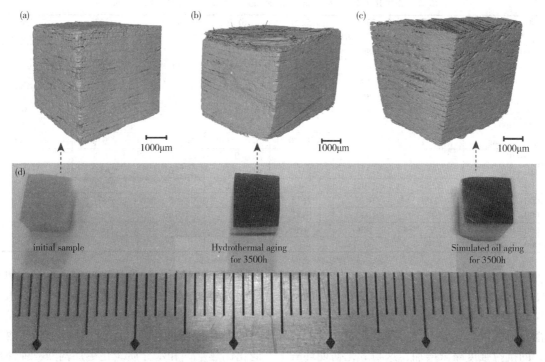

图 9 热老化 GFRP 管 CT 取样与扫描 3D 图

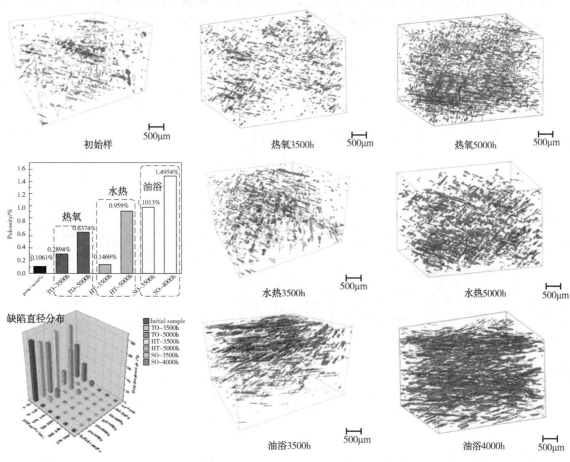

图 10 热老化 GFRP 管内部孔隙缺陷的 CT 扫描图

表 2 热老化 GFRP 管内部孔隙缺陷直径和缺陷体积分布统计表

缺陷数量	4649	5323	5220	6149	1633	7227	2869
缺陷直径/μm	原始	热氧 3500H	热氧 5000H	水热 3500H	水热 5000H	油浴 3500H	油浴 5000H
0~50	4597	4868	4404	5971	393	6932	1685
50~100	47	435	785	168	1120	243	924
100~150	4	19	30	8	103	44	227
150~200	0	0	0	0	12	6	23
200~250	0	0	0	1	3	1	9
250~300	0	0	0	0	1	0	0
缺陷体积/μm³	原始	热氧 3500H	热氧 5000H	水热 3500H	水热 5000H	油浴 3500H	油浴 5000H
0~10	0	0	0	0	0	0	0
10~100	1955	1231	0	2138	6	2634	1
100~1000	2380	1742	0	3254	5	3784	3
1000~10000	279	1039	1615	637	5	587	578
10000~100000	32	1030	3121	116	746	203	1313
100000~1000000	2	280	476	3	824	18	889
1000000~10000000	0	0	7	0	46	0	84

2.4 力学性能评估

图 11 是老化后样品的拉伸应力应变曲线。可以拉伸过程从弹性变形直接发展到断裂，不存在塑性变形阶段与屈服阶段。在老化 50h 时，玻璃钢管的抗拉强度略微增加，这与老化初期环氧树脂的后固化有关。同时，热老化过程中的性能变化是后固化引起的热氧化降解和基质硬化的竞争效应的结果。在图 11.b 中观察到随老化时间的增加，复合材料的拉伸强度明显下降，初始斜率明显增加，在长时间老化后趋于平缓。水介质老化行为比热空气介质剧烈，表现为弹性模量和拉伸强度的快速衰减，而图 11.c 观察到油介质老化比热空气和水介质老化更加剧烈，弹性模型的衰减更加显著。结合微观组织变化可以解释为此时发生了树脂溶解、纤维损伤、界面破坏，力学性能显著下降。

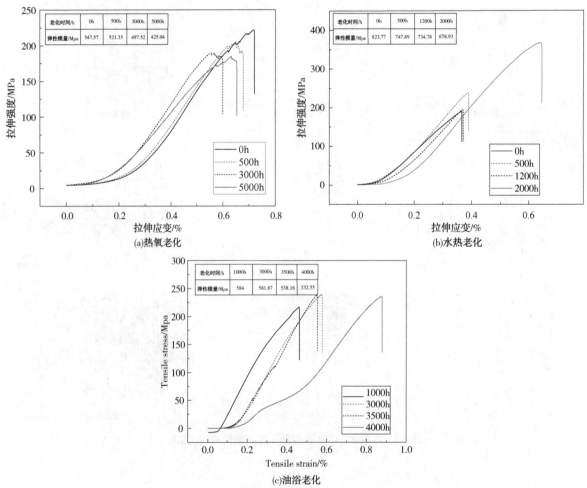

图 11 老化后试样的拉伸应力-应变曲线及强度保持率变化

3 分析与讨论

玻璃钢管道的老化发生在三个层面，一方面是树脂基体的老化，二是纤维损伤，三是界面的脱粘。在初始状态玻璃钢管道内的树脂与纤维界面紧密结合，树脂和纤维没有损伤。当引入高温和氧化性气氛后，由于树脂基体和玻璃纤维的导热性差异，外加高温会在界面处引起热应力，引起树脂基体的收缩变形和纤维截面的破损。在管道截面上观察到树脂收缩孔，在管道内表面观察到纤维暴露。而管道外表层结构为均匀致密的纯树脂，外加高温产生的热应力很小，所以不会在外壁的纯树脂层引发明显的收缩变形和开裂。在研究中发现在 500 小时热氧老化后，玻璃钢管截面的树脂基体开始出现小分子产物溢出形成的微孔，在老化 5000 小时后的树脂基体中发现大量均匀的纳米级微孔，并且通过红外检测到老化样品的成分变化，在高温和氧气的持续作用下，环氧树脂的分子链发生降解。

与热氧老化相比，水热老化的液相介质会进一步促进老化反应。玻璃钢管在水热老化过程中吸湿增重和降解失重相互竞争，但吸湿性更占优势，整体重量持续增加。水介质的存在会促进表面的树脂溶解，形成溶蚀坑，进而介质侵入增强层发生界面破坏，而缺陷的产生又会加速玻璃钢管的吸湿过程。同时，水分子的作用会促进玻璃纤维的水解，氯离子阻碍氢键的形成，纤维和基体界面粘接性能退化。

与水介质的老化相比，油介质将极大促进老化过程。由于油性介质与树脂基体的相似相溶特性，将极大提升了表面树脂层的溶解速度，使得介质进入增强层界面的时间显著缩短，进入增强层后，进一步造成树脂溶解坑和纤维损伤。

4 结论与建议

本论文以油气田环境下使用的玻璃纤维增强复合管作为研究对象，基于时温等效原理设计了热空气、水和油三种介质下的加速老化实验，并对老化之后的样品开展了微观形貌、红外光谱、缺陷扫描以及环向强度测试，对其老化行为和老化机理进行研究。根据目前的实验结果，得到了以下主要结论：

（1）玻璃钢管道老化机理分为三个阶段：①介质从树脂层内表面破损处扩散侵入，产生吸湿增重和树脂降解，此过程主要取决于内表面固有缺陷、温度和介质类型。②介质侵入增强层，在内部界面上造成了局部纤维损伤和基体收缩，界面脱粘，强度快速降低。③介质的进一步扩散使树脂基体发生降解形成小分子气体，小孔洞融合成大空洞，整体强度降低，无法继续使用。

（2）温度是玻璃钢管道性能影响最大的因素，对介质的扩散、树脂溶解和降解等热学过程都有显著影响。

（3）油介质与内表面的树脂溶解，会极大提升介质侵入增强层的速度，因此含油介质比水介质具有更强烈的老化效应。

因此在油气田环境中使用玻璃纤维管时，应该充分重视管材老化的影响，提出如下几点建议：

（1）设计阶段应充分考虑温度、压力和介质的影响。设计时应根据地质条件充分预估运行期工况，以确保管材能够承受油气田中的高温和压力变化。此外，还需考虑介质的特性，选择能够抵抗介质侵蚀的材料。

（2）施工阶段，应建立入场质量抽检机制，全面管控产品入场质量；建立完善的施工质量监管体系，避免微小缺陷成为运行阶段老化和缺陷生长的敏感点。

（3）在运行阶段，防止超温超压运行工况是关键。应建立有效的监控系统，实时监测玻璃纤维管的温度和压力，确保其在安全的工作范围内运行。同时，定期对玻璃纤维管进行检查和维护，包括检查管材的完整性、连接部位的密封性和螺纹的磨损情况，及时发现并修复潜在的问题。

参 考 文 献

[1] Alabtah F G, Mahdi E, Khraisheh M. External Corrosion Behavior of Steel/GFRP Composite Pipes in Harsh Conditions [J]. Materials, 2021, 14 (21): 6501.

[2] Reis J M L, Martins F D F, Da Costa Mattos H S. Influence of ageing in the failure pressure of a GFRP pipe used in oil industry [J]. Engineering Failure Analysis, 2017, 71: 120-130.

[3] 李振林；张笑影；刘彤等. 交通载荷作用下埋地含缺陷玻璃钢管道力学行为研究 [J]. 力学与实践, 2024, 46 (01): 64-76.

[4] 张笑影等. 地质沉降作用下埋地玻璃钢管道的力学

行为．油气储运，2024.43(7)：769-777

[5] 荆旸，赵康．油田集输用在役玻璃钢管道性能测试及失效分析[J]．塑料工业，2021，49(03)：118-121+125

[6] 张志坚；宋长久；章建忠等．纤维缠绕张力对玻璃钢制品质量的影响及控制措施[J]．玻璃钢/复合材料，2019(11)：111-114

[7] 朱央炫；陈建中；吕泳等．玻璃钢空心夹层管道及力学性能分析[J]．玻璃钢/复合材料，2018(11)：37-42．

[8] 胡敏；郭强；习向东；陈磊；刘杰；赵飞．玻璃钢管材无损检测方法综述[J]．材料导报，2023，37(S2)：594-598．

[9] 钱熙文．基于剩余强度模型的油气田用玻璃钢管加速老化研究及寿命预测[D]．四川大学，2023．

[10] 刘玉．湿热环境中环氧树脂的老化特性研究[D]．重庆大学，2018．

[11] 杜永，马玉娥．湿热环境下纤维增强树脂基复合材料疲劳性能研究进展[J]．复合材料学报，2022，39(02)：431-445．

[12] 张志坚，张萍，章建忠等．玻纤增强环氧树脂复合材料加速湿热老化性能研究[J]．玻璃纤维，2018(05)：8-13．

[13] Manalo A, Maranan G, Sharma S, et al. Temperature-sensitive mechanical properties of GFRP composites in longitudinal and transverse directions: A comparative study [J]. Composite Structures, 2017, 173: 255-267.

[14] Gu Y, Liu H, Li M, et al. Macro-and micro-interfacial properties of carbon fiber reinforced epoxy resin composite under hygrothermal treatments [J]. Journal of Reinforced Plastics and Composites, 2014, 33(4): 369-379.

[15] Ridzuan M J M, Abdul Majid M S, Afendi M, et al. Moisture absorption and mechanical degradation of hybrid Pennisetum purpureum/glass-epoxy composites [J]. Composite Structures, 2016, 141: 110-116.

[16] Hawileh R A, Abu-Obeidah A, Abdalla J A, et al. Temperature effect on the mechanical properties of carbon, glass and carbon-glass FRP laminates [J]. Construction and Building Materials, 2015, 75: 342-348.

[17] Stocchi A, Pellicano A, Rossi J P, et al. Physical and water aging of glass fiber-reinforced plastic pipes [J]. Composite interfaces, 2006, 13 (8-9): 685-697.

[18] 钱熙文；廖丹丹；谷坛等．95℃热氧老化对玻璃钢管性能的影响[J]．塑料工业，2023，51(08)：98-101+121．

[19] 杜增智；罗顺友；陈思维等．油田用玻璃钢管老化行为与机制的实验研究[J]．塑料工业，2023，51(12)：77-81．

[20] 吴瑞，李岩，于涛．不同种类纤维增强复合材料湿热老化性能对比[J]．复合材料学报，2022，39(09)：4406-4419．

[21] Deniz M E, Karakuzu R. Seawater effect on impact behavior of glass-epoxy composite pipes [J]. Composites Part B: Engineering, 2012, 43(3): 1130-1138.

[22] Alawsi G, Aldajah S, Rahmaan S A. Impact of humidity on the durability of E-glass/polymer composites [J]. Materials & Design, 2009, 30(7): 2506-2512.

[23] 杨成瑞，王豪，李鹏等．环氧树脂及其复合材料高温高压矿化水条件下的老化机理及性能[J]．工程塑料应用，2022，50(04)：97-105．

[24] 时中猛，邹超，周飞宇等．碳纤维增强树脂基复合材料紫外老化机理及寿命预测[J]．压力容器，2022，39(05)：8-15．

[25] Delor-Jestin F, Drouin D, Cheval P Y, et al. Thermal and photochemical ageing of epoxy resin-Influence of curing agents [J]. Polymer Degradation and Stability, 2006, 91(6): 1247-1255.

[26] Zhang X, Wu Y, Wen H, et al. The influence of oxygen on thermal decomposition characteristics of epoxyresins cured by anhydride [J]. Polymer Degradation and Stability, 2018.

[27] 宋海硕．玻纤增强尼龙10T复合材料热氧老化性能及寿命预测研究[D]．贵州大学，2017．

[28] Fan W, Li J L. The Effect of Thermal Aging on Properties of Epoxy Resin [J]. Advanced Materials Research, 2012, 602-604: 798-801.

[29] 李娟子．热氧老化对三维碳/玻璃纤维双马复合材料耐久性的影响[D]．西安工程大学，2020．

[30] Jojibabu P, Ram G D J, Deshpande A P, et al. Effect of carbon nano-filler addition on the degradation of epoxy adhesive joints subjected to hygrothermal aging [J]. Polymer Degradation and Stability, 2017, 140: 84-94.

[31] LAFARIEFRENOT M. Damage mechanisms induced by cyclic ply-stresses in carbon-epoxy laminates: Environmental effects [J]. International Journal of Fatigue, 2006, 28(10): 1202-1216.

[32] Lafarie-Frenot M C, Rouquie S. Influence of oxidative environments on damage in c/epoxy laminates subjected to thermal cycling [J]. Composites Science and Technology, 2004, 64(10-11): 1725-1735.

多井水溶造腔排量优化调配及现场应用

秦 垦[1]　任众鑫[1]　廖友强[2]

（1. 国家管网集团储能技术有限公司；2. 中国科学院武汉岩土力学研究所）

摘　要　对于盐穴储气库造腔工程，下游盐化工厂的消纳能力和接收卤水要求往往决定了盐穴造腔施工设计参数。提高盐穴溶蚀速度和确保卤水浓度是一组相互矛盾的技术指标，为解决淡水注入量、综合盐浓度和循环泵能耗等参数的优化问题，提出了一种盐穴储气库多井腔排量优化分配方法。不同空腔施工方案的能耗损失和空腔速度相差50%以上，造腔施工速度相差约14.4%，证明造腔排量的优化具有重要的工程意义。现场应用表明，在牺牲一定能耗和造腔速率的前提下，满足多井成腔标准速率的混合卤水浓度可提高15.98%，此方法可以有效降低成本并提高盐穴储气库的建设效率。

关键词　盐穴储气库；水溶造腔；流量调节；优化方法；成本与效率

Optimization and Application of Multi-Well Cavity Discharge Volume Allocation for Salt Cavern Gas Storage

Qin Ken[1]　Ren Zhongxin[1]　Liao Youqiang[2]

(1. PipeChina Energy Storage Technology Co., Ltd.; 2. Institute of Rock and Soil Mechanics, Chinese Academy of Sciences)

Abstract　For salt cavern gas storage cavity, the salt chemical plant's capacity to consume and the requirements for receiving brine often determine the design parameters of salt cavern. Increasing theleaching rate and ensuring brine concentration are a pair of contradictory technical indicators. To solve the problems of optimizing parameters such as freshwater injection volume, comprehensive salt concentration, and energy consumption of circulation pumps, a multi-well cavity discharge volume optimization allocation method for salt cavern gas storage is proposed. The energy consumption loss and cavity formation speed of different cavity construction schemes differ by more than 50%, and the cavity formation speed varies by approximately 14.4%, demonstrating the significant engineering significance of optimizing cavity volume. Field applications show that, at the expense of certain energy consumption and cavity formation rate, the mixed brine concentration meeting the standard rate of multi-well cavity formation can be increased by 15.98%. This method can effectively reduce costs and improve the construction efficiency of salt cavern gas storage.

Key words　Salt cavern gas storage; Water-leaching cavity; Flow rate regulation; Optimization method; Cost and efficiency

1 引言

我国能源供需不平衡问题仍然突出，进一步呈现消费重心东倾、生产重心西移的态势。建设大尺度的能源存储库是保障能源供应安全的重大战略需求。地下盐穴储气库是能源储存的主要方式之一，也是我国基础储气设施的重要组成部分。相比于其他类型的储气库设施，盐穴储气库具有注采效率高、密封性好、垫层气用量少、安全性高等优点。盐穴储气库建设于地下盐岩层中，盐岩的渗透率和孔隙率极低，天然气、空气等气体难以在盐岩层中渗漏，使得盐穴储气库具有良好的密封性。盐岩还具有良好的蠕变性和损伤自愈性，这使得盐穴腔体在注采运行过程中即使发生体积收缩、产生较大的变形，也不会发生明显的力学破坏或形成大规模的裂隙，影响腔体的安全性和密封性。

水溶造腔是盐穴储气库建设的重要组成部分。国外虽然有多年的注水造腔历史，但因其盐岩纯度高、地处沿海地区，具备卤水直接排海的天然优势，其工艺流程、成腔效率等相关理论在金坛储气库适用性较差。我国盐穴储气库地处内陆，水溶造腔产生的卤水需外输卤水处理厂，并有严格的卤水浓度要求，一般要求>295g/L。目前，为满足卤水处理厂对卤水浓度的要求，最常用的方式是进行卤水多次循环，以提高排卤浓度。然而注入的卤水浓度越高，井下的水溶造腔速度就越慢。相关研究表明：二次循环的水溶造腔效率在一次循环的50%以下。因此，面对水溶造腔速度与卤水处理厂的浓度要求的矛盾，需要对造腔的排量进行优化，在满足浓度要求的条件下实现水溶造腔速率的最优。此外，为加快盐穴储气库的建腔效率，现场通常同时进行多井造腔。多井造腔过程中的排量调配优化更为复杂，在兼顾造腔速度和排卤浓度的同时，还需考虑注水能耗。具体体现在：（1）不同腔体距离注水站的距离不同，排量的调配对多井水溶造腔的总能耗影响较大；（2）由于不同腔体的地质条件和所处造腔阶段不同，所以即使在相同的注入浓度下，各腔体的排卤浓度和水溶造腔速度也存在较大差异；（3）相较于单井造腔，多井造腔是一个多目标综合最优解问题，在满足多井的综合排卤浓度满足卤水处理厂要求的同时，还需考虑各腔体的综合造腔速率。然而，目前对于多井造腔过程中的排量调配很大程度依赖于现场工程师的经验以及现场排卤浓度的实测值，往往会造成排卤浓度不达标、造腔效率低和注水能耗大等技术难题，缺乏完善系统的优化理论支撑排量优化从滞后被动向提前主动的重大技术跨越。因此，迫切需要建立完善的能耗和排卤浓度预测模型，以实现排量调配的精确优化。

班凡生等对盐穴储气多腔组合水溶造腔方案进行了深入研究，采用灰色系统理论和物质元素分析方法确定了最优方案，客观评价了盐穴储气水溶腔的各种预选方案。提出了盐穴储气库建腔时间、平均排量、空腔容积、卤水浓度和空腔设计合格率等综合评价指标，采用 AHP 对评价指标的权重系数进行量化，实现造腔方案的综合评价。李淑平等提出了一种基于三个约束条件的最佳造腔参数求解方案：淡水消耗与供给的平衡、排出卤水量与注入淡水量之间的平衡以及混合外排卤水浓度的标准。杨海军等采用罚函数法将型腔注水问题转化为无约束优化问题，以单位型腔体积最低能耗作为注水优化的目标函数，提出了基于 SUMT 外点法的造腔注水优化模型，并提出了迭代求解计算方法。

针对水溶性腔体排放优化的技术局限性，阐明卤素浓度、腔速与注水能耗和排量之间的内在关系，提出了盐穴储气多井腔排放优化分配方法，以期在满足浓度要求的基础上，为多井水溶性腔体排放分配提供最优求解路径盐水处理厂。降低盐穴储气库建设的成本并提高效率。

2 模型构建

实现多井造腔状态参数调整优化的关键在于精确地预测参数调整前后排卤浓度的变化规律并建立相应的排卤浓度预测模型，以实现对未来的造腔过程以及卤水处理的积极动态反馈。因此本节将从水溶造腔排卤浓度预测、能耗预测和综合造腔速率三个方面建立模型。

2.1 排卤浓度预测模型

根据李金龙等建立的模型，根据腔体内盐质量守恒，构建了一个盐穴腔体内的卤水含盐浓度计算模型。其中，正循环条件下的模型表达式为：

$$\rho \frac{\partial V}{\partial t}(1-u)+\left(\frac{\partial V_\mathrm{d}}{\partial t}\right)c_\mathrm{s}-\left(\frac{\partial V}{\partial t}-\frac{\partial V_\mathrm{d}}{\partial t}\right)C=\frac{\partial (VC)}{\partial t} \quad (1)$$

式中，ρ 为盐岩的密度，kg/m^3；V 为控制体的体积，m^3；u 为盐层中的不溶物含量；V_d 是饱和区的体积，m^3；C_s 为饱和卤水的浓度，g/L；t 为时间，h。

其中，等式左端第一项为单位时间内盐岩壁面溶解掉的盐量，第二项为单位时间内产生的不溶物落到控制体下方的死水区饱和卤水中，造成饱和卤水上移到控制体内的盐量；第三项为经外管排出卤水中的盐量。等式右端为单位时间之内控制体内的盐量变化量。

同理，反循环条件下的模型表达式为：

$$\rho \frac{\partial V}{\partial t}(1-u)-\left(Q-\frac{\partial V}{\partial t}\right)C=\frac{\partial (VC)}{\partial t} \quad (2)$$

基于腔体的基本信息，可对上述模型进行数值求解，得出水溶造腔过程中整个腔体内卤水含盐浓度的空间分布特征。排卤管位置处的含盐浓度则为此时的排卤浓度。

综上，多井造腔过程中的综合排卤浓度预测模型可以表示为：

$$C_{\mathrm{opt}} = \frac{\sum_{i=1}^{i=N}(Q_i C_i)}{\sum_{i=1}^{i=N} C_i} \tag{3}$$

式中，C_{opt} 为多井造腔综合排卤浓度，g/L。

2.2 注水能耗预测模型

水溶造腔过程中的能耗主要在于克服卤水在管线以及腔体内的流动阻力，与注水泵的泵压、排量和效率呈正比，其计算方法如下：

$$P_{\mathrm{sum}} = \sum_{i=1}^{N}(\eta \rho_l Q h)_i g = \sum_{i=1}^{N}(\eta p Q)_i \tag{4}$$

式中，P_{sum} 为各腔体水溶造腔过程中的综合能耗，W；ρ_l 为卤水的密度，kg/m³；h 为泵的扬程，m；η 为泵效；p 为腔体注液泵的泵压，MPa，假设卤水在管线内做一维流动，压力分布可以通过下式计算：

$$\frac{dp}{dz} + a\rho_l g \cos\theta + \frac{1}{2}f \frac{\rho_l v_l^2}{D} = 0 \tag{5}$$

式中，z 为管线长度，m；a 为流动方向，与重力加速度方向相同取1，相反则取-1；θ 为管线的水平倾角，度；f 为管线流体流动摩阻系数；D 为管线内径，m；v_l 为管线内流体的流动速度，m/s。

从上式可以看出，决定管线内压力分布的关键在于摩阻系数 f 的计算，摩阻系数与流体在管线内的流态（层流和紊流）、管壁粗糙度相关，对于卤水这种近似牛顿流体，可采用如下模型进行摩阻压降的计算。

$$\begin{cases} f = \dfrac{64}{\mathrm{Re}} & (\mathrm{Re} < 2300) \\ f = 0.06539 \exp\left[-\left(\dfrac{\mathrm{Re}-3516}{1248}\right)^2\right] & (2300 \leq \mathrm{Re} \leq 3400) \\ \dfrac{1}{\sqrt{f}} = -2.34 \lg\left\{\dfrac{\varepsilon}{1.72D} - \dfrac{9.26}{\mathrm{Re}} \lg\left[\left(\dfrac{\varepsilon}{29.36D}\right)^{0.95} + \left(\dfrac{18.35}{\mathrm{Re}}\right)^{1.108}\right]\right\} & (\mathrm{Re} > 3400) \end{cases} \tag{6}$$

式中，Re 为流体的雷诺数；ε 为管壁的粗糙度。

从能耗损耗模型可以看出，决定不同排卤方案能耗差异的最关键因素在于注卤管线的长度和注卤排量，与管线长度呈线性正相关、与排量呈二次正相关。如下图所示，存在一个最优的路径满足水溶造腔阶段的能耗最低。

2.3 综合造腔速度预测模型

造腔速度主要反应在腔体壁面盐岩的溶解，溶解速度方程可以表述为：

$$\omega_0 = \frac{0.977}{\rho}(c_s-c)^{\frac{5}{4}} D^{\frac{3}{4}} \nu^{-\frac{1}{4}} x^{-\frac{1}{4}}(1.70+0.26c_s) \tag{7}$$

式中，ω_0 是盐岩侧溶速率，mm/h；ν 是卤水的运动黏性系数；x 是流动路径。

对整个腔体壁面进行积分可得各个腔体的造腔速度。因此，综合造腔速度可以表示为：

$$\omega_{\mathrm{sum}} = \sum_{i=1}^{i=N}\left(\iint_A \omega_0 dA\right) \tag{8}$$

式中，ω_{sum} 为多井造腔的综合造腔速度，m³/h。

3 排量优化方法及流程

实现盐穴储气库建设降本增效，必须识别能耗关键点，针对注水造腔工艺过程进行优化，避免注水二次循环。在满足造腔量及外输卤水浓度的要求下，建立多井造腔排量调配优化数学模型，进行多井溶腔工艺参数方案优化，保证一次循环多井采卤浓度满足外输要求，提高盐穴储气库建库效率。图1表示多井造腔的排量调配优化方法。该方法主要包括如下步骤：

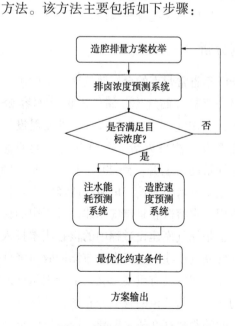

图1 多井造腔排量调配优化方法

① 读取目前金坛盐穴储气库中正在进行造

腔的井，选择参与多井联合造腔的井，并获取这些井的基本参数，包括但不限于：当前注水排量、注入浓度、排卤浓度、循环方式、管柱下深和当前腔体形态；

② 根据注水泵的量程，枚举可能出现的造腔排量分配方案；

$$Q_i = \forall Q \in [0, Q_{\max}] \quad (9)$$

式中，Q_i 为腔体 i 的水溶造腔排量，m^3/h；$Q_{\max}$ 为注液泵的最大排量，m^3/h。

③ 将造腔排量分配方案以及所获取的腔体的基本信息导入水溶造腔排卤浓度预测模型，获取每口井的排卤浓度，计算综合的排卤浓度；

④ 判断当前的浓度值是否满足卤水处理厂的要求，若满足，则跳入下一步，否则跳回第一步，进行下一组枚举的计算；

⑤ 将当前的多井造腔排量调配方案，带入注水能耗预测系统和造腔速度预测系统中，计算当前造腔能耗和造腔速度；

⑥ 判断是否完成所有枚举的计算，若满足，则进行下一步，否则跳回第一步，进行下一组枚举的计算；

⑦ 用户定义最优的约束条件，如在造腔速度处于 10% 的方案中，选择最低的注水能耗，对上述排卤浓度条件的排量调配方案进行对比分析，并选择最优的调配方案输出。

4 实例分析

位于江苏省常州市的金坛储气库，是中国乃至亚洲第一座盐穴储气库，储气库设计库容 26 亿立方米，工作气 17 亿立方米。一期工程投用后，管理者发挥注采灵活、短期吞吐量大的优势，注采气量逐年上升。到 2017 年，年采气量突破 7.3 亿立方米，为长三角地区调峰保供发挥了重要作用。水溶造腔是储气库建设过程中的核心步骤，也是储气库建设过程中的核心成本投入部分。开展储气库建设过程中的节能增效对降低储气库的建设成本具有重要意义。下图表示金坛储气库的注卤站与井位的管道分布图，不同井与注卤站之间的距离存在较大差异，因此导致整个水溶造腔过程中的能耗差异较大。

表 1 为当前 JT-1、JT-2 和 JT-3 的基本参数，包括循环方式、内管深度、外管深度、油垫深度、排量、注卤密度和排卤密度等。其中特别说明的是，卤水的浓度升高会导致密度的升高，因此可以通过实测卤水密度反算排卤浓度。

表 1 所选腔体的基本信息

参数	JT-1	JT-2	JT-3
循环方式	反循环	反循环	反循环
内管深度/m	1050	1104	1135
外管深度/m	1004.78	1084.8	1114.61
油垫深度/m	972.7	1050.3	1087.4
排量/m^3/h	75	80	86
注卤密度	1.01	1.02	1.02
排卤密度	1.197	1.178	1.196

4.1 规律分析

为对比不同排量方案下的注卤能耗和造腔速度。本节将分别从最大造腔速率和最小注卤能耗两个角度讨论多井造腔的排量优化方案。

先将 JT-1 的排量固定在 74.9m^3/h（如图 2 所示，此时的造腔速率最大），进行排量的优化研究。下图表示最大的造腔速率下注卤能耗和造腔速度的变化曲线。从图中可以看出随着 JT-3 排量的逐渐增加，造腔速率呈现先近似线性增加后趋于稳定的趋势，当 JT-3 排量从 75m^3/h 增加到 80m^3/h，造腔速率从 16.91m^3/h 增加到 16.94m^3/h，增幅仅为 0.16%。这主要是排卤浓度的限制，增加 JT-3 井的排量就必须减小 JT-2 井的排量。当 JT-3 井的排量低于 65m^3/h 时，随着 JT-2 排量的增加，注卤能耗近似线性增加，这主要是因为此时排卤浓度尚有余量，即使两者流量同时增加，也能满足排卤浓度的要求，同时造腔速度曲线也可以清晰看出此阶段的造腔速度也显著提高。当 JT-3 井的排量大于 65m^3/h 时，为满足排卤浓度，JT-2 井的排量逐渐减小，注卤能耗呈现先减小后增大的趋势，当 JT-3 井的排量为 77.5m^3/h，此时的能耗最小为 544.73kW。综合分析表明：当 JT-3 井的排量在 75~80m^3/h 这个区间时，能够获得一个较小的能耗以及较大的造腔速率，综合效率最高。

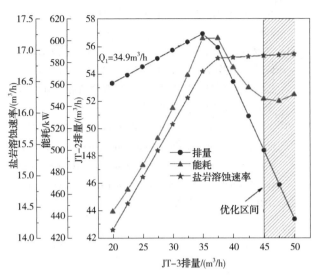

图 2　最大造腔速率下的排量优化

下图表示排量方案最优条件下（如图 3 所示，此时的注卤能耗最小），造腔速度和注卤能耗随 JT-3 井排量的变化曲线。从图中可以看出，随着 JT-3 井排量的逐渐增大，造腔速度呈现先增大后趋于平稳的趋势；而注卤能耗呈现先增大后平缓，随后迅速上升的趋势。因此，可以明显看出当 JT-3 井的排量在 67.5 到 75m³/h 区间时，此时的造腔速度较大但能耗且相对较小。例如，相较于 JT-3 井 80m³/h 的排量，当排量为 70m³/h 时，综合造腔速度仅减小 0.49%，但能耗可减小 7.40%。

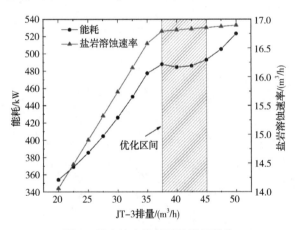

图 3　最小综合能耗下的排量优化

综上所述，多井造腔排量优化是一个极为复杂的过程，在追求最大造腔速度的同时势必会提高造腔能耗。因此，需要根据实际的生产建造需求，确定合适的优化条件。

4.2　选井方案优化

图 4 表示优化前的排卤浓度监测值。可以发现优化前排卤浓度达标率为 75.34%。在牺牲一定能耗和造腔速率的前提下，将多井造腔的综合排卤浓度达标率提高至 91.32%（如图 5）。

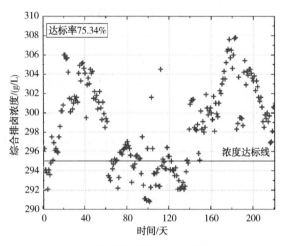

图 4　优化前的排卤浓度监测值

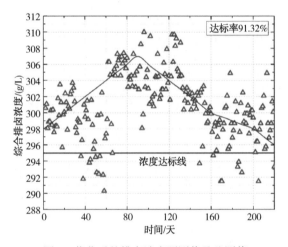

图 5　优化后的排卤浓度预测值及监测值

5　结论

为应对卤水处理厂对排卤浓度的要求并提高多井造腔协同效率，本章提出了以卤水浓度为目标，注卤能耗和造腔速度为约束的多井造腔排量调配优化方法，开发了相应的数字化软件模块，实现了多井造腔排量分配和选井方案对比分析，主要得出以下几点结论：

（1）不同造腔方案间的能耗损失相差 50% 以上，造腔速度相差约 14.4%，证明了开展造腔排量优化具有重要的工程意义。

（2）通过优化各井的排量分配，在牺牲一定能耗和造腔速率的前提下，将多井造腔的综合排卤浓度达标率提高至 91.32%。

（3）处于造腔阶段末期和初期的腔体之间的结合可以互相弥补，能够发挥互相之间的最大优势，实现排卤浓度的达标。

参 考 文 献

[1] 马新华，丁国生. 中国天然气地下储气库[M]. 石油工业出版社，2018，11.

[2] 班凡生. 盐穴储气库水溶建腔优化设计研究[D]. 中国科学院研究生院（渗流流体力学研究所），2008.

[3] 李淑平，刘继芹，齐得山，等. 盐穴储气库造腔管理分析系统设计与应用[J]. 石油化工应用，2016，35(8)：4.

[4] 耿凌俊，李淑平，吴斌，等. 盐穴储气库注水站整体造腔参数优化[J]. 油气储运，2016，35(7)：5.

[5] 杨海军，李龙，李建君. 盐穴储气库造腔工程[M]. 南京大学出版社：2018，04.

[6] Jinlong Li, Xilin Shi, Tongtao Wang, et al. A prediction model of the accumulation shape of insoluble sediments during the leaching of salt cavern for gas storage. Journal of Natural Gas Science & Engineering, 2016, 33: 792-802.

文23储气库完整性协同管理体系

许　锋　苏小健　张思远　杨佳坤　施玉霞　周栋梁　高立超　苗　刚　常　帅

（国家管网集团储能技术有限公司）

摘　要　地下储气库作为天然气行业上中下游、产供储销体系中的关键一环，具有强大的功能定位和市场需求。储气库完整性管理是储气库全生命周期安全生产和高效运行的重要保障。在分析总结文23储气库运行管理方法与技术创新成果的基础上，创新提出了储气库地质体-井-地面设施三位一体完整性协同管理概念、体系及技术内涵。系统集成了文23储气库盖层、断层动态密封性、储层稳定性、封堵井长期封堵有效性、水侵、结盐、出砂等地质体完整性监测评价及管理方法，管柱选材设计、固井质量、环空带压等井完整性监测评价及井控安全管理措施，地面设备设施检修保养、优化改造、节能降耗，数智化平台建设等储气库完整性协同管理能力提升举措，为气藏型储气库的完整性管理和安全生产运行提供了技术支撑和管理借鉴。

关键词　文23储气库；地质体；井；地面设施；完整性管理

Integrity Collaborative Management System Of Wen 23 gas storage

Xu Feng　Su Xiaojian　Zhang Siyuan　Yang Jiakun　Shi Yuxia
Zhou Dongliang　Gao Lichao　Miao Gang　Chang Shuai

（PipeChina Energy Storage Technology Co., Ltd.）

Abstract　As a key link in the upstream, midstream, downstream, production, supply, storage, and sales system of the natural gas industry, underground gas storage facilities have strong functional positioning and market demand. The integrity management of gas storage facilities is an important guarantee for safe production and efficient operation throughout their entire lifecycle. Based on the analysis and summary of the operation and management methods and technological innovation achievements of the Wen 23 gas storage, the concept, system, and technical connotation of the integrated and coordinated management of the geological body well surface facilities of the gas storage facility are innovatively proposed. The system integrates monitoring, evaluation, and management methods for the integrity of geological bodies such as the cap layer, dynamic sealing of faults, reservoir stability, long-term sealing effectiveness of sealed wells, water invasion, salt deposition, and sand production of the Wen 23 gas storage reservoir. It also includes monitoring, evaluation, and management measures for well integrity such as pipe selection design, cementing quality, and annular pressure, as well as well as well as well control safety management measures. The system also includes measures to enhance the collaborative management capabilities of gas storage facilities such as maintenance, optimization, and energy conservation, and the construction of intelligent platforms. This provides technical support and management reference for the integrity management and safe production operation of gas storage reservoirs.

Key words　Wen 23 gas storage; Geological body; Well; Ground facilities; Integrity management

1　引言

国家"十四五"规划对能源结构转型和能源行业发展提出明确要求：进一步加快清洁能源开发利用，推动非化石能源和天然气成为能源消费增量的主体，更大幅度提高清洁能源消费比重。完善能源产供储销体系，加强国内油气勘探开发，加快油气储备设施建设，加快全国干线油气管道建设。地下储气库作为天然气行业上中下游、产供储销体系中的关键一环，具有强大的功能定位和市场需求。

文 23 储气库是国家"十三五"、"十四五"重点工程项目，设计总库容 103.65 亿方，工作气量 40.02 亿方，注气规模 2400 万方/天，采气规模 3900 万方/天，是我国中东部地区库容最大、工作气量最高、调峰能力最强的地下储气库，地处中原腹地，贴近天然气消费需求市场，区位优势明显，承担大华北地区天然气应急调峰、市场保供的重要任务，可为国家管网多条天然气长输管道的平稳运行提供保障，对保障国家能源安全、推进能源结构转型、改善大气环境质量和提高人民生活水平具有重要意义。

文 23 储气库埋藏深，达 3200m，地层温度高，达 120℃，运行压力高，达 38.6MPa，地层水矿化度高，达 30×10^4 ppm，储层厚度大，达 400m，储层低孔低渗，断层发育，断块复杂。注采井、老井利用井、监测井、封堵井多达 155 口，是全国井数最多的储气库；地面工艺复杂，动设备多，对安全生产技术与管理要求严苛。储气库完整性管理是储气库全生命周期安全生产和高效运行的重要保障。为此，在总结提炼文 23 储气库建设与运行管理创新成果的基础上，创新提出了储气库完整性三位一体协同管理概念与技术内涵，首次创建了文 23 储气库地质体-井-地面设施三位一体完整性协同管理体系。

2 文 23 储气库完整性内涵

对于气藏型地下储气库完整性，既要从空间上承接管道与油气井完整性的特点，也要从时间上贯穿储气库全生命周期，确保地质体、井与地面设施功能始终处于安全可靠的服役状态。作为复杂断块砂岩气藏型储气库，文 23 储气库完整性内涵主要包括三个方面：（1）地质体、井与地面设备设施各单元在物理上与功能上是完整的；（2）在设计、建设、运营等全生命周期内始终处于可靠受控状态，防止各环节天然气泄漏事故的发生；（3）不断采取行之有效的完整性管理措施，确保文 23 储气库"注得进、存得住、采得出、输得稳"，在国家管网冬季保供、应急调峰中发挥气脉仓廪、冲峰担当的重要作用。

储气库完整性管理指对所有影响储气库地质体、注采井和地面注采设施三大单元完整性的风险因素进行识别和评价，并综合运用技术、操作和组织管理措施，将储气库运行的天然气泄漏风险水平始终控制在合理和可接受的范围之内。储气库完整性管理是对地质体、注采井和地面设施的一体化管理，是贯穿于储气库全生命周期的全过程管理，是应用技术、操作和组织措施的全方位综合管理。文 23 储气库完整性协同管理体系主要包含地质体-井-地面设备设施三位一体协同管理的内容。

3 文 23 储气库完整性协同管理体系

3.1 地质体完整性管理体系

地质体作为储气库系统的核心单元，其完整性状态是决定建设储气库的先决条件，也是储气库安全生产和高效运行的基本保障。在文 23 储气库多周期注采运行过程中，地质体主要存在盖层完整性失效、断层稳定性失效、封堵井密封性失效、水体侵入、结盐、出砂等风险。因此，地质体完整性管理的目的就是要预防上述六类风险的发生，通过地质体密封性、气水界面动态监测评价，结盐、出砂监测防控，以实现地质体泄漏风险预警，储气库注采平稳运行（表 1）。

表 1 文 23 储气库地质体完整性管理类别及方法

管理类别	管理方法
盖层完整性	持续监测盖层上部采油井日产气量及气质组分
	微地震监测评价
断层稳定性	持续监测断层两侧注采井、监测井压力系统动态变化
	微地震监测评价
封堵井密封性	强化地层、井筒封堵，多级段塞复合注入
	持续监测封堵井井口压力
水体侵入	优化注采井射孔方案，留足避水高度
	气水界面动态监测评价

续表

管理类别	管理方法
结盐	优化注采气方案，合理控制下限压力生产运行
	持续监测采出液氯离子浓度动态变化
	记录分析测井、测试过程中通井状况，及时发现结盐现象及程度
	做好抑盐、溶盐、除盐技术储备
出砂	科学优化开关井制度
	合理控制单井节流阀开度和生产压差

文 23 储气库上覆盖层为文 23 盐，盐层厚度 100-600m，岩性纯，厚度大，展布范围广，封闭性好(图 1)。通过持续监测分析盖层上部油藏采油井日产气量及气质组分，跟踪评价盖层封闭完整性。

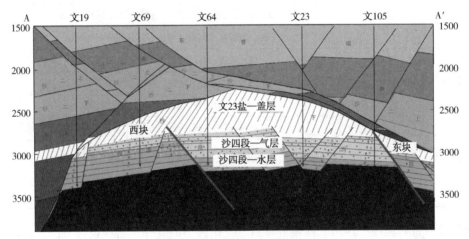

图 1　文 23 储气库过文 19-文 105 井气藏剖面图

通过井震结合、小层对比，综合分析断层特征，识别文 23 主块与边块断层两侧四种地层接触关系(图 1)；对比分析断层两侧注采井、监测井压力系统动态变化特征，结合微地震监测成果，综合评价分块断层稳定性及动态密封性。

针对气田废弃井井况差、地层亏空、堵剂突进、气库密封性要求高的难题，开展封堵剂体系室内研究、评价与现场试验，结合"强化地层封堵、井筒封堵并重、多级段塞复合注入"的封堵理念及工艺技术，实现了地层、管外及井筒全面高质量密封，46 口封堵井均未出现起压现象。

岩芯分析化验结果表明：储层段岩芯渗透率受速敏、酸敏、碱敏影响较弱，水敏影响中等，盐敏影响较强。气水相渗、气水互驱实验结果表明：水体侵入易导致气相渗透率下降，易出现水锁、结盐问题，降低有效库容动用程度；为防止边底水侵入，优化了注采井射孔方案，留足了避水高度，降低了水锁、结盐风险。根据各单井测井解释结果确定气水界面，优选气水界面附近的注采井定期开展含气饱和度测井，持续监测评价气水界面动态变化特征，通过优化注采运行降低水侵风险。

建立了储层和气井结盐模拟方法，揭示了结盐机理和主控因素，预测了地层结盐临界压力、温度、结盐量及井底结盐范围，制定了结盐防控措施。一是优化注采气方案，最大程度控制下限压力在设计范围内生产运行，防止或抑制地层及井筒结盐，降低结盐对储气库安全运行的影响。二是加大采出液氯离子浓度变化监测力度和测井、测试过程中通井状况记录分析，及时发现井筒结盐现象及程度。三是开展抑盐剂、溶盐剂优选实验及机械除盐措施优选及工艺研究，形成抑盐、溶盐、除盐技术储备。

在严格执行上级调度令指定日采气量的同时，科学优化开关井制度，合理控制单井节流阀开度和生产压差，防止地层出砂。

3.2 井完整性管理体系

储气库井完整性管理从时间跨度上应贯穿其设计、建设、运营等全过程。每一个过程的主要目标是对储存流体的有效控制与监测。文23储气库井完整性管理体系主要包括生产套管、注采管柱、固井结构的完整性管理。由于储气库具有短期内"强注强采"的生产特点，套管、油管柱材质及结构应适应注采循环、载荷交变的运行工况，并根据地下岩层岩性选择抗剪切、抗腐蚀材料。优化改进固井结构及固井水泥浆配方体系，确保地层-水泥环-套管的胶结质量、管柱丝扣密封性良好（表2）。

表2 文23储气库井筒完整性管理类别及方法

管理类别	管理方法
生产套管、注采管柱	多周次交变载荷条件下管柱拉伸/压缩强度评价
	多周次循环丝扣气密封评价
	气液两相动态腐蚀和应力腐蚀评价
	建立巨厚盐层地质条件下的管柱选材和设计方法
	油、套管柱选材设计与物资技术规格书的编制
固井质量	优化改进固井结构及固井水泥浆配方体系
	增大套管环空间隙，提高套管居中度
	采用高效前置液清洁井筒，大排量替浆
环空带压	识别分析潜在泄漏途径，开展环空带压风险评估
	持续监测环空液面，及时加注环空保护液
	制定井控分级管理制度，完善"一井一策"
	设计并建设丛式井场多功能辅助流程

在设计阶段，针对温度、压力、流量等参数变化引起的载荷交变，开展了多周次交变载荷条件下管柱拉伸/压缩强度试验及多周次循环丝扣气密封试验，开展了在CO_2、H_2环境下气液两相动态腐蚀和应力腐蚀开裂试验。针对盐层蠕变对管柱抗剪切性能的影响，建立了巨厚盐层地质条件下的管柱选材和设计方法。完善了注采管柱设计方法（图2），创新形成了地下储气库套管技术条件规范，指导了92口新钻井的油、套管柱选材设计与物资技术规格书的编制，取得了显著的现场应用效果。

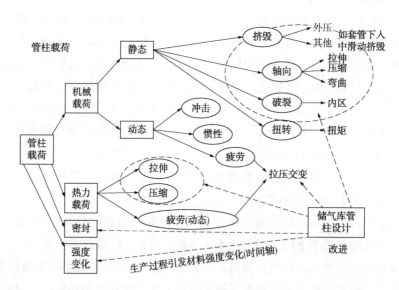

图2 考虑拉压交变载荷和密封性的储气库管柱设计技术流程

在建设期，主要是延长固井结构的寿命，确保地层-水泥环-套管的胶结质量、管柱丝扣密封性良好。为提高新钻井膏盐层固井质量，实施 2 口取芯井对膏盐层取芯，开展盐岩物化性质及其对固井水泥浆稠化、流变、强度、失水等性能的影响评价，以及工作液对盐岩浸泡、冲刷实验等室内实验研究，优化改进固井水泥浆配方体系；设计上增大套管环空间隙，应用软件模拟扶正器最佳位置，提高套管居中度，施工中开展地层承压试验，采用高效前置液清洁井筒，大排量替浆等措施，显著提高了文 23 储气库井膏盐层固井施工质量，在文 96、文 13 西、卫 11 储气库得到了广泛应用。

在运营期，主要是针对环空带压井的井控管理风险，开展了井筒完整性管理研究，识别分析潜在泄漏途径，建立了两级井屏障模型，开展了井屏障元素分解评价及风险评估。通过分析环空液面、压力恢复测试资料，研究确定环空带压原因及泄漏位置，制定了井控分级管理制度，完善了"一井一策"及泄压辅助流程管控措施。考虑环空泄压、保护液加注、应急压井等功能需要，同时受井池限制，注采井两侧无法直接连接压井管线，针对文 23 储气库生产运行工况，设计并建设了丛式井场多功能辅助流程（图 3），实现了环空泄压、保护液加注、应急压井等目的。组织环空带压井放压，单井最高降压达 23.86MPa，取得了较好的降压效果。

图 3 文 23 储气库丛式井场多功能辅助流程

3.3 地面设施完整性管理体系

与常规气田开发相比，储气库具有"大吞大吐、注采交替、高压运行"的特点，注采两套系统交替运行，注采气量波动范围大，对地面工程布局优化和设备管理运行灵活性要求高。文 23 储气库地面设施完整性管理体系主要包括采气树/井口装置完整性检测评价及管理，地面系统设备设施检修保养，注采流程、计量、增压、脱水系统优化改造、节能降耗，数字化、智能化平台建设及数据管理、实时监控、风险预警等（表 3）。

表 3 文 23 储气库地面设施完整性管理类别及方法

管理类别	管理方法
采气树/井口装置	定期开展天然气泄漏检测，做好维护保养
	推行单井 IPOS 巡检，提高巡护质量，及时发现并处理隐患
	高低压导阀更换改造，杜绝发生导阀泄漏事件
计量系统	丛式井场 π 型弯改造，提高单井计量精准度
注采流程	注采站内联络线增加调节阀门，注采转换期节约放空气量
	优化改造电伴热，降低冰堵事件发生率
	实施防雷接地隐患治理，提升地面设施防雷效果和运行可靠性
增压系统	针对进口机组气缸磨损问题，完成国产缸套选材及结构设计
	研发高压活塞环、高压大排量气阀等零部件制造工艺，并形成适配方案，降低故障频次，提高机组运行效率，实现国产替代
	开展增压系统能效评价，制定节能降耗措施，降低生产成本
脱水系统	开展脱水系统检修及优化改造，采气期减少三甘醇损失
	加装气液分离器撬，减少三甘醇污染，改善脱水处理效果
	开展脱水系统能效评价，制定节能降耗措施，降低生产成本
自控系统	优化上位机流程图、报警信息，提升自控系统安全性和稳定性
	构建数字化、智能化储气库监控运行管理体系，建立实体间各要素关联关系，建立生产运行管理系统、设备实施监测及应急管理安全监测体系，实现地下地上一体化安全生产运营管理。

定期开展采气树/井口装置天然气泄漏检测，做好维护保养，消除泄漏隐患。推行单井IPOS巡检，不断提高单井巡护质量，及时发现并处理隐患。完成单井高低压导阀更换改造，改造后效果良好，未发生导阀泄漏事件。

为解决采气期单井流量计受节流阀节流后噪声干扰，出现不计量或计量跳变等问题，开展了丛式井场π型弯改造（图4），提高了单井计量的精准度，确保了储气库生产数据的准确性。

图4 文23储气库丛式井场π型弯改造后流程

针对注采转换期因平压造成球阀磨损及放空浪费气量的问题，开展注采工艺流程优化改造，在注采站内联络线增加调节阀门，节约注采转换期放空气量。

针对投产初期采气期间发现的电伴热功率不足，漏电保护功能设计缺失的问题，实施了隐患改造，显著降低了冰堵事件的发生率，确保了冬季保供期间安全平稳采气。

针对一期工程地面设备防雷接地检测和检查中发现的接地不规范、防雷失效、浪涌保护未分级设置等隐患问题，实施了防雷接地隐患治理工程，对部分丛式井场、增注站、注采站接地装置进行了优化改造，显著提升了雷雨季节地面设施的防雷效果，全面提升了地面设施的运行可靠性。

针对进口压缩机组存在的气缸不同程度磨损问题，开展气缸缸套材料及结构研究，应用有限元仿真分析方法，完成国产缸套选材及结构设计。研发了高压活塞环、高压大排量气阀等零部件制造工艺，并形成了适配方案。成功实现了压缩机关键部件国产化，有效解决了进口压缩机组缸套、活塞等零部件异常磨损问题，机组运行情况持续向好，故障频次降低了52%，平均无故障运行时间由改造前的122小时提升至目前的7099小时。

利用能效分析平衡方程，建立了文23注采站地面工艺系统的能效分析模型及计算评价方法。计算得出：增压系统、脱水系统能量传输效率分别为99.23%、98.84%，能源利用率分别为61.43%、55.69%。构建了文23注采站地面系统适应性评价指标体系，计算得出：增压系统、脱水系统能效评价值分别为81.26、67.77。基于压缩机热力复算理论，利用数值模拟和Java编程技术，研发了往复式压缩机性能模拟程序（图5），实现了不同工况条件下的压缩机热力参数模拟计算，提出了四项增压系统节能降耗优化措施：提高一级进气压力，降低二级排气压力，降低二级进气温度，优化压缩机辅助系统（提高中间冷却器的换热性能，清除冷却器管束沉积物，尽可能减少设备内外泄漏、降低阀门节流能损）。

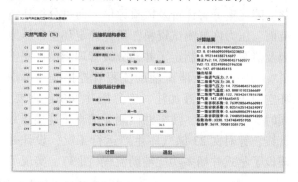

图5 文23储气库往复式压缩机热力复算程序

针对吸收塔三甘醇泄漏、冷凝水外排泵运行不稳等问题，开展了脱水系统检修及优化改造，采气期减少三甘醇损失。脱水系统加装气液分离器撬：由一台叶片分离器和一台聚结分离器组成，主要用于气液分离。介质先进入叶片分离器进行一次气液分离，降低介质含液量，再进入聚结分离器进行二次聚结分离，减少了三甘醇污染，改善了脱水处理效果，确保了文23储气库3112万方/天冲峰能力的发挥。

文23注采站脱水系统采用了贫、富液先换热再闪蒸和汽提再生工艺，具有自身优点，但分析认为该流程仍然存在贫、富液换热效果不理想、高压富液的压力能未得到有效利用、重沸器热效率较低等问题。利用HYSYS软件，模拟文23注采站脱水系统运行工况，定量分析运行参数对脱水系统能耗的影响，选取影响能耗的关键参数作为决策变量，建立脱水系统能耗最优化模型。利用HYSYS优化器，选取混合法（MIX）、序列二次型(SQP)算法进行求解，提出了四项脱水系统节能降耗优化措施：提高过滤器气液分离效率，改善三甘醇贫富液换热效果，高压三甘醇富液的压力能利用，提高重沸器热效率(采用超强燃烧器、利用辐射室强化传热技术)。

针对上位机报警信息描述不准确，流程图设置不规范，报警分类不清晰等问题，梳理 DCS 系统点数 3099 处，修改点数 727 处；优化流程图 63 幅；梳理 SCADA 系统参数点 2375 项，优化报警信息 786 处，自控系统的安全性和稳定性得到大幅提升。

构建了地质体-井-地面设备设施三位一体，含七大模块，28 项管理功能的信息化支持平台，文 23 智能储气库一体化管控平台(图 6)。通过全要素实体管理、全方位安全监测、科学化注采设计、智能化生产运行，实现生产运行、动态监测、设计优化、智能控制一体化管理与决策，为储气库全生命周期的安全生产和高效运行提供信息技术支持。建立了文 23 储气库实体编码体系及实体间的关联关系，集成了储气库建设与运行期地质、井筒、地面设备设施各类结构化数据、图形文档、实时数据、视频监控数据以及三维模型数据，建立了地质构造、井筒工程、地面工程、气藏管理、生产管理、压缩机管理、安全环保七大专题的数据结构。建立了地质体-井-地面设备设施三位一体仿真模型，构建了包括气藏、压缩机、生产、安全环保和运行分析的生产运行管理系统，建立了设备设施监测及应急管理安全监测体系，实现了地下地上一体化安全生产运营管理。

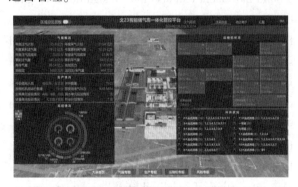

图 6　文 23 智能储气库一体化管控平台

4　结论

储气库全生命周期完整性管理是保障天然气地下储气库本质安全的有效手段，并向着技术体系化、标准规范化、数字智能化方向发展。针对文 23 储气库地质体、井下、地面工艺特点，实际运行工况及完整性管理难点，系统总结集成了文 23 储气库盖层完整性、断层稳定性、封堵井密封性、水体侵入、结盐、出砂等地质体完整性监测评价及管理方法，管柱选材设计、固井质量、环空带压等井筒完整性评价及井控安全管理措施，地面设备设施检修保养、优化改造、节能降耗，数字化、智能化平台建设等储气库完整性协同管理能力提升举措，确保了文 23 储气库各注采周期的安全生产和高效运行，为类似储气库的三位一体完整性协同管理提供了成功经验和典型示范，推广应用前景广阔。

参 考 文 献

[1] 马增辉，徐长峰，王明锋，等.气藏型地下储气库完整性管理体系及风险识别[J].中国安全生产科学技术，2022，18(3)：76-81.

[2] 魏东吼，董绍华，梁伟，等.地下储气库完整性管理体系及相关技术应用研究[J].油气储运，2015，34(2)：115-121.

[3] 罗金恒，李丽锋，王建军，等.气藏型储气库完整性技术研究进展[J].石油管材与仪器，2019，5(2)：1-7.

[4] 马新华，郑得文，申瑞臣，等.中国复杂地质条件气藏型储气库建库关键技术与实践[J].石油勘探与开发，2018，45(3)：489-499.

[5] 孙建华，周军，彭井宏，等.文 23 储气库增压系统性能模拟及优化研究[J].西南石油大学学报(自然科学版)，2022，44(2)：123-134.

[6] 周军，肖瑶，孙建华，等.储气库地面脱水系统能耗优化方法比选[J].天然气化工—C1 化学与化工，2022，47(2)：129-134.

Numerical study on cathodic protection effect of corrosion defects on pipelines under constant load

Dongxu Sun　Bo Wang　Lei Li　Yang Yu　Yi Ji

(Key Laboratory of Oil & Gas Storage and Transportation, College of Petroleum Engineering, Liaoning Petrochemical University, Fushun)

Abstract　In this work, the effectiveness of cathodic protection on corrosion defects of X100 pipelines under constant load was studied experimentally and numerically. A finite element model was designed to simulate the local potential and the distribution of anode/cathode current density inside the defect. The results show that the potential changes are relatively small after applying stress compared to when no stress is applied. When stress is applied, the rate of increase in anode dissolution current density is faster than that without stress, due to the concentration of stress at the bottom of defect, the current density increases significantly near the bottom of defect; before and after applying stress, there is no significant change in cathode current density, but due to stress concentration, there is a trend of cathode current density increasing within a certain distance near the bottom of the defect.

Keywords　cathodic protection; stress corrosion; finite element analysis; crevice; defect

Mechanical damage from manufacturing and construction, third party factors, ground movement and corrosion issues negatively impact pipeline safety and reliability. Especially the corrosion, as it is one of the main reasons for the reduced reliability of pipeline structures and the increased risk of failure. When pipelines are subjected to corrosion, the wall thickness of the pipeline decreases. This loss directly reduces the strength and stiffness of the pipeline, thereby affecting its load-bearing capacity and seismic resistance. If the corrosion process is not controlled, it can eventually lead to pipeline rupture or leakage accidents. Typically, corrosion results in several types of defects on the pipeline wall, such as macroscopic holes caused by widespread corrosion, corrosion pit caused by localized corrosion, and cracks caused by stress corrosion cracking (SCC). Compared to macroscopic corrosion holes, corrosion pits and cracks, due to their greater geometric depth, rapid propagation, and difficulty in detection, typically pose higher risks to pipelines.

Corrosion-induced loss of wall thickness reduces the structural reliability of pipelines, making them more susceptible to external loads such as earthquakes and water pressure, and increasing the risk of leakage or rupture. Stress corrosion can cause embrittlement of pipeline materials, leading to the formation of cracks under relatively low-stress conditions. This embrittlement may result in cracks that do not significantly reduce the wall thickness initially, but over prolonged use, the cracks will continue to propagate and eventually lead to pipeline rupture when it over prolonged use. In addition to the hoop stress generated by internal working pressure, ground movement can also induce significant longitudinal strain on the pipeline. In fact, in addition to the hoop stress applied on the pipeline and soil strain, there are typically localized stress and strain increments on the pipeline surface, such as corrosion defects and mechanical dents. These surface irregularities can lead to localized plastic deformation. The internal pressure at corrosion defect locations and the strain from soil contribute to a complex interaction of localized stress or stress concentration, which in turn impacts defect propagation. It has been confirmed that applied stress significantly enhances the pipeline corrosion. Plastic deformation will considerably increase mechanical electrochemical effects. Moreover, plastic deformation enhances both anodic and cathodic reactions, with a more pronounced effect on the anodic reaction. Cathodic protection (CP) is typically used in conjunction with coatings and is often considered the most effective corrosion prevention method. However, cathodic protection effectiveness may be reduced at corrosion pits and cracks. It has been observed through practical construction that after several years of applying cathodic protection, a significant number of corrosion defects, such as pitting and cracking, appeared on the external surface of the pipeline. Although cathodic protection design can be conducted through numerical methods, but there is still no model available to quantify the performance of cathodic protection under various geometric corrosion defects on pipelines. This experimental difficulty is typically due to the inability of measuring probes or electrodes to penetrate the environment inside pits and cracks. In addition, the effectiveness of cathodic protection on corrosion defects such as pitting and cracking has not received sufficient attention, particularly when considering the effect of stress on corrosion. The corrosion behavior of defects beneath detached coatings is currently not thoroughly researched. Therefore, the available references or works that can be cited in this field are very limited.

In this work, a solid mechanics model is incorporated on top of cathodic protection to investigate the mechanical-electrochemical effects of corrosion under stress at defects in high-strength pipeline steel exposed to NS_4 solution with a near-neutral pH. The reliability of the model was validated through various mechanical and corrosion measurements. Simulation parameters included stress distribution, corrosion potential, as well as anodic and cathodic current densities as functions of defect size and longitudinal tensile strain. It is anticipated that this study will provide a sufficiently reliable method for simulating and predicting localized corrosion in pipelines under complex stress and strain conditions, and offer recommendations for industry risk assessment and integrity management.

1 Model

1.1 geometric model

The specimen was cut axially from wall plates of X100 pipeline, which chemical compositions areas shown in Table 1. The specimen was machined into rectangular solids, each containing five cylindrical holes of different diameters to simulate corrosion defects on the pipeline wall. The schematic diagram of the geometric model is shown in Figure 1. The simulation employed NS_4 solution with a near-neutral pH. The solution is prepared with 0.483g/L $NaHCO_3$, 0.122g/L KCl, 0.181g/L $CaCl_2 \cdot H_2O$, and 0.131g/L $MgSO_4 \cdot 7H_2O$. The initial conditions for the finite element analysis adopted various electrochemical corrosion parameters derived from polarization curves of the steel, including corrosion potential, corrosion current density, Tafel slope, and exchange current density.

The schematic diagram of the geometric model is shown in Figure 1. Five cylindrical defects were positioned on the simulated specimen, with defect data as depicted in Figure 1.

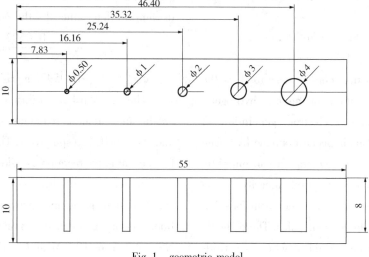

Fig. 1 geometric model

Table 1 chemical compositions

C	Mn	S	Si	P	Ni	Cr	Mo	V	Cu	Al
0.07	1.76	0.005	0.1	0.018	0.154	0.016	0.2	0.005	0.243	0.027

1.2 Conditions

Since all energy pipelines transport pressurized fluids, most of the stress on the pipeline is primarily from internal pressure. A solid stress simulation was performed on the pipeline, considering elastic-plastic behavior. An isotropic hardening model was selected, with the hardening function σryhard defined as follows.

$$\sigma_{\text{yhard}} = \sigma_{\exp}(\varepsilon_{\text{eff}}) - \sigma_{\text{ys}} = \sigma_{\exp}\left(\varepsilon_{\text{p}} + \frac{\sigma_{\text{e}}}{E}\right) - \sigma_{\text{ys}}$$

where, $\sigma_{\exp}$ is derived from the stress-strain curve of X100 steel, ε_{eff} represents the total effective strain, σ_{ys} is the yield strength of the steel, which is 800MPa, ε_{p} denotes the plastic strain, σ_{e} is the effective stress, E is Young's modulus at 207000MPa, and σ_{e}/E is the elastic strain. Elastic-plastic modeling employs the Von Mises yield criterion, applying different tensile strains longitudinally on the pipeline. It is understood that the applied tensile strain is equivalent to the total strain. Local strains may vary due to the presence of defects.

1.3 Reaction equation

The electrochemical anodic and cathodic reactions of X100 pipeline steel in deoxygenated, near-neutral pH NS_4 solution involve iron dissolution as the anodic reaction and hydrogen evolution as the cathodic reaction.

Anodic reaction: $Fe \longrightarrow Fe^{2+} + 2e$

Cathodic reaction: $H^+ + e \longrightarrow H$

The cathodic reaction here is described as the single-electron generation of atomic hydrogen, rather than the double-electron generation of hydrogen molecules. This is because atomic hydrogen may be generated, and its penetration can embrittle the steel and compromise structural integrity. However, some atomic hydrogen does not penetrate but recombines to form hydrogen molecules. The cathodic reduction of hydrogen ions in the reaction is considered and simulated here, without distinguishing the destination of the generated atomic hydrogen. Furthermore, in this study, it is assumed that the primary anodic reaction is the dissolution of iron in the steel, along with the dissolution of alloying elements with iron. Additionally, the pipeline steel is in an active dissolution state in the test solution. Modeling of steel corrosion relies on the description of iron dissolution. Corrosion of the steel is activation-controlled. Pipeline steels, including X100 steel, are in an active dissolution state in NS_4 solution. The electrode kinetics descriptions of the anodic and cathodic reactions are provided when applying cathodic protection potential.

$$i_a = i_{0,a} \exp\left(\frac{\eta_a}{b_a}\right)$$

$$i_c = i_{0,c} \exp\left(\frac{\eta_c}{b_c}\right)$$

$$\eta = \varphi - \varphi_{\text{eq}}$$

where, subscripts a and c respectively refer to the anodic and cathodic reactions. i represents the charge transfer current density of the electrochemical reaction, i_0 represents the exchange current density, φ is the electrode potential, φ_{eq} is the equilibrium electrode potential, η is the overpotential, and b is the Tafel slope. b_a and b_c are the anodic and cathodic Tafel slopes respectively. The equilibrium potentials for the dissolution of steel and the reduction of hydrogen are determined by the Nernst equation.

$$\varphi_{a,\text{eq}} = \varphi_{a,\text{eq}}^0 + \frac{0.0592}{2} \log[Fe^{2+}]$$

$$\varphi_{c,\text{eq}} = \varphi_{c,\text{eq}}^0 + 0.0592 \log[H^+] = -0.0592\text{pH}$$

The standard equilibrium potentials for the anodic and cathodic reactions are −0.409V (SHE) and 0V (SHE) respectively. The standard potential for the cathodic reaction is taken from the cathodic portion reaction of hydrogen reduction, but the standard state of atomic hydrogen is not clearly defined. Assuming a concentration of ferrous ions in the NS_4 solution of 10^{-6} M and a pH of 6.8, the equilibrium potentials for the anodic and cathodic reactions are calculated to be $-0.86V$ (SCE) and $-0.645V$ (SCE) respectively. The exchange current densities for the anodic and cathodic reactions are determined from the measured polarization curves. The electrochemical corrosion parameters are shown in table 2.

Table 2 Electrochemical corrosion parameters

electrochemical reaction	$\log i_0 (A/cm^2)$	Equilibrium potential (V, SCE)	Tafel slope (V/decade)	$i_0 (A/cm^2)$
Cathodic hydrogen evolution	-5.837	-0.645	-0.207	1.457×10^{-6}
anodic oxidation	-6.628	-0.86	0.118	2.353×10^{-7}

1.3.1 Anodic reaction

Gutman proposed the effect of elastic and plastic deformation on the equilibrium potential of the anodic reaction.

Elastic deformation: $\Delta\varphi_{a,ep}^{e} = -\dfrac{\Delta P V_m}{zF}$

Plastic deformation: $\Delta\varphi_{a,ep}^{p} = -\dfrac{TR}{zF}\ln\left(\dfrac{\nu\alpha}{N_0}\varepsilon_p + 1\right)$

Continuous elastic-plastic tension: $\Delta\varphi_{a,ep} = \Delta\varphi_{a,ep}^{0} - \dfrac{\Delta P_m V_m}{zF} - \dfrac{TR}{zF}\ln\left(\dfrac{\nu\alpha}{N_0}\varepsilon_p + 1\right)$

where $\varphi_{a,eq}^{e}$ and $\varphi_{a,eq}^{p}$ represent the displacement of the equilibrium potential of the anodic reaction under elastic and plastic deformation, respectively. $\varphi_{a,eq}^{0}$ is the standard equilibrium potential of the anodic reaction. ΔP is the overpotential equal to one-third of the uniaxial tensile stress (Pa), ΔP_m is the overpotential equal to one-third of the elastic deformation limit yield strength of steel, V_m is the molar volume of steel (7.13×10^{-6} m^3/mol), z is the charge number ($z=2$), R is the ideal gas constant (8.314J/mol K), $T = 298.15K$, F is the Faraday constant ($F = 96485$C/mol), v is the orientation factor ($v = 0.45$), α is 1.67×10^{11} cm^{-2}, N_0 is the initial density of dislocations before plastic deformation (1×10^{8} cm^{-2}), and ε_p is the plastic strain obtained through mechanical elasto-plastic simulation.

1.3.2 Cathodic reaction

Gutman illustrated through Evans diagrams that plastic deformation can lead to the redistribution of electrochemical heterogeneity and an increase in the cathodic reaction area, thereby enhancing hydrogen evolution in deaerated neutral pH solutions. Additionally, the increase in slip deformation, microcracks, and surface defects during plastic deformation can lower the activation energy for hydrogen evolution. This describes the mechanoelectrochemical effects on the cathodic reaction.

$$i_c = i_{0,c} \times 10^{\frac{\sigma_{Mises} V_m}{6F(-b_c)}}$$

where $i_{0,c}$ is the exchange current density for hydrogen evolution on X100 steel in the absence of external stress, σ_{mises} is the von mises stress calculated for iron, and b_c is the cathodic tafel slope.

1.3.3 electric field distribution in solution

according to the theory of electric fields, the general equation defining the distribution of the current field in the solution during electrochemical reaction processes is formulated as follows.

$$\nabla i_k = Q_k$$
$$i_k = -\sigma_k \nabla\varphi_k$$

where Q_k represents the general source term with exponent k, where the liquid is used for electrolytes (e.g., NS$_4$ solution) or the solid is used for metal electrodes (e.g., X100 steel electrode), σ_k is the conductivity, and φ_k is the potential. the conductivity of the solution and steel are 0.096S/m and 10^6S/m, respectively.

1.4 Numerical processing

The boundary conditions are set to be electrically isolated, except for the reaction interface, which is set as a free boundary. when the left end of the specimen is fixed, the right end is subjected to prescribed tensile strain as described above. the reaction interface is discretized using a triangular mesh, while the rest of the domain is discretized using a free tetrahedral mesh. a complete mesh consists of 139343 elements. the maximum and minimum element sizes are 5mm and 0.1mm, respectively, with a maximum element growth rate of 1.4. the solver is used for the solution.

2 Model validation

Based on the experimental studies by Xu, the effectiveness of the simulation was demonstrated by comparing the simulated results with the experimental results. The potentials of different defects and the current density of steel dissolution were compared. the comparison between the simulation and experimental results is shown in Figures 2 and 3. it can be observed from the figures that the error between the simulation and experimental results is very small, with only slight deviations at individual locations. This could be attributed to discrepancies between the assumptions or data parameters of the model and the actual conditions.

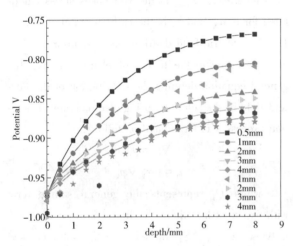

Fig 2 Simulation and experimental potentials

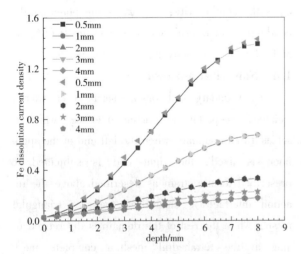

Fig 3 Simulation and experimental dissolution current density

3 Results

3.1 Corrosion defect stress concentration finite element simulation

The distribution of the potential field in the solution and von Mises stress at different depths of corrosion defects under fixed longitudinal tensile strain are shown in Figure 4. It can be observed that stress concentration is significant at the bottom of the specimen, with stress notably lower at the fixed constraint positions compared to the free interface positions. Additionally, stress concentration becomes more pronounced as the width of the defect decreases.

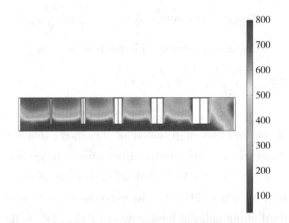

Fig 4 The distribution of the potential field in NS_4 solution and the von Mises stress at corrosion defects of different widths under fixed longitudinal tensile soil strain

3.2 Stress effect on the potential distribution inside defects

The three-dimensional potential distribution at the steel electrode in NS_4 solution at $-1.00V$ (SCE) are shown in Figure 5. Figures 5(a) and 5(b) represent the potential distribution on the steel electrode surface with and without applied stress, respectively, where the electrode surface potential is $-1.00V$ (SCE). As the depth of the defect increases, the potential gradually becomes more positive, with the potential being most positive at the bottom of the defect. Additionally, as the width of the defect increases, the potential at the defect decreases.

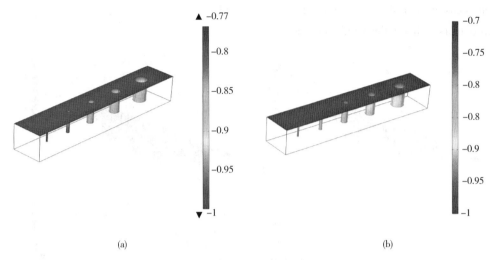

Fig 5 Potential distribution inside defects with and without stress

Figure 6 shows the variation in potential at defect locations under cathodic protection potential of −1.0V (SCE) for the steel, with and without applied stress. From the line graph, it is observed that the potential at the defect opening, i.e., within the range of defect locations of −10 to −8mm and 8 to 10mm, remains relatively constant at around −0.97V (SCE). The potential gradually becomes more positive with increasing defect depth. At the same defect depth, narrower defects exhibit more negative potentials, indicating a greater potential drop along the defect depth for narrower defects. It is visually apparent from Figure 6 that as the defect width increases, the potential at the bottom of the defect tends to stabilize and flatten out.

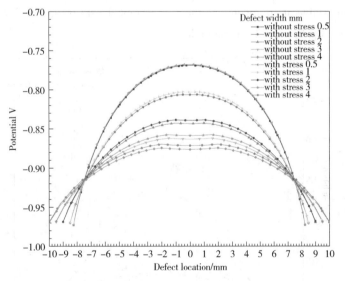

Figure 6 Potential changes at various defect locations with and without stress

3.3 Stress effect on the current distribution inside defects

Figure 7 shows the variation in three-dimensional steel electrode dissolution current density in NS_4 solution at −1.00V (SCE) before and after the application of stress. Figure 7(a) shows the distribution of steel dissolution current density without applied stress. Cathodic protection at −1.00V (SCE) can effectively mitigate steel corrosion outside the defects. However, due to the uneven distribution of cathodic potentials at the defect sites, some defect areas do not receive proper protection. Especially for small defects with diameters of 0.5mm and 1mm, the cathodic protection current cannot fully enter the defects, resulting in cathodic protection shielding at the bottom of the defects. Figure 7(b) shows the distribution of steel dissolution current density after the application of stress. Corrosion in defects with smaller widths is exacerbated.

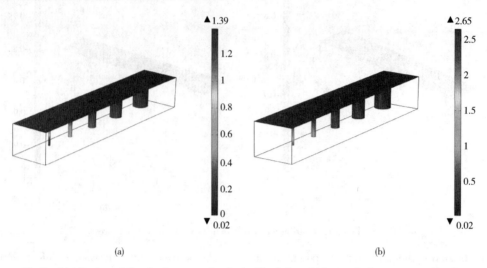

Fig 7　Distribution of dissolved current density inside defects before and after stress application

Figure 8 shows the variation in anodic dissolution current density at defect locations under cathodic protection potential of −1.0V (SCE) before and after the application of stress. As depicted in the line graph, the current density at the defect opening, i.e., at defect locations of −10 to −8mm and 8 to 10mm, remains stable at approximately the same magnitude, around $0\mu A/cm^2$. The magnitude of current density at each defect location changes with increasing defect depth. At the same defect depth, narrower defects exhibit higher current density rates. In the absence of applied stress, the current density at the bottom of the 0.5mm defect reaches $1.39\mu A/cm^2$. After applying stress, the current density spikes to $2.8\mu A/cm^2$.

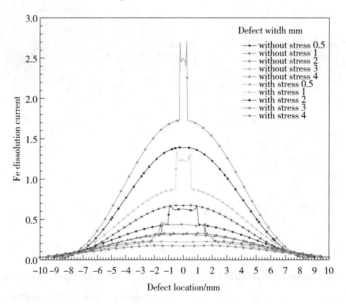

Fig 8　Changes in current density at various defect locations before and after stress application

Figure 9 shows the variation in three-dimensional hydrogen evolution current density when the steel electrode is in NS_4 solution at −1.00V (SCE). Figure 9(a) represents the distribution of hydrogen evolution current density without applied stress. For small defects with diameters of 0.5mm and 1mm, the cathodic protection current cannot fully penetrate the defects, leading to cathodic protection shielding at the bottom of the defects. Figure 9(b) shows the distribution of hydrogen evolution current density after applying stress. It can be visually observed from (a) and (b) that before and after applying stress, the hydrogen evolution current density at each defect location remains almost unchanged, with only slight variations at the bottom of the defects.

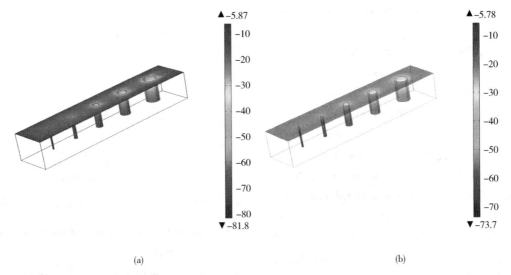

Fig 9 Distribution of hydrogen evolution current density inside defects before and after stress application

Figure 10 shows the variation in hydrogen evolution current density at different defect locations with and without the application of stress under cathodic protection potential. From the line graph, it can be observed that the cathodic current density decreases with increasing defect depth regardless of the presence of stress. At the defect opening where the steel is located, the cathodic current density is approximately $-56\mu A/cm^2$, whereas at the bottom of the 0.5mm defect, the current density is only about $-5.87\mu A/cm^2$. Wider defects allow cathodic current density to reach the bottom of the defect more easily. At the bottom of the defects, the application of stress leads to a noticeable increase in current density, particularly in larger defect locations.

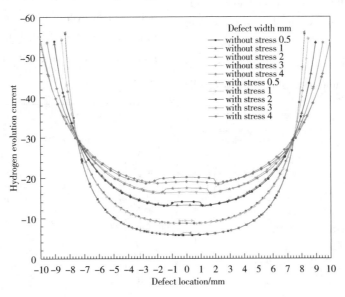

Fig 10 Changes in hydrogen evolution current density at various defect locations before and after stress application

3.4 Numerical simulation of potential at ellipsoidal defects

To better simulate corrosion defects occurring on pipelines in real-life scenarios, ellipsoidal defects were also simulated using software. Here, ellipsoidal defects of 0.1, 0.5, 1, 1.5, 2, 3, 4,

5, 6, and 7mm were uniformly cut out on the simulated specimens. Similar to cylindrical defects, the potential distribution around the ellipsoidal defects changes notably with increasing defect depth and decreasing width.

Both experimental tests and numerical simulations indicate that when the pipeline is at cathodic protection potential, the bottom of the corrosion defect is always associated with the most negative potential, presenting the smallest protected area. The cathodic protection performance at the bottom of the defect is crucial for the service life of the pipeline.

Figure 11 shows the potential distribution under cathodic protection for defects of different widths and depths on X100 steel under unstressed conditions. It can be observed that with increasing defect depth, the potential gradually becomes more positive. Additionally, the wider the defect, the easier the cathodic protection penetrates into the defect. In particular, when the defect width is less than 0.1mm, almost all cathodic protection is shielded, and the potential at the bottom of the defect approaches −0.72V, which is the corrosion potential of X100 steel in NS_4 solution. Even for larger defects, cathodic protection is partially shielded rather than fully penetrated. Cathodic protection can lose its effectiveness partially in narrow and deep defects. And Figure 12 shows the potential distribution under cathodic protection for defects of various widths and depths on X100 steel under applied stress conditions.

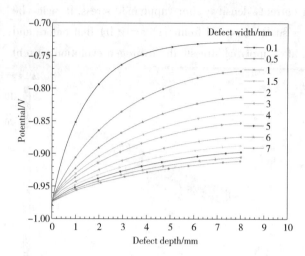

Fig 12 Potential at each defect location with applied stress

3.5 Numerical simulation of current at ellipsoidal defects

Figure 13 shows the anodic dissolution current density at various defect positions under unstressed conditions. It can be observed that the current density at the defect opening is 0, indicating adequate protection of the steel surface. However, as the defect depth increases for defects of various widths, the anodic dissolution current density increases, indicating a decrease in the effectiveness of cathodic protection within the defect. The increase in current density is not significant for wide defects. However, for narrow defects, such as those with widths of 0.1 to 0.5mm, a significant increase in steel dissolution current density is observed.

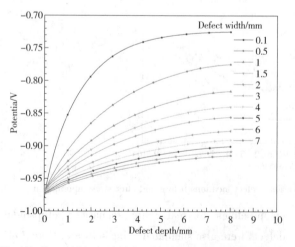

Fig 11 Potential at each defect location without applied stress

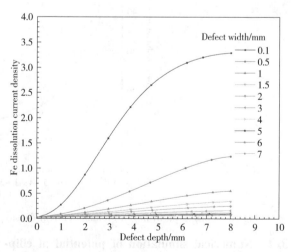

Fig 13 Anodic dissolution current density at various defect locations without applied stress

Figure 14 shows the anodic dissolution current density at various defect positions under applied stress conditions. It can be observed that the current density at the defect opening is zero, indicating that even under stress, the steel surface still receives adequate protection. At the same defect depth position, the smaller the defect width, the greater the anodic dissolution current density. As the defect depth increases, the rate of increase in anodic dissolution current density accelerates. After applying stress, there is a sharp increase in current density near the bottom of the defect, likely due to stress concentration at the defect bottom.

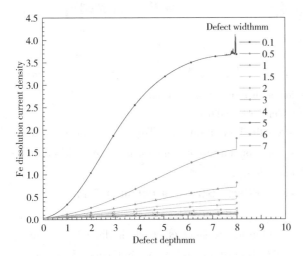

Fig 14 Anodic dissolution current density at each defect location under stress application

Figure 15 illustrates the cathodic current density at various defect positions under unstressed conditions. It can be observed that the cathodic current density decreases with increasing defect depth. At the defect opening, i.e., position 0mm, the cathodic current density is approximately $-58\mu A/cm^2$, while at the bottom of a 0.1mm defect, the current density is only about $-3.5\mu A/cm^2$, indicating that cathodic protection is shielded here, leading to cathodic protection failure. Additionally, the wider the defect, the easier it is for the cathodic current density to reach the bottom of the defect.

Figure 16 shows the cathodic current density at various defect positions under applied stress conditions. At the defect opening, i.e., position 0mm, even with applied stress, the cathodic current density remains around $-58\mu A/cm^2$. At the bottom of a

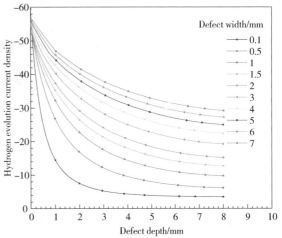

Fig 15 Cathode current density at various defect locations without applied stress

0.1mm defect, the current density is only approximately $-4.1\mu A/cm^2$. Due to stress concentration effects, there is a trend of increasing cathodic current density near the defect bottom within a certain distance.

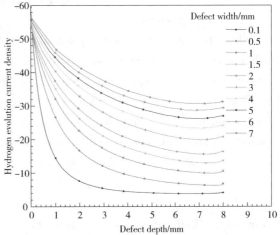

Fig 16 Cathode current density at various defect locations with applied stress

4 Discussion

4.1 Corrosion mechanism of cathodic protection under constant load conditions

Cathodic protection, as a commonly used corrosion prevention measure, can mitigate or prevent stress corrosion of metal materials. Oil pipelines often use cathodic protection and anti-corrosion coating methods to delay corrosion. However, factors such as the insulating performance of anti-corrosion coatings, stray currents, and corrosion defects can lead to under-protection or over-protection in certain areas.

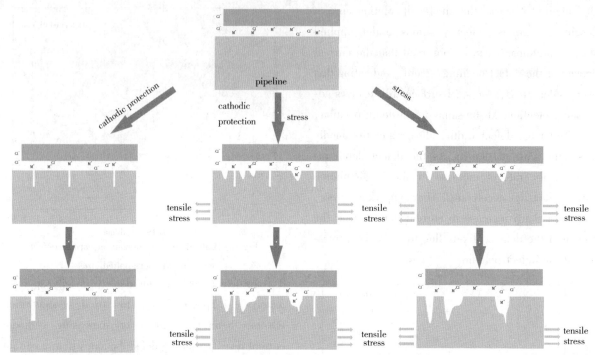

Fig 17 Schematic diagram of possible corrosion mechanisms of samples subjected to cathodic protection under constant load conditions

(1) Under-protection:

When the cathodic protection potential cannot reach a sufficient level of protection, the metal surface potential may be higher than normal. This can lead to intensified metal anodic reactions in the corrosive medium, accelerating its dissolution rate, especially in areas with defects in the sample where stress concentration and stress corrosion risks increase sharply. Weak protection may also cause excessive local current density on the metal surface, leading to localized corrosion and damaging the protective layer on the metal surface, further exacerbating the risks of stress concentration and stress corrosion.

(2) Over-protection:

In high-stress, high-temperature, and high-strength materials, over-protection can result in the generation of large amounts of hydrogen on the surface of the metal structure, leading to phenomena such as hydrogen embrittlement or hydrogen-induced stress cracking. This exacerbates the occurrence of stress corrosion, even causing damage to the metal structure. Similarly, excessively high cathodic protection potential and current density can result in rapid dissolution of the metal structure, leading to a reduction in metal material, lowering its strength and durability. This can cause localized stress concentration, increasing the risk of stress corrosion.

4.2 Effect ofstress on potential field

The influence of applied stress on the potential variation of the steel electrode surface is not clearly visible from the cloud maps in Figures 5(a) and 5(b). This indicates that the effect of applied stress on the potential of the specimen is not significant. From Figure 6, it can be clearly seen that under the action of applied stress, the potential at the defect site shifts more positively compared to when no stress is applied. Additionally, the trend of potential shift increases gradually with the increase in defect width, and when the defect width exceeds 1mm, a significant potential change near the bottom of the defect can be observed.

Comparing the data between the ellipsoidal defect simulation results in Figures 11 and 12, the potential variation is relatively small under applied stress compared to when no stress is applied. This observation is consistent with the simulation results of cylindrical defects.

4.3 Effect of stress on current density

From Figures 7(a) and (b), it is visually apparent that for small defects with diameters of

0.5mm and 1mm, the phenomenon of cathodic protection shielding at the bottom of the defect is more pronounced after the application of stress. The anodic dissolution current density significantly increases, with a more pronounced increase in the current density of steel dissolution at the bottom of the 0.5mm defect, reaching a peak at the bottom of the defect.

Figure 8 shows that due to the influence of stress, the change in current density at the defect location is more significant compared to when no stress is applied. After applying stress, the growth of current density at the same position is noticeable. The smaller the defect width, the more pronounced the effect of stress concentration on the defect, and the current density increases as the defect width decreases. Under stress concentration after stress application, there is a significant current variation at the bottom of the defect due to stress concentration.

Similarly, the comparison of ellipsoidal defect simulations in Figures 13 and 14 confirms this. After applying stress, as the defect depth increases, the rate of increase in anodic dissolution current density is faster compared to when no stress is applied. Smaller defects exhibit higher anodic dissolution currents, also influenced by stress concentration, resulting in a sharp increase in current density near the bottom of the defect.

4.4 Effects of stress on defects of different widths

From the comparison of contour maps in Figures 9(a) and (b), it can be observed that stress application has almost no effect on the cathodic current density. There is no significant difference in the cathodic protection current density at the defect location, while the cathodic protection on the reaction surface is enhanced by approximately 10.9% compared to when no stress is applied.

Figure 10 reveals that regardless of stress application, the cathodic current density decreases with increasing defect depth. After applying stress, smaller defect widths make it more difficult for the cathodic current density to reach the bottom of the defect, resulting in subtle changes in the cathodic current density. Due to stress concentration effects, there is a similar trend of increased current density at the bottom of the defect after stress application, as seen more prominently in Figure 8.

Comparing the data from ellipsoidal defect simulations in Figures 13 and 14, before and after stress application, there is virtually no change in the cathodic current density before reaching the bottom of the defect, with only a slight increasing trend at the bottom due to the influence of stress concentration.

5 Conclusion

A finite element model was established to investigate the mechanical – electrochemical effects of pipeline corrosion through coupled multiphysics simulations of solid mechanics and electrochemical reactions at the reaction interface, building upon cathodic protection. Longitudinal tensile strain was applied to induce local stress on the pipe wall, resulting in significant stress concentration at the bottom of the specimens, with stresses markedly lower at fixed constraint positions compared to free interface locations, and more pronounced stress concentration at locations with smaller defect widths.

During pipeline corrosion prevention using cathodic protection, the distribution of potential and current density at the surface defects of the pipeline becomes non – uniform. This is due to the resistive effect of the solution and the dissipation effect of the current, resulting in a potential drop formed within the defect. Consequently, the applied cathodic protection potential will be shielded at parts located at the bottom of the defect, leading to a reduced corrosion prevention effect of cathodic protection. This effect depends on the geometric shape of the defect and is enhanced with increasing defect depth and decreasing width.

By comparing the changes in potential and current density of specimens before and after applying stress on top of cathodic protection for pipeline corrosion prevention, it was observed that: there is minimal potential change after applying stress compared to before; the rate of increase in anodic dissolution current density is faster after applying stress than without, due to stress concentration at the bottom of the defect, resulting in a noticeable increase in current density near the defect bottom; there is no significant change in cathodic current density before and after stress application, but there is a trend of increasing cathodic current density near the defect bottom within a certain distance due to stress concentration.

References

[1] Liu Chuansen, Li Zhuangzhuang, and Chen Changfeng. "A Review on Stress Corrosion Cracking of Stainless Steel." Surface Technology 49. 03 (2020): 1-13.

[2] Zhao Shuai, et al. "Research Progress on Stress Corrosion Behavior and Protection Techniques of Duplex Stainless Steel." Surface Technology 51. 08 (2022): 123-134.

[3] Saleem, B., et al. "Stress corrosion failure of an X52 grade gas pipeline." Engineering Failure Analysis 46 (2014): 157-165.

[4] Guo Fuqiang, et al. "Stress corrosion behavior and microstructure analysis of Al-Zn-Mg-Cu alloys friction stir welded joints under different aging conditions." Corrosion Science 210. P1(2023).

[5] Cui, C., et al. "Two-dimensional numerical model and fast estimation method for calculating crevice corrosion of cross-sea bridges." Construction and Building Materials 206. MAY 10(2019): 683-693.

[6] Supornpaibul N., et al. "Stress corrosion crack initiation in filler metal 82 in oxygenated high-temperature water." Corrosion Science 214. (2023).

[7] Merwe J. W. van der, et al. "Prediction of stress-corrosion cracking using electrochemical noise measurements: A case study of carbon steels exposed to $H2O-CO-CO_2$ environment." Engineering Failure Analysis 144. (2023).

[8] Bueno, Ahs, E. D. Moreira, and J. Gomes. "Evaluation of stress corrosion cracking and hydrogen embrittlement in an API grade steel." Engineering Failure Analysis 36. 1(2014): 423-431.

[9] Han Yongdian, et al. "Welding heat input for synergistic improvement in toughness and stress corrosion resistance of X65 pipeline steel with pre-strain." Corrosion Science 206. (2022).

[10] Zhang Si, et al. "Effect of Oxide Metallurgy on Inclusions in 125 ksi Grade OCTG Steel with Sulfide Stress Corrosion Resistance." Materials 15. 13(2022).

[11] Xiao, Z. H., et al. "A quantitative phase-field model for crevice corrosion." Computational Materials Science 149(2018): 37-48.

[12] Kennell, G. F., and R. W. Evitts. "Crevice corrosion cathodic reactions and crevice scaling laws." Electrochimica Acta 54. 20(2009): 4696-4703.

[13] Cui, C., et al. "Two-dimensional numerical model and fast estimation method for calculating crevice corrosion of cross-sea bridges." Construction and Building Materials 206. MAY 10(2019): 683-693.

[14] Yadav Rajesh, et al. "Mechanical and stress corrosion cracking behavior of welded 5059H116 alloy." Corrosion Science 206. (2022).

[15] Li Yizhou, et al. "The crevice corrosion behavior of N80 carbon steel in acidic NaCl solution: The effect of $O2$." Materials and Corrosion 73. 2(2021).

[16] Cao Angang, et al. "Study on Stress Corrosion Properties of 1Cr17Ni2 Stainless Steel." Advances in Materials Science and Engineering 2022. (2022).

[17] Chang Litao, et al. "The effect of martensite on stress corrosion crack initiation of austenitic stainless steels in high-temperature hydrogenated water." Corrosion Science 189. (2021).

[18] Serafim Felipe M. F., et al. "Stress corrosion cracking behavior of selected stainless steels in saturated potash brine solution at different temperatures." Corrosion Science 178. (2020).

[19] Lin Shen, et al. "Stress corrosion cracking behavior of laser-MIG hybrid welded 7B05-T5 aluminum alloy." Corrosion Science 165. C(2020).

[20] KkochNim Oh, et al. "A Study on the localized corrosion and repassivation kinetics of Fe-20Cr-x Ni (x = 0~20wt%) stainless steels via electrochemical analysis." Corrosion Science 100. (2015).

[21] Song, F. M. "Theoretical investigation into time and dimension scaling for crevice corrosion." Corrosion Science 57. Apr. (2012): 279-287.

[22] Shidong Wang, et al. "Strain-shock-induced early stage high pH stress corrosion crack initiation and growth of pipeline steels." Corrosion Science 178. (2020).

[23] Anita Toppo, et al. "Effect of Nitrogen on the Intergranular Stress Corrosion Cracking Resistance of 316LN Stainless Steel." Corrosion 76. 6(2020).

[24] E. M. Gutman, Mechanochemistry of Solid Surfaces, World Scientific Publication, Singapore, 1994.

[25] Xu, L. Y., and Y. F. Cheng. "Experimental and numerical studies of effectiveness of cathodic protection at corrosion defects on pipelines." Corrosion Science 78. jan. (2014): 162-171.

盐穴储气库智能造腔预测与设计

王桂九[1]　陈加松[1]　赵廉斌[1]　李金龙[2]　王卓腾[2]

(1. 国家管网集团储能技术有限公司；2. 浙江大学)

摘要　盐穴储气库的水溶建腔设计和形态控制是保证最终储库容量和长期运行安全的重点。水溶建腔是一个非稳态的流-固对流传质过程，目前多通过室内模拟实验、仿真软件模拟进行设计和评价，十分依赖工程设计经验且费时较长。对此，本文提出基于人工神经网络的盐穴储气库水溶建腔快速设计及参数优化方法。利用盐穴储气库水溶造腔模拟软件(SSCLS)开展了1258组水溶建腔模拟，以其设计工艺参数(包括5个阶段的内管/外管深度、油垫深度、持续时间、注水流量和注水量)作为输入，对应的仿真成腔数据结果(有效容量、最大半径、形态)作为输出组成两组数据集，输入都为建腔设计参数，输出分别为腔体体积和最大半径/腔体形态。将数据集划分为训练集和测试集，利用训练集来训练人工神经网络模型，利用测试集进行预测精度检验。最终建成盐穴储气库体积和最大半径预测模型及形态预测模型。体积和最大半径预测模型预测成腔体积的平均绝对百分比误差(MAPE)为1.689%，预测最大半径的MAPE为3.067%，精度满足工程设计需求；形态预测模型预测的测试集在不同深度半径的平均绝对误差(MAE)约1.275m，其预测的一组实地造腔工艺参数，与实际腔体形态的半径MAE为2.83m。这证明神经网络在盐穴储气库水溶建腔这一高度非线性问题的预测方面具有较好的适用性。基于两种预测模型提出两种盐穴储气库建腔参数优化设计方案。首先，利用体积和最大半径预测模型对随机生成的百万组建腔工艺参数进行了预测和优选(耗时<30s)，并用SSCLS对优选出来的参数进行了造腔仿真，所得腔体形态完整规则、盐层利用率高，验证了所提出的参数优化方法的可靠性；然后，利用形态预测模型设计了一个循环程序，该程序能筛选出成腔形态满足目标形态的建腔参数。利用该程序耗时约51分钟从66万组参数中得到一组优化设计参数，其与目标形态偏差度低达0.041，证实了基于形态预测模型的工艺参数定制设计方法是可行且高效的。值得一提的是，本文中的数据集仅限来自于SSCLS软件，且采用了均匀盐层假设，在下一步工作中，可以获取实地数据进行训练，并考虑实际盐岩层的地质特征。

关键词　盐穴储气库；智能造腔设计；人工神经网络；形态预测

Prediction and design of intelligent leaching for salt cavern gas storage

Wang Guijiu[1]　Chen Jiasong[1]　Zhao Lianbin[1]　Li Jinlong[2]　Wang Zhuoteng[2]

(1. PipeChina Energy Storage Technology Co., Ltd.; 2. Zhejiang University)

Abstract　The design and shape control of energy storage salt caverns are key to ensuring the final cavern capacity and long-term operational safety. cavern construction is a non steady state flow solid convective mass transfer process, which is currently designed and evaluated through indoor simulation experiments and simulation software, relying heavily on engineering design experience and taking a long time. This paper proposes a rapid design optimization method for cavern construction based on artificial neural networks. A total of 1258 sets of cavern construction simulations are conducted using the SSCLS. The design process parameters (including inner/outer tube depth, oil pad depth, duration, water injection flow rate, and water injection volume in 5 stages) are used as inputs, and the corresponding simulation cavern results (effective capacity, maximum radius, shape) are used as outputs to form two datasets. The inputs are cavrn design parameters, and the outputs are cavern capacity and maximum radius/cavern shape, respectively. Divide the dataset into a training set and a testing set, use the training set to train the artificial neural network model, and use the testing set to test the prediction accuracy. Finally, a prediction model for the volume and maximum radius of the salt cavern, as well as a shape prediction model, are established. The Mean Absolute Percentage Error (MAPE) of the capacity and maximum radius prediction model for predicting the cavern capacity is 1.689%, and the MAPE for predicting the maximum radius is 3.067%, which meets the accuracy requirements of engineering design; The Mean Absolute Error (MAE) of the test set predicted by the shape prediction model at different depth radii is about 1.275m. The predicted set of on-site cavern construction parameters has a MAE of 2.83m compared to the actual cavern shape. This proves that neural networks have good applicability in predicting the highly nonlinear problem of cavern construction. We propose two optimization design methods for the construction parameters of salt cavern based on two prediction models. Firstly, the capacity and maximum radius prediction model is used to optimize the cavern construction parameters of a randomly generated million data (time<30 seconds), and SSCLS is used to simulate the cavern with the optimized parameters. The obtained cavity shape is complete and regular, and the salt layer utilization rate is high, verifying the reliability of the proposed parameter optimization method; Then, a loop program is designed using the shape prediction model, which can screen the cavern construction parameters that meet the target shape. Using this program, it takes about 51 minutes to obtain a set of optimized design parameters from 660000 sets of parameters, with a deviation of as low as 0.041 from the target shape. This confirms that the optimized design method based on the shape prediction model is feasible and efficient. It is worth mentioning that the dataset in this article is limited to SSCLS software and adopts the assumption of uniform salt layers. In the next step of work, field data can be obtained for training, and the geological characteristics of actual salt rock layers can be considered.

Key words Salt cavern gas storage; Intelligent cavity design; Artificial neural network; Morphological prediction.

1 引言

近年来，随着新能源的发展，地下储气库的建设需求日渐增长。盐穴储气库是一种通过注水溶腔的方式建设在地下沉积盐岩层中的储气库，由于盐岩具有良好的密闭性和损伤自修复等力学特性，在国际上被广泛用于石油、天然气储存。2020年，美国90%以上的战略石油储备在德克萨斯州和路易斯安那州的5个盐穴储库群中，总储存量达1.19亿吨；在天然气储存方面，世界上目前有90多个盐穴地下储气库，总工作气量超过280亿立方米，每天的日采气量约15.6亿立方米，占所有储气库的23%。另外，盐穴储气库还是储氢和压气蓄能的理想场所。水溶建腔指通过一组内外套管向深部盐岩层中注入淡水或不饱和卤水以溶解盐岩，同时排出高浓度卤水。经过数年乃至十数年的溶解循环，逐渐形成数十万方的空腔。水溶建腔的过程，实际上是一个淡卤水与盐岩之间发生的对流传质过程。故可以通过改变注水和排卤的位置、流量、浓度等边界条件，来改变腔体内的卤水流场、浓度场，从而改变腔体边界的溶解速度，来达到控制腔体形态和体积的目的。从盐穴储气库的安全性和经济性要求出发，对水溶建腔过程的优化设计提出了以下三个需求：

a. 通过良好的控制使其最终形成受力合理的接近椭球形的稳定形态，并能够保证盐层顶板的厚度；

b. 能够充分利用盐层资源，在固定的盐层中能够得到较大的成腔体积；

c. 具有更高的成腔效率。

对水溶建腔过程的控制方法包括在不同建腔阶段中调整内外管柱深度、油垫深度、循环方式等，在单个阶段也可以调整注水流量、注水浓度和持续的时间。不同工艺参数组合下的成腔行为难以通过简单的理论进行预测和控制。对此，研究人员广泛通过室内水溶建腔模拟实验、建腔模拟理论与数值方法等手段，对盐穴储气库的成腔机理、优化设计方法开展了大量的研究。

其中，室内模拟实验是认识水溶建腔机理、试验建腔控制方法的重要手段。包括通过室内盐岩溶解实验，认识盐岩的溶解机理，总结其与卤水浓度、卤水流速、界面角度、界面粗糙度、温度等因素的关系，并形成相关经验方程；通过淡卤水射流对流实验，研究注入淡水-排出卤水的流动模式和规律，提出全腔质量平衡、分层对流、浮羽流等降维、降阶的简化流场模型，以求解水溶建腔过程中的非定常三维湍流流场；以及通过开展室内缩尺的全周期水溶建腔模拟实验，通过实时调整和监测注水流量、注水浓度、内管深度、外管深度、油垫深度、循环方式等工艺参数，观察其对腔体形态发展的影响，深化对水溶建腔形状控制机理和方法的认识，或可以针对盐岩水溶建腔理论进行验证。但目前室内相关模拟实验多还只能定性地对建腔过程中的溶解、流动机理进行分析，由于缩尺相似不协调的问题难以解决，难以进行定量描述。

另一方面，建腔模拟数值工具方面，学者们基于盐岩溶解速度经验公式、淡卤水对流简化模型等，编制了包括CAVITY、SALT77、CAVSIM3D、SANSMIC、UBRO、INVDIR、CAVITA、CPSLS、SSCLS、HCLS、TWHSMC等水溶建腔模拟软件，其中SSCLS和HCLS为笔者前期所开发，考虑了中国层状盐岩中不溶杂质的沉积与盐岩溶解的相互影响。这些软件可以根据不同的地质、工艺参数，对水溶建腔过程以时间差分的方法进行计算仿真，求解动态的水溶建腔发展过程，并进行不同程度的数据和可视化的输出。对于特定的建腔工艺参数方案，它们可以定量的仿真和评价；但对于建腔设计方案的优化，需要多次重复工艺参数方案的仿真结果分析进行反馈调整。由于工艺参数的可调节范围非常大，且不同阶段的结果相互影响，这个过程非常耗时且非常依赖操作人员的设计经验。

近年来，由于人工神经网络特有的非线性自适应信息处理能力，基于人工神经网络的机器学习方法在工程设计及优化中得到了广泛的应用。盐穴储气库建腔预测的影响因素多、规律复杂且具有高度的非线性，有望通过神经网络得到新的解答思路，但目前相关研究为空白。

在本文中，提出利用以下基于人工神经网络的水溶建腔设计参数优化方法技术路线，将大幅度提高建腔设计优化效果和效率：

1) 在一定基础规则下随机生成千余组多阶段水溶建腔工艺参数，在本文中，暂仅引入了5个正循环阶段的内管深度、外管深度、阻溶剂深度、持续时间、注水流量；并利用盐穴储气库单井水溶造腔模拟软件SSCLS批量得到这些工艺参数对应的建腔结果，包括最终储库成腔体积、最大半径及腔体形态；

2) 以工艺建腔参数数据作为输入，以成腔体积和最大半径作为输出组建数据集一；以工艺建腔参数数据作为输入，以腔体形态作为输出组建数据集二。使用两个数据集对组建的神经网络模型进行训练，得到根据多阶段水溶建腔工艺参数直接预测成腔体积、平均半径的神经网络模型和根据多阶段水溶建腔工艺参数直接预测成腔形态的神经网络模型。

3) 使用评估参数分别对建立的两个神经网络模型进行预测精度评估。

4) 基于随机生成大量的工艺参数组合，利用上述两个模型对其成腔结果进行预测，对预测结果进行筛选得到优化的建腔设计工艺参数。

2 方法

2.1 数据集建立

2.1.1 体积和最大半径预测模型数据集

通常储库的建设需要经过数个阶段的参数调整，每个阶段中的注水管深度、排卤管深度、阻溶剂深度、注水流量、注水浓度和注水时间等，均会对本阶段的腔体发展产生影响，进而影响下阶段的发展(改变了下阶段的边界条件)乃至最终的腔体形态；在本文中，暂取5个正循环阶段的不同参数组合进行探究；其中，每个阶段的各个工艺参数在以下规则下随机生成：

1) 第一个阶段的注水管深度取为0，其他的深度取与该深度的相对值

2) 最后一个阶段的阻溶剂深度取为80m，以控制算例中最终腔体的高度一致为80m；

3) 各个阶段的注水管深度依次升高；每次升高不低于10m；

图1 部分SSCLS软件模拟的最终腔体形态截面图

4）各个阶段的拍卤管深度比阻溶剂深度低5m，且需高于各个阶段的注水管深度

5）各个阶段的阻溶剂深度依次升高（目前常用阻溶剂为柴油，在腔体发展后降低深度需大量的柴油，增加大量成本），每次升高不低于10m；

6）各个阶段的时间最低为30天，最高为200天，合计共600天（约两年）

7）由于工程上流量和浓度的调整范围较小，故阶段流量均取目前工程上比较常见的60m³/h，注水浓度取为0（纯淡水）；

在上述规则下，最终生成了1258组随机工艺参数组合，每组中包含5个阶段的注水管深度、排卤管深度、油垫深度、持续时间、注水流量和注水浓度；另外，在本文中，盐层中的不溶物含量均被假设为均匀的10%，未考虑不溶夹层的影响；我们将这些参数导入了前期开发的SSCLS软件，经过批量模拟计算，最终得到了1258组模拟腔体及其有效成腔体积、最大半径。这1258组5阶段工艺参数和对应的模拟结果，构成了体积和最大半径预测模型的数据集。

表1 部分体积和最大半径预测模型模拟数据

输入							输出	
阶段1					...	阶段5		
内管深度/m	外管深度/m	阻溶剂深度/m	持续时间/天	注水流量/m³/h	...	...	体积/m³	最大半径/m
0	22	25	41	60	...	...	86369.5	32.90
0	11	14	23	60	...	...	80778.9	27.72
0	21	24	59	60	...	...	75136.8	31.99
0	27	30	36	60	...	...	79405.8	28.45
0	8	21	52	60	...	...	76298.9	27.28

2.1.2 形态预测模型数据集

本节中数据集是在上节的1207组预测盐穴储气库体积和最大半径数据集基础上建立的。

在上述体积和最大半径预测模型输入数据集生成规则的基础之上，将注水流量与注水量合并为注水总量，即第5、6条规则合并为新的第5条：

5）5个阶段的总注水量合计共 60 * 24 * 600m³；

(5) The water injection volume for the 5 stages is 60 * 24 * 600m³ in total.

在上述规则下，生成了1207组随机工艺参数组合，每组中包含5个阶段的注水管深度、排卤管深度、油垫深度和注水量；同样将这些参数导入了前期开发的SSCLS软件，经过批量模拟计算，得到不同高度处的半径，作为形态预测模型的输出数据。输出的半径的高度间隔为5m。这样，我们得到了1207组工艺参数对应的造腔形态，也就是20m、25m、30m、…、75m、79m高度处的半径，作为输出数据集。最低记录高度取20m的原因是盐岩在溶解过程中会在底部产生沉渣，故形成腔体的底部比0m高，所有腔体实际上均高于15m，即该数据集中15m以下的半径全是0，无需预测。这1207组输入数据和输出数据，构成了盐穴储气库形态预测模型的数据集。

表2 部分形态预测模型模拟数据

输入						输出					
阶段1				...	阶段5	某深度处的半径/m					
内管深度/m	外管深度/m	油垫深度/m	注水量/m³			79	75	70	...	25	20
0	14	19	41760	...	...	9.03	15.72	16.067	...	0	0
0	20	25	66240	...	...	6.253	17.283	8.424	...	0	0
0	6	11	28800	...	...	0.608	4.603	8.077	...	11.812	0
0	9	14	30240	...	...	1.216	5.645	18.76	...	0	0
0	5	10	57600	...	...	3.127	20.931	21.8	...	23.363	0

2.2 备选神经网络结构

本文的目标是建立根据盐穴储库建腔工艺参数来预测最终的腔体体积和最大半径的预测模型和预测最终腔体的形态(不同深度处的半径)的预测模型,属于回归问题。其潜在可用的回归神经网络模型有 BP 神经网络、循环神经网络、卷积神经网络等,本文考虑数据的性质和神经网络的契合度,重点选取以下三种神经网络进行训练和筛选。

1) bp 神经网络

BP 神经网络是人工神经网络中常见的网络之一,是一种基于误差反向传播(BP)算法的人工神经网络。其拓扑结构包含一个输入层、几个隐含层和一个输出层,每一层由许多神经元组成,如图 2 所示。BP 神经网络的学习分为前馈和后馈两个阶段。当信号向前传播时,输入层神经元将输入信号传递到隐藏层。隐藏层中信号经过每个神经元的加权和计算后,由其激活函数得到其和,作为输出信号输出到下一层。最终信号到达神经网络的输出层,由输出层运算后输出预测结果。当信号向后传播时,神经网络的预测结果和实际数据之间的偏差从输出层向后传输到输入层,并根据误差调整其内部参数(权值和偏置项)进行学习。BP 神经网络学习的过程就是信号不断向前向后传播的过程,最终使得其达到较好的预测效果。BP 神经网络通过引入激活函数在网络中引入非线性,进一步增强学习能力,使得其适用于逼近非线性映射关系。BP 神经网络已被成功用于风力发电\工程造价\岩土参数等工程问题预测,因此选用 BP 神经网络作为潜在待筛选的神经网络模型之一。

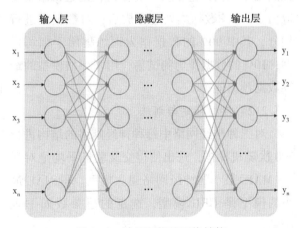

图 2 bp 神经网络的网络结构

2) 长短期记忆网络(LSTM)

LSTM 由 Hochreiter 等人引入,是第一个克服梯度消失和爆炸问题的循环网络体系结构。因为 LSTM 隐藏层细胞之间是有连接的,隐藏层的输入不仅包括输入层或上一隐藏层的输出还包括上一时刻隐藏层的输出(见图 3),所以 LSTM 能处理时间序列问题。其已被广泛用于股票预测,机器翻译等领域。工艺参数预测腔体形态也属于时间序列问题,随着建腔时间的变化,最终的腔体形态也随着时间变化,且后序预测的结果与前序的结果有关。考虑 LSTM 的时间序列处理能力,将其作为潜在待筛选模型之一。

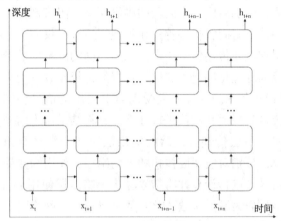

图 3 LSTM 结构

3) 门控循环单元(GRU)

GRU 网络由 Cho 等人提出,主要目的是简化 LSTM,与 LSTM 相比,GRU 宏观结构和其相同,它可以被描述为一种更简单的门控机制。在保持了 LSTM 的效果同时又使其细胞结构更加简单,需要的参数更少,训练时间更短,是一种非常流行循环神经网络。

2.3 模型构建与训练

2.3.1 体积和最大半径预测模型

2.3.1.1 数据处理

前述预测腔体形态和最大半径数据集的输入数据中,注水流量为恒定值,无需纳入训练数据集;另外外管深度取决于阻溶剂深度,为非独立变量,也无需纳入;故最终的输入数据仅包括五个阶段的内管深度、阻溶剂深度和持续时间,共 15 项。输出数据包括有效体积和最大半径 2 项。为提升模型的收敛速度和精度,训练前将输入和输出数据进行归一化处理——即将此数据集中的各列数据点分别映射到[0, 1]区间(最终模型的预测结果还需经过逆归一化处理使之返回本身的取值空间)。

归一化完成后对数据集进行划分,其中90%的数据作为训练集用于训练模型;10%的数据作为测试集,用来评估模型的最终数据预测能力。

2.3.1.2 模型搭建与训练

本节通过训练对不同神经网络的结构超参数和训练超参数进行优化,并通过最终模型在验证集上的平均绝对百分比误差(MAPE)越小模型越精准为原则来确定最高效的超参数,构建出根据多阶段水溶建腔工艺参数预测成腔体积和平均半径的BP神经网络模型。

该BPANN模型的拓扑结构由输入层、隐藏层和输出层构成。其中输入层由15个神经元组成,分别对应五个阶段的内管深度、阻溶剂深度和持续时间,这些神经元不执行运算,仅向下一层神经元传递信号。在输入层后面的隐藏层数量经训练优化确定为3层,各层的神经元数量均为512,其内的神经元接受来自上一层的信号,结合自身权值和阈值进行计算,并将结果作为信号向下一层传递。最后的输出层由2个神经元组成,分别对应建腔体积和平均半径,作用为接受信号,进行计算并将结果输出。隐藏层使用relu函数作为激活函数,此函数可以给神经网络引入非线性因素,使得其具有较强的非线性函数映射能力;输出层不设激活函数。

模型训练采用的超参数,包括损失函数Loss、评估标准Metrics、优化器Optimizer、迭代次数Epochs、批尺寸Batch size、回调函数Callbacks、初始化方法Initializer也根据训练过程进行了优化(如表3)。其中,考虑神经网络的学习率需要随着模型的逐渐收敛而降低以防出现震荡,本模型使用ReduceLROnPlateau函数作为回调函数使得模型的学习率可自动更新,其参数设置为:monitor = val_ loss,factor = 0.2,patience = 5,min_ lr = 0.001;

表3 模型训练采用的超参数

超参数	设置
损失函数	mean_ squared_ error(MSE)
矩阵	mean_ absolute_ error(MAE)
优化器	Adam
训练轮次	200
批尺寸	5
召回函数	ReduceLROnPlateau
初始值	normal

对本节所建立的体积和形态预测模型模型进行训练,图4为模型误差函数值MSE与模型训练轮次之间的关系。MSE的值随着训练轮次的增加而逐渐降低,且降低速率逐渐减小,最后趋于稳定。在第100个训练轮次之后,模型训练集上的MSE基本稳定在0.0003左右,验证集上的MSE基本稳定在0.0033左右,至此可认为模型已经完全收敛。

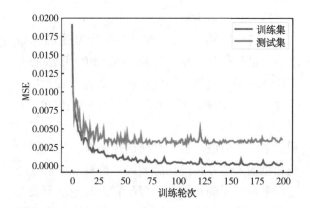

图4 体积和最大半径预测模型在训练集和测试集上的损失函数趋势图

2.3.2 形态预测模型

2.3.2.1 数据准备

前述预测腔体形态工艺参数输入数据中,外管深度取决于油垫深度,为非独立变量,无需纳入训练数据集。故最终的输入数据仅包括五个阶段的内管深度、油垫深度和注水量,共15项。为提升模型的收敛速度和精度,训练前也将输入和输出数据进行归一化处理——即将数据集中的各列数据点分别映射到[0,1]区间(最终模型的预测结果还需经过逆归一化处理使之返回本身的取值空间)。数据集处理完成后,对数据集进行划分,其中90%的数据作为训练集用于训练模型;10%的数据作为测试集,用来评估模型的最终数据预测能力。

2.3.2.2 模型搭建与训练

对三种神经网络中每种神经网络的结构进行200次随机抽样,选出使得交叉验证平均绝对误差(MAE)最小的神经网络建立模型。得MAE最小的模型是GRU模型,选定的超参数如下表所示。

表4　GRU模型采用的超参数

超参数	设置
隐藏层数量	3
隐藏层单元数	250；125；63
输出层单元数	13
初始值	He_normal
隐藏层激活函数	Hard_sigmoid
输出层激活函数	Sigmoid
优化器	Adam
召回函数	ReduceLROnPlateau
批尺寸	100
训练轮次	600

使用训练集对本节建立的预测不同深度处盐穴储气库半径的模型进行训练，再根据模型在测试集上的表现来评估模型。图5反映了模型损失函数值和模型迭代次数之间的关系。由图可见，MSE随着模型的训练轮次增加逐渐降低，最后趋于平缓。最终测试集的MSE稳定在0.004，这说明所建立的形态预测模型已经训练至完全收敛。

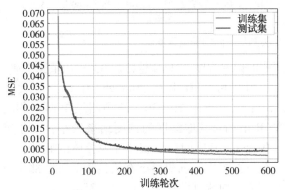

图5　形态预测模型在训练集和测试集上的loss趋势

3 结果

3.1 体积和最大半径预测模型精度评估

我们利用平均绝对百分比误差（MAPE）和相关系数R对体积和最大半径预测模型进行了预测精度评估，其中R是研究模型输出值和实测值之间线性相关程度的量，当R越接近1时，表示模型的参考价值越高[28]。

如图6为模型输出的成腔体积和不同数据集内实测的成腔体积之间的误差和R。可以看到在训练集上模型预测精度较高，模型输出值与实测值的R和MAPE分别为0.998和0.509%。而测试集的R和MAPE则要略微低于训练集，R为0.962，MAPE为1.689%。此预测在成腔体积值的不同区间的准确程度不同：相对而言成腔体积较大的部分，测试集的预测准确度更高、MAPE更小。取其中体积实测值较大的一半数据，其在测试集上的MAPE为0.724%。而成腔体积较小的区间里，个别预测数据与实测数据的偏差较大，其原因可能是在成腔体积较小区间里的数据量偏少使得模型缺乏训练。由于本模型的最终目的是优选成腔体积大的工艺参数，所以相对而言成腔体积较大区间内的预测精确度更为重要，故可以认为本例能够较好地满足对成腔体积的预测准确度要求。

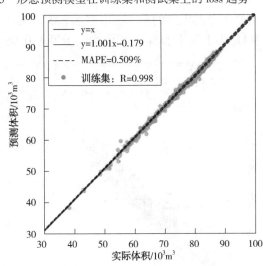
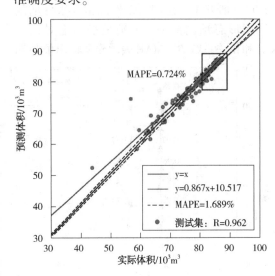

图6　体积输出值与体积实测值之间的偏离程度

如图7为预测的最大半径和不同数据集内实际的最大半径之间的MAPE和R。最大半径的测试集的R为0.892，MAPE为3.067%，整体而言最大半径的预测准确度要低于成腔体积；但3%左右的 MAPE，仍在工程可接受的误差范围之内。

由于每次神经网络初始化的权值和阈值是随机产生的，所以每次训练出的模型对测试集的预测结果有一定差别。为了使测试结果具有普遍性，我们采用 10 次训练来综合统计测试集的表现，每次训练模型前将数据集顺序随机打乱，按 8∶1∶1 重新划分训练集、验证集和测试集。10 次训练中，成腔体积的 MAPE 均值为 1.838%，而最大半径的 MAPE 均值为 3.144%，可以认为此盐穴储气库体积和最大半径预测模型具有根据多阶段水溶建腔工艺参数直接预测成腔体积和最大半径的能力，预测精度能够满足工程实际需求。

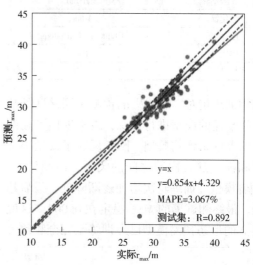

图 7　最大半径输出值与最大半径实测值之间的偏离程度

表 5　重复训练 10 次的平均绝对百分比误差

数据类别	MAPE			
	最大值	最小值	均值	方差
体积	2.042	1.672	1.838	0.017
最大半径	3.452	2.948	3.144	0.032

3.2　形态预测模型精度评估

3.2.1　测试集预测精度评估

在这项研究中，使用两个性能指标来评估盐穴储气库形态预测模型，分别是平均绝对误差（MAE）和相关系数 R。MAE 表示模型输出值与期望值的偏差程度。MAE 越接近 0，输出值越准确。

图 8 可以看出在 30m 处 MEA 为最大值 2.3m；实际上对 20~30m 的预测结果偏差均较大（R 值相对其他深度的预测结果较小，小于等于 0.95），其原因在于该部分为腔体的底部，其形态与不溶物的沉积形态有关，相对其余部位更为复杂，预测难度较大。但 20~25m 处的半径本身比较小，故 MAE 也较小；而 30m 处的 MAE 值则为全部深度的最大值。其余（35~79m）点的 MAE 小于等于 1.5m，R 值均大于 0.96。

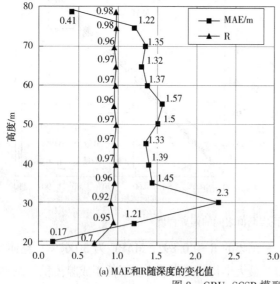

(a) MAE 和 R 随深度的变化值

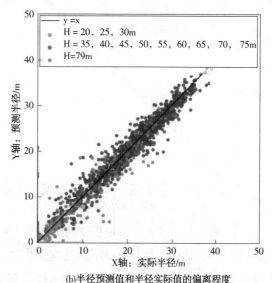

(b) 半径预测值和半径实际值的偏离程度

图 8　GRU-SCSP 模型在测试集上的结果

在测试集(未参与训练)中随机抽取十组数据,将数据和模型对应的预测结果作图,得图9。由图可知,该模型除对第三、四、五组数据预测效果稍差外,对其他组数据均能取得较好的预测效果。至此认为本形态预测模型能够较好地满足对各个深度处半径的预测准确度要求。

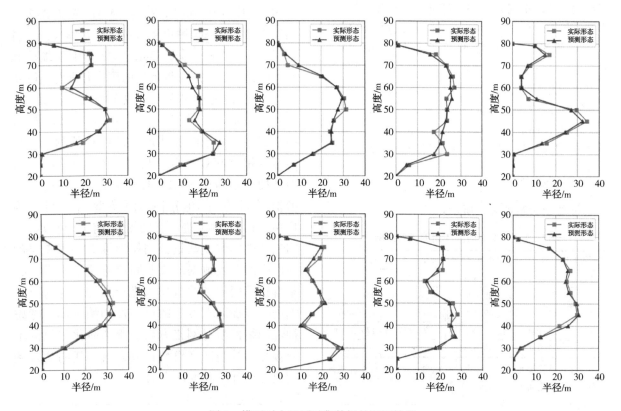

图9 模型对十组测试集数据的预测结果

3.2.2 实地腔体 JT52 形态预测精度评估

通过5.1节中的模型评估,展示了该模型能够较好的预测各个深度处半径(平均误差1.5m),但由于其基础数据集是由数值软件模拟而来,该方法对现场腔体形态的预测能力尚待验证。因此,本节中,我们演示使用该模型对一组实地造腔数据的设计和预测过程。

首先获许一组实地试验工艺参数和其对应的腔体形态。我们选用 JT52 洞室作为本次研究的对象。JT52 洞室是江苏金坛地下地质研究所最早建成的盐洞室之一,建于2003年至2010年,洞的体积约为 186675m³,深度范围为 −1089.5m 至 −1015.4m。整个洞室施工过程共7个阶段,每个阶段包括4个不同工艺参数,包括内外管深度、油垫深度和注水量,因为所提出的形态预测模型目前只支持预测5个阶段的数据,故只取前5阶段数据(见表6)和5阶段结束后的洞穴形态(图10)。

表6 Technological parameters of cavern JT52

阶段	内管深度/m	外管深度/m	油垫深度/m	注水量/m³
1	−1129.2	−1111.9	−1095.2	42134.4
2	−1117.1	−1065.5	−1053.3	107308.8
3	−1113.4	−1065.0	−1051.1	696528
4	−1099.1	−1035.7	−1035.2	316648.8
5	−1085.2	−1025.5	−1025.4	363235.2

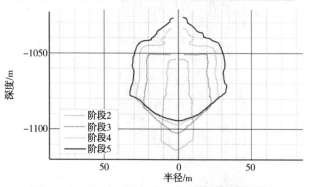

图10 JT52 洞室5个浸出阶段结束后的洞穴形态

由于 JT52 的工艺参数与本文的假设不完全符合,故在模拟之前,需要先按照量纲分析对这些工艺参数进行等比例的缩放:

1. 第一阶段的注水管深度取为0,其他的深度取与该深度的相对值;

2. JT52 的高度(采动区域)为 103.8m 缩放为第三节假设中的 80m，缩放比例为 a=80/103.8。对应的各阶段内外管深度、油垫深度也同时乘以 a 等比例缩放，注水量则乘以 a3，对应的，预测完成后对预测结果中的腔体高度和半径乘以 1/a 进行反缩放；

3. 缩放后的 JT52 的总注水量为 $6.985*10^5 m^3$，第三节假设中为 $8.64*10^5 m^3$，则各阶段注水量仍需进行二次缩放，缩放比例为 x = 8.6/6.98，由于注水量增大之后，腔体总体积等比例增大，而腔体高度不变(油垫保护)，则各处半径增大，故还需之后预测出的腔体半径乘以 $\sqrt{\dfrac{1}{x}}$ 进行反缩放。

缩放后参数如表 7 所示：

表 7 腔体 JT52 的缩放后的工艺参数

阶段	内管深度/m	外管深度/m	油垫深度/m	注水量/m³
1	0	13.333	26.204	43200
2	9.326	49.094	58.497	60774.69
3	12.177	49.480	60.193	374400
4	23.198	72.062	72.447	179335.08
5	33.911	79.923	80	205719.44

备注：外管深度对成腔形状没有影响，故忽略数据生成规则中外管深度比油垫深度低 5m 这一要求。

将形态预测模型预测结果、和实际的声纳测腔结果、以及 SSCLS 的模拟结果进行对比，如图 11 所示。SSCLS 模拟结果与实际形态的半径 MAE 基本在 3m 以内，其 MAE 为 1.292m；相对而言，形态预测模型预测的误差偏大，在各个深度处的 MEA 为 2.83m，可能是由于缩放和反缩放，以及对地层等信息的简化假设与实际不符。但该误差仍在工程可接受范围之内。

本例充分说明，所提出的形态预测模型可用于实际工程中的腔体形态设计和预测，其预测精度相对 SSCLS 低但仍在工程要求范围内，但其拥有远高于 SSCLS 的效率。SSCLS 的单次模拟时间约在 10 分钟，而模型的时间约为 $1.14*10-4$ 秒。进一步的提升方法，可以减少第三节中的假设数量，例如训练更多含夹层的地层中的造腔数据，或非固定阶段数量等，甚至直接采用现场的数据来进行训练，这将有可能进一步提高本方法的可靠性和精确度。考虑神经网络模型对网格方法、流场简化假设的不依赖性，神经网络模型将成为新一代的盐穴储库建腔设计工具发展方向。

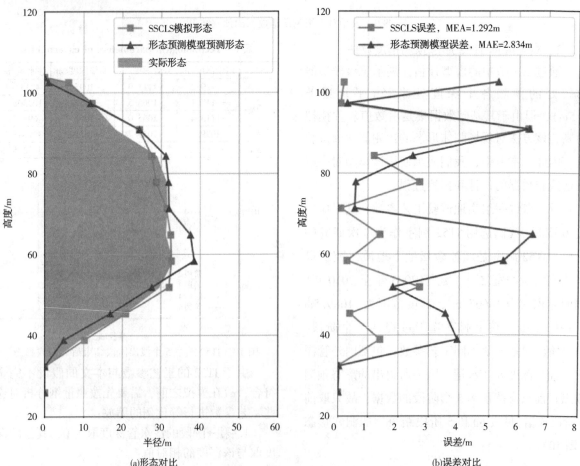

(a)形态对比　　(b)误差对比

图 11 GRU-SCSP 预测半径和 SSCLS 模拟半径与实际半径的对比结果

4 讨论

4.1 基于体积和最大半径预测模型的工艺参数优化设计方法

基于第二小节中所述的工艺参数规则,我们进一步随机生成了一百万组工艺参数组合,利用上述盐穴储气库体积和最大半径预测模型对其成腔结果进行计算预测。一组好的造腔工艺参数设计应兼顾成腔体积大、成腔效率(成腔体积/总时间)高,和盐层利用率(成腔体积/最大半径)高和,由于总时间在本例中是一致的,所以此处工艺参数优选仅需考虑成腔体积和盐层利用率。则对百万组预测结果计算归一化的成腔体积、盐层利用率,并求和、排序。考虑最大半径的预测误差相对偏大,我们得到归一化的体积和利用率最高的100组参数之后,取其中成腔体积最大的10组作为最终的优选参数。最终百万组参数生成耗时1.53s,模型预测耗时27.88s,排序筛选耗时0.40s。以该方法进行造腔工艺参数优选,极大地提高了设计参数初步优化的效率。

针对利用该流程优选出来的工艺参数,开展进一步的基于SSCLS软件造腔模拟,并人工去除模拟结果存在大平顶、形态受力不合理的参数,最终得到4组工艺参数如表8(外管深度取比油垫深度低5m,流量统一取60m^3/h),其软件模拟结果如图12,4组腔体形态均规则、饱满,未出现畸形形态,证明了此造腔参数优选方法的可行性和可靠性。

表8 最终优选出来的4组工艺参数

	阶段	时间/天	内管深度/m	外管深度/m	油垫深度/m	流量/m^3/h
A	1	33	0	11	16	60
	2	164	8	45	50	60
	3	216	12	60	65	60
	4	99	46	70	75	60
	5	88	59	75	80	60
B	1	29	0	7	12	60
	2	133	8	44	49	60
	3	227	13	59	64	60
	4	122	26	69	74	60
	5	89	61	75	80	60
C	1	30	0	14	19	60
	2	100	7	45	50	60
	3	248	16	60	65	60
	4	123	42	68	73	60
	5	99	59	75	80	60
D	1	29	0	7	12	60
	2	133	8	44	49	60
	3	227	13	59	64	60
	4	100	26	69	74	60
	5	111	42	75	80	60

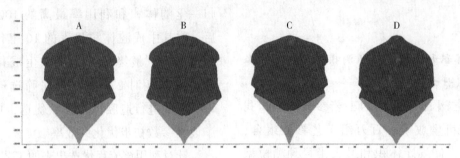

图 12 最终优选出的 4 组工艺参数在 SSCLS 软件中的模拟腔体形态结果

4.2 基于形态预测模型的工艺参数优化设计方法

基于盐穴储气库形态预测模型，我们提出了一个新的盐穴储气库建腔设计优化方法。利用神经网络的高效，我们可以在大量试算的基础上，对造腔工艺参数进行定制化设计，使最终的腔体形态能够满足我们的需求。

如图 13 展示了整个优化设计的流程，该流程分为以下六步：

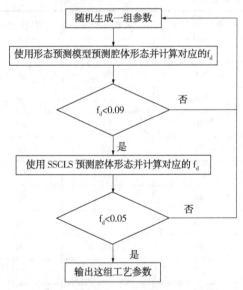

图 13 优化设计流程

拟定理想设计目标形态(如椭球形)；

基于工艺参数生成原则随机生成一组工艺参数；

2. 利用训练好的形态预测模型预测腔体形态；

3. 计算该预测形态与理想形态的偏差度 fd；

4. 若第 4 步中 fd 小于设定的阈值，如 0.09，则利用 SSCLS 对这组工艺参数进行仿真，得到设计腔体形态。否则回到第 2 步。

5. 计算 SSCLS 预测形态与理想形态的 fd；

6. 若第 6 步中得到的 fd 小于设定的阈值，如 0.05，则输出这组工艺参数为优化的工艺参数，循环结束。否则回到第 2 步。

在本节中，将以一个标准的椭球形腔体作为设计目标的建腔参数优化设计过程为例演示该方法。本例中，设计目标为长半径为 30m，短半径为 28m 的椭球。预测形态与椭球的 fd 通过以下公式描述。fd 越小，预测的腔体形态越接近椭球，则其对应造腔工艺参数的参考价值越高。

$$f_d = \frac{\sum_{i=1}^{13} |r'_i - r_i|}{13R_S}$$

其中，r'_i 为长半径 Rl 为 30m，短半径 RS 为 28m(短半径方向和盐穴储气库深度方向一致)，球心在 52m 深度处的椭球在各个深度处的半径；r_i 为盐穴储气库在各个深度处的半径。

表 9 为运行一次优化设计程序得到的建腔设计参数，该程序共循环了 660752 次，得到的建腔设计参数的 fd 低至 0.041。图 14 为此建腔设计参数的 SSCLS 模拟形态和标准椭球形态的对比图，由图可知 SSCLS 模拟形态与标准椭球之间的相似度是极高的。

表 9 优选的建腔设计参数

阶段 1	内管深度/m	0
	油垫深度/m	16
	注水量/m³	36000
阶段 2	内管深度/m	13
	油垫深度/m	50
	注水量/m³	198720
阶段 3	内管深度/m	10
	油垫深度/m	65
	注水量/m³	236160

续表

阶段 4	内管深度/m	21
	油垫深度/m	75
	注水量/m³	133920
阶段 5	内管深度/m	44
	油垫深度/m	80
	注水量/m³	259200

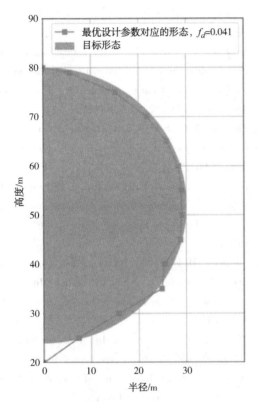

图 14 优选建腔设计参数的 SSCLS 模拟形态与目标形态

以上，验证了我们提出的基于盐穴储气库形态预测模型的盐穴储气库建腔设计优化方法的可行性。这个优化的过程也具有极高的效率，优化过程共耗时约 51min，其中，形态预测模型从 66 万组参数中选出 5 组初步优化参数耗时 79.11s，利用 SSCLS 模拟 5 组得到最终优化参数共耗时 50min。

5 总结

针对传统的室内模拟实验和仿真软件模拟方法对建腔工艺设计参数优选耗时间长且依赖工程设计经验的问题，本文提出了基于神经网络的盐穴储气库水溶建腔设计及参数优化方法，具体工作和主要结论如下。

1) 利用盐穴储气库单井水溶造腔模拟软件 SSCLS 开展水溶造腔模拟，获取了 1258 组造腔设计工艺参数组合（包括 5 个阶段的内管、外管、油垫深度，注水流量，阶段时间和注水量）及对应的模拟成腔体积、最大半径和成腔形态。以五个阶段中每个阶段的内管深度、阻溶剂深度和持续时间为有效输入数据，以成腔体积和最大半径为有效输出数据，组建了训练盐穴储气库体积和最大半径预测模型所需的数据集；以五个阶段中每个阶段的内管深度、油垫深度和注水量为有效输入数据，以腔体在 20m, 25m, ⋯, 75m, 79m 深度处的半径为有效输出数据，组建了训练盐穴储气库形态预测模型所需的数据集。

2) 将数据集划分为训练集和测试集对神经网络模型进行训练和验证。构建了含有 3 层隐藏层、每层 512 个节点的体积和最大半径预测模型和含有 3 层隐藏层、每层节点数分别为 250，125，63 的形态预测模型。

3) 体积和最大半径模型输出的预测值与实际值比较显示，成腔体积在训练集和测试集上的相关系数 R 均高于 0.96、平均绝对百分比误差（MAPE）均低于 2%。其中成腔体积较大的一半数据对造腔设计意义更大，其在测试集上的 MAPE 为 0.724%；最大半径在训练集和测试集上的 R 均高于 0.89，MAPE 均低于 3%。重复 10 次随机划分数据集、训练和预测，成腔体积和最大半径预测的 MAPE 均值分别为 1.838% 和 3.144%，预测精度较高，可以满足设计需求。

4) 形态预测模型输出的预测值与实际值比较显示，测试集中大多数深度的 MAE 小于等于 1.5m。除深度等于 20，30 米的数据外，测试集在其他深度的数据 R 值均大于等于 0.95。为了进一步验证形态预测模型预测效果，在测试集中随机抽取十组数据，将数据和模型对应的预测结果对比，模型预测结果显示该模型对大部分数据均能取得较好的预测效果。应用此形态预测模型预测实地腔体 JT52 的工艺参数，并将模型预测的腔体形态和 SSCLS 模拟形态和实际形态进行对比。SSCLS 模拟结果与实际半径的 MAE 为 1.292m，其单次模拟时间约在 10~30 分钟；相对而言，由于采用了均质盐层假设，形态预测模型预测的误差偏大，在各个深度处的 MEA 为 2.83m，但仍在工程可接受范围之内，其单次模拟时间约为 1.14×10^{-4} 秒。这说明所提出的神经网络预测方法可用于实际工程中的腔体形态设计和预测,且拥有远高于SSCLS的效率。

5）基于体积和最大半径预测模型提出盐穴储气库工艺参数优化设计方法。应用模型对随机生成的一百万组工艺参数进行成腔体积和最大半径预测,并优选出归一化的成腔体积、盐层利用率之和最高的100组参数,考虑成腔体积的预测误差更小,进一步取其中成腔体积最大的10组工艺参数作为最终优选工艺参数,此过程共耗时不到30s。优选工艺参数得到的模拟造腔腔体形态规则、饱满,相同时间内成腔体积大且盐层利用率高。对比传统方法,神经网络模型的时间成本极低,大幅度提高了工艺参数优选的效率。

6）基于形态预测模型提出盐穴储气库工艺参数优化设计方法。以长短轴比为15:14的椭球形为目标成腔形态为例,对随机生成的工艺参数进行形态预测和评估筛选。整个优化设计的流程为：使用模型对随机生成的一组工艺参数进行形态预测并计算预测形态与椭球形态的偏差度fd；判断fd是否小于0.09,若fd小于0.09则进行下一步,否则返回第一步；使用SSCLS软件对这组工艺参数进行成腔模拟,并计算预测形态与椭球形态fd；判断fd是否小于0.05,若fd小于0.05则输出这组工艺参数,循环结束,否则返回第一步。运行一次程序得到的优选工艺参数的fd低达0.041。整个优化设计过程共花费约51分钟,其中形态预测模型从66万组参数中选出5组初步优化参数耗时79.11s。对比传统方法,基于形态预测模型的工艺参数优化设计方法的时间成本极低,大幅度提高了工艺参数优化设计的效率。

目前本文中仅针对均匀盐层、正循环造腔模拟结果构建数据集,更进一步可以考虑反循环工况、考虑不溶夹层,甚至使用现场数据进行模型构建和训练,将有望取代造腔模拟软件直接预测现场腔体成腔体积及成腔形态,成为盐穴储库建设工程造腔设计工具。

参 考 文 献

[1] 杨春和,梁卫国,魏东吼,杨海军. 中国盐岩能源地下储存可行性研究[J]. 岩石力学与工程学报, 2005(24): 4409-4417.

[2] N. Zhang, W. Liu, Y. Zhang, P. F. Shan, X. L. Shi. Microscopic pore structure of surrounding rock for underground strategic petroleum reserve (SPR) caverns in bedded rock salt Energies, 13 (7) (2020), p. 1565.

[3] N. Zhang, L. J. Ma, M. Y. Wang, Q. Y. Zhang, J. Li, P. X. Fan Comprehensive risk evaluation of underground energy storage caverns in bedded rock salt. J Loss Prevent Proc, 45 (2016), pp. 264-276.

[4] 班凡生,高树生,单文文,熊伟. 岩盐储气库水溶建腔岩盐溶蚀模型研究[J]. 辽宁工程技术大学学报, 2005(S2): 102-104.

[5] 班凡生,高树生,单文文,熊伟. 岩盐储气库水溶建腔岩盐溶蚀模型研究[J]. 辽宁工程技术大学学报, 2005(S2): 102-104.

[6] Cristescu N D, Paraschiv I. The optimal shape of rectangular-like caverns[J]. International Journal of Rock Mechanics & Mining Science & Geomechanics Abstracts, 1995, 32(4): 285-300.

[7] Ehgartner B L, Sobolik S R. Analysis of cavern shapes for the strategic petroleum reserve. [J]. Technical Report, 2006.

[8] Durie R W., Jessen F. W. Mechanism of the dissolution of salt in the formation of underground salt cavities [J]. Society of Petroleum Engineers Journal, 1964, 4(2): 183-190.

[9] Kazemi H. Jessen F W. Mechanism of Flow and Controlled Dissolution of Salt in Solution Mining [J]. Society of Petroleum Engineers Journal. 1964, 4(4): 317-328.

[10] Webb, S. W., O'Hern, T. J., Hartenberger, J. D. SPR salt wall leaching experiments in lab-scale vessel-data report[R]. (Office of Scientific & Technical Information Technical Reports, 2010).

[11] Kazemi H. Jessen F W. Mechanism of Flow and Controlled Dissolution of Salt in Solution Mining [J]. Society of Petroleum Engineers Journal. 1964, 4(4): 317-328.

[12] Saberian A. and Von Schonfeldt. Convective mixing of water with brine around the periphery of vertical tube [C]. //Fourth symposium on salt. Houston. Texas, 1973.

[13] Reda D C, Russo A J. Experimental Studies of Salt-Cavity Leaching by Freshwater Injection[J]. Spe Production Engineering, 1986, 1(1): 82-86.

[14] 李龙,屈丹安,李建君,史辉,肖恩山,杨海军. 盐穴储气库反循环造腔的试验研究[A]. 宁夏回族自治区科学技术协会、宁夏社会科学界联合会、共青团宁夏回族自治区委员会、宁夏回族自治区青年联合会. 青年人才与石化产业创新发展——第七届宁夏青年科学家论坛论文集[C]. 宁夏回族自治区科学技术协会、宁夏社会科学界联合会、

共青团宁夏回族自治区委员会、宁夏回族自治区青年联合会:, 2011: 4.

[15] Liu W, Zhang ZX, Chen J, Fan JY, Jiang DY, Daemen JJK, Li YP. Physical simulation of construction and control of two butted-well horizontal cavern energy storage using large molded rock salt Specimens. Energy 2019; 185: 682-694.

[16] Jiang D Y, Yi L, Chen J, et al. Laboratory similarity test relevant to salt cavern construction in interlayer-containing moulded saliferous aggregates specimen[J]. Current Science, 2016, 111(1): 157-167.

[17] Sears G F. Controlled solution mining in massive salt [J]. Society of Petroleum Engineers Journal, 1966, 6(2): 115-125.

[18] Saberian A. Numerical simulation of development of solution-mined storage cavities [J]. Thesis Texas Univ, 1974.

[19] Nolen J S, Hantlemann O, Meister S, et al. Numerical Simulation of the Solution Mining Process [C]//SPE European Spring Meeting. 1974.

[20] Russo A J. Solution mining code for studying axisymmetric salt cavern formation [R]. United States, 1981.

[21] Kunstman A. S. and Urbanczyk K. M. UBRO – A computer model for designing salt cavern leaching process developed at CHEMKOP [C]. //Solution Mining Research Institute Fall Meeting, Paris, 1990.

[22] Chaudan E. INVDIR: A convenient and Efficient Solution Mining Model[C]. //Solution Mining Research Institute Fall Meeting Paris. 1990

[23] M. Guarascio. CAVITA: A Multipurpose Numerical Code for Brine Production Planning and Cavern Design and Control[C]. //Solution Mining Research Institute Fall Meeting Cleveland. 1996, 375-404

[24] 班凡生. 盐穴储气库水溶建腔优化设计研究[D]. 中国科学院研究生院(渗流流体力学研究所), 2008.

[25] Li J, Shi X, Yang C, et al. Mathematical model of salt cavern leaching for gas storage in high-insoluble salt formations [J]. entific Reports, 2018, 8(1): 372.

[26] Li J, Yang C, Shi X, et al. Construction modeling and shape prediction of horizontal salt caverns for gas/oil storage in bedded salt[J]. Journal of Petroleum Science and Engineering, 2020, 190: 107058.

[27] Wan JF, Peng TJ, Shen RC, Jurado MJ. Numerical model and program developmentof TWH salt cavern construction for UGS. J Petro Sci Eng 2019; 179: 930-740.

[28] Wang Z, Chen Q, Wang Z, Xiong J. The investigation into the failure criteria of concrete based on the BP neural network. Eng Fract Mech 2022; 275. https://doi.org/10.1016/j.engfracmech.2022.108835.

[29] Nazari A, Riahi S. Prediction split tensile strength and water permeability of high strength concrete containing TiO2 nanoparticles by artificial neural network and genetic programming. Compos B Eng 2011; 42. https://doi.org/10.1016/j.compositesb.2010.12.004.

[30] Hochreiter S, Schmidhuber J. Long short-term memory. Neural Comput 1997; 9: 1735-80. https://doi.org/10.1162/neco.1997.9.8.1735.

[31] Chung J, Gulcehre C, Cho K, Bengio Y. Gated feedback recurrent neural networks. 32nd International Conference on Machine Learning, ICML 2015, vol. 3, 2015.

[32] Cristescu ND, Paraschiv I. The optimal shape of rectangular-like caverns. International Journal of Rock Mechanics and Mining Sciences And 1995; 32. https://doi.org/10.1016/0148-9062(95)00006-3.

Seismic-hazard risk assessment of long-distance pipelines based on an improved unascertained measurement model

Ying Wu[1]　Qing Peng[1]　Yu Tian[1]　Xiao You[1,2]

(1. School of Civil Engineering and Geomatics, Southwest Petroleum University;
2. Karamay Fucheng Energy Company.)

Abstract　Long-distance pipeline infrastructure is essential for ensuring urban energy supply and economic stability. Pipes safe and stable operations have a direct impact on urban order. Owing to their long lengths, high pressures, and complex environments, pipelines are susceptible to risks from natural disasters, aging, and human interference. Pipeline safety is particularly critical in seismically active areas. The seismic-hazard risk assessment of long-distance pipelines involves complex, dynamic-variable information. Owing to incomplete seismic-hazard data records, conventional methods cannot fully deal with the large amount of uncertain information in seismic hazards and provide quantitative risk assessments. The unascertained measure theory is an effective method for assessing uncertainties and handling uncertain information. Therefore, this study proposes a risk-assessment method for the seismic hazards of long-distance pipelines based on an improved unascertained measurement model. This study introduces the unascertained measure theory and combines finite element analysis (FEA) and the improved analytic hierarchy process (AHP) methods to assess the seismic risk of long-distance pipelines. A long-distance pipeline in Southwest China was used as a case study; the process of application of the method was demonstrated; and the validity and reliability of the method were verified. Risk mitigation measures were proposed based on the final risk assessment results. The results indicated the following: (1) the 18 indicators comprehensively cover the factors influencing the seismic hazard of long-distance pipelines; (2) the application of unascertained measure theory effectively reduces the uncertainty in assessment results caused by incomplete baseline data; (3) the effective integration of the FEA and improved AHP methods better accounts for the significance of indicator variation and enhances the accuracy of indicator weights; (4) case studies demonstrate that this method can assess the risk of target pipelines, identify the most critical seismic factors, and recommend target seismic measures. In conclusion, the methodology of this study provides an effective tool for pipeline companies, offering valuable insights for assessing the safe operation of long-distance pipelines in seismically active regions.

Keywords　Long-distance pipeline; Seismic hazard; Risk assessment; Unascertained measurement theory; Finite element analysis

1　Introduction

Long-distance pipeline infrastructure is critical for energy transport and therefore essential for ensuring the urban energy supply and stable functioning of the national economy (Sun et al., 2024; Chen et al., 2022a; Li et al., 2021). These pipelines transport natural gas from production sites to cities, and their safe and stable operation directly affects urban production and daily life (Xu et al., 2023). However, owing to their long transport distances, high pressures, and complex environments, long-distance pipelines face a range of risks during operation, including natural disasters, aging and wear, and human interference (Chen et al., 2022b; Hu et al., 2022). In seismically active regions, the safety of these pipelines is particularly crucial.

China is a major player in natural gas pipeline

construction, where these pipelines serve as vital arteries in the national energy supply. However, for long-distance pipelines that are located in seismically active regions, unpredictable seismic activity poses a considerable threat, particularly near tectonic plate boundaries (Zhang et al., 2024; Melissianos et al., 2023). These pipelines are characterised by high pressures, high flow rates, and extensive routing. Earthquakes can cause these pipelines to rupture or sustain damage, leading to natural-gas leakages. This gas can easily spread into adjacent underground spaces and cause explosions, which could result in severe casualties and economic losses (Li et al., 2023). For example, the 7.3-magnitude Kobe earthquake in Japan in 1995 ruptured several natural-gas pipelines, leading to~ 459 fires and, substantial property and life losses (Frolova et al., 2017). In 2020, a 5.7-magnitude earthquake in Magna damaged 11 natural-gas pipelines and 468 gas-meter components (Eidinger et al., 2023).

These incidents underscore the significant threat that earthquakes pose to the safe operation of long-distance pipelines and highlight the need to strengthen pre-earthquake prevention and emergency-response capabilities. Therefore, conducting seismic-hazard risk assessments for long-distance pipelines is crucial for ensuring pipeline integrity. Such efforts can mitigate massive economic losses from disasters, enhance the safety of pipeline transportation, and ensure the stability of the energy supply.

In recent years, domestic and international have conducted numerous studies on natural-gas pipeline risk assessments. Xu et al. used a branch of the China-Myanmar natural-gas pipeline as an example and proposed a comprehensive pipeline safety-evaluation method based on the fuzzy fault-tree model (Xu et al., 2023). Bai et al. proposed a Bayesian network (BN) and computational fluid dynamics based quantitative risk assessment of natural-gas explosions in utility tunnels (Bai et al., 2022). Bai et al. proposed a new risk-assessment model integrating a knowledge graph, decision experiment, and evaluation laboratory, and proposed a data-driven approach based on BN to analyse natural-gas pipeline accidents and reduce reliance on experts (Bai et al., 2023). He et al. proposed a nonlinear risk-assessment method based on fuzzy measures and the Choquet integral, which can conduct multi-source and single-source independent evaluations of chemicals (He et al., 2022). Yasir et al. proposed a natural-gas pipeline risk-assessment method based on a hybrid fuzzy BN and expert inspiration (Mahmood et al., 2024). Wu et al. analysed the influence of different factors on the diffusion law of gas leakage and, through exploratory data analysis, identified the factor with the greatest impact on diffusion (Wu et al., 2024). Lu et al. proposed a new quantitative risk-assessment model that included data on the surrounding soil, pipeline-protection layer, cathodic protection, and pipeline wall-thickness readings to guide excavation inspection and maintenance decisions (Lu et al., 2022). Wen et al. proposed a quantitative assessment model for landslide-induced long-distance-pipeline risk based on a combination of the recursive-feature removal and particle-swarm optimisation AdaBoost method, fuzzy-clustering method, and CRITIC method, analysing historical landslide-disaster data along oil and gas pipelines (Wen et al., 2023). Li et al. used the Monte Carlo method to solve the limit state equation of pipeline-instability probability based on the reliability theory, and the results were used to determine the risk level of pipeline instability (Li et al., 2017). Liu et al. used a subset simulation to estimate the probability of corrosion failure in buried natural-gas pipelines (Liu et al., 2019).

From a review of existing studies and literature, the issues can be summarised as follows:

(1) Most of the aforementioned risk assessment studies only considered probabilities, post-failure. However, for pipelines under seismic activity predictive modelling and analysis is necessary, rather than post-event reasoning.

(2) Owing to incomplete data on failure incidents encountered in long-distance pipelines. The

methods mentioned earlier are insufficient for effectively addressing the considerable information uncertainties. Thereby hindering precise quantitative risk assessments for long-distance pipelines. Considering the destructive potential and complexity of seismic-disaster incidents, innovative approaches are required to assess the pipeline risks.

(3) The weighting methods used in certain pipeline risk-indicator systems often emphasise a single aspect and neglect the integration of objective and subjective weights. This lack of holistic consideration of both the degree of change in indicator values and the importance of the indicators has led to inaccuracies in the assessment results.

Numerous scholars have applied themented measure theory proposed by Guangyuan et al. (Guangyuan W, 1990; Liu et al., 1999) to various fields and achieved relatively good evaluation results (Ma et al., 2021; Li et al., 2020; Shu et al., 2023; Ding et al., 2024). The unascertained measure theory (UMT) establishes an unascertained measurement model (UMM) to analyse index factors and confidence-recognition criteria to determine the reliability of simple, pertinent, practical, and reliable index factors. The introduction of an unascertained measure theory into the risk assessment of long-distance pipelines can solve the evaluation difficulty caused by unascertained information in the seismic-hazard risk assessment of long-distance pipelines.

To make a more reasonable determination of index weight, in the process of risk evaluation, an improved analytic hierarchy process (AHP) was developed based on the classical AHP (Ba et al., 2022). However, the AHP requires expert expertise. To improve the accuracy of the weights obtained by this method, this study innovatively proposed carrying out finite element analysis (FEA). With the help of numerical simulation for quantitative indicators and using grey correlation analysis to transform the weights based on its calculation results. To complete the conversion between the grey correlation degree and weight, a conversion formula was proposed. This aided the realisation of quantitative optimisation of the weights that were obtained by the improved AHP. To improve the calculation efficiency of pipeline seismic dynamic-response analysis and improve the calculation efficiency, orthogonal tests were designed to reduce the calculation conditions.

Based on the existing issues identified in prior studies, this study introduced the theory of unascertained measurements. Then proposed a seismic-hazard risk-assessment method for long-distance pipelines grounded in an improved unascertained-measurement model. First, the factors influencing long-distance pipeline damage during earthquakes were analysed from multiple perspectives to establish a seismic-hazard risk-evaluation index system for long-distance pipelines. Next, an improved unascertained measurement model was developed, detailing the procedures for its application, including the application of unascertained measure theory and the calculation of indicator weights. Considering a long-distance pipeline in the Sichuan Province, China, as a case study, the risk level of the evaluated target was determined. Additionally, radar charts were created to identify major risk sources and visually present the risk-assessment results, further validating the applicability and effectiveness of the proposed method. Based on the risk-assessment findings, targeted risk mitigation measures were recommended to ensure the safe operation of long-distance pipelines in seismically active regions.

The remainder of this paper is organised as follows. The first section covers the background challenges, a literature review of similar approaches, and the novelty of the study. The second section elaborates on the methodology adopted in this study, with an emphasis on the complete process and the proposed, improved unascertained measurement model. The third section presents the application of the method, using a long-distance pipeline in Sichuan Province, China, as a case study, including specific operations, results, and analysis. Finally, the fourth section presents conclusions and future research directions.

2 Methodology

Fig. 1 illustrates the proposed framework for the seismic-hazard risk assessment of long-distance pipelines using the improved unascertained measurement model, which consists of three steps. The first step uses inputs from various sources. The second step construct an improved unascertained measurement model. The third step uses the improved unascertained measurement model for risk assessment. The steps are as follows:

Step 1: Identification of pipeline information and risk factors.

Comprehensive information on long-distance pipeline earthquakes was collected. Detailed information included material, operating-environment, and historical-fault or accident data. The purpose of this study was to identify the potential risk factors that affect pipeline safety, integrity, and reliability. These risk factors comprise three parts: pipeline vulnerability, seismic hazards, and disaster losses.

Step 2: Construction of an improved unascertained measurement model.

The construction of an improved unascertained measurement model includes four main parts: construction of a single-indicator unascertained measurement matrix, weight determination, multi-indicator unascertained comprehensive measurement evaluation vector, and risk-level identification. The improved unascertained measurement model can reduce the uncertainty of risk-assessment results caused by incomplete basic data. Thereby improving the accuracy of indicator weights in risk assessment and achieving seismic-hazard risk assessment for long-distance pipelines.

Step 3: Use of an improved unascertained measurement model for risk assessment.

After selecting the pipeline and obtaining the indicator values for the evaluation, a risk assessment was conductedin the evaluation section to obtain the risk level. The use of radar charts to identify risk factors and obtain major risk sources for evaluating pipeline segments. It can provided a basis for developing effective risk-mitigation measures.

The improved unascertained measurement model considers the incompleteness of the basic indicator data and the subjectivity in determining indicator weights. The subjectivity of the weights is reduced by combining subjective and objective methods, and thus, a more accurate assessment of the overall risk is provided. The effectiveness of this method has been validated through examples. The verification results are consistent with reality, indicating that the proposed method has practical operability and reliability. This study can help pipeline companies effectively manage the seismic risk of long-distance pipelines, and accurately identify pipelines with high seismic-risk levels. Thereby actively implement risk-mitigation measures and ensure the safe operation of long-distance pipelines.

2.1 Establishment of a seismic-hazard risk-evaluation-index system

2.1.1 Evaluation-index system

Rational development of an indicator system is critical for the accuracy of risk assessment. An indicator system should first adhere to the principles of systematicity, scientific rigor, feasibility, and hierarchy. In addition, the characteristics of earthquake hazards must be considered for the risk assessment of long-distance pipelines underseismic damage.

Referring to the concept of disaster vulnerability in vulnerability assessments (Wang et al., 2020; Deng et al., 2018), this evaluation system introduced an innovative improvement. Adding earthquake-hazard indicators to the first-level indicator layer, building upon a conventional risk-indicator system. During an earthquake, the likelihood of pipeline failure is jointly determined by the seismic-hazard risk and the inherent conditions of the pipeline. The failure likelihood is divided into seismic risk and pipeline vulnerability, and the consequences of failure are addressed through disaster-loss indicators. Based on this analysis and drawing from methods such as accident-cause analysis, statistical analysis of seismic - damage characteristics, literature review, field investigation, and expert consultation. Thereby an indicator system suited to seismic-hazard risk in long-distance pipelines was established. Finally, the first-level indicators were categorised into three: seismic risk, pipeline vulnerability, and disaster loss, comprising 18 indicators, in total. The indicator system is illustrated in Fig. 2.

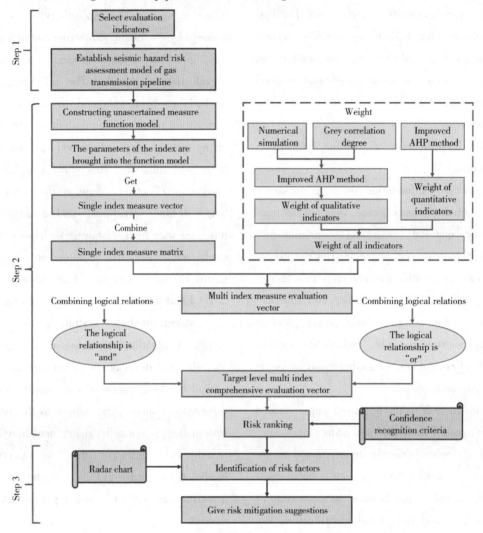

Fig. 1 Framework of the proposed methodology

2.1.2 Classification standard of index system

In this study, the grading standard quantitative method(Dong et al., 2019) was adopted to form the grading standard for the indicators. The basic principle of the grading standard quantisation method, taking the interval method in Subsection 5 as an example, is as follows:

$$x_{ij} = \begin{cases} a, & A_{ij} = Text_1 \\ b, & A_{ij} = Text_2 \\ c, & A_{ij} = Text_3 \\ d, & A_{ij} = Text_4 \\ e, & A_{ij} = Text_5 \end{cases} \quad (1)$$

where a–e is usually taken in the form of 1-3-5-7-9 or 1-2-3-4-5 and A_{ij} is the description of the jth qualitative index state of the ith factor.

The classification standard of each index in this study was derived from relevant statistical data, the literature, national standards, and industry norms. The classification standards of the qualitative indicators are listed in Table 1, and the summary results of all indicators are listed in Table 2.

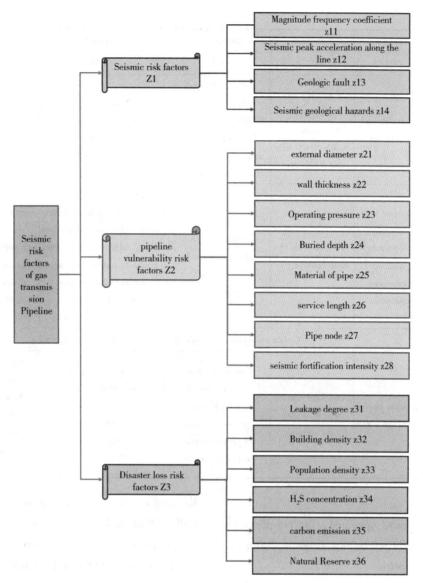

Fig. 2 Seismic-hazard risk-evaluation-index system of long-distance pipelines

Table. 1 Description of the qualitative index-classification standard of seismic-hazard risk-evaluation-index system for long-distance pipelines

Qualitative index	Low risk (C1)	Lower risk (C2)	Medium risk (C3)	High risk (C4)	Higher risk (C5)
Geologic fault z_{13}	No faults in the area	No faults in the near field	Non-near faults in the near field	Near faults in the near field	Pipelines pass through the fault and near the fault field
Seismic geological disaster z_{14}	No geological hazard around the pipeline	Lower-risk geological disasters around the pipeline	Medium-risk geological disasters around the pipeline	More dangerous geological disasters around the pipeline	Dangerous geological disasters around the pipeline
Material of pipe z_{25}	Above X70 (L485)	X60(L415)-X70(L485)	X42(L290)-X56(L390)	L245 or 20#	Below L245
Pipe node z_{27}	No node	With nodes and goodseismic effect	With nodes, and general seismic effect	With nodes, and relatively poor seismic effect	With nodes, and poor seismic effect

· 483 ·

续表

Qualitative index	Low risk (C1)	Lower risk (C2)	Medium risk (C3)	High risk (C4)	Higher risk (C5)
Leakage degree z_{31}	$P \leq 0.3$Pe and $D \leq 219$mm	$P \leq 0.3$Pe and 219mm$<D \leq 323$mm, or 0.3Pe$<P \leq 0.6$Pe and $D \leq 219$mm	$P \leq 0.3$Pe and $D > 323$mm, or 0.3Pe$<P \leq 0.6$Pe and 219mm$<D \leq 323$mm, or 0.6Pe$<P \leq$Pe and $D \leq 219$mm	0.3Pe$<P \leq 0.6$Pe and $D > 323$mm, or 0.6Pe$<P \leq$Pe and 219mm$<D \leq 323$mm	0.6Pe$<P \leq$Pe and $D > 323$mm
Population density z_{33}	Sections not frequently occupied and no permanent residences	Sections with 15 households or less	Sections with more than 15 households and less than 100 households	Sections with 100 households or more including suburban residential areas, commercial areas, industrial areas, and planned-development areas and densely populated areas not qualified for level IV	Sections with buildings with four floors or more (excluding basement floors), generally concentrated, frequent traffic, many underground facilities
Carbon emission z_{35}	Carbon emissions are classified according to the leakage degree				
Natural Reserve z_{36}	No nature reserve around the section	One local nature reserve around the section	One local nature reserve around section and pipeline crosses the experimental area of the reserve	Two or more local nature reserves or one national nature reserve around the section	Several national nature reserves or one National Nature Reserve around the section and the pipeline crosses the experimental area of the reserve

Notes:

1. Unless otherwise specified, the periphery of the pipeline generally refers to a range of 200m on both sides of the pipeline.

2. Regional scope refers to the scope centred on the project site with a radius of no less than 150km. Near-field range refers to the range centred on the project site with a radius of no less than 25km. The near fault is a fault with a fault distance of 20km.

Table 2 Classification standard of seismic-hazard risk-evaluation-index system of long-distance pipeline

Index	Low risk (C1)	Lower risk (C2)	Medium risk (C3)	High risk (C4)	Higher risk (C5)
Magnitude frequency coefficient z_{11}	>1.8 or $=0$	(1.6, 1.8]	(1, 1.6]	(0.7, 1]	(0.44, 0.7]
Seismic attenuation law z_{12}	[0, 0.09g)	[0.09g, 0.19g)	[0.19g, 0.38g)	[0.38g, 0.75g)	$\geq 0.75g$
Geologic fault z_{13}	1	2	3	4	5
Seismic geological disaster z_{14}	1	2	3	4	5
External diameter z_{21}/mm	≥ 406	[323, 406)	[219, 323)	[114, 219)	[0, 114)
Wall thickness z_{22}/mm	≥ 20	[15, 20)	[10, 15)	[5, 10)	[0, 5)
Operating pressure z_{23}/MPa	[0, 0.8)	[0.8, 1.6)	[1.6, 2.5)	[2.5, 4)	≥ 4
Buried depth z_{24}/m	[0, 0.6)	[0.6, 1.0)	[1.0, 1.4)	[1.4, 1.8)	≥ 1.8
Material of pipe z_{25}	1	2	3	4	5
Service life z_{26}/year	[0, 5)	[5, 10)	[10, 15)	[15, 20)	≥ 20

续表

Index	Low risk (C1)	Lower risk (C2)	Medium risk (C3)	High risk (C4)	Higher risk (C5)
Pipe node z_{27}	1	2	3	4	5
Seismic fortification intensity z_{28}	[0, 6)	[6, 7)	[7, 8)	[8, 9)	≥ 9
Leakage degree z_{31}	1	2	3	4	5
Building density z_{32}	[0, 0.1)	[0.1, 0.15)	[0.15, 0.25)	[0.25, 0.35)	≥ 0.35
Population density z_{33}	1	2	3	4	5
H_2S concentration z_{34}/mg · m^{-3}	[0, 6]	(6, 20]	(20, 285]	(285, 720]	> 720
Carbon emission z_{35}	1	2	3	4	5
Natural Reserve z_{36}	1	2	3	4	5

2.2 Construction of single-index unascertained measure matrix

The most widely used and simplest linear unascertained measurement function (Fig. 3) was adopted; it is expressed as follows:

$$\begin{cases} \mu_i(x) = \begin{cases} \dfrac{-x}{a_{i+1}-a_i} + \dfrac{a_{i+1}}{a_{i+1}-a_i} & a_i < x \leq a_{i+1} \\ 0 & x > a_{i+1} \end{cases} \\ \mu_{i+1}(x) = \begin{cases} 0 & x \leq a_i \\ \dfrac{x}{a_{i+1}-a_i} - \dfrac{a_i}{a_{i+1}-a_i} & a_i < x \leq a_{i+1} \end{cases} \end{cases} \quad (2)$$

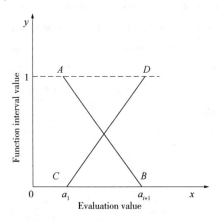

Fig. 3　Linear unascertained measure function diagram

By substituting the grading results for the indicators listed in Table 2 into Eq. (2), the unascertained measure function graphs for each indicator were obtained, as illustrated in Fig. 4.

For each index Z_{ij} of each factor seismic risk Z_1, pipeline vulnerability Z_2, and disaster loss Z_3 in factor space $Z = \{Z_1, Z_2, Z_3\}$. The measure function $\mu(x_{ij} \in C_k)$ ($i = 1, 2, \cdots, n, j = 1, 2, \cdots, m, k = 1, 2, \cdots, p$) corresponding to each index Z_{ij} was constructed according to the evaluation space $C = \{C_1, C_2, \cdots, C_p\}$. Then the structural formula of the linear unascertained measure function, and the single index unascertained measure evaluation matrix were obtained:

$$(\mu_{ijk})_{m \times p} = \begin{pmatrix} \mu_{i11} & \mu_{i12} & \cdots & \mu_{i1p} \\ \mu_{i21} & \mu_{i22} & \cdots & \mu_{i2p} \\ \vdots & \vdots & \ddots & \vdots \\ \mu_{im1} & \mu_{im2} & \cdots & \mu_{imp} \end{pmatrix} \quad (3)$$

Applying Eq. (2), the single-index unascertained measurement and evaluation matrix of seismic risk Z_1, pipeline vulnerability Z_2, and disaster loss Z_3, are:

$$(\mu_{1jk})_{4 \times 5} = \begin{pmatrix} z_{11} \\ z_{12} \\ z_{13} \\ z_{14} \end{pmatrix} = \begin{pmatrix} \mu_{111} & \mu_{112} & \mu_{113} & \mu_{114} & \mu_{115} \\ \mu_{121} & \mu_{122} & \mu_{123} & \mu_{124} & \mu_{125} \\ \mu_{131} & \mu_{132} & \mu_{133} & \mu_{134} & \mu_{135} \\ \mu_{141} & \mu_{142} & \mu_{143} & \mu_{144} & \mu_{145} \end{pmatrix} \quad (4)$$

$$(\mu_{2jk})_{8 \times 5} = \begin{pmatrix} z_{21} \\ z_{22} \\ z_{23} \\ z_{24} \\ z_{25} \\ z_{26} \\ z_{27} \\ z_{28} \end{pmatrix} = \begin{pmatrix} \mu_{211} & \mu_{212} & \mu_{213} & \mu_{214} & \mu_{215} \\ \mu_{221} & \mu_{222} & \mu_{223} & \mu_{224} & \mu_{225} \\ \mu_{231} & \mu_{232} & \mu_{233} & \mu_{234} & \mu_{235} \\ \mu_{241} & \mu_{242} & \mu_{243} & \mu_{244} & \mu_{245} \\ \mu_{251} & \mu_{252} & \mu_{253} & \mu_{254} & \mu_{255} \\ \mu_{261} & \mu_{262} & \mu_{263} & \mu_{264} & \mu_{265} \\ \mu_{271} & \mu_{272} & \mu_{273} & \mu_{274} & \mu_{275} \\ \mu_{281} & \mu_{282} & \mu_{283} & \mu_{284} & \mu_{285} \end{pmatrix} \quad (5)$$

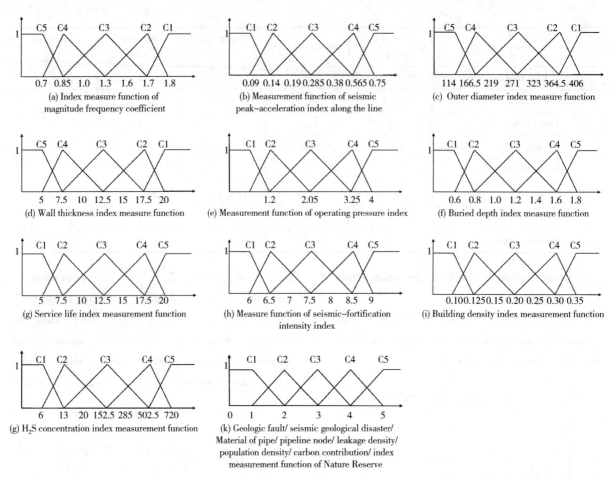

Fig. 4 Unascertained measure function of single index

$$(\mu_{3jk})_{6\times 5} = \begin{pmatrix} z_{31} \\ z_{32} \\ z_{33} \\ z_{34} \\ z_{35} \\ z_{36} \end{pmatrix} = \begin{pmatrix} \mu_{311} & \mu_{312} & \mu_{313} & \mu_{314} & \mu_{315} \\ \mu_{321} & \mu_{322} & \mu_{323} & \mu_{324} & \mu_{325} \\ \mu_{331} & \mu_{332} & \mu_{333} & \mu_{334} & \mu_{335} \\ \mu_{341} & \mu_{342} & \mu_{343} & \mu_{344} & \mu_{345} \\ \mu_{351} & \mu_{352} & \mu_{353} & \mu_{354} & \mu_{355} \\ \mu_{361} & \mu_{362} & \mu_{363} & \mu_{364} & \mu_{365} \end{pmatrix}$$

(6)

2.3 Weight of evaluation indices

In this study, the evaluation indices were classified as qualitative and quantitative evaluation based on the feasibility of numerical simulation (Fig. 5). Among them, the following five indices were quantitative: intensity, outer diameter, wall thickness, internal pressure, and buried depth. For greater objectivity, numerical simulation and grey correlation degree were used to achieve the index-weight optimisation of the aforementioned five quantitative indices (whose index weight was obtained by the AHP). To improve the calculation efficiency of the pipeline seismic dynamic-response analysis, an orthogonal experimental design was conducted.

For the qualitative indices, the weights were determined using the improved AHP thereby improving the accuracy of weight determination in risk assessment.

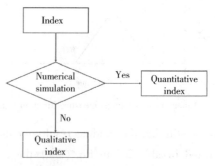

Fig. 5 Classification criteria for qualitative and quantitative indices

2.3.1 Determination of quantitative index weights
2.3.1.1 Orthogonal experimental design

A combined scheme with five factors and five levels was designed using the orthogonal test method. The level settings of each quantitative index are listed in Table 3, and the $L_{25}(5^5)$ orthogonal test is listed in Table 4.

Table 3 Factor level setting

Level	Factor				
	1 Earthquake intensity	2 External diameter mm	3 Wall thickness mm	4 Internal pressure MPa	5 Buried depth mm
1	6	89	4	0.6	400
2	7	159	6	1.2	800
3	8	273	8	2.0	1200
4	9	355	10	3.2	1600
5	10	457	12	5.0	2000

Table 4 Orthogonal experimental design table and the corresponding numerical simulation results

Number	Factor					von Mises stress/Pa
	Earthquake intensity	External diameter/mm	Wall thickness/mm	Internal pressure/MPa	Buried depth/m	
1	6	89	4	0.6	0.4	1.264×10^7
2	6	159	6	1.2	1.6	2.382×10^7
3	6	273	8	2.0	0.8	3.999×10^7
4	6	355	10	3.2	2.0	6.422×10^7
5	6	457	12	5.0	1.2	9.941×10^7
6	7	89	10	5.0	1.6	2.556×10^7
7	7	159	12	0.6	0.8	1.201×10^7
8	7	273	4	1.2	2.0	8.158×10^7
9	7	355	6	2.0	1.2	7.065×10^7
10	7	457	8	3.2	0.4	9.497×10^7
11	8	89	6	3.2	0.8	4.484×10^7
12	8	159	8	5.0	2.0	6.160×10^7
13	8	273	10	0.6	1.2	1.687×10^7
14	8	355	12	1.2	0.4	3.315×10^7
15	8	457	4	2.0	1.6	1.552×10^8
16	9	89	12	2.0	2.0	3.378×10^7
17	9	159	4	3.2	1.2	9.871×10^7
18	9	273	6	5.0	0.4	1.404×10^8
19	9	355	8	0.6	1.6	3.307×10^7
20	9	457	10	1.2	0.8	4.131×10^7
21	10	89	8	1.2	1.2	1.209×10^8
22	10	159	10	2.0	0.4	1.121×10^8
23	10	273	12	3.2	1.6	1.253×10^8
24	10	355	4	5.0	0.8	1.347×10^8
25	10	457	6	0.6	2.0	1.431×10^8

2.3.1.2 Numerical simulation of buried pipeline

The time history analysis method was used to establish the numerical model of a buried pipeline, using finite element software. A typical Wenchuan seismic wave was selected to calculate the seismic dynamic response of each group of working conditions in the orthogonal test table (Fig. 6).

The specific steps follow.

This model is based on the following assumptions.

① The pipeline section under consideration is buried in uniform soil.

② All parts of the pipeline have the same mechanical properties.

③ The influence of temperature on pipeline performance is not considered.

The numerical simulation used a 200-m long L245 steel pipe for the long-distance pipeline (Wu et al., 2020). The other parameters of the pipeline are listed in Table 5. The soil volume was 200m×4m×4m; other parameters of the soil mass are summarised in Table 6.

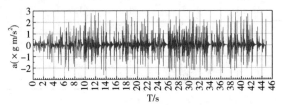

Fig. 6 Wenchuan wave acceleration time history (horizontal axial)

Table 5 Pipe parameters

External diameter/mm	Wall thickness/mm	Buried depth/m	Internal pressure/MPa	Elastic modulus/GPa	Poisson's ratio	Density/(kg/m^3)	Yield strength/MPa
114.3	5.6	2	2	203	0.3	7850	245

Table 6 Soil parameters

Modulus of elasticity/MPa	Poisson's ratio	Density/(kg/m^3)	Friction angle/°
8	0.38	1800	15.8

The results of the pipeline grid sensitivity test are shown in Fig. 7.

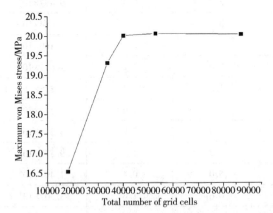

Fig. 7 Grid sensitivity test

The maximum von Mises stress increases significantly with an increase in the number of grids (Fig. 7). When the number of grids increased to 40000, the maximum von Mises stress did not change significantly. To save calculation time and ensure calculation accuracy, 40000 grids were selected in this study. The model and grid divisions are illustrated in Fig. 8.

Fig. 8 Model and grid division

2.3.1.3 Model calculation

Using the finite element software ABAQUS, the dynamic response under different working conditions was simulated according to the orthogonal experimental design table (Table 7). The results are included in Table 7.

Table 7 Orthogonal experimental design table and the corresponding numerical simulation results

Number	Factor					von Mises stress/Pa
	Earthquake intensity	External diameter/mm	Wall thickness/mm	Internal pressure/MPa	Buried depth/m	
1	6	89	4	0.6	0.4	1.264×10^7
2	6	159	6	1.2	1.6	2.382×10^7
3	6	273	8	2.0	0.8	3.999×10^7
4	6	355	10	3.2	2.0	6.422×10^7
5	6	457	12	5.0	1.2	9.941×10^7
6	7	89	10	5.0	1.6	2.556×10^7
7	7	159	12	0.6	0.8	1.201×10^7
8	7	273	4	1.2	2.0	8.158×10^7
9	7	355	6	2.0	1.2	7.065×10^7
10	7	457	8	3.2	0.4	9.497×10^7
11	8	89	6	3.2	0.8	4.484×10^7
12	8	159	8	5.0	2.0	6.160×10^7
13	8	273	10	0.6	1.2	1.687×10^7
14	8	355	12	1.2	0.4	3.315×10^7
15	8	457	4	2.0	1.6	1.552×10^8
16	9	89	12	2.0	2.0	3.378×10^7
17	9	159	4	3.2	1.2	9.871×10^7
18	9	273	6	5.0	0.4	1.404×10^8
19	9	355	8	0.6	1.6	3.307×10^7
20	9	457	457	1.2	0.8	4.131×10^7
21	10	89	8	1.2	1.2	1.209×10^8
22	10	159	10	2.0	0.4	1.121×10^8
23	10	273	12	3.2	1.6	1.253×10^8
24	10	355	4	5.0	0.8	1.347×10^8
25	10	457	6	0.6	2.0	1.431×10^8

2.3.1.4 Grey correlation degree calculation and conversion to weight

1) Calculation of the grey correlation degree γ_i

Finally, the grey correlation degree of the von Mises stress intensity, outer diameter, wall thickness, internal pressure, and buried depth were calculated, and the results are listed in Table 8.

Table 8 Results of grey correlation degree calculation

Grey correlation degree	Index				
	Earthquake intensity	External diameter	Wall thickness	Internal pressure	Buried depth
γ_i	0.266	0.401	0.132	0.389	0.045

2) Conversion to weights ω_i

The grey correlation degree lies in the range $[-1,1]$, and the sum is not always 1. The grey correlation degrees were converted to the corresponding weights using Eq. (7).

$$w_i = \frac{|\gamma_i|}{\sum_{i=1}^{n} |\gamma_i|} \quad (7)$$

The grey correlation degree of earthquake intensity, pipe outer diameter, wall thickness, internal pressure, and buried depth were normalised according to Eq. (7). Then the respective weights of the five indices were calculated. The results are summarised in Table 9.

Table 9 Results of the quantitative index weight calculations

Weight	Index				
	Earthquake intensity	External diameter	Wall thickness	Internal pressure	Buried depth
w_i	21.57%	32.52%	10.71%	31.55%	3.65%

2.3.2 Determination of qualitative index weights

The improved AHP uses the three-scale method to construct a judgment matrix. The calculations do not require a fast, high-precision consistency test. Therefore, the improved AHP was used to determine the weights of the qualitative indicators in this study.

The weights of the other indices in the index system are calculated using the basic principles and steps of the improved AHP. The details follow.

(1) Hierarchical structural model establishment

This study adopted a three-level structural model. The index system illustrated in Fig. 2 provides further details.

(2) Comparison matrix establishment

The actual data of the comparison matrix were based on expert experience and knowledge. An expert questionnaire was designed, optimised, and sent to the experts. After the experts completed the questionnaire, the data was collected and sorted, and the comparison matrix was established.

The following example, used to demonstrate the process of calculation of weights, uses the seismic-risk index Z_1 of the criterion layer.

A comparison matrix, A, of the seismic risk Z_1 index {magnitude frequency coefficient, seismic peak acceleration along the line, fault, and seismic geological hazard} of the criterion layer was built.

$$A = \begin{pmatrix} 1 & 2 & 0 & 0 \\ 0 & 1 & 0 & 0 \\ 2 & 2 & 1 & 2 \\ 2 & 2 & 0 & 1 \end{pmatrix}$$

(3) Importance ranking index calculation

$r_1 = 3$, $r_2 = 1$, $r_3 = 7$, $r_4 = 5$

(4) Judgment matrix construction B

$$B = \begin{pmatrix} 1 & 3 & 1/5 & 1/3 \\ 1/3 & 1 & 1/7 & 1/5 \\ 5 & 7 & 1 & 3 \\ 3 & 5 & 1/3 & 1 \end{pmatrix}$$

(5) Finding the transfer matrix C of the judgment matrix B

$$C = \begin{pmatrix} 0 & 0.4771 & -0.6990 & -0.4771 \\ -0.4771 & 0 & -0.8451 & -0.6990 \\ 0.6990 & 0.8451 & 0 & 0.4771 \\ 0.4771 & 0.6990 & -0.4771 & 0 \end{pmatrix}$$

(6) Finding the optimal transfer matrix D of the transfer matrix C

$$D = \begin{pmatrix} 0 & 0.3306 & -0.6990 & -0.3495 \\ -0.3306 & 0 & -1.0106 & -0.6801 \\ 0.6801 & 1.0106 & 0 & 0.3306 \\ 0.3495 & 0.6801 & -0.3306 & 0 \end{pmatrix}$$

(7) Finding the quasi-optimal uniform matrix B′

$$B' = \begin{pmatrix} 1 & 2.1407 & 0.2089 & 0.4472 \\ 0.4671 & 1 & 0.0976 & 0.2089 \\ 4.7867 & 10.2470 & 1 & 2.1407 \\ 2.2361 & 4.7867 & 0.4671 & 1 \end{pmatrix}$$

(8) Find the eigenvector W of the quasi-optimal uniform matrix B′

1) Finding the element product N_i of each line of the quasi-optimal uniform matrix B′

$N_1 = 0.2$, $N_2 = 0.0095$, $N_3 = 105$, $N_4 = 5$

2) Finding the square root $\overline{W_i}$

$\overline{W_1} = 0.6687$, $\overline{W_2} = 0.3124$,

$\overline{W_3} = 3.2011$, $\overline{W_4} = 1.4953$

3) Normalizing the square root $\overline{W}_i$ to get W_i

$W_1 = 0.1178$, $W_2 = 0.0550$,
$W_3 = 0.5638$, $W_4 = 0.2634$

(9) Calculating the weight w

$$w = \frac{1}{f}\sum_{i=1}^{f} W_i \qquad (8)$$

The other index weights were calculated using MATLAB. The calculation process is not explained further, here. The calculation results are summarised in Table 10.

Table 10 Results of index weight calculation

Criterion layer	Index	Weight
Seismic risk Z_1	Magnitude frequency coefficient z_{11}	$w_{11} = 0.152$
	Seismic attenuation law z_{12}	$w_{12} = 0.287$
	Geologic fault z_{13}	$w_{13} = 0.368$
	Seismic geological disaster z_{14}	$w_{14} = 0.193$
Pipeline vulnerability Z_2	External diameter z_{21}/mm	$w_{21} = 0.130$
	Wall thickness z_{22}/mm	$w_{22} = 0.043$
	Operating pressure z_{23}/MPa	$w_{23} = 0.126$
	Buried depth z_{24}/m	$w_{24} = 0.015$
	Material of pipe z_{25}	$w_{25} = 0.236$
	Service life z_{26}/year	$w_{26} = 0.167$
	Pipe node z_{27}	$w_{27} = 0.197$
	Seismic fortification intensity z_{28}	$w_{28} = 0.086$
Disaster loss Z_3	Leakage degree z_{31}	$w_{31} = 0.067$
	Building density z_{32}	$w_{32} = 0.059$
	Population density z_{33}	$w_{33} = 0.377$
	H_2S concentration z_{34}/mg·m^{-3}	$w_{34} = 0.283$
	Carbon emission z_{35}	$w_{35} = 0.083$
	Natural Reserve z_{36}	$w_{36} = 0.131$

2.4 Construction of a multi-index comprehensive measurement and evaluation vector

(1) When the logical relationship is "or"

For the multi-index measure with the logical relationship "or" between indices in the index layer, the weighted summation method is used for calculations.

Let w_j be the weight of index F_j of evaluation factor Z_i. Then, the multi-index unascertained measure μ_{ik} of any evaluation factor Z_i in the criterion layer satisfies the following equations.

$$0 \leq \mu_{ik} \leq 1 \qquad (9)$$

$$\mu_{ik} = \sum_{j=1}^{m} w_j \mu_{ijk} \qquad (10)$$

Therefore, the multi-index unascertained measure evaluation matrix for any evaluation factor Z_i in the criterion layer is obtained as follows:

$$(\mu_{ik})_{n \times p} = \begin{pmatrix} \mu_{11} & \mu_{12} & \cdots & \mu_{1p} \\ \mu_{21} & \mu_{22} & \cdots & \mu_{2p} \\ \vdots & \vdots & \ddots & \vdots \\ \mu_{n1} & \mu_{n2} & \cdots & \mu_{np} \end{pmatrix} \qquad (11)$$

where $(\mu)_{i1}, \mu_{i2}, \cdots, \mu_{ip})$ is the multi-index measure evaluation vector of Z_i.

If $\mu_k = \mu(Z_i \in C_k)$ indicates that the evaluation factor Z_i belongs to the unascertained measure of the Category C_k, then:

$$\mu_k = \sum_{i=1}^{n} w_i \mu_{ik} \quad k = 1, 2, \cdots, p \quad (12)$$

where w_i is the weight of the Z_i. Obviously, $0 \leq \mu_k \leq 1$, $\sum_{k=1}^{p} \mu_k = 1$ Eq. (12) are called at this time, and $(\mu_1, \mu_2, \cdots, \mu_p)$ is the multi-index comprehensive measurement evaluation vector of the target layer under the logical relationship of "or".

(2) When the logical relationship is "and"

The multi-index comprehensive measure, where the logical relationship between the evaluation factors of the criterion layer is "and", is calculated by normalisation processing after quadrature.

$$\mu_k = \frac{\prod_{j=1}^{m} \mu_{ijk}}{\sum_{i=1}^{n} \prod_{j=1}^{m} \mu_{ijk}} \quad (13)$$

At this time, there is no weight addition relationship between the evaluation factors for the criterion layer. Finally, $(\mu_1, \mu_2, \cdots, \mu_p)$, calculated using Eq. (13), is the multi-index comprehensive measurement evaluation vector of the target layer with the logical relationship "and".

The μ_{1k}, μ_{2k}, and μ_{3k} of the seismic hazard Z_1, pipeline vulnerability Z_2, and disaster loss Z_3 are:

$$\mu_{1k} = \begin{pmatrix} w_{11} \\ w_{12} \\ w_{13} \\ w_{14} \end{pmatrix}^T \cdot \mu_{1jk} = \begin{pmatrix} w_{11} \\ w_{12} \\ w_{13} \\ w_{14} \end{pmatrix}^T \cdot \begin{pmatrix} \mu_{111} & \mu_{112} & \mu_{113} & \mu_{114} & \mu_{115} \\ \mu_{121} & \mu_{122} & \mu_{123} & \mu_{124} & \mu_{125} \\ \mu_{131} & \mu_{132} & \mu_{133} & \mu_{134} & \mu_{135} \\ \mu_{141} & \mu_{142} & \mu_{143} & \mu_{144} & \mu_{145} \end{pmatrix} \quad (14)$$

$$\mu_{2k} = \begin{pmatrix} w_{21} \\ w_{22} \\ w_{23} \\ w_{24} \\ w_{25} \\ w_{26} \\ w_{27} \\ w_{28} \end{pmatrix}^T \cdot \mu_{2jk} = \begin{pmatrix} w_{21} \\ w_{22} \\ w_{23} \\ w_{24} \\ w_{25} \\ w_{26} \\ w_{27} \\ w_{28} \end{pmatrix}^T \cdot \begin{pmatrix} \mu_{211} & \mu_{212} & \mu_{213} & \mu_{214} & \mu_{215} \\ \mu_{221} & \mu_{222} & \mu_{223} & \mu_{224} & \mu_{225} \\ \mu_{231} & \mu_{232} & \mu_{233} & \mu_{234} & \mu_{235} \\ \mu_{241} & \mu_{242} & \mu_{243} & \mu_{244} & \mu_{245} \\ \mu_{251} & \mu_{252} & \mu_{253} & \mu_{254} & \mu_{255} \\ \mu_{261} & \mu_{262} & \mu_{263} & \mu_{264} & \mu_{265} \\ \mu_{271} & \mu_{272} & \mu_{273} & \mu_{274} & \mu_{275} \\ \mu_{281} & \mu_{282} & \mu_{283} & \mu_{284} & \mu_{285} \end{pmatrix} \quad (15)$$

$$\mu_{3k} = \begin{pmatrix} w_{31} \\ w_{32} \\ w_{33} \\ w_{34} \\ w_{35} \\ w_{36} \end{pmatrix}^T \cdot \mu_{3jk} = \begin{pmatrix} w_{31} \\ w_{32} \\ w_{33} \\ w_{34} \\ w_{35} \\ w_{36} \end{pmatrix}^T \cdot \begin{pmatrix} \mu_{311} & \mu_{312} & \mu_{313} & \mu_{314} & \mu_{315} \\ \mu_{321} & \mu_{322} & \mu_{323} & \mu_{324} & \mu_{325} \\ \mu_{331} & \mu_{332} & \mu_{333} & \mu_{334} & \mu_{335} \\ \mu_{341} & \mu_{342} & \mu_{343} & \mu_{344} & \mu_{345} \\ \mu_{351} & \mu_{352} & \mu_{353} & \mu_{354} & \mu_{355} \\ \mu_{361} & \mu_{362} & \mu_{363} & \mu_{364} & \mu_{365} \end{pmatrix} \quad (16)$$

2.5 Risk-level identification

After obtaining the multi-attribute unascertained measure μ_{ik} of X, the confidence criterion is used to identify the risk level of the index. The evaluation set $\{C_1, C_2, \cdots, C_n\}$ is an ordered set, the order relation is $C_1 < C_2 < \cdots < C_n$, and λ is the confidence. The confidence judgment criterion k_0 is:

$$k_0 = \min \left\{ k: \sum_{l=1}^{k} xl \geq; (k = 1, 2, \cdots, p) \right\} \quad (17)$$

Currently, X is considered Grade C_{k_0}, and the confidence level in the engineering application is 0.6 (Feng et al., 2020).

3 Application of the method

3.1 Pipeline overview

The long-distance pipeline (gas-source part) under the jurisdiction of a prefecture-level city in Sichuan Province exceeds 300km. More than 50% of the pipelines are located in mountainous areas with a seismic-fortification intensity of 6 or above, complex topographic and geological conditions, and active fault structures. The gas supply pipeline is under the jurisdiction of a company that undertakes long-distance supply tasks for the life, industry, and transportation needs of more than two million people in the city. Therefore, safe operation and management of gas-supply pipelines in seismically prone areas are significantly important. In this study, the pipeline between two gas-distribution stations was selected for evaluation and application. The pipeline trends are illustrated in Fig. 9. The selected pipeline was 29.179km long, and the designed daily gas supply was 15.0×10^4 m^3. Based on the characteristics of the pipeline and its surrounding environment (pipe diameter, wall thickness, pipe material, service life, population density, seismic fortification intensity, and station yard), it was partitioned into seven evaluation units.

Fig. 9 Pipeline route

3.2 Evaluation-index value

The value of each index of each evaluation unit was determined through the basic pipeline data collected in the early stage of the investigation. The detailed information of the evaluation unit is listed in Table 11. Simultaneously, the specific values were determined according to the above index classification standards (Table 12).

Table 11 Description of basic data of evaluation unit Pipe-section 1 Index

Criterion layer	Index	Explanation
Seismic risk	Magnitude frequency coefficient	0.828
	Seismic attenuation law	0.14-0.15g
	Geologic fault	There are fault zones in the near field
	Seismic geological disaster	No
Pipeline vulnerability	External diameter/mm	159mm
	Wall thickness/mm	6mm(Epoxy coal tar pitch
	Operating pressure/MPa	1.5MPa
	Buried depth/m	0.6~1.0m
	Material of pipe	20
	Service life/year	20 years
	Pipe node	There are nodes and the seismic effect is general
	Seismic fortification intensity	7
Disaster loss	Leakage degree	The pipe diameter is less than 219mm, and the operating pressure accounts for 37.5% of the design pressure
	Building density	0.3
	Population density	4
	H$_2$S concentration/mg · m^{-3}	0mg · m^{-3}
	Carbon emission	Carbon emission is directly related to the leakage degree, and the value refers to the leakage degree
	Natural Reserve	No

Table 12 Summary of evaluation unit Pipe Section 1 index value results

Criterion layer	Index	Result
Seismic risk	Magnitude frequency coefficient	0.828
	Seismic attenuation law	0.14~0.15g
	Geologic fault	3
	Seismic geological disaster	1
Pipeline vulnerability	External diameter/mm	159
	Wall thickness/mm	6
	Operating pressure/MPa	1.5
	Buried depth/m	0.6~1.0
	Material of pipe	3
	Service life/year	20
	Pipe node	3
	Seismic fortification intensity	7
Disaster loss	Leakage degree	2
	Building density	0.3
	Population density	4
	H_2S concentration/mg·m^{-3}	0
	Carbon emission	2
	Natural Reserve	1

3.3 Construction of unascertained measure matrix

The following only shows the detailed calculation process for Pipe Section 1. For the others, Visual Basic is used to program the calculations.

The single index unascertained measure matrix $(\mu_{1jk})_{4\times5}$, $(\mu_{2jk})_{8\times5}$, $(\mu_{3jk})_{6\times5}$ of seismic risk, pipeline vulnerability, and disaster loss is as follows:

$$(\mu_{1jk})_{4\times5} = \begin{pmatrix} z_{11} \\ z_{12} \\ z_{13} \\ z_{14} \end{pmatrix} = \begin{pmatrix} 0 & 0 & 0 & 0.853 & 0.147 \\ 0 & 0.931 & 0.069 & 0 & 0 \\ 0 & 0 & 1 & 0 & 0 \\ 1 & 0 & 0 & 0 & 0 \end{pmatrix}$$

$$(\mu_{2jk})_{8\times5} = \begin{pmatrix} z_{21} \\ z_{22} \\ z_{23} \\ z_{24} \\ z_{25} \\ z_{26} \\ z_{27} \\ z_{28} \end{pmatrix} = \begin{pmatrix} 0 & 0 & 0 & 0.857 & 0.143 \\ 0 & 0 & 0 & 0.4 & 0.6 \\ 0 & 0.647 & 0.353 & 0 & 0 \\ 0 & 0.5 & 0.5 & 0 & 0 \\ 0 & 0 & 1 & 0 & 0 \\ 0 & 0 & 0 & 0 & 1 \\ 0 & 0 & 1 & 0 & 0 \\ 0 & 0.5 & 0.5 & 0 & 0 \end{pmatrix}$$

$$(\mu_{3jk})_{6\times5} = \begin{pmatrix} z_{31} \\ z_{32} \\ z_{33} \\ z_{34} \\ z_{35} \\ z_{36} \end{pmatrix} = \begin{pmatrix} 0 & 1 & 0 & 0 & 0 \\ 0 & 0 & 0 & 1 & 0 \\ 0 & 0 & 0 & 1 & 0 \\ 1 & 0 & 0 & 0 & 0 \\ 0 & 1 & 0 & 0 & 0 \\ 1 & 0 & 0 & 0 & 0 \end{pmatrix}$$

3.4 Construction of multi-index comprehensive measurement and evaluation vector

Based on the weight-calculation results and the single-index unascertained measure matrix of each evaluation factor of the criterion layer (established inSection 3.3). First, we determine the measured vector (seismic risk, pipeline vulnerability, and disaster loss) of the evaluation factor of the criterion layer. Then, calculate the multi-index unascertained measure matrix of the criterion layer using the weighted summation method. It is used to calculate the comprehensive measurement and evaluation vector of the seismic damage risk of long-distance pipelines in the target layer. As follows:

(1) Criteria layer multi-index comprehensive measure evaluation vector

Measure vector μ_{1k} of seismic risk:
$\mu_{1k} = (0.1930 \quad 0.2672 \quad 0.3878 \quad 0.1297 \quad 0.0223)$

Measure vector μ_{2k} of pipeline vulnerability:
$\mu_{2k} = (0 \quad 0.1320 \quad 0.5280 \quad 0.1286 \quad 0.2114)$

Measure vector μ_{3k} of disaster loss:
$\mu_{3k} = (0.4140 \quad 0.1500 \quad 0 \quad 0.4360 \quad 0)$

(2) Target layer multi-index comprehensive measurement evaluation vector

Based on the measurement vector results of seismic risk, pipeline vulnerability, and disaster loss in (1) and the logical relationship of "and" among them. The quadrature normalisation was selected to establish the multi-index comprehensive measurement and evaluation vector of seismic-damage risk of long-distance pipeline in the target layer. The results were as follows:

Quadrature
$\mu' = (0 \quad 0.0053 \quad 0 \quad 0.0073 \quad 0)$

After normalisation, the comprehensive measure vector of the target layer of the evaluation object pipe section 1 is:

$\mu = (0 \quad 0.4206 \quad 0 \quad 0.5794 \quad 0)$

3.5 Risk-level identification

The risk level of Pipe Section 1 was determined using the confidence criterion, and λ was set as 0.6. Therefore, the risk level of Pipe Section 1 is calculated according to risk-level identification, and the specific process follows.

When $\lambda = 0.6$, $\mu_1 + \mu_2 = 0 + 0.4206 < 0.6$ and $\mu_1 + \mu_2 + \mu_3 + \mu_4 = 0 + 0.4206 + 0 + 0.5794 > 0.6$. Therefore, Pipe Section 1 is at high-risk level IV.

3.6 Result analysis and suggestions

Pipeline Section 1 was further analysed to identify the sources of the risk of seismic damage so that targeted risk-mitigation measures could be proposed. Radar diagrams were used to identify the risks associated with different factors.

Taking the pipeline criterion-layer factors as coordinates, we multiplied and added the measurement vector and its risk-level vector elements. The score obtained was the point marked to draw the radar map to complete the risk identification of different factors. The same method was used for the index layer. The radar-mark values of each factor and index were calculated; the radar diagram of the criterion layer is listed in Fig. 10. The radar diagram of the index layer corresponding to each factor is illustrated in Figs. 11–13.

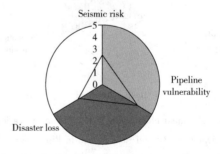

Fig. 10 Radar chart of the risk status of several factors at the criterion layer of Pipe Section 1

It can be observed fromFig. 10 that among the three factors of the criterion layer, the factor causing the high-risk level of Pipe Section 1 is the pipeline vulnerability. From Fig. 12, it can be seen that the three indicators (service life, wall thickness, and outer diameter) have risk levels be-

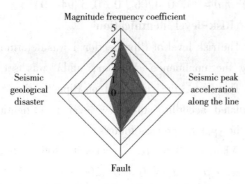

Fig. 11　Radar chart of the risk status of each index of the seismic risk factors in Pipe Section 1

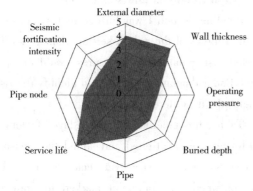

Fig. 12　Radar chart of the risk status of each index of the vulnerability factors in Pipe Section 1

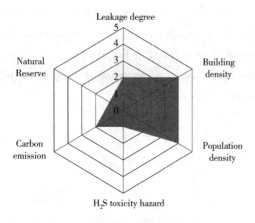

Fig. 13　Radar chart of the risk status of the various indicators of disaster-loss factors in Pipe Section 1

tween 4 and 5 in the radar diagram. These are major sources of pipeline high risk.

The service life of this pipe section is 20 years.

It is a steel pipe with a 159-mm outer diameter and 6-mm wall thickness; this also verifies the correctness of the results obtained from the radar map.

The following risk mitigation measures are proposed for Pipe Section 1:

(1) Tube change

For old pipes that need to be replaced, the new pipes should be larger in diameter, and the wall thickness should be increased as much as possible to meet the engineeringrequirements and economic conditions. Simultaneously, the connection quality between the new and old pipes should be managed well.

(2) No tube replacement

① The pipeline corrosion and other defects must be monitored.

② The inspection frequency of the pipeline must be increased.

③ Proper operation must be ensured in densely populated areas or high-value areas.

4　Conclusion

This study proposed a seismic-hazard risk-assessment method for long-distance pipelines based on an improved unascertained measurement model. The method considers the incompleteness and uncertainty of the basic data, thereby providing more accurate risk-assessment results. The results of this study show that the method is effective, consistent with actual situations, and has practical operability and reliability. This study not only assesses the risk level of pipelines but also identifies the most hazardous factors in pipeline sections, thus enabling the accurate mitigation of risks. This enables pipeline enterprises to effectively manage seismic risks in long-distance pipelines and ensure their safe operation. The following conclusions were drawn from this study.

1) This study develops a seismic-hazard risk-assessment method for long-distance pipelines. The method allows quantitative assessment of complex and difficult to determine seismic hazard information, thus providing a valuable tool for risk management in this field.

2) The most hazardous of the factors of dangerous pipe sections can be identified to provide accurate risk-mitigation suggestions as demonstrated by the evaluation results.

3) The theory of unascertained measure can

maximise the retention of known information about evaluation indicators, enabling them to participate directly in quantitative operations and minimising cumulative errors. The introduction of the theory of unascertained measure into the seismic-hazard risk assessment of long-distance pipelines can effectively mitigate the uncertainty inherent to the assessment results. It is often a consequence of incomplete basic data in the risk-assessment process.

4) In the indicator weight section, orthogonal experiments and finite element analysis (FEA) methods were used to analyse the quantitative indicators. Based on the calculation results, the grey correlation method was used to solve the weights and determine the weights of qualitative indicators by the improved Analytic Hierarchy Process (AHP). This result can achieve optimization of the weights obtained by the improved AHP method, improving the accuracy of weight determination in risk assessment.

5) Radar charts can clearly and intuitively express the risk status of different stages or influencing factors, aiding in identifying major risk sources quickly. The results indicate that the major risk sources of the evaluated pipeline section are service life, outer diameter, and wall thickness. It can provide a basis for developing effective risk-mitigation measures. In addition, if radar charts can be integrated into SCADA systems to achieve real-time online evaluation, the pipeline safety-management capabilities and risk-control levels will be greatly improved.

References

[1] Ba, Z., Wang, Y., Fu, J., & Liang, J. (2022). Corrosion risk assessment model of gas pipeline based on improved AHP and its engineering application. Arab. J. Sci. Eng. 1-19.

[2] Bai, Y., Wu, J., Sun, Y., Cai, J., Cao, J., & Pang, L. (2022). BN & CFD-based quantitative risk assessment of the natural gas explosion in utility tunnels. J. Loss Prev. Process. Ind. 80, 104883.

[3] Bai, Y., Wu, J., Ren, Q., Jiang, Y., & Cai, J. (2023). A BN-based risk assessment model of natural gas pipelines integrating knowledge graph and DEMATEL. Process Saf. Environ. Protect. 171, 640-654.

[4] Chen, Y., Zhang, L., Hu, J., Liu, Z., & Xu, K. (2022a). Emergency response recommendation for long-distance oil and gas pipeline based on an accident case representation model. J. Loss Prev. Process. Ind. 77, 104779.

[5] Chen, Y., Zhang, L., Hu, J., Chen, C., Fan, X., & Li, X. (2022b). An emergency task recommendation model of long-distance oil and gas pipeline based on knowledge graph convolution network. Process Saf. Environ. Protect. 167, 651-661.

[6] Deng, L., Yuan, H., Huang, L., Yan, S., Deng, Q., & Liu, H. (2018, November). Spatial Distribution Analysis and Regional Vulnerability Assessment of Geological Disasters in China. In Proceedings of the 4th ACM SIGSPATIAL International Workshop on Safety and Resilience (pp. 1-4).

[7] Ding, X., Tian, X., & Wang, J. (2024). A comprehensive risk assessment method for hot work in underground mines based on G1-EWM and unascertained measure theory. Sci. Rep. 14(1), 6063.

[8] Dong, S., Yi, X., & Feng, W. (2019). Quantitative evaluation and classification method of the cataclastic texture rock mass based on the structural plane network simulation. Rock Mech. Rock Eng. 52, 1767-1780.

[9] Eidinger, J. M., Maison, B. F., & McDonough, P. W. (2023). Natural gas system performance in Magna M5.7 2020 earthquake. Earthq. Spectra. 39(1), 362-376.

[10] Feng, L., Li, B., Zhang, X., Xu, Y., Liu, N., Zhou, X., & Chen, H. (2020). Capacity allocation optimization of energy storage system considering demand side response. Proc. CESS, 40(S1), 222-231.

[11] Frolova, N. I., Larionov, V. I., Bonnin, J., Sushchev, S. P., Ugarov, A. N., & Kozlov, M. A. (2017). Loss caused by earthquakes: rapid estimates. Nat. Hazards. 88, 63-80.

[12] Guangyuan, W. (1990). Unascertained information and its mathematical treatment. Journal of Harbin University of CE & Architecture, 4, 1-10.

[13] He, Z., Fu, M., & Weng, W. (2022). A nonlinear risk assessment method for chemical clusters based on fuzzy measure and Choquet integral. J. Loss Prev. Process. Ind. 77, 104778.

[14] Hu, J., Chen, C., & Liu, Z. (2022). Early

warning method for overseas natural gas pipeline accidents based on FDOOBN under severe environmental conditions. Process Saf. Environ. Protect. 157, 175-192.

[15] Li, A., Zhang, J., Zhou, N., Li, M., & Zhang, W. (2020). A model for evaluating the production system of an intelligent mine based on unascertained measurement theory. J. Intell. Fuzzy Syst. 38(2), 1865-1875.

[16] Li, X., Penmetsa, P., Liu, J., Hainen, A., & Nambisan, S. (2021). Severity of emergency natural gas distribution pipeline incidents: Application of an integrated spatio-temporal approach fused with text mining. J. Loss Prev. Process. Ind. 69, 104383.

[17] Li, X., Chen, G., Zhu, H., & Zhang, R. (2017). Quantitative risk assessment of submarine pipeline instability. J. Loss Prev. Process. Ind. 45, 108-115.

[18] Li, Y., Qian, X., Zhang, S., Sheng, J., Hou, L., & Yuan, M. (2023). Assessment of gas explosion risk in underground spaces adjacent to a gas pipeline. Tunn. Undergr. Space Technol. 131, 104785.

[19] Liu, A., Chen, K., Huang, X., Chen, J., Zhou, J., & Xu, W. (2019). Corrosion failure probability analysis of buried gas pipelines based on subset simulation. J. Loss Prev. Process. Ind. 57, 25-33.

[20] Liu, K. D., Pang, Y. J., Sun, G. Y., & Yao, L. G. (1999). The unascertained measurement evaluation on a city's environmental quality. Systems engineering theory and practice, 12, 52-58.

[21] Lu, D., Cong, G., & Li, B. (2022). A New Risk Assessment Model to Check Safety Threats to Long-Distance Pipelines. J Press Vess - T ASME, 144(5), 051801.

[22] Ma, L., Zhang, J., Xu, C., Lai, X., Luo, Q., Liu, C., & Li, K. (2021). Comprehensive evaluation of blast casting results based on unascertained measurement and intuitionistic fuzzy set. Shock. Vib. 2021(1), 8864618.

[23] Mahmood, Y., Chen, J., Yodo, N., & Huang, Y. (2024). Optimizing natural gas pipeline risk assessment using hybrid fuzzy bayesian networks and expert elicitation for effective decision-making strategies. Gas Science and Engineering, 125, 205283.

[24] Melissianos, V. E., Danciu, L., Vamvatsikos, D., & Basili, R. (2023). Fault displacement hazard estimation at lifeline-fault crossings: A simplified approach for engineering applications. Bull Earthq. Eng. 21(10), 4821-4849.

[25] Shu, H., Li, N., Dong, L., Luo, Q., & Sabao, A. R. (2023). Thermal humidity risk assessment in high-temperature environment of mines based on uncertainty measurement theory. Case Stud. Therm. Eng. 50, 103401.

[26] Sun, D., Chen, Y., Jiang, J., Li, H., & Li, Q. (2024). Numerical design of composite material crack arrestors for long-distance gas pipelines. Eng. Fract. Mech, 307, 110272.

[27] Wang, Q., Zhang, Q. P., Liu, Y. Y., Tong, L. J., Zhang, Y. Z., Li, X. Y., & Li, J. L. (2020). Characterizing the spatial distribution of typical natural disaster vulnerability in China from 2010 to 2017. Nat. Hazards, 100, 3-15.

[28] Wen, H., Liu, L., Zhang, J., Hu, J., & Huang, X. (2023). A hybrid machine learning model for landslide-oriented risk assessment of long-distance pipelines. J. Environ. Manage. 342, 118177.

[29] Wu, L., Qiao, L., Fan, J., Wen, J., Zhang, Y., & Jar, B. (2024). Investigation on natural gas leakage and diffusion characteristics based on CFD. Gas Science and Engineering, 123, 205238.

[30] Wu, Y., Meng, B., Wang, L., & Qin, G. (2020). Seismic vulnerability analysis of buried polyethylene pipeline based on finite element method. Int. J. Press. Vessels Pip. 187, 104167.

[31] Xu, J., Jiang, F., Xie, Z., & Wang, G. (2023). Risk assessment method for the safe operation of long-distance pipeline stations in high-consequence areas based on fault tree construction: Case study of China-Myanmar natural gas pipeline branch station. ASCE-ASME J. Risk Uncertain. Eng. Syst. Part A Civ. Eng. 9(1), 05022003.

[32] Zhang, D. Y., & Wu, J. Y. (2024). Experimentally validated numerical analyses on the seismic responses of extra-large LNG storage structures. Thin-Walled Struct. 195, 111407.

基于数据中台的油气管道行业
数据治理模式及实施路径探讨

李 梁　姜 辉　张建军　徐加兴　李步伟　刘 峰

（国家管网集团北京智网数科技术有限公司）

摘 要　油气管道数据具有多样性、时序性、空间分布性、专业性和安全性等特点，这些数据特点对传统的数据治理模式提出挑战。本文深入剖析油气储运行业数据特点，基于数据中台理论体系，引入数据挖掘、多模态数据治理等技术，结合数据管理能力成熟度模型，提出了适用于油气管道行业的基于数据中台的创新治理模式与实施路径。基于数据中台的数据治理模式涵盖数据战略、组织机制、标准体系、元数据驱动、数据架构、数据质量、数据安全、资源管控及数据开放共享等多方面，通过中台能力支撑实现数据的"可知、可用、可控"。实施路径分为夯实基础、深化重点领域治理和持续构建运营机制三个阶段。此研究旨在为油气管道行业数据治理提供关键思路与实践指南，进一步推动行业的数字化转型朝着更深层次、更具实效的方向稳步迈进。

关键词　油气管道行业数据治理；数据治理；基于数据中台的数据治理；行业数据治理

Research on the Data Governance Model and Implementation Path of the Oil and Gas Pipeline Industry Based on the Data Middle Platform

Li Liang　Jiang Hui　Zhang Jianjun　Xu Jiaxing　Li Buwei　Liu Feng

（Pipechina Digital Co., Ltd）

Abstract　Oil and gas pipeline data possess characteristics such as diversity, temporality, spatial distribution, specialization, and security, which pose challenges to traditional data governance models. This paper deeply analyzes the data characteristics of the oil and gas storage and transportation industry. Based on the theoretical system of the data middle platform, technologies such as data mining and multi-modal data governance are introduced. Combined with the data management maturity model, an innovative governance model and implementation path based on the data middle platform suitable for the oil and gas pipeline industry are proposed. The data governance model based on the data middle platform covers multiple aspects such as data strategy, organizational mechanism, standard system, metadata-driven, data architecture, data quality, data security, resource control, and data open sharing. Through the support of the middle platform capabilities, the data can be "knowable, available, and controllable". The implementation path is divided into three stages: consolidating the foundation, deepening the governance of key areas, and continuously constructing the operation mechanism. This research aims to provide key ideas and practical guidelines for data governance in the oil and gas pipeline industry, further promoting the digital transformation of the industry to move forward steadily in a deeper and more practical direction.

Key words　Oil and Gas Pipeline Industry Data Governance; Data Governance; Data Governance Based on Data Middle-platform; Industry Data Governance

油气管道作为能源输送的重要基础设施，其安全运行和高效管理对于保障国家能源安全具有重要意义。截止2024年底，我国长输油气管网总里程约19万公里，形成了庞大而复杂的能源输送网络。当前我国油气管道行业大力实施市场化、平台化、科技数字化和管理创新"四大战略"，加快建设"全国一张网"，着力打造智慧互联大管网、构建公平开放大平台、培育创新成长新生态，建设中国特色世界一流能源基础设施。这标志着油气管道行已从"数字化"建设迈向"数智化"的阶段，在此过程中庞大多样的数据对数据治理提出了更严峻挑战。因此，如何通过数据治理使数据真正成为油气管道行业乃至国家的核心资产，成为行业信息化数字化的规划重点。

1 油气管道行业数据治理现状与挑战

1.1 油气管道行业数据特点分析

油气管道行业数据的多样性、时序性、空间分布性、专业性和安全性，为数据治理带来了巨大挑战。

1. 多样性。油气管道行业在"管道建设"、"管道运营"、"管道维护"、"管道设备制造"等程中会产生复杂繁多的数据。例如，一个中等规模的油气管道工程从建设到运行，其涉及的数据不仅包括勘察设计资料、施工焊接参数、地理信息图片、设备参数、运行压力、温度、流量数据、日志文件，还涉及大量的管道沿线视频监控资料等，包含结构化数据、半结构化数据、非结构化数据、时序数据、GIS（时空数据）等多种类型。

2. 时序性。油气管道的运行状态需要实时监测和记录，以确保安全、稳定和高效。因此，为了反映管道的真实状态，需要对数据实时、时序地完成采集、传输和处理。同时，由于油气管道是连续运行的，数据也具有连续性，需要持续不断地进行采集和分析。

3. 空间分布性。油气管道行业具备庞大生产规模、分散的地理位置和极长的产业链条的特点，使数据在空间上分布广泛。例如，西气东输管道干线长达4000公里，沿线设有众多泵站、阀室等设施，因此其产生的数据也广泛分布在广阔的空间范围内。

4. 专业性。油气管道行业涉及复杂的专业知识，如地质、工程、机械、化学等多个专业领域。在管道建设时，需要根据不同地质条件进行设计和施工，涉及到复杂的地质勘探数据解读；在运行过程中，需要依赖化学专业知识分析油气成分变化对管道的影响。

5. 安全性。油气管道行业数据涉及国家战略资源储备、涉及千万家日常生活，同时也具备极高的商业价值，对国家能源经济安全和能源供应稳定具有直接影响，因此具有很高的安全性要求。

1.2 油气管道行业数据治理的目标和挑战

油气管道行业数据治理目标是通过构建和实施数据治理体系，确保数据可知、可用、可控，使数据真正成为油气管道行业乃至国家的核心资产。"数据可知"强调对数据的全面了解和清晰认知；"数据可用"聚焦于提供标准化的、清洁的数据；"数据可控"主要涉及对数据的安全性和基础资源管控。

然而数据治理推进过程中面临诸多挑战，诸如数据管理责任不清晰、数据标准不统一、数据质量不可靠、数据安全需要加强等，这些挑战导致数据在使用时面临诸如数据采集困难、有价值的数据无法被有效利用、数据孤岛、数据应用不足等问题。

为应对这些挑战，实现总体目标，油气管道行业本着科学性、可行性、创新性原则，探索和设计了基于数据中台的数据治理模式。

2 基于数据中台的数据治理模式

2.1 油气管道行业数据中台

数据中台是处于前台应用与后台数据存储之间，将数据进行资产化管理进而统一提供服务的平台，它能提升数据利用效率、促进业务创新、增强数据治理能力。数据中台是一套持续不断把数据变成资产并服务于业务的机制，其核心能力包括汇聚整合、提纯加工、服务可视化、价值变现。因此，数据中台贯穿了数据资源管理全生命周期，为数据治理提供了可靠的治理环境。

油气管道行业数据中台（图1）充分考虑油气储运行业特点，借鉴DCMM和TOGAF方法论，以数据架构、数据管理、数据应用为牵引，构建和夯实数据底座，提供数据湖仓平台、数据开发治理平台和数据资源共享平台等能力，加速实现数据资产化、资产价值化、价值生态化，推进数据赋能业务。

图 1 油气管道行业数据中台架构

数据治理与数据中台关联密切，相辅相成。一方面，数据治理是保障数据资产高质量和高价值的基础；另一方面，数据中台能够拓展数据治理的范围，进而达成对数据治理场景的全面覆盖。

2.2 基于数据中台的数据治理模式

《"十四五"大数据产业发展规划》要求围绕数据全生命周期，提高数据质量，打造分类科学、分级准确、管理有序的数据治理体系。油气管道行响应国家号召，在"十四五"期间将数据治理作为重点工作内容，为数字化转型提供核心能力支撑。

油气管道行业数据治理模式依托于中台提供的数据治理能力，以数据中台数据及其上下游数据作为治理内容，实现行业数据的"可知、可用、可控"，促使行业核心数据复用，指标数据打通。数据治理体系核心将以数据战略、数据治理组织、数据治理工具作为牵引和支撑，以数据标准体系作为指导，以元数据管理实现数据可知，以数据架构和数据质量管理实现数据可用，以数据安全和资源治理实现数据可控，以数据共享和开放管理作为价值落实点，全面构建形成油气管道行业数据治理模式。

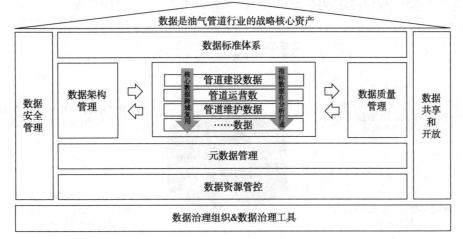

图 2 数据治理体系总框架

2.2.1 以数据战略和组织机制作为牵引

油气管道行业数据战略包括战略解码、业界标杆分析、数据治理成熟度评估和现状问题调研与差距分析四方面内容。通过内外部调研明确数据作为油气管道行业的核心资产的战略目标。

油气管道行业数据治理和数据管控工作的组织机制涵盖了一系列管理制度与管理流程，具体包括为之配套的组织机构设置、角色定义、岗位职责及要求、资源配置、培训宣传以及监督考核等方面。

2.2.2 以数据管理标准体系作为指导

遵循数字经济发展规律，以促进数据"供得出、流得动、用得好、保安全"为主线，遵循顶层设计、协同推进、问题导向、务实有效，应用牵引、鼓励创新，立足国内、开放合作的基本原则，建立油气管道行业数据标准体系，指导数据领域标准的制修订和协调配套。

数据标准体系构建有助于提升数据质量、促进数据共享与集成、提高数据管理效率、降低管理成本、推动实现行业内数据的互联互通。

2.3.3 以元数据作为驱动

元数据是指定义和描述其他数据的数据。根据行业特点油气管道行业元数据分为四类：(1)业务元数据，包括数据目录、业务对象、业务指标等。(2)技术元数据，包括信息系统、库、表、字段、GIS(时空数据)、非机构化数据等。(3)操作元数据，包括访问日志、作业调度等。(4)管理元数据，包括组织架构、人员、权限等。

油气管道行业数据治理采用元数据驱动理念，通过元数据驱动数据目录构建知识谱系和数据特征、驱动数据采集和集成形成数据汇聚和数据地图、驱动数据质量和数据标提升质量和标准化、驱动数据应用形成数据服务和流通管理。元数据驱动理念实现业内数据的认识和感知。

2.2.4 以数据架构作为基础

基于TOGAF框架，借鉴业内外优秀实践，油气管道行业将数据架构管理作为基础工作，以满足行业数据治理的需求。油气管道行业的数据架构主要包括数据资产目录、数据模型、数据分布、数据标准。数据架构是以结构化的方式描述在业务运作和管理决策中所需要的各类信息及其关系的一套整体组件规范。一方面可以有效地支持管理层的策略制定，另一方面可以准确地指导IT项目的实施。针对非机构化数据、GIS数据、时序数据也应该具备相应的数据架构信息。

通过数据架构各模块的联通，据此推进数据架构在各IT系统的落地管控，形成落标报告，确保数据设计阶段、开发阶段、上线阶段的标准化统一，提升数据架构治理水平，同时数据架构管理也将确保进入数据中台的数据具备高度标准化特点，大大提升数据分析的易用性。

2.2.5 以数据质量作为管理核心

数据质量是数据治理的核心目标之一，油气管网行业数据治理的核心目的之一是提供可用的数据。"数据可用"体现在多方面的数据质量管控。在数据用于决策时，保证数据的时效性，在数据综合分析应用时，保证数据的完整性，在数据逻辑关系中，保证数据的一致性。通过参考国家标准并结合自身行业情况，油气管道行业创新了数据质量的管理方法论。通过数据质量"六性"（图3）对数据进行评估。同时对于行业中非机构化数据、GIS数据、时序数据，数据质量的评估在六性的基础上进行扩充，例如：可理解性、波动性、可访问性等。

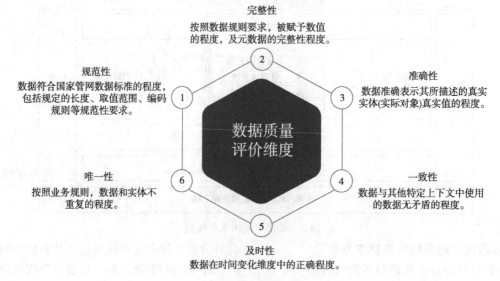

图3 油气管道行业数据质量六性

数据质量需要对准业务目标,识别数据质量风险,制定数据质量衡量标准和改进机制,以确保数据质量可控与持续提升,实现问题闭环管理,全面提升数据质量。

2.2.6 以数据安全和资源管控作为可靠保障

数据安全是指通过采取必要措施,确保数据处于有效保护和合法利用的状态,以及具备保障持续安全状态的能力。油气管道行业需要针对各类数据的采集、使用、传输、共享、存储、销毁等数据处理活动,明确并和落实数据安全保护等要求,通过实施技术保障、经费保障等一系列措施,确保行业数据的安全。

数据资源是指数据存储资源以及数据加工、分析、应用等过程的计算资源。油气管道行业需要针对数据存储设施和数据算力设施进行系统性管理,确保数据的存算信息、存算过程可追溯。

2.2.7 以数据共享和开放管理作为价值落实

数据共享和开放管理在油气管道行业数据治理中扮演着价值转化枢纽的角色,数据共享和开放指行业数据对内进行共享,对外进行开放。其中,数据共享的管理旨在构建数据需求和服务字典,实现的高效流通与复用,同时共享过程中,建立反馈与验证机制,不断完善数据质量,进一步提升数据的可用性;数据开放管理核心在于发展数据要素市场,积极促进数据资产的交易和流通,形成完整闭环的数据资产和数据产品认定、评估、交易流通的管理机制。

2.2.8 以数据中台作为工具支撑

数据中台提供了数据治理的基础能力,包括:(1)元数据。通过对四类元数据的盘点和采集整合,并经过元数据治理后,形成数据资产目录,提供数据血缘、资产关联、数据地图和元数据分析等能力。(2)数据架构。提供数据架构设计、发布、终止的全过程管理能力。通过数据架构的设计和落标,可以确保数据设计阶段、开发阶段、上线阶段的标准化统一。(3)数据质量。为数据的开发及使用提供全套的数据质量解决方案,构建数据质量PDCA闭环管理机制,提供从质量规则设置、质量规则执行、质量结果分析管理到问题改进等数据质量管理全流程的工作环节所需能力。(4)数据安全。提供数据加密、行级和列级的细粒度权限控制等多种安全防护手段,满足行业内外数据共享安全需求,确保数据安全可控共享(5)存算资源管控。支持对所有数据资源进行统计管理,包括资源浏览量、资源申请量、资源调用量。

3 数据治理实施路径

数据治理作为油气管道行业战略工作之一,具有复杂性和长期性的特点,需要进行总体规划和分步实施。

3.1 夯实基础

夯实基础阶段需完成数据治理的基础建设工作,实现覆盖全域的数据架构及安全管理,以确保数据治理全面效推行。这一阶段要完成行业数据治理顶层规划建设,明确治理标准体系、组织流程及运作机制,并使其贯穿"管道建设"、"管道运营"、"管道维护"、"管道设备制造"、"安全与环境管理"、"技术研发与创新"等领域进行落地。在顶层规划的指引下,围绕各领域开展数据架构设计和数据资产盘点工作,形成覆盖全领域的数据资产目录,并加强数据架构的审核和管理,确保数据架构的有效性和落地实施;同时进行数据安全建设,开展数据安全评估,制定全面的安全策略,落实数据访问控制与认证授权。此外需要搭建数据治理工具平台,以满足可持续的治理需求。

3.2 深化重点领域的数据治理

在上一阶段的基础上,针对"管道建设"、"管道运营"、"管道维护"等行业重点域进行全面深入的治理落地,为数据分析和数据流通提供有效支撑。本阶段需加强对相关业务领域数据架构和数据质量的识别和管理,形成数据架构落标报告和数据质量报告,并推动问题识别和问题改进,提升数据质量。同时依托上阶段数据资源盘点成果,增加与地质、工程、化学等专业相关的元数据分类和管理,建立元数据仓库,通过元数据驱动数据资产汇聚,形成数据血缘和数据地图,支撑油气管道数据分析中,能够快速溯源,促进数据可见、可查询、可理解。另外在本阶段需要开展数据要素流通相关的研究和拓展,建立油气管道行业内部的数据资产化管理与经营体系,包括数据资产认定、确权、价值评估、资产处置和流通等,推动油气管道行业与相关行业(如能源市场、物流运输等)的数据协同创新。此外,本阶段可以借鉴 GB/T 36073—2018《数据管理能力成熟度评估模型》围绕数据管理的8个能力域、28个能力项定期进行治理效果评估,确保数据治理工作处于可量化级以上。

3.3 持续构建智能数据治理运营机制

本阶段除了全面推进其他领域数据治理外，还将进一步探索主动数据治理运营机制。建立全域数据资产共享模式，打通横纵向数据脉络，构建油气管道行业知识谱系，推动行业数据广泛共享，助力数据价值挖掘，同时拉通和构建行业数据标准，持续推动数据标准化进程。另一方面推动实现数据资产的广泛经营和流通，并实现数据安全溯源，形成事前，事中，事后多方位的数据安全监控和审计。此外将结合AI能力构建智能数据治理平台，通过智能数据治理优化数据治理策略，形成良性循环。在此基础上，构建具备油气管道行业特色的数据治理能力成熟度评估体系，验证数据治理效果，提升总体数据管理能力。

4 结语

数字化转型过程中，数据治理是基础。基于数据中台的油气管道行业数据治理模式及实施路径整合了数据战略、组织架构、标准体系、元数据驱动、数据架构、质量控制、安全保障及资源管理与共享等关键要素，形成有机整体，旨在实现数据的可知、可用、可控，提升数据资产价值；同时实施路径分阶段推进，确保治理工作可持续发展。该治理模式和实施路径有助于推动行业整体数据治理水平的提升，进而保障国家能源基础设施的安全、高效运行。未来，随着技术的不断进步和数智中台的演进，新的智能数据治理模式将在油气管道行业发挥更加重要的作用。

参 考 文 献

[1] 人民日报海外版. 横跨东西、纵贯南北、覆盖全国——油气电全国网络正越织越密[新闻报道]. (2024-10-09)[2024-10-09]. http://www.sasac.gov.cn/n2588025/n2588139/c31841657/content.html

[2] 付登坡，江敏，任寅等数据中台[M]. 北京：机械工业出版社，2020：22。

[3] 工信部信息技术发展司.《"十四五"大数据产业发展规划》解读[N]. 中国电子报，2021-12-03(008).

[4][8] 国家发展改革委、国家数据局、中央网信办、工业和信息化部、财政部、国家标准委. 国家数据标准体系建设指南，2024：6。

[5] 国家质量监督检验检疫总局，中国国家标准化管理委员会. 信息技术元数据注册系统（MDR）第1部分：框架：GB/T18391.1-2009[S]. 定义3.2.16。

[6] 翁瑜卿. 基于标准化的油气能源企业数据治理研究[J]，2022：2。

[7] 中华人民共和国全国人民代表大会.《中华人民共和国数据安全法》(2021年)

[9] 国家质量监督检验检疫总局，中国国家标准化管理委员会. 数据管理能力成熟度评估模型：GB/T36073—2018[S]. 北京：中国标准出版社，2018：48.

Seismic vulnerability analysis of natural gas distribution station

Wu Ying Tian Yu Meng Bojie

(School of Civil Engineering and Geomatics, Southwest Petroleum University)

Abstract With the continuous promotion of low-carbon clean energy, the gas supply system has developed rapidly, and the pipeline mileage has shown a significant upward trend. The natural gas distribution station is an important part of the gas supply system and the gas transmission pipeline system. The process flow of the gas distribution station is complex, there are manypipeline equipment, and it shows its vulnerability in the face of earthquake disasters. And also the station with the most serious consequences after the earthquake. The explosion, fire hazard and other accidents of the gas distribution station will lead to huge economic losses and serious social impact. However, most of the current studies rarely consider the seismic vulnerability analysis of gas distribution station, which left a great danger and risk for seismic damage prevention of natural gas supply system. In this study, a method which can evaluating the vulnerability of urban natural gas distribution stations to earthquake is proposed to improve the theory of seismic vulnerability analysis. The finite element model of the pressure regulating system of the gas distribution station is established, the incremental dynamic analysis is carried out, the vulnerability curve is drawn. Obtaining the probability of the gas distribution station exceeding the limit state under different seismic peak ground accelerations. The whole research process can form a set of seismic vulnerability assessment methods suitable for urban gas distribution stations, which has important reference significance for disaster prevention and mitigation planning and policy formulation of urban gas distribution stations.

Keywords Gas distribution stations, Seismic vulnerability, Risk quantification, Incremental dynamic analysis.

1 Introduction

Lifeline system is generally considered to include power system, transportation system, communication system, heating and gas supply system, etc. The urban natural gas distribution station is an important part of the gas supply system and the gas transmission pipeline system. The gas distribution station is characterized by complex process flow, numerous pipeline equipment and valves. Process pipeline rupture, valve damage, equipment failure and other consequences are prone to occur during high-intensity earthquakes, which seriously affect the stability of downstream gas supply. In case of explosion, fire and other accidents, huge economic losses and social impacts will be caused. In the 1964 Niigata earthquake in Japan, the fire caused by the rupture of the gas storage tank in a station lasted for more than half a month, burning more than 80 oil tanks and large areas, resulting in almost paralysis of the refinery in Niigata, causing serious economic losses and casualties. In the 1994 Los Angeles earthquake, the gas pipeline network was seriously damaged, more than 30 high and medium pressure pipelines were damaged, more than 700 low pressure pipelines were damaged, 151000 users stopped gas, and the gas system leaked up to 150000, causing a large number of fires and explosions. In 1995, a major earthquake occurred in Kobe, Japan. The earthquake caused the rupture of the gas pipeline, which led to gas leakage. In case of a fire source, it immediately caused a raging fire. There were 459 fires, with a burning area of tens of thousands of square meters, causing a large number of casualties. In the 2008 Wenchuan Earthquake in China, 70% of cities and counties in Sichuan Province were affected by the earthquake, and large areas of pipelines were damaged, which brought huge economic losses to the country and serious impact on people's production and life. In the face of random and extremely destructive earthquake disasters, urban natural gas distribution stations bear huge risks.

In order to protect urban natural gas distribution stations, scholars have conducted a series of related studies. In 2004, Carl E. Jaske and others learned from the pipeline integrity management model, carried out the research on asset integrity management technology based on API 1160, and proposed the basic concept and specific implementation steps of oil and gas transmission station integrity management. In 2010, William V. Harper et al. used a reliability based approach to assess the risk of station facilities. In 2012, Susan Urra et al. established a risk assessment model of human factors from the aspects of design, construction, operation, maintenance, etc. based on human factors in the pipeline industry. In 2015, R Sivaprakasam et al. proposed a method for reliability analysis of complex engineering systems using fault tree analysis and data uncertainty/imprecision information, and carried out quantitative risk assessment on LPG filling stations. In 2019, Hye Ri Gye proposed a quantitative risk assessment of a high-pressure hydrogen fuel station in urban areas with a large population and crowded equipment. The research results show that the main risks are the leakage of the tube trailer and distributor and the potential explosion of the tube trailer. In 2021, Meng graded the risk of gas distribution stations from the perspective of failure probability.

In 2019, Leoni et al. proposed an improved risk-based maintenance (RBM) method by taking the Italian gas pressure regulating and metering station as an example, and used Bayesian networks to model risks and related uncertainties. In terms of seismic vulnerability, mature theories and methods of structural seismic vulnerability have also been formed. In 2017, Y L Chen defined the connotation of the urban vulnerability of oil and gas pipelines. Based on the analysis of the transmission performance of oil and gas structural vulnerability, he analyzed the indicators that affect the vulnerability of oil and gas pipelines and built a mechanism framework for the urban vulnerability of oil and gas pipelines from three aspects: environment, structure and response. In 2018, He Si et al. designed and completed 24 groups of 72 reciprocating loading tests with diameter, water pressure, wall thickness and loading mode as variables for building water supply pipes such as PPR and galvanized steel pipes commonly used in new and old buildings in China, and summarized the damage characteristics and seismic vulnerability of various types of pipes according to the test results. In 2020, Zou Rong proposed an expert evaluation method for seismic vulnerability of urban gas polyethylene buried pipe network based on cloud theory, which is of great significance for ensuring the safe operation of urban gas pipelines, sustainable supply of energy and the safety of people's lives and property. In 2019, Zhang and Wu proposed a seismic vulnerability analysis method for reinforced concrete (RC) bridges based on Kriging model. It aims to reduce the calculation effect when Monte Carlo technology is used to establish the structural vulnerability curve. In 2018, Qiu et al. carried out the seismic vulnerability analysis of the rocky mountain tunnel. The UDM method is used to generate experimental samples considering the variability of variables. Different values of tunnel depth, tunnel size, rock mass and lining thickness are considered. The probabilistic seismic demand model is established, the dynamic finite element analysis is conducted, and the vulnerability curve is established based on the probabilistic seismic demand model. The rationality of this method is verified. In 2020, Ratiranjan et al. will assess the seismic vulnerability of Banda Aceh through the combination of multi-level decision-making method, analytic hierarchy process and geographic information system method. Banda Yaqi City is close to the Sumatran fault in North Sumatra. The vulnerability index includes three aspects: social vulnerability index, structural vulnerability index and geotechnical vulnerability index. In 2021, Wang established the seismic vulnerability evaluation system of buried pipelines in mountainous areas from the perspective of logical relationship combined with borda method. In 2022, N. S. analyzed the existing earthquake disaster model and consequence model in the United States to encourage more systematic quantification of natural gas pipeline damage models. Dong established a seismic vulnerability analysis model from the perspective of pulse characteristics of ground motion. Xiao proposed an interdependent lifeline function to study seismic vulnerability from the perspective of two-way dependence of lifeline. S. M. proposed a seismic vulnerability assessment method for the collapse of existing buildings, and compared the vulnerability indicators. G. K studied the risk identification and vulnerability of the support structure overhead pipeline and analyzed the consequences of the accident. X. U proposed a mixed uncertainty seismic vulnerability assessment method for underground structures considering seismic demand and capacity. H. U proposed a new seismic vulnerability analysis framework for probabilistic seismic performance and risk assessment of slopes.

However, based on the above research, the author found the following two problems:

(1) The risk study of natural gas stations is relatively mature, and relevant evaluation systems and standards have been formed. The research on seismic risk assessment has also been carried out in a small amount, but it is not perfect and mature enough;

(2) At present, the research on seismic vulnerability of different structures in the field of seismic resistance is relatively mature, forming four different types of vulnerability assessmentmethods, but the research on seismic vulnerability of urban natural gas distribution stations is still lacking;

It can be seen from the above that the current theoretical methods for seismic vulnerability research have been formed, and have been well applied in many fields, which play an important role. Therefore, it is necessary to absorb, digest and summarize these excellent methods to form a set of seismic vulnerability assessment methods suitable for urban gas distribution stations to guide the disaster prevention and mitigation planning of urban gas distribution stations and the formulation of policies and measures.

In order to comprehensively analyze the seismic vulnerability of gas distribution stations from the perspective of theory and numerical simulation, the second section deduces the formula of seismic vulnerability of gas distribution stations and shows the basic framework of seismic vulnerability assessment of gas distribution stations. The third section establishes the model. In the fourth section, the dynamic incremental analysis is carried out by ground motion amplitude modulation. Section 5 gives seismic fragility curves. Finally, the conclusion is summarized (Section 6).

2 Seismic vulnerability theory

Seismic vulnerability analysis is considered as an effective means to assess the degree of structural damage after an earthquake, usually in the form of a vulnerability curve. The seismic vulnerability curve defines the conditional probability of the structure reaching or exceeding the preset damage state under a specific seismic intensity, which has been successfully applied to the seismic performance evaluation of different structures [3~5]. There are usually four vulnerability assessment methods: empirical vulnerability, expert judgment vulnerability, analytic vulnerability and mixed vulnerability.

Vulnerability analysis is a method with certain evaluation accuracy and moderate evaluation difficulty. The analytical vulnerability method is selected in this paper. Compared with the direct method of directly obtaining the failure probability of the structure under earthquake action, the indirect method of analytical vulnerability transforms the study of seismic vulnerability into a classical reliability problem: $F(x) = P[S > R \mid IM = x]$, where R and S are the seismic forces and responses of the structure, $F(x)$ is seismic vulnerability. In the PBEE (Performance Based Earthquake Engineering) probability framework proposed by PEER (The Pacific Earthquake Engineering Research Center), seismic vulnerability is usually expressed as:

$$F(x) = \psi\left[\frac{\ln m_{D|IM} - \ln m_C}{\sqrt{\beta_{D|IM}^2 + \beta_C^2}}\right] \quad (1)$$

In the formula, $m_{D|IM}$ and $\beta_{D|IM}$ are the conditional median and logarithmic standard deviation of seismic demand D under the action of ground motion intensity IM, m_C and β_C are the median and logarithmic standard deviation of structural seismic capacity.

According to Formula (1), seismic vulnerability analysis is divided into probabilistic seismic demand analysis and probabilistic seismic capacity analysis, which also correspond to the probabilistic seismic demand model and probabilistic seismic capacity model in the PBEE probabilistic decision-making framework proposed by PEER.

Seismic vulnerability is generally summarized as the conditional probability that the structure reaches or exceeds the set damage state or performance level under a given ground motion intensity. The seismic vulnerability analysis based on probabilistic seismic demand model mainly includes four parts: ① selecting ground motion records; ② Select reasonable ground motion intensity index; ③ Define the damage state of the structure; ④ The incremental dynamic analysis (IDA) method is used to establish the probabilistic seismic demand model and then fit the vulnerability curve. The framework of seismic vulnerability analysis of pressure regulating system in gas distribution station is shown in figure 1.

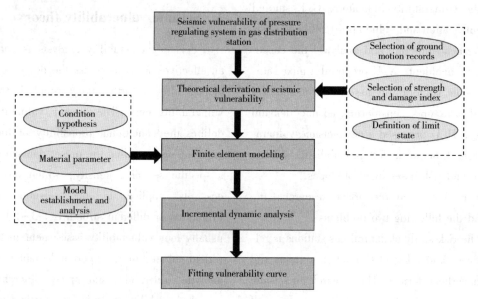

Fig. 1 Frame diagram of seismic vulnerability analysis of pressure regulating system in gas distribution station

2.1 Derivation of vulnerability theoretical formula

2.1.1 General vulnerability function derivation

In all kinds of vulnerability distribution, the logarithmic normal distribution model is proved to be a good fit by experiments. Although the model has some shortcomings, such as the low tail region is not conservative, it is widely used because of its convenient application:

$$f_{R_g}(x) = \Phi\left[\frac{\ln x - \mu_{R_g}}{\beta_{R_g}}\right] \quad (2)$$

Where, $\Phi[\cdot]$ is the probability distribution function (PDF) of standard normal random variables; μ_{R_g} represents the logarithmic mean of Rg; β_{R_g} represents the dispersion. Rg is generalized resistance.

Since there is a relationship between the median A_{R_g} and the logarithmic mean μ_{R_g} of the lognormal distribution random variable Rg:

$$\mu_{R_g} = \ln A_{R_g} \quad (3)$$

Substituting Eq. (3) into Eq. (2), the expression of seismic vulnerability function can be obtained:

$$f_{R_g}(x) = \Phi\left[\frac{\ln x - \ln A_{R_g}}{\beta_{R_g}}\right] \quad (4)$$

Where, A_{R_g} is the median value of the lognormal

distribution random variable Rg, which corresponds to the quantile on the seismic fragility curve with a failure probability of 50%.

The logarithmic normal distribution model is used to derive the vulnerability function. Because of the generality of Formula (2), different parameters need to be substituted into different models. The engineering demand parameter EDP, the damage measure DM, and the decision variable DV are all regarded as random variables. Assuming that they obey the lognormal distribution, the demand vulnerability function, the capacity vulnerability function and the loss vulnerability function can be obtained from the formula (2) as shown in (5)~(7).

$$G_{EDP \mid IM}(edp \mid im) \triangleq f_{R_{demand}}(im) = P[EDP \geqslant edp \mid im] = 1 - \Phi\left[\frac{\ln\left(\frac{edp}{A_{EDP \mid IM}}\right)}{\beta_{EDP \mid IM}}\right] \quad (5)$$

$$G_{DM \mid EDP}(dm \mid edp) \triangleq f_{R_{capacity}}(edp) = P[DM \geqslant dm \mid edp] = 1 - \Phi\left[\frac{\ln\left(\frac{edp}{A_{DM \mid EDP}}\right)}{\beta_{DM \mid EDP}}\right] \quad (6)$$

$$G_{DV \mid DM}(dv \mid dm) \triangleq f_{R_{loss}}(dm) = P[DV \geqslant dv \mid dm] = 1 - \Phi\left[\frac{\ln\left(\frac{edp}{A_{DV \mid DM}}\right)}{\beta_{DV \mid DM}}\right] \quad (7)$$

Where, $G(x \mid y)$ is the complementary cumulative distribution function (CCDF) of random variable X under the condition of $Y = y$. im、edp、dm is the value of corresponding variable IM, EDP and DM; $X \mid Y$ is the variable X under the given variable Y. Where, $X = EDP$、DM、DV, $Y = IM$、EDP、DM, $A \mid_{X \mid Y}$、$\beta \mid_{X \mid Y}$ is the median and deviation of the corresponding variable $X \mid Y$.

However, in (5) ~ (7), the relationship between the seismic intensity parameter IM and the damage measure DM and the decision variable DV cannot be directly obtained. In general, it is impossible to effectively quantify the relationship between seismic vulnerability and damage or decision. Therefore, the convolution of formula (5) and (6) is used to obtain (8) as follows:

$$G_{DM \mid IM}(dm \mid im) \triangleq f_{R_{damage}}(im) = P[ls_{damage} \mid im]$$
$$= P[DM \geqslant dm \mid im]$$
$$= \Phi\left[\frac{\ln\left(\frac{m_{EDP \mid IM}}{m_{DM \mid EDP}}\right)}{\sqrt{\beta^2_{EDP \mid IM} + \beta^2_{DM \mid EDP}}}\right] \quad (8)$$

In the formula, ls_{damage} is the value of the damage limit state LS_{damage}, that is, a limit state, which can be called the damage vulnerability function, $\beta_{X \mid Y}$ is the logarithmic standard deviation of damage vulnerability.

Similarly, the decision vulnerability function (9) can also be obtained by convolving formulas (6) and (7):

$$G_{DV \mid IM}(dv \mid im) \triangleq f_{R_{decision}}(im) = P[ls_{decision} \mid im]$$
$$= P[DV \geqslant dv \mid im]$$
$$= \Phi\left[\frac{\ln\left(\frac{m_{DM \mid EDP}}{m_{DV \mid DM}}\right)}{\sqrt{\beta^2_{DM \mid EDP} + \beta^2_{DV \mid DM}}}\right] \quad (9)$$

If the strong nonlinear behavior such as structural collapse is not considered, assuming that the log-linear relationship between the median value of the variable and the adjacent variables is satisfied during the analysis process, the relationship between the variables IM, EDP, DM, and DV can be represented by (10) to (12):

$$\ln(m_{EDP}) = A + B\ln(IM) \quad (10)$$
$$\ln(m_{DM}) = C + D\ln(EDP) \quad (11)$$
$$\ln(m_{DV}) = E + F\ln(DM) \quad (12)$$

Where, A, B, C, D, E and F are unknown parameters, which can be obtained by regression analysis.

The formulas (5), (8) and (9) are transformed into the form similar to (4), which correspond to the exceedance probabilities of EDP, DM and DV for a given ground motion intensity, respectively.

2.1.2 Theoretical derivation of seismic vulnerability of gas distribution station

The seismic vulnerability curve of pipeline reflects the conditional probability of pipeline damage exceeding the defined pipeline damage index when the pipeline is subjected to different seismic intensities. By formula (13):

$$P_f = P(\sigma_c/\sigma_{max} < 1) \quad (13)$$

In the formula, P_f represents the conditional probability of the structural damage index σ_c corresponding to the structural state point defined by the pipeline damage index σ_{max} under earthquake, σ_c represents structural state damage index, σ_{max} represents the pipeline damage index, rewrite formula (13) to (14)

$$P_f = P(\ln\sigma_c - \ln\sigma_{max} < 0) \quad (14)$$

Let $X = \ln\sigma_c - \ln\sigma_{max}$, because $\ln\sigma_c$ and $\ln\sigma_{max}$ are independent random variables and both obey normal distribution, then X also obeys normal distribution. The mean is $\mu_X = \mu_{\sigma_c} - \mu_{\sigma_{max}}$ and the standard deviation is $\theta_z = \sqrt{\theta_{\sigma_c}^2 + \theta_{\sigma_{max}}^2}$. The formula of pipeline failure probability is:

$$P_f = P(X < 0) = \int_{-\infty}^{0} f(X)\,dX = \int_{-\infty}^{0} \frac{1}{\theta_X\sqrt{2\pi}}\exp\left[-\frac{1}{2}\left(\frac{X-\mu_X}{\theta_X}\right)^2\right]dX \quad (15)$$

To simplify the calculation, $N(\mu_X, \theta_X)$ is transformed into standard normal distribution function $N(0, 1)$. Let $x = \frac{X-\mu_X}{\theta_X}$, $dX = \theta_X dx$, $X = \mu_X + \theta_X \cdot x < 0$, $x < \frac{\mu_X}{\theta_X}$. Thus the formula can be further written as:

$$P_f = P\left(x < -\frac{\mu_Z}{\theta_Z}\right) = \int_{-\infty}^{-\frac{\mu_Z}{\theta_Z}} \frac{1}{\sqrt{2\pi}}\exp\left[-\frac{1}{2}(x)^2\right]dx = \Phi\left(-\frac{\ln\sigma_c - \ln\sigma_{max}}{\sqrt{\theta_{\sigma_c}^2 + \theta_{\sigma_{max}}^2}}\right) \quad (16)$$

In this article, a specific failure probability can be written as:

$$P_f = \Phi\left(-\frac{\ln\left(\frac{\sigma_c}{\sigma_{max}}\right)}{\sqrt{\theta_{\sigma_c}^2 + \theta_{\sigma_{max}}^2}}\right) = \Phi\left(\frac{\ln\left(\frac{\sigma_{max}}{\sigma_c}\right)}{\sqrt{\theta_{\sigma_c}^2 + \theta_{\sigma_{max}}^2}}\right) = \Phi\left(\frac{\ln(\alpha(PGA)^\beta/\sigma_c)}{\sqrt{\theta_{\sigma_c}^2 + \theta_{\sigma_{max}}^2}}\right) \quad (17)$$

2.2 Selection of ground motion records

Ground motion is an important input parameter in dynamic time history analysis, which can be understood as external accidental load borne by the structure. Different ground motions have different effects on the results. Reasonable selection of ground motions is a strong guarantee for the accuracy of dynamic time history analysis results. The Pacific EarthquakeCenter (PEER) of the United States combed the ground motion parameters after earthquakes in various countries, and systematically collated them, and built a ground motion database based on the measured ground motion data, which is convenient for scholars to conduct seismic time history analysis and selection.

The preliminary screening results of actual strong earthquake records selected in this study are shown in Table1.

Table1 PEER website actual strong earthquake records first screening results show

Number	Time	Event	Seismic station	D5-95/s	Mag	Vs30/(m/s)
1	1952	Kern County	Taft Lincoln School	30.3	7.36	385.43
2	1971	San Fernando	Santa Felita Dam (Outlet)	23.6	6.61	389.0
3	1979	ImperialValley-06	Cerro Prieto	36.4	6.53	471.53
4	1980	Irpinia, Italy-01	Bisaccia	27.0	6.9	496.46
5	1980	Irpinia, Italy-01	Calitri	24.2	6.9	455.93
6	1980	Irpinia, Italy-02	Bisaccia	22.0	6.2	496.46
7	1984	Morgan Hill	Gilroy Array #3	20.4	6.19	349.85

Plot seven seismic records, as shown in Figure 2. The figure contains three directions in the seismic record: horizontal 1, horizontal 2, and vertical. They refer to Horizontal-1 Acc. Filename, Horizontal-2 Acc. Filename, and Vertical Acc. Filename in the seismic record respectively.

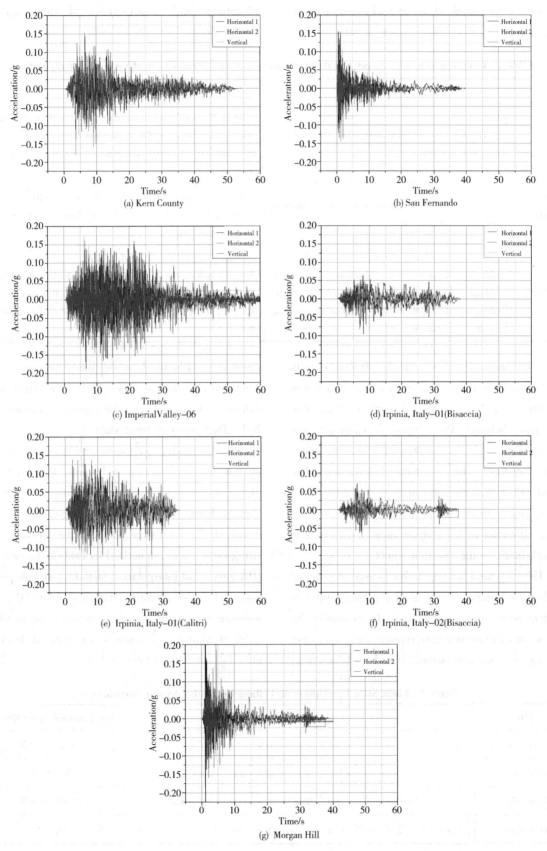

Fig. 2 Ground motion time history curve

2.3 Selection of ground motion intensity index and damage index

The ground motion intensity index is used to measure the magnitude of the ground motion intensity. The IDA method often uses ground motion intensity indexes such as Peak Ground Acceleration (PGA), Spectral Acceleration (Sa), Peak Ground Velocity (PGV), etc. Table2 compares the advantages and disadvantages of three different ground motion intensity indexes.

Table 2 Comparison of advantages and disadvantages of three different ground motion intensity indices

	Advantage	Disadvantage
Sa	IDA curves exhibit low dispersion and are suitable for most monomer structures	When the high-order modal effect ($T<T1$) dominated by the frequency component of $T \neq T1$ and the system period extension effect caused by nonlinear deformation, this index will produce great variability.
PGV	High efficiency and practicality for most structures compared to other metrics	When PGV is selected as the strength index, the IDA curve corresponding to the damage index will have high discreteness in the whole strength range
PGA	Widest range of use, high technology maturity. Suitable for complex station structure, with strong adequacy, in the calculation of full probability structure failure probability, do not need to rely on the remaining conditional probability	It mainly reflects the amplitude characteristics of the local high frequency components of seismic waves, and the high frequency components do not play a key role in the seismic response of the structure, which will lead to the discreteness of the calculation results.

In summary, for station structures with complex structures, Sa is adopted as the ground motion intensity index, which will result in a large discrete type. Although PGV has high efficiency and practicability, it is far less than PGA in sufficiency, which makes the calculated results less reliable. PGA is more applicable as the strength index, which is in line with the actual situation of the station. Therefore, this paper uses PGA seismic strength index as the research object.

The structural damage index is used to evaluate whether the structure is in good condition and how the damage is, expressed in DM. The evaluation indicators of pipeline structure damage include: pipeline stress, pipeline strain, etc. In engineering, stress is usually used as the index to evaluate the structural damage, so this study selects stress as the damage index to evaluate the pipeline structure.

2.4 Definition of limit state

In this study, the fourth strength theory is used as the failure criterion to define four states of pipelines under earthquake: basically intact, slightly damaged, moderately damaged and severely damaged. The limit state points of the pressure regulating system are summarized in Table3. According to IDA curve and interpolation method, we can get the limit state ground motion intensity index IM of the pressure regulating system structure, as shown in Table 4. Material parameters are derived from the material test data in Section 3.2.

Table 3 Limit State Description of Voltage Regulating System Structure

State	Degree of pipeline damage	Judgement index (MPa)
Basically intact	Pipe not damaged	$\sigma_M < 141.336$
Slight damage	Pipeline may be partially damaged without leakage	$141.336 \leqslant \sigma_M \leqslant 235.56$
Moderate damage	Pipe partially damaged, minor leak, serviceable after repair	$235.56 < \sigma_M \leqslant 392.6$
Severe damage	Pipeline severely damaged, a large number of leaks, need to be replaced in time	$\sigma_M > 392.6$

Table 4 Limit state ground motion intensity index of surge system structure

State	Maximum structural stress interval/MPa	PGA/(m/s^2)	PGA/g
Basically intact	[0, 141.336)	[0, 63.51)	[0, 6.48)
Slight damage	[141.336, 235.56]	[63.51, 94.8]	[6.48, 9.67]
Moderate damage	(235.56, 392.6]	(94.8, 128.370]	(9.67, 13.09]
Severe damage	(392.6, +∞)	(128.37, +∞)	(13.09, +∞)

3 Finite element modeling

This paper selects the most risky pressure regulating system in gas distribution station as the modeling objects object, but the proposed method and analysis results are applicable to each system of natural gas distribution station. Our team has done similar research as a basis for support.

3.1 Condition hypothesis

(1) Without considering the internal structure of the pressure regulator, emergency shut-off valve and ball valve, it is simplified as a volume block with a certain mass, but the volume block is still a solid unit model.

(2) The connection mode of pipeline is welding, without considering pipeline defects, welding quality defects, etc.

(3) Flange connection is also simplified as welding.

(4) The influence of piping and instrumentation equipment other than the pressure regulating system on the system time history analysis is not considered.

3.2 Material parameters

3.2.1 Mechanical model test of tubing

In this section, the material mechanics test of No. 20 steel pipe used in the gas distribution station was carried out, and its constitutive curve was obtained after measuring its mechanical properties. The stress-strain relationship of the pipe was measured by a universal testing machine under the specified conditions by using the strip sample cut from No. 20 steel pipe. The sample and structure size are shown in Fig. 3.

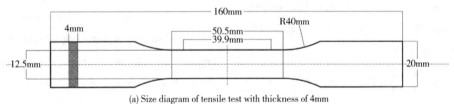

(a) Size diagram of tensile test with thickness of 4mm

(b) Size diagram of tensile test with thickness of 6mm

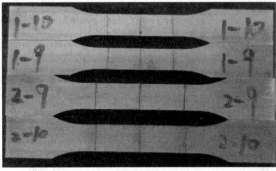

(c) Tensile specimen object

Fig. 3 Tensile specimen size and object

In this study, MTS809.25 fatigue testing machine was used to carry out the tensile test of the pipe specimen. The structure of the instrument is shown in Fig. 4.

Fig. 4 MTS809.25 Fatigue testing machine

In practice, the size of the specimen is not constant, and the real strain is related to the load and the cross-sectional area of the specimen at a certain time during the tensile process [37]. Assuming that the volume of the specimen is constant during the tensile process, the relationship between the true stress and the true strain can be obtained by equations (18) and (19).

$$\sigma_T = \frac{F}{A} = \frac{Fl}{A_0 l_0} = \frac{F}{A_0}(1+\varepsilon) = \sigma(1+\varepsilon) \quad (18)$$

$$\varepsilon_T = \int_{l_0}^{l} \frac{1}{l} dL = \ln\frac{l}{l_0} = \ln\frac{l_0 + \Delta l}{l_0} = \ln(1+\varepsilon) \quad (19)$$

In the formula, F is the tensile force of the sample, A is the instantaneous cross-sectional area of the sample, and A_0 is the original cross-sectional area of the sample. l is the instantaneous length of the sample, l_0 is the original length of the sample, Δl is the sample length deformation.

According to Eq. (18) and Eq. (19), the material tensile test results are transformed to obtain the true stress-strain curve and engineering stress-strain curve of No.20 steel pipe specimen, as shown in Fig. 6.

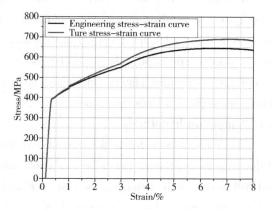

Fig. 6 Stress-strain curve of typical 20# steel sample

It can be seen from the diagram that in the elastic stage, because the deformation of the cross section is small, the true stress-strain curve and the engineering stress-strain curve basically coincide. After entering the plastic stage, the true stress-strain curve will be significantly higher than the engineering stress-strain curve. This is because the material undergoes plastic deformation, the cross-section deformation increases, and the corresponding stress will increase accordingly.

According to the mechanical test, the specific material parameters are shown in Table 5.

Table 5 20# steel linear elastic stage modeling parameters

Material	Elastic modulus/Pa	Poisson ratio	Density/(kg/m³)	σ_s/MPa
20#steel	205.679	0.3	7850	392.6

Table 6 Modeling parameters of support structural steel

Material	Elastic modulus/GPa	Poisson ratio	Density/(kg/m³)	σ_s/MPa
structural steel	200	0.3	7850	250

3.3 Modelling

(1) Geometric model

The pipeline layout, pipe diameter, wall thickness and equipment model of the pressure regulating system are basically the same, so one of the pipelines is selected for modeling, and the model size is shown in Figure 7. The total length of the model is 2412mm, the pipe size before reducing is 89.4mm (outer diameter), and the pipe size after reducing is 57.4mm; The support model is shown in Figure 8. The support is 60mm high and its outer wall is tangent to the outer wall of the pipe.

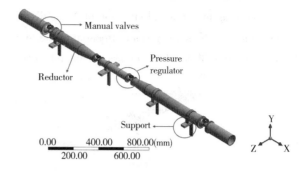

Fig. 7 finite element model of equipment pipeline in pressure regulating area

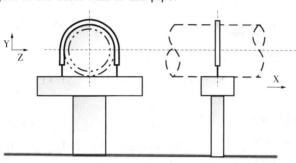

Fig. 8 Pipe rack constraint form

(2) Gridding

This model adopts solid element "solid186", which is a high-order 3D-20 node solid element, defined by 20 nodes, each node has 3 degrees of freedom, and the element supports plastic, hyperelastic, creep, large deflection and large strain operations. Compared with hexahedral mesh, tetrahedral mesh requires more cell nodes to obtain the same result accuracy, so it will consume more CPU computing time and more data storage space. At the same time, the transient dynamic analysis needs uniform mesh size, hexahedral mesh is still the first choice. However, in this model, due to the complex structure and shape of the four valve equipment and the poor quality of hexahedron mesh generation, tetrahedron element mesh generation is adopted; The rest of the model is a hexahedral element grid, and the division results are shown in Figure 9.

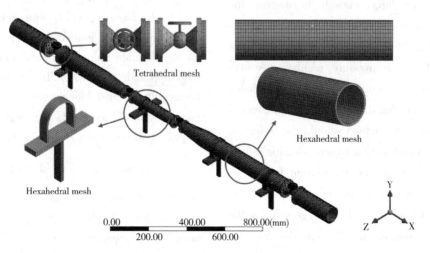

Fig. 9 Pressure regulating system model grid division

(3) Constraint setting

According to the actual situation of A gas distribution station, the six degrees of freedom of the four bottom surfaces of the bracket are completely constrained. This model constrains the translational degree of freedom of the x-axis at both ends of the pipeline and the rotational degree of freedom of the y and z axes. Constraint settings are shown in Figure 10.

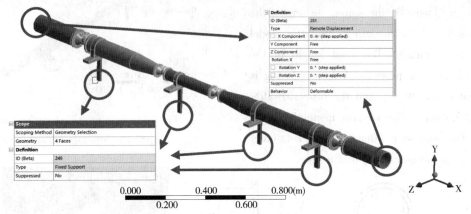

Fig. 10 Pressure regulating system model constraint setting

4 Incremental Dynamic Analysis

Incremental dynamic analysis is a method based on nonlinear time history analysis. It can be used to calculate the process of pipeline structure from elastic to elastic-plastic and finally collapse under different intensity earthquakes. In order to accurately describe the relationship between seismic response of pipeline structure and ground motion intensity, this paper uses incremental dynamic analysis method to analyze the structure.

4.1 Amplitude modulation of ground motion

When the earthquake magnitude is too small and the energy released is not enough, it is difficult to destroy the structure in the calculation process. So let the energy be large enough to destroy the structure by amplitude modulation. Before the amplitude modulation, ANSYS workbench is used to analyze the modal of the pressure regulating system model. For structures with deterministic periods, Liu. [38] suggested that the $Sa(T1)$ seismic record amplitude modulation method should be selected. Each different ground motion record is multiplied by the corresponding amplitude modulation factor, so that the response spectrum value of each ground motion record at $T1$ is equal to the average value of the group of seismic records at $T1$, and the response spectrum curve after amplitude modulation is obtained.

$$Sa(T1)_{geomean} = \sqrt[n]{Sa(T1)_1 \cdot Sa(T1)_2 \cdots Sa(T1)_n} \quad (20)$$

$$SF_1 = Sa(T1)_{geomean}/Sa(T1)_i \quad (21)$$

In the formula, $Sa(T1)_{geomean}$ is the geometric mean of the response spectrum acceleration of a set of seismic records at the basic period $T1$; SF_i is the scaling factor of the response spectrum; $Sa(T1)_i$ is the acceleration value of each ground motion response spectrum at the basic period $T1$.

The calculation of Sa before and after the first amplitude modulation is calculated according to formula (22), and the i time amplitude modulation process is shown in formula (23)~(24).

$$PGA_2 = SF_1 \cdot PGA_1 \quad (22)$$

$$PGA_{i+1} = SF_i \cdot PGA_1 \quad (23)$$

$$SF_i = \zeta_i \cdot SF_{i-1} \quad (24)$$

In the formula, PGA_1 represents the initial acceleration value of ground motion, PGA_i represents the acceleration value of ground motion after the i time amplitude modulation. In this study, a total of 10 amplitude modulations were carried out, so i takes 2~10, ζ_i represents the amplitude modulation amplification coefficient, and the value is $\zeta_i = \zeta_1 \cdot \frac{\sigma_s}{a \cdot \sigma_k}$, where σ_s is the yield strength of the material, σ_k is the maximum von Mises stress calculated at the k time, k takes 1~9, ζ_1 takes 1, a is the amplification accuracy coefficient, which characterizes the amplitude modulation degree. Because this study has 10 amplitude modulations, a takes 10.

Through Eq. (24), the amplitude modulation factor of ground motion can be determined, and then the ground motion after amplitude modulation can be calculated according to Eq. (23).

4.2 IDA curve

IDA curve is the amplitude modulation of seven selected ground motion records; then the amplitude-modulated ground motion time history is loaded onto the pipeline structure for transient time history analysis, and the maximum von Mises stress response value of the pipeline structure under any ground motion is obtained. Finally, the IDA curve is drawn with the maximum von Mises stress as X axis and PGA as Y axis. The IDA curve is shown in Figure 11.

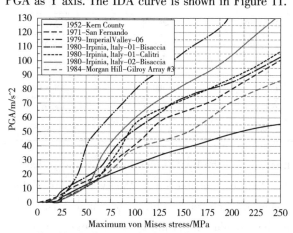

Fig. 11 IDA curve of voltage regulating system

Through the incremental dynamic analysis, the IDA curves of the system model under different ground motion can be obtained. The IDA curve is shown in Figure11.

The connection of points on the curve is B-spline interpolation. Due to the large discrete type of IDA curve, this paper chooses 50 % quantile line (Median line) to determine the limit state point and draw the vulnerability curve. If the extreme value line is selected, the vulnerability curve results may be too large or too small, which will have a certain impact on the final seismic measures. Therefore, this paper chose the median curve as shown in Figure 7 red bold lines.

5 Establishing probabilistic seismic demand model and fitting vulnerability curve

5.1 Seismic probabilistic demand model

The seismic vulnerability function formula derived from Section 2.1, see Formula (6), in which there are three undetermined coefficients $\sqrt{\theta_{\sigma_c}^2 + \theta_{\sigma_{max}}^2}$, α and β.

$$P_f = \Phi\left(-\frac{\ln\left(\frac{\sigma_c}{\sigma_{max}}\right)}{\sqrt{\theta_{\sigma_c}^2 + \theta_{\sigma_{max}}^2}}\right) = \Phi\left(\frac{\ln\left(\frac{\sigma_{max}}{\sigma_c}\right)}{\sqrt{\theta_{\sigma_c}^2 + \theta_{\sigma_{max}}^2}}\right) = \Phi\left(\frac{\ln(\alpha(PGA)^\beta/\sigma_c)}{\sqrt{\theta_{\sigma_c}^2 + \theta_{\sigma_{max}}^2}}\right) \quad (17)$$

$\ln(PGA)$ and $\ln(\sigma_{max})$ are calculated and substituted into Formula (14) for regression analysis. Formula (25) is as follows.

$$\ln \sigma_{max} = a + b \cdot \ln(PGA) \quad (25)$$

The data of regression analysis are shown in table7.

Table 7 Structural seismic demand model regression analysis parameters of surge tank

$\ln(PGA)$	$\ln(\sigma_{max})$	$\ln(PGA)$	$\ln(\sigma_{max})$	$\ln(PGA)$	$\ln(\sigma_{max})$
-0.67531	3.012933	4.167573	4.870254	4.13493	4.705775
0.712459	3.01578	4.319577	4.973349	4.282123	5.012733
2.098876	3.517913	4.424966	5.299567	4.436917	5.237803
3.48514	4.741361	4.548769	5.414677	4.556667	5.409662
3.70829	5.107157	4.660037	5.546739	4.705016	5.547051
3.942494	5.31598	-0.77252	3.190229	-0.15202	3.199
4.135934	5.752971	0.613779	3.207896	1.639734	3.498022
1.271445	3.018667	1.709102	3.219915	2.33292	3.925334
2.370057	3.486457	2.402249	3.491952	3.026019	4.082272

续表

ln(PGA)	ln(σ_{max})	ln(PGA)	ln(σ_{max})	ln(PGA)	ln(σ_{max})
2.657739	3.881254	3.095397	3.802654	3.513365	4.172077
3.063204	4.244387	3.788544	3.914759	3.820254	4.375939
3.350886	4.320151	3.954239	4.178349	4.135407	4.611053
3.756351	4.668333	4.176769	4.358707	4.326184	4.887311
4.016113	4.79165	4.285648	4.523483	4.460318	5.051489
4.11933	4.887231	4.437355	4.660224	4.550249	5.22389
4.256237	5.195869	4.528138	4.833532	4.792927	5.422303
4.449487	5.351669	4.706372	5.053593	4.87168	5.503745
4.854917	5.767195	4.793176	5.228646	1.778032	3.58018
0.651961	3.020669	4.90909	5.301512	2.694357	4.108247
2.038229	3.047993	4.945492	5.418941	3.38747	4.48255
2.443737	3.510052	0.323387	3.200508	3.649255	4.512419
2.731441	3.69511	1.709718	3.491952	3.857525	5.014627
3.136798	4.176846	2.625972	4.080584	4.016113	5.125392
3.55689	4.293387	3.031485	4.202302	4.136446	5.218083
3.75298	4.437461	3.724609	4.485373	4.298455	5.30039
3.942378	4.580023	3.96999	4.548759	4.479081	5.54966

Through regression analysis, the regression curve of seismic probabilistic demand model of surge system is shown in Fig. 12.

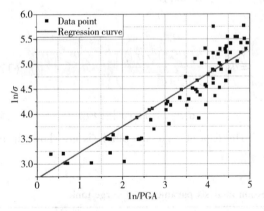

Fig. 12 Seismic probability demand model curve of pressure regulating system

From the regression analysis, $a = 2.70891$, $b = 0.52087$ in Formula (25) can be obtained, that is, the original formula can be rewritten as Formula (26) as shown below. $\alpha = e^a = 15.0129$ and $\beta = b = 0.52087$ in Eq. (17) can also be calculated. At the same time, $\sqrt{\theta_{\sigma_c}^2 + \theta_{\sigma_{max}}^2}$ can be calculated from (27), where n is the number of samples, and its value is calculated to be 0.0752. Substituting it into Eqs. (17), the final vulnerability function, that is, the failure probability of the voltage regulation system in each limit state, is shown in Eqs. (28). In the formula, σ_c is the capacity parameter of different limit state structures.

$$\ln\sigma_{max} = 0.52087 \cdot \ln(PGA) + 2.70891 \quad (26)$$

$$\sqrt{\theta_{\sigma_c}^2 + \theta_{\sigma_{max}}^2} = \sqrt{\frac{1}{n-2}\sum_{i=1}^{n}(\ln(\sigma_{max}) - a - b\ln(PGA))^2} \quad (27)$$

$$P_f = \Phi\left(\frac{\ln(15.0129(PGA)0.52087/\sigma_c)}{0.0752}\right) \quad (28)$$

5.2 Seismic vulnerability curve

The seismic vulnerability curve of the pressure regulating system can be drawn from Formula (27) and Formula (28), as shown in Figure 9.

It can be seen from the figure that the exceedance probability of the pressure regulating system under the three limit states when subjected to different peak ground motions. ① When PGA is 0.5g, the probability of the pressure regulating system exceeding the basic integrity is 0.68, the probability of exceeding the moderate failure is 0.53, and the probability of exceeding the severe failure is

0.41; ② When PGA is 1.0g, the probability of the pressure regulating system exceeding the basic integrity is 0.8, the probability of exceeding the moderate failure is 0.67, and the probability of exceeding the severe failure is 0.55; ③ When PGA is 2.0g, the probability of the pressure regulating system exceeding the basic integrity is 0.89, the probability of exceeding the moderate failure is 0.78, and the probability of exceeding the severe failure is 0.7.

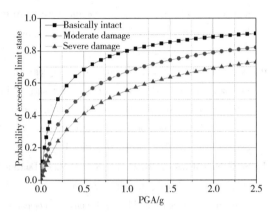

Fig. 13 Seismic vulnerability curve of surge system

6 Conclusion

Based on the seismic theory of performance-based seismic engineering, this paper proposes and deduces the pipeline vulnerability analysis theory for urban natural gas distribution stations, introduces the commonly used ground motion record selection methods, and determines 7 ground motion records according to the selection rules. PGA is selected as the ground motion intensity index and the pipeline stress as the damage index of the pipeline structure. The definition standard of the limit state of the pipelinestructure is given. The selected seven ground motions are amplitude - modulated to achieve sufficient strength and loaded onto the pipeline for incremental dynamic analysis. Because of the large discrete type, the median line of the 7 curves is selected as the appropriate IDA curve. The selected data are calculated and processed to obtain the values of all undetermined coefficients in Formula (17), so as to obtain the final seismic vulnerability function (28) of gas distribution station. Finally, the seismic vulnerability curve is drawn by using the obtained vulnerability function (28), and the pressure regulating system of A gas distribution station is analyzed. The following conclusions were reached.

It can be seen from the incremental dynamic analysis and the curve drawn that the mechanical response of the pipeline of the pressure regulating system shows great dispersion for the ground motion with different spectral characteristics. According to the vulnerability curve, 1) when PGA is 0.5g, the probability of the pressure regulating system pipeline exceeding the basic integrity is 0.68, the probability of exceeding the moderate damage is 0.53, and the probability of exceeding the severe damage is 0.41; 2) When PGA is 1.0g, the probability of exceeding the basic integrity of the pressure regulating system pipeline is 0.8, the probability of exceeding the moderate damage is 0.67, and the probability of exceeding the severe damage is 0.55; 3) When PGA is 2.0g, the probability of exceeding the basic integrity of the pressure regulating system pipeline is 0.89, the probability of exceeding the moderate damage is 0.78, and the probability of exceeding the severe damage is 0.7.

References

[1] Wang Daqing. Study on Quantitative Risk Assessment Method of Gas Transmission Stations Considering Uncertainty [D]. Cheng Du, Southwest Petroleum University, 2014.

[2] Liu Zhi. Analysis on the Estimation of Earthquake Damage and Losses of Heating System [D]. Institute of Engineering Mechanics, China Earthquake Administration, 2011.

[3] Sun Shaoping. Seismic damage and analysis of water supply pipeline in Kobe earthquake [J]. SPECIAL STRUCTURES, 1997(02): 51-55.

[4] Hou Xiangqin, Zhang Peng, Duan Yonghong, etc.. Seismic Reliability Analysis of Urban Gas Pipeline Network [C]//2010 Academic Conference of Sichuan Mechanics Society, Journal of Sichuan University (Engineering Science Edition), 2010: 46-49.

[5] Jaske C E, Lopez-Garrity A. Integrity assessment methods adapted for stations, term-inals [J]. Oil and Gas Journal, 2004, 102(47): 54-58.

[6] Harper. W V, Stucki D J, Shie T M, et al. Reliability Based Facility Risk Asse-ssment[C]. 2010 8th Inter-

national Pipeline Conference. American Society of Mechanical Engineers, 2010: 247-253.

[7] Urra S, Green J. Human Factor Modelling in the Pipeline Industry[C]. 2012 9th Inter-national Pipeline Conference. American Society of Mechanical Engineers, 2012: 589-594.

[8] Rajakarunakaran S, Kumar A M, Prabhu. A V. Applications of fuzzy faulty tree analysis and expert elicitation for evaluation of risks in LPG refuelling station[J]. Journal of Loss Prevention in the Process Industries, 2015, 33: 109-123.

[9] Seo H G, Kwon S, Bach Q V, Ha D, Lee C J. Quantitative risk assessment of an urban hydrogen refueling station[J]. International Journal of Hydrogen Energy, 2019, 44: 1288-1298.

[10] Meng Bojie. Wu Ying. Chen Liangruo. Analysis of seismic event sequence in gas distribution station based on improved risk matrix method[J]. Journal of Safety Science and Technology, 2021, 17(03): 110-116.

[11] Leoni L, Toroody A B, Carlo F D, Paltrinieri N. Developing a risk-based maintenance model for a Natural Gas Regulating and Metering Station using Bayesian Network[J]. Journal of Loss Prevention in the Process Industries, 2019, 57: 17-24.

[12] Chen, Y. L., Han, L.. The Vulnerability Formation Mechanism and Control Strategy of the Oil and Gas Pipeline City[J]. IOP Conference Series: Earth and Environmental Science, 2017, Vol. 104: 012009

[13] Zou Rong. Seismic vulnerability assessment method of urban gas polyethylene pipeline based on cloud theory [D]. Cheng Du, Southwest Petroleum University, 2020.

[14] Zhang Y, Wu G. Seismic Vulnerability Analysis of RC Bridges Based on Kriging Model[J]. Journal of Earthquake Engineering, 2019, 23(2): 242-260.

[15] Qiu W, Huang G, Zhou H, et al. Seismic Vulnerability Analysis of Rock Mountain Tunnel[J]. International journal of geomechanics, 2018, 18(3): 1-16.

[16] Ratiranjan J, Biswajeet P, Ghassan B. Earthquake vulnerability assessment in Northern Sumatra province by using a multi-criteria decision-making model[J]. International Journal of Disaster Risk Reduction, 2020, 46: 10-18.

[17] Wang Wenyue, Wu Ying, You Xiao, Chen Lang. Seismic vulnerability assessment of mountainous gas pipelines based on improved Borda method and attribute recognition[J]. Oil & Gas Storage and Transportation, 2021, 11: 1265-1271.

[18] N. Simon Kwong, Ph. D., M. ASCE; Kishor S. Jaiswal, Ph. D., P. E., M. ASCE; Jack W. Baker, Ph. D., M. ASCE; Nicolas Luco, Ph. D., A. M. ASCE; Kristin A. Ludwig, Ph. D.; and Vasey J. Stephens. Earthquake Risk of Gas Pipelines in the Conterminous United States and Its Sources of Uncertainty [J]. ASCE - ASME Journal of Risk and Uncertainty in Engineering Systems, Part A: Civil Engineering, 2022, Vol. 8; No. 1.

[19] Dong Jinqi, Zheng Shansuo, Xie Xiaokui, Yang Feng, Che Shunli, Liu Xiaohang. Seismic Vulnerability Analysis of Buried Pipelines Considering the Influence of Near-fault Impulsive Ground Motions [J/OL]. Engineering Mechanics, 1-13[2022-11-24].

[20] Xiao, Yuanhao; Zhao, Xudong; Wu, Yipeng; Chen, Zhilong; Gong, Huadong; Zhu, Lihong; Liu, Ying. Seismic resilience assessment of urban interdependent lifeline networks. Reliability Engineering & System Safety, 2022, Vol. 218, Part B.

[21] Sebastiano Marasco; Ali Zamani Noori; Marco Domaneschi; Gian Paolo Cimellaro. Seismic vulnerability assessment indices for buildings: Proposals, comparisons and methodologies at collapse limit states[J]. International Journal of Disaster Risk Reduction, 2021, Vol. 63: 102466.

[22] George Karagiannakis; Luigi Di Sarno; Amos Necci; Elisabeth Krausmann. Seismic risk assessment of supporting structures and process piping for accident prevention in chemical facilities[J]. International Journal of Disaster Risk Reduction, 2022, Vol. 69: 102748.

[23] Minze Xu, Chunyi Cui, Jingtong Zhao, Chengshun Xu, Kun Meng, A nove approach to seismic fragility evaluation of underground structures considering hybrid epistemic uncertainties of both seismic demand and capacity, [J] Tunnelling and Underground Space Technology, 2025, vol156, 106278.

[24] Ruilin Xiao, Wenguang Liu, Derui Kong, Failure criterion and seismic fragility evaluation of isolated nuclear power plant piping system using BP neural network, [J] Engineering Failure Analysis, 2024, Vol165: 108790.

[25] Shinozuka M, Feng M Q, Kim H K, et al. Nonlinear Static Procedure for Fragility Curve Development[J]. Journal of Engineering Mechanics, 2000, 126(12): 1287-1295.

[26] Rossetto T, Elnashai A. A new analytical procedure for the derivation of displacement-based vulnerability curves for populations of RC structures [J].

Engineering Structures, 2005, 27(3): 397-409.

[27] Jeong S H, Elnashai A S. Probabilistic fragility analysis parameterized by fundamental response quantities [J]. Engineering Structures, 2007, 29 (6): 1238-1251.

[28] Erberik M A. Fragility-based assessment of typical mid-rise and low-rise RC buildings in Turkey [J]. Engineering Structures, 2008, 30(5): 1360-1374.

[29] Pan Feng. Global Probabilistic Seismic Demand Analysis of Reinforced Concrete Frame Structures [D]. Harbin Institute of Technology, 2007.

[30] Cornell C A, Jalayer F, Hamburger R O, et al. Probabilistic Basis for 2000 SAC Federal Emergency Management Agency Steel Moment Frame Guidelines [J]. Journal of Structural Engineerin, 2002, 128 (4): 526-533.

[31] Ellingwood B R. Earthquake Risk Assessment of Building Structures [J]. Reliability Engineering & System Safety, 2001, 74(3): 251-262.

[32] Lu Dagang, Yu Xiaohui, Song Pengyan, et al. Simplified Vulnerability Analysis Method for Optimal Fortification Level Decision and Life Cycle Optimization Design of Seismic Structures. [J]. Earthquake Engineering and Engineering Dynamics, 2009, 29(4), 23-32.

[33] Lu Dagang, Yu Xiaohui. Application study of probabilistic seismic risk assessment based on analytical functions of seismic fragility [J]. Journal of Building Structures, 2013, 34(10): 41-48.

[34] Mackie K, Mackie K, Wong J M et al. Integrated Probabilistic Performance - Based E - valuation of Benchmark Concrete Bridges [M]. Berkeley: Pacific Earthquake Engineering Research Center, College of Engineering, University of California, 2008.

[35] Ibarra L F, Krawinkler H. Global Collapse of Frame Structures under Seismic Excita-tions [M]. Berkeley: Pacific Earthquake Engineering Research Center, College of Engineering, University of California, 2005.

[36] Baker J W, Cornell C A. Vector - Valued Ground Motion Intensity Measures for Proba-bilistic Seismic Demand Analysis [M]. Berkeley: Pacific Earthquake Engineering Research Center, College of Engineering, University of California, 2006.

[37] Li Linya. Study on safety evaluation method of X80 pipeline dent based on ductile damage [D]. Cheng Du, Southwest Petroleum University, 2020.

[38] Liu Hongbo, Sun Weixuan, Zhang Yuxin, et al. Experimental study on the scaling method for ground motions for frame structures [J]. Earthquake Engineering and Engineering Dynamics, 2019, 39(01), 80-87.

基于 GA-PSO 混合算法和 XGBoost 算法的城市燃气管道风险评价

彭善碧　王鸿扬

（西南石油大学）

摘　要　【目的】随着城市燃气管道规模的日益增大，燃气管道事故频繁发生，严重威胁到居民的安全。有效管理城市燃气管道并降低事故风险已成为一项重要挑战，而风险评估在应对这些挑战中发挥着至关重要的作用。【方法】为此，本文基于遗传算法（Genetic Algorithm, GA）、粒子群优化（Particle Swarm Optimization, PSO）算法和极梯度提升（eXtreme Gradient Boosting, XGBoost）算法提出了一种新型城市燃气管道风险评价模型，即 GA-PSO-XGBoost 模型。该模型采用 GA-PSO 混合算法来优化 XGBoost 的超参数，从而进行城市燃气管道风险评价。【结果】实验结果显示，GA-PSO-XGBoost 模型的均方误差（Mean Square Error, MSE）、平均绝对误差（Mean Absolute Error, MAE）和决定系数（R-square, R^2）分别为 4.1264、1.4939 和 99.1146%，表示该模型具有较高的精度。此外，特征重要性分析表明，使用寿命和员工素质是影响风险评价的两个最重要指标。【结论】因此，本研究表明，GA-PSO-XGBoost 模型在城市燃气管道风险评价中具有出色的性能和实用性。

关键词　城市燃气管道；风险评价；GA-PSO-XGBoost；特征重要性

Risk assessment of urban gas pipeline based on GA-PSO hybrid algorithm and XGBoost algorithm

Peng Shanbi　Wang Hongyang

(Southwest Petroleum University)

Abstract　[Objective] With the continuous expansion of the total length of urban gas pipelines, urban gas pipeline accidents occur frequently, which makes the personal safety of residents under great threat. It is challenging to manage urban gas pipelines and reduce the risk of pipeline accidents effectively. Risk assessment plays a critical role in addressing these challenges. [Methods] To that end, this paper proposes a novel model based on genetic algorithm (GA), particle swarm optimization (PSO), and extreme gradient boosting (XGBoost) to assess the risk of urban gas pipelines, or "GA-PSO-XGBoost". A GA-PSO hybrid algorithm is employed to optimize the hyperparameters of the XGBoost algorithm for assessing the risk of urban gas pipelines. [Results] Then, the risk of urban gas pipelines in a city is analyzed based on the failure data, and the mean square error (MSE), mean absolute error (MAE) and r-square (R^2) of the GA-PSO-XGBoost model are 4.1264, 1.4939 and 99.1146%, respectively. The results show that the GA-PSO-XGBoost model exhibits superior precision. Finally, the feature importance is calculated, and the service life and staff quality are the two most significant possibility indexes. [Conclusion] According to this study, the GA-PSO-XGBoost model has an excellent performance and practicality in the risk assessment of urban gas pipelines.

Key words　Urban gas pipeline; Risk assessment; GA-PSO-XGBoost; Feature importance

1　引言

随着中国迅速发展，城市燃气普及率不断增长。在国家环保意识日益强化的背景下，天然气已成为我国主要的城镇燃气能源。因此，对城市燃气管道进行风险评价至关重要，以确保居民的生命和财产安全以及燃气系统的稳定运行。风险评价能够揭示潜在的安全隐患，降低事故风险，优化系统运营，精确配置资源，指导决策制定。

风险评价应用于燃气管道已有几十年的历史，从长输管道逐渐普及到城市燃气管道。目前，城市燃气管道的风险评价方法已经较为丰富，主要包括：定性评价方法、半定量评价方法和定量评价方法。巴振宁等人采用基于三标度法改进层次分析法（Analytic Hierarchy Process，AHP）确定各指标权重，并使用模糊综合评价法建立了腐蚀风险评价模型。与传统的AHP-模糊综合评价模型相比，该模型不需要进行一致性检验，并且精度更高。曾小康等提出了一种基于AHP-熵权法的城市燃气管道风险评价模型。通过建立包含105个底层评价因子的燃气管道风险评价指标体系，为燃气管道风险的预警和管理提供了依据，并运用模糊综合评价法和风险分析矩阵评估燃气管道的风险等级。张鹏等人采用灰色理论与决策试验法（DEMATEL）以及模糊综合评价，对燃气管道进行风险评价。该方法有效解决了各因素相互作用会影响评价结果的问题，并能够有效评估燃气管道风险。Liang等人基于模糊TOPSIS模型和云推理的方法，评估城市PE天然气管道的风险。采用风险因素分析和模糊TOPSIS分析法建立风险指标体系，并使用云推理呈现评估结果。案例分析验证了该方法的有效性。Chen等人基于变权理论和云理论评估管道风险，能够更准确地分析管道的风险等级和关键风险因素。实例分析验证了该方法的有效性，有助于为风险控制和修复提供依据。单克基于失效数据和修正因子对燃气管道进行风险评价。将修正因子分为定量、半定量和定性对基本失效概率进行修正和实时更新以降低主观依赖性和提高准确性。随着时代的进步，大数据技术与机器学习方法的结合为燃气管道风险评价方法的研究提供了新思路。王新颖等人将最小二乘支持向量机（LS-SVM）应用到燃气管道风险评价中，利用样本数据进行训练和测试后，建立LS-SVM预测风险评价模型。2019年，陈毓飞等人利用长短期记忆（LSTM）建立模型，预测了燃气管道施工损坏的概率。Li等人采用支持向量机（SVM）和人工神经网络（ANN）对城市埋地燃气管网的脆弱性进行评估，结果表明SVM优于ANN，其MSE和对称平均绝对百分比误差（SMAPE）分别为0.0274%和0.79%。2020年，Wang和Li提出了一种基于聚类算法的风险评估方法。2021年，熊威建立了以关键指标为输入值，燃气管道风险的概率和综合风险值作为输出值的三层BP神经网络。经过验证，表明了该模型具有良好的适用性。2023年，Bai等人使用贝叶斯网络（BN）评估天然气管道风险。Zhang和An建立了基于径向基函数和反向传播神经网络的径向基函数神经网络（RBFNN）模型，用于管道故障概率的预评估，误差接近3%。

回顾以上历史文献可以发现，与传统的风险评估方法相比，机器学习方法由于其优越的性能，在燃气管道风险评价领域中应用越来越广泛。在此背景下，越来越多的学者开始研究、开发和应用各种机器学习算法。陈天奇博士于2016年提出了一种高效、灵活和轻便的XGBoost算法。为了快速、准确和客观的评价城市燃气管道的风险状况，所以本文选择XGBoost作为基础风险评估模型。然而，XGBoost算法需要优化的超参数较多并且传统的超参数优化方法费时费力，因此，初步选择粒子群算法来优化XGBoost的超参数，但粒子群算法容易陷入局部最优解。因此，在粒子群算法的基础上融合遗传算法以增强粒子群算法的全局搜索能力。最终建立GA-PSO-XGBoost城市燃气管道风险评价模型。

2 方法

2.1 风险评价

城市燃气管道风险评价的目的是根据构建的风险评价模型，按照风险指标计算出管道风险值。随后，根据管道风险值判断管道的风险状况，以便及时采取措施，减轻城市燃气管道事故可能带来的不利后果。图1展示了利用机器学习模型进行风险评估的过程。在图1中，输入特征代表风险指标值，而输出标签则是机器学习模型生成的管道风险值。因此，本文将城市燃气管道风险评价定义为通过风险评价机器学习模型，利用管道数据和其他相关信息综合评价潜在风险的过程。这一过程不仅能提高风险评价的准确性和客观性,还能为制定有针对性的风险缓解策略提供科学依据。

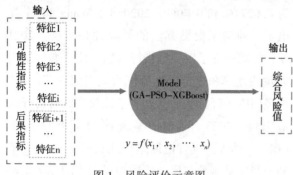

图1 风险评价示意图

2.2 相关性分析

为了解数据样本中是否具有信息重复的冗余特征,需要计算每两个特征之间的相关性。本文采用皮尔逊(Pearson)相关系数度量两个特征之间的相关性,Pearson 相关系数计算公式如下:

$$c = \frac{\sum_{i=1}^{n}(X_i - \overline{X})(Y_i - \overline{Y})}{\sqrt{\sum_{i=1}^{n}(X_i - \overline{X})^2}\sqrt{\sum_{i=1}^{n}(Y_i - \overline{Y})^2}} \quad (1)$$

Pearson 相关系数介于 -1 和 1 之间,如果相关性系数接近于零,则表示其两个变量之间没有线性相关性,否则,相关性系数绝对值越大,则两个变量之间的相关性越强。由于冗余特征对预测模型没有贡献,还会增加模型训练时间,因此,应对冗余特征进行删减。

2.3 XGBoost 算法

近年来,XGBoost 被广泛应用于各领域的回归和分类问题。该算法是一种基于梯度提升决策树(GBDT)改进的集成学习算法,当其应用于回归问题时,会连续添加新的回归树,并通过新生成的 CART 树拟合先前模型的残差。表1详细列出了 XGBoost 算法的 8 个超参数。

表1 XGBoost 算法的超参数

名称	类型	取值范围	意义
num_round	Integer	$(0, +\infty)$	梯度提升树的数量
max_depth	Integer	$(0, +\infty)$	树的最大深度
eta	floating-point	$[0, 1]$	学习率
gamma	floating-point	$[0, +\infty)$	复杂度的惩罚项
alpha	floating-point	$[0, +\infty)$	L1 正则项的参数
ambda	floating-point	$[0, +\infty)$	L2 正则项的参数
subsample	floating-point	$(0, 1]$	样本随机抽样比例
colsample_bytree	floating-point	$(0, 1]$	每次生成树时随机抽样特征比例

XGBoost 的目标函数包含损失函数和正则化项两部分,损失函数表示预测值与真实值的误差,正则化项用于控制模型复杂度并防止过拟合。XGBoost 目标函数推导如下:

$$Obj = \sum_{i=1}^{n} l(y_i, \hat{y}_i) + \sum_{k=1}^{K} \Omega(f_k) \quad (2)$$

其中,$\hat{y}_i$ 为预测值,y_i 为真实值,f_k 表示树结构。

$$\Omega(f) = \gamma T + \frac{1}{2}\lambda \|\omega\|^2 \quad (3)$$

其中,γ、λ 为惩罚系数。

通过泰勒二阶展开式对目标函数进行简化并移除常数项。

$$Obj = \sum_{i=1}^{n}\left[g_i f_t(x_i) + \frac{1}{2}h_i f_t^2(x_i)\right] + \Omega(f_t) \quad (4)$$

其中,$g_i = \partial_{\hat{y}^{(t-1)}} l(y_i, \hat{y}^{(t-1)})$ 和 $h_i = \partial^2_{\hat{y}^{(t-1)}} l(y_i, \hat{y}^{(t-1)})$ 分别为损失函数的一、二阶偏导。

对简化后的目标函数进行求解得到树的每个叶结点输出值的计算公式,以及对树结构的评估函数。定义叶子样本集合 $I_j = \{i \mid q(x_i) = j\}$,定义 $G_j = \sum_{i \in I_j} g_i$,$H_j = \sum_{i \in I_j} h_i$,目标函数改写为:

$$\begin{aligned} Obj &= \sum_{i=1}^{n}\left[g_i f_t(x_i) + \frac{1}{2}h_i f_t^2(x_i)\right] + \Omega(f_t) \\ &= \sum_{i=1}^{n}\left[g_i \omega_{q(x_i)} + \frac{1}{2}h_i \omega_{q(x_i)}^2\right] + \gamma T + \frac{1}{2}\lambda \sum_{j=1}^{T}\omega_j^2 \\ &= \sum_{j=1}^{T}\left[\left(\sum_{i \in I_j}g_i\right)\omega_j + \frac{1}{2}\left(\sum_{i \in I_j}h_i + \lambda\right)\omega_j^2\right] + \gamma T \\ &= \sum_{j=1}^{T}\left[G_j \omega_j + \frac{1}{2}(H_j + \lambda)\omega_j^2\right] + \gamma T \quad (5) \end{aligned}$$

假定树结构 $q(x_i)$ 是固定的,每个叶节点的最优权重为 $\omega_j = -\frac{G_j}{H_j + \lambda}$;将其带入公式(4),最终目标函数为:

$$Obj = -\frac{1}{2}\sum_{j=1}^{T}\frac{G_j^2}{H_j + \lambda} + \gamma T \quad (6)$$

2.4 GA-PSO 混合优化算法

GA 的核心思想是通过模拟自然界中的进化过程，使用适应函数对候选解进行评估，并通过遗传操作（选择、交叉、变异）产生新的候选解，最终找到最优解。PSO 算法是一种受到鸟群或鱼群等群体行为启发的优化算法，它通过不断调整粒子的速度和位置，寻找最优解。该算法具有较快的收敛速度和较好的全局搜索能力，但对于复杂问题，容易陷入局部最优解。

本文提出了一种 GA-PSO 混合算法，以克服 GA 和 PSO 各自的局限性，并兼顾全局搜索能力和收敛速度。在该算法中，PSO 被用来增强 GA 的搜索能力，以加速向最优解的收敛。而 GA 则用来保持种群的多样性，防止 PSO 陷于局部最优解。如果要实施 GA-PSO 混合算法，首先要定义几个函数。这些函数包括适合度函数、初始化种群函数、更新速度和位置函数以及遗传操作函数。在本研究中，GA-PSO 混合算法的具体步骤如下：

步骤1：根据种群大小和粒子维数设置以下参数后初始化种群：交叉概率、变异概率、自我认知因子、社会认知因子、迭代次数、惯性权重等。GA-PSO 中的惯性权重会影响其搜索效果。惯性权重越小，局部搜索越容易；惯性权重越大，全局搜索越容易。本文引入了可变惯性权重，以动态适应不同阶段的要求，在迭代次数较少时强调全局搜索能力，在迭代次数增加时确保更好的局部搜索能力。计算公式如下：

$$w=w_1-(w_1-w_2)\frac{t}{max\_iter} \quad (7)$$

其中，w_1 为最大惯性权重；w_2 为最小惯性权重；t 为当前循环次数；max_iter 为最大迭代次数。

步骤2：计算粒子的适应度值，并记录粒子自身的最优解和当前的全局最优解。

步骤3：根据公式（8）和公式（9）更新粒子的速度和位置。

$$v_{id}^{k+1}=wv_{id}^k+c_1r_1(p_{id,pbest}^k-x_{id}^k)+c_2r_2(p_{d,gbest}^k-x_{id}^k) \quad (8)$$

$$x_{id}^{k+1}=x_{id}^k+v_{id}^{k+1} \quad (9)$$

步骤4：对完成更新的粒子进行选择、交叉和变异操作，以提高种群多样性。

步骤5：更新粒子的适应度值，并重新记录粒子自身的最优解和当前的全局最优解。

步骤6：如果满足结束条件，则输出最优粒子及其适应度值；否则，返回步骤3。

2.5 GA-PSO-XGBoost 模型

基于 GA-PSO 混合算法和 XGBoost 算法建立 GA-PSO-XGBoost 风险评价模型。在该模型中，XGBoost 算法计算管道风险值，而 GA-PSO 混合算法则优化 XGBoost 的超参数，以提高其性能。基于 GA-PSO-XGBoost 模型的城市燃气风险评估流程如图2所示。GA-PSO-XGBoost 模型的具体步骤如下：

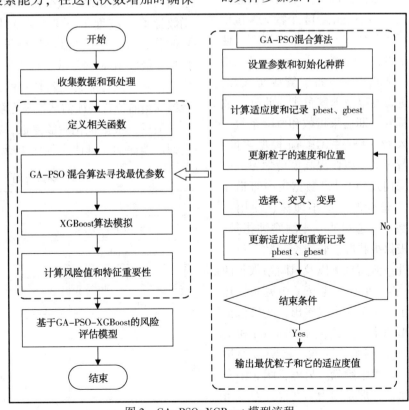

图2 GA-PSO-XGBoost 模型流程

步骤 1：收集数据并进行预处理。

步骤 2：使用 GA-PSO 混合算法优化 XGBoost 的超参数，具体步骤见第 2.4 节。

步骤 3：将最优超参数输入 XGBoost 并运行程序。

步骤 4：计算城市燃气管道的风险值和特征重要性。

步骤 5：输出结果并结束程序。

2.6 度量指标

本文使用风险系数预测值计算 MSE、MAE 和 R^2 三种指标度量预测精度和性能，公式如下：

$$MSE = \frac{1}{n}\sum_{i=1}^{n}(y_i - \hat{y}_i)^2 \quad (10)$$

$$MAE = \frac{1}{n}\sum_{i=1}^{n}|y_i - \hat{y}_i| \quad (11)$$

$$R^2 = 1 - \frac{\sum_{i=1}^{n}(y_i - \hat{y}_i)^2}{\sum_{i=1}^{n}(y_i - \bar{y})^2} \quad (12)$$

其中，y_i 表示管道风险的实际值；$\hat{y}_i$ 表示管道风险的预测值。

3 试验

3.1 数据处理

本研究所采用的数据集源自参考文献，总计包含 681 组数据。每组数据涵盖 14 个特征，包括管道压力、公称直径、管道壁厚、最小埋深、使用年限、周边商业指数、管道材质、土壤腐蚀指数、防腐层类型、阴极保护状况、管线交叉数、员工素质、介质危害性以及泄露危害。此外，数据集还包括两个标签，即后果指数和风险系数。

在将数据输入 GA-PSO-XGBoost 模型之前，应对冗余特征进行删减，以此提高模型学习的效率。因此，本文采用式（1）计算数据集中管道压力、公称直径、管道壁厚、最小埋深、使用年限、周边商业指数、管道材质、土壤腐蚀指数、防腐层类型、阴极保护状况、管线交叉数、员工素质、介质危害性、泄露危害以及风险系数每两个之间的 Pearson 相关系数，建立了相关系数热力图，如图 3 所示。从图 3 可以看出，阴极保护状况和防腐层类型两个特征之间的 Pearson 相关系数为 0.96，呈强正相关。因此，将阴极保护状况从以下的训练过程中删除。另外，图 3 中介质危害性无相关系数，那是因为本文仅针对城市燃气管道进行风险评价，管道介质都为天然气，介质危害性一致，所以，将介质危害性也从以下的训练过程中删除。

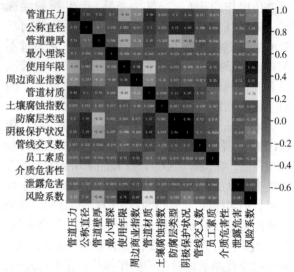

图 3 原始数据相关系数热力图

3.2 实现细节

首先，将预处理后的数据集随机分为训练集和测试集，其中 80% 分配给训练集，20% 分配给测试集。训练集用于模型训练，测试集用于验证模型的性能。

然后，使用 GA-PSO 混合优化算法，在训练集上对 XGBoost 模型的参数进行优化。在参数空间内，随机生成一组初始参数，并通过适应度函数计算适应度值，参数空间见表 2。本文以 MSE 作为适应度值，以评估粒子的优劣，最终选择具有最低 MSE 的参数组合为最优解。在每次迭代中，所有粒子都会更新自身的速度和位置，然后根据当前位置的评估结果来更新全局最优解。

最后，根据优化后的参数，在测试集上进行模型性能评估。具体步骤如 4 所示。

表 2 参数空间

参数	优化范围
num_round	[50, 1000]
max_depth	[3, 10]
eta	[0.01, 0.3]
gamma	[0, 5]
alpha	[1, 10]
ambda	[0, 10]
subsample	[0.5, 1]
colsample_bytree	[0.5, 1]

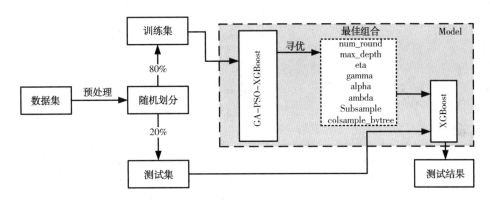

图 4 建立风险评价模型步骤

表 3 GA-PSO 混合优化算法参数设置值

名称	含义	数值
pop_size	种群规模,即粒子数。	500
Max_iter	最大迭代次数。	100
w_1	最大惯性权重。	0.95
w_2	最小惯性权重。	0.7
c_1	个体学习因子。	1.5
c_2	群体学习因子。	2.5
elites	选择精英粒子数。	5
crossover_rate	交叉概率。	0.9
mutation_rate	变异概率。	0.5

本文采用 Python 语言,基于 Jupyter Notebook 编写 GA-PSO-XGBoost 模型。在该模型中,将表 3 中参数的值输入 GA-PSO 混合优化算法,利用其优化表 1 中的超参数。GA-PSO 算法的全局最优适应度值变化曲线如图 5 所示。在图 5 中,横坐标是迭代次数,纵坐标是全局最优适应度值。从图 5 可以看出,适应度值随迭代次数的增加而不断减小。大约经过 20 次迭代后,全局最优适应度值收敛于 2.6042,种群趋于稳定。经过 GA-PSO 混合优化算法的迭代计算后,得到 max_depth、eta、gamma、subsample、lambda、alpha、colsample_bytree、和 num_round 的最优值分别为 3、0.3、0.15、1.00、2.18、0.52、1.00 和 544。

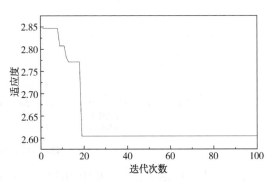

图 5 GA-PSO 算法优化中的适应度变化曲线

4 结果与讨论

4.1 GA-PSO-XGBoost 风险评价模型

当 XGBoost 算法的 max_depth、eta、gamma、subsample、lambda、alpha、colsample_bytree、和 num_round 参数分别为 3、0.3、0.15、1.00、、2.18、0.52、1.00 和 544 时,成功建立了用于城市燃气管道风险评价的 GA-PSO-XGBoost 模型。然后,利用测试集对 GA-PSO-XGBoost 风险评价模型进行测试,得到的结果如图 6、图 7 和图 8 所示。

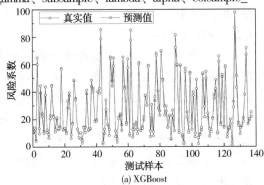
(a) XGBoost

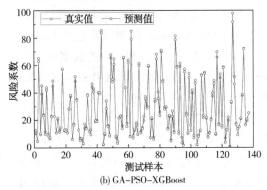
(b) GA-PSO-XGBoost

图 6 XGBoost 参数优化前后风险评价性能表现

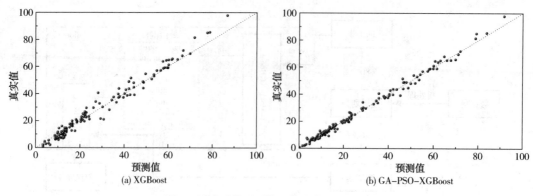

图 7 风险系数预测值和真实值的交叉图

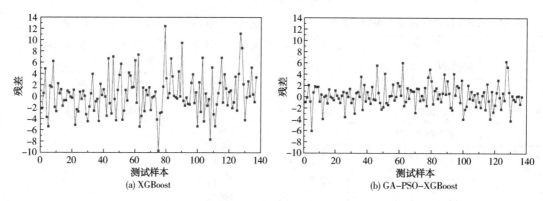

图 8 风险系数预测值和真实值之间的残差图

在图 6 中，横轴代表测试集中的样本，纵轴代表风险值。红色折线代表实际风险值，蓝色折线代表预测值。图 6 显示了测试集中 137 个数据集的预测结果。图 6(a) 和 (b) 分别展示了 XGBoost 和 GA-PSO-XGBoost 的预测风险值和实际风险值的对比。如图 6(a) 所示，在优化超参数之前，XGBoost 的风险预测精度较低。而在优化超参数之后，GA-PSO-XGBoost 可以实现预测值折线与实际值折线之间非常出色地拟合，并可以准确预测大部分风险值数据点，只有少数几个点存在明显误差。因此，超参数优化后的风险预测准确性明显提高。

在图 7 中，横轴为风险预测值，纵轴为风险实际值。图 7(a) 和 (b) 分别为实际风险值与 XGBoost 和 GA-PSO-XGBoost 预测风险值之间的交叉图。与图 8(a) 相比，图 8(b) 中的数据点更集中在 $y=x$ 的两侧。可以看出，GA-PSO-XGBoost 的管道风险预测值与实际值更为吻合。

在图 8 中，横轴表示测试集中的样本，纵轴表示真实值和预测值之间的残差。图 8(a) 和 (a) 分别显示了参数优化前后风险系数真实值与预测值之间的残差。如图 8(a) 和 (a) 所示，在使用 XGBoost 模型预测或使用 GA-PSO-XGBoost 模型预测时，某些样本的预测结果与其他样本相比误差较大。因此，图 8(a) 和 (a) 中存在一些共同异常数据点。子图 (a) 中，残差绝对值最大为 12.4291，子图 (a) 中，残差绝对值最大为 6.3134。如图 8 所示，(b) 中的残差点相较于 (a) 中的残差点，更加集中于 $y=0$ 这条直线两侧。由此可见，优化超参数后，风险预测结果的误差明显减小。

基于相同的评估样本，在优化超参数之前，预测结果的 MSE、MAE 和 R2 值分别为 12.9534、2.6639 和 97.2206%。超参数优化后，这些值重新计算后分别为 4.1264、1.4939 和 99.1146%。

4.2 特征重要性

特征重要性分析可以提高模型的可解释性，并为风险管理提供决策支持。所有输入特征的重要性都是通过 XGBoost 算法中包含的特征重要性计算器工具计算得出的。该工具通过计算每个特征在分割所有节点时的使用次数来确定其重要性得分。根据验证结果，输出每个特征的重要性得分并进行排序，如图 9 所示。

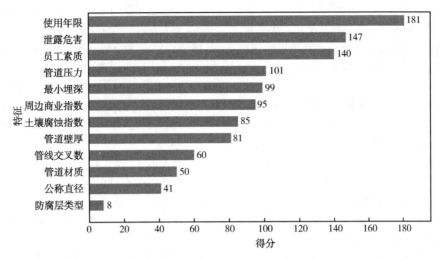

图9 特征重要性排序

这些特征的重要性分数表示单个特征对整个模型的重要性，而不是特征之间的相对权重。不过，XGBoost 模型能够学习它们之间的差异。从如图 9 所示，在数据集的 12 个输入特征中较为重要的是使用年限、泄露危害和员工素质，重要性分数分别为 181、147 和 140。针对于高使用年限的管道，基本上都是钢制管道。大部分钢制管道已运行几十年，各方面性能都在全面下降，潜在风险大，容易出现缺陷并发生事故。天然气是一种易燃易爆的气体，一旦泄漏并与氧气混合，会形成可燃气体。它燃烧温度高，燃烧速度快，如果遭受到点火，火势很难控制，容易造成大面积火灾和可能引发爆炸事故。员工素质的高低对管道的安装、运营管理和维修具有重大影响。员工素质不足会极大增加燃气管道事故风险。

使用年限、泄露危害和员工素质中使用年限和员工素质是事故可能性指标，如果不引起重视会增加事故发生概率；泄露危害是事故后果指标，由介质天然气决定。要降低管道事故风险，首先要着重对高使用年限的管道进行监管，提高巡检频率，及时维修。其次，要经常对员工进行知识技能培训和考核。最后，要时刻注意管道的运行压力、建管时的埋深、土壤酸碱性等方面

5 结论

城市燃气管道作为一个城市的血管，是重要的基础设施。在本文中，使用 GA-PSO-XGBoost 模型对燃气管道进行风险评价。根据获得的风险评价结果，这项工作的主要结论如下：

（1）Pearson 相关性分析结果热力图表明，阴极保护状况和防腐层类型之间的相关系数为 0.96，介质危害性是无用特征。从数据集删减阴极保护状况和质危害性有助于训练风险评价模型。

（2）与参数优化前的 XGBoost 相比，经过 GA-PSO 算法优化参数后，指标 MES、MAE 和 R^2 分别为 4.1264、1.4939 和 99.1146%，模型的性能有了明显的提升。本文建立的 GA-PSO-XGBoost 城市燃气管道风险评价模型上具有可观的精度和实用性。

（3）12 个特征的重要性分数表明，最重要的特征是使用年限，它的重要性分数为 181。这意味着，我们应该更加注重城市中使用年限高的钢制管道的检查和维护。

（4）本文创造性地将 GA-PSO-XGBoost 引入城市燃气管道风险评价过程中，希望这一想法可以为城市燃气管道风险评价以后的机器学习方法提供有价值的参考。

在这项工作中，由于数据采集方面的挑战，风险评价模型仅针对一个城市的燃气管道进行训练和测试。因此下一步的研究重点是收集更多的数据样本，并测试各种算法模型，以创建更适用于不同环境的城市燃气管道风险评估模型。

参 考 文 献

[1] 巴振宁，匡田，梁建文，梁凯，王鸣铄，王智恺. 城市燃气管网运行安全风险评估研究进展[J]. 消防科学与技术，2020，39(04)：533-537.

[2] 巴振宁，韩亚鑫，梁建文. 基于改进 AHP 和模糊综合评价法的燃气管道腐蚀风险评价[J]. 安全与环境

学报, 2018, 18(6): 2103-2109.

[3] 曾小康, 冯阳, 赖文庆, 汤彬坤, 吴涛, 伏喜斌, 等. 基于 AHP-熵权法的城市燃气管道风险评价[J]. 中国安全生产科学技术, 2021, 17(5): 130-135.

[4] 张鹏, 杨宗强, 沈豪, 游亚菲, 潘灏航. 灰色理论和 DEMATEL 在城市燃气管道风险评价中的运用[J]. 安全与环境学报, 2023, 23(9): 3036-3043.

[5] LIANG X B, MA W F, REN J J, DANG W, WANG K, NIE H L, et al. An integrated risk assessment methodology based on fuzzy TOPSIS and cloud inference for urban polyethylene gas pipelines[J]. Journal of Cleaner Production, 2022, 376: 134332.

[6] CHEN Y N, XIE S Y, TIAN Z G. Risk assessment of buried gas pipelines based on improved cloud-variable weight theory[J]. Reliability Engineering & System Safety, 2022, 221: 108374.

[7] 单克. 基于数据统计及情景模拟的燃气管道定量风险评价方法研究[D]. 北京: 中国石油大学(北京), 2021.

[8] 王新颖, 宋兴帅, 杨泰旺, 陈海群, 王凯全. LS-SVM 模型在城市燃气管道风险评估中的应用[J]. 消防科学与技术, 2017, 36(11): 1598-1601.

[9] 陈毓飞, 金跃辉, 杨谈. 一种基于 LSTM 的燃气管道施工破坏风险预测模型[J]. 网络新媒体技术, 2019, 8(1): 24-29.

[10] LI F, WANG W H, Xu J, Yi J, WANG Q S. Comparative study on vulnerability assessment for urban buried gas pipeline network based on SVM and ANN methods[J]. Process Safety and Environmental Protection, 2019, 122: 23-32.

[11] WANG Z F, LI S Z. Data-driven risk assessment on urban pipeline network based on a cluster model[J]. Reliability Engineering & System Safety, 2020, 196: 106781.

[12] 熊威. 基于 BP 神经网络的 N 市燃气管道风险管理[D]. 南昌: 南昌大学, 2021.

[13] BAI Y P, WU J S, REN Q R, JIANG Y, CAI J T. A BN-based risk assessment model of natural gas pipelines integrating knowledge graph and DEMATEL[J]. Process Safety and Environmental Protection, 2023, 171: 640-654.

[14] ZHANG X F, AN J Y. A new pre-assessment model for failure-probability-based-planning by neural network[J]. Journal of Loss Prevention in the Process Industries, 2023, 81.

[15] SU X Y, SHANG S W, XU Z H, QIAN H, PAN X L. Assessment of Dependent Performance Shaping Factors in SPAR-H Based on Pearson Correlation Coefficient[J]. CMES - Computer Modeling in Engineering & Sciences, 2024, 138(2): 1813-1826.

[16] WANG L, WU C Z, TANG L B, ZHANG W G, LACASSE S, LIU H L, et al. Efficient reliability analysis of earth dam slope stability using extreme gradient boosting method[J]. Acta Geotechnica, 2020, 15: 3135-3150.

[17] 舒漫. XGBoost 算法在成都市燃气负荷预测分析中的应用[D]. 成都: 成都理工大学, 2020.

[18] LIU W, CHEN Z X, HU Y. XGBoost algorithm-based prediction of safety assessment for pipelines[J]. International Journal of Pressure Vessels and Piping, 2022, 197: 104655.

[19] CHEN T Q, GUESTRIN C. Xgboost: A scalable tree boosting system[C]. San Francisco: Proceedings of the 22nd ACM SIGKDD International Conference on Knowledge Discovery and Data Mining, 2016: 785-794.

[20] LIN C. An adaptive genetic algorithm based on population diversity strategy[C]. Guilin: 2009 Third International Conference on Genetic and Evolutionary Computing, 2009: 93-96.

[21] JAIN M, SAIHJPAL V, SINGH N, SINGH S B. An overview of variants and advancements of PSO algorithm[J]. Applied Sciences, 2022, 12(17): 8392.

[22] 李红亚, 彭昱忠, 邓楚燕, 龚道庆. GA 与 PSO 的混合研究综述[J]. 计算机工程与应用, 2018, 54(2): 20-28+39.

[23] ANAND A, SUGANTHI L. Hybrid GA-PSO optimization of artificial neural network for forecasting electricity demand[J]. Energies, 2018, 11(4): 728.

[24] BI K X, QIU T. An intelligent SVM modeling process for crude oil properties prediction based on a hybrid GA-PSO method[J]. Chinese Journal of Chemical Engineering, 2019, 27(8): 1888-1894.

A Panoramic Visualization Platform for Petroleum Pipeline Supply Chain Procurement: Enhancing Transparency, Efficiency, and Data-Driven Decision-Making

Han Bing

(Software Development Engineer National Petroleum and Natural Gas Pipeline Network Group Co., Ltd)

Abstract The petroleum pipeline industry's procurement operations are inherently complex, involving multiple stakeholders, stringent compliance requirements, and geographically dispersed suppliers. Traditional procurement practices often rely on fragmented information sources, manual oversight, and limited data integration, resulting in inefficiencies and reduced transparency. This study aims to address these challenges by developing a panoramic visualization platform that centralizes procurement data, streamlines supplier collaboration, enhances quality assurance, and supports informed, strategic decision-making. By doing so, we intend to improve procurement efficiency, optimize resource allocation, and ultimately contribute to more resilient and reliable petroleum pipeline supply chains.

We employed a design science research approach, starting with a comprehensive needs assessment of the petroleum pipeline procurement landscape. Based on identified requirements, we conceptualized and developed a modular software architecture integrating ERP data, supplier management functionalities, and automated workflows for procurement planning, supplier evaluation, contract management, and order approvals. A supplier production coordination module enables real-time data sharing of production progress, material preparation status, and quality metrics. Data analytics and visualization components—such as procurement dashboards, supplier performance analytics, cost forecasting models, and deviation alerts—were implemented to facilitate data-driven decision-making. Stringent security and compliance features, including role-based access controls, encryption protocols, and audit logs, ensure data integrity and adherence to industry standards. The platform's effectiveness was evaluated through pilot testing, scenario simulations, user feedback, and performance metrics (e.g., order cycle times, supplier response rates, data accuracy).

The platform successfully consolidated previously fragmented procurement data, providing stakeholders with a unified view of orders, supplier qualifications, contract terms, and operational metrics. This integration significantly enhanced process transparency and reduced manual data consolidation efforts. Users reported improved collaboration with suppliers, as real-time production progress and material readiness data minimized last-minute delays and enhanced predictability. Data-driven analytics allowed procurement managers to identify cost variations, forecast future demands, and detect potential bottlenecks, leading to more informed negotiation strategies and proactive risk mitigation. The system's role-based access and audit logs supported stringent compliance requirements, while automated workflows reduced administrative overhead and expedited order approvals. Overall, initial tests demonstrated shorter procurement cycle times, higher supplier reliability, and increased user satisfaction, underscoring the platform's practical value.

This research confirms that an integrated, data-centric visualization platform can substantially improve the efficiency, transparency, and strategic rigor of petroleum pipeline procurement processes. By centralizing data, automating key workflows, and fostering closer buyer-supplier collaboration, the system enables more agile and evidence-based decision-making. Moreover, robust security measures and compliance functions ensure that the platform aligns with regulatory standards and corporate governance policies. While the solution is tailored to the petroleum pipeline sector, its architectural principles, data integration strategies, and analytical capabilities have broader applicability to other industries facing complex supply chain procurement challenges. Future work may include incorporating advanced machine learning models for supplier risk assessment, integrating sustainability metrics, and expanding mobile accessibility, thus continually refining and strengthening the platform's capacity to support dynamic, high-stakes procurement environments.

Keywords Petroleum Pipeline; Supply Chain Procurement; Panoramic Visualization Platform; Data-Driven Decision-Making; Supplier Collaboration; Quality Control; Compliance Management; Role-Based Access Control

Introduction

In the contemporary petroleum industry, efficiency and transparency within supply chain operations have emerged as paramount considerations. The petroleum pipeline sector, in particular, presents a range of complex procurement challenges that extend far beyond the simple exchange of goods and services. These challenges are reflected in multi-tier supplier relationships, fluctuating raw material costs, strict compliance requirements, diverse geographical operations, and rapidly changing market demands. The procurement process in petroleum pipeline projects involves multiple stakeholders—including procurement managers, suppliers, subcontractors, quality inspectors, financial controllers, and regulatory auditors—each with distinct informational needs and decision-making criteria. Achieving a high degree of visibility, control, and agility in these processes is not a trivial endeavor.

Against this backdrop, the development and deployment of an integrated software platform that supports full-spectrum procurement management and collaboration can offer a transformative solution. In recent years, significant advances have been made in digital procurement solutions, including integration with enterprise resource planning (ERP) systems, deployment of supplier relationship management tools, incorporation of data analytics and visualization dashboards, and the application of intelligent forecasting models. However, the petroleum pipeline context, with its long project lifecycles, stringent safety and quality requirements, and the necessity of secure, transparent procurement transactions, has continued to require a tailored, domain-specific solution.

This paper introduces a comprehensive petroleum pipeline supply chain procurement panoramic visualization platform. The platform is designed to streamline procurement workflows, enhance visibility across the value chain, and facilitate real-time decision-making. It integrates functionalities including procurement dashboards, supplier production coordination modules, and holistic management features—all tied together with robust data security and flexible integration capabilities. By doing so, it offers procurement managers and stakeholders a unified digital environment that centralizes procurement data, automates key workflows, ensures compliance with internal policies and industry standards, and supports strategic decision-making through analytics and visualizations.

The primary objectives of this research are fourfold. First, we aim to design a conceptual and technical framework for a procurement visualization platform that meets the unique requirements of petroleum pipeline supply chains. Second, we seek to implement functionalities—ranging from procurement planning, contract management, supplier evaluation, and real-time production progress tracking—that collectively support end-to-end procurement processes. Third, we intend to demonstrate the platform's ability to integrate with existing enterprise systems (e.g., ERP) and external e-commerce platforms. Finally, we will evaluate the effectiveness of this platform by analyzing improvements in procurement transparency, decision-making accuracy, supplier collaboration, and overall operational efficiency.

Following this introduction, the paper is organized into three further sections. The second section outlines the technical thinking and methods applied, detailing the architecture, data models, user roles, and system components developed. The third section presents results and effects, discussing improvements in key performance indicators, supplier management processes, collaboration outcomes, data-driven decision - making, and system usability. The final section draws conclusions about the platform's significance, its potential to serve as a model for other sectors, and areas of future research and development.

Technical Approach and Research Methods

1 System Conceptualization and Architecture Design

The petroleum pipeline procurement panorama platform was conceptualized to address the complexity and fragmentation often encountered in large-scale energy supply chains. The platform's architecture emphasizes modularity, scalability, interoperability, and security. Each module corresponds to a specific aspect of the procurement process—ranging from procurement planning, supplier evaluation, contract management, and order approval to supplier production coordination, quality control, and data visualization.

Previous studies have highlighted the importance of integrating ERP systems with procurement platforms to enhance data consistency and streamline supply chain operations. At the foundational layer, the platform integrates with the enterprise's existing ERP system to synchronize procurement, inventory, and financial data. Standard APIs facilitate communication with external data sources, such as third-party e-commerce platforms supplying raw materials or finished components. The system's data layer centralizes procurement - related information (e.g., supplier data, order histories, pricing records, contract details, quality metrics), ensuring consistency and accessibility. Above this data layer, application services implement domain-specific business logic: generating procurement plans, managing supplier relationships, monitoring budgets, and performing real - time analytics.

The user interface layer provides access points for various stakeholders—procurement managers, supplier administrators, quality inspectors, financial officers—according to role-based permissions. The interface is designed for intuitive navigation, with a left - side functional menu that segregates procurement dashboards, supplier collaboration interfaces, and integrated management consoles. On top of this architecture, a robust security framework ensures data confidentiality, integrity, and availability through SSL/TLS encryption, multi-factor authentication, operation logging, and periodic security audits.

2 Functionality Modules

2.1 Procurement Dashboard:

The procurement dashboard presents a comprehensive overview of key performance and business indicators. It includes metrics such as procurement concentration (i.e., degree of supplier consolidation), direct purchase rates, online procurement ratios, and essential timelines. Historical data queries enable users to explore past procurement events filtered by organizational unit or date range. Integrations with third-party e-commerce orders provide insights into external procurement activities. By consolidating these data points in interactive charts and tables, the dashboard empowers procurement managers to quickly assess procurement efficiency, identify performance gaps, and respond proactively.

2.2 Supplier Production Coordination:

Effective buyer-supplier collaboration is crucial for mitigating supply chain disruptions and enhancing operational efficiency. This module aims to reduce information asymmetry and foster real-time collaboration between buyers and suppliers. Supply plans, generated from confirmed contracts and orders, outline the required materials, quantities, and delivery deadlines. Suppliers can upload or manually enter production progress updates, enabling procurement managers to track real - time manufacturing status, anticipate potential delays, and coordinate mitigating actions. Materials preparation (or kitting) processes are streamlined by allowing suppliers to upload, edit, and confirm material readiness data. Quality control measures are strengthened through integrated quality assurance workflows, allowing for prompt reporting and rectification of non-conformance issues. The module also supports electronic label creation, exporting, and printing (e.g., via QR codes), facilitating efficient on-site materials management and logistics tracking.

2.3 Integrated Management and Visualization:

Advanced visualization tools play a pivotal role in enabling decision-makers to interpret complex supply chain data and address operational challenges effectively. A comprehensive "panoramic" view is achieved through large-screen visualization tools that aggregate procurement activities, order execution statuses, supplier performance ratings, and inventory data into a unified, intuitive display. This top-level visualization helps senior executives and decision-makers rapidly identify trends, anomalies, and areas requiring immediate attention. The module incorporates intelligent analytical models to forecast procurement demands, evaluate supplier rankings based on timeliness and accuracy, detect pending order deviations, and provide cost and trend analyses. Automated alerts, notifications, and drill-down capabilities ensure that stakeholders can move seamlessly from a high-level overview to granular details.

3 Data Management and Analytics

Effective data management underpins a platform's analytical capabilities. The system maintains a historical repository of procurement orders, supplier credentials, contract documents, and transaction logs. This historical data facilitates the calculation of performance indicators—such as procurement concentration, bidding frequencies, direct purchase ratios—and informs the benchmarking of current operations against historical trends.

Additionally, the platform integrates advanced analytics for forecasting procurement demands, identifying cost overruns, and uncovering supplier risk patterns. Predictive models leverage time-series forecasting to anticipate future procurement volumes based on historical consumption, lead times, and seasonal variations. Anomaly detection algorithms alert users to potential operational disruptions, such as unusually slow production progress or significant price deviations from baseline levels. These insights form the basis for proactive decision-making, enabling procurement managers to adjust inventory strategies, renegotiate contracts, or reallocate resources.

4 Security and Compliance

Security and compliance are vital in an environment where sensitive data (such as contract amounts, supplier ratings, and procurement budgets) is frequently accessed and shared. The platform's design includes stringent user authentication, role-based access control, encryption of data in transit, and meticulous operation logging to ensure traceability. Sensitive fields are masked or only accessible by authorized individuals, minimizing the risk of data leakage. Regular security audits and vulnerability scans are conducted, and patches or remedial measures are promptly implemented.

Compliance with industry standards—whether corporate governance policies, financial regulations, or quality and safety standards for petroleum pipelines—is integrated into system workflows. Approval processes can be configured to require sign-offs from finance, legal, or engineering departments before large orders become effective. Audit logs, accessible to authorized auditors, document every data query, modification, or approval, supporting both internal and external compliance verifications.

5 Integration with External Systems and Data Formats

To reduce data silos and redundant inputs, the platform supports integration with ERP systems and external e-commerce interfaces. This ensures automatic updates of supplier catalogs, pricing, and order confirmations. Bid solicitations can be managed within the system, and responses from suppliers can be compared and analyzed. Contract data can be aligned with corresponding procurement orders, and payment information synchronized with financial modules.

Data import and export capabilities, in formats such as CSV, Excel, and PDF, support interoperability and reporting. Seamless data exchange and compatibility with standard file formats are critical for enhancing cross-system collaboration in supply chain management. Standard APIs allow integration with business intelligence tools (e.g., Tableau, Power BI) for more specialized analyses. The platform can also operate on various operating systems (Windows, Linux, Mac), enhancing its accessibility and adaptability. Mobile access is enabled to ensure that users can track orders, approve requests, and view analytics on the go.

6 Research and Evaluation Methodology

The design science methodology is particularly effective for addressing complex, real-world problems through iterative development and evaluation. From a research perspective, we adopted a design science methodology. First, we analyzed current procurement pain points in the petroleum pipeline sector. Next, we designed a conceptual architecture to address these challenges, focusing on key functionalities (procurement dashboards, supplier collaboration, integrated analysis). We iteratively refined the system based on stakeholder feedback, pilot testing, and performance metrics gathered in a controlled environment. Qualitative user feedback was collected through interviews and surveys, while quantitative data—such as order processing times, supplier response lead times, and user adoption rates—were monitored to measure improvements.

Throughout the pilot phase, test scenarios were constructed to emulate typical procurement events: annual budget allocations, new supplier onboarding, contract negotiations, material production and inspection, and emergency order placement. Process metrics, system response times, data accuracy, and user satisfaction were recorded and analyzed. These assessments guided incremental adjustments to the platform and informed future feature enhancements.

Results and Effects

1. Enhanced Procurement Transparency and Control

One of the platform's primary objectives was to provide real-time transparency into procurement operations. Post-implementation observations revealed that procurement managers gained immediate access to up-to-date order statuses, supplier performance metrics, and budget consumption rates. This visibility significantly reduced the time and effort needed to consolidate information from disparate sources. Historical data queries allowed for rapid retrieval of past orders, supplier contracts, and performance metrics, which previously required manual lookups in disconnected databases or spreadsheets.

Effective procurement visualization tools are essential for identifying supply chain risks and optimizing decision-making processes. The procurement dashboard's visual indicators offered clear, comprehensible insights into organizational purchasing patterns. For instance, the concentration of procurement activities with a single supplier or region could be instantly identified. This facilitated more informed risk management decisions—such as diversifying supplier bases or re-negotiating contracts. By centralizing all procurement-related data and presenting it in a user-friendly format, decision-makers could identify bottlenecks, reduce procurement cycle times, and ensure alignment with strategic procurement objectives.

2. Improved Supplier Relationship and Production Coordination

By enabling suppliers to upload production progress data and manage materials readiness information, the platform strengthened buyer-supplier collaboration. Suppliers were no longer treated as information black boxes; instead, they became active participants in shared workflows. This real-time exchange fostered trust, reduced uncertainties, and minimized last-minute surprises in the supply chain. Procurement managers reported fewer urgent interventions needed to expedite delayed shipments, as potential delays were detected earlier. Suppliers, for their part, appreciated the clarity of requirements and expectations provided by the digital supply plans.

The electronic labeling and QR code generation functionalities improved on-site material handling and logistics tracking. Warehousing and logistics personnel could scan labels to instantly view material specifications, associated orders, and supplier credentials. This reduced misplacement risks, improved inventory counts, and accelerated inbound inspection processes. Over time, better coordination and clearer visibility contributed to stronger supplier relationships, reinforced by data-driven performance evaluations and transparent performance reporting.

3. Strategic Decision-Making Through Analytics

Analytical capabilities and forecasting tools integrated into the platform enabled procurement teams and management to make more strategic, data-driven decisions. Monthly and annual procurement amount trends displayed in dynamic charts helped financial controllers compare actual expenditures against budgets. Procurement trend predictions guided adjustments to inventory policies, ensuring materials availability while avoiding overstocking.

Supplier performance ratings—incorporating timeliness, accuracy, cost-effectiveness, and quality metrics—enabled objective supplier evaluations. With quantifiable performance metrics, contract managers could negotiate better terms, allocate larger volumes to reliable suppliers, and reduce relationships with underperforming ones. This data-driven supplier optimization process led to more resilient supply networks, better pricing leverage, and overall cost containment.

Real-time alerts and predictive models addressing potential order deviations improved risk management. Early warnings about potential late deliveries or cost overruns empowered managers to mitigate these issues proactively. Such proactive measures not only saved time and money but also maintained stable project timelines and safeguarded the integrity of the pipeline construction and maintenance activities.

4. Efficiency Gains in Procurement Workflows

The automation of procurement planning, order generation, and supplier collaboration reduced manual interventions and error rates. Previously, procurement staff spent significant amounts of time entering data into spreadsheets, verifying supplier qualifications, checking inventory levels, and reconciling purchase orders. With automated workflows, orders were created based on predefined thresholds, integrated budgets, and approved suppliers. Supplier selection and qualification processes were accelerated through integrated vendor management tools. These automations freed procurement professionals to focus on strategic activities—such as evaluating new suppliers, analyzing long-term procurement strategies, and pursuing cost-saving initiatives—instead of routine administrative tasks.

5. Compliance and Security Outcomes

The platform's stringent access controls, multifactor authentication, and encrypted data transmissions mitigated data security risks. Users reported confidence in the platform's security, a critical factor given the sensitivity of procurement and contract data. Compliance audits were simplified due to comprehensive logging of every system interaction. Auditors could trace order approvals, supplier evaluations, and contract modifications to authorized individuals, with timestamps and rationale clearly recorded. This level of transparency reduced compliance burdens, facilitated smooth audits, and lowered the risk of financial or reputational repercussions.

6. User Adoption and Satisfaction

In pilot tests, user feedback indicated that the system's intuitive interface, role-based navigation, and centralized data access improved the overall user experience. Procurement managers appreciated how easily they could customize performance dashboards. Suppliers, who initially expressed reservations about new data entry responsibilities, became more receptive after understanding the mutual benefits of transparency and more predictable order flows. Over time, user satisfaction rose as the platform became integrated into daily workflows, providing consistent value and reducing overall complexity.

7. Scalability and Future Integration Potential

The system's modular architecture and compatibility with multiple platforms ensured that additional functionalities, data sources, and user roles could be integrated as procurement needs evolved. This scalability is particularly valuable in the petroleum pipeline industry, where new projects, expanding geographies, and evolving regulatory requirements demand flexible digital infrastructures. Plans are already underway to incorporate deeper inventory analytics, maintenance scheduling integrations, and advanced artificial intelligence algorithms for supplier risk scoring and cost anomaly detection.

Conclusion

This paper presented a comprehensive petroleum pipeline supply chain procurement panoramic visualization platform designed to enhance efficiency, transparency, and strategic decision-making in a complex procurement environment. The platform combines procurement dashboards, supplier production coordination tools, integrated management features, and robust security measures to address the unique challenges inherent in petroleum pipeline procurement operations.

The introduction established the pressing need for enhanced visibility and centralized data management in the petroleum pipeline sector. Complex procurement activities, multiple stakeholder interests, and stringent compliance standards created conditions ripe for a digital solution that could streamline workflows, reduce information asymmetries, and promote informed decision-making.

The technical approach and research methods sections detailed how the platform's modular architecture, data integration capabilities, analytics tools, and security frameworks were conceptualized and implemented. By delineating procurement dashboard functionalities, supplier production coordination, comprehensive data management, analytics integration, and interoperability with existing ERP and external e-commerce systems, we demonstrated a holistic approach to procurement digitalization. This holistic design addressed the entire procurement lifecycle—from planning and supplier qualification to contract management, order execution, quality assurance, and final delivery—while maintaining adherence to strict data security, compliance, and auditability requirements.

In the results and effects section, we highlighted tangible improvements observed in procurement operations following the platform's deployment. Enhanced transparency allowed stakeholders to access real-time data on orders, suppliers, and budgets, reducing delays and guesswork. Supplier collaboration modules led to improved relationships and smoother production coordination, mitigating last-minute changes and disruptions. Data-driven analytics informed better strategic decisions, including supplier optimization, cost management, and risk mitigation. Workflow automation reduced manual efforts and error rates, enabling procurement staff to focus on higher-value tasks. Compliance and security measures supported reliable, audit-ready records and reduced the risk of data breaches or non-compliance. Overall, users expressed higher satisfaction and confidence in the system's capabilities, and the platform's scalable architecture positions it well to adapt to future requirements.

In conclusion, the petroleum pipeline procurement panoramic visualization platform represents a significant advancement in supply chain digitalization and integrated procurement management. By centralizing data, fostering buyer-supplier collaboration, and leveraging advanced analytics, it significantly contributes to the transparency, reliability, and strategic rigor of procurement processes. For the petroleum industry—and potentially other sectors with similarly complex supply chains—this platform offers a blueprint for integrating data-driven decision-making into the procurement lifecycle.

Looking ahead, further enhancements could include incorporating advanced machine learning algorithms for supplier risk analysis, predictive maintenance scheduling for pipeline components, and more sophisticated inventory optimization models. Additional integration with environmental and sustainability metrics would reflect the growing emphasis on responsible sourcing and corporate social responsibility. The continuous evolution of data analytics, cloud infrastructure, and intelligent automation tools ensures that such a platform will remain at the forefront of effective supply chain management practices. In sum, the research confirms that an integrated, data-centric approach to procurement can generate substantial operational, financial, and strategic benefits, setting a new standard for digital transformation in the petroleum pipeline industry.

References

[1] Alvarez-Rodríguez J M, Labra-Gayo J M, Ordoñez de Pablos P. New trends on e-Procurement applying semantic technologies: Current status and future challenges[J]. Computers in Industry, 2014, 65(5): 800-820.

[2] Saputelli L, Duran J, Rivas F, Casas E, et al. Success Cases and Lessons Learned After 20 Years of Oilfield Digitalization Efforts[C]. New Orleans: SPE Annual Technical Conference and Exhibition, 2024: SPE-220932-MS.

[3] Syed Z A, Dapaah E, Mapfaza G, et al. Enhancing supply chain resilience with cloud-based ERP systems [J]. IRE Journals, 2024, 8(2): 106-128.

[4] Mwesiumo D, Nujen B B, Buvik A. Driving collaborative supply risk mitigation in buyer-supplier relationships[C]. Supply Chain Forum: An International Journal. 2021, 22(4): 347-359.

[5] Al-Kassab J, Ouertani Z M, Schiuma G, et al. Information visualization to support management decisions [J]. International Journal of Information Technology & Decision Making, 2014, 13(02): 407-428.

[6] Arunachalam D, Niraj K, Kawalek J P, Understanding big data analytics capabilities in supply chain management: Unravelling the issues, challenges and implications for practice[J]. Transportation Research Part E: Logistics and Transportation Review, 2018, 114: 416-436.

[7] Kharfan M, Chan V W K, Efendigil T F. A data-driven forecasting approach for newly launched seasonal products by leveraging machine-learning approaches [J]. Annals of Operations Research, 2021, 303: 159-174.

[8] Bachlechner D, Thalmann S, Maier R. Security and compliance challenges in complex IT outsourcing arrangements: A multi-stakeholder perspective[J]. Computers & Security, 2014, 40: 38-59.

[9] Yang L, Ni S T, Wang Y, et al. Interoperability of the Metaverse: A Digital Ecosystem Perspective Review [J]. arXiv,

[10] Baskerville R, Pries-Heje J, Venable J. Soft design science methodology[C]. Philadelphia: The 4th International Conference on Design Science Research in Information Systems and Technolog, 2009: 1-11.

[11] Odutola A. Advanced procurement analytics: Building a model for improved decision making and cost efficiency within global supply chains[J]. International Journal of Scientific and Management Research, 2022, 5(1): 273-286.

[12] Adebisi E O, Alao O O, Ojo S O. Assessment of early warning signs predisposing building projects to failure in Nigeria[J]. Journal of Engineering, Design and Technology, 2020, 18(6): 1403-1423.

[13] Flechsig C, Anslinger F, Lasch R. Robotic Process Automation in purchasing and supply management: A multiple case study on potentials, barriers, and implementation[J]. Journal of Purchasing and Supply Management, 2022, 28(1).

[14] Prestidge K L. Digital Transformation in the Oil and Gas Industry: Challenges and Potential Solutions[D]. Massachusetts: Massachusetts Institute of Technology, 2022. URL: https://hdl.handle.net/1721.1/143178

[15] Pasupuleti V, Thuraka, B, Kodete C S, Malisetty S. Enhancing supply chain agility and sustainability through machine learning: Optimization techniques for logistics and inventory management[J]. Logistics, 2024, 8(3): 73.

设计参数对硬岩储气库稳定性影响规律研究

周小松[1,2]　黄康康[2]　闫 磊[2]　王颖蛟[2]　刘 卫[2]

(1. 西安理工大学岩土工程研究所；2. 机械工业勘察设计研究院有限公司)

摘　要　围岩等级、埋深、洞径是影响人工开挖洞室围岩稳定性的主要设计参数。采用MIDAS GTS NX软件建立模型，从应力、位移、塑性区多个角度研究了围岩等级、埋深以及洞径对洞室围岩稳定性的影响规律。研究表明：储气压力为16MPa时，衬砌的顶部和底部均出现受拉破坏；随着围岩等级的降低，洞室围岩的位移量和塑性区面积不断增大；随埋深的增加，在开挖完成后，各关键点位移不断增大。在运行期间，各关键点先向外部发生位移，位移量逐渐减小，之后向部发生位移，位移量逐渐增大，洞室围岩塑性区面积不断减小；随洞径的增加，各关键点位移和洞室围岩塑性区面积不断增大。经过分析得到，储气库布置在围岩等级为Ⅰ~Ⅱ级、埋深为300~600m、洞径为5~25m范围时，具有较好的稳定性。

关键词　洞室围岩；围岩等级；埋深；洞径；稳定性

Study on the influence of design parameters on the stability of hard rock gas storage

Zhou Xiaosong[1,2]　Huang Kangkang[2]　Yan Lei[2]　Wang Yingjiao[2]　Liu Wei[2]

(1. Xi'an University of Technology Institute of Geotechnical Engineering;
2. Mechanical Industry Survey and Design Research Institute Co., Ltd.)

Abstract　The surrounding rock grade, buried depth and hole diameter are the main design parameters that affect the stability of the surrounding rock of the artificial excavation chamber. Using MIDAS GTS NX software to establish a model and studies the influence of surrounding rock grade, burial depth, and tunnel diameter on the stability of tunnel surrounding rock from multiple perspectives of stress, displacement, and plastic zone. The study shows that when the gas storage pressure is 16MPa, the top and bottom of the displacement of the surrounding rock and the plastic area of the surrounding rock are increasing. With the increase of buried depth, the displacement of each key point is increasing after the excavation. During operation, key points first shift outward, with a gradual decrease in displacement, and then shift inward, with an increase in displacement and a continuous decrease in the plastic zone area of the surrounding rock of the tunnel; As the diameter of the tunnel increases, the displacement of each key point and the plastic zone area of the surrounding rock of the tunnel continuously increase. After analysis, it is found that the gas storage layout has good stability when the surrounding rock grade is Ⅰ~Ⅱ grade, the buried depth is 300~600m, and the hole diameter is 5~25m.

Key words　Cave surrounding rock; Surrounding rock grade; Buried depth; Hole diameter; Stability

压缩空气储能是一种将电能转化为空气压力势能和储热介质热能进行储存，在需要时又将其转换为电能的长时储能技术。由于大型压缩空气储能电站需要较大的储气空间，而地上储气罐价格昂贵，现阶段在建、已投产项目多采用地下储气库，即地下盐穴、人工开挖洞室、废弃矿洞等3种形式。人工开挖洞室具有不受特定地质条件限制、存储压力大、气密条件好、循环效率高的特点，使其成为在压缩空气储能电站储气库建造时优先考虑的方案。特别是越来越多的化工、钢铁冶金等用电大户，每天近千万度电的外购成本，成为其沉重的负担。在用户侧建设人工开挖洞室压缩空气储能电站，可以充分利用当地的波谷电价差、共享储能容量补贴、电网服务等收益，极大降低企业用电成本，提高企业的生成竞争力。

目前，国外已有两座正在商业化运行的大型压缩空气储能电站，分别是德国Huntoyf 290MW级电站、美国Mcintosh 110MW级电站。国内压缩空气储能电站也发展迅速，相继建成了芜湖0.5MW级电站、金坛60MW级电站、张北100MW级电站等，其中张北压缩空气储能电站利用硬岩储气库进行储气。硬岩储气库洞室稳定性是压缩空气储能系统安全运行的关键，因此国内外学者对人工开挖储气库洞室进行了大量的研究。Zimmels等利用FLAC软件研究了隧道式洞室在不同气压、间距下围岩塑性区的分布状态，并对储气库容量进行了计算。Perazzelli等根据岩体变形破坏准则，研究了内衬式岩洞在不同地质条件下开挖的可能性。Pornhasem等对含有孔洞的岩石样品施压空气压力，通过物理试验的方法研究了围岩在空气压力作用下的破坏特征。王其宽采用ABAQUS软件，以关键点的位移以及围岩塑性区为评价指标，研究了洞室间距、埋深及内径对围岩变形和塑性区的影响规律。夏才初等利用ABAQUS、COMSOL有限元软件，研究了不同截面形状、尺寸、埋深的洞室在10MPa气压作用下围岩受到的应力、变形以及塑性区发展状况。朱荣华通过建立不同围岩条件、高跨比、埋深的洞室数值模型，研究了开挖后围岩的稳定性特征，得到了洞室围岩的变化规律。蒋中明等基于FLAC3D软件平台，研究了储气库截面形式、洞室埋深和运行压力对储气库围岩累计损伤特征的影响。由以上研究可知，储气库运行压力、围岩等级、埋深、洞径对其自身稳定性有着重要的影响。因此本文采用MIDAS GTS NX软件在花岗岩岩体中建立容积为18万 m^3 的圆形隧道式储气库二维模型，分析衬砌厚度为0.5m、混凝土等级为C40、储气压力为16MPa时，不同围岩等级、埋深、洞径条件下，洞室围岩的受力特征以及塑性区变化状态，从而获得合理的设计参数，为将来地下储气库的工程建设提供参考。

1 储气库有限元模型

1.1 数值模型建立

圆形隧道式储气库的长度远大于洞径，为计算简便，将其简化为二维模型。为了避免边界效应影响，保证洞室距模型上下、左右边界的距离大于3倍洞径，模型尺寸设置为400×800m。采用"四边形+三角形"组合形式单元进行网格划分，以洞室为中心向四周扩展，网格逐渐稀疏，最大单元尺寸控制为5m，单元总数为28658个。模型的初始应力为自重应力，左右边界以及底部边界受到法向位移约束，上部边界不受任何约束。模型示意图如图1所示。

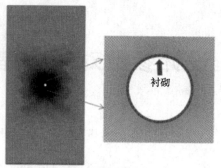

图1 模型示意图

1.2 力学模型及其参数

在数值模拟过程中，将围岩假定为理想弹塑性材料，采用摩尔-库仑屈服准则。剪切破坏与拉伸破坏函数分别为：

$$F_s = \sigma_1 - N_\varphi \sigma_3 + 2c\sqrt{N_\varphi} \quad (1)$$
$$F_t = \sigma_3 - \sigma_t \quad (2)$$

式中，σ_1 为最大主应力，MPa；$N_\varphi = (1+\sin\varphi)/(1-\sin\varphi)$，$\varphi$ 为岩石的内摩擦角（°）；σ_3 为最小主应力，MPa；c 为岩石的黏聚力，MPa；σ_t 为岩石的抗拉强度，MPa。

围岩的剪切破坏势函数采用不相关联流动法则，拉伸破坏势函数采用相关联流动法则，其表达式分别为：

$$g_s = \sigma_1 - \sigma_3 N_\psi \quad (3)$$
$$g_t = -\sigma_3 \quad (4)$$

式中，$N_\psi = (1+\sin\psi)/(1-\sin\psi)$，$\psi$ 为岩石的膨胀角（°）。

衬砌假定为弹性材料，采用C40混凝土，厚度为0.5m。材料参数取值参考《工程岩体分级标准》(GB/T 50218—2014)和《混凝土结构设计规范》(GB 50010—2010)，详见表1。

1.3 数值模拟方案

在模型计算过程中，主要考虑围岩等级、埋深、洞径三个影响因素。围岩级别分别设置为Ⅰ级、Ⅱ级、Ⅲ级，埋深分别设置为100m、200m、300m、400m、500m、600m、700m，洞径分别设置为5m、15m、25m、35m、45m，各工况下圆形隧道式储气库数值模拟方案表2。计算时，首先初始地应力平衡，位移清零；然后开挖洞室，使得应力重分布；最后，浇筑混凝土，施加16MPa储气压力。

表1 材料参数表

材料类别		弹性模量 E/GPa	泊松比 μ	重度 γ/kN·m^{-3}	黏聚力 c/MPa	内摩擦角 φ/°	抗拉强度/MPa
围岩	Ⅰ级	33	0.15	27	2.1	60	8
	Ⅱ级	20	0.2	26	1.5	50	2
	Ⅲ级	10	0.25	24	1	40	1.5
C40混凝土		32.5	0.2	24.4			1.71

表 2　数值模拟方案

影响因素	圆形隧道式储气库	基本方案
围岩等级	Ⅰ级、Ⅱ级、Ⅲ级	Ⅱ级
埋深	100m、200m、300m、400m、500m、600m、700m	400m
洞径	5m、15m、25m、35m、45m	15m

2　计算结果分析

2.1　围岩等级对储气库稳定性的影响

储气压力为 16MPa、埋深为 400m、洞径为 15m 的圆形隧道式储气库在不同围岩等级条件下，衬砌受到的第一主应力云图如图 2 所示。

从图中可以看到，Ⅰ级、Ⅱ级、Ⅲ级围岩条件下，衬砌受力状态分布相同，均处于受拉状态，沿拱腰向拱顶、拱底方向，衬砌受到的拉应力逐渐增大。衬砌拱腰处受到的拉应力分别为 12.39MPa、24.80MPa、53.02MPa，衬砌受到的最大拉应力分别为 37.45MPa、49.52MPa、73.04MPa。随着围岩等级的降低，衬砌受到的主应力逐渐增加，且衬砌受到的拉应力远大于混凝土的抗拉强度，表明衬砌会产生整体性张拉破坏。

不同围岩等级条件下，洞室在开挖完成后和储气压力为 16MPa 时，围岩受到的最大拉应力曲线图如图 3 所示。从图中可以看到，不同工况下，围岩受到的最大拉应力均小于其抗拉强度，表明围岩不会出现张拉破坏。

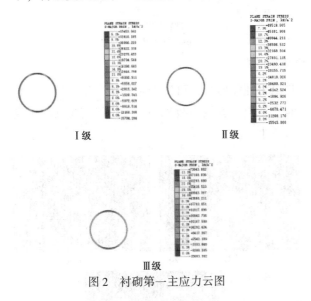

图 2　衬砌第一主应力云图

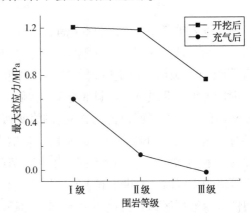

图 3　围岩最大拉应力曲线图

表 3　各关键点位移

	围岩等级	拱顶位移/mm	左侧拱腰位移/mm	拱底位移/mm	右侧拱腰位移/mm
开挖后	Ⅰ级	-4.84	5.30	4.89	-5.34
	Ⅱ级	-7.61	9.75	7.65	-9.78
	Ⅲ级	-15.83	49.38	15.53	-49.63
运行期间	Ⅰ级	-0.72	0.20	1.35	0.20
	Ⅱ级	-0.41	1.44	1.40	-1.47
	Ⅲ级	1.75	35.27	3.23	-35.41

选取储气库拱顶、拱底、左右两侧拱腰四个关键点，对其位移进行分析。统计在不同阶段不同围岩等级条件下各点的位移如表 3 所示，并绘制出各点位移与围岩等级的关系曲线图，如图 4 所示。可知：洞室开挖完成后，拱顶发生沉降，拱底向上隆起，左右两侧拱腰向内发生位移。随着围岩等级的降低，各点的位移呈先缓慢增加后显著增加的趋势，左右两侧拱腰的位移量始终大于拱顶、拱底的位移。围岩等级从Ⅰ级降低到Ⅱ级，拱顶、左侧拱腰、拱底、右侧拱腰的位移分别增加了 2.77mm、4.45mm、2.76mm、4.44mm；围岩等级从Ⅱ级降低到Ⅲ级，四点的位移分别增加了 8.22mm、39.63mm、7.88mm、39.85mm。与Ⅰ级、Ⅱ级围岩条件相比，Ⅲ级围岩条件下洞室的位移量显著大。当储气压力为 16MPa 时，洞室变形受到抑制。与开挖完成后相比，围岩等

级为Ⅰ级时,四点的位移分别减小了4.12mm、5.5mm、3.54mm、5.54mm;围岩等级为Ⅱ级时,四点的位移分别减小了7.2mm、8.31mm、6.25mm、8.31mm;围岩等级为Ⅲ级时,四点的位移分别减小了17.58mm、14.11mm、12.3mm、14.22mm。从开挖完成到储气压力为16MPa时,围岩等级为Ⅲ级的洞室发生的位移变化量最大。

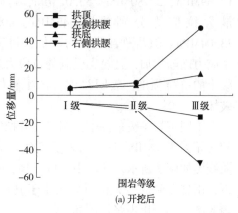
(a) 开挖后

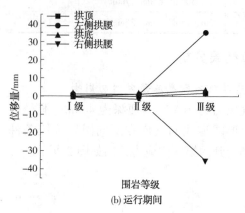
(b) 运行期间

图4 关键点位移曲线图

不同围岩等级条件下,洞室塑性区分布状况如图5所示。从图中可知:在洞室开挖完成后,围岩等级为Ⅰ级时,洞室围岩基本不产生塑性区;围岩等级为Ⅱ级时,在洞室的左右两侧产生部分塑性区,塑性区面积占比5%;围岩等级为Ⅲ级时,洞室左右两侧的塑性区面积显著增大,塑性区面积为18%。在16MPa压力运行期间,围岩等级为Ⅰ级时,在拱顶和拱底区域产生小部分塑性区,塑型区面积占比约为5.5%;围岩等级为Ⅱ级、Ⅲ级时,洞室产生的塑性区面积逐渐增大,塑性区面积占比分别为9%、20%。分析可知,洞室在开挖完成后,围岩等级越差,其自稳能力越弱,产生的塑性变形越大。储气压力为16MPa时,空气压力大于自重压力,导致在拱顶和拱底区域产生新的塑性区,Ⅲ级围岩产生的塑性区远大于Ⅰ级、Ⅱ级围岩。

2.2 埋深对储气库稳定性的影响

储气压力为16MPa、围岩等级为Ⅱ级、洞径为15m的圆形隧道式储气库在不同埋深条件下,衬砌受到的第一主应力云图如图6所示。可以看到,洞室在不同埋深下,衬砌受力状态分布相同,拉应力均分布在衬砌的顶部和底部区域;受到的最大拉应力分别为54.15MPa、53.97MPa、51.71MPa、49.52MPa、47.30MPa、44.86MPa、42.52MPa,远大于混凝土的抗拉强度,表明在衬砌的顶部和底部会产生张拉破坏。随着洞室埋深的增加,衬砌受到的最大拉应力逐渐减小,这是因为埋深增加,衬砌受到的围岩压力增大,与储气压力的差值逐渐减小,使得衬砌受到的拉应力减小。

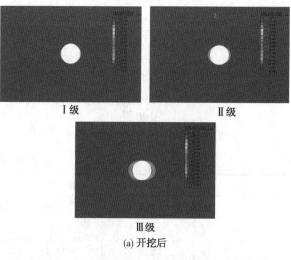
(a) 开挖后

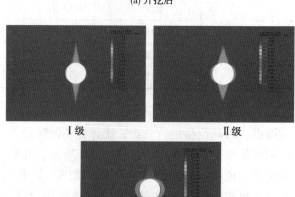
(b) 运行期间

图5 围岩塑性区云图

图7为不同埋深条件下,洞室开挖完成后和储气压力为16MPa时,围岩受到的最大拉应力曲线图。可以看到,埋深在100m至600m时,围岩受到的最大拉应力均小于其抗拉强度;埋深为700mm时,洞室开挖后围岩受到的最大拉应力为4.11MPa,大于其抗拉强度2MPa,表明此时的洞室围岩会出现张拉破坏。

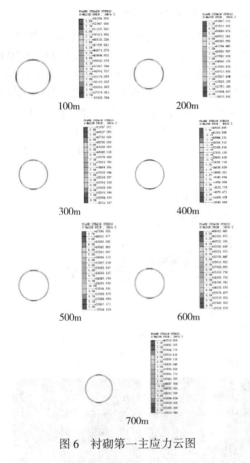

图6 衬砌第一主应力云图

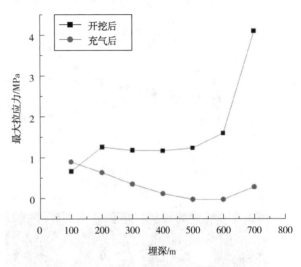

图7 围岩最大拉应力曲线图

表4 各关键点位移

	埋深/m	拱顶位移/mm	左侧拱腰位移/mm	拱底位移/mm	右侧拱腰位移/mm
开挖后	100	-1.40	0.25	2.23	-0.25
	200	-3.34	0.422	3.95	-0.402
	300	-5.42	3.76	5.76	-3.7
	400	-7.61	9.75	7.65	-9.78
	500	-9.91	19.1	9.63	-19.2
	600	-12.31	31.01	11.65	-31.01
	700	-14.89	44.3	13.35	-44.45
运行期间	100	8.45	-10.96	-5.63	10.97
	200	4.66	-9.16	-2.90	9.17
	300	2.05	-5.05	-0.69	5.04
	400	-0.41	1.44	1.40	-1.47
	500	-2.81	11.18	3.44	-11.21
	600	-5.28	23.39	5.51	-23.38
	700	-7.81	36.86	7.30	-37.05

在不同阶段不同埋深下,洞室各关键点的位移如表4所示,各关键点的位移与埋深的关系曲线图如图8所示。结合图表可知,随着洞室埋深的增加,拱顶的沉降量呈线性增大,拱底的隆起量呈线性增大,左右拱腰处向内发生的位移量呈指数增加。埋深在100m至400m范围时,拱顶、拱底的位移大于左右拱腰的位移;埋深大于400m时,拱顶、拱底的位移小于左右拱腰的位移。储气压力为16MPa时,洞室变形受到内压的作用,埋深从100m增加至300m时,拱顶向上拱起,拱底向下沉降,左右拱腰向外发生位移,各点位移量随埋深的增加逐渐减小。随着埋深的增加,围岩的自重应力不断增大,洞室埋深在400m时,自重应力与储气压力大致处于平衡状态,四点的位移分别为-0.41mm、1.44mm、1.40mm、-1.47mm。埋深从400m增加至700m时,自重应力大于储气压力,且压力差不断增大,使得各点向内发生的位移逐渐增加。

不同埋深条件下，洞室塑性区分布如图9所示，从图中可以看到，洞室从100m增加至700m，在开挖完成后，洞室围岩产生的塑性区面积占比分别为0、1%、4%、5%、7%、8.5%、10.5%；储气压力为16MPa时，洞室围岩产生的塑性区面积分别为32%、24%、10%、9%、10%、10%、11%；塑性区面积占比分别增加了32%、23%、6%、4%、3%、2.5%、0.5%。通过分析可知，洞室埋深越大，其所承受的自重应力越大，产生的变形量越大，导致在开挖完成后围岩产生的塑性区面积逐渐增大。洞室储气压力为16MPa时，当其埋深为100m、200m时，其所承受的自重应力远小于运行压力，导致围岩产生大量的塑性区；之后随埋深的增加，其所承受的自重应力由略小于储气压力增加至远大于储气压力，储气压力对围岩的作用影响减小，使得围岩产生的新的塑性区面积逐渐减小。

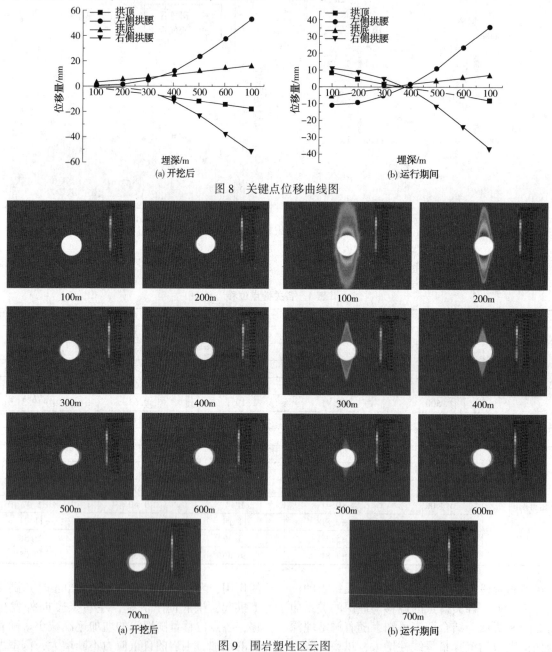

图8 关键点位移曲线图

图9 围岩塑性区云图

2.3 洞径对储气库稳定性的影响

储气压力为16MPa、围岩等级为Ⅱ级、埋深为400m的圆形隧道式储气库在不同洞径条件下，衬砌受到的第一主应力云图如图10所示。衬砌的顶部和底部区域受到拉应力作用，洞径从5m增加到45m时，衬砌受到的最大拉应力分别为30.13MPa、49.52MPa、58.80MPa、63.50MPa、65.58MPa。衬砌最大拉应力随洞径的增加而增大，且远大于混凝土的抗拉强度，表明在衬砌的顶部和底部区域会产生张拉破坏。

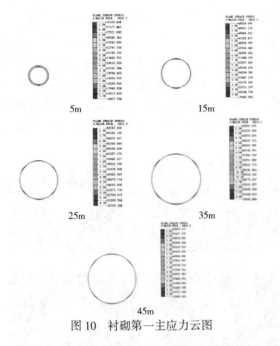

图 10 衬砌第一主应力云图

不同洞径条件下，洞室在开挖完成后和储气压力为 16MPa 时，围岩受到的最大拉应力曲线如图 11 所示。围岩受到的最大拉应力均小于其抗拉强度，表明洞室围岩不会出现张拉破坏。

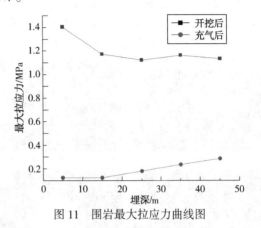

图 11 围岩最大拉应力曲线图

表 5 各关键点位移

	洞径/m	拱顶位移/mm	左侧拱腰位移/mm	拱底位移/mm	右侧拱腰位移/mm
开挖后	5	-2.72	1.82	2.73	-1.76
	15	-7.61	9.75	7.65	-9.78
	25	-12.48	18.47	12.49	-18.47
	35	-17.40	26.43	17.26	-26.43
	45	-22.39	34.55	21.94	-34.56
运行期间	5	-0.93	-0.14	1.01	0.19
	15	-0.41	1.44	1.40	-1.47
	25	0.85	3.43	1.92	-3.46
	35	2.64	5.10	2.71	-5.10
	45	4.97	7.41	3.65	-7.34

不同阶段不同洞径条件下，洞室各关键点的位移如表 5 所示，各关键点的位移与洞径的关系曲线图，如图 12 所示。在洞室开挖完成后，各关键点向洞室内部发生位移，且位移量随洞径的增加线性增大；左右拱腰处的位移始终大于拱顶、拱底的位移。储气压力为 16MPa 时，各关键点位移与开挖完成后相比，当洞径为 5m 时，拱顶、左侧拱腰、拱底、右侧拱腰的位移分别减小了 1.79mm、1.96mm、1.72mm、1.95mm；当洞径为 25m 时，四点的位移分别减小了 13.33mm、15.04mm、10.57mm、15.01mm；当洞径为 45m 时，四点的位移分别减小了 27.36mm、27.14mm、18.29mm、27.22mm。可以得到，洞径越大，从开挖完成到储气压力为 16MPa 时，洞室围岩产生的位移变化越大。

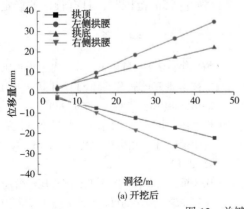
(a) 开挖后

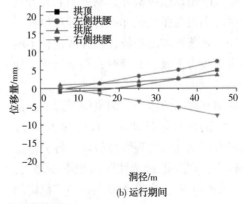
(b) 运行期间

图 12 关键点位移曲线图

不同洞径条件下,洞室塑性区分布云图如图13所示。在洞室开挖完成后,洞室的左右拱腰区域产生塑性区,洞径从5m增加到45m,洞室围岩塑性区面积占比分别为1.5%、5%、12%、17.5%、23%,呈逐渐增大的趋势。储气压力为16MPa时,在围岩拱顶和拱底处产生新的塑性区,产生的塑性区面积占比分别为2.5%、9%、17%、30%、37%,与洞室开挖后相比,塑性区面积增加了1%、4%、5%、12.5%、14%,洞径为35m、45m时,塑性区面积变化量远大于其它洞径。这是因为洞径越大,洞室越容易发生变形,从而产生的塑性区面积越大。由以上分析可知,当洞径在5m至25m范围时,洞室处于较稳定状态。

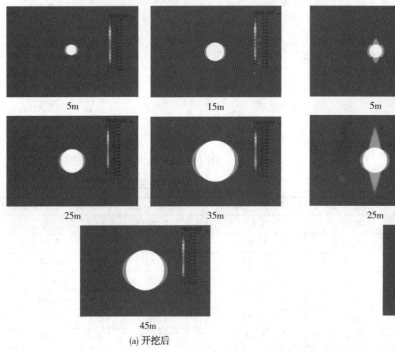
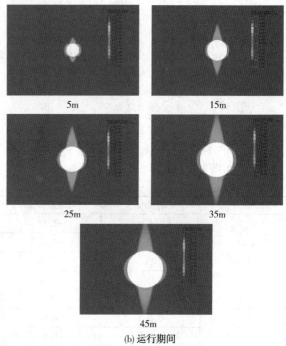

(a) 开挖后　　　　　　　　　　(b) 运行期间

图13　围岩塑性区云图

3 结论

通过利用MIDAS GTS NX软件建立二维圆形隧道式储气库模型,从衬砌第一主应力、洞室围岩最大拉应力、关键点位移以及塑性区的角度,研究了开挖后以及储气压力为16MPa时不同围岩等级、埋深、洞径条件对储气库洞室围岩稳定性的影响。得到以下结论:

(1)随围岩等级的降低,洞室中衬砌的顶部和底部在储气时出现张拉破坏;洞室围岩在开挖后和储气时均未出现张拉破坏;各关键点位移呈先缓慢增加后显著增加的趋势;塑性区面积不断增大。围岩等级为Ⅰ级、Ⅱ级时,洞室处于较稳定状态。

(2)随洞室埋深的增加,洞室中衬砌的顶部和底部在储气时出现张拉破坏;仅埋深为700m的洞室的围岩在储气时出现张拉破坏;各关键点位移不断减小;在开挖后塑性区面积不断增大,在储气时,塑性区面积先快速减小后趋于稳定的趋势。埋深为100m、200m时易出现塑性剪切破坏。

(3)随洞室洞径的增加,洞室中衬砌和洞室围岩均未出现张拉破坏;各关键点位移呈线性增加,塑性区面积不断增大。洞径为35m、45m时,洞室围岩变形量较大,产生的塑性区面积较大,易发生剪切破坏。

(4)根据计算结果,储气库储气压力为16MPa时,布置在围岩等级为Ⅰ~Ⅱ级,埋深为300~600m,洞径为5~25m范围时,具有较好的稳定性。

参 考 文 献

[1] 梅生伟,李瑞,陈来军,等. 先进绝热压缩空气储能技术研究进展及展望[J]. 中国电机工程学报, 2018, 38(10): 2893-2907+3140.

[2] 梅生伟,薛小代,陈来军. 压缩空气储能技术及其应用探讨[J]. 南方电网技术, 2016, 10(03): 11-15+31+3.

[3] 徐新桥,杨春和,李银平. 国外压气蓄能发电技术及其在湖北应用的可行性研究[J]. 岩石力学与工程

学报，2006(S2)：3987-3992.

[4] 郭丁彰，尹钊，周学志，等．压缩空气储能系统储气装置研究现状与发展趋势[J]．储能科学与技术，2021，10(05)：1486-1493.

[5] 张筱萍，王金昌，杨森，等．内衬岩洞储气库（LRC）应用前景分析[J]．青岛理工大学学报，2009，30(06)：107-111.

[6] 付兴，袁光杰，金根泰，等．盐穴压缩空气储能库建设现状及工程难点分析[J]．中国井矿盐，2017，48(03)：14-18.

[7] 王富强，王汉斌，武明鑫，等．压缩空气储能技术与发展[J]．水力发电，2022，48(11)：10-15.

[8] Zimmels Y, Kirzhner F, Krasovitski B. Design Criteria for Compressed Air Storage in Hard Rock[J]. Energy & Environment, 2002, 13(6).

[9] Perazzelli P, Anagnostou G. Design issues for compressed air energy storage in sealed underground cavities [J]. Journal of Rock Mechanics and Geotechnical Engineering, 2016, 8(03): 314-328.

[10] Jongpradist P, Tunsakul J, Kongkitkul W, et al. High internal pressure induced fracture patterns in rock masses surrounding caverns: Experimental study using physical model tests [J]. Engineering Geology, 2015, 197.

[11] 王其宽，张彬，王汉勋，等．内衬式高压储气库群布局参数优化及稳定性分析[J]．工程地质学报，2020，28(05)：1123-1131.

[12] 夏才初，张平阳，周舒威，等．大规模压气储能洞室稳定性和洞周应变分析[J]．岩土力学，2014，35(05)：1391-1398.

[13] 夏才初，赵海斌，梅松华，等．埋深对压气储能内衬洞室稳定性影响的定量分析[J]．绍兴文理学院学报(自然科学)，2016，36(03)：1-7.

[14] 周舒威，夏才初，张平阳，等．地下压气储能圆形内衬洞室内压和温度引起应力计算[J]．岩土工程学报，2014，36(11)：2025-2035.

[15] 周瑜，夏才初，周舒威，等．压气储能内衬洞室高分子密封层的气密与力学特性[J]．岩石力学与工程学报，2018，37(12)：2685-2696.

[16] 朱荣强，丁浩，江星宏，等．地下储气库洞室结构形式对围岩稳定性的影响[J]．油气储运，2022，41(09)：1052-1060.

[17] 蒋中明，秦双专，唐栋．压气储能地下储气库围岩累积损伤特性数值研究[J]．岩土工程学报，2020，42(02)：230-238.

[18] 蒋中明，唐栋，李鹏，等．压气储能地下储气库选型选址研究[J]．南方能源建设，2019，6(03)：6-16.

[19] 蒋中明，李小刚，万发，等．压气储能遂昌地下储气库结构应力变形特性数值研究[J]．长沙理工大学学报(自然科学版)，2021，18(03)：79-86.

[20] 蒋中明，黄毓成，刘澜婷，等．平江浅埋地下储气实验库力学响应数值分析[J]．水利水电科技进展，2019，39(06)：37-43.

基于 PSO-SVR 和熵权-TOPSIS 的双金属复合管焊接残余应力预测及优化

彭星煜　蒋海洋　冯梁俊　祝星语

（西南石油大学石油与天然气工程学院）

摘　要　【目的】金属管道焊接过程中难免会产生残余应力，故需要对其焊接工艺参数进行优化，但油气领域现有研究中，采用优化模型算法与实验相结合的研究相对较少，且未对优化结果进行进一步的评价筛选，有必要对其预测及优化算法进行升级改进。【方法】为了得到 Incoloy825/L360QS 复合管的最优焊接参数，首先采用 SYSWELD 仿真软件建立模型并设计正交试验，通过数值仿真获取样本数据。其次，采用粒子群优化支持向量回归算法（PSO-SVR）预测模型以及多目标灰狼算法（MOGWO）对其焊接残余应力及变形量进行预测优化，在此基础上进一步引入熵权-TOPSIS 综合评价法，对优化结果进行进一步评分筛选，获得其最优焊接参数。最后，采用最优焊接参数进行焊接获取试验样本，并通过一系列的试验，对优化后复合管的力学性能及耐蚀性能进行验证。【结果】最终结果表明：①相较于常用的 GA-BP 预测模型，PSO-SVR 预测模型的预测准确性及计算效率更优；②在 MOGWO 优化结果的基础上，进一步引入熵权-TOPSIS 综合评价法能准确获得一组最优焊接参数，相较于优化前，该参数下残余应力及变形量分别有效降低了 7.42% 以及 7.14%；③通过实验验证，优化后的双金属复合管具有更优异的力学性能以及耐蚀性能。【结论】该预测及优化模型具有可观的准确性及计算效率，获得的最优焊接参数能有效降低焊接残余应力及变形量，研究成果可为双金属复合管焊接参数的优化提供更多理论支撑。

关键词　双金属复合管；残余应力；焊接参数；PSO-SVR；MOGWO

Prediction and optimization of welding residual stress in bimetallic composite pipes based on PSO-SVR and entropy weight TOPSIS

Peng Xingyu　Jiang Haiyang　Feng Liangjun　Zhu Xingyu

(College of Petroleum and Natural Gas Engineering, Southwest Petroleum University)

Abstract　[Objective] During the welding process of metal pipelines, residual stresses are inevitably generated, necessitating the optimization of welding process parameters. However, existing research in the oil and gas sector has relatively few studies that integrate optimization model algorithms with experimental investigations, and the optimization results have not been thoroughly evaluated and refined. Therefore, it is essential to enhance and improve the prediction and optimization algorithms. [Methods] In order to determine the optimal welding parameters for Incoloy 825/L360QS composite pipes, a model was first established using SYSWARD simulation software, and orthogonal experiments were designed. Sample data were obtained through numerical simulation. Subsequently, the particle swarm optimization support vector regression algorithm (PSO-SVR) prediction model and multi-objective grey wolf algorithm (MOGWO) were employed to predict and optimize the welding residual stress and deformation. Building upon this, the entropy weight TOPSIS comprehensive evaluation method was introduced to score and select the optimization results, thereby identifying the optimal welding parameters. Finally, the optimal welding parameters were utilized in welding, and a series of experiments were conducted to verify the mechanical and corrosion resistance properties of the optimized composite pipe. [Results] The following results are obtained. Firstly, in comparison to the commonly utilized GA-BP prediction model, the PSO-SVR prediction model demonstrates superior prediction accuracy and computational efficiency. Secondly, based on the optimization results of MOGWO, the entropy weight TOPSIS comprehensive evaluation method is subsequently introduced to accurately obtain a set of optimal welding parameters. Compared to the pre-optimization results, the residual stress and deformation under these parameters are effectively reduced by 7.42% and 7.14%, respectively. Thirdly, through experimental verification, the optimized bimetallic composite pipe has better mechanical properties and corrosion resistance. [Conclusion] The prediction and optimization model demonstrates significant accuracy and computational efficiency, and the optimal welding parameters obtained can effectively reduce residual stress and deformation in welding. The findings of this research can offer substantial theoretical support for the optimization of welding parameters in bimetallic composite pipes.

Key words　Bimetallic composite tube；Residual stress；Weld parameter；PSO-SVR；MOGWO

在中国深入开发高含硫气田的背景下，Incoloy825/L360QS复合管由于其优良的综合性能，在高腐蚀性油气田的应用中得到了广泛应用。然而在硫化氢环境下，应力腐蚀开裂形成裂纹时的阈值应力较低，双金属复合管在裂纹出现前通常不显现变形，导致许多事故在无预警的情况下突发，进而影响管道的可靠性和安全性。为了进一步降低焊接残余应力，提高双金属复合管的质量，常对其结构参数以及焊接参数进行优化研究。目前，针对双金属复合管残余应力的研究往往都是通过传统实验方式进行的，这无疑会消耗大量的人力物力资源。因此，如何建立一种具有精确性以及泛用性的残余应力预测模型已经成为油气领域的重点研究内容。

随着计算机科技和机器学习理论的快速发展，焊接领域的研究趋势逐渐从传统的有限元模拟和实验验证转向了机器学习方法的应用。2020年王祺等以立管用X65管线钢作为研究对象，采用生死单元法模拟焊料填料填充过程，并建立Kriging近似模型对残余应力进行预测。2022年Zhang等采用PSO-SVR机器学习模型，对交通载荷作用下X65埋地管道的轴向应力进行预测，探讨了车辆质量、埋深等五个因素对管道应力的影响。2023年Rissaki D. K以奥氏体不锈钢管环焊缝为研究对象，开发了两个人工神经网络（ANN）集成模型，对其厚度方向的轴向及环向残余应力进行预测。2023年Peng等提出一种将BP神经网络与连续球压痕法相结合的预测模型，在局部位置可非破坏性地获得无应力压痕曲线，并通过有限元分析及实验对其进行验证。2024年李成文等提出一种GA-BP预测模型，对316L平板对接焊的残余应力及变形进行预测，研究表明，GA-BP模型的预测误差小于3%，相较于BP模型预测更为精确。以上学者虽然对残余应力的预测进行了相关研究，但对于预测结果大多都只能得到分布范围或优化解的集合而不能确定一种最优方案，且针对复合管焊接参数对残余应力影响的研究较少。研究结果发现，在焊接残余应力预测结果的基础上，引入机器优化算法以及综合评价法，能大幅提高优化效率并准确获得最优优化结果。

对于上述问题，笔者考虑传统神经网络和支持向量回归算法的局限性，以Incoloy825/L360QS复合管为研究对象，利用SYSWELD仿真软件获得仿真数据，采用粒子群算法优化支持向量回归（PSO-SVR）作为预测焊接残余应力的工具，获得残余应力预测结果，并通过多目标灰狼算法得到一组优化方案，在此基础上进一步引入熵权-TOPSIS综合评价法对模拟结果进行优化，从而解决机器学习算法只能获得优化解集的问题，得到复合管的最优焊接工艺参数，为Incoloy825/L360QS复合管的焊接残余应力的预测及优化提供建议和理论依据。

1 复合管有限元模型的建立

1.1 实验材料

本文以 Incoloy 825/L360QS 机械复合管为研究对象，外基管材料为 L360QS 碳钢，内衬管材料为 Incoloy825 镍基合金，试件几何尺寸为 φ100×(7+3)mm，坡口角度 60°，采用 TIG 焊，填充金属为 ERNiCrMo-3 焊丝，焊接接头示意图如图 1，其焊接工艺参数范围见表 1。

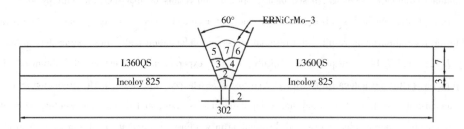

图 1 焊接接头示意图

复合管焊接接头的焊接工艺选择为钨极氩弧焊，具体的工艺参数如表 1 所示。此外，焊接时选用对接焊接接头，采用 V 形坡口，坡口角度为 60°，坡口间隙为 2mm，接头的示意图如图 1 所示。复合管的热输入采用式(1)进行计算。

$$Q = \eta UI/v \quad (1)$$

式中，Q 为焊接热输入；η 为焊接热效率，取 0.7；U 为电弧电压；I 为焊接电流；v 为焊接速度。

表 1 焊接工艺参数

焊道	焊接电流/A	焊接电压/V	焊接速度/(mm/s)	层间温度/℃
打底	110	10	1.25	200
过渡	125	15	1.25	200
填充	125	15	1.25	200
盖面	125	15	1.25	200

焊接过程涉及利用热源迅速升温后快速冷却的局部区域，为了保证仿真结果的准确性，需要对材料的热力学性能进行精确定义(图 2)。

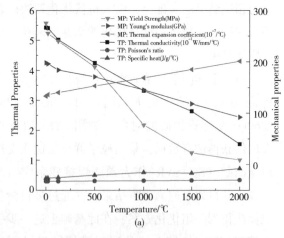
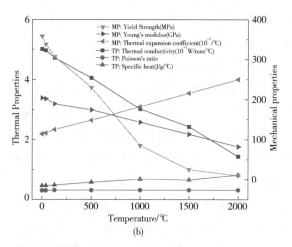

图 2 Incoloy825 (a)以及 L360QS (b)的热力学性能

1.2 有限元模型的建立

使用前处理软件 Visual Mesh 对 Incoloy 825/L360QS 复合管焊接接头进行建模，其中模型尺寸定为 φ300×(7+3)mm。复合管在模拟软件中的 X 和 Y 轴具有对称分布的特性，故只需建立 1/4 的复合管模型即可达到对应效果。在网格的划分上，考虑到焊接过程中影响范围主要集中在焊缝和热影响区附近，因此在这两块区域上网格划分细密，而影响较小的两端划分稀疏。在模拟软件中，通常采用确定三个点限制不同方向上的位移约束来确保焊件在焊接过程中不产生移动，因此为防止 Incoloy 825/L360QS 复合管模型的位移，本文设置了三个点，分别显示其 XYZ、XY 和 Y 方向上的位移。

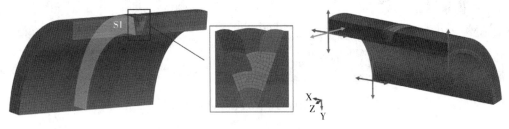

图 3 有限元网格及约束条件

为确保研究结果的独立性于计算网格的影响，本研究对有限元模型执行了网格无关性验证。采用的初步几何模型经过三级网格密度划分为粗网格、中密度网格及细网格，它们的网格单元数分别为 125687、210745 和 305826。通过对比在不同网格密度下获得的焊缝基层和衬层的等效应力，探究网格细化程度对仿真准确性的影响，计算结果如图 4 所示。

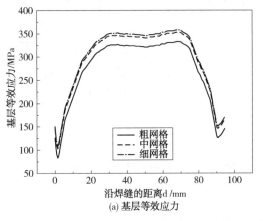
(a) 基层等效应力

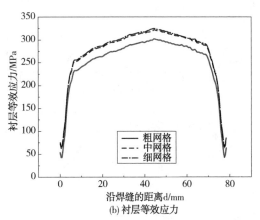
(b) 衬层等效应力

图 4 Incoloy 825/L360QS 复合管模型网格无关性验证

结果表明，从粗网格过渡到中等密度网格，焊接残余应力的估算值出现了明显的变化。然而，从中等密度网格细化到细网格时，变化率降至小于 3%，表明已经达到了网格无关性的要求。对于本研究的目标和几何条件，中等密度网格提供了足够的精度，同时保持了计算效率。因此，本研究选择采用 210745 个网格单元的模型进行残余应力的仿真模拟。

1.3 残余应力实验验证

依据 GB/T 31310—2014 的相关规定，使用 ASM3.0H 型应力应变仪，通过盲孔法检验实际焊接后残余应力大小，而对于复合管内部及衬层残余应力盲孔法无法直接测量，故采用超声波法测量，将实验结果与模拟结果进行对比，即可验证 SYSWELD 模拟的效度。其中，盲孔法的计算公式如下：

$$\sigma_1 = \frac{E}{4A}(\varepsilon_1 + \varepsilon_2) - \frac{\sqrt{2}E}{4B}\sqrt{(\varepsilon_1 - \varepsilon_2)^2 + (\varepsilon_3 - \varepsilon_2)^2} \tag{2}$$

$$\sigma_2 = \frac{E}{4A}(\varepsilon_1 + \varepsilon_2) - \frac{\sqrt{2}E}{4B}\sqrt{(\varepsilon_1 - \varepsilon_2)^2 + (\varepsilon_3 - \varepsilon_2)^2} \tag{3}$$

式中，$\sigma_{1,2}$ 为轴向、环向残余应力；E 为杨氏模量；A、B 为应变释放系数，采用出厂标定默认值，A 取 -0.072，B 取 -0.151；ε_1、ε_2、ε_3 为释放的应变。

图 5 为残余应力测量过程及应变片分布示意

图；图6为试验轴向应力及环向应力与模拟值的对比结果。经过中央截面和焊缝侧上基层和衬层的实测值与模拟值的对比发现，两者之间具有高度的一致性，平均误差小于5%，从而验证了所建立的模型的准确性和可靠性。

图5 残余应力检测过程及测量点分布

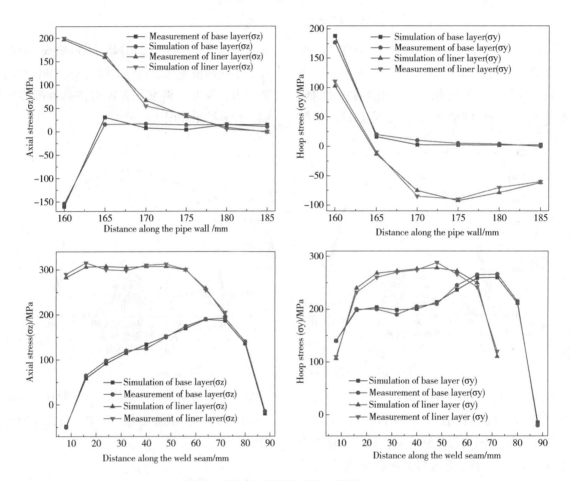

图6 检测值与模拟值的应力结果对比

2 双金属复合管残余应力的预测及优化

2.1 试验数据

因机器学习需要的样本较多，考虑到试验费用问题，无法通过试验获得，故利用数值仿真方法获取足够的样本。打底焊的焊接电流和电压与其他焊层不同，因此确定试验的自变量为打底焊电流、打底焊电压、其他焊电流、其他焊电压、焊接速度和层间温度这6个变量。在这些变量中，电压考虑三个水平，而其余因素各有五个水平。打底焊电压设定在10~12V范围内，其他焊电压设定在13~15V范围内；打底焊电流则介于90~110A，其他焊电流在120~140A；焊接速度设定在1~2mm/s；层间温度设置为100~300℃。为丰富样本数据，试验涵盖了72组数据情况，实验方案及最大残余应力和变形量的仿真结果如表2所示。

表 2 实验方案及仿真结果

试验编号	打底焊电流/A	打底焊电压/V	其他焊电流/A	其他焊电压/V	焊接速度/(mm/s)	层间温度/℃	最大残余应力/MPa	变形量/mm
1	105	10	125	14	1.75	250	484.58	2.81
2	100	11	130	15	2	300	531.78	3.05
3	110	12	125	14	1	150	453.18	1.23
4	95	10	135	14	1.75	100	485.51	1.76
5	100	11	130	13	2	100	474.81	2.01
6	110	11	130	15	2	150	472.27	1.52
7	90	11	140	15	1.25	150	518.75	1.92
8	100	10	140	13	1	250	478.15	1.79
9	95	10	140	14	1.5	150	491.18	1.46
10	90	12	120	14	1.75	300	556.13	3.21
…	…	…	…	…	…	…	…	…
61	110	12	130	15	1.5	100	461.63	1.28
62	105	11	135	15	1	150	468.93	1.86
63	95	11	140	14	1	250	510.36	1.86
64	100	11	120	14	1	150	473.95	1.43
65	110	11	140	14	2	200	474.3	1.79
66	100	10	125	14	1.5	300	504.76	2.31
67	110	12	140	13	1.75	150	459.73	1.95
68	90	11	120	15	1.5	250	524.26	2.45
69	105	10	135	14	1.5	300	490.68	3.21
70	105	10	135	13	1.25	300	477.38	2.84
71	90	10	120	13	1.75	100	490.51	1.83
72	90	11	130	13	1.25	100	490.7	0.88

2.2 PSO-SVR 预测残余应力

在表 2 样本数据的基础上，采用 PSO-SVR 模型对焊接残余应力及变形量进行预测，图 7 为 PSO-SVR 模型的流程图。在支持向量回归（SVR）模型的构建过程中，指出了其中的关键未知变量惩罚参数 C 的作用，尤其在处理非线性问题时显得尤为关键。除了惩罚参数 C 之外，核函数的选取及其参数的设定也在模型的性能优化中扮演了重要角色，因此确定这两者的关键参数，对 SVR 模型的准确度至关重要。此外，由于焊接残余应力的研究是非线性问题，故后续采用 RBF 核函数进行计算。

将 PSO 算法初始化，设置种群和迭代次数为 100，学习因子 Ⅰ 取 1.5，学习因子 Ⅱ 取 2.5，惯性权重取 1，确立 c 和 g 为调优变量。利用 PSO 的快速收敛性，初步确定参数范围为 [2^{-10}, 2^{10}]，搜索步长为 0.05，通过网格搜索进行精细调优。

（1）最大 Von-mises 残余应力

图 8 为残余应力网格搜索结果与预测模型适

应度曲线示意图。经过小步长搜索后,最终的取值为 $c=2$, $g=0.125$,将得到的 $(c,g)_{best}$ 带入到支持向量机函数中,以此建立模型。将 $(c,g)_{best}$ 代入支持向量机后,通过构建优化的 SVR 模型,采用五折交叉法进行模型验证,并对预测集进行测试。由 PSO-SVR 优化过程的适应度变化曲线图可以看出,当迭代至第 36 代时,支持向量机的适应度不再下降,达到最优值。

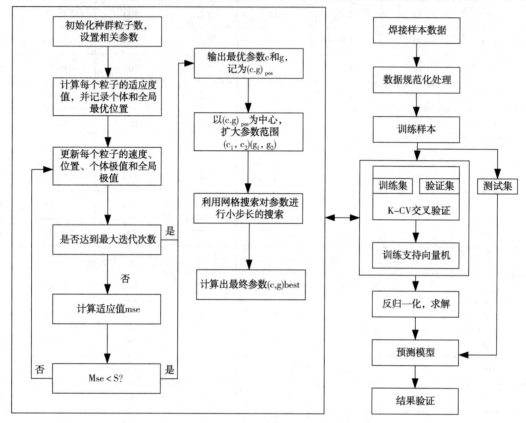

图 7　PSO-SVR 模型流程图

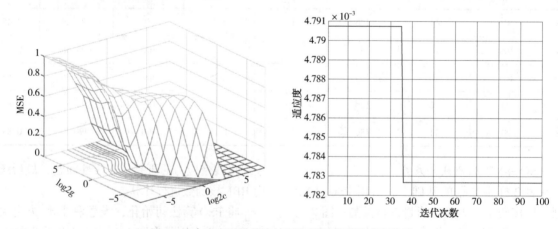

图 8　残余应力网格搜索结果与预测模型适应度曲线

（2）最大变形量

图 9 为变形量的网格搜索结果与预测模型适应度曲线示意图。经过小步长搜索后,最终的取值为 $c=1$, $g=0.25$,以此建立预测模型。同样运用五折交叉法对训练数据集进行验证,最后将预测样本代入算法中进行检验。由 PSO-SVR 预测时的变形量适应度曲线图可得,当迭代次数为 19 代后,适应度趋于稳定不再下降,此时粒子群优化支持向量机的参数达到最佳。

将通过 MATLAB 仿真实验的 PSO-SVR 预测

模型对样本进行焊接残余应力的预测，并将其预测结果与 GA-BP 预测模型进行对比，二者的对比结果如下所示。

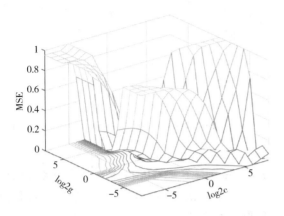 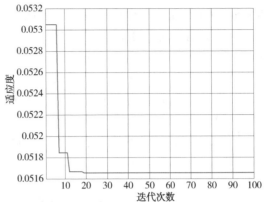

图 9 变形量的网格搜索结果与预测模型适应度曲线

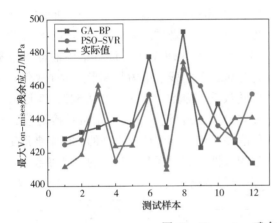 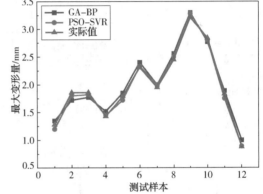

图 10 Von-mises 残余应力及最大变形量对比

表 3 模型计算时间对比

计算时间/s	残余应力/s	变形/s	平均/s
GA-BP	24.96	25.11	25.035
PSO-SVR	23.6	22.9	23.25

对于 Von-mises 残余应力，GA-BP 的差值最大为 27.19MPa，最小为 12.51MPa，平均为 19.36MPa。PSO-SVR 的差值最大为 19.32MPa，最小为 0.24MPa，平均为 9.16MPa。从预测误差百分比上来看，GA-BP 的最大误差为 6.20%，最小为 3.24%，平均为 4.44%；PSO-SVR 的最大误差为 4.38%，最小为 0.05%，平均为 2.12%；预测准确度上，PSO-SVR 相较于 GA-BP 性能提升了 52.25%。

对于最大变形量，PSO-SVR 模型预测的值与实际值最大差值为 0.08mm，而 GA-BP 的差值为 0.14mm，数据的波动上来看 PSO-SVR 模型要优于 GA-BP。而预测误差对比分析发现，PSO-SVR 综合误差均小于 GA-BP，最小值仅为 0.87%，平均为 2.63%。因此，在变形量的预测上 PSO-SVR 模型有着较高的准确度。

此外，对比两者的平均计算时间，PSO-SVR 耗时 23.25s，GA-BP 耗时 25.035s，PSO-SVR 在计算时间上性能提升 7.1%。综上所述，PSO-SVR 预测模型无论在预测准确性，还是在计算效率上均优于 GA-BP。

2.3 基于多目标灰狼算法的残余应力优化

在 PSO-SVR 模型预测结果的基础上，本文采用 MOGWO 算法对预测数据进行优化，在固定的种群大小下通过迭代搜索策略,搜寻特定的最小化函数值,图11为MOGWO焊接工艺参数优化流程图。将灰狼优化算法应用于Incoloy825/L360QS复合管焊接过程的参数优化分析时,考虑到需要优化的n组工艺参数,每一组由六个变量构成:两个焊接电压(U_1、U_2)、两个焊接电流(I_1、I_2)、焊接速度(S)及层间温度(T),形成了一个6n维的优化问题,其优化目标向量$X_P = \{U_{1i}, U_{2i}, I_{1i}, I_{2i}, S_i, T\}$,$i=1, 2, \cdots, n$定义为寻找最优解的灰狼个体所追求的目标。

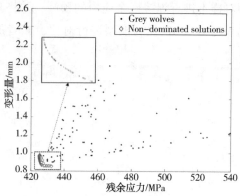

图12 残余应力和变形的寻优结果

2.4 基于熵权-TOPSIS算法的残余应力优化

为了解决MOGWO算法无法确定最优解的问题,本文选用熵权-TOPSIS模型结合了熵权法确定评价指标权重和TOPSIS技术对方案进行排序,对MOGWO的结果进行进一步的优化,以提供科学且客观的决策支持,具体步骤流程如图13所示。对MOGWO最优解集中的样本数据执行分析,确定残余应力与变形这两个关键性能指标的权重(残余应力权重为0.542,变形权重为0.458),通过此方法,计算相对接近度,再进行降序排列,其中评分最高的数据组被选定为最佳焊接性能的指标及其相应的焊接参数。焊接性能指标的综合评分降序排名(前10)见表4。

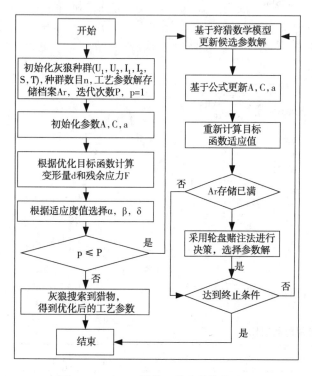

图11 MOGWO焊接工艺参数优化流程

在执行灰狼优化算法过程中,参数设置包括:种群规模100,最大迭代次数100,处理参数的维度为6,引入10个网格点来提高搜索精度,同时设定初始和调节更新系数为2。通过执行100轮迭代计算,可获得优化结果数据集(图12),进一步对该数据集进行反归一化处理,即可得到关于残余应力与变形的Pareto前沿最优解集。

由图12可明显看出在寻求变形和残余应力的优化过程中,并未获得两者同时达到最小化的方案,随着残余应力的增加,变形值呈现轻微下降趋势。因此,在选择工艺参数时还需综合考虑解的优劣进行评估。

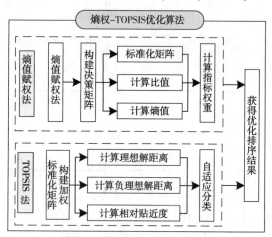

图13 熵权-TOPSIS流程图

由表4可得,焊接性能指标(残余应力和变形)综合评分最高的是第91组数据,其中最大残余应力为429.9992MPa,最大变形为0.776323mm,即打底焊电流为93.159A,打底焊电压为10.251V,其它焊层电流为124.664A,其它焊层电压为13.037V,焊接速度为1.00mm/s,层间温度为100.08℃为Incoloy825/L360QS复合管焊接过程中最优参数。

表4 熵权-TOPSIS 评价计算结果

编号	打底焊电流/A	打底焊电压/V	其它焊电流/A	其它焊电压/V	焊接速度/(mm/s)	层间温度/T	残余应力/MPa	变形/mm	相对接近度	排序
91	93.159	10.251	124.664	13.037	1.00	100.08	430.9992	0.776323	0.68247	1
36	103.316	10.659	126.148	13.009	1.03	100.96	430.6737	0.784157	0.674147	2
35	93.713	10.754	126.921	13.053	1.02	100	428.5922	0.821888	0.670389	3
71	101.225	10.203	125.499	13.051	1.03	100	431.4834	0.776323	0.660553	4
76	102.897	10.138	125.660	13.088	1.00	100	431.7077	0.782497	0.653291	5
31	93.464	10.716	126.749	13.041	1.00	100	429.3221	0.827566	0.650905	6
60	101.034	10.821	125.014	13.088	1.05	100	432.0339	0.776323	0.648705	7
30	93.359	10	126.642	13.015	1.03	100	429.628	0.827568	0.645191	8
4	107.850	10	123.224	13.038	1.01	100	428.2447	0.839883	0.642144	9
97	104.452	10.049	125.359	13.043	1.01	100	432.3405	0.776323	0.636317	10

3 优化结果的验证

3.1 数值模拟验证

以前文得出的复合管最优焊接工艺参数，带入 SYSWELD 仿真软件进行模拟计算，即可得到优化后 Incoloy825/L360QS 双金属复合管的 Von-Mises 残余应力云图(图14)以及整体焊接变形云图(图15)。分析可得，基层和衬层的高应力范围对比优化前明显减小，横截面过渡层附近的焊接残余应力有所降低，双金属复合管整体的最大 Von-Mises 残余应力从优化前 464.45MPa 减小到 429.99MPa，有效降低了 7.42%；整体变形量从 0.84mm 减小到 0.78mm，有效降低了 7.14%。

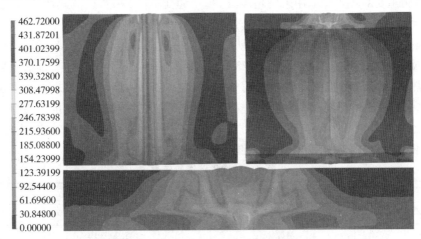

图14 优化后的 Von-Mises 残余应力云图

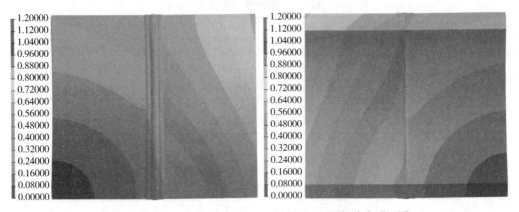

图15 优化后复合管基层(左)及衬层(右)的焊接变形云图

表 8 实际焊接结果对比

参数	打底焊电流/A	打底焊电压/V	其它焊电流/A	其它焊电压/V	焊接速度/(mm/s)	层间温度/T	残余应力/MPa	变形/mm
优化前	110	10	125	15	1	200	460.14	0.88
优化后	93.159	10.251	124.664	13.037	1.009	100.08	431.21	0.82
优化率	/	/	/	/	/	/	6.29%	6.82%

3.2 力学性能验证

以优化后的工艺参数制备双金属复合管(图16),并对其力学性能进行检测。

图 16 焊接试验样品

(1) 无损检测

对于无损检测主要对环焊缝采用采用超声波检测以及磁粉检测。进行超声波检测时,选用 MS380 型超声波探伤仪并搭配 5P 9×9 K2 探头,探伤灵敏度为 N5 刻槽+6dB,依据 SY/T 6423-2013 对焊缝及热影响区进行检测;进行磁粉检测时,使用 CJZ-212E 磁粉探伤仪和磁轭法进行磁化,确保磁化强度不低于 45N,表面照度 ≥ 1000μW/cm²,依据 ASTM E709-2021 进行检测。

通过上述实验可以得出,优化前后样品均未产生缺陷。

(2) 工艺评价

从环焊缝上获取实验试样,依据相关试验标准,分别对其开展拉伸、刻槽锤断、面弯背弯试验,对应的试验结果如表6至表8所示。结果表明,优化后的试样均满足相关标准的要求,且在抗拉强度及断裂载荷的表现均优于优化前的试样。

表 6 拉伸性能试验结果

编号	平均抗拉强度/MPa	平均断裂载荷/kN	断裂位置
优化前	629.08	38.00	热影响区
优化后	741.62	45.90	热影响区

表 7 刻槽锤断试验结果

编号	试验结果	标准要求
优化前	未发现超标缺陷	每个刻槽锋试样的焊透和熔合完整,气孔最大直径不超过 1.6mm,总面积不超 2%;夹渣高度不超 0.8mm,长度限制在钢管公称壁厚的一半且不超过 3mm,夹渣间距至少为 13mm。
优化后	未发现超标缺陷	

表 8 弯曲试验结果

编号	试验结果	GB/T 31032—2014 标准要求
优化前	未发现裂纹	拉伸弯曲测试中,焊缝及熔合区域的缺陷尺寸限制在公称壁厚的一半以内,最大不超过 3.0mm。
优化后	未发现裂纹	

(3) 夏比冲击

采用 ZBC2752-B 冲击试验机，依据相关标准对焊缝及热影响区的试样进行冲击韧性的测试，结果如表9所示。分析可得，检测试样均具有较高的冲击吸收能，具有良好的冲击韧性，将优化前后的冲击韧性进行对比不难看出，优化后焊缝及热影响区的冲击韧性明显高于优化前。

表9 夏比冲击试验结果

编号	焊缝平均冲击吸收能/J	热影响区平均吸收能/J
优化前	146.0	126.5
优化后	178.5	158.5

(4) 硬度分布

考虑到 Incoloy825/L360QS 复合管焊接接头在特定环境下的应力腐蚀问题，焊缝硬度的控制至关重要，在酸性高盐工况下，推荐其硬度值应控制在不超过 345HV，同时镍基合金焊接后热影响区的硬度应不超过母材和熔覆金属的最大硬度。

取焊接接头试样，依据 GB/T 4340.1—2009 开展硬度测试，依据实际测量硬度值并绘制硬度云图。试验中发现。图18分别为优化前和优化后工艺下的焊接接头硬度云图，分析表明，不同工艺参数下焊接接头的硬度分布呈现相似的梯度分布模式，并且，优化前后获得的焊接接头的硬度值均低于 345HV，均符合相关标准要求。

图17 夏比冲击试验

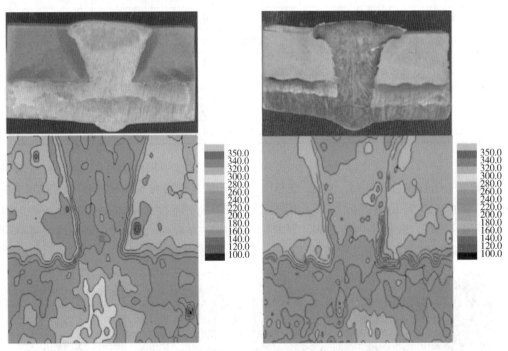

图18 优化前(左)及优化后(后)的环焊缝硬度分布云图

3.3 耐蚀性能的验证

Incoloy825/L360QS 复合管服役于酸性环境下，对管道具有一定的腐蚀，故需要对优化后的环焊缝进行耐蚀性评价，本研究主要从晶间腐蚀、点腐蚀、均匀腐蚀以及抗硫化物应力开裂四个方面检测其耐蚀性能。

(1) 晶间腐蚀

图19(a)为晶间腐蚀过程示意图。取焊接接头样本进行晶间腐蚀试验，按 ASTM G28 A法，在标准硫酸铁-硫酸溶液中微沸状态下腐蚀 120h 后进行称重和腐蚀速率计算。试验前后的形貌如图19(b)以及图19(c)所示，晶间腐蚀试验结果

如表13所示，通过计算可得，焊接工艺优化前后下的晶间腐蚀平均腐蚀速率分别为0.46mm/a、0.45mm/a，2种双金属复合管焊接接头的晶间腐蚀速率均小于0.5mm/a，故耐晶间腐蚀能力满足工艺要求，此外工艺参数优化后的焊接接头晶间腐蚀速率略低于优化前，但两者相差不大。

(a)

(b)

(c)

图19 晶间腐蚀过程及腐蚀前后形貌对比图

表13 晶间腐蚀试验结果

试片编号	试样前重/g	试样后重/g	长/cm	宽/cm	厚/cm	表面积/cm²	腐蚀速率/(mm/a)	平均值/(mm/a)
优化前	7.7695	7.7062	4.996	0.996	0.188	12.2050	0.47	0.46
	7.6492	7.5863	4.998	0.995	0.187	12.1874	0.44	
	7.8332	7.7621	5.002	0.997	0.189	12.2416	0.48	
优化后	7.9088	7.8465	5.004	0.99	0.192	12.2096	0.45	0.45
	7.9084	7.8450	4.996	0.992	0.191	12.1995	0.42	
	7.9125	7.8462	5.002	0.996	0.191	12.2552	0.48	

（2）点腐蚀

进行焊接接头的点腐蚀试验时，遵循ASTM G48 A标准，于22±2℃环境中进行了72小时的测试。试验完毕后通过20倍显微镜观察，试样表面没有出现点腐蚀迹象。试验过程如图20(a)所示，试验结果见表11。由试验结果可知，优化前后的焊接接头的试验样本，点腐蚀的平均腐蚀速率分别为0.41g/m²、0.31g/m²，均小于标准规定的4g/m²，且优化后的腐蚀速率有效降低了24.39%。

(a) 实验前

(b) 实验后

(c) 形貌对比

图20 点腐蚀试验过程

表 11 点腐蚀试验结果

试片编号	试样前重/g	试样后重/g	失重/g	长/mm	宽/mm	厚/mm	面积/cm²	腐蚀速率/(g/m²)	平均值/(g/m²)
优化前	16.1635	16.1631	0.0004	49.95	19.91	1.99	22.6705	0.1764	0.41
	16.0364	16.0354	0.001	49.91	19.96	1.96	22.6630	0.4412	
	16.0652	16.0638	0.0014	49.94	19.97	1.95	22.6725	0.6175	
优化后	15.9918	15.9906	0.0012	49.95	19.98	1.92	22.6453	0.5299	0.31
	15.9741	15.974	1E-04	50.03	19.97	1.91	22.6560	0.0441	
	15.9579	15.9571	0.0008	49.97	19.96	1.93	22.6473	0.3532	

（3）均匀腐蚀

使用高温高压釜，根据 ASTM G111-1997(2018)指南进行了模拟环境下的均匀腐蚀浸泡测试，如图 21(a)。试验总压 9.9MPa，H_2S 分压 1.5MPa，CO_2 分压 0.85MPa，模拟溶液中 Cl^- 含量 10000mg/L，HCO_3^- 含量 20000mg/L，SO_4^{2-} 含量 100mg/L，浸泡时间 720 小时。试验结果见表 12 所示，优化前后焊接工艺下，均匀腐蚀速率分别为 0.0002305mm/a、0.0002197mm/a，2 种试样的均匀腐蚀速率均符合要求，但两者差距不大。

(a) 实验前 (b) 实验后 (c) 形貌对比

图 21 均匀腐蚀试验过程

表 12 均匀腐蚀试验结果

编号	试样前重/g	试样后重/g	失重/g	长/mm	宽/mm	厚/mm	面积/cm²	均匀腐蚀速率/(mm/a)	平均值/(mm/a)
优化后	5.8466	5.8465	0.0001	40.03	9.93	1.88	9.8285	0.0001521	0.0002197
	7.5261	7.5259	0.0002	39.82	10.07	2.40	10.4145	0.0002873	
优化前	7.4229	7.4228	0.0001	40.04	9.88	2.38	10.2881	0.0001453	0.0002305
	5.2443	5.2441	0.0002	39.09	9.92	1.75	9.4708	0.0003156	

（4）抗硫化物应力开裂

取焊接接头四点弯试样，依照试验要求进行硫化物应力腐蚀开裂试验，试验条件模拟生产实际工况，设计总压 9.9MPa，H_2S 分压 1.5MPa，CO_2 分压 0.85MPa，模拟溶液中 Cl^- 含量 10000mg/L，HCO_3^- 含量 20000mg/L，SO_4^{2-} 含量 100mg/L，试验中保持加载应力，试验时间 720 小时(30 天)。图 22 为两种试样腐蚀后的形貌图，在试样表面均无明显表面裂纹产生，均满足标准要求。

3.4 焊缝组织的验证

在优化后的试样制备过程中，首先将焊接区域切割成宽约 30mm 的样本。初步采用 180 目粗砂纸进行打磨，随后转至 600 目砂纸以提升平滑度，最终使用 1000 目细砂纸进一步细化表面。抛光步骤采用绒布，以实现镜面般的光滑效果。完成表面处理后，应用显示剂于焊缝截面，进而通过电子显微镜对焊缝微观结构进行详细观察，如图 23 所示。

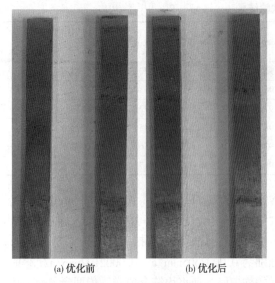

(a) 优化前　　(b) 优化后

图 22　应力腐蚀试验

图 23(a)为焊缝组织微观结构示意图,分析可得其结构为柱状的奥氏体组织,此外还可以看到奥氏体基体上分布着白色的点状物,为合金化合物;图 23(b)所示为熔合线附近组织分布,可观察到熔合线靠近焊缝侧存在晶粒明显较大的奥氏体区,具有一定的宽度,且此区域的奥氏体基体上并未分布有白色的合金化合物。由于合金化合物一般具有较高的硬度,其在奥氏体基体上的弥散分布,使得焊缝的硬度明显高于熔合线周围。经过优化的焊接工艺参数所得试样,在金相检验中没有发现裂纹、疏松、淬硬马氏体、过烧组织或其它缺陷,证明焊接质量符合标准。

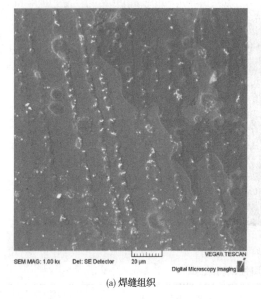
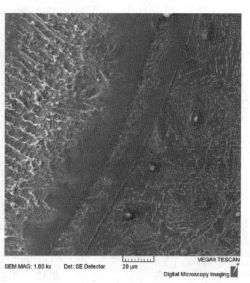

(a) 焊缝组织　　(b) 熔合线附近组织

图 23　环焊缝组织的微观结构

4　结论

本文针对 Incoloy825/L360QS 复合管开展了数值模拟与实验相结合的试验设计,通过机器学习算法对其焊接残余应力及最大变形量进行了预测及优化,提出了一种针对双金属复合管残余应力的预测优化模型,得到了以下结论:

(1)采用 PSO-SVR 组合算法,构建了一套双金属复合管焊接残余应力及变形量的预测模型并与此前常用的 GA-BP 模型进行对比,结果表明,PSO-SVR 模型在预测准确度及计算效率方面均优于 GA-BP 模型;

(2)在上述预测结果的基础上,采用 MOGWO 优化算法对预测结果进行优化,并引入熵权-TOPSIS 综合评价法对其进行进一步的评价筛选,获得最优焊接参数,从而解决机器学习算法无法精确得到一组最优解的问题;通过优化模型,得到其最优焊接参数为:打底焊电流 93.159A,打底焊电压 10.251V,其它焊层电流 124.664A,其它焊层电压 13.037V,焊接速度 1.00mm/s,层间温度 100.08℃。优化后最大残余应力为 429.9992MPa,最大变形为 0.776323mm,相较于优化前分别有效降低了 7.42%以及 7.14%;

(3)采用最优焊接参数进行焊接,并对焊接后的试样进行一系列的试验,对其力学性能、耐蚀性等基本性能进行验证,结果表明优化后的双金属复合管具有优异的力学性能和耐蚀性能,且焊缝组织稳定。

参 考 文 献

[1] 邓华波,刘思远,王喜春,邓宇波,徐鹏.2205双相钢管道环缝焊接温度场与应力场的数值模拟分析[J/OL].热加工工艺,1-6[2024-12-28].

[2] 任阳,刘永良,汪洋,曾云帆,刘建,张少刚,等.埋地双金属复合管焊缝区域腐蚀风险及阴保效果评价[J].腐蚀与防护,2024,45(08):42-49.

[3] LiZ, Guo Y, Dong H, et al. Numerical prediction and control of deformation and residual stress in double-sided arc welding of large pressure hull[J]. Measurement, 2025, 242(PB):115955-115955.

[4] 钟洋,毛汀,黄洪发,曾德智,罗涛,杨廷加,等.双金属复合管环焊缝耐蚀性能评价方法[J].石油与天然气化工,2024,53(01):104-109.

[5] 张杰,胡特.弯曲载荷下双金属机械复合管内衬层屈曲失效机理[J].船舶力学,2024,28(02):283-293.

[6] 王祺,张洪伟,王小涵,樊恒明,王琳.立管环焊缝残余应力Kriging近似预测模型研究[J].机械科学与技术,2020,39(11):1656-1661.

[7] DongZ, Xiaoben L, Yue Y, et al. Field experiment and numerical investigation on the mechanical response of buried pipeline under traffic load[J]. Engineering Failure Analysis, 2022, 142.

[8] D. K. R, P. G. B, G. -C. V, et al. Residual stress prediction of arc welded austenitic pipes with artificial neural network ensemble using experimental data[J]. International Journal of Pressure Vessels and Piping, 2023, 204.

[9] WeiP, Wenchun J, Bin Y, et al. An indentation method for measuring welding residual stress: Estimation of stress-free indentation curve using BP neural network prediction model[J]. International Journal of Pressure Vessels and Piping, 2023, 206.

[10] 李成文,吉海标,闫朝辉,刘志宏,马建国,王锐,等.基于GA-BP神经网络的316L多层多道焊残余应力和变形预测[J].焊接学报,2024,45(05):20-28.

[11] 王丽博,周金华,王宗园.基于GA-BP-FA算法的多轴铣削残余应力优化[J].组合机床与自动化加工技术,2024,(06):175-180.

[12] 曾权,李鑫,王克鲁,鲁世强,刘杰,黄文杰,等.基于GA-BP和PSO-BP神经网络的SLM GH3625高温合金残余应力预测研究[J].塑性工程学报,2024,31(03):193-199.

[13] WuW, Chen K, Tsotsas E. Prediction of particle mixing in rotary drums by a DEM data-driven PSO-SVR model[J]. Powder Technology, 2024, 434119365-.

[14] 葛华,黄海滨,蒋毅,张硕,曹宇光,司伟山.X80管道环缝焊接残余应力数值模拟[J].焊接,2021,(12):17-23+64.

[15] Z. Gao, X. Shao, P. Jiang, et al. Parameters optimization of hybrid fiber laser-arc butt welding on 316L stainless steel using Kriging model and GA[J]. Optics & Laser Technology, 2016, 83:153-162.

[16] 穆晨光,陈曦,邰冠华,邢金华.基于PSO-SVR的巴克豪森效应特征信号预测环轧件残余应力[J].南昌航空大学学报(自然科学版),2024,38(02):1-9.

[17] 汪骥,毛远,李瑞,刘玉君.基于支持向量机的高强钢薄板对接焊残余应力分布预测方法研究[J].船舶力学,2020,24(10):1294-1301.

[18] WangX, Gong J, Zhao Y, et al. Prediction of Residual Stress Distributions in Welded Sections of P92 Pipes with Small Diameter and Thick Wall based on 3D Finite Element Simulation[J]. High Temperature Materials and Processes, 2015, 34(3):227-236.

[19] WangY, Zhao Z, Ding W, et al. An online prediction method of three-dimensional machining residual stress field based on IncepU-net[J]. Measurement, 2025, 242(PA):115794-115794.

[20] 金翔羽.基于改进灰狼算法优化支持向量机的焊接接头疲劳寿命预测[D].大连交通大学,2023.

[21] YangL, Yue Y, Jiaqi W, et al. Thermal error modeling of servo axis based on optimized LSSVM with gray wolf optimizer algorithm[J]. Case Studies in Thermal Engineering, 2024, 53103858-.

[22] 贺康,董玉森,王力哲,曾菲,钱益涵,李慧丽.基于熵权TOPSIS模型的海岸线登陆点评估[J/OL].地质论评,1-10[2024-12-28].2024.11.015.

[23] Zhang QX, Cheng L Q, Sun W, et al. Research on a TOPSIS energy efficiency evaluation system for crude oil gathering and transportation systems based on a GA-BP neural network[J]. Petroleum Science, 2024, 21(1):621-640.

[24] 李昂,曾一达,李智勇,贺荣,郭正华,陈玉华,等.双相钢与铝合金异种金属激光焊接技术研究进展[J].材料导报,2024,38(22):218-226.

[25] 周任远,朱丽慧.750℃时效后Inconel 740H焊接接头的显微组织及性能[J].材料热处理学报,2024,45(03):218-226.

油田含硫天然气小口径管道内检测技术与案例分析

张佳[1,2]　薛文明[3]　周智勇[1,2]　秦林[1,2]　李潮浪[1,2]　孙明楠[1,2]　文绍牧[4]

(1. 中国石油西南油气田公司安全环保与技术监督研究院；2. 国家能源高含硫气藏开采研发中心；
3. 国家管网集团广西分公司；4. 中国石油西南油气田公司)

摘　要　小口径管道作为油气田地面集输系统的关键组成部分，因其管径小、输送环境复杂，长期以来缺乏高效的检测技术。特别是在川渝地区，天然气管道的输送介质成分和地理环境更加复杂，含硫气体(如H_2S)使得管道更易发生腐蚀，这对管道的安全运行构成了严峻挑战。因此，开展含硫天然气管道检测具有重要意义。管道内检测(In-line Inspection, ILI)技术作为保障管道安全的核心手段，已被广泛认可，其能够精确识别并定位管道缺陷。漏磁检测技术在油气管道内检测中应用广泛且成熟；电磁涡流内检测技术近年来也取得了显著进展，该技术具有良好的通过性，尤其在小口径、低压管道中对微小缺陷和裂纹的识别能力较强。小口径管道长期积聚的污物造成管道内径发生较大变化，增加了内检测难度。本文以某油气田含硫小口径管道内检测为案例，分析了漏磁和电磁涡流检测技术在现场的适应性、优势与局限性，并通过对比分析实际检测结果的差异，为未来类似管道的内检测提供宝贵经验和参考。

关键词　小口径管道；漏磁；电磁涡流；结构优化；数据分析

ILI technology and field case analysis of small-diameter pipelines of sulfur natural gas

Zhang Jia[1,2]　Xue Wenming[3]　Zhou Zhiyong[1,2]　Qin Lin[1,2]
Li Chaolang[1,2]　Sun Mingnan[1,2]　Wen Shaomu[4]

(1. Safety, Environment and Technology Supervision Research Institute of PetroChina Southwest Oil and Gas Field Company;
2. China National Energy R&D Center of High Sulfur Gas Exploitation;
3. The National Oil and Gas Pipeline Network Group Co., LTD. Guangxi branch;
4. PetroChina Southwest Oil and Gas Field Company; 5. China Special Equipment Inspection and Research Institute)

Abstract　Small-diameter pipelines, as a critical component of the surface gathering system in oil & gas fields, have long lacked efficient detection technologies due to their small-diameter and complex operating environment. This issue is particularly prominent in the Sichuan-Chongqing region, where the composition of the transported gas and the geographical environment are more complex. The presence of sulfurous gases (such as H_2S) makes the pipelines more susceptible to corrosion, posing a significant challenge to the safe operation of the pipelines. Therefore, conducting in-line inspection (ILI) for sulfur-containing natural gas pipelines is of great importance. ILI technology, as a core method for ensuring pipeline safety, has been widely recognized for its ability to precisely identify and locate defects. Magnetic flux leakage (MFL) ILI technology has been extensively and maturely applied in oil & gas pipeline ILI, while electromagnetic eddy current (ECT) technology has also made significant progress in recent years. This technology demonstrates good through-put capability, particularly for detecting small defects and cracks in small-diameter, low-pressure, and low-flow pipelines. However, most small-diameter pipelines have not undergone cleaning operations, and the long-term accumulation of debris causes significant changes in the ILI diameter, which further increases the difficulty of ILI. This paper uses the ILI of a sulfur-containing small-diameter pipeline in an oil & gas field as a case study to analyze the field adaptability, advantages, and limitations of MFL and ECT ILI technologies. By comparing and analyzing the differences in actual ILI results, this study provides valuable insights and references for future ILI of similar pipelines.

Key words　Small diameter pipeline; Magnetic flux leakage; Electromagnetic eddy current; Structure optimization; Data analysis

四川盆地是中国天然气工业的发源地，已发现的27个含油气层系中有13个高含硫化氢，中国硫化氢含量超过30克/立方米的高含硫气藏中有90%集中在四川盆地，四川盆地已探明高含硫天然气储量约9200亿立方米，占全国天然气探明储量的九分之一。高含硫气田的集输管道敷设环境复杂，面临较大的高程差、弯头多等挑战，腐蚀情况未知，给腐蚀检测带来较高的难度。集输管道主要采用L360QS抗硫碳钢，管径范围为DN100至DN300。随着气田生产年限的增加，产气量逐步下降、产水量上升，流速降低、携液能力减弱，这些因素使得腐蚀问题愈加严峻。因此，开展适应于含硫气田集输小口径管道的内检测技术研究显得尤为重要。某油气公司管道总数为4211条，总长度24107.23公里，其中小口径天然气管道2386条，总长度7906.9公里，占管道总数的56.7%、占管道总长度的35.4%，如图1所示。小口径管道具有管径小、管壁薄、变径频繁、曲率半径小等特点，且内部积垢严重，且大部分管道未施加内涂层（国内外多数200mm以下的小口径支线管道未进行内涂层处理），这些特点使得小口径管道的内检测面临更高的技术挑战。

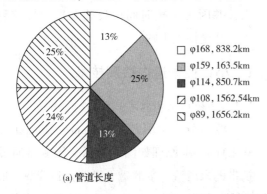
(a) 管道长度

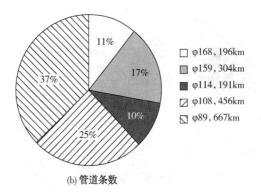
(b) 管道条数

图1　某气田小口径管道现状

小口径管道内检测是保障油气输送和安全运行的重要环节，但与大口径管道相比，其面临的挑战更大。小口径管道内部环境复杂，长期积累的油污、硫沉积物和腐蚀产物使得检测难度加大。此外，小口径管道常因低压、低流速导致流动性差，沉积物堆积及管道堵塞成为常见问题，进一步影响内检测的精度和可靠性。另一方面，小口径管道通常缺乏收发球装置，常规的清管和内检测作业变得更加困难。同时，焊缝焊瘤、椭圆变形、错边及斜接管节等结构性缺陷在小口径管道中更为突出（如图2所示），这些问题增加了内检测的复杂性和技术难度。此外，由于检测器的运行不稳定性，控制内检测器的速度是天然气集输管道内检测成功的关键。速度控制单元与备用安全装置可搭载于智能检测器上，适用于大排量高流速的输气管线，确保设备在不影响管道正常流量的情况下，保持在预定的运行速度范围。然而，小口径管道由于空间有限，往往无法安装速度控制单元，进一步加大了检测的技术难度。

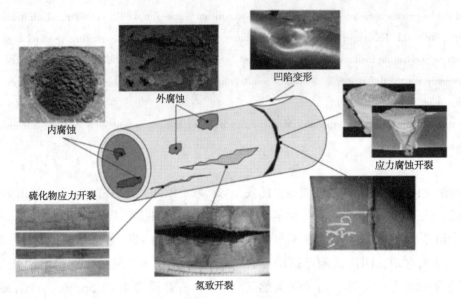

图 2 小口径管道实际工况

图 3 是 EGIG 对输气管道失效原因的统计图。可以看出，腐蚀占比 25.73%，约 1/4；其次为外部干扰/第三方损坏，占比 22.81%；然后是土体移动/地质灾害，占比 19.3%；然后是施工缺陷/材料失效，占比 17.54%；原因未知占比 14.04%。

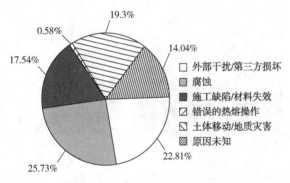

图 3 输气管道失效原因统计图
（来源：12th EGIG report）

随着管道完整性管理的推广，管道内检测技术被广泛应用于老旧管道的安全检测，特别是在存在缺陷的管道中。国内对输油管道内检测技术的研究较为早期，但输气管道的内检测技术研究仅在近年来逐渐受到关注。与液态原油管道相比，气体管道面临更大的挑战，主要由于气体的高压缩比、快速流动及显著的压力变化，这使得内检测器的运行速度更不稳定，尤其在经过弯头和复杂地势时，容易发生堵塞或速度波动，导致无法准确采集检测数据。此外，内检测器过快的运行速度会降低漏磁信号的准确性，甚至可能无法获得缺陷信号，影响检测结果。目前，国内研究主要集中在大管径、高压力的输气管道检测器，而对于油田集输系统中的小口径管道的检测研究较为薄弱。在小口径管道中，弯头可能导致停球蓄能，造成内检测器以超高速弹射，严重时可能损坏探头，并因无法短时间内磁化管壁而导致数据丢失，甚至检测失败。

1 小口径管道内检测技术

1.1 漏磁内检测技术

管道漏磁检测技术（Magnetic Flux Leakage，MFL）的基本原理如图 4 所示。检测器通过自身携带的永磁铁产生的磁力线耦合到管壁，形成一个完整的磁回路，使管壁达到磁饱和状态。在没有缺陷的情况下，磁力线均匀分布在管壁内；若管道存在缺陷，管壁横截面将发生局部减薄，磁力线在该区域发生变形，部分磁力线穿出管壁，形成漏磁场。漏磁信号通过紧贴管壁的霍尔传感器探头检测到，并转化为感应信号，经过滤波、放大和模数转换等处理后，数据被记录到检测器存储器中。小口径漏磁检测器的结构如图 5 所示。检测工作完成后，数据可通过电子包节下载，利用传感器在三个方向（轴向 B_x、径向 B_y、周向 B_z）采集的信号进行定性和定量分析，从而识别缺陷的特征，如图 6 所示。一般在漏磁数据中，轴向和径向信号的变化较为显著，而周向信号则相对较弱。为了区分管道内外壁的缺陷，需要结合 ID/OD（内/外壁）信号进行识别。通过分析内外壁的信号特性，能有效区分不同类型的缺陷。无特征点时，ID/OD 信号通常表现为直线或内凹；当存在特征点时，信号则表现为明显的凸起波峰。

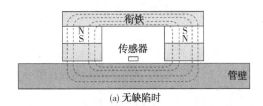
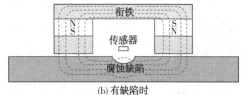

图 4　漏磁检测原理图

图 5　小口径管道漏磁检测器

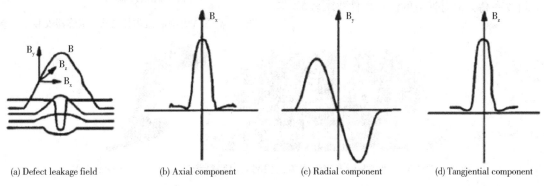

图 6　缺陷处的磁感应强度及磁场分量信号

漏磁检测技术广泛应用于铁磁性管道的表面及近表面缺陷检测。其主要优点包括：高灵敏度、快速检测、高性价比，能够对缺陷的尺寸和形态进行初步量化分析，并且对管道的清洁度要求较低。然而，漏磁检测技术仅适用于铁磁材料，无法检测形状复杂的物体；此外，检测器自身较为沉重，操作上不够灵活；由于要求管道达到磁化饱和，且管道壁厚存在一定限制，导致某些特殊管道无法进行有效检测。在小口径管道内进行智能检测时，为确保检测数据的准确性，通常需要增加磁化构件的重量，但这一措施却增加了检测器发生卡堵的风险。因此，小口径漏磁检测器的研发面临着两个主要挑战：一方面是检测器的设计优化，另一方面是保证数据质量的稳定性。具体技术难点包括如何突破传统漏磁检测技术的局限，尤其是在狭窄弯道和复杂地形下的通过性和速度控制问题，这些因素都直接影响了数据的准确性和完整性。因此，针对小口径管道的漏磁检测器必须在提高磁化能力的同时，解决通过性和卡堵风险，确保高效、准确的检测结果。

1.2　电磁涡流内检测技术

电磁涡流内检测技术（Eddy Current Testing，ECT）通过交变电流产生的交变磁场实现对管道的检测。当检测线圈接近被检测材料表面时，交变磁场会在材料表面诱发涡流，产生的涡流磁场与检测线圈发生电磁感应作用，从而在线圈中感生电压。由于缺陷的存在，会影响涡流的强度和分布，进而导致感应电压的变化，最终使缺陷得到检测和定位，检测原理图如图 7 所示。电磁涡流内检测技术的显著优势在于其良好的通过性和较低的卡堵风险，尤其适用于小口径、低压力、低排量的管道检测，且不易受到管道内污垢或沉积物的影响。该技术能够有效检测和量化管道内壁的浅表面和体积缺陷，广泛应用于钛钢、不锈钢、铸铁、铝、合金钢及内衬不锈钢复合管等多种材质的管道检测。此外，电磁涡流内检测在没有收发球装置的情况下也能进行有效检测，为小口径、复杂管道的检测提供了重要技术支持。

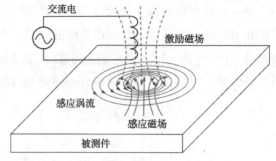

图 7 电磁涡流检测原理图

电磁涡流检测技术在工程应用中主要分为清管式和几何检测器式两种形式，如图 8 和 9 所示。主要区别在于应用方式和适用管道类型。清管式检测器使用柔性探头，能够适应管道的复杂几何形状，如弯头和异径管，特别适合小口径管道的检测，精度高，但相对效率较低，适用性更强。几何检测器式检测器则适用于大直径、长距离管道，检测效率高，但对管道的清管条件要求较高。

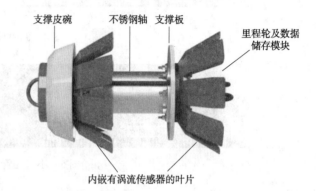

图 8 清管式电磁涡流内检测器(传感器安装于皮碗中)

图 9 ROSEN 几何检测器式电磁涡流内检测器(传感器安装在几何机械探头壁)

1.3 内检测技术对比

在管道内检测技术的选择过程中，不同的检测方法因其独特的技术特点和适用范围，展现出各自的优势与局限性。为全面评估漏磁检测技术与电磁涡流检测技术(包括清管式和变形探头式)，本文进行了系统对比分析。通过对各技术在适用场景、检测精度、适应性等关键性能指标的深入剖析。下表 1 和 2 汇总了传统漏磁和电磁涡流内检测技术的主要优缺点，以期为实际应用提供科学的决策支持，帮助在复杂的管道作业环境中选择最合适的检测手段。

表 1 检测技术适应性对比

检测技术	优点	缺点	适应性
漏磁检测技术	方法简单、方便使用、费用低； 检测灵敏度高，检测速度快，成本低能够形成缺陷尺寸形态的初步量化； 对管内清洁度要求低； 适应于小壁厚管道(<12mm)。	仅适用于铁磁材料检测； 由于磁饱和管壁的要求，管道的最大壁厚有限； 内检测设备沉重。	适应于铁磁性的油气管道表面及近表面腐蚀、点蚀检测； 适应速度 0.5~5m/s； 借助辅助探头可检测管道内外缺陷。
涡流检测技术	检测无需耦合介质，灵敏度高； 基本不受管径影响； 设备相对较轻，可通过清管阀进行发送及接收。	被检测物需为导电材料； 对被测物表面状态要求较高； 难以准确区分缺陷种类； 只能检测内缺陷，渗透能力有限。	具备电功能的油气管道表面及近表面裂纹和腐蚀缺陷； 适应速度 0.1~8m/s。

表 2 检测技术对管道条件要求的对比

管道条件要求	漏磁	清管器式电磁涡流	几何检测器式电磁涡流
推荐管输压力要求	2MPa	0.8MPa	1.6MPa
最小背压要求	0.2MPa	0.2MPa	0.2MPa
推荐运行速度	0.5m/s~5m/s	0.3m/s~7m/s	0.1m/s~5m/s
收发球筒最小长度	发球筒大小头距最近球阀间距应>(1.5m+检测器长度) 168检测器为例：大于3.5m	发球筒大小头距最近球阀间距>检测器长度 168检测器为例：大于0.5m	发球筒大小头距最近球阀间距应>(1.5m+检测器长度) 168检测器为例：大于3.5m
	收球筒大小头距最近球阀间距>(0.5m+2*检测器长度) 168检测器为例：大于4.5m	收球筒大小头距最近球阀间距>检测器长度 168检测器为例：大于0.5m	收球筒大小头距最近球阀间距应>检测器长度 168检测器为例：大于4.5m
最小弯头曲率半径	1.5D	1.5D	1.5D
管道清洁度要求	当清出杂质质量小于5kg或连续两次清管清出杂质质量相当且满足检测要求时。	通过泡沫或清管球清管，即可开展电磁涡流内检测作业	当清出杂质质量小于5kg或连续两次清管清出杂质质量相当且满足检测要求时。

2 小口径检测器卡堵原因分析及解决措施

小口径管道内检测器的卡堵问题是影响检测效率和质量的关键瓶颈之一。由于小口径管道的空间狭窄、弯头复杂以及管道内部可能积聚沉积物等障碍，内检测器在运行过程中易出现卡堵现象，进而导致检测任务无法顺利完成。因此，深入分析卡堵的根本原因，并提出相应的解决措施，显得尤为重要。从卡堵的主要原因出发，探讨如何通过优化检测器的设计、改进操作方法以及引入先进技术手段，提高小口径管道内检测器的通过性和运行稳定性，进而保障管道内检测工作的顺利实施。

2.1 卡堵原因分析

（1）驱动皮碗硬度、过盈量大，摩擦系数大，导致发球困难；

（2）现场发球筒为板材卷曲焊接制作，与传统无缝管相比，一是存在焊缝降低了内部通径；二是焊缝可能存在根部余高，增加了通过难度；

（3）未曾考虑高含硫气田作业环境，检测器在含硫管道中出现钢刷、皮碗、线缆、舱体等腐蚀；

（4）整体检测器直管段的缩颈能力小，导致检测器的通过性低，如遇小弯头、厚壁收发球筒、以及管道阀门未完全打开情况的适应能力差；

（5）电池节和机芯节舱体设计余量较大，导致部分空间浪费，进而降低了通过性；

（6）变形探头稀疏，导致部分小缺陷不能完全识别；弹簧拉力不够，转动不流畅。

2.2 解决措施与改进方案

（1）整体机械骨架小型化设计，关键部件采用钛合金抗硫材料设计；

（2）调整皮碗过盈量，降低皮碗硬度。支撑皮碗采用带V型槽的蝶形皮碗设计，增加检测器通过性；

（3）驱动节增加辅助密封皮碗，同时减少蝶皮碗，增加检测器通过性；

（4）通过重新设计电子包结构大幅减小机芯节直径，减小电子包和电池舱体积，提高通过能力；

（5）优化励磁节的体积和质量，以保证在饱和磁化强度大于1.8T的条件下永磁体质量最小，分别优选磁能积最高和饱和磁感应强度最大的永磁体和铁芯材料。

3 应用案例

3.1 检测器改造

针对 φ168 漏磁检测器进行了全面的结构与性能优化，相较于原有设计，改进后的检测器在管道适配性、体积和重量等关键技术参数上均实现了显著突破。为优化磁路设计，选择了磁能积最高的烧结钕铁硼与具有较高饱和磁感应强度和磁致伸缩系数的铁钴钒软磁合金作为永磁体和铁芯材料。这一材料选择有效减小了永磁体体积，降低了整体重量，使得检测器更适用于低压输气管道检测，并能够满足高清漏磁检测对磁化强度的高要求。当检测器进入管道弯头时，滑动摩擦会转化为静摩擦，导致启动时需要更大的推力来克服静摩擦并恢复运行。考虑到摩擦阻力系数及法向力是影响静摩擦的重要因素，本次改造重点降低了检测器的摩擦阻力系数与法向力。这一改进不仅保持了导向作用，还有效减少了启动时的静摩擦力，从而降低了启动加速度，减少了对设备的冲击与损坏风险，进一步提升了检测器的稳定性与通过性。

3.2 基本情况

将改造后的内检测器应用于某天然气管道，输送介质为天然气，管道外径168mm、壁厚10mm、长度6.5km，运行压力7MPa，运行输量$25×10^4 m^3/d$，弯头曲率半径3D，管线材质L245NB无缝钢管。为保证检测器平稳运行，在现有压力7.5MPa下检测时输量宜控制在$22 \sim 33×10^4 m^3/d$，运行时间约为0.6~0.9h，当前输量为$25×10^4 m^3/d$，速度约为2.4m/s，运行时间约为0.75h，满足检测需求，建议检测时继续维持输量在$25×10^4 m^3/d$，管道的基本信息如表3所示。

表3 待检测管道基本信息

设计压力/MPa	9.0
设计输量/($10^4 m^3/d$)	30
硫化氢含量/(g/m^3)	1.53
运行压力/MPa	7
运行输量/($10^4 m^3/d$)	25
外径/mm	168
壁厚/mm	10
长度/km	6.5
弯头曲率半径/D	3
管线材质	L245NB
制管方式	无缝

3.3 检测过程

在管道内检测实施过程中，依据属地实际情况对管道进行了初步处理，以确保检测顺利进行并提高数据采集的准确性。首先，为控制检测器运行速度，现场对气量进行了调节，确保数据采集的完整性和准确性；同时，为保证检测质量，作业前通过清管器清除了管道内的积水及气井返排杂质。接着，先后进行了普通清管和测径清管，以有效清除管道内的污物与障碍物，同时对管道内径尺寸和变形状况进行了初步评估。在清管工作完成并通过现场评估后，发送了几何检测器，进一步确认了管道几何形状和关键部位的通行情况。经计算，测径板最大变形量为16.2%，几何检测器最大通过量为20%，且测径分析结果显示未发现限制检测器通过的障碍物，结果如图10所示。

(a) 检测前　　　(b) 检测后

图10 测径清管

运行期间未发生憋压或卡阻现象，清管器结构完好，符合漏磁内检测器发送要求。强化检测现场风险管控，沿途加密定标点，采用"专人监听+设备监测"方式多渠道监控检测设备运行情况，确保检测设备运行风险受控。漏磁内检测器于上午11:49从发球站发出，并于12:33到达收球站，历时44分钟，均速约1.69m/s。收球后，经现场技术人员检查，漏磁内检测器外观完好，无机械损伤，所有组件齐全，探头、转接盒及数据线均无损坏。数据下载后，检测结果显示内检测数据完整，无数据丢失或损坏，数据质量清晰且完好。整个检测过程严格按照操作规范执行，确保了检测数据的准确性和完整性，为后续管道评估与维护提供了可靠的技术支持，运行前后的漏磁检测器如图11所示。

(a) 检测前

(b) 检测后

图 11 φ168 小口径漏磁内检测器

3.4 数据分析及开挖验证

为了提高检测结果的可靠性与准确性，本研究采用了涡流检测与漏磁检测数据的对比验证方法。通过在相同管段上同时开展涡流检测和漏磁检测，能够有效地验证这两种检测技术在实际应用中的一致性和互补性。涡流检测作为一种非接触式的电磁检测技术，具有较高的灵敏度，尤其适用于细小缺陷的检测；而漏磁检测则在检测管道内壁的磁性缺陷方面具有良好的效果。通过对比两种技术在相同缺陷位置的检测数据，可以深入分析它们在不同类型缺陷、管道状态和检测条件下的性能差异，从而验证漏磁检测数据的准确性与全面性。对 1、2、3、9、10 等 5 处开挖的数据进行对比分析，结果如表 4 所示。

表 4 开挖验证点

序号	数据类型	特征	钟点	内外部	深度/%t	长度/mm	宽度/mm	距焊缝距离/m
1	内检测信息	金属损失	11：58	外部	21.1	33	557	5.658
	开挖信息	测试接头	12：00	外部	/	/	/	5.358
2	内检测信息	金属损失	9：12	内部	21.1	25	60	1.9
	开挖信息	金属损失	9：30	内部	19.7	75	60	2.2
3	内检测信息	金属损失	9：55	内部	15.5	44	15	2.124
	开挖信息	金属损失	9：30	内部	15.5	224	65	2.06
9	涡流检测数据	金属损失	10：24	内部	5.6	32	15	11.19
	漏磁检测数据	金属损失	3：17	内部	17.4	98	52	11.31
	开挖信息	凹陷	3：30	外部	/	42	25	11.29
10	涡流检测数据	金属损失	11：08	内部	12.7	106	30	11.35
	漏磁检测数据	金属损失	7：02	内部	6.0	69	35	11.20
	开挖信息	划伤	7：00	外部	20.4	60	15	11.29

1 号缺陷为漏磁单独发现缺陷，数据显示为 21.1%金属损失，实测未发现缺陷，往上游走 300mm 有测试接头。

图 12 开挖验证结果（1 号缺陷）

2 号缺陷为漏磁单独发现缺陷，数据显示为 21.1%金属损失，实测 19.7%，长度偏差超过标准要求。3 号缺陷为涡流单独发现缺陷，数据显示为 21.1%金属损失，实测 15.49%，长度和距焊缝距离偏差超过标准要求。

图 13 开挖验证结果（2 号缺陷）

9 号缺陷为漏磁涡流共同发现缺陷，数据均显示为内部金属损失，实测为外部凹陷，缺陷类型均判断错误，涡流在时钟点位的偏差超过标准要求。

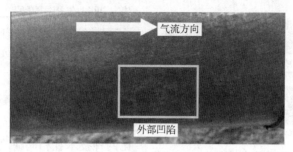

图 14　开挖验证结果(9 号缺陷)

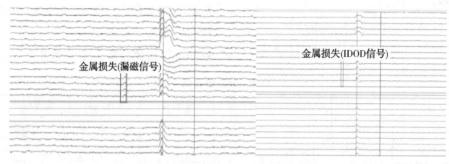

图 15　漏磁和 IDOD 信号(9 号缺陷)

10 号缺陷为漏磁涡流共同发现缺陷，数据均显示为内部金属损失，实测为外部划痕，缺陷内外位置均判断错误，涡流在时钟点位和长度的偏差，漏磁在深度和宽度的偏差超过标准要求。

(a) 金属损失缺陷

(b) 外部划痕

图 16　开挖验证结果(10 号缺陷)

从 9 和 10 号点验证为外部缺陷，漏磁为内部。漏磁一般会把内部误判为外部，很少把外部误判为内部。现在看来，涡流会把小变形误判为内部缺陷。10 号缺陷为漏磁涡流共同发现缺陷，数据均显示为内部金属损失，实测为外部划痕，缺陷内外位置均判断错误，涡流在时钟点位和长度的偏差，漏磁在深度和宽度的偏差超过标准要求。10 号缺陷所在管节，除漏磁和涡流共同发现的缺陷外，还有 1 处外部缺陷，为漏磁与涡流均未发现，实测为 7.04% 外部缺陷。

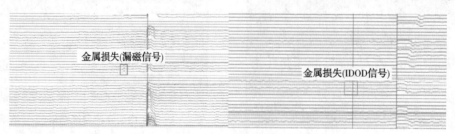

图 17　漏磁和 IDOD 信号(10 号缺陷)

4　结论

本研究探讨了小口径管道内检测面临的技术难，并对小口径漏磁和电磁涡流检测技术在现场应用中的数据进行了对比分析，本文通过对比两种技术在实际检测中的表现，选取了 5 处典型位置进行了开挖验证。验证结果表明，漏磁和涡流检测器对管道缺陷的识别精准，且在缺陷的长度、宽度和深度方面的误差均满足工程要求。在漏磁和涡流数据对比的开挖验证中，检测器成功识别了各类缺陷，并展现出较高的检测精度。然而，管道开挖过程中也发现定位点存在一定偏差，提示在内检测器设计中仍需进一步优化，尤其是在降低启动压差和保障检测器平稳运行方面。总体而言，本文所提出的内检测器的改进与应用有效填补了小口径天然气集输管道内检测技术的空白。该技术的成功应用为管道运行维护人员提供了更为全面的管体状态评估手段，有助于及时识别潜在的安全隐患，从而保障天然气集输管道的安全稳定运行。

参 考 文 献

[1] ZHANG J, WEN M, LIN D, GAO J, QIN L, LIU C, et al. Failure analysis of local effusion corrosion in small diameter gas pipeline: Experiment and numerical[J]. Engineering Failure Analysis, 2024, 161(2024), 108300.

[2] Maxfield K. MFL Inspection of Small Diameter, Previously Unpiggable, Pipelines-Lessons Learned[C]. 34th Pipeline Pigging and Integrity Management Conference, Houston, USA, 2022.

[3] ZHANG J, SUN M N, QIN L, LIN D, LIU C, LI J, et al. In-line inspection methods and tools for oil and gas pipeline: a review[J]. International Journal of Pressure Vessels and Piping, 2024, 214(2024), 105409.

[4] DESFA, ENAGAS, S A, Energinet, EUSTREAM, FGSZ, Fluxys, et al. European Gas Pipeline Incident Data Group (EGIG). 12th report of the European gas pipeline incident data group[J]. 2024.

[5] 赵番. 油气管道漏磁内检测三维有限元动态仿真分析与验证[J]. 压力容器, 2024, 41(09): 77-86.

[6] 吕坦, 孟祥吉, 闻亚星, 王锋, 陈金忠, 马义来. 长输油气管道漏磁内检测信号识别与分析[J]. 无损检测, 2024, 46(12): 58-65.

[7] 唐建华, 张鑫, 刘金海, 刘海超, 卢进. 基于复合骨干网络的漏磁小缺陷信号检测方法[J]. 电子测量与仪器学报, 2024, 38(10): 69-77.

[8] 姜伟光, 何战友, 刘小齐, 陆红军, 李佩. 小口径管道涡流内腐蚀检测技术的优化与应用[J]. 化工安全与环境, 2024, 37(12): 109-113.

[9] 宋汉成, 李攀登, 张理飞. 管道涡流内检测平面线圈传感器设计及性能分析[J]. 无损检测, 2024, 46(05): 1-5.

[10] 李睿. 油气管道内检测技术与数据分析方法发展现状及展望[J]. 油气储运, 2024, 43(03): 241-256.

[11] 金魏峰, 孙静, 章卫文, 孙凡, 方钊峰, 董志明, 等. 智能内检测技术在城镇燃气管道的应用[J]. 煤气与热力, 2024, 44(03): 43-46.

[12] 张行, 富宽, 陈铭浩, 李睿, 石新娜. 压差式多节串联管道机器人越障时动力学演化规律及减振分析[J]. 机械工程学报, 2024, 60(08): 348-359.

[13] 廉正, 刘斌, 刘桐, 武梓涵, 杨理践. 基于双磁场管道复合型缺陷应力信号提取方法研究[J]. 仪器仪表学报, 2023, 44(03): 107-118.

[14] 付双成, 戴朝磊, 张庆保, 陈金忠, 辛佳兴. 直径168mm管道被动喷射式皮碗结构设计及清洗性能仿真[J]. 科学技术与工程, 2023, 23(03): 1048-1055.

[15] 梁守才, 孙皓, 孙超, 马焱, 刘万强. 小曲率半径低压输气管道内检测器研制与应用[J]. 油气储运, 2023, 42(11): 1261-1266.

复杂地质条件下水平定向钻技术在长输
油气管道施工中的应用

曹子建　胡乾彬

（国家管网集团工程技术创新有限公司）

摘　要　随着油气资源的应用对社会发展愈发重要，长输油气管道技术也愈发成熟。其中，水平定向钻技术是长输油气管道中最为常用的技术之一。然而，随着更多的油气管道建成，施工过程中遇到的问题也越来越多，为解决长输油气管道施工过程中岩溶地貌条件下出现的岩层破碎、泥浆偏离问题，研究提出了一系列技术优化方案。首先，对现有钻具进行优化，特别是在管道施工的特殊地质位置，采用锥形PDC钻头与高强度钻杆组合，以提高钻进的稳定性和耐久性。此外，引入智能导向技术，通过地磁传感器与惯性导航单元的结合，实现对钻头在复杂地层中的三维坐标模拟和实时路径修正，确保钻进路径的精准性，减少路径偏差。为应对泥浆偏离和孔壁坍塌问题，研究还引入自动化泥浆控制系统，通过实时监控和自动调整泥浆的流速、粘度和密度，确保泥浆在复杂地质条件下的稳定性和有效性，成功解决了泥浆流失、压力不稳等问题。在施工中，这些优化措施显著提高了施工质量与效率，具体表现为：钻具磨损问题得到有效控制，钻进路径偏差显著减小，泥浆流失率和孔壁坍塌率大幅下降，施工效率提升了30%，施工成本降低了20%。研究结果表明，这些优化方案在复杂地质条件下的长输油气管道施工中具有重要的应用价值，减少非计划停工时间的同时保证项目施工安全，极大程度提高了水平定向钻在施工过程中的准确性、稳定性，为未来类似工程提供了可靠的技术参考。

关键词　复杂地质；水平定向钻技术；长输油气管道；管道施工

Application of Horizontal Directional Drilling Technology in Long-distance Oil and Gas Pipeline Construction under Complex Geological Conditions

Cao Zijian　Hu Qianbin

(Pipechina Engineering Technology Innovation Co., Ltd.)

Abstract　As the application of oil and gas resources becomes more and more important to social development, the technology of long-distance oil and gas pipelines is becoming more and more mature. Among them, horizontal directional drilling technology is one of the most commonly used technologies in long-distance oil and gas pipelines. However, as more oil and gas pipelines are built, more and more problems are encountered in the construction process. This study in order to solve the challenges of rock fragmentation and mud deviation in long-distance oil and gas pipeline construction under karst geomorphology by proposing key technical optimizations. First of all, optimize existing drilling tools, especially in the special geological location of pipeline construction, using a combination of conical PDC bits and high-strength drill pipes are used to enhance the stability and durability of drilling. In addition, intelligent guidance technology is introduced to ensure the accuracy of drilling paths and reducing path deviations, combining geomagnetic sensors with an inertial navigation unit, ensures accurate 3D path correction of drill bits in complex formations. To address the issues of mud deviation and hole wall collapse, the study also introduced an automated mud control system, this automated mud control system monitors and adjusts mud flow, viscosity, and density in real-time, effectively managing mud loss and hole wall collapse. In the construction, these optimization measures have significantly improved the construction quality and efficiency. Specifically, the drilling tool wear problem has been effectively controlled, the drilling path deviation has been significantly reduced, the mud loss rate and the hole wall collapse rate have been significantly reduced, and a 30% increase in construction efficiency, along with a 20% reduction in costs. The research results show that these optimization schemes have important application value in the construction of long-distance oil and gas pipelines under complex geological conditions, reduce the unplanned downtime and ensure the construction safety of the project, greatly improved the accuracy and stability of horizontal directional drilling during the construction process and provide a reliable technical reference for similar projects in the future.

Key words Complex geology; Horizontal directional drilling technology; Long-distance oil and gas pipeline; Pipeline construction

复杂地质条件下的长输油气管道施工因其特殊的岩溶地貌结构面临诸多技术挑战。岩层破碎、溶洞和裂隙带的发育导致传统开挖方法难以保证管道的稳定性与安全性。为了克服这些难题，采用先进的水平定向钻技术，以期在复杂地质环境中实现精确、高效的钻进操作。研究聚焦于通过优化钻具、增强泥浆性能及智能导向技术的应用，系统性地提升钻进精度和施工效率。此次设计方案能够有效减少施工过程中的不确定性因素，降低施工成本提升项目整体安全性与可行性，具有一定现实意义，推动相关领域技术的进步。

1 工程概况

某长输油气管道施工项目在岩溶地貌复杂区域开展施工，地质特征主要表现为地下溶洞、裂隙发育、岩层不稳定性高，施工难度较大。该项目线路全长135.8km，地表起伏不定，地下多层岩石夹杂溶洞和裂隙带，导致传统开挖方式难以保障管道的安全铺设与稳定。为此，项目选择水平定向钻技术以规避地表和地下不稳定地质带对施工的影响。施工深度35~46m，需穿越多个溶洞和断裂带，岩层破碎和泥浆偏离等问题在施工中频繁出现，对导向精度和施工安全构成威胁。

2 水平定向钻技术在长输油气管道施工中的应用

2.1 前期准备

针对复杂地质条件下开展长输油气管道施工，此次选用GSSI的SIR 4000系统地质雷达设备，具有高分辨率和深度穿透能力。经过详细勘察，发现该区域主要由中厚层石灰岩构成，岩层倾角为16°~23°，溶洞深度在28~43m之间，最大溶洞直径为11.8m，周边岩层较软，存在塌陷风险。地下水位距地表约24.7m，流速在每小时4.8~14.3m之间，水中碳酸盐矿物含量较高，并含有硫化氢气体，可能稀释泥浆并影响孔壁的稳定性。基于此选择X80级高强度钢管，并进行三层PE防腐处理，外壁涂覆高密度聚乙烯和陶瓷耐磨层，以抵御腐蚀和磨损。考虑到溶洞和裂隙带的复杂性，选择适用于中深度钻进的Vermeer D100x140水平定向钻机，具有较强的扭矩与拉力，最大扭矩达到18800Nm，最大拉力440kN。

2.2 施工阶段

2.2.1 导向孔钻进

基于前期的地质勘察数据，确定导向孔的钻进路径，重点避开溶洞和主要裂隙带整个过程如图1所示。Vermeer D100x140钻机被部署，初始阶段使用PDC钻头切入地层。通过高精度地磁测量系统对钻头进行实时定位，并校准钻进角度和方向，确保钻头沿预定路径前进。启动PDC钻头，通过旋转和推力结合的方式进行钻进，实时监测钻头位置和倾角，确保导向精度。为应对复杂地质条件，钻进过程中密切关注岩层变化，调整钻具参数，避免因岩层硬度突变导致钻头偏移或磨损。

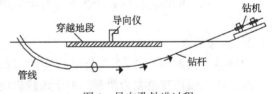

图1 导向孔钻进过程

2.2.2 扩孔与回拖

在扩孔与回拖阶段，采用逐步加大钻头直径的方式对导向孔进行扩孔。使用直径为300mm的锥形滚刀扩孔器，逐步扩大孔径，每次扩孔深度控制在1~2m，确保孔径均匀。扩孔过程中，保持扩孔器的旋转和推进速度协调,避免因岩层硬度变化引发偏差,并通过注入高粘度膨润土泥浆稳固孔壁,防止坍塌。扩孔完成后,立即进行回拖作业,管道连接至滚动轴承设计的回拖头,通过液压系统控制回拖力,回拖速度保持在每小时 8~12m,实时监测管道张力和扭矩,确保均匀受力,特别是在穿越溶洞或软弱地层时,适当降低速度并增加泥浆注入,以减小阻力和防止孔道坍塌。

2.2.3 管道铺设与连接

采用连续管段铺设技术,将预先焊接好的管段通过液压回拖系统逐步拉入孔道,回拖速度控制在 0.5~11.0m/min,使用滚动支架支撑管道,减少摩擦。穿越溶洞与软弱地层时需降低速度并增加泥浆注入量,保证管壁稳定。管道回拖完成后,采用自动焊接机进行现场焊接。焊接前,打磨管端,确保接合面洁净。焊接时,控制电流在 200~250A,焊接速度为 30~50mm/s,确保焊缝强度,焊后进行超声波探伤检测,补焊缺陷位置。焊接检测合格后,对焊缝进行防腐处理,使用热缩套管或环氧树脂涂覆,确保焊缝与管道主体一致,整体无损伤、变形情况,即确认合格。

3 施工过程问题及解决方案

3.1 施工问题

3.1.1 岩层破碎

此次长输油气管道施工过程中,钻进至距离起点约 1.5 公里处时,钻头进入受溶蚀作用严重的石灰岩层。该区域岩层破碎,内部裂隙密集且分布不规则,导致孔壁的稳定性极差。钻头在推进时,频繁出现孔壁塌陷现象,导致钻头偏移,无法保持既定的钻进路径。此外,在同一区域的多个断层交汇处,破碎岩层的结构松散,钻具多次卡滞,严重影响施工进度,钻头和钻杆的磨损程度严重。

3.1.2 泥浆偏离

在穿越距离起点约 2~2.3 公里之间的溶洞区段时,泥浆偏离现象尤为严重。由于该段岩层裂隙和溶洞发育,泥浆无法在钻孔内形成有效的压力系统,大量泥浆通过裂隙流失至孔外,导致孔壁失去支撑,发生局部坍塌。同时,泥浆的过度流失还导致钻屑无法顺利排出,孔道多次出现堵塞现象,迫使施工频繁暂停以清理孔道,施工效率降低,造成一定物资和时间成本。

3.2 应对措施

3.2.1 钻具优化

为降低当前施工过程中岩层破损问题对钻具的伤害,施工队决定在破碎岩层区域施工位置更换锥形 PDC 钻头,具有更强的耐磨性与抗冲击性,能够在松散岩层中提供稳定钻进效果。为防止钻具卡滞,应使用高强度钻杆采用短段钻进策略,每段推进约 2~3m 后进行暂停和稳定性检查,以避免孔壁进一步坍塌。对于频繁卡滞的区域,可考虑采用反循环钻进方式,通过反向流动清理钻屑,在钻进过程中实时清除孔内碎石和松散岩屑,减少钻具与孔壁之间的摩擦和阻力,从而降低钻具卡滞的风险。

针对钻具部分,还引入机器辅助模式确保实时监测钻进情况并优化钻进参数。钻具配备高精度扭矩传感器、推力传感器和振动传感器,实时监测钻进过程中的关键参数,如扭矩变化、推力大小和孔壁振动情况。数据通过无线模块传输至地面控制中心。地面控制中心运行基于机器学习的算法,对传感器数据进行实时分析。当检测到扭矩增加或振动异常时,系统识别出钻头遇到硬岩或孔壁不稳定的情况。具体原理步骤如下:

(1) 数据输入与特征提取:从传感器获取的实时数据,包括扭矩 $T(t)$、推力 $F(t)$、振动 $V(t)$ 等关键参数,作为输入特征进入机器学习模型,如下:

$$X(t) = [T(t), F(t), V(t)] \quad (1)$$

(2) 异常检测算法:基于历史数据进行异常值检测,当扭矩值超出正常范围 T,系统会识别为硬岩或碎石层。识别过程为:

$$|A(t) - \mu_A| > k\sigma_A \quad (2)$$

式中,μ_A 和 σ_A 分别表示均值和标准差;k 表示系数。

整个钻具运行过程依靠反馈控制理论和比例控制理论,通过实时监测关键参数并根据反馈调整系统操作,当检测到某个参数超出预设值,会自动调整,以保证系统迅速响应异常变化,从而维持钻具运行的稳定性。

3.2.2 增强泥浆性能

针对 2~2.3km 溶洞区段的泥浆偏离问题,首先调整泥浆配方,增加膨润土和聚合物的比例,使泥浆粘度提高至 45~55s(用标准漏斗法测量),以确保其能更好地封堵裂隙。同时,增加泥浆的密度至 1.2~1.4g/cm^3,以增强孔壁支撑力。在穿越溶洞时，采用分段注入高粘度泥浆的方法，每推进5m即停止钻进，注入泥浆形成泥浆屏障，并通过压力测试确认泥浆的封堵效果，避免泥浆大规模流失。对于泥浆流失严重的区域，采用膨胀型封堵剂混合泥浆，填充裂隙和小型溶洞。

为进一步增强泥浆性能，研究提出引入自动化泥浆管理系统，实时传感器安装在钻孔内外，持续监测泥浆的密度、粘度、流速和压力等关键参数。中央控制单元通过分析传感器数据，识别泥浆性能的变化，如粘度下降或密度不足。系统自动调整泥浆配方，增加膨润土或聚合物添加剂的比例，以增强泥浆的携屑能力和孔壁支撑力。同时，控制单元自动调节泥浆流速和压力，确保泥浆在破碎岩层和溶洞区段的稳定性。比如，当检测到泥浆压力下降时，系统立即提高泥浆泵的功率，增加泥浆流量，以补偿泥浆流失。对于振动参数超过阈值的情况，系统利用增强泥浆黏度的方式来稳定孔壁，其调整公式为：

$$\eta_n = \eta_c \times \left(1 + \gamma \times \frac{V(t) - V}{V}\right) \quad (3)$$

式中，$V(t)$表示实时监测到的振动参数，反映钻孔内孔壁的振动情况；V表示振动参数阈值；η_c表示当前泥浆黏度；η_n表示调整后泥浆黏度；γ表示比例系数，决定黏度调整幅度。

此环节中，系统通过增加泥浆黏度增强孔壁支撑能力，从而确保泥浆性能能够适应实时的钻进环境。

3.2.3 引入智能导向技术

针对复杂地质条件下的长输油气管道施工中，水平定向钻技术应引入智能导向功能。在施工开始前，智能导向系统通过安装在钻头内的高精度地磁传感器和惯性导航单元（IMU）实时监测钻头的位置和姿态，捕捉钻头在三维空间中的坐标、方位角、俯仰角等，通过无线通信传输至地面控制中心。地面控制中心利用预先规划的钻进路径，与实时传感器数据进行比对。系统通过计算钻头的偏差量确定钻头位置与目标路径误差，公式为：

$$\Delta d(t) = \sqrt{(x_{real} - x_{target})^2 + (y_{real} - y_{target})^2} \quad (4)$$

式中，$\Delta d(t)$表示t时刻中钻头实际位置与目标路径之间的偏差距离；x_{real}、y_{real}分别表示钻头在t时刻的实际横纵坐标，x_{target}、y_{target}表示钻头在时刻t预定路径上的目标横纵坐标。

此过程系统将判断钻头已经偏离预定路径，需要进行路径修正。

3.3 成效分析

在此次长输油气管道施工中，针对岩层破碎和泥浆偏离问题设计优化方案，对比优化前后的施工情况，结果如表1所示。

表1 结果分析

指标	优化前	优化后
钻头卡滞次数(每100米)	5次	≤1次
钻具磨损率	高	低
路径偏差(毫米)	20毫米	<5毫米
泥浆流失率	15%	3%
孔壁坍塌率	高	低
施工效率提升	-	30%

通过对比优化前后的施工数据，结果表明，应用优化措施后，钻具的卡滞频率显著降低，钻具磨损情况也得到了有效控制。路径偏差明显减少，表明智能导向技术提升了钻进精度。泥浆流失率和孔壁坍塌率也显著下降，证明强化泥浆性能和自动化泥浆管理系统有效改善泥浆控制效果。

4 结语

研究通过对钻具、泥浆管理和智能导向技术的优化，成功应对复杂地质条件下的施工挑战。在实际施工中引入先进技术，能够有效缓解钻具磨损问题，避免泥浆大量流失，同时提升孔壁稳定性，减少非计划停工时间。同时，利用智能导向系统的路径修正与自适应调节功能，保证钻头在复杂地层中的精准定位。

参 考 文 献

[1] 陈登旭.球墨铸铁管水平定向钻技术在供水工程中的应用[J].价值工程,2024,43(08):101-103.

[2] 姜宇飞,刘鹏,成磊,等.海对海水平定向钻跨越航道铺设海底管道施工工艺研究[J].石油和化工设备,2024,27(03):105-109.

[3] 宋础.非开挖水平定向钻在海缆穿堤工程中的应用[J].中国水运,2023,(03):121-123.

[4] 王国炜,顾士国,刘琳,等.油气输送管道水平定向钻下穿公路涉路工程技术评价研究[J].路基工程,2021,(04):169-172.

[5] 冒乃兵.石油天然气管道水平定向钻穿越工程质量控制研究[J].石化技术,2020,27(02):44+49.

油气管道智能阴极保护技术应用及发展

刘红波　王　钰　高秀宝　郭春雷　高　晓

（中海油石化工程有限公司）

摘　要　我国正处于油气管道大发展时期，使得传统阴极保护管控过程中存在的效率低、运维成本高、人工巡检难、管控能力不足、数据全面性及规范性不足等问题逐渐暴露。智能化、数字化的发展给阴极保护开拓了新的发展思路，本文综述了当前阴极保护智能化终端及管理平台的应用状况，从阴极保护智能化管理需求出发，分析了不同终端类型及其功能参数的适用性，并提出发展方向；从阴极保护技术管理和工作管理融合角度，分析了现有软件系统功能及其发展方向。从目前阴极保护智能化应用状况看，智能恒电位仪、多功能阴极保护智能测试桩、智能腐蚀记录仪等设备的规模化应用，实现了阴极保护、杂散电流、土壤环境等参数的自动化采集，其未来发展方向是实现多元化、集成化采集，提升终端可靠性。通过建立智能阴极保护管理平台，嵌入数据分析模型、数值模拟模型、可视化、数字化双生等方面，初步实现数据自动诊断分析、终端设备远传远控、阴极保护智能控制等特色功能的探索性应用。未来智能阴极保护系统应结合大数据分析、人工智能等先进技术，进一步提升系统感知分析及综合性预判能力，且应从柔性化工作管理角度融合多方资源，实现阴极保护的一体化管控及优化运行能力。智能阴极保护还通过运用全方位的智能感知、数据智能分析评估、智能控制和终端智能管理，提升阴极保护管理的效率和科学性，助力国家石油天然气管道阴极保护水平跃升，为管道整体管理实现数字化改造奠定了坚实基础。

关键词　油气储运；阴极保护；智能化；数字化

Application and Development of Intelligent Cathodic Protection Technology

Liu Hongbo　Wang Yu　Gao Xiubao　Guo Chunlei　Gao Xiao

(CNOOC Petrochemical Engineering Co., Ltd)

Abstract　China is currently in a period of significant development in its oil and gas pipeline infrastructure. This rapid expansion has gradually exposed several challenges within traditional cathodic protection (CP) processes, including low operational efficiency, high maintenance costs, difficulties with manual inspections, limited control capabilities, and inadequate data comprehensiveness and standardization. The advent of intelligent and digital technologies has opened up new avenues for addressing these issues and enhancing the effectiveness of CP systems. The current application status of intelligent cathodic protection terminals and management platforms has significantly improved the efficiency and reliability of CP systems. These terminals come in various types, each tailored to meet specific requirements based on the nature of the infrastructure being protected. For instance, smart potentiometers, multifunctional CP intelligent test stations, and smart corrosion recorders are now widely used in the industry. Smart potentiometers are capable of automatically collecting and transmitting data related to CP potential, stray currents, and soil environmental conditions. They are designed to operate in harsh environments, ensuring reliable and accurate data collection. Multifunctional CP intelligent test stations provide comprehensive monitoring capabilities, integrating data from multiple sources and sensors to offer a holistic view of the CP system's performance. They can detect anomalies and trigger alerts in real-time, allowing for prompt corrective actions. Smart corrosion recorders continuously track corrosion rates and other key parameters, offering valuable insights into pipeline conditions and facilitating proactive maintenance strategies. The advancement of these intelligent terminals aims to achieve comprehensive and integrated data collection, enhance terminal reliability, and improve the overall functionality of cathodic protection systems. All activities involved in the operation of all kinds of CP are mainly carried out by the core of the system, which is called intelligent CP management platform. By integrating data from various sources, this platform provides a unified interface for monitoring, analysis, and control. Key features include automated data diagnosis and analysis, where the platform employs advanced algorithms and machine learning techniques to analyze data, identify trends, and diagnose potential issues before they become critical. Remote transmission and control of terminal devices allow operators to remotely access and control CP terminals, adjusting settings and performing maintenance tasks without the need for physical presence at the site. Intelligent control of cathodic protection optimizes CP output based on real-time data, ensuring that the system operates efficiently and effectively, while minimizing energy consumption and reducing environmental impact. Visualization and digital twinning technologies enable operators to visualize the entire CP system and simulate various scenarios, facilitating better decision-making and planning. The future of intelligent cathodic protection systems lies in the integration of cutting-edge technologies such as big data analytics and artificial intelligence. These technologies will further enhance the system's perception, analysis, and predictive capabilities, enabling more accurate and timely decision-making. Big data analytics will play a crucial role in extracting meaningful insights from vast amounts of data collected by CP systems. Techniques like machine learning and deep learning can be employed to detect patterns, predict future trends, and identify potential issues proactively. Deep learning technologies such as natural language processing, neural networks are used to automate complicated jobs, decrease human error, and improve system performance. AI-powered systems can adapt to changing conditions, learn from past data, and continuously improve their accuracy over time. From a work management perspective, intelligent CP systems will integrate multiple resources, including personnel, materials, and equipment, to achieve integrated control and optimized operations. This includes scheduling maintenance tasks, allocating resources efficiently, and streamlining workflows to enhance productivity. In summary, intelligent cathodic protection leverages comprehensive smart sensing, intelligent data analysis and evaluation, smart control, and terminal intelligent management to improve the efficiency and scientific rigor of CP management. This not only elevates the level of CP for national oil and gas pipelines but also lays a solid foundation for the digital transformation of pipeline integrity management. As the technology continues to evolve, it is expected to play a pivotal role in shaping the future of the oil and gas industry, ensuring the longevity and safety of critical infrastructure.

Key words Oil and gas storage and transportation; Cathodic protection; Intelligentization; Digitization

当前中国经济社会的高速发展，伴随油气能源需求量也逐年攀升，为解决石油天然气供给运输问题，长输管道得到了大规模应用。但石油天然气的远距离运输面临诸多困难，由于管道多为金属材料，服役过程中长期暴露在复杂的自然环境中，很容易受到腐蚀问题的影响。腐蚀不仅会造成管道壁厚减薄、穿孔外泄，还可能造成严重的事故，如火灾、爆炸等，严重威胁着人民群众的生命财产安全。需要对管道采取腐蚀控制手段，以减缓埋设管道的腐蚀作用。其中，采用阴极保护外防腐层的方式，是目前国际上公认的对外腐蚀控制措施中最经济、最实用、最可靠的一种。强制电流阴极保护作为外腐蚀控制的重要补充措施，已在长输管道及输油气站场中得到广泛的应用。

但在目前的国内油气管道维护工作中，阴极保护系统的日常巡检、模式切换、参数调整等工作还需人工现场操作，不仅导致阴极保护工作效率低、运维成本高、管控能力不足，且阴保数据全面性及规范性不足等问题也逐渐暴露。智能化、数字化的发展，为解决实际工作中存在的问题，开启了阴极保护的发展新思路。智能阴极保护技术具有显著优势，可实现管道实时监控和自动调节保护电流，提高保护效率，降低维护成本，提高安全性，提供管理的智能化和一体化应用。

因此，进行智能阴极保护技术的相关研究，对解决管道腐蚀问题，保障管道安全稳定运行，具有重要的现实意义。不仅如此，随着技术的进步，未来的管道也必然更加智能化，将确保管道运行更加安全可靠，同时也将朝着自动监控、自动控制的方向，提升阴极保护的管理效能并促进技术革新。

1 阴极保护异常分析及解决措施

阴极保护主要在油气站场与长输线路管道工程中被大量应用，但在实际工程中仍存在大量的问题与难点：

大量的地埋管、仪器设备和接地网络存在于现役的石油天然气站场中，它们共同组成了一个庞大的金属结构网。鉴于管网与金属设施在地下错综复杂的布局，它们对阴极保护电流产生了显著的屏蔽效应与干扰，导致阴极保护效果不理想，频繁出现保护不充分、过度保护乃至局部失效的现象。为解决这一问题，地下部分采取了防腐层与阴极保护联合应用的防腐策略。在线路管线中，由于阴保站保护间隔较长，难以进行上下游恒电位仪的联调联控，造成管线每一段的阴极保护水平参差不齐；而且线路阴极保护只有一个单独的阴保间恒电位仪控制点，而且保护范围往往是一段较长的管线，控制点的电位并不能代表整个管线的阴极保护状况，恒电位仪的输出欠妥是由单一的电位来控制的。且常在线路站场绝缘接头外侧设置阴极保护控制点，在线路站场有区域阴极保护时，常因其对绝缘接头外侧产生干扰，致使控制点测量电位出现误差，线路恒定电位仪输出不能满足阴极保护的需要，在线路站场有区域阴极保护的情况下，经常出现阴极保护控制点的故障；由于线路延伸距离远，且沿途易遭受杂散电流的侵扰，这使得管道的保护电位出现波动。然而，当前线路所采用的恒定电位仪运行模式并不具备根据沿线电位变化动态调整其输出的能力，因此难以有效减轻杂散电流带来的干扰影响。这一限制是基于沿线电位实际变动情况而凸显出来的。

在油气管道的维护工作中，阴极保护系统的监检测项目仍大范围使用传统方法，问题与不足主要体现在以下几个方面：

（1）无法实现实时监测

目前，已应用的阴极保护系统对管道腐蚀情况的监测检测手段相对较少，主要依靠定期人工巡检，恒电位仪的运行参数也依赖人工抄报，很容易出现抄报错误的情况，人工巡检的方式存在很大的全局局限性和盲区，极易造成漏检问题，无法做到对阴极保护状态的全面实时监控，因此，目前已采用的保目前已应用的阴极保护系统，由于巡检人员的技能水平所限，导致人工巡检误差较多，监控管道腐蚀情况，且巡检工作效率低，运行维护成本较高；不能第一时间通过传统的监督检测手段发现阴极保护系统的故障和异常，容易造成不能保证阴极保护系统持续稳定运行的经济损失和安全事故，从而延长了故障的不正常处理时间。

（2）无法实现终端设备远传远控

油气管道的位置大多是丛林、山地、荒漠戈壁等腐蚀环境恶劣的偏远地区，而且油气管道的里程跨度大，土壤含水率、含盐率、氧浓度存在差异，腐蚀形态千差万别；偏远地区油气管道实施人工巡检维护有一定难度，维护时可能会出现危及人身安全的情况。

（3）监测内容单一

油气管道的跨度较大，且自然环境复杂多变，可能出现多种腐蚀现象，例如局部腐蚀、微生物腐蚀以及杂散电流干扰引发的腐蚀等。仅仅依赖恒电位仪的运行参数和阴极保护电位来评估管道的腐蚀抑制情况往往不够全面。人们更加需要深入了解管道的实际腐蚀状况以及与其腐蚀相关的环境因素。因此，腐蚀速率监测、杂散电流干扰监测、排流防护措施监测、土壤环境监测等多元化、集成化的阴极保护监测装置需求迫切。

针对这些问题，对阴极保护进行智能化改造具有重要意义，阴极保护智能控制技术可应用于油气站场区域的阴极保护和管道线路的阴极保护，现有的站场与线路所面临的阴极保护问题，能够得到有效的解决。智能控制技术不仅能够实时监测阴极保护系统并进行多点监测，还能根据智能算法动态调整阴保电流。它具备多项优势，包括误差小、精度高以及显著降低能源消耗和人工成本等，具体体现在以下几个方面：

（1）全面且实时的监控：利用智能控制技术，对电位、电流密度等多个关键位置及其参数进行实时监测，实现阴极保护的全覆盖监控，能够及时发现并解决潜在问题。

（2）高效的自动调节与反馈：智能控制技术依据实时监控数据自动调整，确保精准防护。一旦监测到问题，能够迅速响应，自动调整阴极保护参数，使管道始终处于最佳保护电位，从而提高保护效率，降低能源消耗。同时，该系统能够实现区域状态的自动调节与控制，确保各区域均处于最佳状态。

（3）远程管理：智能化控制技术赋予远程监控与管理能力，通过 Internet 实现专业技术人员对该系统的远程访问与监控。监控时效提高，人工巡检、维护频率大幅降低，运维成本也随之降低，监控时效提高，维护成本也随之降低。

（4）故障预测与维修：基于实时监测数据，智能控制技术具备故障预测与预防性维修的能力，能够提前洞察潜在问题。通过智能诊断手段，该技术可辅助专业技术人员识别站场内的屏蔽与干扰现象，从而迅速采取应对措施，有效避免保护不足或过度保护的情况，确保阴极保护系统的稳定运行。通过智能诊断阴极防护系统负载负荷。

阴极保护智能控制的应用总体提升阴极保护管理以及现场工作的便利性，提高工作的质量和效率，同时充分的发挥已有成果的功效，将阴极保护智慧化提升到更高的层次和水平，更好的为安全高效运行提供保障。随着信息技术、通信技术及工业物联网技术的普及，智能控制技术也将迎来大力推广。

2 智能阴极防护终端和控制工艺

2.1 阴极保护智能测试桩

阴极保护智能采集仪（智能测试桩）是一种重要的阴极保护监测设备，具有自动采集、传输数据、在线诊断与评估、实时监控与预警等功能特点。传统的阴极保护采集仪通常需要手动操作，如设置测量参数、启动测量等。操作者需具备专业知识和经验，确保准确、安全地计量。相对于传统的采集仪，智能采集仪可以不通过传统的人工采集，而是通过采集被保护结构的阴保信息，如通、断电电位，自然电位，交流杂散电流、交流感应电压，等，通过数字化、智能化的方式进行阴极保护数据采集。人力物力投入到对管道阴极防护效果的监控上，能够有效降低使用量。然后通过无线传输的方式发送到远程监控中心，远程监控中心接受数据后对数据进行处理分析，最终将经过处理分析后的数据呈现在系统客户端平台上，图1为智能采集仪的原理；通过该装置的应用，能够实时监控并有效管理金属结构阴极保护状态，确保金属结构的安全运行。

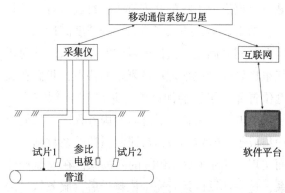

图 1 智能采集仪系统图

智能采集仪被广泛应用于长距离、大范围的长输油气管道、储罐等金属结构的阴极保护监测中。通过该设备的应用，实现了对金属结构阴极保护状态的实时监测和有效管理，确保了金属结构的安全运行。具体应用场景与效果如下：

（1）长输油气管道

能实时监测预警管道阴极保护状态，对潜在腐蚀问题能及时发现，及时处理，延长管道寿命。特别是采集仪能对管路通过人迹罕至或需严格控制的管段、杂散电流干扰严重的管段、穿越管段等进行及时准确的数据采集、传输和分析控制。

（2）储罐

可监测罐体阴极防护电位，确保其处于最佳防护状态；通过对保护状态的异常报警，防止储罐因腐蚀发生泄漏事故，减少安全事故的发生和经济损失。

（3）其他金属结构

如桥梁、船舶等金属结构也可应用阴极保护智能采集仪进行监测和保护，以确保结构物在安全状态的服役。

综上所述，阴极保护智能采集仪是一种具有高精度测量、低功耗设计、高抗干扰能力和易于维护等优点的阴极保护智能监测设备。其应用不仅提高了阴极保护系统的可靠性和稳定性，还降低了管理和运行维护的经济成本，已在阴极保护工程中规模化应用。牟春霖以振弦式感应器为基础，自主研制的智能监控采集仪，将传感器数据的自动化采集上传、人工采集数据对比分析，结果显示，智能监控采集仪数据与人工采集资料相比，通过运用先进技术，显著简化了数据上传与分析的流程。将智能监控采集器所获取的数据与石城市轨道交通工程的基坑监测项目进行对比分析后，结果显示，该采集器的数据精度高达0.4HZ以上，充分满足了监控项目的实际需求。针对川气东送管道沿线地理环境复杂的特点，我们在关键位置部署了智能测试桩，并在RTU阀室安装了智能电位采集器。通过后台服务程序，我们成功集成了GIS空间数据与遥测遥控关系数据，从而实现了基于GIS/GPRS的阴极保护在线监控体系，这一体系的运行完全依赖于后台服务软件的支撑。实现阴极保护系统运行参数异常保护的集中监视管理，智能阴保电位采集仪在阴保系统广泛应用于投入使用；资料可靠性得到了保证；使资料传送的及时性、可控性得到了提高，阴极防护系统的管理、运维费用大大降低；加快了工作的效率，提高了工作质量，用更加方便的日常阴保管理，为用户阴极保护系统的管理以及判断阴极保护的成效，智能化、规范化以及办公管理无纸化的提高方便了办事效率。

2.2 阴极保护智能桩扩展功能

阴极保护多功能智能测试桩是广泛应用于石油、天然气、化工等行业的埋地管道阴极保护系统的智能化监测设备。其中智能采集仪是智能测试桩的核心设备，用来负责采集各类阴保参数，除了对阴极保护参数进行监测外，多功能智能测试桩还可以实现腐蚀速率监测、环境综合参数监测、防盗泄露监测、绝缘接头监测等多种测试功能，且所有功能都为模块化设计，可按实际需求集成融合，可选择性增减。如图2所示为智能测试桩装置结构图，在通过通讯模块向智能阴极保护平台进行数据通信的同时，主控芯片依靠模块化传感器对管道和周围环境进行监测，从而实现多种测试功能。

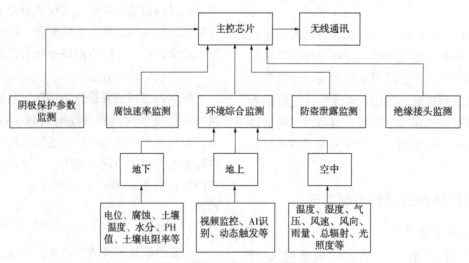

图2 智能测试桩装置结构图

现阶段，国内各厂商智能桩产品技术参数相差不大、基本相同，但在产品应用数量、结构功能优化、数据平台对接、现场调试维护等方面有所差异，青岛雅合与北京安科智能桩具有一定优势；国外智能桩的技术需求、设计思路和实现手段与国内基本相同，与国内产品最大技术差异在通信方式方向。

在当前各应用单位使用的智能桩中，主要品牌也为青岛雅合与北京安科两家，两者占总体数量的80%以上。每台生产厂家生产的智能测试桩均具有采集通电位、断电位和交流电压的功能，直流电流密度、交流电流密度、腐蚀速率、自然电位、湿度、温度等参数采集功能可根据需要增加，如：各厂家生产的智能检测桩均能满足要求，可按要求增加。在腐蚀速率监测方面，厂家选用ER探头采集腐蚀速率居多，但增加腐蚀速率监测功能的设备费用是普通采集设备的2-3倍，建议在腐蚀控制的关键位置、腐蚀控制效果不明显不确定的位置、以及杂散电流混合干扰严重管段位置安装带有腐蚀监测模块的智能测试桩。

截止至2023年，阴极保护智能桩有效运行率超过98%，并以每年超过5000套智能测试桩的建设速度快速增长。在智能桩密度在北京地区管道达到最大，覆盖率约每2.5km安装1处，随着智能桩数量快速增加，其运行维护工作也受到了挑战。经前期调研，智能桩设备常见的故障维修的主要集中在电池欠电、通讯故障、易被第三方破坏等几个方面。在智能桩的发展过程中，各个厂商针对这些问题进行了针对性的更新和升级，包括高性能电池的升级、太阳能电池的引用、电池的可自主更换、智能桩所处区域设备待机功耗的优化、设备频次和数据上传方式的优化、电池的可自主更换等，针对这些问题，各个厂商都进行了针对性的更新和升级；提供多种通讯方式及通讯故障自检功能，并设置通讯异常保护，在数据无法上传的情况下，存储装置可在一定时间内暂时保存测试数据；仪器为智能桩加上振动感应器和定位器，当发生振动或位移倾斜时，可向管理平台发送坐标定位和信息。

随着技术进步和市场需求的动态变化，油气管材智能化发展已成为必然趋势，而装备可靠性高、质量性能更好、后期维护保养更便捷、功能拓展更丰富的智能桩产品更能在智能试验桩大规模投资建设的浪潮中占领市场，搭上油气管材智能化发展的快车，智能试验桩是其中不可或缺的一环；油气管道智能化发展已成必然趋势，桩类产品智能化试验目前桩类相关标准规范在智能化检测方面较少，内容多为产品技术要求，涉及桩类相关标准规范在安装、检测、故障排查、设备维护等方面较少，目前桩类相关标准规范在智能化检测方面较少，涉及产品技术要求较高，目前桩类相关标准规范在智能化检测方面较少未来，制定智能测试桩在运维方向的相关标准规范，对油气管道智能化工程建设与安全运行具有重大意义。

2.3 便携式智能数据记录仪

智能数据记录仪是一款专门针对阴极保护系统进行巡检的设备，能够对阴极保护系统的运行状况进行高效、准确的监控和记录，有力保障了阴极保护系统的维护及管理。图3为某公司智能记录仪硬件体系架构图。硬件系统包含主板和内置防雷单元。主板主要有液晶显示屏单元、GNSS定位单元、蓝牙通讯单元、电源管理单元、单片机单元；内置防雷单元，对管道内的感应雷起到保护作用，可实现雷击电流的释放。

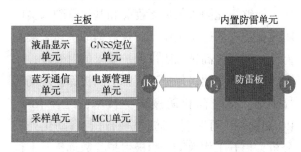

图3 某企业智能DATCR硬件系统架构

目前国内主流销售国外阴极保护资料记录仪主要有美国TR公司的DL-1资料记录机、加拿大CorTalk公司的UDL资料记录机，加拿大CALL-Tech公司的CorrReaderPro电位记录机和SmartStreamII智能数据记录仪；国内主流销售的智能数据记录器主要有天津嘉信技术工程公司的MDL多功能技术数据记录机、北京安科腐蚀技术有限公司的Corrlog-1C阴保电位记录仪、西安雷迪仪器有限公司的PKS型阴极保护真实电位(极化电位)、西安雷迪仪表有限公司的SNB型阴极保护真电位检测记录仪，这些数据记录器都能有效记录埋地管道阴极保护的管地电位和交流电压，在国内管道无损检测公司、中石油中石化管道、燃气公司等应用非常广泛，目前国内主流销售的智能数据记录机主要有天津嘉信技术工程公司的MDL多功能技术数据记录仪、北京安科腐蚀技术有限公司的SNB型阴极保护真电位检测记录仪。

其中进口设备虽然在采集参数、采集功能方面相对较全面，但其产品价格昂贵、操作不便、维护困难，国内产品方面，其产品应用规模较小，产品性能不一，无法满足阴极保护日常管理检测对采样参数及采样功能多样化的需求，随着科技的发展，巡检设备的智能化、便携化已成为仪器设备发展的重要趋势，在长输管线阴极保护系统巡检工作中，亟需功能全面、价格适中、操作简便、功耗低、适应现场恶劣环境的便携式巡检设备。随身携带的资料记录器,小巧、轻便,方便了去实地查房的用户随身携带;智能化的设计让它拥有了用户可以轻松操作的特点,界面友好,只需按提示操作即可实现。

便携式智能阴保巡检仪作为阴极保护系统巡检的重要工具,未来研发的阴极保护智能数据记录仪,将引入无线传输技术和智能诊断维护功能,能够简化阴极保护工程师和基层管理人员的阴极保护现场巡检工作,实现数据的快速采集和上传,同时提高感知阴极保护效果的时效性,同时,还可简化阴极保护工程师的阴极保护现场巡检工作,也可简化阴极防护技术,配合 APP、PC 端软件,可对阴极防护资料进行自动化处理分析,提升工作效能。阴极保护参数远传及上位机系统的自动统计功能,也能简化工作人员数据整理上报工作,大大降低了人员工作量,提升了阴极保护工作质量为阴极保护的维护和管理提供更加全面和高效的支持。

2.4 智能排流监测仪

目前,市场上各种排流设备相对齐全,排流功能完善,埋地金属结构的腐蚀保护绕不开也必须解决杂散电流干扰;但是,对于安装在现场的排流设备,其运行状况往往依赖于测试员现场实测,缺乏与物联网相结合的应用,因此,一款智能排流监测仪的出现,具备了对排流设备工作参数进行远程监控传输,将有效降低排流设施运行维护的工作量,成为目前国内急需的具备排流设备工作参数的智能排流监测仪,而这一点,也将成为目前我国再配合相应的数据可视化软件,对于干扰源对保护对象的影响变化有很好的帮助,可以及时调整排流策略,使保护效果达到最佳。

智能排流监测仪在普通监测仪的基础上,通过安装固态去耦合器/极性排流器/嵌位式排流器/接地排流装置等,使产品增加了排流器交直流排流电流采集(外置分流器)、地床电位/电压采集(通断前后)、全线远控开断排流地床、去极化电位采集等功能,能够得到管道更真实的数据和状态,智能排流监测仪工作原理图如图4所示。在常规情况下为闭合状态,用于导通管道和地床,在有测量需求的情况下,对此直流接触器可通过驱动单元实现通断控制,对其进行排前、排后阴保数据的采集。

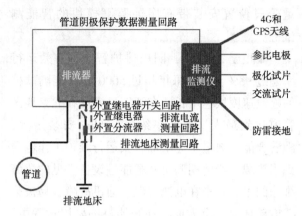

图 4 排水监控智能仪原理图

在工程项目中,智能排流检测仪因其高效、精确的监测与控制能力而得到广泛应用。特别是在地铁等轨道交通系统中,杂散电流的管理与排流是确保系统安全、稳定运行的关键环节。张洪健提出的集杂散电流与排流监测于一体的智能排流系统方案,为这一领域带来了创新性的解决方案。该方案提出了以大功率开关器件 IGBT 为核心的新型智能排流装置。WEB 监控系统是在 ASP.NETMVC 框架下设计的,实现了系统实时显示监控数据、维护设备故障信息等功能;马国栋在昆明地铁沿线管道杂散电流整治工程中,燃气管网杂散电流排监一体化管理系统是结合杂散电流干扰状况建立起来的,应用杂散电流排监一体化装置对管道阴极保护效果、杂散电流干扰状况、排流设施有效性等进行在线监测和评价,指导燃气管网日常运行和维护保养,并结合杂散电流的干扰状况,对管道出现的风险要及时排除。经试验,效果比较理想;汪理通过实施一系列创新措施,对智能导通柜、排流柜以及钢轨电位限制装置进行了全面的智能化升级。这一改造不仅优化了原有单一设备的监控控制算法,还成功地将南京地铁的杂散电流排流监控整合为一个高效的系统。该一体化系统有效遏制了杂散电流向地铁外部环境的扩散,相较于老旧设备,显著减少了其潜在的不良影响,并几乎消除了这些影响。这一系列改进确保了地铁系统的安全、稳定运行,并为未来的智能化管理奠定了坚实基础。

2.5 多路智能恒电输出装置

在阴极保护系统中,恒电位仪的应用范围是非常广泛的,所以恒电位仪的应用范围是非常广泛的，作为强制电流阴极保护系统的核心设备。在其工作状态和工作参数上可以直接反映阴极保护系统的运行状态和保护效果。但是在实际油气管道的维护工作中，传统的恒电位仪会存在以下问题：

（1）传统恒电位仪控制模式现存问题

a. 对测点的管路极化电位，不能准确反应通电电位的数据；

b、恒电位仪的"恒电位"方式因动态直流杂散电流干扰而不能正常运行；

（2）面向站场区域阴极保护的多路输出传统恒电位仪在应用中现存问题：

a. 传统恒电位仪的各种输出之间的相互干扰是恒电位模式下不能运行的；

b. 恒电位仪的人工调节过程复杂，要求管理人员水平必须要高；

c. 传统恒电位仪监测点数量少，无法全面掌控站场阴极保护；

但是智能多路输出恒电位仪可以同时输出实现自动控制阴保电流输出，一般作为电源使用外加电流阴极保护，原理架构图如图5所示，是实现智能阴极保护的基础设备，在整个智能阴保护系统中至关重要，是解决目前传统恒电位仪存在问题的主要措施，也是传统恒电位仪升级的首要方向。

Smart多路输出恒电位拥有先进的智能和自动化，其应用可在石油天然气实际站场区域带来强大的阴极保护优势，具体功能优势如下：

（1）恒电位仪断电电位控制

采用试片通断法，连续测量管道的断电电位，恒定电位仪的输出按测量的断电电位调整，使管道的断电电位达到控制目标(预置断电电位)。

（2）定时获取运行参数，智能多路恒定输出

在优化后的人机交互界面中，实现了对输出电压、输出电流及参比电位等关键数据的实时展示。为了增强数据的管理与追踪能力，我们在上位机配置软件中创新性地加入了日志记录功能，该功能能够自动记录并显示这些输出参数的详细资料。这样，用户不仅可以在人机交互界面上即时查看到这些重要数据，还能通过查阅软件日志，获得参数的历史记录，从而更全面地了解系统的运行状态。这一改进既避免了信息的重复展示，又提升了数据的可访问性和可追溯性。

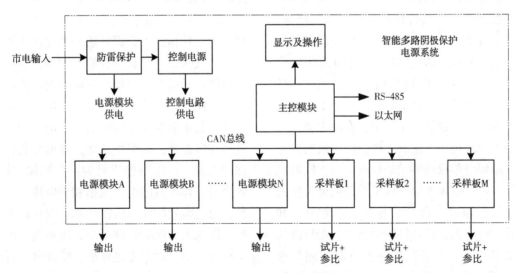

图5 智能多路恒电位仪电源系统架构图

（3）具有系统时钟功能

为系统中提供精确的系统时间，提高了智能恒电位仪的时钟同步性。

（4）通过先进的人机交互界面，用户可以灵活配置智能恒电位仪的多种工作参数。这包括但不限于设定其工作状态(如激活或待机)、选择具体的工作模式、规划开机测试的开始与结束时间、安排日常的开机与关机时段，以及调整输出状态参数等。此界面设计旨在提供一个直观且全面的控制面板，使用户能够轻松根据实际需求调整恒电位仪的运行配置，同时避免了对同一信息的重复设置与操作。

（5）支持两种调节方式：手动调节，自动调节。

在正常作业状态下，智能恒电位仪默认运行于自动调节模式。该模式下，一旦外部干扰导致电位发生变化，智能恒电位仪会根据预设的参数自动调整其输出。系统能够实时采集电位反馈，并据此动态调整输出功率，确保系统维持稳定运行状态。若自动调节功能发生故障，系统会自动切换至手动调节模式，以无缝衔接保障智能恒电位仪的持续运行。值得注意的是，在自动调节过程中，系统会持续监控并适时调整至最佳状态，确保智能恒电位仪的稳定性和可靠性，避免任何中断，保障整体系统的顺畅运作。

（6）具有远程资料控制功能。

为了满足多样化的现场环境需求，系统提供了灵活的沟通方式选项，使用户能够根据实际场景和需求，轻松选择最适合自己的交流途径。

（7）具有数据存储功能。

智能恒电位仪配备有专门的存储机制，用于安全保存系统组态数据，确保在系统复位后这些数据不会丢失。在资料上传流程中，系统会同步保存所收集的信息。面对网络通信不稳定的情况，系统会暂存收集到的资料，待网络连接恢复良好后再自动上传，确保资料上传的连续性。此外，该机制还能主动执行物资整理与储备任务，提升整体管理效率。八、内部集成同步接通功能

运用卫星授时模块，实现了时间的精准获取，进而显著提升了阴极保护系统的时间同步性能。高桂飞在某管道现场测试中，采用了断电电位作为控制基准。测试结果显示，在动态直流杂散电流的干扰环境下，配合土壤管道测量的断电过程，新型恒电位仪能够精确控制电位、实时调整且运行稳定，这极大地增强了线路阴极保护的效果。史汉宸通过组合GPS模块、控制电路和内部负载，研制出负载切换的恒电位，通过现场应用，可实现多条管道阴极保护电源的同步通断、GPS授时准确、数据可靠、通用性强等优点，通过组合GPS模块、控制电路和内部负载，研制出负载切换的恒电位；杨文乐研制了恒电位仪，该仪表支持全线同步电位通断，远程自动调节了全线贯通监控功能，并输出了恒电位仪；杨文乐成功研发了一款恒电位仪系统，该系统创新性地实现了全线同步通断的远程自动监控与调整功能，并有效增强了电平仪器的输出性能。经过严格试验验证，这款智能恒电位仪展现出了卓越的电位控制精度，其输出误差被精准地控制在1%以内，同时完全满足了设计要求的同步通断与通讯功能，确保了实验中的电位精确度达到很高水平。尤为值得一提的是，该恒电位仪具备远程自动调节能力，能够在管道遭遇欠保护或过保护状况时及时响应。目前，这款高性能产品已在深圳中石化管道项目中成功应用，并取得了显著成效。作为阴极保护系统的核心装置，智能恒电位仪通过对保护电流输出的精确控制，实现对金属结构物的有效保护；该恒电位仪系统完美契合了阴极保护领域对于智能化、网格化、数字化建设的高标准要求，有效弥补了市场上传统恒电位仪在数据上传、远程监控及同步通断测试功能上的不足。其应用显著提升了阴极保护系统的整体稳定性、可靠性及保护效果，同时，通过优化能耗，大幅降低了系统的运行成本，为用户带来了更为经济、高效的保护解决方案，促进了阴极保护技术的创新和发展。智能恒电位仪将在未来阴极保护系统中继续发挥重要作用，引领物联网、大数据、人工智能等技术的持续进步和发展。

2.6 阴极保护智能控制技术

阴极保护智能控制是一种基于电化学腐蚀原理、无线传输、大数据算法控制先进阴极保护腐蚀防控技术，它结合了智能监测与控制技术，可以对被保护的金属结构实施精确的阴极保护。

智能控制技术（SmartControlTechnology）是通过监测技术对金属结构的电位、电流等关键参数进行实时监测，并根据这些参数的变化自动调整保护电流，具体调控流程为：在被保护结构中，各个电位监控点的电位数据和恒电位仪的运行参数，通过高效的无线传输技术，以预设的频率实时发送至阴极保护管理系统。该系统内的智能自动控制模块，运用先进算法，综合分析各监控点的断电电位信息及恒电位仪的当前运行状态，精确计算出所需的恒电位仪输出调整量。随后，这些调整指令被迅速发送至对应的恒电位仪，指导其精准调整输出参数。这一过程确保了金属结构持续获得最优化的保护状态，同时，通过精细调控电流输出，实现了能源的合理分配与高效利用。使恒电位仪如图6所示的智能阴极保护控制机制也能有效降低阴极保护带来的能源消耗。

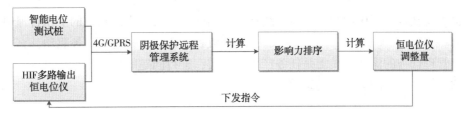

图 6　阴极保护智能控制机理

近年来，油气管道领域见证了智能控制技术的广泛研究、应用与推广，这一技术涵盖了通信技术、信息技术及工业物联网技术等多个方面。高桂飞等人开发了一种创新的区域阴极保护智能控制技术，该技术能依据站内多个监控点的实时断电电位数据，对恒电位仪的输出进行动态调整，确保各监控点的电位均满足阴极保护标准。通过智能控制算法，该技术实现了站场区域阴极保护的智能均衡控制。

智能控制技术的应用显著提升了区域阴极保护的效能，不仅有助于油气管道的安全高效运行，还降低了管控的复杂性和人力成本。与此同时，立冰等人则结合了防腐层缺陷检测技术与基于腐蚀电化学原理的控制技术，通过智能算法使被保护管道达到最优保护电位，确保管道电位符合国家标准的保护要求。这一成果通过人工模拟的区域阴极保护现场实验得到了验证。

高媛开展燃气智能化示范工程改造，将云计算、物联网、阴极保护智能终端等先进IT软硬件技术集成，实现燃气管网标准化、流程化、电子化、智能化的全过程管理，以北京一家燃气公司的智能燃气管网改造建设为例；贺琦在山东能源勃中海A场址提出了应用效果实验验证的智能阴极保护系统策略。测试结果显示，在海上风电设备的运行稳定性和可靠性方面，智能保护策略能够得到有效的改善。具体体现在降低设备的腐蚀速率、实时监测海上风电桩基的阴极保护参数、实现智能管理的可视化、提升设备的运行效率、延长使用寿命以及减少维护费用。

2.7　智能阴极保护系统

智能化阴极保护系统是以"打造智慧互联大管网"战略目标为基础，围绕阴极防护技术管理及工作管理的要求，通过首页、统计分析、日常管理、设备管控、基础资料及系统管理等六个基础功能单元，开发集监控、控制、管理、数据查看、辅助诊断分析于一体的智能化管理平台，围绕阴极保护技术管理和工作管理的要求，共计6个基础功能单元。

它促进阴极防护技术管理与工作管理的融合，使阴极防护整体提升工作质量和效率，更好地为安全高效运行提供保障，从传统的人工管理向智能管理模式的跨越。阴极保护管理是管道各项专业管理任务之一，包含技术管理和工作管理两个方面。其中，技术管理是根本，工作管理是保障。

（1）技术管理：以全面达标为导向，由数据驱动，实现细化、深化、融合、先进的目标。

智能阴极保护系统管理对象包括管线、站场阀室、防腐层以及阴极保护相关的设备设施。管理的数据包括完成阴极保护、防腐层和相关设备、技术管理所需的所有技术参数，具体工作内容如图7所示。

数据来源包括建设阶段数据（阴极保护设计、施工及验收数据）、日常运行参数（含智能终端采集数据）、检测数据、维修维护数据、整改施工数据、基础台账等涵盖管道运行全寿命周期的阴极保护数据。

（2）工作管理：以解决问题为导向，由事件驱动，实现全面、闭环、聚焦、高效的目标。

智能阴极保护系统管理对象主要来自于公司阴极保护管理程序中规定的工作任务、任务产生的问题处理及工作任务执行情况等。根据管理对象不同可分为防腐层管理、检测检验任务、设备设施管理、杂散电流干扰管理、日常运维测试管理、故障及隐患治理等。根据任务的生成的周期不同可将任务定义为两类：手动任务及自动任务。手动任务是指任务开始和结束时间不确定，需要人为制定；而自动任务是指根据管理文件要求的定期需要进行工作，具体内容如图8所示。

（3）技术管理与工作管理相融合

实际工作中，技术管理与工作管理密不可

分,技术管理需要工作管理提供数据来源保障,而工作管理也需要技术管理中的数据分析、统计结果作为驱动。

所有工作管理任务的流转本质上是任务产生的技术数据的流转,任务的关闭也是以数据的填报、审核完成作为控制节点。而技术管理中发现的问题,又会生成相应的工作管理中的问题处理流程,流程由相应的任务驱动,直至问题得到处理,进行问题闭环。

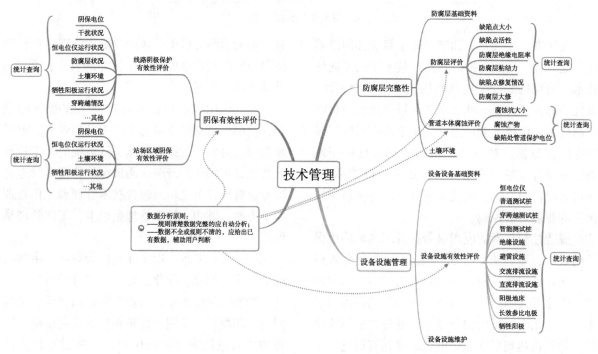

图 7 技术管理工作内容树状图

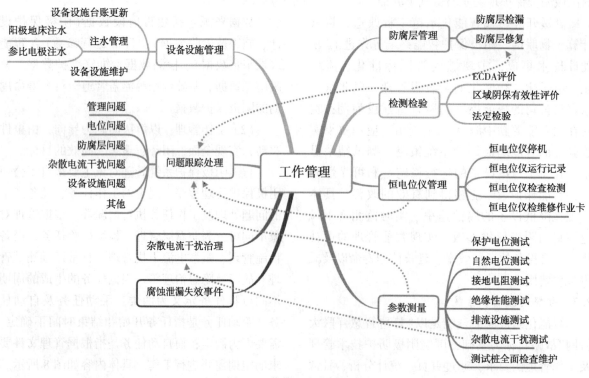

图 8 工作管理内容树状图

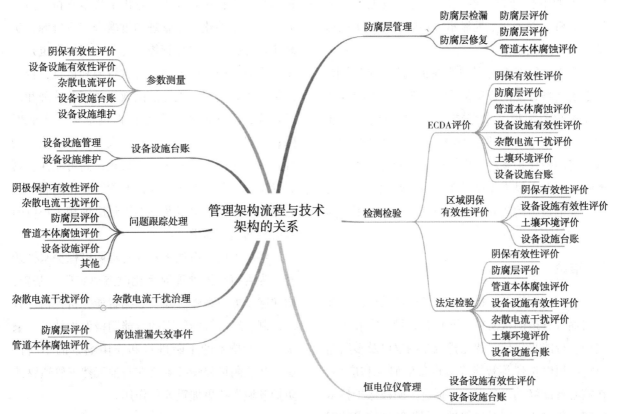

图9 技术管理与工作管理融合应用树状图

智能阴极保护系统的研制与应用对于提高防腐效率、降低维护成本、提升安全性、促进智能化发展以及适应多种环境等方面都具有重要意义，下面是智能阴极保护系统应用案例。

徐浩深入分析了管道事故案例，并据此设计了一套系统化的油气管道阴极保护监控数据管理软件及其实操流程。该软件旨在针对油田现场服务的油气管道，进行剩余寿命的可预测性计算和评估。这一创新不仅推动了管道寿命预测技术的发展，还在管道完整性研究领域取得了显著的进步；佰添以电指纹法（FSM）为基础，结合ZigBee数据无线传输技术和使用LabVIEW开发的上位机软件，设计出一套能够实时监控管道腐蚀情况的监控系统，并在管道工程中试验使用；李丹涛设计的油气管道在线监控系统方案，依托超低功耗MSP430单片机遥测终端为核心，全面支撑油气管道的阴极保护服务。该系统实现了实时监控功能，确保数据的有效存储与分析，并通过数据可视化手段提供预警信息，从而实现对油气管道状态的全面掌控；刘昊天基于NBIOT设计的分布式阴极保护智能监控系统，利用NBIOT物联数据传输升级GSM和GPRS通信方式，融合分布式监控与微处理器技术，我们成功制定了一套针对埋地油气管道阴极保护参数的动态监测方案，实现了参数的实时追踪与监控。

梁锦基于C/S架构，精心设计了包含节点软硬件、上位机控制软件及点对点无线远传协议的完整方案，开发出无线传感网络阴极保护数据采集管理系统。该系统通过深入研究串口通信、平台架构、数据库管理及权限控制等关键环节，实现了石油管道阴极保护数据的高效采集、传输、总结与管理。

在山东济华港润公司的105公里长输管线工程中，潘尔璟设计的智能阴极保护管理系统得以应用，该系统极大地促进了管道安全运行的智能化、信息化与统一化管理，确保了阴极防护系统为管道安全提供坚实保障。

李春雨在花格输油管道试验中，引入了智能阴极保护系统，采用GPRS与北斗卫星通信双模数据传输，创新性地使用长效参比电极填充物隔离土壤离子污染并保湿，结合定期浇水维护，选用多孔陶瓷材质以减少渗漏，有效抑制了长效参比电极的电位漂移。该系统自动完成测量、上传、分析、警示及数据显示等功能，提升了阴保管理的智能化水平。

李光达针对海上风电系统的腐蚀控制需求，开发了新型智能阴极保护系统。该系统不仅与阴极保护数值模拟技术相结合，优化辅助阳极布置，还实现了远程监控保护状态与远程操作控制，将阴极保护系统全面智能化。同时，该系统能够监控海上风力发电设备的多种腐蚀要素，为风力发电设施的防腐时效性与有效性提供了更为坚实的保障。

3 结论

智能化阴极保护技术在站场区域阴极保护和线路油气管道阴极防护方面均有大量研究和工程应用，智能采集仪、智能排流监测仪以及多路输出智能恒电位仪等智能终端设备的研发和应用，在核心智能终端设备方面，促进了阴极保护技术的智能化进程，同时也促进了阴极防护技术的智能化和应用，并在核心智能终端设备、智能排流监测仪、多路输出智能恒电位仪等多个方面进行了深入的研究和应用，促进了阴极保护技术在核心智能终端的应用。智能采集仪能够自动采集、传输数据，实现在线诊断与评估、监控与预警，有效提高了阴极保护状态的监测精度和时效性。智能排流监控仪实现了交直流杂散电流排流与采集监测一体化的地床电位等关键参数，使管道腐蚀防护得到更全面的数据支持。而智能恒电位仪则通过精确控制保护电流输出，更加有效地保护金属结构物，为阴极保护夯实基础，实现智能化、网格化、数字化。

在智能阴极保护技术的发展过程中，智能化和数字化技术的融合应用发挥了关键作用，未来通过物联网、大数据、人工智能等先进技术的引入，智能阴极保护技术将继续向更高层次发展，以实现更全面的感知和智能分析评估管道状态。一方面，随着大数据分析和人工智能技术的不断进步，智能阴极保护系统的感知分析能力会更强，预测能力也会更全面。通过深入挖掘和分析历史资料，可以为管道的维护管理、预测管道腐蚀趋势和潜在风险提供更精准的决策支持。另一方面，从柔性化工作管理角度融合多方资源，实现阴极保护的一体化管控及优化运行能力也是未来的发展方向。通过整合各方资源和信息，智能阴极保护系统将实现更加高效、协同的管理和运维模式，为管道的安全运行提供更加坚实的保障；同时，针对油气管道跨度大、自然环境复杂等特点，应研发多元化、集成化、高可靠性、具有异常诊断分析的阴极保护监测装置，以满足对管道实际腐蚀状况以及影响管道腐蚀相关环境因素的全面监测需求。

将助力国家石油天然气管道阴极保护水平跃升，为管道整体管理数字化改造奠定坚实基础，构建起全方位智能感知、数据智能分析评估、智能控制、终端智能管理的智能阴极保护技术。未来，随着技术的不断进步和应用场景的不断扩展，智能阴极保护技术将在石油天然气管道腐蚀防护领域发挥更加重要的作用。

参 考 文 献

[1] 牟春霖. 智能监测采集仪在城轨基坑监测中的应用[J]. 2023(05)：36-41.

[2] 薛光. 管道工程智能测试桩和阴极保护监测系统[J]. 2011(06)：63-65.

[3] 张洪健. 地铁杂散电流智能排流系统研究[D]. 中国矿业大学. 2023.

[4] 马国栋. 城市燃气杂散电流排监一体化系统的应用[C]. 中国燃气运营与安全研讨会（第十一届）暨中国土木工程学会燃气分会2021年学术年会. 2021.

[5] 汪理. 杂散电流一体化智能综合监控分析系统的研究[J]. 计算机测量与控制. 2020(28)：81-85

[6] 高桂飞. 一种新型数控高频开关恒电位仪的研制与应用[J] 腐蚀与防护. 2023(44)：82-86.

[7] 史汉宸. 输气管道阴极保护恒电位仪改造研究. [D]西南石油大学. 2018.

[8] 杨文乐. 埋地管道智能恒电位仪的研制[D]. 河北工业大学. 2020.

[9] 高桂飞. 站场区域阴极保护智能控制技术[J]. 材料保护. 2022(55)：212-217.

[10] 李冰. 用于地下金属管网的智能控制阴极保护装置[J]. 用于地下金属管网的智能控制阴极保护装置, 2008, 6(6)：464-467.

[11] 高媛. 智能技术在城市燃气输配管网系统的应用研究[D]. 北京建筑大学. 2017.

[12] 贺琦. 海上风电阴极保护的智能化应用研究[J]. 材料保护. 2024，38(8)：116-121.

[13] 徐浩. 油气管道运行监测系统研究与实现[D]. 西安石油大学. 2023.

[14] 白添. 管道内腐蚀状态监测技术研究[D]. 吉林建筑大学. 2021.

[15] 李丹涛. 基于阴极保护的油气管道在线监测系统设计与实现[D]. 郑州大学. 2022.

[16] 刘昊天. 埋地油气管道阴极保护智能监测系统的研究及应用[D]. 西安石油大学. 2023.

[17] 梁锦. 基于无线传感器网络的阴极保护数据采集管理系统[D]. 北京邮电大学. 2013.

[18] 潘尔璟. 智能化、信息化阴保系统对高压燃气管网腐蚀防护的重要意义[C]. 2022年第五届燃气安全交流研讨会论文集(上册). 2022.

[19] 李春雨，何鹏程，耿立娟等. 智能阴极保护系统在青海油田的应用与评价[J]. 石油化工腐蚀与防护，2022，39(6)：60-64.

[20] 李光达. 新型海上风电智能阴极保护系统研究与应用[J]. 全面腐蚀控制. 2023，37(10)：42-45.

国内外油气行业管材标准现状及发展趋势分析

田 灿 崔绍华

(国家石油天然气管网集团有限公司)

摘 要 随着全球经济增长和工业化进程加速,世界各国对石油天然气等能源的需求量逐年递增,推动了油气钻采和储运产业的快速发展。管材是油气钻采和储运的关键装备之一,主要包括钻杆、套管、油管和管线管。石油和天然气工业专用管材标准,规定了油气行业专用管材的生产制造及使用规范,具有严谨性、实用性和持续更新的特点,为保证产品质量、提高操作安全、克服技术壁垒、促进贸易便利化提供了基础。在此基础上石油天然气行业内的工程设计、建造、试验、运行等方面能具有统一性及规范性,还可以以文件的形式说明不推荐、不适合、非规范的事项。基于此,为了营造稳定、高效、安全的石油天然气产业发展氛围,探析石油天然气行业专用管材标准应用现状、存在的问题以及发展趋势具有非常重要的意义。本文首先对现行的国内外石油和天然气工业专用管材标准分类和特征进行了梳理,分析了现行的不同级别油气管材标准规范。然后从标准体系、用户应用以及生产制造商三个方面论述了标准在实施过程中的难点和问题,认为部分标准缺乏统一性,标准条款含糊不清,缺乏可操作性;制造商和用户对标准缺乏全面的理解,在恶劣环境下对标准的选择存在不一致现象;制造商的标准管理部门缺乏经验,标准质量管理体系尚不健全。最后,根据标准应用的实践经验,提出了油气管材标准未来的发展方向,尤其在氢和二氧化碳等新能源领域方面提出了建立健全相关管材标准的建议。

关键词 油气行业;管材;标准;现状;趋势;新能源

Application Status and Development Tendency for the Standards of Tubular Goods for Oil and Gas Industries

Tian Can   Cui Shaohua

(China Oil & Gas Pipeline Network Corporation)

Abstract With the acceleration of global economic growth and industrialization, the demand for energy such as oil and gas is increasing year by year all around the world. This situation has promoted the rapid development of oil and gas drilling, production, transportation and storage industry. Tubular goods are part of the key equipment in oil and gas drilling, production, transportation and storage industry, which mainly include drill pipe, casing, tubing and pipeline. The standard stipulates the manufacture, construction and operation specifications for tubular goods of oil and gas industry. It has the characteristics of rigor, practicality and continuous update. And this type of standards provides a basis for ensuring product quality, improving operation safety, overcoming technical barriers and promoting trade facilitation. On this basis, the engineering design, construction, test and operation and other aspects of the oil and gas industry can be unified and standardized. At the same time the unrecommended, unsuitable and non-standard items can be explained in the form of documents. Based on this, in order to create a stable, efficient and safe atmosphere for the development of oil and gas industry, it is of great significance to analyze the application status, existing problems and development trend of tubular goods standards for oil and gas industry. In this paper, firstly the classification and characteristics of the current domestic and foreign tubular goods standards for oil and gas industry are analyzed, and the current standard specifications for different kinds and levels are shown. Then the difficulties and problems in the implementation of the standards are discussed based on three aspects: the standard system, the user application and the manufacturer. The analysis results show that some standards are lack of uniformity, and the standard clauses are ambiguous and lack of operability. Manufacturers and users lack a comprehensive understanding of standards, and there is inconsistency in the selection of standards in harsh environments such as ultra deep well and corrosion surroundings with H_2S. The manufacturer's standards management department lacks mature experience, and the standard quality management system is not yet perfect. Finally, according to the practical experience of standard application, the future development direction of tubular goods standards is put forward. Especially in the field of new energy area such as hydrogen and carbon dioxide, suggestions on establishing and optimizing relevant tubular goods standards are put forward.

Key words Oil and Gas Industry; Tubular Goods; Standards; Application Status; Development tendency; New Energy Industry

标准是指在科学、技术和实践经验的综合成果基础上，对活动或其结果规定共同的和重复使用的规则、导则或特性的文件，它以在一定范围内获得最佳秩序、促进最佳社会效益为主要目的，标准文件需经协商一致制定并经一个公认机构的批准发布。相应地制定、发布和实施标准的过程被称为标准化。近年来，随着工业化、国际化水平的飞速发展，标准尤其是国际通用标准在提高操作安全、保证产品质量、克服技术壁垒和贸易便利化等方面发挥着越来越重要的作用和价值，被认为是技术和经济发展的关键。

随着全球经济增长和工业化的迅速发展，世界范围内石油和天然气的需求呈持续增长的趋势。各国政府和各大石油公司都在不断推动世界各地的石油勘探和生产活动。油气管材包括石油专用管材(OCTG)和长输管线管材，这些材料是构成油气勘探、钻采、储运等生产活动用关键装备的重要组成部分。OCTG 主要由套管、油管和钻杆(包括钻杆、加重钻杆、钻铤、方钻杆等)组成，根据其特定的服役环境(如高温、高压或强腐蚀等)承受载荷。长输管线管材是指用于长距离输送石油、天然气以及氢等新能源介质类流体的钢管。根据制造方法，长输管线管材可分为无缝(SMLS)管和焊接管两大类。其中，焊接管主要包括螺旋埋弧焊(SSAW)管、直缝埋弧焊(LSAW)管和电阻焊(ERW)管。近年来，随着油气钻采向深海、深地、高腐蚀等非常规环境等方向发展，以及氢、二氧化碳等新能源相关介质长距离输送需求的增加，对油气管材的技术创新提出了更多的要求，石油天然气工业中也不断涌现出各种新型石油装备，如非金属管、复合/内衬管、钛基/镍基合金管等。

油气管材装备的持续更新进步，对相应的标准也提出了不断更新优化的要求，以保障管材制造规程的规范以及服役过程的安全。本文简要介绍了油气管材标准的现状和特点，详细讨论了标准实施过程中出现的的问题。在此基础上，对油气管材标准未来的发展方向提出了建议，尤其在氢和二氧化碳等新能源领域方面提出了建立健全相关管材标准的倡议。

1 油气管材标准的现状

按照应用范围可将油气管材标准分为四个主要层次。①国际标准，由公认的国际组织制定，如 ISO 11960 石油和天然气工业-用于井套管或管道的钢管、ISO 3183 石油和天然气工业-管道钢管等；②国家/地区标准，在中国由全国石油天然气标准化技术委员会提出并归口管理，如 GB/T 19830 石油天然气工业-油气井套管或油管用钢管、GB/T 9711 石油天然气工业 管线输送系统用钢管等；③行业/协会/团体标准，由油气管道行业协会等组织提出并归口管理，如 SY-T6194 石油天然气工业油气井套管或油管用钢管、SY/T 6601 耐腐蚀合金管线管等；④企业标准，由油气管道材料设计、制造、施工与应用等企业制定的，在企业内部的技术要求、管理要求和工作要求，是企业组织生产、经营活动的依据，如 Q/SY 1513 油气输送管道用管材通用技术条件、Q/SYGD 0503 中俄东线天然气管道工程技术规范等。

在这里需要特别指出的是，一些具有较高的国际影响力的行业协会、团体等机构发布的标准文件，也在国际上通用，如欧洲标准化委员会发布的一系列 EN 标准和美国石油协会发布一系列 API 标准。ISO 和 API 标准作为最重要的基石已在全球石油管材领域广泛实施。事实上，许多 ISO 标准都是基于类似的 API 规范，API 也采用并共同发布了许多 ISO 标准，同样 API 与 ANSI（美国国家标准学会）、ISO 与 NACE（美国国家腐蚀工程师协会）、ISO 与 CEN（欧洲标准化委员会）之间也存在着密切的合作，最大限度地获得了兼容性，确保了行业效益。甚至由于其完备成熟的条款和广泛的应用范围，已经作为通用的基础性标准文件。例如，由国际标准组织发布的最新版本 ISO 3183：2019 石油和天然气工业-管道钢管（第四版）删除了与 API SPEC 5L—2018 相同的内容，将 API SPEC 5L-2108 作为引用标准。

油气管材企业标准多在现有的国际或国家标准基础上进行了更为严格的规定。企业标准在国际贸易活动中受到广泛关注，因为它们不仅在企业内部使用，而且在其商业伙伴中使用，作为贸易过程的重要参考条款。同时，越来越多的油气行业领先企业，如埃克森美孚、壳牌、沙特阿美和中国的中石油、国家管网等，在油气管材的生产和应用实践中不断制定和完善自己的企业标准作为国际/国家标准的补充。

随着中国石油工业国际化的发展，标准已被广泛认为是能源石化行业提高技术和经济竞争力的重要手段。中国石油工业标准化委员会（CPSC）的许多活动都在加强与国际领先的标准化组织，如 ISO 和 API 的合作。因此，中国近一半的国家标准（GB）和石油行业标准（SY）采用了国际标准、有实力的行业标准和先进的企业标准。

表 1 展示了目前常用的石油管材标准。其中，GOST 系列标准是由欧亚标准化理事会（EASC）制定的一套技术标准，欧亚标准化委员会（EASC）是一个在独立国家联合体（CIS）主持下运作的区域标准化组织，GOST 标准在包括俄罗斯在内的所有独联体国家使用，而 GOST R 标准通常在俄罗斯联邦境内有效。ASME 标准是由美国机械工程师协会制定，CGA 由欧洲压缩气体协会制定，CSA 标准由加拿大标准协会制定，AIGA 标准由亚洲工业气体协会制定，IPS 标准为伊朗石油标准；AS 是澳大利亚标准的缩写。

虽然 API 起源于美国，但它是全球石油和天然气行业使用的标准中最受尊敬和最有影响力的贡献者之一，特别是在管材方面，事实上从国家/地区到行业或企业级别的许多标准都是根据 API 标准制定的。API 标准因其对所有相关方的广泛开放性和实际行业实施的高度可行性而获得广泛的国际认可。然而，当前的 API 管材规范只确保管材满足最低要求，对于实际工程项目或应用环境的一般需求或特殊应用场景，行业或企业标准均进行了不同程度的修改和补充，如对钢质管材的合金成分提出更窄的控制要求，尤其是 P、S、O、N、B 等杂质/气体元素；或者对管材的力学、耐蚀等性能提出了更高的要求，例如低温环境下的夏比冲击功和 CTOD 值、高温环境下的耐腐蚀性能等。可以说，行业或企业级别的油气管材标准规范的条款规定较 API/ISO 国际通用标准更为严格，但其适用范围相对较窄。

表 1 油气管材标准统计表

产品类别	国际标准	国家/地区标准	行业标准	具有国际影响力的权威团体标准
油套管	ISO 11960 ISO 13680	GB/T 19830 GB/T 9253.2 GB/T 23802 GOST R 53366 GOST R 53365	SY/T 6194	API SPEC 5CT API SPEC 5B API SPEC 5CRA
连续油管	-	-	SY/T 6895	API SPEC 5ST
钻杆	ISO 11961 ISO 15546	GB/T 20659 GOST 631	SY/T 5561 SY/T 5290	API SPEC 5DP

续表

产品类别	国际标准	国家/地区标准	行业标准	具有国际影响力的权威团体标准
钻铤、加重钻杆、方钻杆	ISO 10424	GB/T 29166 GB/T 20659 GB/T 41343	SY/T 5144 SY/T 5146 SY/T 5561 SY/T 6765 SY/T 6509 SY/T 6857 SY/T 6896	API SPEC 7-1 API SPEC 7-2
长输管线管	ISO 3183	GB/T 9711 CSA Z245.1 IPS-M-PI-190	SY/T 5038 SY/T 6601 SY/T 6700 SY/T 6623	API SPEC 5L API SPEC 5LC API SPEC 5LCP API SPEC 5LD
玻璃纤维增强塑料管	ISO 14692	GB 15558.1 GB/T 13663 AS 14692	SY/T 6267 SY/T 6266	API SPEC 15HR API SPEC 15LR
氢能输送专用管	—	氢气储输管道用钢管(起草) 氢气输送和存储管道用钢板和钢带(起草) 输氢用钢质热煨弯管(起草) 输氢用钢质管件(起草)	T/CAS 851	ASME B31.12 CGA G-5.6 AIGA 033/14
CO_2 输送专用管	ISO 27913	GB/T 42797	SH/T 3202	DNVGL-RP-F104

另外，面对着环境危机和能源危机的双重挑战，在"双碳"目标的推动下，氢气和二氧化碳的输送和存储已成为近年来的热点。高性能管材是实现氢气和二氧化碳的输送的重要方式和基本保障。

国际上氢气输送技术及输氢管材方面开始研究较早，目前已形成了较为成熟和健全的氢气输送标准体系，公布了一系列的输氢管道相关标准规范。如欧洲压缩气体协会的 CGA G-5.6 (R2013) 氢气管道系统，美国机械工程师协会的 ASME B31.12 氢用管道系统和管道和亚洲工业气体协会的 AIGA 033/14 氢气管道系统等。而国内由于起步较晚，尚未发布适用于埋地长距离输氢管道的标准规范，但国内的相关企业和科研机构已经对氢损伤机理、纯氢输送管材开发等相关理论和技术开展了研究工作，中石油、中石化、国家管网以及管材制造企业已经启动了氢气储输管道用钢管、氢气输送和存储管道用钢板和钢带、输氢用钢质热煨弯管、输氢用钢质管件等纯氢输送相关管材的国家标准编制工作。

关于 CO_2 输送专用管材，目前尚无系统的标准规范。国外由挪威船级社 DNV 发布了 DNVGL-RP-F104 二氧化碳管道的设计和运行，由 ISO 发布了 ISO 27913 二氧化碳捕集、输送和地质封存—管道输送系统，但该两项标准仅规定了 CO_2 管道的设计、运行和管理准则，对于管材的成分、性能等未做明确要求。国内同样如此，仅公布了 SH/T 3202 二氧化碳输送管道工程设计标准以及 GB/T 42797 二氧化碳捕集、输送和地质封存管道输送系统，局限于二氧化碳输送管道的设计、建设和运营等方面。

2 油气管材标准中存在的问题

本部分从标准条款、标准体系、用户应用和制造商等方面讨论了标准实施中的问题。

2.1 标准条款

通过对国内西气东输三线、西气东输四线等重大天然气长输管道工程中 D1219×18.4mm 规格 X80 直缝埋弧焊管的冶金成分和力学性能实际情况进行了大批量的统计分析。发现虽然这些工程采购的钢管均执行 API SPEC 5L 管线管规范、GB/T 9711 石油天然气工业管线输送系统用钢管和国家石油天然气管网集团设计与工程建设准则 DEC-NGP-S-PL-003 输气管道工程钢管通用技术规格书等相关标准规范的要求，但因钢板/钢管生产制造企业装备水平、技术路线的差异，造成了同一管道工程用管材的冶金成分和力学性能分布具有很大的分散性，如图1和图2所示。

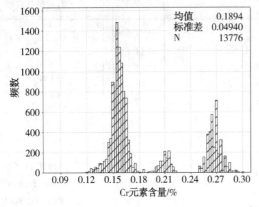 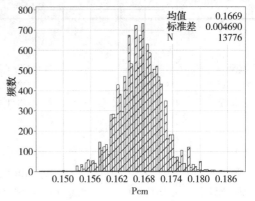

图1 D1219×18.4mm X80 管线钢 Cr 元素和碳当量 Pcm 分布图

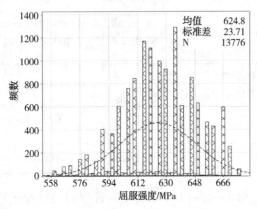 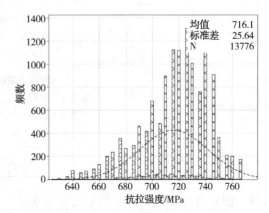

图2 D1219×18.4mm X80 管线钢强度分布图

从图示的统计结果可以看出，由于冶金成分和生产制造工艺不同，导致同种尺寸规格的 X80 钢管强度具有很大的分散性，而一般情况下管道施工过程中环焊一般采用相同的焊材和焊接工艺，因此环焊缝的强度理论上处于一个较为稳定的范围。这就导致了部分环焊缝存在与母材低强匹配的问题，在承受较高的外部载荷时，环焊缝成为了变形的薄弱环节，存在失效风险。

所以，对同一管道工程或相同级别和尺寸规格的钢管，国际通用标准和国家标准存在对合金成分和力学性能规定范围较宽的问题。相关标准规范尤其是企业规范中应该提出更高的要求，即对冶金成分和力学性能的波动范围提出更窄的控制要求。

2.2 标准体系

（1）不分标准之间缺乏一致性。目前，国内油气管材行业除了广泛使用 API 标准外，还同时存在大量的其他级别的标准，包括国家标准、国际标准、行业标准甚至工程项目标准规范，如中俄东线管道工程、西气东输三线/四线管道工程标准规范等。这些标准之间或其中包含的要求之间存在一些差异和不一致，这种差异导致了关于哪个标准优先于另一个标准的混淆，可能会给标准用户带来额外的成本或损失。此外，部分技术内容存在重复或重叠问题，而没有为某些共性关键问题制定具体的规则，阻碍了标准化和行业的发展。因此，标准制定的组织和单位必须考虑到各种标准的统一和协调，从而减少或消除混淆和不便。

（2）某些特殊产品缺乏统一性。近年来随着石油开采的不断深入，油井越来越深，井况越来越复杂，定向井、水平井和海上深水井的数量也在显著增加。复杂油井的特点是腐蚀条件严重，包括高温、高 CO_2 或 H_2S 分压和高 Cl^- 浓度，因此传统管材因无法承受如此恶劣的环境而变得不满足油气开采需求。因此，市场对先进的高级别产品有着强烈的市场需求，如气密螺纹套管和油管、高扭矩钻杆、抗酸钻杆、耐腐蚀套管、高抗挤毁套管、深井超深井套管、非标准尺寸套管和油管等，这些产品与传统 API 标准产品相比具有更好的性能。对于传统管材，API 标准基本上规定了从产品设计、制造、测试和检验、性能要求到工程应用的所有方面以及现场使用的推荐做法，形成了一个集成系统。随着先进技术和产品的不断涌现，人们已经认识到现行标准的局限性，不分高端管材的性能远远超过 API 标准的要求。然而却缺乏相应的标准来规定相关管材的工艺、材料、测试和检查以及施工流程，可能会导致严重的质量问题和安全风险。总之，迫切需要标准制定组织制定并发布新的产品标准或修订现有标准，以跟上市场要求和新技术的步伐。

（3）某些标准术语含糊不清，缺乏可操作性。考虑到制造商或用户的实际情况等原因，一些标准条款中存在有意和无意的歧义。尽管会使标准实施的灵活性变大，但不可避免的是，这种模糊性很容易引起用户和制造商之间的纠纷。例如，在 API SPEC 5CT 套管和油管规范中，PSL2 级套管和油管必须进行螺纹喷砂处理，但没有具体的技术要求和验收标准，如表面粗糙度和轮廓深度。事实上，如果标准包含模糊的规范，无论实施过程多么周密，实施之间的互操作性问题仍然可能出现。除了上述的歧义问题外，某些标准术语缺乏可操作性也是一个不容忽视的突出问题。例如，所有 API 套管和油管的具体储存、运输和现场操作应按照 API RP 5C1 套管和油管保养和使用的推荐规程中概述的要求确定。然而，考虑到现有的基础设施和复杂的现场条件，很难从概念要求过渡到实践中的标准实施。

2.3 用户应用

（1）对标准缺乏全面的理解。尽管 API 标准取得了巨大的成功并得到了广泛的应用，但在实践中仍然存在对这些标准理解不全面的情况。一些用户相信 API 标准已经走向全球，并在石油和天然气行业的所有领域得到使用，因此它们代表了行业的最高技术水平。然而，越来越多的用户认识到了当前 API 标准的局限性并在某些特定的服务环境中使用其他公认的标准，但由于用户理解不全面或不到位，仍会产生混淆。例如，一位用户没有完全理解 NACE 标准，当他们需要在酸性条件下进行材料选择时，他只向制造商询问符合 NACE 标准的耐酸性油管，但并没有说明是哪一种 NACE 标准。然而，NACE International 是一个全球性的权威组织，主要处理任何形式的腐蚀。与 API 类似，NACE International 在收集实验室实验数据和现场经验的基础上，发布了涵盖标准实践、试验方法和材料要求的多种标准，如 NACE MR0175、NACE MR0103、NACE TM0177 和 NACE TM0284。具体而言，MR0175 主要规定了酸性腐蚀服役环境下油气专用管材的选材标准，MR0103 是腐蚀性环境中硫化物应力腐蚀开裂的指南，TM0177 为金属在 H_2S 环境中应力腐蚀开裂的试验方法指导，TM0284 则侧重于管道和压力容器钢的抗氢致开裂性评价实验方法。标准之间存在较大差异，供应商不能代替用户做决策。在这种情况下，用户有必要对标准有一个全面的了解，并深入了解工程实践。

（2）油气管线服役环境愈加复杂，人们认识到 API 标准已不能完全满足实际要求，需要提出更专业化和规范化的专用标准。随着使用环境的日益复杂，理应使用更严格的标准条款要求，但必须谨慎。目前油气管材的采购过程常采用 API 标准和补充技术要求相结合的方式，补充技术要求是基于 API 标准为具体项目合同文件"定制"的，但侧重于限制性更强或大大超出 API 标准的问题。例如，为了提高质量，通常在补充技术要求中提到测试频率、检查水平、检测方法和具体的外观要求。决定一个具体的补充技术标准需要对不断变化的技术和广泛的实践经验有很高的了解。尽管用户在安全性提升等方面受益于使用更具限制性的标准，但是这些特殊要求应该对制造过程中的额外成本和制造难度增加负责。实际上对于特定应用场景，性能的提高应始终与成本的增加和制造难度相权衡。一般来说，建议仅在应用场景需要时使用更严格的标准，否则就没有必要浪费资源。

（3）将标准作为技术壁垒。在对贸易、经济和程序问题的几项限制中，技术性贸易壁垒受到了极大的关注。技术壁垒包括标准、技术条例、经认可的实验室对产品的测试检查和认证制度。标准的制定和采用可能并不总是能够达到扩大网络、增加总体利益和促进公平竞争的目的，它也被一些公司战略性地用作竞争工具。显然，这种技术壁垒使供应商很难、甚至不可能与另一个国家进行贸易。例如，出口油气管材产品到俄罗斯时，通常要求管材必须符合 GOST 标准。虽然 GOST 标准与 API 标准大致相同，但它们在尺寸和连接特性方面存在一些关键区别。具体而言，除了 GOST 标准，在俄罗斯 Rosneft Oil 采购和招标过程中需要同时参考 Ty 规范。Ty 规范是俄罗斯石油公司为具体项目或合同制定的补充技术规范，供应商或国外制造商难以获得和执行，因此在实践中给行业带来了不良的信息不对称和技术障碍。

2.4 制造商

随着全球化进程的发展，国际标准对于希望在国际市场上合规的制造商变得越来越重要。由于标准不仅可以合理化制造工艺，还可以降低制造成本，因此合规化已成为大多数制造商产品开发和营销的主要义务。制造商努力通过实施国际标准来制造具有竞争力的产品，但许多实施过程的失败往往源于对标准的理解不足和缺乏生产或管理部门的经验。作为一家负责任的先进制造商，标准合规性应该是日常业务运营中根深蒂固的一部分。此外，制造商应该意识到，当他们的员工接受培训并有资格根据行业标准履行职责时，他们可以为公司的成功做出贡献。因此，制造业必须建立一套适合其产品及顾客不断变化需求的、持续有效的品质管理系统。

3 油气管材标准发展的几点建议

（1）在通用标准的基础上，针对特定工程项目或服役环境，对油气管材制定补充技术要求。尤其是针对长输管材的需求量较大的产品相关标准，应该重视对产品金属原材料冶金成分和生产制造工艺的限制，以降低产品服役性能的分散性，提高管道工程完整性和安全性。

（2）通过国际、国家和地区层面各种标准的统一和协调，可以减少产品和服务交流的潜在障碍，从而推进全球贸易目标。这是标准制定者在编制和修订标准时时需重点考虑的内容，以达到减少重复工作，并最有效地利用可用资源的目的。此外，还应加快制定特殊恶劣环境下管材专用新标准，如超深井用管材规范、特殊接头套管和油管的选择评价和检验规范等，进一步完善标准，促进行业优化。

（3）对现有标准和新制定的标准，需要进一步宣传贯彻。它可以帮助制造商和供应商确定哪种解决方案最能满足当前的要求，并最终利用机会促进行业专业技术的发展。

（4）建议国家或公司级别的标准用户参与国际标准制定过程。持续参与国际标准制定活动对于信息和情报收集至关重要。它将确保国内行业不仅学习和采用国际标准，而且还支持基于中国经验和最佳工程实践的国际标准的发展。

（5）建议尽快推进国内氢气和二氧化碳等新能源介质输送用管材相关的选材、检验评价方法、生产制造工艺和服役性能要求等标准规范的制定和实施，以响应"双碳"战略的推进和推进新能源产业的发展。

参 考 文 献

[1] 毛浓召，祝少华，宋海辉. 国内外油气输送管道钢管标准发展现状与研究[J]. 中国标准化，2024，（14）：62-66.

[2] 曹燕，谭笑，刘冰，常晓然，郑素丽. 我国油气管道标准国际化的现状、瓶颈与突破路径研究[J]. 中国标准化，2023，（17）：65-70.

[3] 高进伟，崔绍华，曹杉，时文. 石油管材产品标准应用现状及发展建议[J]. 石油管材与仪，2018，4（02）：1-4+8.

[4] 李为卫，谢萍，杨明，吴锦强. 油气输送管道用钢管标准的发展历程及趋势[J]. 油气储运，2019，38（06）：601-608.

[5] Matthew L. Nelson, Michael J. Shaw, William Qualls. 2005. Interorganizational system standards development in vertical industries[J]. Electronic Markets. 2005. 15：378-392.

[6] International Association of Oil & Gas Producers. Catalogue of International Standards Used in the Petroleum and Natural Gas Industries, 2012. Report No. 362.

[7] 徐晓峰，李为卫，秦长毅，方伟，徐婷. 石油管材标准体系优化研究[J]. 石油管材与仪器，2015，1（06）：77-81.

[8] 牛爱军，毕宗岳，韦奉，黄晓辉，刘斌，席敏敏. 我国新能源输送用管材研究进展及发展趋势[J]. 焊管，2024，47（10）：16-24.

[9] 毕宗岳，余晗，鲜林云，王维东，赵博. 连续管技术研究现状与发展趋势[J]. 焊管，2023，46（07）：1-13.

[10] 刘翠伟，裴业斌，韩辉，周慧，张睿，李玉星，等. 氢能产业链及储运技术研究现状与发展趋势[J]. 油气储运，2022，41（05）：498-514.

[11] 程德宝，蔺建刚，魏乃腾，于雯霞，党春蕾，田华峰. 氢气输送管道技术发展现状[J]. 油气储运，2024，43（06）：624-631.

[12] 代志健，吴仲文，王振生，罗天宝，刘春，王婷，等. 输氢管道用钢板和直缝埋弧焊管的研发[J].

钢管，2024，53(02)：39-45.

[13] 伍其兵，张行，张萌，张笑影，董绍华. 基于知识图谱的掺氢天然气管输研究现状与演进趋势[J]. 油气储运，2022，41(12)：1380-1394.

[14] 张对红，李玉星. 中国超临界 CO_2 管道输送技术进展及展望[J]. 油气储运，2024，43(05)：481-491.

[15] 陈嘉琦，蒲明，李育天，张斌，王晓峰，孙骥，等. 国内外 CO_2 管道设计标准对比分析[J]. 油气与新能源，2023，35(01)：94-100.

[16] 李鹤，赖兴涛. 超临界二氧化碳输送用钢管选材分析[J]. 焊管，2024，47(10)：86-90.

[17] 黄维和，李玉星，陈朋超. 碳中和愿景下中国二氧化碳管道发展战略[J]. 天然气工业，2023，43(7)：1-9.

[18] 赵与越，陈小伟，李轶鹏，王斌，王学仕. 二氧化碳输送管道技术研究进展[J]. 焊管，2024，47(06)：1-6+16.

[19] Shao Bing, Yan Xiangzhen, Yang Xiujuan, Wang Tongtao, Li Gensheng. Applicable Material Selection for Oil Country Tubular Goods (OCTG) in Sour Conditions and Software Development [M]//ICPTT 2011: Sustainable Solutions For Water, Sewer, Gas, And Oil Pipelines. 2011：179-187.

[20] Sykes AO. Product standards for internationally integrated goods markets[M]. Brookings Institution Press, 1995.

[21] 徐婷，丁飞，张华，吴浩，方伟，李茹，等. 非API 石油专用管标准体系构建及展望[J]. 石油管材与仪器，2021，7(03)：37-40.

[22] Bush D R, Brown J C, Lewis K R. Introduction to NACE standard MR0103[J]. Hydrocarbon processing (International ed.), 2004, 83(11)：73-77.

[23] 熊庆人，张永红，刘文红，李炎华，许晓锋，韩新利. 油气钻采用耐蚀合金无缝管材标准的发展分析[J]. 钢管，2023，52(02)：80-83.

天然气管道投产及运行工艺计算与软件开发

姜新慧[1]　何国玺[1]　孙　勇[2]　杨　洋[3]　钟瀚宇[1]　廖柯熹[1]

(1. 西南石油大学油气藏地质及开发工程国家重点实验室；2. 国家管网集团北方管道有限责任公司；
3. 西南石油大学地球科学与技术学院)

摘　要　【目的】天然气是中国新型能源体系建设中不可或缺的主体能源，天然气管道工程建设有条不紊的向前推进。天然气管道投产工作是管道建设到正常运行必不可少的关键步骤，但天然气管道投产过程的计算分散、繁琐，亟需开发拥有自主知识产权天然气管道投产全过程的计算软件。[方法]基于状态方程、流动模型、水力模型等理论模型，采用隐式差分法、特征线法、迭代法等方法，建立了适用于不同结构形式的管网系统投产仿真模型。形成耦合经验模型、现场实测数据的阀门开度-流量修正关系。采用C/S架构、MySQL数据库系统、模块结构、分散开发统一集成等技术，开发了天然气流体包计算、氮气与天然气用量、气头追踪、水合物抑制剂加注量、压力异常数据诊断等核心功能。实现组态化管网可视化、在线仿真、局域网(或专网)内共享、数据灵活查询等功能，完成了投产过程在线仿真软件的开发，实现了复杂天然气管网投产仿真核心技术自主可控。[结果]依托研发的天然气管道投产及运行工艺计算软件，对需投产的MX天然气管道进行实例应用。以MX天然气投产管道为对象进行组态化管网建模，将软件模拟仿真结果与现场实测数据对比。发现投产过程中天然气瞬时流量、阶段流量、水合物生成温度等参数最大相对误差小于8%，满足投产工作需求，天然气管道投产及运行工艺计算软件具备仿真计算能力。[结论]所研发的天然气管道投产及运行工艺计算软件能够满足天然气管道高效灵活、安全稳定的投产工作，可为国家数字化、智慧化、现代化发展与转型提供专业的仿真模拟计算服务，对维护国家能源信息安全具有重要意义。

关键词　天然气管道；投产；仿真计算；软件开发

Gas pipeline commissioning and operation process calculation and software development

Jiang Xinhui[1]　He Guoxi[1]　Sun Yong[2]　Yang Yang[3]　Zhong Hanyu[1]　Liao Kexi[1]

(1. State Key Laboratory of Oil and Gas Reservoir Geology and Development Engineering, Southwest Petroleum University;
2. National Pipeline Network Group Northern Pipeline Co., Ltd;
3. School of Geoscience and Technology, Southwest Petroleum University)

Abstract　[Purpose] Natural gas is an indispensable main energy source in the construction of China's new energy system, and the construction of natural gas pipeline projects is moving forward in an orderly manner. The putting into operation of natural gas pipeline is an essential step from pipeline construction to normal operation. However, the calculation process of natural gas pipeline production is scattered and complicated, and it is urgent to develop calculation software with independent intellectual property rights for the entire process of natural gas pipeline production. [Method] Based on the theoretical models such as state equation, flow model and hydraulic model, implicit difference method, eigenline method, iterative method and other methods are used to establish the production simulation model applicable to different structural forms of pipe network systems. The valve openness-flow correction relationship based on a coupled empirical model and on-site measured data is established. Using C/S architecture,

MySQL database system, modular structure, decentralized development and unified integration technologies, we have developed core functions such as natural gas fluid package calculation, nitrogen and natural gas dosage, gas head tracking, hydrate inhibitor refill volume, and pressure anomaly data diagnosis. The development of online simulation software for the commissioning process has been completed by realizing the functions of visualization of pipeline network in configuration, online simulation, sharing in LAN (or private networks), flexible data query, etc., and the core technology of complex natural gas pipeline network production simulation has been realized to be independently controllable. [Result] Relying on the developed natural gas pipeline commissioning and operation process calculation software, the application of MX natural gas pipeline to be put into operation is carried out. Build a configured pipeline network model based on the MX natural gas production pipeline, and compare the software simulation results with the on-site measured data. It is found that the maximum relative error of parameters such as instantaneous flow rate, stage flow rate and hydrate generation temperature during operation is less than 8%, which meets the requirements of the commissioning work. The natural gas pipeline production and operation process calculation software has the capability of simulation calculation. [Conclusion] The developed natural gas pipeline production and operation process calculation software can guarantee the efficient, flexible, safe and stable commissioning work of natural gas pipelines, and can provide professional simulation and calculation services for national digitization, intelligent and modernization development and transformation, which is of great significance for maintaining national energy information security.

Key words Natural gas pipeline; Production process; Simulation calculation; Software development

"十四五"规划提出要加快建设天然气主干管道，完善油气互联互通网络。天然气管道的投产是使用天然气将建好管道中的空气置换出来并投入正常使用的过程。同时，随着中国天然气管网系统规模不断增大、结构日趋复杂、智慧化投产的要求逐渐提高。

我国已积累了涩宁兰管道、陕京管道、西气东输管道、忠武管道、川气东送、中俄东线等多条天然气管道投产案例。在现有的天然气管道投产方式中，使用氮气隔离的"气推气"置换方法具有受地形限制小、氮气损失量小、能有效分隔天然气和空气等优势，应用最为广泛。孙勇等以中俄东线天然气管道长岭—永清段为例，进行投产过程、投产技术叙述。叶恒等研究了投产过程中不同氮气封存量、置换速度对置换过程氮气的分布规律影响，并得到中俄东线最佳置换速度与最优注氮管容比，但并未跟踪天然气—氮气混气段的变化规律。刘建武等分别进行了二氧化碳、超临界二氧化碳管道投产置换过程的模拟分析。目前 LedaFlow、Fluent 等国外商业软件被用来模拟计算投产中的置换过程。同时，相关研究人员还对输气管道站场氮气与天然气置换节点、高落差管道投产调压、注醇量、气体节流温降变化、阀门组低温性能等方面进行了研究。但天然气管道投产的全过程以及全过程软件的开发均未见报道。与此同时，国外软件受其特殊版权与核心技术保护机制的影响，存在采购使用维护费用高、信息安全与深层次开发深受局限的问题。如何实现天然气管道投产过程灵活可靠、准确调配、智能安全运行是投产工作面临的关键问题，同时也对投产过程仿真软件的适用规模、计算精度与速度也提出了响应的要求。

为了适应快速化投产的需要，提高天然气投产效率以及安全性。本文基于天然气管道投产全过程，抓住国家和行业的需求，基于状态方程、能量守恒等理论模型，融合大数据、智能算法等新兴技术，开发拥有自主知识产权的天然气管道投产全过程在线仿真模拟计算软件，适用于新建管网投产过程。该系统的开发有利于提高投产方案的编制效率，对管道安全稳定、高效投产运行以及国家信息安全具有重要意义。

1 研究思路和方法

天然气管道的安全投产方案的制定主要参照 GB/T 35068《油气管道运行规范》、SY/T 6233《天然气管道试运投产规范》、SY/T 5922《天然气管道运行规范》、Q/SY GDJ 0356《天然气管道试运投产技术规范》等相关标准中的投产工艺及流程(图1)。将针对投产过程的流体包计算、气头追踪、异常诊断等理论展开叙述。

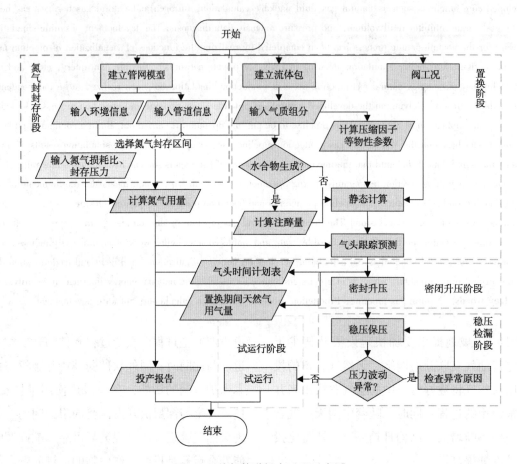

图 1 天然气管道投产过程示意图

1.1 氮气封存阶段

氮气封存是天然气置换过程中的一个重要阶段。天然气与空气直接接触十分容易在局部形成爆炸性混合气体。封存的氮气可以用于隔绝天然气与空气直接接触，降低爆炸的风险。同时，氮气在封存过程中能够将灰尘等杂质吹除，使管道或设备内部保持相对清洁的环境。

1.2 置换阶段

置换阶段是天然气投产作用核心环节，涉及流体包计算、水合物抑制剂用量计算、流量控制、气头追踪等过程。

(1) 流体包计算

天然气管道投产过程气质的压缩因子、密度等基础参数是投产过程的基础，对于用气量计算、气体追踪、异常诊断等至关重要。通过气质组分、温压条件、状态方程计算得到天然气的压缩因子、焓熵、焦耳-汤姆逊系数等物性参数。

(2) 水合物抑制剂加注

输气管道内含有液态水(气体温度低于水露点温度)时，温度和压力一旦达到水合物生成条件，就会有水合物生成。天然气投产过程是一个动态变化的过程，管道内温度和压力是不断变换的。因此判定管道内何时何处形成水合物，即应首先确定管道内任一点的温度、压力是否满足所输气体的水合物生成条件。当存在水合物生成风险时，需添加抑制剂防止水合物生成。天然气投产过程中一般采用甲醇，其中甲醇的加注量也与管道内含水量密切相关。本文通过管道进、出口天然气水露点换算得到进、出口天然气含水量，从而得到管道内部含水量，即可计算出需要加注的甲醇量。

(3) 阀门流量反算

天然气管道投产作业通过阀门开度控制流通气体流量。气体流量是进行气头追踪、混气段气量的基础。根据阀门两端的压降，阀门流量特性分固有流量特性和工作流量特性。固有流量特性是阀门两端压降恒定时的流量特性，亦称为理想流量特性。工作流量特性是在工作状态下(压降变化)阀门的流量特性，阀门出厂提供的流量特性为固有流量特性。阀门流量特性曲线分为直线、等百分比(对数)、抛物线及快开四种。

本文基于流量系数法，结合型号为 FCV-A800、公称直径为 DN600、压力等级为 Class 900 的电动调节阀现场测试的流量数据，对现有阀门-流量公式进行修正，得到了天然气管道投产过程中的阀门开度-流量关系(图2)。

(4) 气头追踪预测

天然气置换基本内容为单向投产和双向投产。单向投产即是置换气体从首站注入，从末站流出。双向投产即是注入置换气体的作业点数不只是一个，但气体的最终流向方向一致。无论哪种置换方式，置换过程中管道内存在空气段、氮气与空气混气段、氮气段、氮气和天然气混气段以及天然气段5段气体段。两种气体界面的判断依据、检测仪器和要求进行规定(图3)。

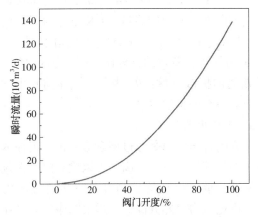

图 2 阀门开度-流量关系图

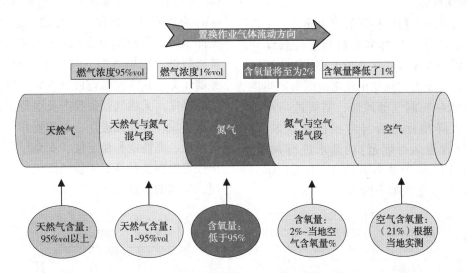

图 3 置换过程示意图

在置换过程中，为了保证较小的混气量，特别重要的是防止气体流态出现"层流化"。按照《天然气管道运行规范》中的相关条款要求，同时也为了保证安全，给站场置换人员留有足够的时间完成检测、操作和指挥等工作，置换时天然气的推进速度应不大于5m/s(18km/h)，一般为3~5m/s。氮气的推进速度不得低于0.6m/s。天然气置换过程涉及流动、扩散传质等多个过程，遵守连续性方程、动量守恒方程、能量守恒方程和组分输运方程，依据实际的情况考虑气体扩散方程，最终建立的数学模型分别为：

$$\frac{\partial \rho}{\partial t}+\frac{\partial \rho v}{\partial x}=0 \quad (1)$$

$$\frac{\partial (\rho v)}{\partial t}+\frac{\partial (\rho v^2)}{\partial x}=-g\rho\sin\theta-\frac{\partial P}{\partial x}-\frac{\lambda}{D}\frac{v^2}{2}\rho \quad (2)$$

$$\frac{\partial}{\partial t}\left[\left(h-\frac{P}{\rho}+\frac{MA^2}{2A^2\rho^2}\right)\rho\right]+\frac{1}{A}\frac{\partial}{\partial x}\left[\left(h+\frac{MA^2}{2A^2\rho^2}\right)M\right]+\frac{4K(T-T_0)}{D}+\frac{Mg\sin\theta}{A}=0 \quad (3)$$

$$\frac{\partial c_A}{\partial t}+v_x\frac{\partial c_A}{\partial x}=D'_{AB}\frac{\partial^2 c_A}{\partial x^2} \quad (4)$$

$$p=p(\rho,T) \quad (5)$$

$$h=h(\rho,T) \quad (6)$$

式中，x 为管道任意位置到注气点的距离，m；g 为重力加速度，m/s²；λ 为水力摩阻系数；θ 为管道与水平面间的倾角，rad；Q 为单位质量的气体向外界放出的热量，J/Kg；u 为气体的内能，J/Kg；v_x 为气体在 x 方向上的速度分量，m/s；c_A 为气体 A 所占的百分比；D'_{AB} 为两种气体 A、B 间的扩散系数；R_A 为两种气体间发生化学反应而引起的扩散通量。

1.3 密闭升压阶段

密闭升压阶段是通过向管道或设备内注入天然气并使其压力逐步升高，可以检查整个系统的完整性；也可以让管道和设备逐步适应压力变化，避免压力突然升高产生过大的应力。升压过程可以根据投产压力选择一级升压和多级升压。

1.4 稳压检漏阶段

异常压力诊断是稳压检漏的重要环节，是为投产动态过程的压降指标异常数据进行自动判断。稳压过程中，压力变化处于正常范围内则证明管道正常；若压力变化超出规定范围则证明该区域管道可能存在漏气等风险，需要进一步排查。其中，压降指标是否异常根据标准 SY/T 5922《天然气管道运行规范》的要求进行判断，稳压保压阶段的压降速率满足 Δp 小于 $[\Delta p]$，Δp 和 $[\Delta p]$ 的计算公式如下所示：

$$\Delta p = \left(1-\frac{(p_2+0.101325)T_1}{(p_1+0.101325)T_2}\right)\times 100\% \quad (7)$$

$$[\Delta p]=\frac{5}{DN}\times 100\% \quad (8)$$

式中，Δp 为实际压降，MP；T_1 是稳压开始时温度，K；T_2 是稳压结束时温度，K；p_1 是稳压开始时压力，MPa；p_2 是稳压结束时压力，MPa；$[\Delta p]$ 是允许压降率；DN 是公称直径，mm。

2 软件开发

结合天然气管道投产智能化的业务需求，以上述理论、数学模型为基础，设计了天然气管道投产及运行工艺计算软件系统构架（图4），采用隐式差分法、特征线法、迭代法等方法进行求解，完成了投产过程在线仿真软件的开发。天然气管道投产及运行工艺计算软件通过模块结构、分散开发统一集成等方法进行实现。客户端采用 C#（基础计算、UI 模块开发）的模式开发成 Windows 桌面程序，数据统一存储到远端 MySQL 数据库系统，客户端可在局域网（或专网）内共享基础数据、成果数据。利用 C/S 模式可降低服务器压力，在客户端完成计算后，数据存储到远端数据库。天然气管道投产及运行工艺计算软件的总体设计遵循先进性、可配置和可管理性、可扩展性、高性能和稳定性、安全、保密性、易维护性等原则，并具有完善的组织与管理方式及直观、生动的信息展示。

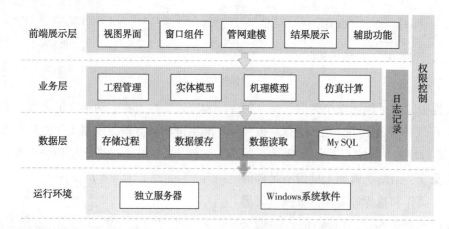

图 4 系统架构示意图

2.1 前端展示层

本软件界面借鉴国内外仿真软件视图界面，并设计开发了组态化管网建模界面，可根据不同投产管道节点设置与里程高程情况进行管网建模与展示，操作便捷。计算结果实时同步传输至各用户，数据结果可视化。

2.2 业务层

业务层实现在线仿真的功能。基于天然气管

道投产过程中的理论与模型，通过调用各种算法模型来完成天然气管道投产过程整体的评价业务逻辑，并利用利用经验模型和机器学习算法对模型、机理进行实现。本文开发了基于状态方程、流动模型、水力模型等理论模型的天然气流体包计算、氮气与天然气用量、气头追踪、水合物抑制剂加注量、压力异常数据诊断等核心功能。

2.3 数据层

数据层为数据提供持久化以及灵活查询的支撑，并对数据进行标准化归约。针对天然气管道投产过程中数据量计算大、数据传输实时的特点，设计了历史投产资料数据库、投产标准与管理要求库、气液及管材物性参数库、专家经验库、风险记录和应急措施库五部分静态数据库、流体物性数据库、阀室节点工况库、仿真结果库三部分动态数据库。并通过局域网内在线数据库的形式实现数据的灵活调用。

2.4 运行环境

天然气管道投产及运行工艺计算软件拥有独立的服务器，安装于Windows操作系统，具有足够的CPU核心、内存和存储容量，可以满足多用户同时使用的需求。

3 应用实例与验证

MX天然气管道线路总长321.5km，全线设计压力10MPa，管径为D914mm、D1016mm，采用L485M钢管。此次天然气投产管线共计站场7座，线路截断阀室15座(图5)。投产期间采用调控中心指挥、现场操作的控制方式。

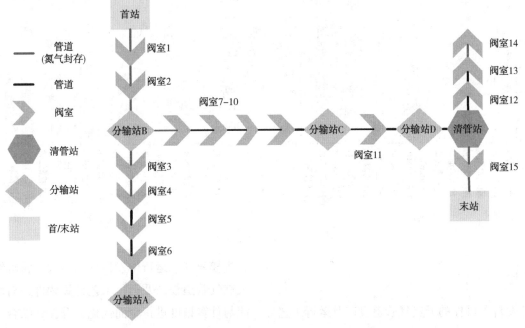

图5 MX投产管道示意图

投产置换方式采取气推气的方式投产，即管道投产期间先用氮气置换空气，再用天然气置换氮气，使天然气和空气完全隔离。注氮采用液氮加热泵车注氮的方式，注氮及氮气封存区域(图5中绿色管道)为首站-阀室7、分输站B-阀室4、清管站-末站、阀室13-阀室14。本次投产所用天然气的气源为LNG来气，来气压力约为7.5MPa，考虑节流效应，采用两级调压方式。

投产过程中用于观测管道压力的基准点为分输站C。按照1.0MPa、3.0MPa、5.0MPa、7.5MPa(干线压力)建立升压阶梯，分别进行稳压检漏。在分输站C压力升至5.0MPa之前，投产采用调控中心统一指挥、现场操作的控制方式；在分输站C压力升至5.0MPa之后，阀室干线截断阀门的关闭、开启操作由调控中心远程控制，现场人员进行监护。

由于新建输气管道干燥后管道内可能会存有一定量的液态水，升压期间天然气水露点随着压力上升会大幅上升。采用软件计算不同压力条件下的水合物生成温度，并与现场采样的实验结果对比，得到误差(图5)。由此可见，在一定压力下的水合物生成温度软件计算值的相对偏差为2.26%，最大不超过5%，计算结果可现场应用。为了防止管道内产生冰堵，在首站设置1个注醇点，阀室1-阀室14间管道置换、升压至1.0MPa前同步向管道注醇。

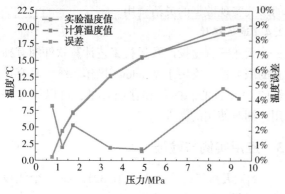

图6 水合物生成温度对比及误差图

MX管道投产分为三个阶段实施。第一阶段：完成首站-阀室1的天然气置换、升压至3MPa、稳压检漏工作。第二阶段：完成阀室1-阀室11、分输站B-分输站A、清管站-末站的天然气置换、升压至3MPa、稳压检漏工作。第三阶段：完成全线升压至5MPa、稳压检漏24h、升压至7.5MPa、稳压检漏24h的工作，最后进行72h试运行。将各阶段的现场实际流量与输气量与软件计算结果进行对比（图7），计算结果符合实际运行情况，软件计算误差小于8%，可用于现场应用。

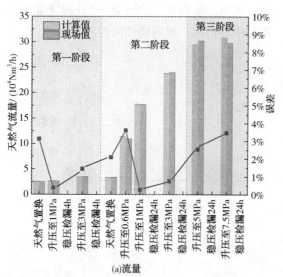

(a)流量

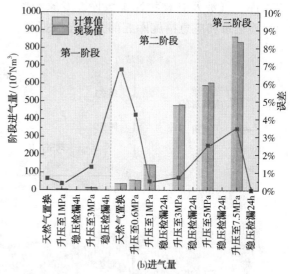

(b)进气量

图7 阶段气量对比及误差图

4 结论

开发自主可控的天然气管道投产及运行工艺计算软件是适应快速化投产、推进投产全过程工作智能化的迫切需求，对于提高天然气投产效率和确保信息安全性具有重大意义。基于天然气管道投产全过程，介绍了流体包计算、气头追踪等理论模型，修正了阀门开度-流量关系，基于组态化管网可视化，论述了天然气管道投产及运行工艺计算软件的系统架构、数据管理与功能实现。通过MX天然气管道的实例应用，验证了该软件的计算精度符合要求，可用于天然气管道投产工作。随着天然气管道的不断建成，应着力构建与现代化、智能化相适配，可用于模拟投产全过程的具有自主知识产权的技术，在天然气管网大规模发展、运行中发挥作用。未来将继续完善天然气管道投产及运行工艺计算软件，对软件性能与计算精度进行全面优化提升，并结合数字孪生技术，实现投产过程场景化，以期为中国天然气管网的投产运行与管理提供技术支持。

参 考 文 献

[1] 蒲丽珠. 天然气管道投产过程中混气规律的研究[D]. 西南石油大学, 2014.

[2] CUI M, WU C. Principles of and tips for nitrogen displacement in gas pipeline commissioning[J]. Natural Gas Industry B, 2015, 2(2-3): 263-269.

[3] 孙勇, 赵国辉, 游泽彬, 李秋娟, 崔茂林, 刘朝阳. 中俄东线中段天然气管道投产技术探讨[J]. 油气储运, 2022, 41(11): 1312-1318.

[4] 叶恒, 李光越, 刘钊, 张博越, 刘家乐. 中俄东线大口径输气管道的投产气体运移规律及注氮量优化

[J]. 天然气工业, 2020, 40(9): 123-130.

[5] 刘建武, 李毅, 龚霁昱, 张千昌, 胡其会. 二氧化碳长输管道投产置换过程模拟分析[J]. 科学技术与工程, 2021, 21(28): 12109-12116.

[6] 李欣泽, 王德中, 张海帆, 刘双全, 朱涛, 邢晓凯. 新疆油田超临界二氧化碳管道投产置换过程温压及相变特性[J]. 新疆石油天然气, 2024, 20(4): 77-86.

[7] 朱浩宇, 李长俊, 贾文龙, 沐峻丞. 基于LedaFlow的液相乙烷输送管道投产过程数值模拟[J]. 油气储运, 2023, 42(3): 335-342.

[8] 张振永, 孟献强, 孙学军, 周亚薇, 张金源. 中俄东线站场工艺管道用高钢级低温钢管韧性指标[J]. 油气储运, 2018, 37(4): 435-442.

[9] 程玉峰. 保障中俄东线天然气管道长期安全运行的若干技术思考[J]. 油气储运, 2020, 39(1): 1-8.

[10] 陈思杭. 起伏大落差液体管道投产过程中气相动态运移和演化规律研究[D]. 中国石油大学(北京), 2023.

[11] KHOSRAVI A, MACHADO L, NUNES R O. Estimation of density and compressibility factor of natural gas using artificialintelligence approach[J]. Journal of Petroleum Science and Engineering, 2018, 168: 201-216.

[12] MOISSEYEVA Y, SAITOVA A, STROKIN S. Calculating densities and viscosities of natural gas with a high content of C2+ to predict two-phase liquid-gas flow pattern[J]. Petroleum, 2023, 9(4): 579-591.

[13] AL GHAFRI S Z S, JIAO F Y, HUGHES T J, ARAMI-NIYA A, YANG X X, Siahvashi A, et al. Natural gas density measurements and the impact of accuracy on process design[J]. Fuel, 2021, 304: 121395.

[14] FARZANEH-GORD M, FARSIANI M, KHOSRAVI A, ARABKOOHSAR A, DASHTI F. A novel method for calculating natural gas density based on Joule Thomson coefficient[J]. Journal of Natural Gas Science and Engineering, 2015, 26: 1018-1029.

[15] MARIĆ I. A procedure for the calculation of the natural gas molar heat capacity, the isentropic exponent, and the Joule-Thomson coefficient[J]. Flow Measurement and Instrumentation, 2007, 18(1): 18-26.

[16] 苑伟民. 新的理想气体焓、熵计算公式[J]. 石油工程建设, 2024, 50(2): 1-5+16.

[17] ÖZCAN K, BOZTEPE A, TARCAN E, TASDEMIRCI Ç. Measuring the enthalpy of combustion of methane gas and establishing the model function to perform comprehensive uncertainty calculations[J]. The Journal of Chemical Thermodynamics, 2024, 189: 107196.

[18] PAKRAVESH A, ZAREI H. Prediction of Joule-Thomson coefficients and inversion curves of natural gas by various equations of state[J]. Cryogenics, 2021, 118: 103350.

[19] 刘春花, 郝忠献, 刘新福, 何鸿铭, 魏松波, 周超. 致密气水合物相平衡动态模型及其热力学特性[J]. 石油机械, 2021, 49(9): 85-91.

[20] FANG S Q, ZHANG X Y, ZHANG J Y, CHANG C, Li P, BAI J. Evaluation on the natural gas hydrate formation process[J]. Chinese Journal of Chemical Engineering, 2020, 28(3): 881-888.

[21] KE W, CHEN D. A short review on natural gas hydrate, kinetic hydrate inhibitors and inhibitor synergists[J]. Chinese Journal of Chemical Engineering, 2019, 27(9): 2049-2061.

[22] ZUO L L, JIN A T, CHEN Q, DONG Q L, LI Y X, ZHAO S R. Simulation on venting process and valve opening control method for gas trunk pipelines[J]. Petroleum Science, 2022, 19(6): 3016-3028.

[23] 王剑, 范宜霖, 雷艳, 刘贵超, 张继伟, 黄健等. 阀门流量特性综合测试系统的研制[J]. 流体机械, 2023, 51(2): 28-32+77.

[24] 史晓娟, 王磊, 汪学明, 宋传智. 基于滑模扰动观测器的阀门开度预测控制[J]. 流体机械, 2024, 52(0): 64-70.

[25] LIN Z H, LI J Y, JIN Z J, QIAN J Y. Fluid dynamic analysis of liquefied natural gas flow through a cryogenic ball valve in liquefied natural gas receiving stations[J]. Energy, 2021, 226: 120376.

[26] 张楠, 胡其会, 李玉星. 起伏输气管道投产置换混气过程机理研究[J]. 油气田地面工程, 2019, 38(S1): 56-60.

[27] Li P, SHEN J, LYU P, DONG C L, CHEN T. Architecture Design of Protocol Controller Based on Traffic-Driven Software Defined Interconnection[J]. Chinese Journal of Electronics, 2024, 33(2): 362-370.

[28] ZAKERI-NASRABADI M, PARSA S, JAFARE S. Measuring and improving software testability at the design level[J]. Information and Software Technology, 2024, 174: 107511.

基于 GWO-BP 算法的页岩气集输管道内腐蚀速率预测

祝星语　彭星煜　冯梁俊　蒋海洋

（西南石油大学石油与天然气工程学院）

摘　要　针对页岩气集输管道的内腐蚀，提出了一种基于 GWO-BP 组合模型的腐蚀速率预测算法。本研究系统分析了 SRB 和 CO_2 共存环境下温度、pH、CO_2 分压、SRB 数量、SO_4^{2-} 浓度等因素对腐蚀速率的影响，运用反向传播（BP）神经网络建立预测模型，运用灰狼算法优化算法（GWO）优化了神经网络权值和阈值的初始值，在模型建立的过程中不断优化提升模型的预测精度，采用所建模型对另一条相邻管道进行预测并开挖验证。研究结果表明：流速、SRB 数量和 CO_2 分压为页岩气集输管道腐蚀的主要影响因素。基于 GWO-BP 算法的模型预测的内腐蚀速率与实测值的最大相对误差减小至 6.18%，对页岩气集输管道内腐蚀速率进行预测具有一定的准确性，该模型可为预防和控制管道腐蚀提供参考。

关键词　页岩气集输管道；内腐蚀速率；遗传算法；灰狼算法；均方根误差

Corrosion Rate Prediction in Shale Gas Gathering Pipelines Based on GWO-BP Algorithm

Zhu Xingyu　Peng Xingyu　Feng Liangjun　Jiang Haiyang

(College of Petroleum and Natural Gas Engineering, Southwest Petroleum University)

Abstract　A corrosion rate prediction algorithm based on a combined GWO-BP model is proposed for the internal corrosion of shale gas gathering pipelines. In this study, the effects of temperature, pH, partial pressure of CO_2, SRB quantity, SO_4^{2-} concentration and other factors on the corrosion rate under the coexistence of SRB and CO_2 were systematically analysed, and the prediction model was established by using a back propagation (BP) neural network, and the Grey Wolf Algorithm Optimization algorithm (GWO) was used to optimize the initial values of the neural network weights and thresholds, and the prediction accuracy was continuously optimized to improve the prediction of the model in the course of the model building process. The model was continuously optimised to improve the model prediction accuracy during the model building process, and another adjacent pipeline was predicted and excavated for verification using the constructed model. The results show that the flow velocity, the number of SRBs and the partial pressure of CO_2 are the main factors influencing the corrosion of shale gas gathering pipelines. The maximum relative error between the predicted and measured internal corrosion rate of the model based on GWO-BP algorithm is reduced to 6.18%, which is an accurate prediction of the internal corrosion rate of the shale gas gathering pipeline, and the model can be used as a reference for the prevention and control of pipeline corrosion.

Key words　Shale gas gathering pipeline; Internal corrosion rate; Genetic algorithm; Grey wolf algorithm; Root mean square error

我国探明的页岩气主要分布在四川盆地，根据已探明的页岩气储量规模和产能建设规划，未来四川盆地页岩气产量将继续大幅跃升。管道失效会造成严重的经济损失和灾难性的后果，腐蚀是导致管道失效的主要原因之一。页岩气集输管道中含有 CO_2、Cl^- 等腐蚀性介质，其以凝析水为载体，易在管道低洼处或上坡段积聚，压裂液、钻井泥浆和蓄水都可能将有害的微生物如 SRB 引入页岩气藏，加剧页岩气集输管道的腐蚀，造成管道内腐蚀甚至穿孔，引发安全事故并造成财产损失。威荣页岩气田在投运后三个月内就发生管道泄漏，2021 年泄漏事件达到 128 起，泄露位置主要位于阀门和管道，其中管道泄漏部位主要包括弯头、三通和焊缝；川渝页岩气田某区块站内四年间累计穿孔 1000 余次，其中集气站的穿孔部位几种发生在水平管段和弯管处。页岩气集输管道内腐蚀问题日益严重。

目前石油和天然气行业存在多种 CO_2 或 SRB 腐蚀速率预测模型。2009 年，顾停月等人基于生物阴极催化硫酸盐还原机理(BCSR)，考虑生物膜-金属界面的生化反应开发了 SRB 腐蚀机理模型，结合 Butler-Volmer 方程和质量平衡方程式可计算出无钝化金属的点蚀速率和最大凹坑深度。2013 年，Fatah 等人基于非生物化学和线性极化电阻(LPR)方法开发了一个 SRB 腐蚀的经验模型以预测微生物腐蚀速率；A Chamkalani 等人提出了一种以 pH、流速、温度和 CO_2 分压为输入，腐蚀速率为输出的神经网络模型，以系统地预测注水管道的腐蚀速率。2014 年，郭少强等人结合 De Waard 模型建立了修正后的 CO_2 腐蚀速率预测模型，用于预测油气管线顶部 CO_2 的腐蚀速率，经过实验证明模型的精确性良好。2022 年，郑度奎等人提出了基于自适应改进的人工鱼群算法优化的广义回归神经网络(IAFSA-GRNN)模型预测 X65 管线钢的 CO_2 腐蚀速率，经实例验证表明预测结果的平均相对误差为 6.90%。页岩气管道的内腐蚀问题日益严重，开展页岩气集输管道内腐蚀预测研究具有重大意义。

灰狼优化算法(Grey Wolf Optimizer, GWO) 是一种基于灰狼群体行为的群体智能优化算法，该算法具有操作简单，参数调整较少以及能较好地平衡全局搜索和局部搜索等优点。采用 GWO 算法优化 BP 神经网络，不仅可以提高收敛速度，减少训练时间，还可以优化网络结构减少 BP 神经网络过拟合的现象。因此，本文考虑以灰狼算法优化的神经网络(GWO-BP)技术与腐蚀机理模型并联的修正方式，以期有效提高 SRB 和 CO_2 共存环境下的内腐蚀机理模型的预测精度。

1 实验

1.1 实验材料制备

采用 L245N 钢作为该实验的实验材料。参考《金属材料均匀腐蚀全浸实验方法》(GB10124-88)中的要求对 L245N 钢做试验前处理：将 L245N 钢切割为 10mm×5mm×3mm 规格，用砂纸逐级打磨至 2000#后，采用游标卡尺测量钢片对应的长宽高，随后将钢片浸泡在丙酮溶液去除油污，接着将钢片浸泡在浓度分别为 25%、50%、75%、100% 的无水乙醇中各 15min 脱水，最后将挂片干燥后称重并记录。

试片的化学成分如表 1 所示。

表 1 试片的化学成分(wt%)

元素	C	Si	Mn	P	S	V	Nb	Ti	Fe
含量	0.23	0.35	1.15	0.02	0.01	0.0068	0.0009	0.022	余量

实验所用水样为页岩气田的 1#、2# 管道分离器的采出液，其组分见表 2。

表 2 实验水样组分

水样组分	Na^+	K^+	Ca^{2+}	Mg^{2+}	Cl^-	SO_4^{2-}	HCO_3^-	pH
水样含量/(mg/L)	6000	200	400	60	7000	200	400	7.0~8.0

1.2 实验方法

1.2.1 动态腐蚀正交实验

取室内配制的 800mL 模拟水置于高压釜，将试片固定在挂片架上，半浸泡悬挂于水样中，安装密封组件，向釜体内通入 1.5h 的 N_2 除氧后，按照实验设计泵入 CO_2，剩余压力由 N_2 补充，使釜内压力提高至实验所需压力，启动加热程序加热至实验所需温度。当釜内温度和压力都达到实验需求时，启动转子，调节至所需转速，记录实验开始时间，周期为 14d，每组动态腐蚀实验各做 3 个平行实验。为保证结果的可靠性，每次试验均使用三份样品。

正交实验因素水平设计如表 3。

表 3 动态正交实验因素水平

水平	因素			
	流速/(m/s)	SRB 数量/(个/mL)	压力/MPa	CO_2 分压/MPa
1	0	0	0	0
2	0.4	1000	4.0	0.1
3	2.0	7.5×10⁴	6.0	0.03

1.2.2 表面分析

利用电子扫描电镜对实验 7d 和 14d 清洗腐蚀产物膜前后的 L245N 钢进行腐蚀产物的微观形貌分析;利用能谱分析仪和 X 射线衍射仪对实验 14d 的 L245N 钢进行腐蚀产物元素成分和腐蚀产物组成分析。

1.3 实验结果分析

1.3.1 腐蚀影响因素分析

动态腐蚀正交实验结果如图 1 所示。

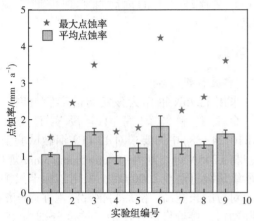

图 1 动态腐蚀挂片正交实验点蚀速率结果

动态实验条件下试片的平均点蚀速率范围为 0.8343mm/a~2.1379mm/a,动态正交实验的条件下平均点蚀速率也均处于高等级。SRB 为 75000 个/mL 时,试片的点蚀速率均较高,分别为 4.2236mm/a(流速 0.4m/s、压力 0MPa、CO_2 分压 0.1MPa)、3.5979mm/a(流速 2.0m/s、压力 4MPa、CO_2 分压 0MPa)和 3.4936mm/a(流速 0m/s、压力 6MPa、CO_2 分压 0.03MPa)。低流速会帮助加快有机物从溶液向金属表面扩散的速率,从而促进 SRB 在金属表面附着,形成生物膜加剧腐蚀;但流速过高时,因流速产生的剪切应力会抵消甚至超过 SRB 吸附在金属表面的作用力,使得 SRB 无法固着在金属表面产生点蚀效应[102-103]。经点蚀速率的实验结果推测 CO_2 与 SRB 共同存在时会促进试片的点蚀速率。

方差分析(ANOVA,全称为 Analysis of Variance)是一种统计方法,用于比较三个或更多组数据的均值是否存在显著异异。它通过分析数据的方差来判断这些均值之间的差异是否超过随机误差所能解释的范围。采用 SPSS 软件进行方差分析,动态腐蚀正交实验方差分析结果见表 4。

表 4 动态腐蚀正交实验方差分析结果

差异源	III 类平方和	自由度	均方	F 值	显著性	置信度
v	0.027	2	0.013	278.237	0.000	* * *
n	0.023	2	0.011	236.514	0.000	* * *
P	0.001	2	0.000	9.2	0.060	* *
P_{CO_2}	0.004	2	0.002	42.286	0.004	* * *
误差	0.001	0	0.001	—	—	—

根据表 4 的分析结果,在 SRB/CO_2 共存的环境下,具有显著性的腐蚀因素为流速、SRB 数量和 CO_2 分压,压强显著性较小。因此选择流速、SRB 数量和 CO_2 分压为页岩气集输管道腐蚀主控因素,并将其作为腐蚀机理建模的主要参数。

1.3.2 腐蚀形貌及腐蚀产物成分分析

对实验 7d 和 14d 清洗腐蚀产物膜前后的 L245N 钢进行腐蚀产物的微观形貌进行分析,结果如图 2 和图 3 所示。

(a) 实验7d　　(b) 实验14d

图 2 挂片清洗腐蚀产物膜前试片的 SEM 图

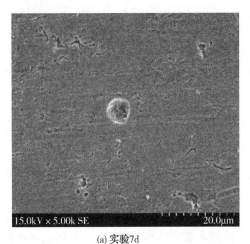
(a) 实验7d

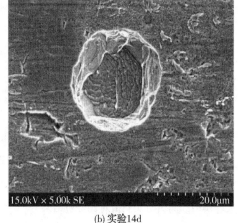
(b) 实验14d

图 3　挂片清洗腐蚀产物膜后试片的 SEM 图

实验结果显示，挂片 7d 与 14d 清洗前试片表层明显附着大量的微生物与沉积物。挂片 7d 时，从图 2(a) 可见，试片表面附着了大量条状的 SRB，且膜层与 SRB 的络合结构较为致密，挂片 14d 时，从图 2(b) 可看出试片表面的膜层较为疏松，颗粒粗大且分布不均；清洗腐蚀产物膜后，挂片 7d 与 14d 的试片表面均表存在明显的点蚀坑，对比图 3(a) 和图 3(b)，挂片 14d 的点蚀坑明显大于 7d 的点蚀坑。

对实验 14 天后的 L245N 钢进行腐蚀产物（图 2(b) 红色标注处）进行 EDS 与 XRD 分析，分析结果如图 4 和图 5 所示。

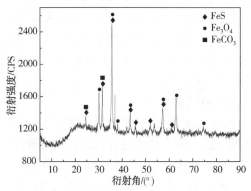

图 5　实验 14d 后试片腐蚀产物 XRD 图

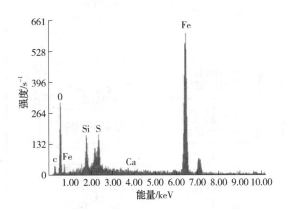

图 4　实验 14d 后试片腐蚀产物 EDS 图

图 4 为试片表面腐蚀产物的能谱图，腐蚀产物成分主要为 Fe、S、O 等元素，还有少量 Ca、C、Si 等元素。试样表面碳含量的增加的原因可能是 CO_2 腐蚀和 SRB 生物膜的产生。结合图 5 的 XRD 检测结果，试片表面腐蚀产物为 FeS、$FeCO_3$ 与 Fe_3O_4，这也说明在模拟溶液中试片发生了 SRB 以及 CO_2 的腐蚀。

2　基于 GWO-BP 算法修正的内腐蚀机理模型

2.1　内腐蚀机理模型建立

考虑 H_2CO_3 及 H_2O 还原和 BCSR 理论，基于电荷转移与传质理论，建立 SRB 与 CO_2 共存环境下的内腐蚀速率预测机理模型，其表述如下：

$$V_{corr} = \frac{M_{Fe} i_{Fe}}{n F \rho_{Fe}} \tag{1}$$

式中，M_{Fe} 为铁的原子量，取 0.056kg/mol；n 为参与反应的电子数，取 2；ρ_{Fe} 为铁的密度，取 7870kg/cm³；i_{Fe} 为阳极腐蚀电流密度，A/cm²；F 为法拉第常数，取 96485.34C/mol；V_{corr} 为腐蚀速率，mm/a。

选取某页岩气田 1# 和 2# 页岩气集输管道近三年的 56 处开挖点为模型验证样本，点蚀速率通过开挖后的实测壁厚损失计算得出，温度、pH、CO_2 分压、SRB 数量、SO_4^{2-} 浓度均为现场取样数据，计算机理模型预测点蚀速率与实测点蚀速率的相对误差，并绘制误差图。

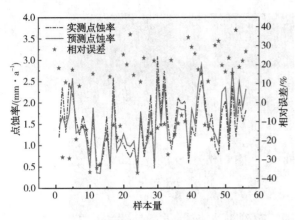

图6 机理模型预测值与实测值对比分析结果

从图6中可以看出,使用该模型预测页岩气管道腐蚀,整体相对误差普遍偏大,最大相对误差达到38.33%。显然,该模型在现场并不适用。因此为了提高机理模型的精度,引入 GWO-BP 模型建立机理模型的误差模型,以提升机理模型的整体精确性。

2.2 GWO-BP 算法

GWO 算法是基于灰狼的位置信息来对 BP 神经网络的权值和阈值进行优化。当灰狼位置不断地发生变化时,BP 神经网络算法的权值和阈值也不断更新,灰狼的最优位置即为 BP 神经网络寻求的最优解,具体算法步骤如下文所示。

(1)确定 BP 神经网络的输入和输出样本。利用 Matlab 将不同量纲下的样本数据均转化至 $[-1,1]$ 之间,从而消除量纲影响。

$$y = (y_{max} - y_{min}) \frac{x - x_{min}}{x_{max} - x_{min}} + y_{min} \quad (2)$$

式中,x 为归一化前的值,y 为归一化前后的值;x_{max} 为输入数据的最大值;x_{min} 为输入数据的最小值;y_{min} 取-1,y_{max} 取1。

选取经验公式和实际训练相结合的方法来确定隐含层节点数。

$$c = \sqrt{m+n} + a \quad (3)$$

其中,c 为隐含层节点数、n 为输入层节点数、m 为输出层节点数,a 为 1~10 之间的常数。

(2)GWO 算法种群初始化。随机初始化灰狼种群位置 $X_z(z=1,2,\cdots,Z,Z$ 为灰狼种群数),每个灰狼个体包含一组权重 $\alpha_i(\alpha_1+\alpha_2+\cdots+\alpha_m=m)$,初始化参数 a、A、C。

(3)构造适应度函数。归一化后的每个输入变量乘上对应的权重作为新的输入参数,并对 BP 神经网络进行训练,利用模型预测输出数据与期望输出数据计算所得的 R^2 构造适应度函数,R^2 的计算公式为式(4)所示,其中 $\hat{y}$ 为实际值的平均值。

$$R^2 = 1 - \frac{\sum_{k=1}^{n}(\hat{y}_k - y_k)^2}{\sum_{k=1}^{n}(\hat{y}_k - \bar{y})^2} \quad (4)$$

(4)对种群中的每个灰狼个体执行 BP 神经网络训练,并计算每个灰狼个体的适应度,从中选出适应度最高的 3 个灰狼,作为当前的最优解 X_α、次优解 X_β、第三优解 X_δ。

(5)更新 ω 狼个体的位置和参数 a、A、C。根据步骤(3)所述重新构造新的 BP 神经网络,并对网络进行训练,重新计算每个灰狼个体的适应度,更新 X_α、X_β、X_δ。

(6)判断 GWO 算法是否达到最大迭代次数。若是,则停止迭代,并输出最优结果 X_α;反之,则重复执行步骤(3)~步骤(5),直至达到最大迭代次数。

(7)利用 GWO 算法得到的最优输入变量的权重再次训练 BP 神经网络,可得到训练精度最高的 BP 神经网络模型,并将其作为最终模型。

GWO 算法优化 BP 神经网络模型流程如下图所示。

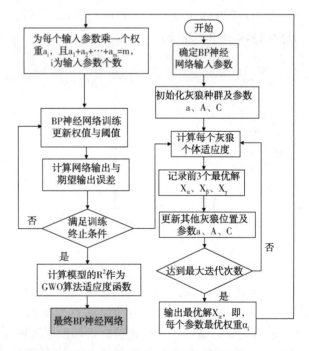

图7 GWO 优化 BP 神经网络过程

2.3 基于 GWO-BP 算法建立的误差模型

根据实验结果分析,BP 神经网络为 3 层神经网络模型,输入节点为 5 个,输出层为 1 个,隐含层节点 c 为 3~12 之间的常数。将不同隐含层节点代入模型进行实验，拟采用 L-M 算法训练神经网络，选择前 46 组数据作为训练集，后 8 组数据作为测试集。不同隐含层模型的预测结果与均方根误差结果如表 5 和图 8 所示。

表 5 不同隐含层节点数 BP 模型的预测值与实测值对比

预测值/%										实际值/%
3	4	5	6	7	8	9	10	11	12	
35.52	28.32	23.12	25.19	25.19	24.89	30.24	30.62	24.25	24.50	32.47
13.25	12.95	25.63	10.76	9.76	10.12	26.21	13.76	25.98	28.23	19.67
25.69	25.02	19.35	25.84	23.02	8.39	11.69	8.61	22.65	9.24	16.26
29.06	28.25	26.39	19.06	18.01	34.21	32.34	31.59	30.32	14.01	23.60
23.35	25.01	20.39	12.77	5.85	13.28	23.44	21.98	7.53	23.85	15.55
27.54	29.36	28.45	30.59	34.68	40.49	29.04	25.12	30.70	45.28	38.33
11.76	13.01	10.09	26.88	17.65	16.01	10.92	11.15	7.38	5.39	18.66
30.17	16.58	29.56	16.82	19.65	28.17	29.73	31.61	15.01	14.23	22.08

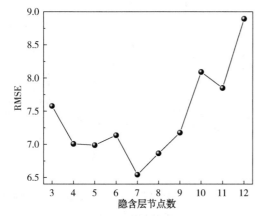

图 8 不同隐含层节点数的 RMSE 对比

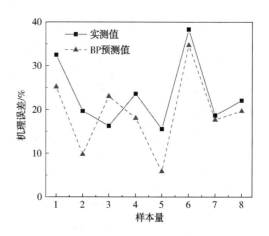

图 9 BP 模型预测值与实际值对比图

当隐含层节点数为 7 的时候，BP 神经网络具有最小的误差和最优的预测结果。因此最终确定以隐含层节点数为 7，L-M 算法训练的 BP 神经网络模型对 46 组训练样本进行训练，8 组样本作为验证。网络模型训练过程的拟合情况用相关系数 R 表示，R 越接近 1 表明网络拟合情况越好。将此条件下模型的实测值和预测值进行对比，结果如图 9 所示。

根据 BP 神经网络训练结果分析，BP 神经网络在最佳训练条件下的相关系数 R 的值为 0.82，拟合效果尚有不足。因此，再根据 GWO 算法对 BP 神经网络进行优化，从而更好地满足训练需求。

通过 BP 神经网络模型的构建以及对重要仿真参数的优选，可知 GWO-BP 模型的输入层为 5 个变量，输出层为 1 个变量，隐含层节点数为 7 个，模型结构为 5-7-1，GWO-BP 模型的网络结构如图 10 所示。

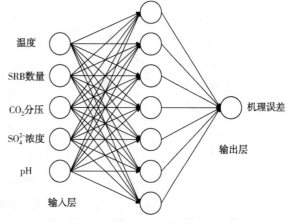

图 10 GWO-BP 网络模型结构

设置 GWO 算法初始狼群规模为 30，迭代次数为 200。利用 Matlab 软件对 GWO-BP 模型进行仿真预测，得到预测结果以及回归系数，部分代码见附录所示，仿真结果见图 11 所示。

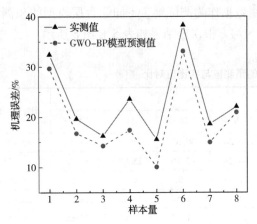

图 11 GWO-BP 模型预测值与实际值对比图

GWO-BP 神经网络模型仿真过程中，模型对机理误差的预测逐渐逼近实测值，且预测值与实测值之间误差较小，GWO-BP 模型具有更加良好的适用性。从图 12 也可见 GWO-BP 模型训练集的拟合系数也逼近于 1，拟合效果优秀。

2.4 基于 GWO-BP 算法建立的误差模型修正机理模型的验证

采用机理模型以及 GWO-BP 算法修正后的机理模型，对 8 组测试样本进行点蚀速率预测。表 6 列出了两种模型的预测结果及其与实测点蚀速率之间的相对误差，并绘制了两种模型预测结果与实测值的相对误差对比图，见图 13 所示。

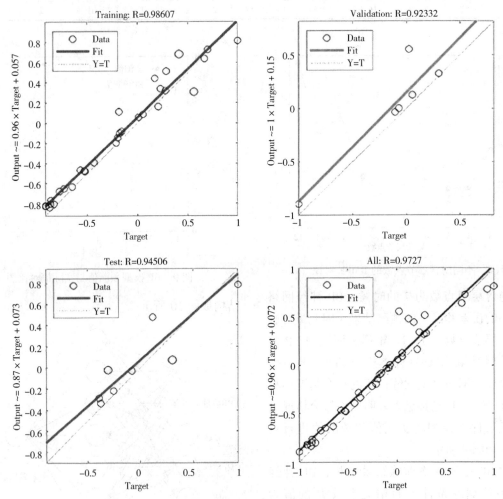

图 12 GWO-BP 模型训练回归图

表 6 两种模型的预测结果对比

样本编号	实测值/(mm/a)	机理模型		GWO-BP 算法修正后的机理模型	
		点蚀速率/(mm/a)	相对误差/%	点蚀速率/(mm/a)	相对误差/%
1	0.77	1.02	32.47	0.79	2.80
2	1.83	2.19	19.67	1.88	2.92

续表

样本编号	实测值/(mm/a)	机理模型		GWO-BP算法修正后的机理模型	
		点蚀速率/(mm/a)	相对误差/%	点蚀速率/(mm/a)	相对误差/%
3	2.03	2.36	16.26	2.07	2.00
4	0.89	1.10	23.60	0.95	6.18
5	2.38	2.75	15.55	2.51	5.48
6	1.20	1.66	38.33	1.26	5.19
7	2.09	2.48	18.66	2.17	3.69
8	1.54	1.88	22.08	1.56	1.13

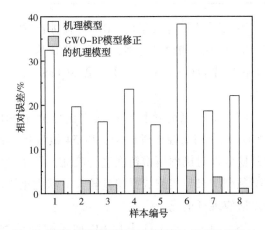

图13 两种模型预测值与实测值的相对误差对比

由表6和图13可以看出，采用GWO-BP算法在机理模型计算结果基础上进行机理误差的反馈修正之后，GWO-BP算法修正后的机理模型预测的点蚀速率与实测开挖数据的最大相对误差为6.18%，较之机理模型预测结果有了大幅度地降低，既说明了GWO-BP模型能够对机理误差进行较好地预测，又验证了GWO-BP模型修正机理模型方法的有效性。

3 工程实例

以某页岩气田的3#集输管道为例，点蚀速率通过实测壁厚损失计算得出，温度、pH、CO_2分压、SRB数量、SO_4^{2-}浓度以靠近开挖点分离器检测的数据为代表。该页岩气集输管道长度为1.82km，管材为L245N，管道外径为273mm，壁厚为5.6mm。近三年中，入口压力范围为2.19~3.45MPa，CO_2分压在0.052~0.273MPa之间，入口温度为17~42℃。图14为3#页岩气集输管道实际沿程走向，表7为3#页岩气集输管道5处开挖点开挖结果。

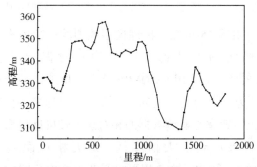

图14 3#页岩气集输管道里程图

表7 3#页岩气集输管道腐蚀数据及实测点蚀速率

序号	温度/℃	CO_2分压/MPa	SRB数量/(个/mL)	SO_4^{2-}浓度/(mg/L)	pH	实测点蚀速率/(mm/a)
1	29	0.07	250	36.69	6.98	0.55
2	32	0.03	300	58.34	6.75	0.52
3	26	0.06	1500	95.32	6.65	0.79
4	27	0.08	750	76.83	6.70	0.68
5	36	0.04	140	44.16	7.30	0.49

将开挖得到的5组数据分别带入机理模型和GWO-BP模型中预测点蚀速率，得到的实际点蚀速率与预测点蚀速率如表8所示，两种模型预测结果与实测值的相对误差对比如图15所示。

表8 两种模型预测测试样本的结果对比

样本编号	实测值/(mm/a)	机理模型		GWO-BP算法修正的机理模型	
		点蚀速率/(mm/a)	相对误差/%	点蚀速率/(mm/a)	相对误差/%
1	0.55	0.71	29.09	0.57	3.27
2	0.52	0.7	34.62	0.56	7.46
3	0.79	1.05	32.91	0.83	4.83
4	0.68	0.92	35.44	0.71	3.81
5	0.49	0.64	30.61	0.50	2.92

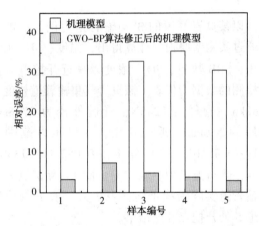

图15 两种模型预测结果与实测值的相对误差对比

由图15可知，机理模型预测的点蚀速率与现场实测点蚀速率的平均相对误差达到32.53%，而采用GWO-BP算法在机理模型计算结果基础上进行误差修正，修正后机理模型预测的点蚀速率与实测数据的相对误差较机理模型预测结果得到了大幅降低，最大相对误差为7.46%，该误差在工程可接受范围之内。明本文使用的GWO-BP模型修正机理模型的方法可以准确预测由SRB与CO_2共同作用引起的腐蚀。并且该模型计算出的点蚀速率略高于实际点蚀率，这有利于现场人员提前预估腐蚀发展，确保页岩气集输管道安全可靠运行。

4 结论

（1）通过正交实验分析，得到了在SRB/CO_2共存的环境下，影响页岩气集输管道腐蚀的主要因素为浸泡方式、温度、pH、流速、SRB数量以及CO_2分压；在SRB/CO_2共存的腐蚀极端环境下，通过SEM检测发现实验7d和14d的试片表层明显附着大量的SRB与腐蚀产物，且7d时的腐蚀产物含量更多；根据XRD检测发现14d后试片表面的腐蚀产物主要为FeS、$FeCO_3$和Fe_3O_4。

（2）基于电荷转移和传质理论建立了SRB与CO_2耦合的腐蚀机理模型，以54组现场实测的开挖数据对机理模型精度进行验证，发现机理模型预测结果与实测数据相差较大，需优化该模型以适应现场情况。

（3）引入GWO算法和BP神经网络模型，对内腐蚀机理模型进行了训练。运用建立的模型对另一条页岩气管道的腐蚀速率进行预测，并开挖验证。5组开挖点有4组都在95%的预测区间内，模型具有良好的准确性，能够运用于后续的腐蚀评价中。

参 考 文 献

[1] 高芸,王蓓,胡迤丹,等.2023年中国天然气发展述评及2024年展望[J].天然气工业,2024,44(02):166-177.

[2] 徐军,李世勇,陈世波.川南页岩气地面流程测试管线刺漏原因及机理分析[J].钻采工艺,2021,44(06):83-87.

[3] 王腾,严小勇,张伟,等.威荣页岩气田腐蚀机理与管控对策研究[J].石化技术,2022,29(12):123-125.

[4] 王月,袁曦,王彦然,等.页岩气田细菌腐蚀与控制技术研究及应用[J].石油与天然气化工,2021,50(05):75-78.

[5] Gu T, Zhao K, Nesic S. A New Mechanistic Model for Mic Based on A Biocatalytic Cathodic Sulfate Reduction Theory[J]. Nace International Corrosion Conference, 2009.

[6] Fatah M C, Ismail M C, Wahjoedi B A. Empirical equation of sulphate reducing bacteria (SRB) corrosion based on abiotic chemistry approach [J]. Anti-Corrosion Methods and Materials, 2013(4): 206-212.

[7] Chamkalani A, Nareh'ei M A, Chamkalani R, Zargari M H, Dehestani-Ardakani M R, Farzam M. Soft computing method for prediction of CO_2 corrosion in flow lines based on neural network approach [J]. Chemical engineering communications, 2013, 200(6): 731-747.

[8] 郭少强, 朱海山, 柳歆, 路民旭, 许立宁, 覃慧敏. 油气输送管线顶部 CO_2 腐蚀预测模型[J]. 腐蚀与防护, 2014, 35(05): 469-472.

[9] 郑度奎, 程远鹏, 李昊燃, 何天隆. IAFSA-GRNN 在油田集输管道 CO_2 腐蚀速率预测中的应用[J]. 中国安全科学学报, 2022, 32(01): 110-117.

[10] 李宁, 李刚, 邓中亮. 改进灰狼算法在土壤墒情监测预测系统中的应用[J]. 计算机应用, 2017, 37(04): 1202-1206.

[11] 赵雪霏, 时景光, 谢湘海. 基于正交试验的电阻焊焊接参数方差分析[J]. 现代商贸工业, 2021, 42(19): 153-155.

气田在线仿真平台架构及仿真模型探讨

李长俊　廖钰朋　贾文龙　杨帆　黄巧竟

（西南石油大学石油与天然气工程学院）

摘　要　天然气是重要的能源和化工原料，93%通过管道输送，保障管道安全对于化工生产意义重大。近年来，我国加大了凝析气、含硫气和页岩气田的开发，集输管道建设持续向西部高山峡谷等恶劣环境挺进，在复杂介质、复杂环境的协同作用下，集输管道内极易出现多相流、硫沉积、水合物等异常工况，给集输管道的高效输送、稳定运行、风险识别与防控带来严峻挑战，迫切需要先进的管理技术与手段。管道智能化技术能够弥补传统管理中的不足与短板，有效提升气田的安全高效管理水平，而在线仿真作为管道智能化的基础，可为管道流动保障问题提供策略支持。在气田集输管道多相流在线仿真技术方面，作者团队进行了一系列尝试，开发了具有在线仿真、运行参数预警、流动保障分析、虚拟计量等功能的气田在线仿真平台。基于质量守恒、动量守恒、能量守恒原理及多场景下的源项模型，建立了适用于任意集输管道形式、任意工况条件下的多相流相变模型。将气田在线仿真平台分别应用于龙王庙气田、川西北地区超高压含硫气田和伴随硫颗粒、水合物颗粒相变的气固、气液固集输管道，展示了该平台的计算精度及对复杂流动场景的适应能力。未来，该平台可进一步拓展至气田集输系统全生命周期内的优化与决策，以及多介质融合输送管网的流动仿真，为管道智能化技术提供有力支撑。

关键词　气田；在线仿真；数学模型；多相流；集输管道

The architecture of the online simulation platform for gas fields and discussion on simulation models

Li Changjun　Liao Yupeng　Jia Wenlong　Yang Fan　Huang Qiaojing

(Petroleum Engineering School, Southwest Petroleum University)

Abstract　Natural gas is a crucial energy source and chemical feedstock, with 93% transported via pipelines, making pipeline safety vital for chemical production. In recent years, China has intensified the development of condensate gas fields, sulfur-containing gas fields, and shale gas fields, with pipeline construction advancing into challenging environments such as mountainous and canyon regions in the west. Under the combined effects of complex media and environments, pipelines are prone to issues like multiphase flow, sulfur deposition, and hydrate formation, posing significant challenges to efficient transportation, stable operation, and risk management. Intelligent pipeline technology addresses these challenges by enhancing gas field safety and efficiency, with online simulation providing strategic support for flow assurance. The author's team developed an online simulation platform with functions such as simulation, parameter warnings, flow analysis, and virtual metering. Using principles of mass, momentum, and energy conservation, they established a multiphase flow phase-transition model adaptable to any pipeline configuration and condition. Applied to fields like Longwangmiao and ultra-high-pressure sulfur-containing pipelines in Sichuan, the platform demonstrates accuracy and adaptability. Future expansion could optimize lifecycle decisions and simulate multi-medium networks, supporting intelligent pipeline technologies.

Key words　Gas field; Online simulation; Mathematical model; Multiphase flow; Gathering and transportation pipeline

习近平总书记指出，能源保障和安全事关国计民生，是须臾不可忽视的"国之大者"。"双碳"背景下，国家大力倡导低碳绿能，天然气作为优质高效的绿色能源，在化石能源主导阶段，可发挥"补位+调峰"的作用，在非化石能源主导阶段，可发挥"调峰+减碳"作用。《"十四五"现代能源体系规划》提出，需要加大力度推动天然气增储上产，确保对外依存底线。天然气是重要的能源和化工原料，93%通过管道输送，保障管道安全对于化工生产意义重大。近年来，我国加大了凝析气、含硫气、页岩气田的开发，集输管道建设持续向西部高山峡谷等恶劣环境挺进，造就了高凝析（如迪那2气田）、高压力（如克拉2气田、川西北九龙山气田）、高含硫（如普光气田、龙王庙气田）、多相流（如川南页岩气田）等国内外罕见的复杂环境集输管道，在复杂介质、多相流动等多因素耦合作用下，集输管道易出现硫沉积、水合物，引发腐蚀泄漏、超压爆燃等突发性事故，造成气田运输通道中断。面对日益复杂化、精细化的气田集输管道系统，其高效输送、稳定运行、风险识别与防控面临严峻挑战，迫切需要先进的管理技术与手段。

《关于加快推进能源数字化智能化发展的若干意见》提出，要推进智能管道建设，为气田的可持续发展提供有力保障。气田集输管道多相流动在线仿真技术基于现场仪表的数采信息，可实现集输管道全线压力、温度、流量、持液率及其他重要参数的实时模拟，量化表征管道内多相流体的流动及变化过程，为管道硫沉积、水合物等流动保障问题提供策略支持，从而大大提高工作效率和管理水平。经过数十年的发展，国外的气田集输管道多相流动在线仿真技术已经基本成熟，并已形成了多套仿真软件产品，中国使用该类软件产品一直受国外制约，每年要缴纳高额的使用授权费用。

十四五以来，国家高度重视工业软件自主化开发，工信部《工业重点行业领域设备更新和技术改造指南》指出，要围绕石油等关系经济命脉和国计民生的行业领域，推动工业软件等更新换代。在气田集输管道多相流在线仿真技术领域，中国在自主发展能力的形成和保持方面还有所欠缺，核心技术、核心软件产品面临"卡脖子"难题。在中美贸易战的大背景下，国外软件随时面临授权终止的风险，并且，国外软件受其特殊版权与核心技术保护机制的影响，难以融入中国自主构建的智慧管网体系。从国家急迫需要和长远需求出发，必须要掌握气田集输管道多相流在线仿真关键技术，研发自主、成熟的软件产品，搭建气田在线仿真平台，从而有效提升气田的安全高效运行水平，为建成智能气田，建设智慧气田打好基础。

1 国内外发展动态

气田集输管道多相流在线仿真技术是气田在线仿真平台的核心，也是气田管道智能化建设的基础，该技术一般集成了图形化组态、实时数据通信、在线数字滤波、大数据存储、在线仿真、参数评价与修正以及大数据可视化等功能，多以软件产品的形式直接出现。自上世纪80年代起，国外就开始了这类软件的研发，经过长时间的应用积累，形成了一系列商用的多相流仿真软件，如用于稳态仿真的PIPEPHASE、PIPESIM等，用于瞬态仿真的OLGA、LEDAFLOW以及用于在线仿真的OLGA online、ISIS、WPM等。

其中，OLGA软件是开发最早、应用最广的仿真软件，已被国内外许多气田作为可行性研究、工程设计和安全运行的依据。OLGA软件的控制方程基于扩展的双流体模型，共包含6个控制方程，包括3个质量守恒方程（气相、液相或液膜、液滴），2个动量守恒方程（含液滴的气体/烃类流体、液相或液膜），以及1个混合能量守恒方程，基于有限体积法与交错网格技术，通过SIMPLE算法求解控制方程，从而获得待求的未知变量。OLGA软件可以实现气田集输管道多相流动、蜡沉积、水合物、段塞流等多种工况的仿真模拟，在大多数工况下，OLGA软件的仿真结果与实际结果对比，具有较好的匹配度。

虽然，OLGA软件在多数情况下表现良好，但在复杂多相流动中，为了获得更高精度模拟结果和更详细的流动信息，OLGA软件有时不能满足需要。例如，在进行流型判别时，OLGA使用的是"最小滑移准则"，并非基于控制方程的计算结果，因而不能预测流型的转变过程。此外，OLGA在预测段塞频率时，也存在不足，并且，由于OLGA只使用了单一的混合能量方程，当各相流体之间热物性差异过大时，温度计算可能不准确。为了推出更精确的多相流动仿真软件，挪威康士伯公司联合TOTAL开发了LEDAFLOW软件。在LEDAFLOW软件的控制模型中，可能存在多达9个场，包括3个连续场和6个分散场。每一个管道控制体（Cell）中包含16个控制方程，允许相间的热力学不平衡,并且还可以追踪每一类物质在每个场内的浓度变化。新推出的LEDAFLOW软件对段塞流的预测效果较OLGA软件更好,在温度、压力、持液率等流动参数计算中展现出与OLGA软件相同或者超越OLGA软件的性能。目前,OLGA软件和LEDAFLOW软件均已在我国海上、陆上气田中得到了广泛应用。

在国内,各高校、科研院所主要针对单相流管网进行自主软件的开发,并取得了不错的成果,例如西南石油大学的天然气管网系统在线仿真软件、西安石油大学的PNS软件、北京石油化工学院的CloudLPS软件等。在气田集输管道多相流在线仿真技术方面,国内的研究起步较晚,但发展很快。1992年,西安交通大学就建立了动力工程多相流国家重点实验室,各石油高校、科研院所投入了大量的人力物力进行多相流研究,也相继推出了自己的软件,例如西安交通大学的油气两相混输稳态计算程序STPHD软件、西南石油大学的NGLPES软件以及中国石油大学(北京)的MPF软件等。但与国外相比,国产多相流软件在仿真规模、仿真速度、仿真精度、稳定性等方面还存在较大的提升空间。

长期以来,西南石油大学一直从事开展管网仿真模型与算法的相关研究,并逐步实现了气体、液体、两相流和复杂流体仿真理论的构建、仿真模块的开发与整合,形成了适用于任意复杂介质、复杂管网的高性能气田在线仿真平台,服务于气田的设计、调度、安全运行与管理。本文详细论述了气田在线仿真平台的架构搭建、气田集输管道多相流仿真建模、开发及相关应用实例,以期为国产化气田在线仿真平台的开发提供参考。

2 气田在线仿真平台架构

在线仿真是计算机系统通过实现虚拟现实技术,模拟管道在不同输送条件和控制下水力热力参数分布的动态变化过程。根据气田集输管道系统智能运行需求,设计并开发了气田在线仿真平台(图1)。气田在线仿真平台基于B/S架构,采用Java EE企业级开发平台开发,从用户界面层级、数据库层级、核心数学模型层级、高效数值算法层级及软件与SCADA系统实时数据通讯层级进行详细设计,可实现气田集输管道的在线仿真、运行参数预警、流动保障分析、虚拟计量等功能。

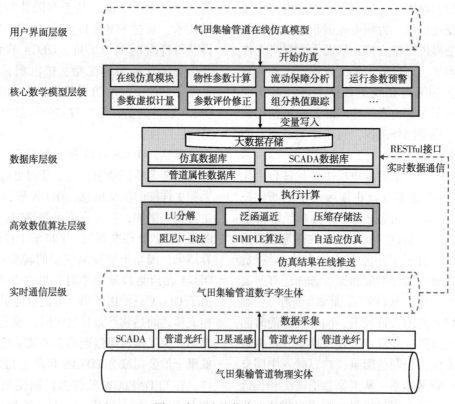

图1 气田在线仿真平台架构

（1）用户界面层级

基于 Microsoft Visual C#和 .Net Framework 4.5软件开发平台，开发了基于实时绘图技术的软件组态界面。该层级是向所有用户开放的层级，在界面上，软件具有全中文的图形化用户组态界面，含有管道和主要设备的图形库工具栏。用户通过工具栏中管道和设备的组态，形成各种不同的管道设计方案或运行方案，建立相应的管道系统仿真模型，并通过对话框和设备属性参数、边界条件、运行工况进行定义，实现气田集输管道水力、热力工况的仿真，流动保障参数的分析。同时，用户可以进行组态、属性参数与边界条件的定义，结果查看和输出。

（2）核心数学模型层级

基于流体力学、热力学理论，开发了正常运行、紧急关断等各种工况的水力热力计算模块。在线仿真平台中的所有水力、热力参数计算都在此层级完成。仿真平台具备气田集输管道不同工况的模拟功能，根据用户建立的组态模型和仿真需求，自驱动后台仿真核心，得到计算结果。

（3）数据库层级

基于时序数据库分布式大数据平台，以及在线仿真的输入输出参数需要，开发了包括管道属性数据库、油气物性数据库、环境参数数据库、仿真结果数据库在内的在线仿真平台核心数据库。

（4）高效数值算法层级

该层级可实现油气物性参数、数学模型及设备参数的求解计算。针对各工况数学模型的特点，以数值分析方法为基础，采用高阶非线性方程组的高效、快速求解技术、大型稀疏矩阵高效压缩存储算法、大型稀疏矩阵自适应仿真算法及多CPU、多核、多线程、并行及GPU编程技术、基于OPC协议的通讯技术、接口技术、数据共享等先进技术开展仿真模型的求解，大幅提高运算速度。

（5）与 SCADA 系统实时数据通讯层级

依托时序数据库分布式大数据平台及数据流传输服务，支撑在线仿真平台与 SCADA 系统等外部系统之间的数据交互。采用 SCADA 实时数据诊断和智能整定技术、在线仿真模型自适应和自学习技术、在线仿真系统分布式、多用户共享展示与应用技术等先进技术完善了在线仿真平台。

3 气田集输管道系统仿真模型

3.1 管道多相流相变模型

气田集输管道系统是由管道、阀门、分离器等多元件、多节点构成的密闭的、统一的水动力系统，对其压力、温度、流量、持液率分布，以及气液相变、硫沉积、水合物等生成情况进行模拟，是实现集输系统安全、经济设计的基础与前提。为了准确描述气田集输管道中各流体相的流动与相变过程，将每一种流体特定形式(连续和分散)描述为一个场，如图2所示。

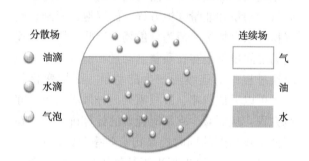

图2 多相流动的连续场和分散场

考虑场与场之间的质量、动量和能量传递，可建立气田集输管道的多相流相变模型，包含各个场的质量守恒方程、动量守恒方程和能量守恒方程：

$$\frac{\partial \alpha_k \rho_k}{\partial t} + \frac{\partial}{\partial x}(\alpha_k \rho_k w_k) = \sum_{i \neq k} \dot{m}_{ki} \quad (1)$$

$$\frac{\partial}{\partial t}(\alpha_k \rho_k w_k) + \frac{\partial}{\partial x}(\alpha_k \rho_k w_k w_k) = -\alpha_k \frac{\partial P}{\partial x} - \alpha_k \rho_k g \sin\theta - \alpha_k \rho_k g \cos\theta \frac{\partial h}{\partial x} + \sum_{i \neq k} \tau_{ki} S_{ki} - \tau_{kw} S_{kw} + \sum_{i \neq k} \dot{m}_{ki} w_{ki} \quad (2)$$

$$\frac{\partial}{\partial t}\left\{\rho_k \alpha_k \left(h_k + \frac{w_k^2}{2}\right)\right\} + \frac{\partial}{\partial x}\left\{\rho_k \alpha_k w_k \left(h_k + \frac{w_k^2}{2}\right)\right\} + \sum_{i \neq k} q_{ki} S_{ki}$$

$$= -\rho_k \alpha_k w_k g \sin\theta - \frac{\partial (P \alpha_k w_k)}{\partial x} + \sum_{i \neq k} \tau_{ki} S_{ki} u_{ki} - q_{kw} S_{kw} + \sum_{i \neq k} \dot{m}_{ki}\left(h_{ki} + \frac{w_{ki}^2}{2}\right) \quad (3)$$

式中，k 为不同的场；x 为管道位置，m；t 为时间，s；α_k 为场 k 的体积分率；ρ_k 为场 k 的

密度，kg/m³；w_k 为场 k 的平均速度，m/s；w_{ki} 为场 k 与其他场相接触界面的平均速度，m/s；$\dot{m}_{ki}$ 为场 k 从其他场中获得的净质量流量，kg/(m³·s)；P 为压力，Pa；g 为重力加速度，m²/s；θ 为管道倾角，rad；h 为液面高度，m；τ_{ki} 为场 k 与其他场的相间剪切应力，N/m²；τ_{kw} 为场 k 与管壁之间的剪切应力，N/m²；S_{ki} 为场 k 与其他场相接触界面的长度，m；S_{kw} 为场 k 的湿周，m；h_k 为场 k 的比焓，J/kg；h_{ki} 为场 k 与其他场相接触界面传递的比焓，J/kg；q_{ki} 为场 k 和其它场间的热量传递速率，J/(m²·s)；q_{kw} 为场 k 和管壁的热流通量传递速率，J/(m²·s)。

除上述守恒方程外，还需要构建体积相分率的约束方程，保证各个场的体积相分率之和为1：

$$\sum \alpha_k = 1 \quad (4)$$

气田集输管道中，一般将气相、油相和水相视为连续场，连续场流动时的质量、动量和能量变化较大，因而针对不同的连续场，必须分别列写质量、动量和能量守恒方程；而气泡、液滴、硫沉积、水合物等一般为分散场，这些分散场与连续场之间虽然也存在质量、动量和能量交换，但交换过程中对连续场产生的动量、能量变化较小，可以近似忽略，故只需列出分散场的质量守恒方程，从而减少求解时耗费的计算机资源，在保证求解精度的同时提高仿真效率。经简化后，可通过16个守恒方程来表征气田集输管道中的绝大部分流动工况，如图3所示。

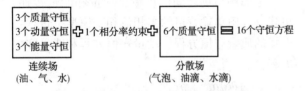

图3　简化后的多相流相变模型

要求解该方程组，必须要补充3个连续场流体的状态方程和表征分散场流体热物性关系的表达式，以及确定各场之间的质量传递、动量传递和能量传递，使模型封闭，该方程组才能被解出。求解时，基于有限体积法和交错网格离散简化后的控制方程，采用SIMPLE算法进行迭代求解。

式(1)~式(4)还可写为更通用化的形式：

$$\frac{\partial \mathbf{U}(x,t)}{\partial t} + \frac{\partial \mathbf{F}(\mathbf{U}(x,t))}{\partial x} + \mathbf{B}(\mathbf{U}(x,t))\frac{\partial \mathbf{U}(x,t)}{\partial x} = \mathbf{S}(\mathbf{U}(x,t)) \quad (5)$$

式中，$\mathbf{U}(x,t)$ 为各场的解向量；$\partial \mathbf{U}(x,t)/\partial t$ 代表各场的瞬态项；$\partial \mathbf{F}(\mathbf{U}(x,t))/\partial x$ 代表各场的保守项；$\mathbf{B}(\mathbf{U}(x,t))\times\partial \mathbf{U}(x,t)/\partial t$ 代表各场非保守项的矩阵向量积；$\mathbf{S}(\mathbf{U}(x,t))$ 为源项。

在不同流动场景中，式(5)左端的表示形式都是相似的，它代表了流体力学中统一的偏微分方程形式；而式(5)右端的源项 $\mathbf{S}$，它代表了流动过程中所受的外力变化(液滴与气泡的夹带/聚并)、热力变化(蒸发、冷凝、融化和凝固)和化学变化(溶解、扩散)等。由于连续场和分散场之间不断转化和场界面变化的复杂性，不同流动场景下源项 $\mathbf{S}$ 的差异性可能较大，因此需要有针对性的进行源项 $\mathbf{S}$ 的构建，保证流动参数求解的准确性。

3.2 多场景源项模型

(1) 复杂天然气大压差节流

含 CO_2、H_2S、凝液等复杂介质的高压/超高压天然气进入气田集输系统时，首先就需要进行节流调压，大压差节流处温度骤降，一般有液体析出，导致节流过程中出现非平衡多相流动。基于等焓节流原理，通过改进Lee-Kesler状态方程的混合规则，提升焓值计算精度，建立了考虑液滴非平衡凝结相变的大压差节流工艺参数计算方法，准确预测节流压力、温度变化。大压差节流过程中的传质源项可表示为：

$$S_m = \frac{4\pi}{3}\rho_1 r_c^3 J + 4\pi r^2 \rho_v \rho_1 N \frac{dr}{dt} \quad (6)$$

式中，ρ_1 为液相密度，kg/m³；ρ_v 为气相密度，kg/m³；r 为液滴半径，m；r_c 为临界液滴半径，m；J 为成核速率，1/(m³·s)；N 为液滴数密度，1/kg。

(2) 气田集输管道硫沉积

高含硫天然气集输管道中单质硫沉积是一个涉及化学反应、气固相变、结晶过程以及气固两相流动的复杂动态现象。在此过程中，管道内的压力、温度、气体组分等不断变化的参数，驱动着单质硫的生成、析出、形核及团聚生长。同时，硫颗粒的生长过程会影响局部气固相比例和天然气组分浓度，进一步作用于管道内的压力、温度和流速等气固两相流动参数。这些参数之间的动态变化与相互耦合，使得管道内部形成了一种伴随硫颗粒团聚生长的复杂气固两相流动过程。只有将多相流相变模型与单质硫的生成机理、硫颗粒形核与生长机理、硫颗粒的团聚与凝聚机理相结合，构建准确描述单质硫沉积过程的质量、动量和能量传递源项，才能实现集输管道中硫沉积和沉积位置的准确预测。基于元素硫的溶解-析出-结晶-运移-沉积机制，结合元素硫溶解度模型，修正了气固多相流模型中的传质源项。改进后的传质模型如下，

$$S_m = \frac{M_{S_8}}{V_g \cdot V^S_{S_8}} \times \eta \times (Q \times y_{S_8} - Q_x \times y_{S_{8x}}) \quad (7)$$

$$y_{S_8} = y^{phy}_{S_8} + y^{chem}_{S_8} \quad (8)$$

$$y^{phy}_{S_8} = \frac{\phi^{sat}_{S_8} P^{sat}_{S_8}}{\phi^V_{S_8} P} \exp\frac{V^S_{S_8}(P - P^{sat}_{S_8})}{RT} \quad (9)$$

$$y^{chem}_{S_8} = y^V_{H_2S_9} = \frac{(y^V_{H_2S}\phi^V_{H_2S})(y^{phy}_{S_8}\phi^V_{S_8})}{\phi^V_{H_2S_9}}$$

$$\exp\left[-\frac{\mu^\Theta_{H_2S_9} - \mu^\Theta_{S_8} - \mu^\Theta_{H_2S}}{RT}\right] \quad (10)$$

式中，M_{S_8} 为单质硫的摩尔质量，kg/mol；V_g 为气相的体积，m³；$V^S_{S_8}$ 为 S_8 的摩尔体积，m³/mol；η 为硫颗粒的沉积率；Q 为管道的体积流量，m³/s；Q_x 为管道 x 处的体积流量，m³/s；y_{S_8} 为元素硫的总溶解度；$y_{S_{8x}}$ 为管道 x 处元素硫的总溶解度；$y^{phy}_{S_8}$ 为单质硫的物理溶解度；$y^{chem}_{S_8}$ 为单质硫的化学溶解度；$y^V_{H_2S_9}$ 为气相中 H_2S_9 的摩尔分数，其值等于 S_8 的化学溶解度；$\phi^{sat}_{S_8}$ 为饱和硫蒸气的逸度系数，可视为 1；$\phi^V_{S_8}$ 为气相 S_8 的逸度系数；$\phi^V_{H_2S_9}$ 为气相中 H_2S_9 的逸度系数 $\phi^V_{H_2S}$ 为气相中 H_2S 的逸度系数；$\mu^\Theta_{H_2S_9}$、$\mu^\Theta_{S_8}$ 和 $\mu^\Theta_{H_2S}$ 分别为 H_2S_9、S_8 和 H_2S 的标准化学势；$P^{sat}_{S_8}$ 为 S_8 的饱和蒸气压，Pa；R 为气体常数，J/(kg·K)。

（3）气田集输管道水合物生成与运移

水合物堵塞是气田集输系统广泛面临的安全生产问题（图4），确定水合物的生成压力、温度条件对于集输系统的设计有重要意义。同时，受沿线不断变化的温度和压力条件影响，管道中的水合物颗粒会出现生长或分解的现象，这种水合物颗粒的相变与管道流动参数的动态耦合作用，使得对水合物动力学参数和管道流动参数的预测和分析变得非常困难。

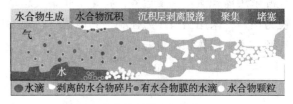

图 4 水合物的形成与堵塞过程

考虑水合物颗粒的生长和分解与管道状态参量的耦合作用，基于守恒原理，建立了伴随水合物颗粒相变的气液固三相管流数学模型。水合物生长和分解过程中的传质源项可表示为，

水合物生长的传质源项：

$$S_m = -\frac{M_h}{V_l \cdot V_w} k_m A_{g-l}(C_{g-w} - C_{w-h}) \quad (11)$$

式中，g、w、l 和 h 分别代表气相、水相、液相和水合物相；V_l 为液相体积，m³；V_w 为水相的摩尔体积，m³/mol；M_h 为水合物的摩尔质量，kg/mol；k_m 为气体传质系数，m/s；A_{g-l} 为气液界面面积，m²；C_{g-w} 为系统温度和压力条件下气水两相平衡时气体的溶解度；C_{w-h} 为系统温度和压力条件下水-水合物两相平衡时气体在水相中的溶解度。

水合物分解的传质源项：

$$S_m = M_h K A_s(f_s - f) \quad (12)$$

式中，K 为水合物分解动力学总速率常数，mol/(MPa·s·m²)；A_s 为参与分解的水合物颗粒表面积，m²/m³；f_s 为水合物固体颗粒与液相界面处甲烷的逸度，MPa；f 为系统温度压力条件下气相中甲烷的逸度，MPa。

上述多相流相变模型与多场景下的源项模型对输送介质、管道结构没有特殊的要求，因此可以适用于任意结构形式的气田集输管道，构成了气田在线仿真平台的核心。

4 应用实例

4.1 龙王庙气田仿真模型

以简化的龙王庙气田集输管网为研究对象（图5），共包含 6 个主要集气站，12 个气源，11 条主要管线。采用气田在线仿真平台建立仿真模型，进行稳态和动态仿真，检验软件的计算精度。

（1）稳态仿真

稳态仿真边界条件及部分计算设定见表 1。

气源采用体积流量温度边界,将龙王庙集气总站视为用户,采用出口压力控制,其他集气站视为气源,也采用体积流量温度边界。

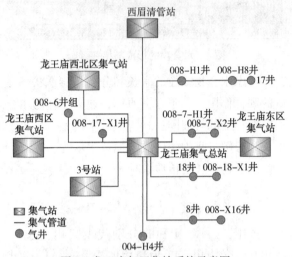

图5 龙王庙气田集输系统示意图

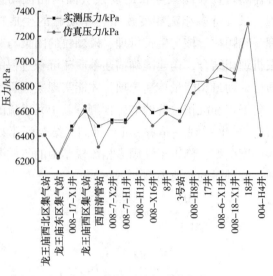

图6 龙王庙气田集输系统部分井及集气站压力

表1 稳态仿真边界条件及部分计算设定

元件类型	气源	集气站
边界类型	体积流量温度	体积流量温度
摩阻系数	Colebrook-White	
状态方程	PR	

针对图5所述的集输管网,软件模拟计算得到的气源、集气站的压力数据如图6所示。由图6可知,计算结果符合实际运行情况。

(2)动态仿真

在稳态仿真的基础上,进行动态仿真,改变008-7-H1井的流量,设置最小时间步长为1min,最大时间步长为10min。仿真模拟得到的008-7-H1井和相邻的008-7-X2井的压力、流量变化情况见图7。可见,当008-7-H1井流量增加时,其压力呈现相同的变化趋势,同时,处于相邻管段的008-7-X2井的压力变化趋势和008-7-H1相同,符合实际运行规律。

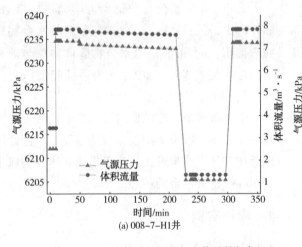

(a) 008-7-H1井

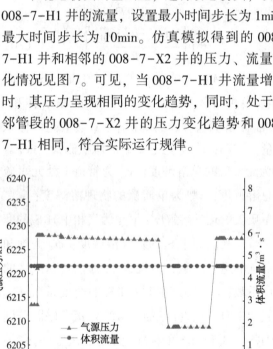

(a) 008-7-X2井

图7 008-7-H1井及其相邻008-7-X2井6h内流量、压力变化曲线

4.2 川西北地区超高压含硫天然气节流仿真

川西北气田开发潜力巨大,但由于地面多级复杂的节流过程和超高压含酸气引起的频繁水合物堵塞,严重降低了其生产效率。考虑到水合物的防治,合理设计节流温度和节流压力,优化地面生产工艺是解决这一问题的关键。通过气田在线仿真平台的节流计算模块,实现了该地区8口气井节流温度的精确预测(图8)。同时,将超高压酸性气井L004-X1和L016-H1的地面节流技术从5级优化为了3级(图9),提高了该气田的生产运营水平,为川西北盆地深层海相碳酸盐岩气藏开发前评价提供了支撑。

4.3 高含硫天然气与硫颗粒气固两相管流仿真

高含硫天然气集输系统中的硫沉积会降低集输系统的输送效率、诱发设备腐蚀，甚至造成安全生产事故。采用气田在线仿真平台对节流阀后的硫颗粒沉积情况进行模拟，模拟结果如图 10 所示。经过节流阀后，硫颗粒在孔板出口处发生了显著的团聚，在与实物图的对比中可以看出，模拟的团聚位置与实际团聚位置非常接近，间接说明了硫沉积源项模型的准确性。

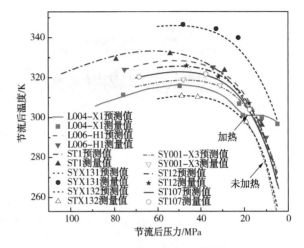

图 8 地面节流温降预测结果

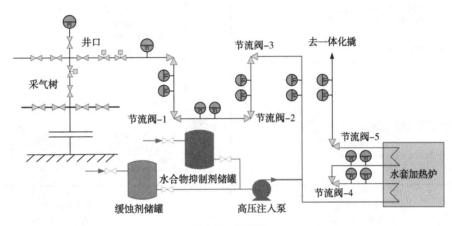

(a) 优化前-五级节流

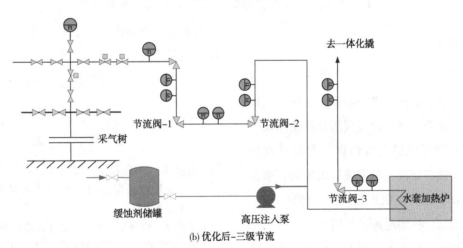

(b) 优化后-三级节流

图 9 L004-X1 井地面节流工艺

4.4 伴随水合物颗粒相变的气液固三相管流仿真

在气田集输管道中，不稳定的温度和压力可能使水合物颗粒出现生长或分解等现象。为了描述伴随水合物颗粒相变的气液固三相流动，基于 Joshi 等人气水两相体系水合物生成流动实验数据，采用气田在线仿真平台进行建模求解，模拟结果如图 11 所示。模拟值与实际值的平均相对误差为 16.4%，最大不超过 40%，模型较为准确，可以实现对水合物输送管道气液固三相流动的模拟预测。

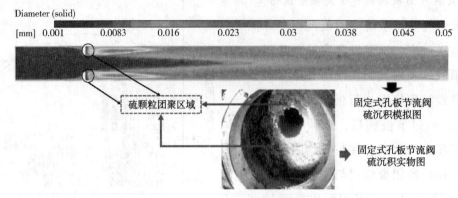

图 10 固定式孔板节流阀硫颗粒团聚模拟结果与实际工况的对比

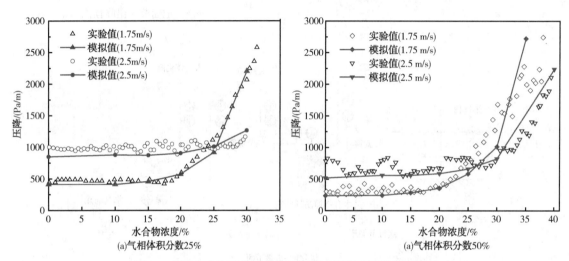

图 11 伴随水合物颗粒相变的气液固三相管流压降模拟结果

5 结论与建议

开发具有自主知识产权的气田集输管道在线仿真平台是掌握多相流仿真核心技术、确保气田集输系统安全高效运行、推进气田智能化建设的迫切需求。介绍了气田在线仿真平台的开发架构，包含5大核心层级，可实现气田集输管道系统的在线仿真、运行参数预警、流动保障分析、虚拟计量等功能，详细论述了气田集输管道多相流相变模型及多场景下源项模型的构建方法，形成了气田在线仿真平台的核心。通过龙王庙气田、川西北高含硫天然气集输管道节流、伴随硫颗粒、水合物颗粒相变的气固两相流和气液固三相流的实际应用，验证了该平台的计算精度及对复杂流动场景的适应能力，为自主化气田在线仿真平台的开发奠定了基础。未来，该平台可进一步拓展至气田集输系统全生命周期内的仿真、优化、预测及决策应用，以及多介质融合输送管网的流动仿真，为管道智能化发展提供有力的技术支撑。

参 考 文 献

[1] 余晓钟,刘梦薇,邱湖森,等.我国天然气产业高质量发展路径探讨[J].中国工程科学,2024,26(04):63-71.

[2] 王雅菲,周淑慧,李广.基于天然气视角的《"十四五"现代能源体系规划》解读[J].国际石油经济,2022,30(04):1-10.

[3] 李长俊,宇波,张对红,等.油气综合立体调运关键技术现状与趋势[J].前瞻科技,2024,3(02):39-49.

[4] 姚茂堂,刘举,袁学芳,等.高温高压凝析气藏井筒结垢及除垢研究[J].石油与天然气化工,2020,49(04):73-77.

[5] 杨海军,李勇,唐雁刚,等.塔里木盆地克深气田成藏条件及勘探开发关键技术[J].石油学报,

2021,42(03):399-414.

[6] 曾婷婷,李长俊,贾文龙.九龙山异常高压气田地面节流工艺研究[J].油气田地面工程,2016,35(03):59-61.

[7] 李士伦,杜建芬,郭平,等.对高含硫气田开发的几点建议[J].天然气工业,2007,27(02):137-140+163-164.

[8] 田华,鲁雪松,张水昌,等.四川盆地安岳气田龙王庙组气藏多层盖层天然气扩散量[J].石油学报,2022,43(03):355-363+398.

[9] 马新华,谢军.川南地区页岩气勘探开发进展及发展前景[J].石油勘探与开发,2018,45(01):161-169.

[10] 钱建华,牛彻,杜威.管道智能化管理的发展趋势及展望[J].油气储运,2021,40(02):121-130.

[11] 关于加快推进能源数字化智能化发展的若干意见[J].大众用电,2023,38(04):11-14.

[12] 陈宏举,王靖怡,康琦,等.多相管流模拟软件MPF与OLGA和LedaFlow预测能力对比[J].中国海上油气,2022,34(06):168-176.

[13] 工信部:印发《工业重点行业领域设备更新和技术改造指南》[J].中国设备工程,2024,40(19):1.

[14] 李长俊,张员瑞,贾文龙,等.大型天然气管网系统在线仿真方法及软件开发[J].油气储运,2022,41(06):723-731.

[15] 康琦,吴海浩,张若晨,等.面向智能油田的集输管网工艺模拟软件研制[J].油气储运,2021,40(03):277-286.

[16] 袁玉.煤层气井生产系统流动模型研究与应用[D].北京:中国石油大学(北京),2020.

[17] Berg K, Davalath J. Field Application of Idun Production Measurement System[C]. Houston: Offshore Technology Conference, 2002, OTC-14007.

[18] Theuveny B, Kosmala A, Sagar R K. Real-Time Production-A Virtual Dream or Reality? The Case of Remote Surveillance of ESP and Multiphase Metering[C]. San Antonio: SPE Annular Technical Conference, 2006, SPE-102351.

[19] Haouche M, Tessier A, Deffous Y et al. Smart metering: An Online Application of Data Validation and Reconciliation Approach[C]. Utrecht: SPE intelligent Energy Conference and Exhibition, 2012, SPE-149908.

[20] Kjølaas J, De Leebeeck A, Johansen S T. Simulation of hydrodynamic slug flow using the LedaFlow slug capturing model[C]. Cannes: BHR International Conference on Multiphase Production Technology, 2013, BHR-2013-H1.

[21] 杨可嘉.多相流模拟软件LedaFlow与OLGA的对比分析研究[D].北京:中国石油大学(北京),2018.

[22] 钱东良,李长俊,廖柯熹,等.水改气海底管道输送高压天然气可行性研究[J].中国安全生产科学技术,2014,10(10):173-178.

[23] 陈海宏,李清平,姚海元,等.流花16-2油田海底管道清管方案优化研究[J].中国海上油气,2022,34(02):188-193.

[24] 史静怡,樊建春,武胜男,等.深水井筒天然气水合物形成预测及风险评价[J].油气储运,2020,39(09):988-996.

[25] 李长俊,刘浠尧,贾文龙,等.大落差天然气管道清管冲击分析[J].油气田地面工程,2015,34(12):46-48.

[26] 宋尚飞,史博会,兰文萍,等.多相混输管道水合物流动的LedaFlow软件模拟[J].油气储运,2019,38(06):655-661.

[27] 宋尚飞,史博会,霍小倩,等.基于LedaFlow的天然气凝析液管道瞬变过程模拟[J].油气储运,2019,38(04):404-411.

[28] 朱浩宇,李长俊,贾文龙,等.基于LedaFlow的液相乙烷输送管道投产过程数值模拟[J].油气储运,2023,42(03):335-342.

[29] 王雪媛,陈文峰,鞠朋朋,等.基于LedaFlow的深水海管蜡沉积模拟分析[J].天然气与石油,2022,40(06):25-29.

[30] 亓佳宁,樊迪,张明思,等.气液混输管道组分跟踪及相变量预测[J].油气储运,2023,42(02):215-222.

[31] 王寿喜,邓传忠,陈传胜,等.天然气管网在线仿真理论与实践[J].油气储运,2022,41(03):241-255.

[32] 王军防,宇波,李亚平,等.国产化液体管道云仿真软件CloudLPS的应用[J].油气储运,2023,42(12):1419-1434.

[33] 王树众,梁志鹏,林宗虎,等.油气两相混输的稳态计算程序——STPHD[J].油田地面工程,1997,(06):6-10+81.

[34] 贾文龙.天然气液烃输送管网仿真理论与技术研究[D].成都:西南石油大学,2014.

[35] Loilier P. Numerical simulation of two-phase gas-liquid flows in inclined and vertical pipelines[D]. Cranfield: Cranfield University, 2006.

[36] Jia W, Yang F, Li C, et al. A unified thermodynamic framework to compute the hydrate formation conditions of acidic gas/water/alcohol/electrolyte mixtures up to 186.2MPa[J]. Energy, 2021, 230: 120735.

[37] Li C, Zhang C, Li Z, et al. Numerical study on the condensation characteristics of natural gas in the throttle valve[J]. Journal of Natural Gas Science and Engineering, 2022, 104: 104689.

[38] Huang J, Liu G, Fan S, et al. Numerical simulation of sulfur particle agglomeration at bends of high sulfur natural gas gathering pipelines based on Euler-PBM coupling [J]. Scientific Reports, 2024, 14 (1): 19190.

[39] 黄婷. 深水天然气水合物输送管道气液固三相流基础理论研究[D]. 成都：西南石油大学, 2018.

[40] Luo Z, Liu Q, Yang F, et al. Research and Application of Surface Throttling Technology for Ultra-High-Pressure Sour Natural Gas Wells in Northwestern Sichuan Basin[J]. Energies, 2022, 15(22): 8641.

[41] Joshi S V. Experimental investigation and modeling of gas hydrate formation in high water cut producing oil pipelines [D]. Golden：Colorado School of Mines, 2012.

Application of Composite Materials in Pipeline Repair and Optimization

Casey Whalen　Matthew Green

[(Henkel)CSNRI]

Abstract　n pipeline repair engineering, the selection of materials is crucial for optimizing performance, cost, and durability. Traditional isotropic materials, such as metals, have simple and predictable design processes, but for specific applications, they may not always be the most efficient (or economical) solution. This article explores the potential applications of fiber-reinforced polymer (FRP) composites, which exhibit anisotropic behavior and allow for customized material properties to be optimized for specific stress scenarios. The article presents three case studies demonstrating the benefits of using composite materials for pipeline repair and optimization. These case studies highlight the versatility of composite materials in addressing various pipeline defects and environmental challenges by adjusting resin, fiber orientation, and fabric structure. Each case study examined specific design methods and related test results. Finally, this article emphasizes the importance of intentional design, testing, and certification in utilizing the unique properties of composite materials to meet specific requirements of pipeline infrastructure, ensuring long-term integrity and safety.

Key words　Pipeline repair; Fiber-reinforced polymer; Customization; Corrosion defects; Pipeline depression; Circumferential weld repair

Introduction - Research background and purpose

In the field of pipeline engineering, material selection plays a crucial role in determining the performance, cost, and durability of the infrastructure. A keycomponent of material selection includes the directionality of the material's physical properties. The standard characteristic of many materials, especially metals, is their isotropic nature which means that these materials have uniform properties in all directions, which simplifies the engineering design process. For example, once the best type of metal composition is determined for a pipeline, calculating the necessary thicknesses or other design factors can be a relatively straightforward task. Engineers often perform simple comparisons between various stress scenarios, such as the pressure-induced hoop stress and the bending loads that the pipe may encounter. The highest required thickness, whether it is driven by hoop stress or bending loads, becomes the minimum required thickness for the pipe design. This thickness determines the pipe's weight, cost, inspection protocols, and overall structural integrity. The isotropic behavior of materials leads to a linear, predictable design process.

However, while this simplicity may seem advantageous, it does not always result in the most efficient solution for all applications. Engineers are often tasked with balancing multiple factors, such as strength, durability, weight, and cost. In many cases, isotropic materials, although reliable and easy, may not be the best option when seeking the optimal performance or efficiency for specific pipeline applications.

In contrast to isotropic materials, fiber-reinforced polymers (FRPs), or composite materials, exhibit anisotropic behavior, meaning that their properties differ depending on the direction in which they are loaded. This introduces additional complexity into the design process, but also presents the opportunity for high levels of optimization. Anisotropic materials offer engineers the ability to tailor their material properties in specific directions, making them ideal for applications whereload directionality is consistent such as in pressure loading or bending in a pipe.

This paper explores how composite materials can be utilized in pipeline repair and optimization, focusing on three distinct repair scenarios. By leveraging the customizable properties of composites, significant improvements in pipeline longevity, cost efficiency, and performance can be achieved. The objective is to illustrate how composites can be tailored to meet the unique challenges posed by different pipeline defects and environmental conditions. Each case study in this paper highlights the benefits and design considerations involved in selecting composite materials for repair, demonstrating how their versatility enables optimized solutions that go beyond traditional methods.

Fiber-Reinforced Polymers: Anisotropic Materials with Customizable Properties

Beforedelving into the specific repair scenarios, it is important to define the fundamental concept of composite materials and their anisotropic nature. Composite materials, such as carbon fiber or glass fiber reinforced with resin, are made by layering multiple plies or fabric sheets that have been impregnated with a resin system. These materials differ significantly from metals because their properties can be engineered by varying the orientation of the fibers within each ply, the type of fiber used, and the type of resin that encapsulates the fiber. The layering technique is particularly advantageous because it allows the material's performance to be customized according to the specific stresses the pipeline will face. For instance, by adjusting the orientation of the fiber fabric (i.e., at 0°, 45°, or 90° angles relative to the pipe's axis), the material can be designed to better resist tensile, compressive, or shear stresses in different directions. This level of customization makes composite materials an ideal candidate for specialized applications where performance optimization is a priority.

In contrast to the example provided at the beginning of the paper, a composite system might only require four plies of composite materials, or layers, to address hoop stress in a pipeline, but twelve layers for bending stress. By strategically changing the fiber orientation and the resin system, the composite material could be optimized to require only eight layers to address both stress scenarios while reducing both the weight and the cost of the system. This ability to fine-tune material properties for specific stress scenarios enables composite systems to outperform traditional materials in various applications.

Case Study 1: General Corrosion Repair in Natural Gas Pipelines

The first case study to be examined will focus on hoop stresses due to general corrosion on a natural gas pipeline. Based on the defect assessment, the only injurious stress load is hoop stress with a potential burst-type failure. However, additional practical design considerations, including design life validation, also play a role in product selection. The repair that will be discussed here was installed in live field service and removed after 25 years for testing.

Product Design and Methodology

For general corrosion on a natural gas line, the primary threat of failure isthe reduction in the pipe's capacity to handle internal pressure. This comparison is done by comparing the pipe's response to the pressure called hoop stress against the pipe's specified minimum yield stress (SMYS). If only internal pressure is considered as an input load, the hoop stress of a pipe can be calculated using Barlow's Formula:

$$P = \frac{2\sigma_h t}{D} \rightarrow \sigma_h = \frac{PD}{2t}$$

Where P = internal pressure, σ_h = hoop stress, t = pipe wall thickness, D = outside diameter.

In the case of external corrosion, the remaining pipe wall thickness is decreasedas the steel is consumed in the corrosion process. A simplified approach to assessing the damage caused by this corrosion would be to replace the term 't', which is normally the nominal wall thickness, with the remaining wall thickness. The use of remaining wall thickness would indicate the estimated hoop stress currently in the pipe when solving for σ_h. This value could be compared to the pipe's SMYS to determine if a repair is needed. Additional assessment methods such as ASME B31.G provide additional methods for a more accurate assessment of the damage incurred on the pipe.

When considering the axial loads only due to internal pressure, the stress equation changes to be:

$$\sigma_a = \frac{PD}{4t}$$

Immediately apparent is that the axial stress load is half of the hoop stress load when only considering the effects of internal pressure. With the metallic pipe being isotropic, this means that the hoop stress will be the first point of failure in this scenario. With this information, a composite repair can be designed to focus specifically on hoop stress reinforcement as no other external load scenarios are expected or are of concern. The composite repair system utilized in this repair was the Clock Spring® repair system which is a uni-directional fiberglass system with all fabric aligned in the pipe's circumferential orientation. This allows for maximum reinforcement of hoop stress. When a composite material is applied to the outside of a pipe and any defect spots are filled in with an appropriately stiff material, Barlow's formula can be rearranged to account for a stress distributed across the pipe and the composite. The following equation can be used as a starting point to estimate the remaining stress in the pipe substrate at the defect area:

$$\sigma_h = \frac{1}{t_{rem}} \left(\frac{PD}{2} - \sigma_c t_c \right)$$

Where the terms t_{rem} = remaining wall thickness, σ_c = composite hoop stress, and t_c = composite repair thickness are introduced. Assumptions including installation pressure, and more are required for a detailed analysis, however, the important piece of information to focus on is the term σ_c. This term, being the composite repairs stress loading oriented in the circumferential direction of the pipe, directly determines the amount of potential stress relief given to the pipe. While utilizing a term such as short-term tensile stress would provide an estimated maximum capacity, this approach is not utilized for long-term pipeline repairs.

Intended as a 'permanent' repair, per requirements from relevant standards and regulations, a long-term stress, or equivalent strain times modulus, must be tested for in relation to overall design life. Composite materials, as most materials, will suffer from material creep when exposed to a continuous stress. Fortunately, a method of testing exists to estimate the reduced stress capacity based on duration. Completion of this test and proper implementation allows one to utilize a design stress for the composite. With the addition of safety factors introduced as well, a composite thickness can be calculated to ensure that the stress in the composite is below a maximum design load while simultaneously ensuring that the stress in the pipe stays within operating parameters.

Relevant to this case study, the Clock Spring repair system was tested to determine a design stress for 50 years. In this test, the Clock Spring was exposed to multiple loads over 10,000 hours. Plotting the time to failure against the stress loads applied to the composite test samples, an extrapolation can be drawn which ultimately suggests that for a 50-year design, a maximum composite stress load of 20,000 psi should be considered. This is shown in the figure below. Considering the inclusion of safety factors, however, it is likely that this design stress, when used to determine repair thickness, is significantly higher than the actual stresses the repair system will experience, implying that the actual failure life will far exceed design expectations.

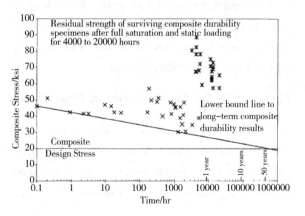

Case Study Results

In late 2020, a natural gas pipeline provided an opportunity to conduct testing and validation on a Clock Spring repair which they had installed and has been operating in service since 1995. The section of pipeline was being replaced and a section which had the repairs installed was provided to perform testing to evaluate the overall condition of the repairs. To the authors' knowledge, this is the oldest composite repair in continuous operation to be removed from service for this type of testing.

The pipe section of note, which was removed from one of their assets in the northern United States and sent to CSNRI for testing, is a 24″ (610mm) diameter natural gas pipeline operating at 750 psi (51.7 bar). The length of pipe had a total of 9 Clock Spring repairs over multiple corrosion defects (the worst of which was approximately 12″ [305mm] long and 63% deep) that were installed in 1995 and have been in operation since.

Once received by CSNRI, the pipe section was fitted with end caps and other required pieces to conduct a full-scale pressure test. The testing protocol followed was to gradually increase pressure to 1.5 times the MAOP of the pipeline (970psi [66.9bar]), which works out to 1455psi (100.3bar) and hold for 5minutes, followed by pressurization until rupture. After a successful hold at the target value, the pressure was then increased until final failure pressure was achieved. This occurred at 2180psi (150.3bar) and, as can be seen in the pictures, occurred in the non-defect section of the pipe outside of the repair area. This shows that a high degree of confidence can be placed in the 25-year-old installed Clock Spring repairs as they were able to continue to maintain a structural repair over the defect section which made it stronger than the pipe itself even after their targeted service life.

This result indicates that a properly designed repair considering only internal pressure and built for hoop stress reinforcement can survive for at least 25 years in service and still provide sufficient reinforcement to cause failure to occur outside of the repair in a forced over-pressurization event.

Case Study 2: Dent Repair in Pipe Spools

The second case study considers a pipe spool with a significant dent. In this scenario, the largest potential threat to system failure is the accelerated fatigue in aggressive cyclic environments due to the increased stresses imposed by the dent shape. An optimized repair for this scenario will need to focus on multi-directional stress states and have a design objective of reducing, or counteracting, the increased substrate stress. To best demonstrate this material's capacity, a test was performed that measured strain in the dent of an unreinforced sample against one that was reinforced with a carbon-fiber epoxy repair system.

Product Design and Methodology

For dented pipelines, the assumption of hoop

stress induced bursting is insufficient for a design basis. Instead, when a dented pipe undergoes a pressure cycle, multidirectional stresses are present that may cause fatigue cracking or other damage to unidirectional composite systems if the dent is severe enough with high pressure cycling. Additionally, the primary mode of failure shifts to fatigue in the dented region of the pipe as each cycle sees a potentially large increase in its stress response due to its non-ideal geometry. This increase in stress can be summarized by a term known as the Stress Concentration Factor or SCF. This term is a ratio of the peak stress in a defect area compared to the stress of pristine, undamaged pipe at the same input load or pressure.

This increased stress that occurs during pressure cycling results in a significant reduction in the amount of remaining fatigue life for the component. However, in an over-pressurization event, it has been demonstrated that dents have only a small portion of the capacity removed, unlike large scale corrosion. Therefore, the goal of a composite here is not burst, or strength, reinforcement but must be to reduce the stress that the damaged pipe sees per cycle. Doing so will increase its remaining fatigue life and allow the pipe to continue to operate for significantly longer.

With this change in objective in mind, a few alterations can be made to a composite repair system to allow for a more targeted design approach. First, the introduction of carbon fiber as opposed to glass fiber allows for a significant increase in the composite's modulus of elasticity. This is relevant because a higher modulus means that the same load can be absorbed without requiring the material to strain as much. When combined with the metallic pipe suffering a dent, this directly relates to a reduction of cyclic stress or strain that the pipe experiences minimizing fatigue damage. By minimizing the amount of potentially plastic deformation over time, a carbon fiber composite can significantly improve the performance of a dented pipe.

Additionally, instead of using uni-directional fabric, a bi-directional woven fabric was utilized with heavy preference in the hoop direction. The expectation is that for dents, the majority of fatigue will still occur due to hoop stresses and therefore a preference for that direction should be given. However, enough axial reinforcement should be provided to ensure that other loads are accounted for, and that premature failure of the composite system's resin does not occur.

Lastly, instead of using a pre-cured system such as Clock Spring, the resin matrix of the Atlas repair system utilizes asite-applied epoxy. This allows the composite material to be applied over the dented area while still in a moldable, pre-cure condition therefore allowing the composite to conform around the non-ideal shape of the dent. Doing this allows for maximum fit and optimizes the amount of support the composite can directly give the pipe while also minimizing the chance for human error during application.

Case Study Results

To test the effectiveness of composite repair systems on dents, a joint industry program (JIP) was launched and completed in 2015 that ultimately showed a composites ability to reduce the dents effective SCF value. In this study, named DV-CIP, multiple pipes were artificially dented and rerounded to simulate potential real-life applications. All pipes were comprised of 24" (610mm) diameter, API 5L Grade X42 with a wall thickness of 0.25" (6.35mm). All pipes underwent an artificial indention of 15% of its diameter and was then rerounded to an average of 4.8% of its diameter.

Of direct interest to this paper are three specific sample pipes:

1) A baseline pipe was dented and left unrepaired.

2) A pipe was identically dented and repaired with the Atlas repair system.

3) A pipe was dented then had 40% of its thickness machined out to simulate an extreme case of an interacting dent-corrosion defect.

All samples were then pressurized and cycled to failure or a runout condition near 100,000 cycles. The table below shows a summary of the three discussed pipespulled from the DV-CIP report.

Sample	Description	Repair Thickness	Measured SCF (1000cycle)	Cycles to Failure	Comments
UR-PD-24-1	Unrepaired	-	4.20	23,512	
PW-PD-24-1	Repaired Dent	0.364"	0.87	101,151+	Runout reached
PW-DC-24-1	Repaired Dent + Wall Loss	0.552"	0.92	101,152+	Runout reached

Therepair thicknesses were calculated based on expected stress loading with a notable increase in the thickness when considering the additional 40% wall loss scenario. However, in both cases of repaired dents, the measured change in strain per cycle was less than the pristine pipe outside of the dented area. This resulted in a measured SCF of less than 1.00 which clearly demonstrates the potential for composite materials to reduce the stress loading in the base pipe which reduces fatigue effects. This can further be seen by looking at the Cycle to Failure column which shows a 4x increase in fatigue life and is only limited to that because runout conditions were reached and testing ended.

Case Study 3: Girth Weld Reinforcement in Geohazard Environments

The third and final case study explores the use of composite materials for reinforcing girth welds in geohazard environments. This case study demonstrates the versatility of composite materials, as a simple change in fabric orientation can turn a repair system designed for one set of stresses into a solution for entirely different loading conditions. It emphasizes the potential of composites to be tailored to a wide range of environmental challenges, providing enhanced protection where traditional materials might fall short. In this case study, the composite repair system was applied to damaged girth welds, and a focused testing program was conducted to measure its effectiveness in reducing the risk of early weld failure. The results of these tests demonstrated that the composite repair significantly reduced the strain on the welds, providing a durable solution for pipelines located in geohazard zones.

Product Design and Methodology

Girth welds, which are the joints that connect segments of a pipeline, are often vulnerable to failure due to external forces such as ground movement, soil settlement, or seismic activity if there is a manufacturing flaw or defect within the girth weld. In these scenarios, the welds are subject to stresses that can lead to early failure if not adequately reinforced. Of primary concern in this defect scenario are bending loads and axial tension loads that can be caused by ground movement. This combination acting on a weakened girth weld results in concerns in both the hoop and axial direction. However, in most cases, the defining failure mode will be due to axial loads induced by some combination of internal pressure, pure axial tension, and tension induced from bending.

By modifying the orientation of the carbon fiber

fabric used in the composite system, the material can be specifically engineered to address the unique stresses experienced at the girth welds. For this scenario, the aforementioned Atlas repair system was modified to create a new repair system named Atlas UA. This system utilizes the same resin but has a modified carbon fiber fabric architecture that consists entirely of unidirectional reinforcement in the transverse direction. When wrapped circumferentially around the pipe per standard installation techniques, this results in reinforcement aligned with the axial direction of the pipe.

By combining the Atlas bi-directional, but hoop focused, system with the Atlas UA axially aligned system, a composite repair can be engineered to match the anticipated requirements in both hoop and axial directions. In scenarios with little axial loading concern, the proposed repair system may be entirely comprised of the Atlas bi-directional system. In scenarios where there is little hoop stress concern in the girth weld, the final repair system may be primarily Atlas UA with several planned layers of Atlas to provide cyclic hoop reinforcement and kink resistance.

Lastly, a taper of the repair system on the edges can be critical when repairing axial dominated loads. Without a taper, a stress discontinuity exists at the pipe-to-repair interface, leading to potentially premature buckle failure under bending loads. Due to this and the general higher thickness requirements for axial concerns, a new specified installation technique was implemented utilizing a dual-system installation method. This installation technique involves alternating between the Atlas and Atlas UA layers at specified intervals while building a step-wise taper. The resulting taper can be seen below in the buckled pipe image.

Case Study Results

During 2022 and 2023, multiple coupons were tested that included machined crack-like defects in the girth weld or incompletewelding in the girth weld. These samples were then either loaded in a 4-point bend frame or pipe-scale tensile frame. The samples then saw a variety of pressure related considerations including installs at high pressure followed by testing at low pressure to simulate worst-case scenarios in terms of axial loading. A close up of a simulated girth weld defect and the resulting pipe failure due to bending is shown below:

In bending, the composite repair easily, and repeatably, minimized strain in the girth weld to values near 0.1% axial strain while eventually causing a buckle in the base pipe, which measured near 2.5% axial strain prior to buckling. For comparison, unreinforced bending loads on the same defect saw the girth weld experience near 2.3% axial strain prior to a rupture-type failure while the base pipe only strained near 0.4% axially.

The results in bending clearly show that aproperly designed and installed composite repair can reinforce severely damaged girth welds and change geohazard induced failure modes from potentially brittle girth weld failures to ductile buckle failures outside of the reinforcement area. This can be significant as this change in failure mode could take a potential release situation and instead allow for ductile pipe to move as needed while keeping fluid contained and operational until remediation can occur following a large-scale geohazard event.

In addition to bending loads, pure axial loading was considered. It should be noted that a pure axial load scenario is expected to be a very rare occurrence as most geohazard events are more likely to induce significant load through bending whichwould engage the composite repair in a very different manner. When looking at pure axial loading, the test focuses entirely on the bond between the pipe and composite. In this scenario, sufficient axial loading will cause the pipe to eventually begin to yield and neck which rapidly decreases the pipes outside diameter until separation occurs with the composite.

Knowing this would be the primary failure mode, several different scenarios were considered including varying install pressure and test pressureas well as cycling either pressure or axial loads. The results are shown below. The main takeaway from the information below is that in pure axial tension, the axial strain in the base pipe increased at least 6-times compared to an unreinforced sample. While the unreinforced pipe was seeing 1.0% or higher axial strain, the girth weld experienced approximately 10-times less strain before failure.

Sample#	Loading Condition	Internal Pressure	Total Axial Load/kips	Maximum Measured Girth Weld Strain/%	Maximum Measured Base Pipe Strain[1]/%
3	Unreinforced Axial tension	Yes	500	0.132	0.16
4	Reinforced Axial tension ($P_{install}$ = 0psig)	Yes	634	0.11	1.81
5	Reinforced Axial tension ($P_{install}$ = 64% SMYS)	Yes	613	0.11	1.37
6	Reinforced Axial tension ($P_{install}$ = 0psig)	No	531	0.10	1.00
7	Reinforced Axial tension ($P_{install}$ = 64% SMYS)	No	499	0.15	1.44

Overall, this test series indicates that awell-engineered and strategically implemented composite reinforcement designed to resist axial loading can be successful in reducing the stress or strain seen due to axial or bending loads.

Conclusions

Considering these case studies, as well as historical data and experience, it can be seen that composite materials allow for a wide range of solutions. By altering the composite's resin, fabric type or architecture, and possibly combining multiple systems, a composite repair can be designed and optimized for many specific scenarios. The first case study demonstrated an easy-to-install, pre-cured repair system entirely focused on reinforcing against hoop stress. Additionally, it was discussed how longevity of the repair system can be considered, providing justification for a permanent, long-term repair.

The second case studyshows how changing the resin and fabric of a composite allows for better optimization for non-ideal scenarios. In this case, utilizing the carbon-fiber/epoxy Atlas repair system, a much more defect-specific repair can be accomplished that now considers long-term cyclic fatigue and stress reduction in the base pipe as opposed to only considering long-term burst reinforcement.

Lastly, by looking at the axial loads induced by geohazard type events, it was demonstrated that introducing a new fabric architecture and changing the installation parameters allowed for a more targeted axial stress reinforcement. By combining the Atlas and Atlas UA systems in an engineered, designed way, various types of loading scenarios, including axial tension and bending can be designed for in addition to hoop stress.

Ultimately, composites materials have long made a case for their usefulness in industry based on their ability for customization and diversity. This paper hopes to have demonstrated the same potentialregarding the use of composite materials for pipeline repair or reinforcement. With the right materials, the right engineering, and agreed-upon design expectations, composite materials can provide highly optimized and targeted repair scenarios for many different defects that exist. As pipeline infrastructure continues to age and face new challenges, the ability to harness these unique properties of composites will be essential for maintaining the integrity, safety, and cost-effectiveness of pipeline systems worldwide.

It should be noted that the overall message of this paper is one thatdiscusses the importance of intentional design and methodology supported by testing and is not intended to imply that any composite repair system can be used for every defect. Instead, it is of high importance that specified repair systems only be used for the scenarios they have been designed and tested for. There is no single system or design that will solve all problems; rather the benefit of composite materials is their ability to be customized, and therefore optimized, when designed by qualified individuals.